Chemistry of Spices

This page intentionally left blank

Chemistry of Spices

Edited by

Villupanoor A. Parthasarathy

Indian Institute of Spices Research
Calicut, Kerala, India

Bhageerathy Chempakam

Indian Institute of Spices Research
Calicut, Kerala, India

and

T. John Zachariah

Indian Institute of Spices Research
Calicut, Kerala, India

www.cabi.org

CABI is a trading name of CAB International

CABI Head Office
Nosworthy Way
Wallingford
Oxfordshire OX10 8DE
UK

Tel: +44 (0)1491 832111
Fax: +44 (0)1491 833508
Email: cabi@cabi.org
Web site: www.cabi.org

CABI
745 Atlantic Avenue
8th Floor
Boston, MA 02111
USA

Tel: +1 617 682-9015
Email: cabi-nao@cabi.org

A catalogue record for this book is available from the British Library, London, UK

Library of Congress Cataloging-in-Publication Data
Chemistry of spices / [edited by] V.A. Parthasarathy, B. Chempakam, T. John Zachariah.
p. cm.
Includes bibliographical references and index.
ISBN 978-1-84593-405-7 (alk. paper)
1. Spices--Analysis. 2. Spice plants--Composition. I. Parthasarathy, V.A. II Chempakam, B., Dr. III. Zachariah, T. John. IV. Title.

SB305.C44 2008
641.3'383—dc22

2007043551

ISBN-13: 9781-84593-405-7

First published 2008
Transferred to print on demand 2016

Printed and bound in the UK by Marston Book Services Ltd, Didcot, Oxon.

Contents

Contributors

Indian Institute of Spices Research, Calicut – 673 012, Kerala, India
Phone: 0091 – 0495 – 2731410, Fax: 0091 – 0495 – 2730294
E-mail: mail@spices.res.in, Web site: www.spices.res.in

Dr V.A. Parthasarathy, Director
E-mail: parthasarathy@spices.res.in

Division of Crop Production & PHT

Dr B. Chempakam, Principal Scientist & Head
E-mail: chembakam@spices.res.in

Dr T. John Zachariah, Senior Scientist – Biochemistry
E-mail: john@spices.res.in

Dr N.K. Leela, Senior Scientist – Organic Chemistry
E-mail: leela@spices.res.in

Dr K.S. Krishnamurthy, Senior Scientist – Plant Physiology
E-mail: kskrishnamoorthy@spices.res.in

Dr Shamina Azeez, Senior Scientist – Biochemistry
E-mail: shamina@spices.res.in

Gobinath, P. E-mail: gobinath_bio@yahoo.com

Balaji, S. E-mail: bio_balaji@yahoo.co.in

Sapna, V.P. E-mail: sapnajeesh@yahoo.co.in

Shafeekh, K.M. E-mail: sheficlt@yahoo.com

Sindhu, S. E-mail: intelnidhu@yahoo.com

Vipin, T.M. E-mail: vip_kkd@yahoo.co.in

This page intentionally left blank

Preface

Spices are woven into the history of nations. The desire to possess and monopolize the spice trade has, in the past, compelled many a navigator to find new routes to spice-producing nations. In the late 13th century, Marco Polo's exploration of Asia established Venice as the most important trade port. Venice remained prosperous until about 1498. Portuguese explorer Vasco de Gama sailed around Africa's Cape of Good Hope to reach Calicut, India. He returned with pepper, cinnamon, ginger and jewels, and also deals for the Portuguese to continue trade with India.

Spices impart aroma, colour and taste to food preparations and sometimes mask undesirable odours. The volatile oils from spices give the aroma and the oleoresins impart the taste. There is a growing interest in the theoretical and practical aspects of the inner biosynthetic mechanisms of the active principles in spices, as well as in the relationship between the biological activity and the chemical structure of these secondary metabolites. The antioxidant properties of herbs and spices are of particular interest in view of the impact of oxidative modification of low-density lipoprotein cholesterol in the development of atherosclerosis. A range of bioactive compounds in herbs and spices has been studied for anticarcinogenic properties in animals, but the challenge lies in integrating this knowledge to ascertain whether these effects can be observed in humans, and within defined cuisines. Research on the structure activity relationships in spice components has become an exciting field since these compounds play a major role in the culinary, industrial and pharmacological fields.

Hence, we have attempted to compile all available information on the chemistry of spice crops such as black pepper, cardamom (small), cardamom (large), ginger, turmeric, cinnamon and cassia, clove, nutmeg and mace, coriander, cumin, fennel, fenugreek, paprika, vanilla, ajowan, star anise, aniseed, garcinia, tamarind, parsley, celery, curry leaf and bay leaf. To edit this book, we have used the current Indian expertise on spices and we have made every effort to collate all available information so that the book will be useful to researchers, industrialists and postgraduate students of agriculture, horticulture and phytochemistry. It will also be a very useful resource book for spice traders and processors. We are grateful to CABI for giving us the opportunity to edit this book and we are indebted to Ms Sarah Hulbert of CABI Head Office for her immense help in getting the book into final shape. She has answered an array of e-mails and strings of questions to help us in this venture and we thank her for her patience and assistance.

We appreciate the help rendered by Mr A. Sudhakaran, artist-cum-photographer of IISR, Calicut, Kerala, for designing the cover page. The help given by Ms T.V. Sandhya in typesetting the manuscript is gratefully acknowledged. We also thank the Director of the Indian Institute of Spices Research, Calicut, India, for providing photographs of the spices.

V.A. Parthasarathy
B. Chempakam
T.J. Zachariah

1 Introduction

V.A. Parthasarathy, B. Chempakam and T. John Zachariah

Spices and herbs have played a dramatic role in civilization and in the history of nations. The delightful flavour and pungency of spices make them indispensable in the preparation of palatable dishes. In addition, they are reputed to possess several medicinal and pharmacological properties and hence find position in the preparation of a number of medicines.

1.1. Historical Perspective

Many maritime routes were developed to India and China with an ultimate desire to develop a spice route. In the late 13th century, Marco Polo's exploration of Asia established Venice as the most important trade port. Venice remained prosperous until about 1498. The Portuguese explorer, Vasco de Gama, sailed around Africa's Cape of Good Hope to reach Calicut, India. He returned with pepper, cinnamon, ginger and jewels, and also deals for the Portuguese to continue trade with India.

Rosengarten (1969) has presented a very interesting history of spices. In 1492, Christopher Columbus arrived in America while searching for a direct western route to the Spice Islands. Though he did not find the Spice Islands, Columbus brought allspice, vanilla and red peppers from the West Indies back to his Spanish supporters. Conflict developed over who would dominate this prosperous trade. Wars over the Indonesian Spice Islands broke out between the expanding European nations and continued for about 200 years, between the 15th and 17th centuries.

In 1780, the Dutch and English fought a war over the spice trade and the Dutch lost all spice trading centres. The Americans began their entry into the world spice race in 1672 (ASTA, 1960).

From the beginning of history, the strongest nations have controlled the spice trade. The same is true today; the USA is now the world's major spice buyer, followed by Germany, Japan and France.

In short, the trade in spices, usually carried out along the many historic spice routes, has been one of the most important commercial activities throughout ancient and modern times. The importance placed on spices is reflected by economic developments that began early in many ancient civilizations, where spices found applications in food preservation, cooking and traditional medicine.

Asia still grows most of the spices that once ruled the trade, including cinnamon, pepper, nutmeg, clove and ginger. However, more and more spices are being planted in the Western hemisphere, along with a wide variety of herbs and aromatic seeds. Brazil is a major supplier of pepper. Guatemala is a leading producer of cardamom. Grenada grows nutmeg and ginger, and allspice is grown in Jamaica. Nicaragua, El Salvador and the USA grow sesame seed. Europe and the USA produce many herbs and Canada grows several aromatic seeds.

1.2. Global Spice Trade

The major markets in the global spice trade are the USA, the European Union, Japan, Singapore, Saudi Arabia and Malaysia. The principal supplying countries are China, India, Madagascar, Indonesia, Vietnam, Brazil, Spain, Guatemala and Sri Lanka. During the review period from 2000 to 2004, the value of spice imports increased by an average of 1.9% per year and the volume increased by 5.9%. World trade in spices in 2004 consisted of 1.547 million t, valued at US$2.97 billion. An annual average rate of 7% was seen in the global import volume of spices in the period 2000–2002, whereas the import values decreased by 5% annually. This was attributed to the dramatic decrease in the value of whole pepper during 2000/01 by about 40% and a further 18% in 2002/03 (Table 1.1).

Higher market prices for major commodities such as paprika, vanilla, ginger, bay leaves and spice mixtures resulted in an upward value trend by 4.6% from 2003 to 2004, with a stabilized import volume. There was a growing trend towards the trade of processed spices, which fetched higher prices. The increasing demand for value-added processing of spices, such as capsicum and ginger, offers business opportunities for the food and extraction industries in international markets (International Trade Centre, 2006).

World import for black pepper achieved only minor increases in volume during 2000–2004. On average, 260,000 t of black pepper is imported yearly into the global market. While growth in volume trade rose marginally, import values for *whole pepper* declined steeply by 54% from US$854 million to US$394 million in that period, resulting in lower world prices for pepper. Vietnam, Indonesia, Brazil, Malaysia and India are the major producers and exporters of black pepper. With an export volume of 96,113 t, valued at US$136.6 million in 2004, Vietnam is the world's largest exporter in the black pepper trade.

In the case of ginger, Japan is the number one importer in the world. Japan's imports of ginger reached more than 100,000 t, valued at US$126 million, which accounted for 50% of the country's total spice imports in 2004. The principal supplier of quality ginger to the Japanese market is China, with exports exceeding 70,000 t, valued at US$93 million, followed by Thailand with 26,000 t.

Vanilla is the second most expensive spice after saffron because its production is very labour-intensive. The world market for vanilla is highly concentrated in the USA, France and Germany. In 2004, US imports of vanilla amounted to US$205 million, followed by France and Germany (US$44 million and US$36 million, respectively). These importing countries represent 72.5% of the world vanilla trade.

As an average, import values of nutmeg, mace and cardamom decreased by 7% annually, whereas volumes recorded a slight increase over 2000–2004. Imports of cardamom made up 60% and nutmeg and mace 40% of the total import value of US$204 million in 2004.

International trade in mixed spices (curcuma, turmeric and curry powder, laurel leaves, curry paste, dill and fenugreek seeds) grew by 5% and 11% in volume and value terms, respectively, in 2003/04. The main importing countries were the USA, Belgium, Germany, the Netherlands and the UK. India supplied 14% of the total import value of this spice category to the US and UK markets in 2004.

Table 1.2 shows the exports and market shares of the leading spice producing countries during 2000–2004. These major exporters account for a value share of more than 55% in the 2004 world import trade of spices. In terms of export competitiveness, China has emerged as the principal exporter. Its export share increased sharply in 2003/04 to 13.2%, up from 9.7%, surpassing India

Table 1.1. World imports of different spices.

	Quantity (thousand t)					Value (US$ million)				
Spice category	2000	2001	2002	2003	2004	2000	2001	2002	2003	2004
Pepper, whole	216.1	228.9	246.6	228.8	237.0	854.3	492.3	402.4	425.1	394.6
Pepper, crushed/ground	23.7	22.1	27.4	30.5	32.4	95.0	72.1	75.4	92.3	99.5
Total pepper	239.8	251.0	274.0	259.3	269.4	949.3	564.4	477.8	517.4	494.1
Capsicum	230.7	273.1	324.8	350.1	371.0	370.6	426.1	453.5	492.0	590.4
Vanilla	4.3	4.4	6.8	5.0	3.5	108.2	240.7	308.5	535.9	394.9
Cinnamon, whole	73.4	68.3	78.4	70.4	75.2	108.6	108.1	106.5	100.1	105.6
Cinnamon, crushed/ground	9.8	10.1	13.4	13.0	13.2	16.7	16.2	20.2	20.6	22.6
Total cinnamon	83.2	78.4	91.8	83.4	88.4	125.3	124.3	126.7	120.7	128.2
Cloves, whole and stems	50.3	53.1	29.5	50.3	43.9	148.2	148.2	124.1	101.2	115.9
Nutmeg, mace, cardamom	42.2	41.9	46.3	50.1	47.5	279.9	279.9	236.9	215.6	204.4
Spice seeds	201.2	186.4	207.0	213.8	220.3	207.8	207.8	207.0	201.3	207.5
Ginger (except preserved)	213.7	234.1	236.2	313.8	284.1	206.6	206.6	143.1	177.9	305.3
Thyme, saffron, bay leaves	15.3	17.9	18.3	20.1	20.6	77.9	77.9	80.0	95.9	106.9
Other spice mixtures	173.5	249.2	202.0	189.5	198.4	292.7	292.7	321.6	383.3	427.3
Total spice imports	1254.0	1389.6	1436.7	1535.4	1547.2	2766.5	2766.5	2479.2	2841.2	2973.9

Source: International Trade Centre (2006).

Table 1.2. Main spice-exporting countries by commodity; value and percentage share, 2004.

Spice category	Import value (US$ thousand)	First	%	Second	%	Third	%
Pepper, whole	394,560	Vietnam	32.6	Indonesia	17.5	Brazil	16.7
Pepper, crushed/ground	99,536	Germany	18.2	India	14.8	Vietnam	8.0
Capsicum	590,420	China	23.8	India	15.9	Spain	9.3
Vanilla	394,928	Madagascar	51.8	Indonesia	12.2	Papua New Guinea	8.9
Cinnamon, whole	105,580	Sri Lanka	45.0	Indonesia	21.1	China	19.9
Cinnamon, crushed/ground	22,594	Indonesia	28.7	Brazil	14.8	Netherlands	11.1
Cloves, whole and stems	115,869	Madagascar	30.4	Sri Lanka	17.3	Tanzania, U.R.	12.5
Nutmeg, mace, cardamom	204,383	Guatemala	38.8	Indonesia	24.1	Nepal	5.7
Spice seeds	207,526	India	18.2	Syria Arab Rep.	14.7	Turkey	8.7
Ginger (except preserved)	305,321	China	64.3	Thailand	12.3	Brazil	3.3
Thyme, saffron, bay leaves	105,896	Iran Islam Rep.	29.3	Spain	25.0	Turkey	12.0
Spices n.e.s. mixtures	427,268	Germany	15.9	India	13.9	Netherlands	6.9

Note: n.e.s. = not elsewhere specified.

Table 1.3. Main spice-importing countries by commodity; value and percentage share, 2004.

Spice category	Import value (US$ thousand)	First	%	Second	%	Third	%
Pepper	494,096	USA	23.1	Germany	10.9	Netherlands	5.3
Capsicum	590,420	USA	23.6	Malaysia	7.6	Germany	7.1
Vanilla	394,928	USA	51.9	France	11.3	Germany	9.3
Cinnamon	128,174	Mexico	21.0	USA	16.9	India	6.0
Cloves	115,869	Singapore	46.3	India	23.7	Malaysia	7.1
Nutmeg, mace, cardamom	204,383	Saudi Arabia	25.0	India	8.0	Netherlands	8.0
Spice seeds	207,526	USA	11.1	Germany	8.4	Malaysia	6.5
Ginger (except preserved)	305,321	Japan	41.2	USA	12.1	Pakistan	6.2
Thyme, saffron, bay leaves	105,896	Spain	20.2	USA	13.9	Italy	8.0
Spices n.e.s. mixtures	427,266	USA	13.0	Belgium	7.8	Germany	6.8

Note: n.e.s. = not elsewhere specified.

with 8.6%, followed by Madagascar 8.2%, Indonesia 7.3%, Vietnam 5.1%, Brazil 4.1%, Spain 3.1%, Guatemala and Sri Lanka 2.8%. Table 1.3 shows the rankings of the top three exporting countries of individual spices to international markets.

Developing countries, including least developed countries, supply about 55% of spices to global markets. The USA, the European Union, Japan and Singapore are among the major markets, accounting for about 64% of the world import share of spices. Germany, the Netherlands and Singapore are significant re-exporters in the spice trade.

Apart from competing for markets, developing country producers and exporters face many challenges, including that of quality issues. Spice exports are subject to strict quality standards for food safety set by the American Spice Trade Association (ASTA) and the European Spice Association (ESA). Demand is growing for high quality and processed spices. This trend for value-added products offers new business opportunities in the spice trade.

Global production of spices

Table 1.4 gives the major spice-producing areas in the world, while Table 1.5 shows the area and production of important spices in the world. Compared with many other field and horticultural crops, area and production of spices is limited. The FAO database gives the area and production of a limited number of spices only. Spices were cultivated in an area of 7587.02 thousand ha, with a production of 31,859.69 thousand t during 2005. The world export of spices during 2005 was 3592.48 thousand t and import was 3454.40 thousand t (Anon., 2007).

1.3. Major Compounds in Spices

Spices impart aroma, colour and taste to food preparations and sometimes mask undesirable odours. Volatile oils give the aroma, and oleoresins impart the taste. Aroma compounds play a significant role in the production of flavourants, which are used in the food industry to flavour, improve and increase the appeal of their products. They are classified by functional groups, e.g. alcohols, aldehydes, amines, esters, ethers, ketones, terpenes, thiols and other miscellaneous compounds. In spices, the volatile oils constitute these components (Zachariah, 1995; Menon, 2000).

In black pepper, caryophyllene-rich oils possess sweet floral odours, whereas oils

Table 1.4. Spice-producing areas.

Spices	Botanical name	Edible part(s)	Major source/origin
Ajowan	*Trachyspermum ammi* (L.) Sprague	Seed	Persia and India
Aniseed	*Pimpinella anisum* L.	Fruit	Mexico, The Netherlands, Spain
Basil	*Ocimum basilicum* L.	Sweet, leaf	France, Hungary, USA, Serbia and Montenegro
Bay leaf	*Laurus nobilis* L.	Leaf	Turkey, USA, Portugal
Cardamom	*Elettaria cardamomum* White et Mason	Fruit	India, Guatemala
Large cardamom	*Amomum subulatum* Roxb.	Fruit	India, Nepal, China
Cassia	*Cinnamomum cassia* (L.) Presl	Stem, bark	China, Indonesia, South Vietnam
Celery	*Apium graveolens* L.	Fruit	France, India
Chilli	*Capsicum frutescens* L.	Fruit	Ethiopia, India, Japan, Kenya, Mexico, Nigeria, Pakistan, Tanzania, USA
Cinnamon	*Cinnamomum verum* syn. *C. Zeylanicum*	Stem, bark	Sri Lanka, India
Clove	*Syzygium aromaticum* (L.) Merr. et Perry	Buds	Indonesia, Malaysia, Tanzania
Coriander	*Coriandrum sativum* L.	Fruit	Argentina, India, Morocco, Romania, Spain, Serbia and Montenegro
Cumin	*Cuminum cyminum* L.	Fruit	India, Iran, Lebanon
Curry leaf	*Murraya koenigii* Spreng	Leaf	India, Burma
Dill	*Anethum graveolens* L.	Fruit	India
Fennel	*Foeniculum vulgare* Mill.	Fruit	Argentina, Bulgaria, Germany, Greece, India, Lebanon
Fenugreek	*Trigonella foenum-graecum* L.	Fruit	India
Garcinia	*Garcinia cambogia*	Fruit	India, Sri Lanka
Garlic	*Allium sativum* L.	Bulb/clove	Argentina, India
Ginger	*Zingiber officinale* Rosc.	Rhizome	India, Jamaica, Nigeria, Sierra Leone
Mint	*Mentha piperita* L.	Leaf/terminal shoot	Bulgaria, Egypt, France, Germany, Greece, Morocco, Romania, Russia, UK
Mustard	*Brassica nigra* (L.) Koch	Seed	Canada, Denmark, Ethiopia, UK, India
Nutmeg	*Myristica fragrans* Houtt.	Aril/seed kernel	Grenada, Indonesia, India
Onion	*Allium cepa* L.	Bulb	Argentina, Romania, India
Oregano	*Origanum vulgare* L.	Leaf	Greece, Mexico
Paprika	*Capsicum annuum* L.	Fruit	Bulgaria, Hungary, Morocco, Portugal, Spain, Serbia and Montenegro
Parsley	*Petroselinum crispum* (Mill) Nyman ex A.W. Hill	Leaf	Belgium, Canada, France, Germany, Hungary
Black pepper	*Piper nigrum* L.	Fruit	Brazil, India, Indonesia, Malaysia, Sri Lanka, Vietnam

Continued

Table 1.4. *Continued*

Spices	Botanical name	Edible part(s)	Major source/origin
Poppy	*Papaver somniferum* L.	Seed	The Netherlands, Poland, Romania, Turkey, Russia
Rosemary	*Rosmarinus officinalis* L.	Leaf, terminal shoot	France, Spain, USA, Serbia and Montenegro
Saffron	*Crocus sativus* L.	Pistil of flower	Spain
Sage	*Salvia officinalis* L.	Leaf	Albania, Serbia and Montenegro
Star anise	*Illicium verum* Hooker fil.	Fruit	China, North Vietnam
Tamarind	*Tamarindus indica* L.	Fruit	Indonesia, Vietnam
Thyme	*Thymus vulgaris* L.	Leaf	France, Spain
Turmeric	*Curcuma longa* L.	Rhizome	China, Honduras, India, Indonesia, Jamaica
Vanilla	*Vanilla planifolia* Andrews	Fruit/beans	Indonesia, Madagascar, Mexico, India

Source: cookingsecrets.org/herbs-spices/spice-producing-areas.

Table 1.5. Area and production of important spices in the world.

Spice(s)	Area (thousand ha)	Production (thousand t)
Anise, badian, fennel, coriander	661.16	467.86
Chillies and peppers (dry)	2,004.81	2,662.73
Chillies and peppers (green)	1,725.54	24,803.01
Cinnamon (canella)	176.98	134.8
Cloves	466.08	145.18
Ginger	338.9	1,119.74
Nutmeg, mace and cardamom	222.89	74.02
Pepper (*Piper* sp.)	473.55	407.41
Vanilla	76.44	10.36
Other spices	1,440.67	2,034.58
Total	7,587.02	31,859.69

Source: FAO database (2007).

with high pinene content give turpentine-like off-odours (Lewis *et al.*, 1969). The major compounds in fresh pepper are *trans*-linalool oxide and α-terpineol, whereas dry black pepper oil contains α- and β-pinenes, d-limonene and β-caryophyllene as major components.

In cardamom, the oil has very little mono- or sesquiterpenic hydrocarbons and is dominated by oxygenated compounds, all of which are potential aroma compounds. While many of the identified compounds (alcohols, esters and aldehydes) are commonly found in many spice oils (or even volatiles of many different foods), the dominance of the ether, 1,8-cineole, and the esters, α-terpinyl and linalyl acetates in the composition make the cardamom volatiles a unique combination (Lewis *et al.*, 1966; Salzer, 1975; Korikanthimath *et al.*, 1997).

Ginger owes its characteristic organoleptic properties to two classes of constituents: the odour and the flavour of ginger are determined by the constituents of its steam-volatile oil, while the pungency is determined by non-steam-volatile components, known as the gingerols. The steam-volatile oil comprises mainly of sesquiterpene hydrocarbons, monoterpene hydrocarbons and oxygenated monoterpenes (Purseglove *et al.*, 1981). The monoterpene constituents are believed to be the most important contributors to the aroma of ginger and are more abundant in the natural oil of the fresh ('green') rhizome than in the essential oil distilled from dried ginger. Oxygenated sesquiterpenes are relatively minor constituents of the volatile oil, but appear to be significant contributors to its flavour properties. The major sesquiterpene hydrocarbon constituent of ginger oil is (–)-α-zingiberene. Australian ginger oil has a reputation for possessing a particular 'lemony' aroma, due to its high content of the isomers, neral and geranial, often collectively referred to as citral (Wohlmuth *et al.*, 2006).

Cinnamon possesses a delicate, spicy aroma, which is attributed to its volatile oil. Volatile components are present in all parts of cinnamon and cassia. They can be classified broadly into monoterpenes, sesquiterpenes and phenylpropenes (Senanayake, 1997). The oil from the stem bark contains 75% cinnamaldehdyde and 5% cinnamyl acetate, which contribute to the flavour (Angmor *et al.*, 1972; Wijesekera, 1978; Krishnamoorthy *et al.*, 1996).

The minor constituents like methyl amyl ketone, methylsalicylate, etc., are responsible for the characteristic pleasant odour of cloves. The oil is dominated by eugenol (70–85%), eugenyl acetate (15%) and *β*-caryophyllene (5–12%), which together make up 99% of the oil. *β*-Caryophyllene, which was earlier thought of as an artefact of distillation, was first reported as a constituent of the bud oil by Walter (1972).

The volatile oil of nutmeg constitutes the compounds: monoterpene hydrocarbons, 61–88%; oxygenated monoterpenes, i.e. monoterpene alcohols, monoterpene esters; aromatic ethers; sesquiterpenes, aromatic monoterpenes, alkenes, organic acids and miscellaneous compounds. Depending on the type, its flavour can vary from a sweetly spicy to a heavier taste. The oil has a clove-like, spicy, sweet, bitter taste with a terpeny, camphor-like aroma.

Among the seed spices, cumin fruits have a distinctive bitter flavour and strong, warm aroma due to their abundant essential oil content. Of this, 40–65% is cuminaldehyde (4-isopropylbenzaldehyde), the major constituent and important aroma compound, as also the bitterness compound reported in cumin. The odour is best described as penetrating, irritating, fatty and overpowering, curry-like, heavy, spicy, warm and peristent, even after drying out (Weiss, 2002). The characteristic flavour of cumin is probably due to dihydrocuminaldehyde and monoterpenes.

In the mature fruit of fennel, up to 95% of the essential oil is located in the fruit, greater amounts being found in the fully ripe fruit. Hydrodistillation yields 1.5–3.5%. Generally, anethole and fenchone are found more in the waxy and ripe fruits than in the stems and leaves (Akgül, 1986; Kruger and Hammer, 1999). Anethole has flavouring properties and is distinctly sweet, being 13 times sweeter than sugar.

As for coriander, in the unripe fruits and the vegetative parts of the plant, aliphatic aldehydes predominate in the steam-volatile oil and are responsible for the peculiar aroma. On ripening, the fruits acquire a more pleasant and sweet odour and the major constituent of the volatile oil is the monoterpene alcohol, linalool. Sotolon (also known as sotolone, caramel furanone, sugar lactone and fenugreek lactone) is a lactone and an extremely powerful aroma compound and is the major aroma and flavour component of fenugreek seeds (Mazza *et al.*, 2002).

Among the leafy spices, 45 aroma volatiles of desert parsley have been identified, with the major constituents as myristicin, apiole, *β*-phellandrene, *p*-mentha-1,3,8-triene and 4-isopropenyl-1-methylbenzene (MacLeod *et al.*, 1985). Among these, apiole in particular has a desirable parsley odour character. The leaf stems of celery show three main constituents of volatiles, e.g. apiole (about 23%), 3-butylphthalide (about 22%) and sedanolide (about 24%). The last two possess a strong characteristic celery aroma (MacLeod *et al.*, 1988). Limonene (40.5%), β-selinene (16.3%), *cis*-ocimene (12.5%) and β-caryophyllene (10.5%) are some of the volatile oil constituents present in celery leaves from Nigeria (Ehiabhi *et al.*, 2003).

The curry leaf plant is highly valued for its characteristic aroma and medicinal value (Philip, 1981). A number of leaf essential oil constituents and carbazole alkaloids have been extracted from the plant (Mallavarapu *et al.*, 1999). There are a large number of oxygenated mono- and sesquiterpenes present, e.g. *cis*-ocimene (34.1%), α-pinene (19.1%), γ-terpinene (6.7%) and β-caryophyllene (9.5%), which appear to be responsible for the intense odour associated with the stalk and flower parts of curry leaves (Onayade and Adebajo, 2000). In fresh bay leaves, 1, 8-cineole is the major component, together with α-terpinyl acetate, sabinene, α-pinene, β-pinene, β-elemene, α-terpineol, linalool and eugenol (Kilic *et al.*, 2004).

The major chemical constituents in spices are tabulated in Table 1.6.

1.4. Value Addition and New Product Development

Farm-level processing operations are the most important unit operations for value addition and product diversification of spices. It is essential that these operations ensure proper conservation of the basic qualities like aroma, flavour, pungency, colour, etc. Each of these operations enhances the quality of the produce and the value of the spice. The clean raw materials form the basis for diversified value-added products.

The first spice oil and oleoresin industry was started in 1930 in India at Calicut by a private entrepreneur. Extracts of ginger were manufactured during the Second World War. The major oils are from black pepper, cardamom, chilli seed, capsicum, paprika, clove, nutmeg, mace, cinnamon, cassia, kokkam, galangal, juniper and peppermint (Guenther, 1950). Pepper oil, ginger oil, celery seed oil, kokkam oil and peppermint are the major oils exported from India. Oleoresins exported are from black pepper, cardamom, chillies, capsicum, paprika, ginger, turmeric, white pepper, coriander, cumin, celery, fennel, fenugreek, mustard seed, garlic, clove, nutmeg, mace, cinnamon, cassia, tamarind, galangal, rosemary and curry powder oleoresins. Table 1.7 lists the value-added products from major spices.

1.5. Pharmacological aspects

Chemopreventive and anticancerous

Recent advances in our understanding at the cellular and molecular levels of carcinogenesis have led to the development of a promising new strategy for cancer prevention, that is, chemoprevention. Chemoprevention is defined as the use of specific chemical substances – natural or synthetic, or their mixtures – to suppress, retard or reverse the process of carcinogenesis. It is one of the novel approaches of controlling cancer alternative to therapy, which has some limitations and drawbacks in the treatment of patients (Stoner and Mukhtar, 1995; Khafif *et al.*, 1998; Kawamori *et al.*, 1999; Bush *et al.*, 2001; Jung *et al.*, 2005).

The chemopreventive and bioprotectant property of curcumin in turmeric increases cancer cells' sensitivity to certain drugs commonly used to combat cancer, rendering chemotherapy more effective. It also possesses strong antimicrobial and antioxidant activity and may slow down other serious brain diseases like multiple sclerosis and Alzheimer's disease (Lim *et al.*, 2001). The specific inhibition of HIV-1 integrase by curcumin suggests strategies for developing antiviral drugs based on curcumin as the lead compound for the development of inhibitors of HIV-1 integrase (Li *et al.*, 1993). The effect of polyacetylenes in celery leaves towards human cancer cells, their human bioavailability and their ability to reduce tumour formation in a mammalian *in vivo* model indicates that they may also provide benefits for health (Christensen and Brandt, 2006).

In star anise, the presence of a prenyl moiety in the phenylpropanoids plays an important role in antitumour-promoting activity. Hence, the prenylated phenylpropanoids might be valuable as a potential cancer chemopreventive agent (Padmashree *et al.*, 2007).

Table 1.6. Major chemical constituents in spices.

Spice crop (botanical name)	Compound and structure

Black pepper (*Piper nigrum* L.)
Piperine, *β*-caryophyllene, chavicine

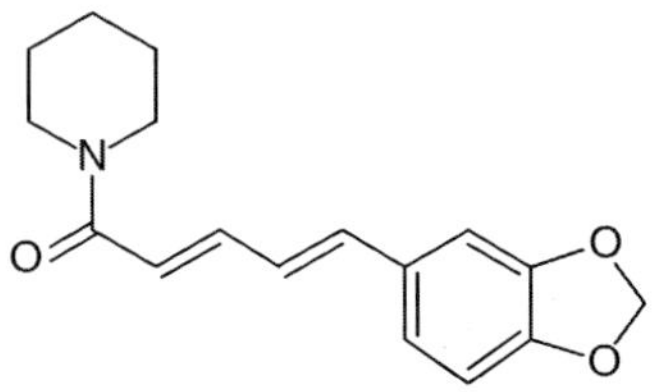
Piperine

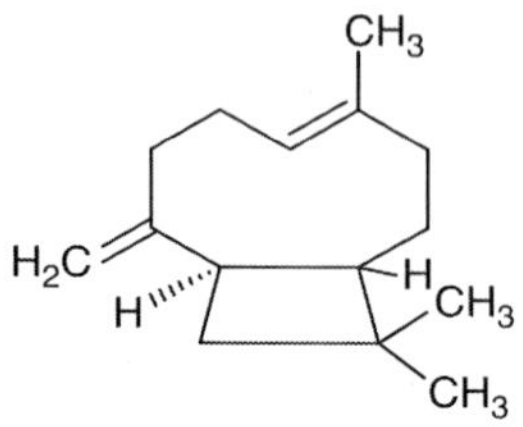
β-Caryophyllene

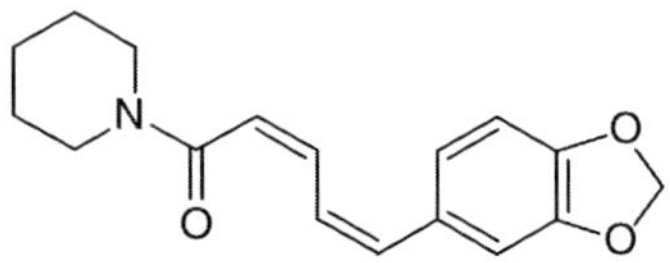
Chavicine

Small cardamom (*Elettaria cardamomum* Maton) and large cardamom (*Amomum subulatum* Roxburgh)
1,8-cineole, *α*-terpinyl acetate

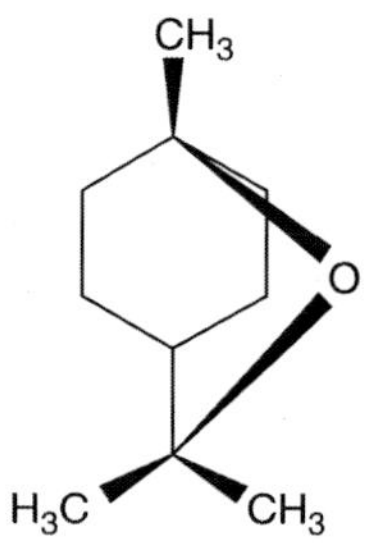
1,8-cineole

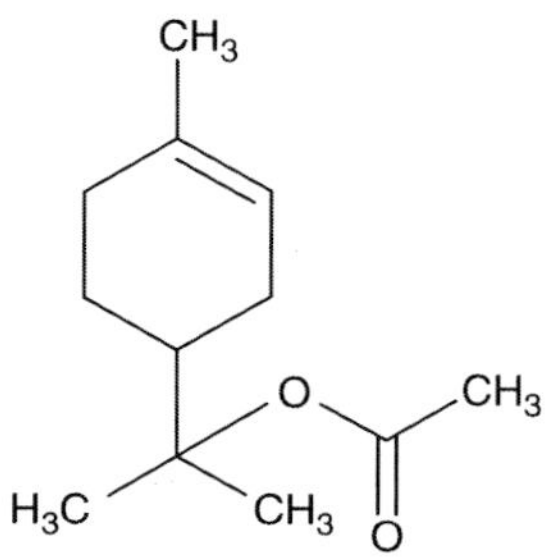
α-Terpinyl acetate

Ginger (*Zingiber officinale* Rosc.)
Gingerol, shogoal, citral, zingiberene, *ar*-curcumene

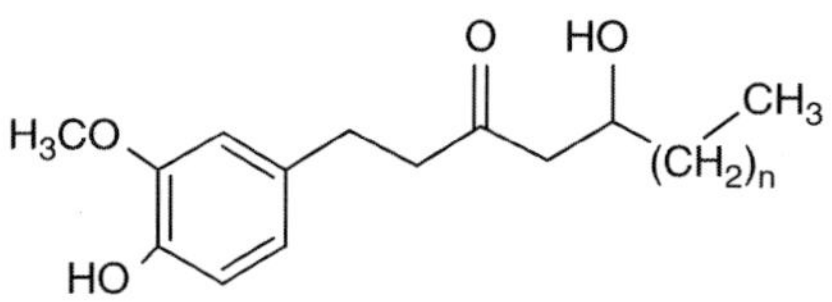
Gingerol

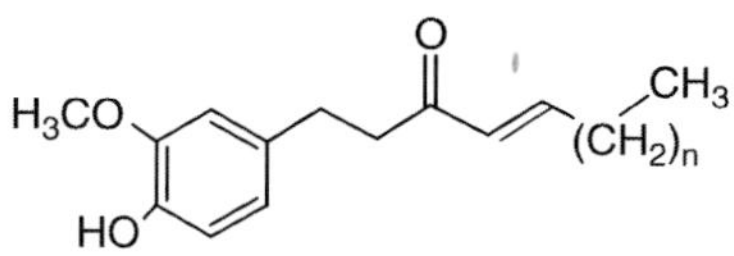
Shogaol

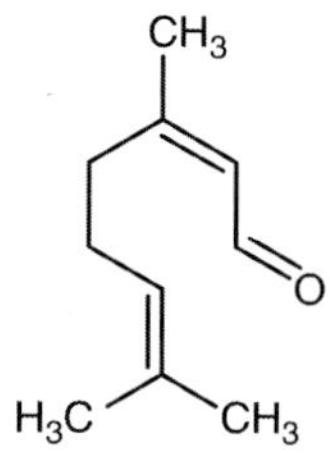
Citral

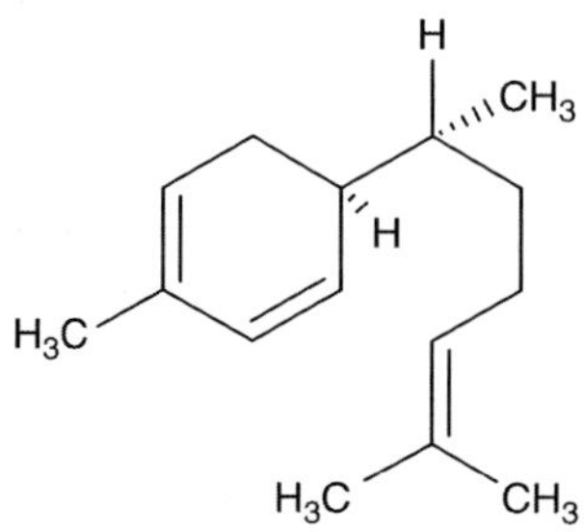
(–)Zingiberene

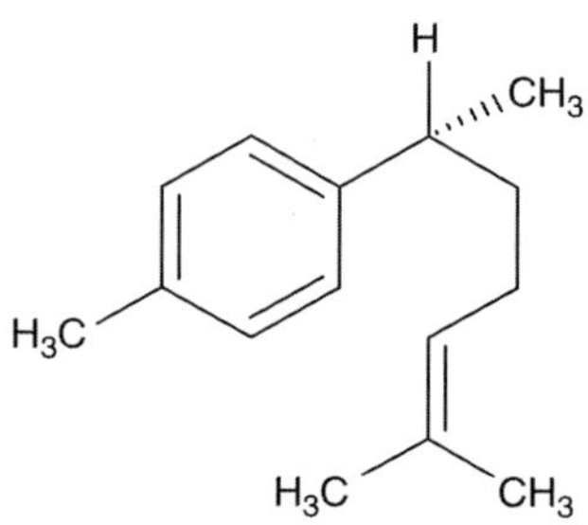
ar-Curcumene

Continued

Table 1.6. *Continued*

Spice crop (botanical name)	Compound and structure

Turmeric (*Curcuma longa* L.)
ar-Turmerone, curcumin, demethoxy curcumin, *bis*-demethoxy curcumin

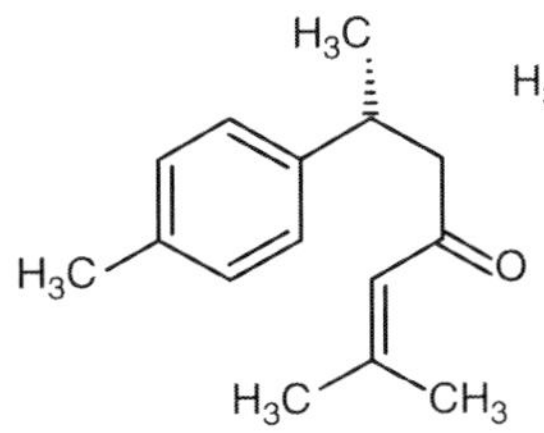
ar-Turmerone

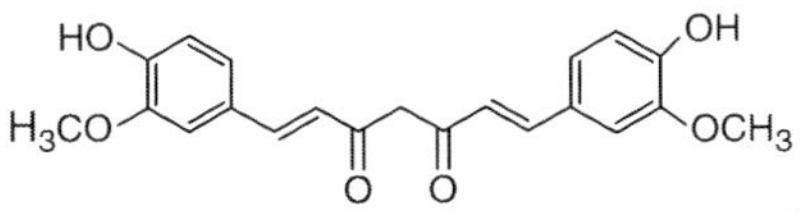
Curcumin

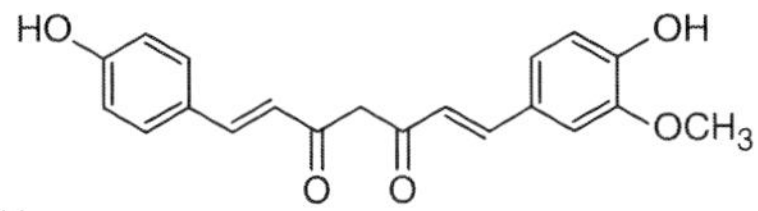
Demethoxycurcumin

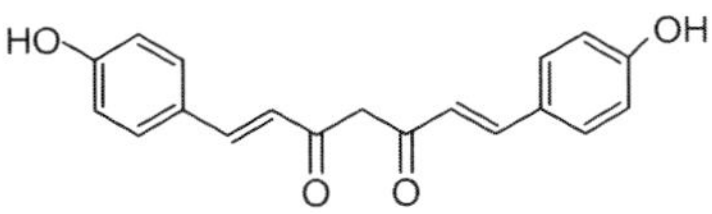
Bisdemethoxycurcumin

Cinnamon (*Cinnamomum verum* syn. *C. Zeylanicum*) and Cassia (*Cinnamomum cassia* (L.) Presl)
Eugenol, benzyl benzoate, cinnamaldehyde

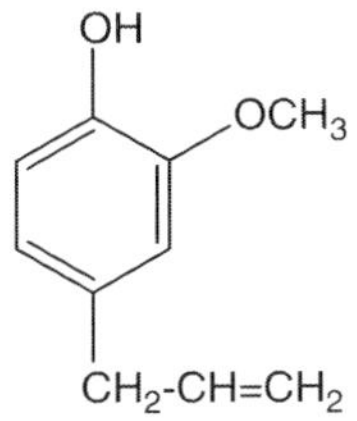
Eugenol

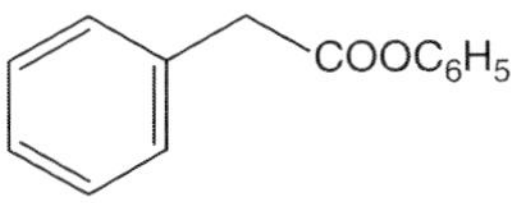
Benzyl benzoate

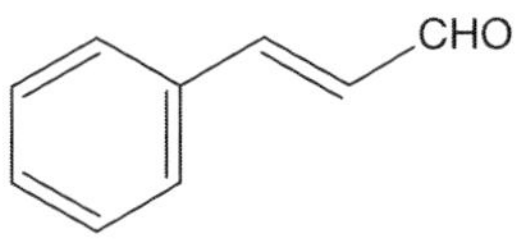
Cinnamaldehyde

Clove (*Syzygium aromaticum* (L.) Merr. et Perry)
Eugenol, eugenyl acetate

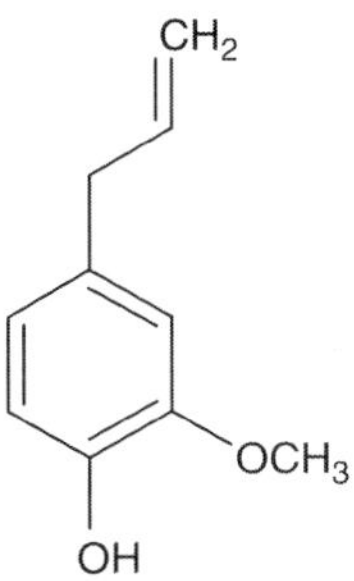
Eugenol

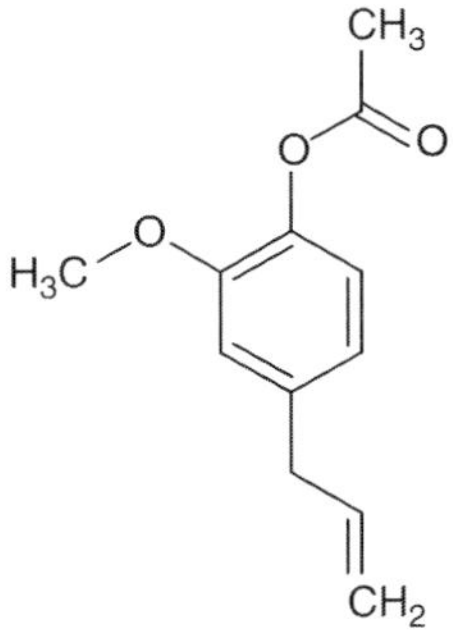
Eugenyl acetate

Continued

Table 1.6. *Continued*

Spice crop (botanical name)	Compound and structure

Nutmeg and mace (*Myristica fragrans* Houtt)
Myristicin, elemicin

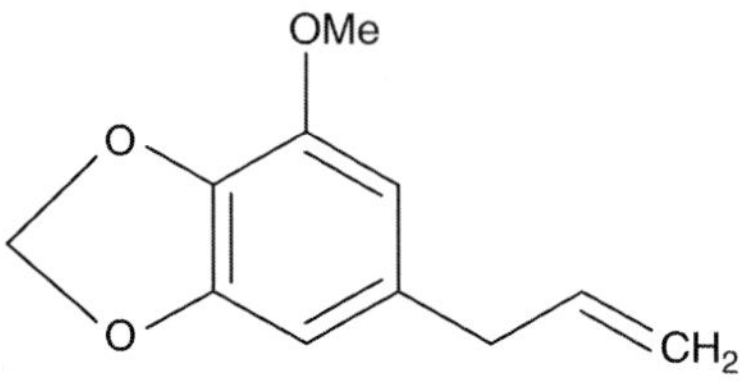

Myristicin

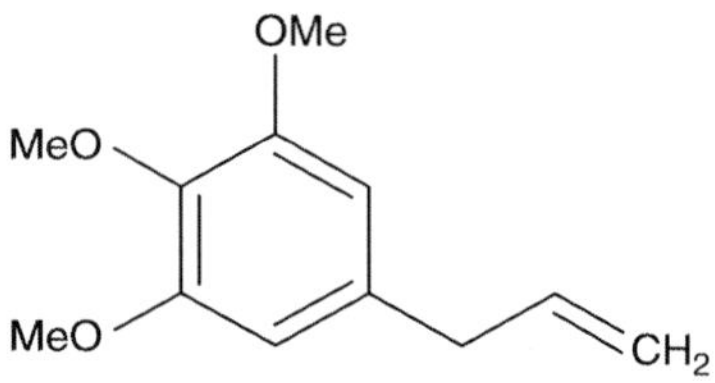

Elemicin

Coriander (*Coriandrum sativum* L.)
Linalool

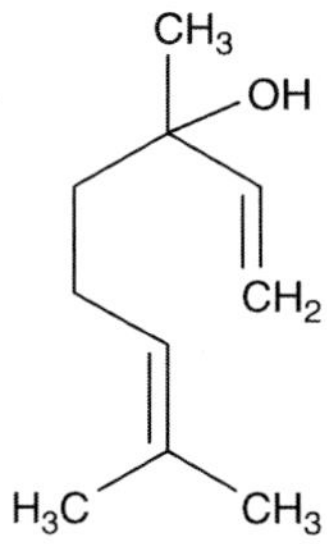

Linalool

Cumin (*Cuminum cyminum* L.)
Cuminaldehyde, β-pinene, *cis*-β-farnesene

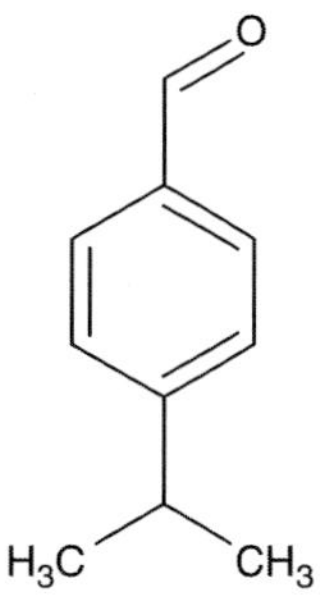

Cuminaldehyde

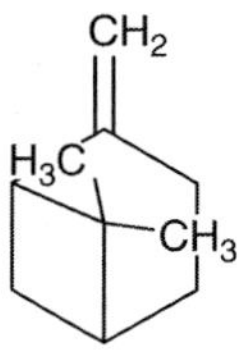

β-Pinene

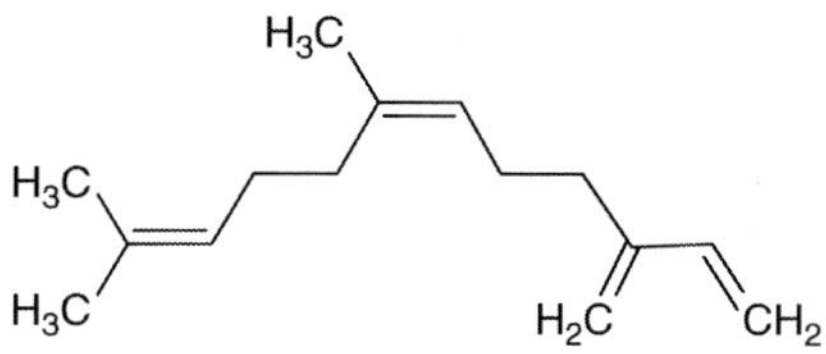

cis-β-Farnesene

Continued

Table 1.6. *Continued*

Spice crop (botanical name)	Compound and structure

Fennel (*Foeniculum vulgare* Mill.)
Anethole, estragol

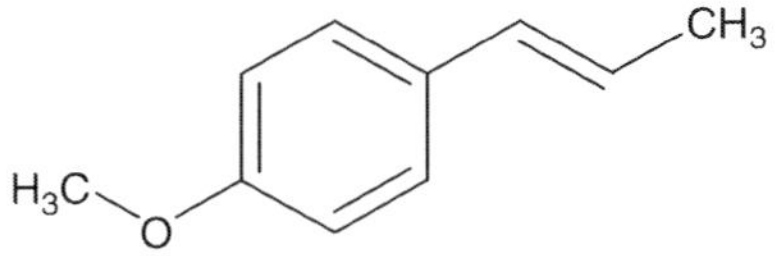

(*E*)-Anethole

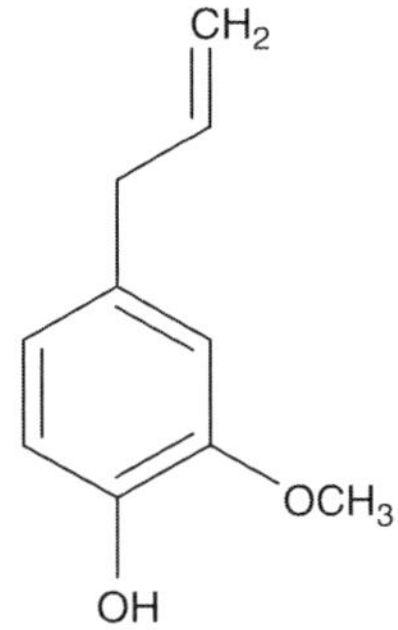

Estragol (methyl chavicol)

Fenugreek (*Trigonella foenum-graecum* L.)
Diosgenin

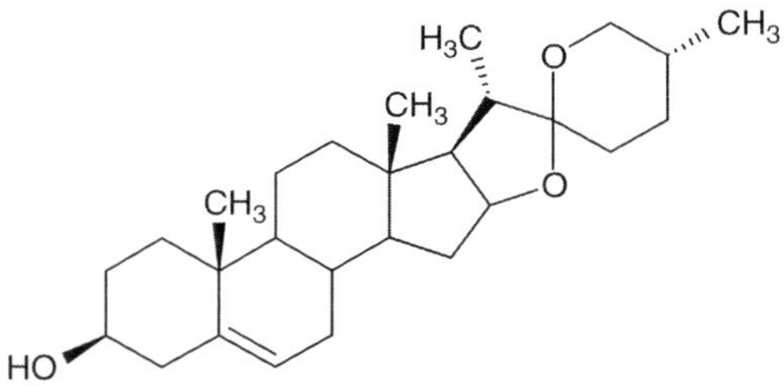

Diosgenin

Paprika (*Capsicum annum* L.)
Capsanthin, capsorubin

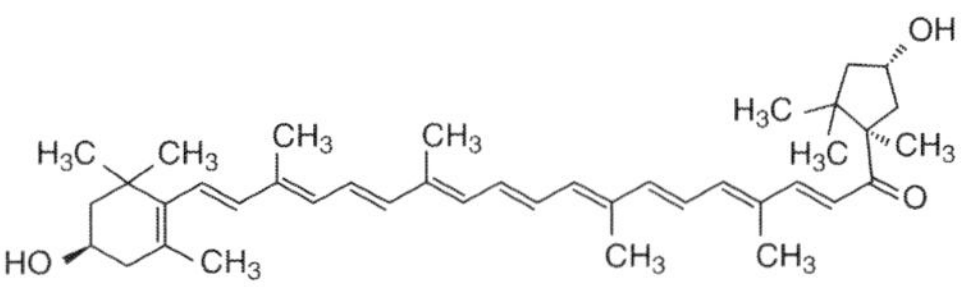

Capsanthin

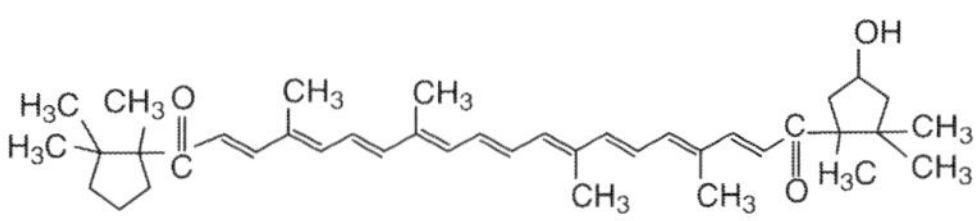

Capsorubin

Vanilla (*Vanilla planifolia* Andrews)
Vanillin

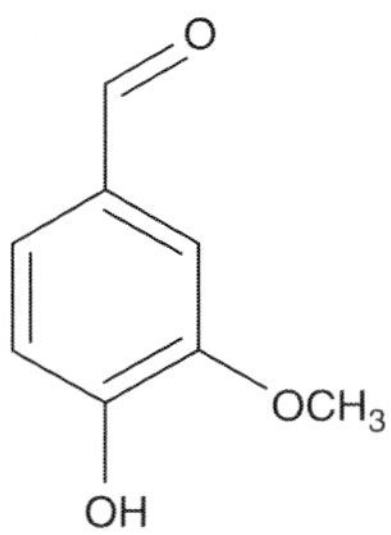

Vanillin

Continued

Table 1.6. *Continued*

Spice crop (botanical name)	Compound and structure

Ajowan (*Trachyspermum ammi* (L.) Sprague)
Thymol, γ-terpenene

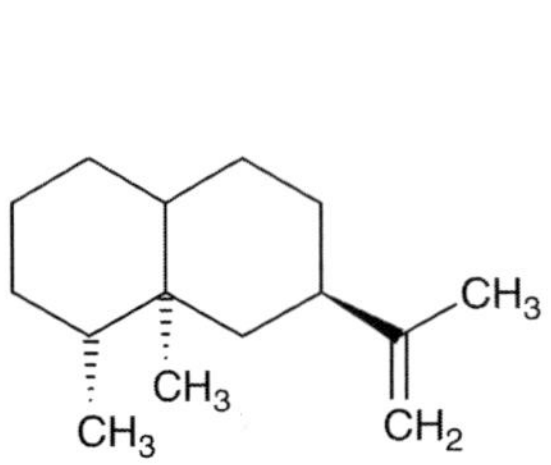

Valencene

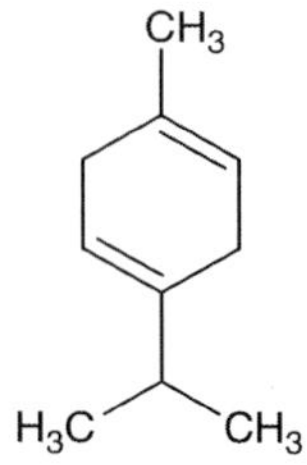

γ-Terpenene

Star anise (*Illicium verum* Hooker fil.)
(*E*)-Anethole

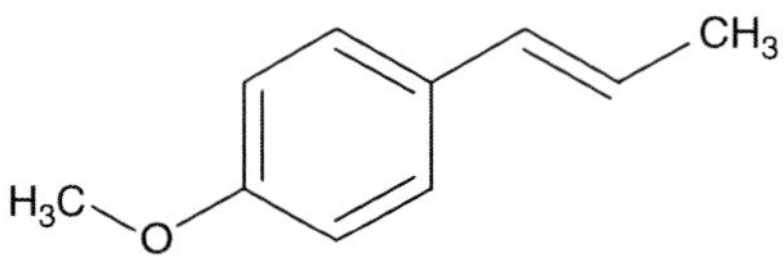

(*E*)-Anethole

Aniseed (*Pimpinella anisum* L.)
(*E*)-Anethole, anisaldehyde

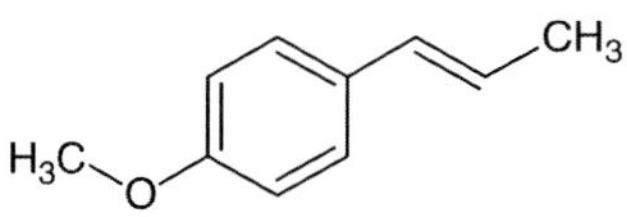

(*E*)-Anethole

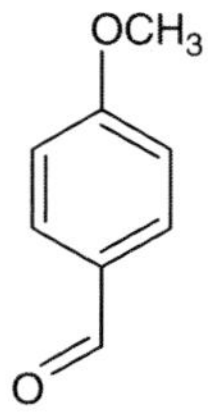

Anisaldehyde

Garcinia (*Garcinia cambogia*)
α-Humelene, valencene, β-caryophyllene

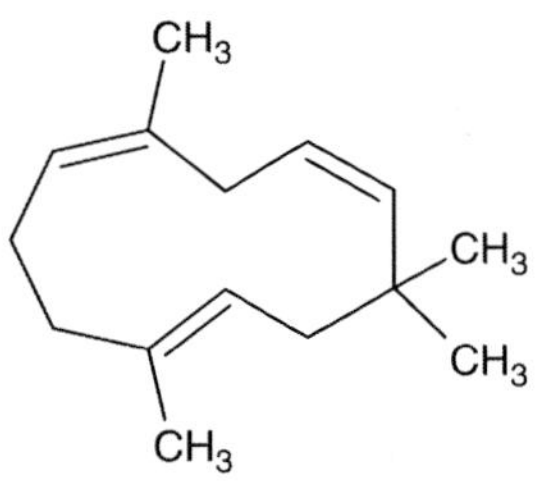

α–Humulene

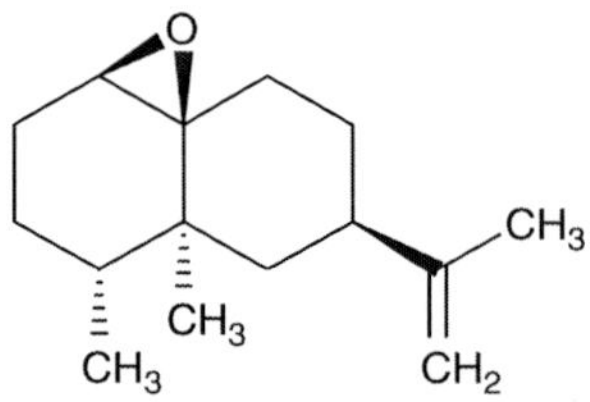

Valencene

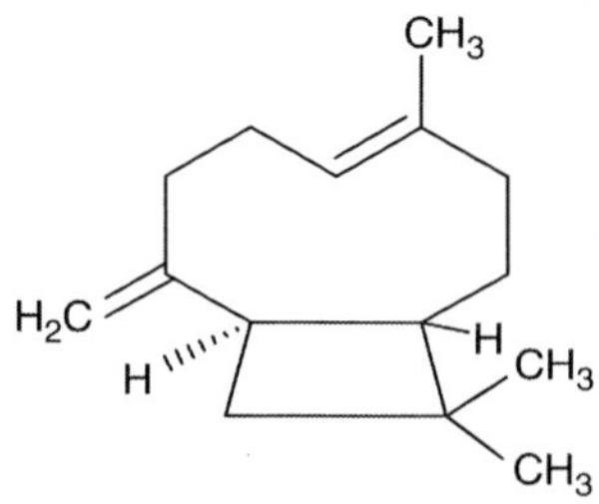

β-Caryophyllene

Continued

Table 1.6. *Continued*

Spice crop (botanical name)	Compound and structure

Tamarind (*Tamarindus indica* L.)
Furfural, 2-phenyl acetaldehyde

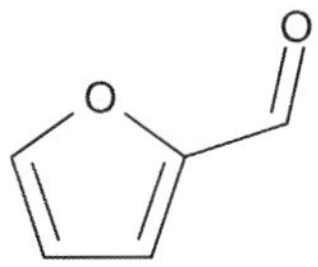

Furfural

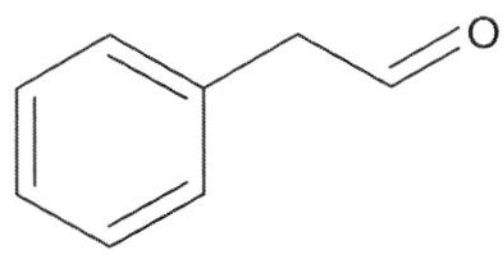

2-Phenylacetaldehyde

Parsley (*Petroselinum crispum* (Mill) Nyman ex A.W. Hill)
1,3,8-*p*-Menthatriene, *β*-phellandrene, myristicin

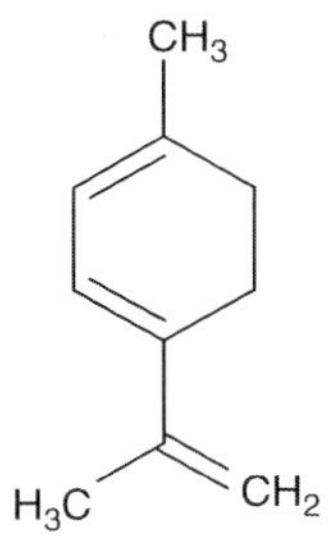

1,3,8-*p*-menthatriene

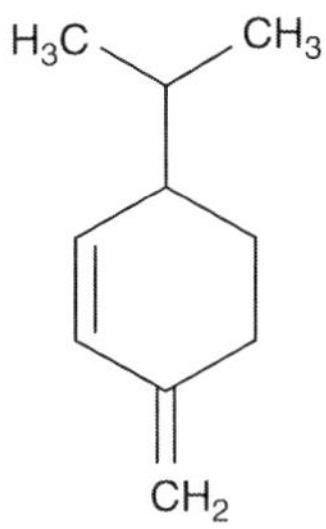

β-Phellandrene

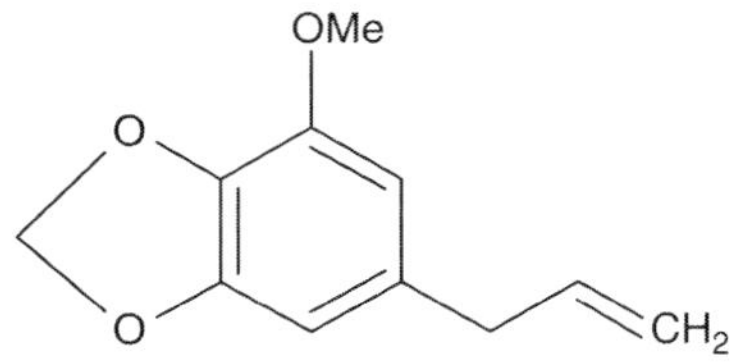

Myristicin

Celery (*Apium graveolens* L.)
Myrcene, limonene, *α*-pinene

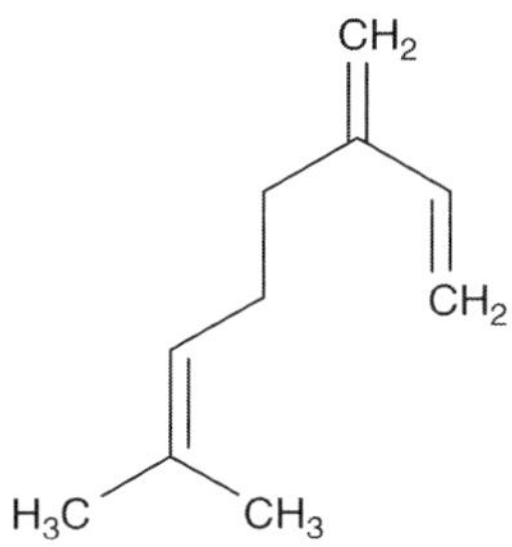

Myrcene

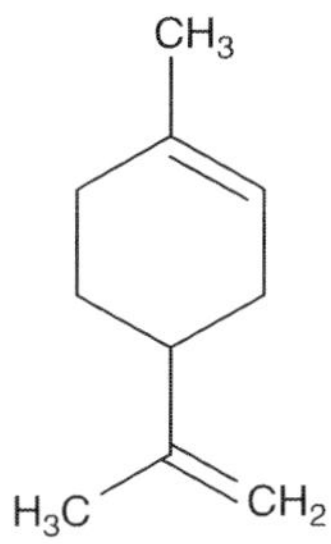

-(–) Limonene

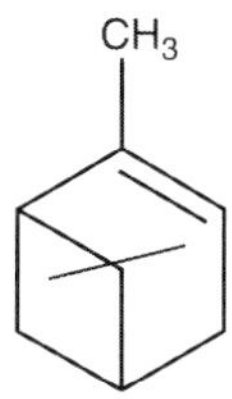

α-Pinene

Bay leaf (*Laurus nobilis* L.)
1,8-Cineole, linalool, *α*-terpinyl acetate, methyl eugenol

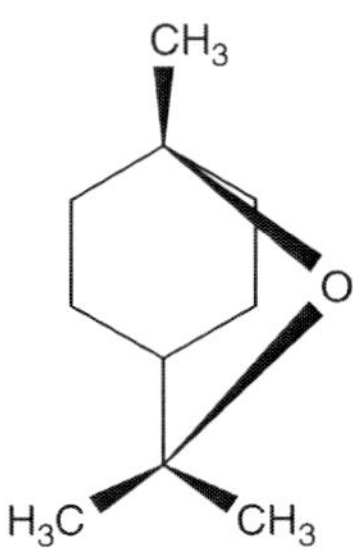

1,8-Cineole

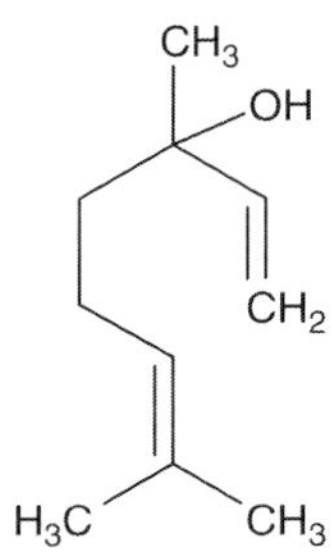

Linalool

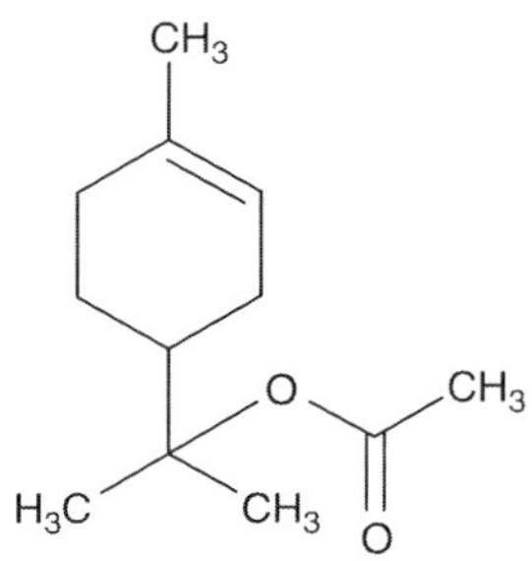

α-Terpinyl acetate

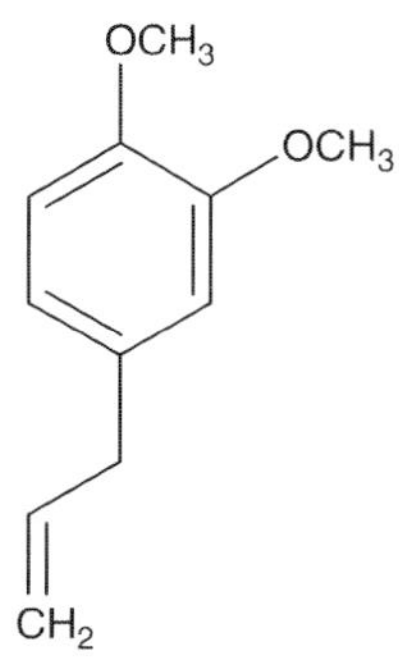

Methyl eugenol

Continued

Table 1.6. *Continued*

Spice crop (botanical name)	Compound and structure

Curry leaf (*Murraya koenigii* Spreng.)
Murrayacine, koenigine, α-pinene, β-phellandrene

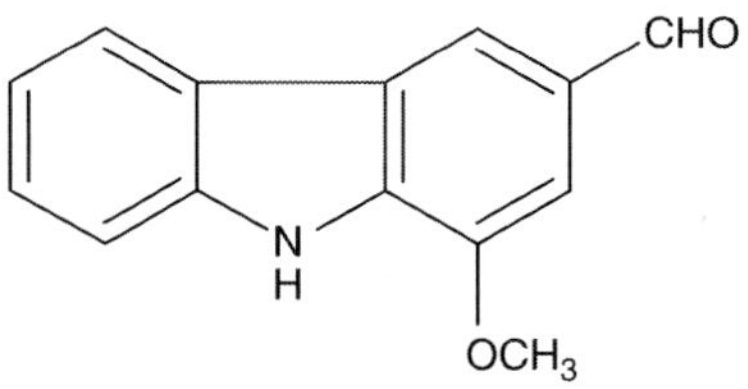

Murrayacine

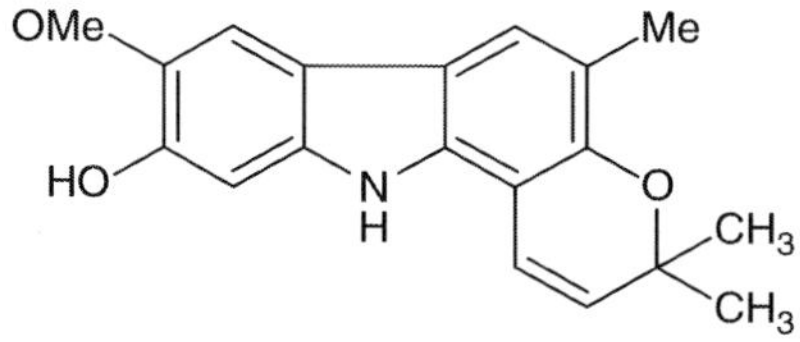

Koenigine

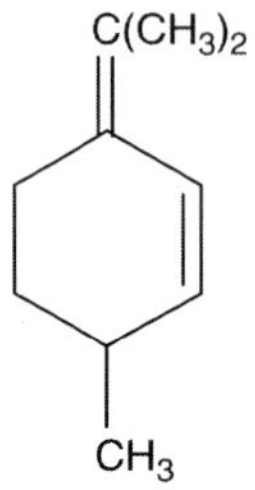

β-Phellandrene

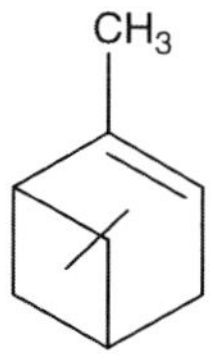

α-Pinene

Table 1.7. Value-added products from major spices.

Spices	Product
Black pepper	Dehydrated green pepper, freeze-dried green pepper, frozen green pepper, white pepper, green pepper in brine, pepper oil, pepper oleoresin, ground pepper, organic pepper, sterile pepper, canned tender green pepper
Cardamom (small)	Green cardamom, cardamom oil, cardamom oleoresin
Cardamom (large)	Oil, oleoresin
Ginger	Ginger oil, oleoresin, candy, preserves, vitaminized effervescent ginger powder, plain effervescent powder, starch from spent ginger, wine, beer, medicinal beverages, encapsulated ginger oil, dehydrated ginger
Turmeric	Curcuminoids, dehydrated turmeric powder, oil, oleoresin

Antioxidant

Tamarind is used traditionally as an astringent, anti-inflammatory and antidiuretic agent, and a laxative, carminative and digestive agent (Sudjaroen *et al.*, 2005; Siddhuraju, 2007). As for garcinia, the major flavouring compound is (–)-hydroxycitric acid, which is emerging as an antiobesity factor (Greenwood *et al.*, 1981; Rao and Sakariah, 1988; Jena *et al.*, 2002). However, more evidence needs to be compiled to prove its potential satisfactorily.

Apart from culinary uses, parsley is known for its anticancer, antioxidant, diuretic and laxative properties. Photosensitizing, toxic furocoumarines, including psoralen, bergaptene and isoimperatorin, have been found in parsley roots, which can induce dermatitis (Peterson *et al.*, 2006).

As a remedy for bird flu

Star anise is the industrial source of shikimic acid, a primary ingredient used to create the antiflu drug, Tamiflu, which is regarded as the most promising drug to mitigate the severity of the bird flu H5N1 strain of virus (Goodman, 2005). Currently, Tamiflu is the only drug available which may reduce the severity of bird flu (also known as avian flu).

As a bioenhancer

Piperine (1-piperoyl piperidine) in black pepper is shown to possess bioavailability-enhancing activity with various structurally and therapeutically diverse drugs. This property of piperine may be attributed to increased absorption, which may be due to alteration in membrane lipid dynamics and a change in the conformation of enzymes in the intestine (Khajuria *et al.*, 2002).

Antimicrobial

Clove bud oil has various biological activities, such as antibacterial, antifungal, antioxidant and insecticidal properties. The high level of eugenol present in the essential oil imparts strong biological and antimicrobial activity (Raghavenra *et al.*, 2006).

Curry leaves have been studied for their antifungal activity against three plant-pathogenic fungi, i.e. *Rhizoctonia solani*, *R. bataticola* [*Macrophomina phaseolina*] and *Helminthosporium oryzae* [*Cochliobolus miyabeanus*] (Ray and Srivastava, 2006).

Insecticidal

The volatile oil from cardamom is a potential grain protectant by killing various life stages of the stored-product insects attacking wheat, e.g. *Tetropium castaneum* and *Sitophilus zeamais Motschulsky*, via contact and fumigant action (Huang *et al.*, 2000). Cinnamaldehyde in cinnamon has strong insecticidal activity against *Acanthoscelides oblectus* and antifeedant activity against *Ceratitis capitata*, a pest causing damage to fruit crops.

Nutmeg oil also possesses strong antibacterial, antifungal and insecticidal properties. Myristicin, which imparts hallucinogenic properties, is also reported to be an effective insecticide, while the lignin types of the constituents in the nut are anticarcinogenic (Narasimhan and Dhake, 2006). Larvicidal properties, against second stage larvae of *Toxocara canis*, are also reported in mace (Nakamura *et al.*, 1988).

Curry leaves have also been proven to be effective against *Rhizopus stolonifer* [*R. stolonifer var. stolonifer*] and *Gloeosporium psidii* [*Colletotrichum coccodes*] infecting guava (Dwivedi *et al.*, 2002). Bay leaf has been used as a herbal medicine and has pharmaceutical activity which includes antibacterial, antifungal, antidiabetes and anti-inflammatory effects (Guynot *et al.*, 2003).

1.6. Conclusion

Spices produce a vast and diverse assortment of organic compounds, the great majority of which do not appear to participate directly in growth and development. These substances, traditionally referred to as secondary metabolites, assume great significance. Although noted for the complexity of chemical structures and biosynthetic pathways, the volatile and non-volatile natural products are perceived generally as biologically insignificant.

Secondary metabolites in spices have been a fertile area for chemical investigation for many years, driving the development of both analytical chemistry and of new synthetic reactions and methodologies. In recent years, there has been an emphasis on secondary metabolites in relation to dietary components, which may have a considerable impact on human health. The majority of herbs and spices constitute important bioactive secondary metabolites which possess versatile pharmacological and medicinal properties. The structure–activity relationship of these compounds is an exciting field, where molecular biology and nanotechnology can definitely play a symbiotic role.

References

Akgül, A. (1986) Studies on the essential oils from Turkish fennel seeds. In: Brunke, E.J. (ed.) *Progress in Essential Oil Research*. W. de Gruyter and Co., New York, pp. 487–489.

American Spice Traders' Association (1960) *A History of Spices*. Bernard L. Lewis, Inc., New York.

Angmor, J.E., Dicks, D.M., Evans, W.C. and Sandra, D.K. (1972) Studies on *Cinnamomum zeylanicum* Part 1. The essential oil components of *C. zeylanicum* Nees grown in Ghana. *Planta Medica* 21, 416–420.

Anon. (2007) *Arecanut and Spices Database*. Premaja, P. and Manoj Kumar, K. (eds), Directorate of Arecanut and Spices Development, India, p. 110.

Bush, J.A., Cheung, K.J.J. and Li, G. (2001) Curcumin induces apoptosis in human melanoma cells through a Fas receptor/caspase-8 pathway independent of p53. *Experimental Cell Research* 271(2), 305–314.

Christensen, L.P. and Brandt, K. (2006) Bioactive polyacetylenes in food plants of the Apiaceae family: occurrence, bioactivity and analysis. *Journal of Pharmaceutical and Biomedical Analysis* 41(3), 683–693.

Cookingsecrets.org/herbs-spices/spice-producing-areas.

Dwivedi, B.K., Pandey, R.C., Pandey, G., Pant, H.L. and Logani, R. (2002) Evaluation of angiospermic plant extracts against *Rhizopus stolonifer* and *Gloeosporium psidii* fungi of guava. *Bioved* 13(1/2), 129–134.

Ehiabhi, O.S., Edet, U.U., Walker, T.M., Schmidt, J.M., Setzer, W.N., Ogunwande, I.A., Essien, E. and Ekundayo, O. (2003) Constituents of essential oils of *Apium graveolens* L., *Allium cepa* L., and *Voacanga africana* Staph. from Nigeria. *Journal of Essential Oil Bearing Plants* 9(2), 126–132.

FAO (2007) FAO database.

Goodman, P.S. (2005) Star Rises in Fight Against Bird Flu. Demand for a Chinese Fruit Skyrockets. *Washington Post Foreign Service*, 18 November, p. D01.

Greenwood, M.R., Cleary, M.P., Gruen, R., Blasé, D., Stern, J.S., Triscari, J. and Sullivan, A.C. (1981) Effect of (–)-hyroxycitrate on development of obesity in the Zucker obese rat. *American Journal of Physiology* 240, E72–E78.

Guenther, E. (1950) *The Essential Oils*, Volume IV. Van Nostrand Publishing Co., New York, pp. 396–437.

Guynot, M.E., Ramos, A.J., Seto, L., Purroy, P., Sanchis, V. and Marin, S. (2003) Antifungal activity of volatile compounds generated by essential oils against fungi commonly causing deterioration of bakery products. *Journal of Applied Microbiology* 94(5), 893–899.

Huang, Y., Lam, S.L. and Ho, S.H. (2000) Bioactivities of essential oil from *Elletaria cardamomum* (L.) Maton to Sitophilus zeamais Motschulsky and Tribolium castaneum (Herbst). *Journal of Stored Products Research* 36, 107–117.

International Trade Centre UNCTAD/WTO (2006) *World Markets in the Spice Trade – 2000–2004*. ITC, Geneva, vi + 111 p.

Jena, B.S., Jayaprakasha, G.K., Singh, R.P. and Sakariah, K.K. (2002) Chemistry and biochemistry of (–)-hydroxycitric acid from *garcinia*. *Journal of Agricultural and Food Chemistry* 50, 10–22.

Jung, E.M., Lim, J.H., Lee, T.J., Park, J., Choi, K.S. and Kwon, T.K. (2005) Curcumin sensitizes tumor necrosis factor-related apoptosis-inducing ligand (TRAIL)-induced apoptosis through reactive oxygen species-mediated up-regulation of death receptor 5 (DR5). *Cancer Biology* 26(11), 1905–1913.

Kawamori, T., Lubet, R., Vernon, E.S., Kelloff, G.J., Kaskey, R.B., Rao, V.R. and Reddy, B.S. (1999) Chemopreventive effect of curcumin, a naturally occurring anti-inflammatory agent, during the promotion/progression stages of colon cancer. *Cancer Research* 59, 597–601.

Khafif, A., Schantz, S.P., Chou, T.C., Edelstein, D. and Sacks, P.G. (1998) Quantitation of chemopreventive synergism between (–)-epigallocatechin-3-gallate and curcumin in normal, premalignant and malignant human oral epithelial cells. *Carcinogenesis* 19, 419–424.

Khajuria, A., Thusu, N. and Zutshi, U. (2002) Piperine modulates permeability characteristics of intestine by inducing alterations in membrane dynamics: influence on brush border membrane fluidity, ultrastructure and enzyme kinetics. *Phytomedicine* 9(3), 224–231.

Kilic, A., Hafizoglu, H., Kollmannsberger, H. and Nitz, S. (2004) Volatile constituents and key odorants in leaves, buds, flowers, and fruits of *Laurus nobilis* L. *Journal of Agricultural and Food Chemistry* 52(6), 1601–1606.

Korikanthimath, V.S., Ravindra, M. and Zachariah, T.J. (1997) Variation in yield and quality characters of cardamom clones. *Journal of Medicinal and Aromatic Plant Sciences* 19(4), 1024–1027.

Krishnamoorthy, B., Rema, J., Zachariah, T.J., Jose, A. and Gopalam, A. (1996) Navasree and nithyasree – two high yielding and high quality cinnamon (*Cinnamomum verum* – Bercht & Presl.). *Journal of Spices and Aromatic Crops* 5(1), 28–33.

Kruger, H. and Hammer, K. (1999) Chemotypes of fennel. *Journal of Essential Oil Research* 11(1), 79–82.

Lewis, Y.S., Nambudiri, E.S. and Phillip, T. (1966) Composition of cardamom oils. *Perfumery and Essential Oil Records* 57, 623.

Lewis, Y.S., Nambudiri, E.S. and Krishnamurthy, N. (1969) Composition of pepper oil. *Perfumery and Essential Oil Records* 60, 259–262.

Li, C.J., Zhang, L.J., Dezube, B.J. and Crumpacker, C.S. (1993) Three inhibitors of human type 1 immunodeficiency virus long terminal repeat directed gene expression and virus replication. *Proceedings of the National Academy of Science (USA)* 90, 1839.

Lim, G.P., Chu, T., Yang, F., Beech, W., Frantschy, S.A. and Cole, G.M. (2001) The curry spice curcumin reduces oxidative damage of amyloid pathology in an Alzheimer transgenic mouse. *Journal of Neurosciences* 21, 8370–8377.

MacLeod, A.J., Snyder, C.H. and Subramanian, G. (1985) Volatile aroma constituents of parsley leaves. *Phytochemistry* 24(11), 2623–2627.

MacLeod, A.J., MacLeod, G. and Subramanian, G. (1988) Volatile aroma constituents of celery. *Phytochemistry* 27(2), 373–375.

Mallavarapu, G.R., Ramesh, S., Syamasundar, K.V. and Chandrasekhara, R.S. (1999) Composition of Indian curry leaf oil. *Journal of Essential Oil Research* 11, 176–178.

Mazza, G., Tommaso, D. di and Foti, S. (2002) Volatile constituents of Sicilian fenugreek (*Trigonella foenum-graecum*) seeds. *Sciences des Aliments* 22(3), 249–264.

Menon, A.N. (2000) The aromatic compounds of pepper. *Journal of Medicinal and Aromatic Plant Sciences* 22(2/3), 185–190.

Nakamura, N., Kiuchi, F., Tsuda, Y. and Kondo, K. (1988) Studies on crude drugs effective on visceral larva migrans. V. The larvicidal principle in mace (aril of *Myristica fragrans*). *Chemical and Pharmacological Bulletin* 36(7), 2685.

Narasimhan, B. and Dhake, A.S. (2006) Antibacterial principles from *Myristica fragrans* seeds. *Journal of Medicinal Food* 9(3), 395–399.

Onayade, O.A. and Adebajo, A.C. (2000) Composition of the leaf volatile oil of *Murraya koenigii* growing in Nigeria. *Journal of Herbs, Spices and Medicinal Plants* 7(4), 59–66.

Padmashree, A., Roopa, N., Semwal, A.D., Sharma, G.K., Agathian, G. and Bawa, A.S. (2007) Star-anise (*Illicium verum*) and black caraway (*Carum nigrum*) as natural antioxidants. *Food Chemistry* 104(1), 59–66.

Peterson, S., Lampe, J.W., Bammler, T.K., Gross-Steinmeyer, K. and Eaton, D.L. (2006) Apiaceous vegetable constituents inhibit human cytochrome P-450 1A2 (hCYP1A2) activity and hCYP1A2-mediated mutagenicity of a.atoxin B1. *Food and Chemical Toxicology* 44, 1474–1484.

Philip, J. (1981) Curry leaf and its uses. *World Crops* 33, 125–127.

Purseglove, J.W., Brown, E.G., Green, C.L. and Robbins, S.R.J. (1981) *Spices*, Volume 2. Longman, New York, pp. 447–531.

Raghavenra, H., Diwakr, B.T., Lokesh, B.R. and Naidu, K.A. (2006) Eugenol the active principle from cloves inhibits 5-lipoxygenase activity and leukotriene-C4 in human PMNL cells. *Prostaglandins, Leukotrienes and Essential Fatty Acids* 74, 23–27.

Rao, R.N. and Sakariah, K.K. (1988) Lipid-lowering and antiobesity effect of (–)-hydroxycitric acid. *Nutrition Research* 8, 209–212.

Ray, D.P. and Srivastava, S. (2006) Curry leaf (*Murraya koenigii*): the aromatic biopesticide. *Journal of Interacademicia* 10, 231–235.

Rosengarten, F. Jr. (1969) *The Book of Spices*. Jove Publications, Inc., New York, pp. 23–96.

Salzer, U.J. (1975) Analytical evaluation of seasoning extracts (oleoresins) and essential oils from seasonings. I. *International Flavours and Food Additives* 6(3), 151–157.

Senanayake, U.M. (1997) The nature, description and biosynthesis of volatile oils of *Cinnamomum* spp. PhD thesis, University of New South Wales, Kensington, Australia.

Siddhuraju, P. (2007) Antioxidant activity of polyphenolic compounds extracted from defatted raw and dry heated *Tamarindus indica* seed coat. *Food Science and Technology* 40(6), 982–990.

Stoner, G.D. and Mukhtar, H. (1995) Polyphenols as cancer chemopreventive agents. *Cell Biochemical Supplement* 22, 169–180.

Sudjaroen, Y., Haubner, R., Würtele, G., Hull, W.E., Erben, G., Spiegelhalder, B., Changbumrung, S., Bartsch, H. and Owen, R.W. (2005) Isolation and structure elucidation of phenolic antioxidants from Tamarind (*Tamarindus indica* L.) seeds and pericarp. *Food and Chemical Toxicology* 43(11), 1673–1682.

Walter, R.H. (1972) *β*-Caryophyllene in native clove bud oil. *Phytochemistry* 211(1), 405–406.

Weiss, E.A. (2002) Umbelliferae. In: Weiss, E.A. (ed.) *Spice Crops*. CAB International, Wallingford, UK, pp. 261–268.

Wijesekera, R.O.B. (1978) The chemistry and technology of cinnamon. *CRC Critical Reviews in Food Science and Nutrition* 10, 1–30.

Wohlmuth, H., Smith, M.K., Brooks, L.O., Myers, S.P. and Leach, D.N. (2006) Essential oil composition of diploid and tetraploid clones of ginger (*Zingiber officinale* Roscoe) grown in Australia. *Journal of Agricultural and Food Chemistry* 54(4), 1414–1419.

Zachariah, T.J. (1995) Essential oil and its major constituents in selected black pepper accessions. *Plant Physiology and Biochemistry New Delhi* 22(2), 151–153.

2 Black Pepper

T. John Zachariah and V.A. Parthasarathy

2.1. Introduction

Black pepper (*Piper nigrum*) belongs to the family Piperaceae. It is cultivated for its fruit, which is usually dried and used as a spice and seasoning. The same fruit is also used to produce white pepper and green pepper. Black pepper is native to South India, where it is cultivated extensively, and also to some other tropical regions. The fruit, known as peppercorn when dried, is a small drupe, 5 mm in diameter, dark red when fully mature, containing a single seed.

Dried, ground pepper, and its variants, is one of the most common spices in European cuisine, having been known and prized since antiquity for both its flavour and its use as a medicine. The spiciness of black pepper is due to the chemical, piperine. Ground black peppercorn, usually referred to simply as 'pepper', may be found on nearly every dinner table in some parts of the world, often alongside table salt. Black pepper, also nicknamed as 'black gold' and the 'king of spices', is the most important and widely consumed spice in the world. Compared with many other spices, properly dried black pepper (~ moisture content 8–10%) can be stored in airtight containers for many years without losing its taste and aroma.

The word 'pepper' is derived from the Sanskrit *pippali*, via the Latin *piper* and Old English *pipor*. The Latin word is also the source of German *pfeffer*, French *poivre*, Dutch *peper* and other similar forms. 'Pepper' was used in a figurative sense, meaning 'spirit' or 'energy', at least as far back as the 1840s.

The average total export from the different producing countries is about 138,000 t. India, Indonesia, Malaysia, Sri Lanka, Vietnam and Brazil are some of the major producing countries (Peter, 2000).

Peppercorns are, by monetary value, the most widely traded spice in the world, accounting for about 20% of all spice imports. The price of pepper can be volatile and this figure fluctuates a great deal year by year. The International Pepper Exchange is located in Kochi, India.

Recently, Vietnam has become the world's largest producer and exporter of pepper (116,000 t in 2006). Other major producers include Indonesia (67,000 t), India (65,000 t), Brazil (35,000 t), Malaysia (22,000 t), Sri Lanka (12,750 t), Thailand and China. Vietnam dominates the export market, exporting almost the entire produce.

The International Trade Centre (ITC), Geneva, put the latest trade in spices at 400,000–450,000 t, valued at US$1.5–2.5 billion annually. Black pepper accounts for about 35% of the world trade in spices. Table 2.1 gives an approximate estimate of world production and export of black pepper (Ravindran, 2000).

(eds V.A. Parthasarathy, B. Chempakam and T.J. Zachariah)

Table 2.1. World production and export of black pepper (t).

Country	Production	Export
Brazil	25,400	23,300
India	58,600	31,800
Indonesia	47,700	41,600
Malaysia	21,500	21,600
Sri Lanka	4,400	3,700
Thailand	8,600	2,300
IPC countries	166,200	124,300
Other countries	30,600	19,200
World	196,800	143,500

Source: International Pepper Community.

2.2. Botany and Uses

Botany

The black pepper of commerce is the matured dried fruits (berries) of the tropical, perennial plant *P. nigrum* L. of the *Piperaceae* family. The common black pepper is found extensively in the evergreen forests of Western Ghats and adjoining areas, almost from sea level up to an elevation of 1300 m. It is a perennial climber, climbing by means of ivy-like roots which adhere to the support tree.

The sessile, small white flowers are borne in pendulous, dense, slender spikes of about 50 blossoms each. The berry-like fruits, or peppercorns, are round, about 0.5–1.0 cm in diameter and contain a single seed. They become yellowish-red at maturity and bear a single seed. Spike length varies greatly, based on cultivar. The young berries are green, whitish green or light purple, while mature ones are green, pale purple or pale yellow and change to red on ripening. Wild forms are usually dioecious, while cultivated ones are bisexual (Ahlert *et al.*, 1998).

The cultivars of black pepper may have originated from the wild varieties through domestication and selection. Over one hundred cultivars are known, but many of them are becoming extinct due to various reasons, such as the devastation of pepper cultivation by diseases like foot rot and slow decline, replacement of traditional cultivars by a few high-yielding varieties, etc. Cultivar diversity is richest in the Indian state of Kerala, followed by the state of Karnataka. Most of the cultivars are bisexual forms, unlike their wild counterparts. Some of the popular Indian black pepper cultivars and their features are illustrated in Table 2.2. Once, there were specific cultivars of black pepper identified with major growing tracts. However, during the turn of the present century, extensive plantations of tea, cardamom and coffee were established in the hilly tracts of Western Ghats and there was a lot of human migration from the plains to these hills, mainly in search of land and work, bringing with it, among other things, pepper cultivars. Such human activities influenced the selective spread of certain high-yielding cultivars and they became very popular in all pepper-growing tracts (Ravindran *et al.*, 2000a).

Black pepper has multiple uses in the processed food industry, in kitchens, in perfumery, in traditional medicine and even in beauty care. Pepper is valued for its pungency and flavour, which is attributed by the alkaloid piperine and the volatile oil (Ravindran *et al.*, 2000b).

Uses

Black pepper oil can be used to help in the treatment of pain relief, rheumatism, chills, flu, colds, exhaustion, muscular aches, physical and emotional coldness, fevers, as a nerve tonic and to increase circulation. Furthermore, it increases the flow of saliva, stimulates appetite, encourages peristalsis, tones the colon muscles and is a general digestive tonic (Pruthi, 1993).

Products from pepper

Black pepper, matured dehydrated green pepper and tender green pepper are processed for various end products. The various

Table 2.2. Black pepper cultivars popular in India.

Sl No.	Cultivar	Features
1	Aimpiriyan	Performs well in plains and hilly regions, not suitable for heavy shaded areas, late maturing and high yielder.
2	Arakkulam Munda	Early variety, medium yield and quality.
3	Arimulaku	Small ovate leaves, lobes unequal, small fruits and spikes, early maturity, poor yield, medium quality.
4	Balankotta	Tolerant to shade, performs well as mixed crop in arecanut gardens, large leaves, medium yield, bold fruit, medium quality.
5	Cheppukulamundi	Ovate, cordate leaves, medium long spikes, setting moderate, medium yield and quality.
6	Cheriyakaniakkadan	Small lanceolate leaves, tips acuminate, spikes short, fruits small, early maturity, poor yielder, quality medium.
7	Cholamundi	Small, lanceolate leaves, spikes medium, setting often poor, fruits small, medium quality, predominantly female.
8	Chumala	Medium ovate leaves, spikes short to medium, good fruit set, fruits medium, medium yield and quality.
9	Doddigae	A cultivar grown in Karnataka state, leaves ovate, poor yielder.
10	Jeerakamundi	Small, lanceolate leaves, spikes small, setting poor, spiking intensity high.
11	Kalluvally	A hardy cultivar, minutely hairy in nature, medium yield and quality.
12	Karimkotta	A common hardy cultivar of Malabar, poor yielder.
13	Karimunda	Originally from south Kerala, now very popular throughout Kerala, tolerant to shade, performs well as a mixed crop, widely adaptable, good yielder, medium quality.
14	Karimundi	Medium-long ovate leaves, spikes medium, setting moderate, yield medium.
15	Karivilanchi	Medium ovate leaves, predominantly female, fruit bold, oblong, medium quality, poor yielder.
16	Kottanadan	Performs well in plains and hilly regions up to 700–800 m MSL, widely adapted and high yielding, high quality.
17	Kurimalai	Performs well as intercrop in coconut and arecanut gardens, not suitable for plains, good yielder, medium quality.
18	Kuriyalmundi	Elliptic to lanceolate leaves, good spiking, spikes very short, 5–6 cm, curved or twisted, fruits very small, setting good, poor yielder.
19	Kuthiravally	A stable yielder, long spikes, good setting, high quality.
20	Malamundi	Leaves ovate with round base, spikes medium long, peduncle small, flowers bisexual and female almost in equal proportion, fruits medium, good setting.
21	Malligesara	One of the most popular cultivars in the Karnataka, especially in the Malanad (hilly tracts) of Uttara Kannada and Shimoga districts, good yielder, two types of Malligesara are commonly recognized – Karimalligesara and Bilimalligesara – moderate yielder, medium quality.
22	Mundi	Leaves ovate, spikes short to medium, fruit set moderate, fruits medium, quality medium.
23	Narayakkodi	Common in all pepper-growing tracts, said to be field tolerant to *Phytophthora* foot rot, medium yield and quality.
24	Nedumchola	Leaves are smallest among the cultivars, ovate to obovate, base round, spikes very short, 4–6 cm, berries very small, slightly obovate, poor yielder and characteristically small-statured vine.
25	Neelamundi	Reported to be field-tolerant to foot rot, suitable for high-elevation areas, moderate yielder, medium quality.
26	Perambramunda	Resembles Neelamundi, berries bold, medium-long spikes, medium yield and quality.

Continued

Table 2.2. *Continued*

Sl No.	Cultivar	Features
27	Perumkodi	Leaves ovate to ovate-elliptic, spikes medium, setting poor, fruits bold, quality medium and alternate bearer.
28	Poonjaran munda	Leaves broadly ovate, base cordate, long spikes, moderate yielder and alternate bearer.
29	Thulamundi	Leaves ovate, base round, spikes medium in length, flowers (male, female and bisexual-mixed), alternate bearer, poor yield, quality medium.
30	Thevanmundi	Leaves moderately large, ovate, spikes medium, setting good, good spiking, berries medium oblong, good yield, quality medium.
31	Thommankodi	A vigorous cultivar, leaves ovate to widely ovate in the main stem, medium large in lateral, spikes long (13–14 cm) setting good, fruits medium, globose, closely resembles *Kuthiravally*, good yielder and quality.
32	Uddaghere	A popular and high-yielding cultivar from the Uttara Kannada and Shimoga districts of Karnataka, good yield, moderate quality.
33	Uthirankotta	Predominantly female, poor yield.
34	Vadakkan	A natural triploid, vigorous vine, leaves ovate to ovate elliptic, long petiole, spikes medium, setting poor, fruit very bold, medium quality, spikes light purplish.
35	Valiakaniakkadan	Spikes medium to long, berries bold, medium yielder, alternate bearer.
36	Vattamundi	Vigourous vine, leaves medium, widely ovate, spikes medium, setting moderate, berries bold, round, medium yield and quality.
37	Vellanamban	Tolerant to drought, medium yield and quality.
38	Velliyaranmunda	Leaves large, ovate, base often oblique, or round, interveinal region raised dorsally, spikes medium long, fruits medium, round, medium yield and quality.

products prepared are as follows (Dhas and Korikanthimath, 2003):

1. Green pepper-based products.
2. Black pepper-based products.
3. White pepper-based products.
4. Miscellaneous products.

GREEN PEPPER-BASED PRODUCTS Canned green pepper; green pepper in brine; bulk-packaged green pepper in brine; cured green pepper; frozen green pepper; freeze-dried green pepper; dehydrated green pepper; green pepper pickle; mixed green pepper pickle; green pepper sauce; and green pepper-flavoured products.

BLACK PEPPER-BASED PRODUCTS Whole black pepper; sterilized black pepper; ground black pepper; cryoground black pepper powder; pepper oil; oleoresin; microencapsulated spice flavour.

WHITE PEPPER-BASED PRODUCTS White pepper whole; white pepper powder.

MISCELLANEOUS PRODUCTS Curry powder-spice blends; pepper-flavoured products; pepper extract; preservative; pepper oil; pepper oleoresin; lemon pepper; garlic pepper; sauces; paste; etc.

PEPPER BY-PRODUCTS Light pepper; pepper hulls; pepper pinheads.

PEPPER-FLAVOURED PRODUCTS Pepper mayonnaise; pepper tofu; pepper cookies; candy and perfume (Dhas and Korikanthimath, 2003).

2.3. General Composition

There are two main components of black and white pepper: the volatile oil and pungent compounds. The volatile oil level in black pepper is usually higher than in white pepper. The hull of pepper contains fibre and some essential oil. Black pepper contains about 2.0–2.6% volatile oil and about 6–13% oleoresin. The nutritional composition of black pepper is given in Table 2.3.

The pungency of black pepper (*P. nigrum* L.) was attributed initially to the presence of piperine only, the structure of which is *trans,trans*-5-(3,4-methylenedi-oxyphenyl)-2,4-pentadienoic acid piperidide. Further investigations into the pungency of this spice by several workers led to the discovery that materials other than piperine also contributed to its pungency (Traxler, 1971).

Chun *et al.* (2002) found that 88% of the polysaccharide of black pepper berries was glucose, followed by galactose, arabinose, galacturonic acid and rhamnose in smaller proportions. Zachariah *et al.* (2005) evaluated major black pepper cultivars for oil, oleoresin and piperine, and the details are given in Table 2.4. The accumulation of these constituents tends to vary during maturation. Purseglove *et al.* (1981) demonstrated the variation in major constituents during maturation in two black pepper cultivars (Table 2.5).

Table 2.3. Nutritional composition of black pepper per 100 g.

Composition	ASTA requirement
Water (g)	8.0
Food energy (Kcal)	400.0
Protein (g)	10.0
Fat (g)	10.2
Carbohydrates (g)	66.5
Ash (g)	4.6
Calcium (g)	0.4
Phosphorus (mg)	160.0
Sodium (mg)	10.0
Potassium (mg)	1200.0
Iron (mg)	17.0
Thiamine (mg)	0.07
Riboflavin (mg)	0.210
Niacin (mg)	0.8
Ascorbic acid (mg)	ND
Vitamin A activity	19.0

Note: ASTA = American Spice Trade Association, ND = not detected.
Source: Tainter and Grenis (1993).

Table 2.4. Levels of oil, oleoresin and piperine content of common black pepper cultivars.

Cultivar	Oil (%)	Oleoresin (%)	Piperine (%)
Panniyur-1	3.2	8.9	2.8
Panniyur-2	4.4	12.7	3.6
Panniyur-3	4.0	11.2	3.6
Panniyur-4	2.8	9.7	3.5
Panniyur-5	2.4	8.2	3.6
Pournami	3.6	10.9	3.6
Panchami	3.2	8.7	3.0
Sreekara	3.2	9.8	3.8
Poonjaranmunda	2.8	10.2	3.5
Kuthiravally	3.2	9.5	3.0
Balankotta	2.4	8.5	3.0

Source: Zachariah *et al.* (2005).

Blackening of pepper

Apart from the major quality attributes such as pungency and aroma, the appearance with respect to colour (brown/black) is of importance for the use of black pepper as a spice in the whole or ground form. Since phenols are known to contribute to browning/blackening of finished peppercorns, the nature and distribution of phenolic compounds are very important. Blackening of fresh green pepper is due to enzymatic oxidation of (3,4-dihydroxy phenyl) ethanol glycoside by an *o*-diphenol oxidase (PPO) present in the fresh fruit. Bandyopadhyay *et al.* (1990) reported that conversion of green pepper to black pepper by the drying process was accompanied by a 75% decrease in total phenolic content and a complete loss of *o*-diphenol oxidase oxidizable phenolic fraction, which suggested a major role for enzymatic phenolic oxidation during pepper blackening. They had characterized 3,4-dihydroxy-6-(*N*-ethylamino) benzamide as the substrate for *o*-diphenol oxidase.

Table 2.5. Changes in the chemical composition of Indian black pepper cultivars during maturation.

Cultivar	Karimunda					Panniyur-1				
Stage of maturation	Months after fruit setting					Months after fruit setting				
Components	3.0	4.5	5.5	6.5	7.0	3.0	4.5	5.5	6.5	7.0
Volatile oil (%)	6.8	10.4	8.2	4.4	3.6	6.4	7.6	6.3	2.8	2.0
NVEE	10.3	9.7	8.6	7.5	7.4	8.7	8.8	8.7	8.1	7.8
Piperine	1.9	2.4	2.4	3.1	3.1	1.9	2.6	2.7	3.1	3.5
Starch	2.6	4.9	6.2	15.3	15.3	2.5	3.7	5.1	10.2	16.8

Note: NVEE - non-volatile ether extract.
Source: Purseglove *et al.* (1981).

2.4. Chemistry

Volatiles

The aroma of black pepper is contributed mainly by the volatile oil, which varies between 2 and 5% in the berries.

Produced by steam distillation from the black peppercorns, the essential oil is water-white to pale olive in colour, with a warm, spicy (peppery), fresh aroma. It has a middle note and blends well with rose, rosemary, marjoram, frankincense, olibanum, sandalwood and lavender; however, it should be used in small amounts only (Borges *et al.*, 2003).

A promising technology for the extraction of black pepper essential oil using liquid carbon dioxide was described by Ferreira *et al.* (1993). This technology has now become very popular in the spice and aromatic crops industry. Composition of the extracted oil was obtained by chromatographic analysis and solubility of the oil was determined using the dynamic method of extraction. About 70% of the total oil was extracted during the constant rate period (Ferreira *et al.*, 1999). In the extraction of black pepper essential oil with supercritical CO_2, it was observed that the fluid phase concentration of the soluble components began to decrease after a portion of the solute had been removed (decreasing extraction rate period). This effect may be explained by the combination of increased solute–fluid mass transfer resistance and a decrease in the 'effective' length of the fixed bed, due to exhaustion of the extract in the solid substratum in the direction of the flow. The fixed bed extraction of black pepper essential oil using supercritical carbon dioxide was modelled by the extended Lack's plug flow model developed by Sovová (Sovová's model) (Ferreira and Meireles, 2002).

The impetus to the characterization of constituents of essential oils began with the advent of gas chromatography. A combination of methods like vacuum distillation, column chromatography, thin-layer chromatography, UV, IR, NMR, GC and MS was employed by later investigators to separate and identify the constituents. Hyphenated techniques like GC-IR, GC-MS, etc., speed up the identification process. High-resolution capillary column GC coupled to MS or IR, along with the availability of IR and MS spectral libraries, made identification of known compounds easier. The developments adopting data on relative retention indices of Kovats indices and usage of capillary columns of 50 m in length was of tremendous use. This led to the identification of over 135 compounds consisting of monoterpenoids, sesquiterpenoids, aliphatic, aromatic and miscellaneous-nature compounds (Shiratsuchi *et al.*, 1993; Zhao and Cranston, 1995; Korány and Amtmann, 1998; Narayanan, 2000; Roessner *et al.*, 2000).

Constituents of black pepper oil

Earlier workers established the presence of α-pinene, β-pinene, 1-α-phellandrene, dl-limonene, piperonal, dihydrocarveol, a compound melting at 161°C, β-caryophyllene and a piperidine complex from the essential oil obtained by steam distillation of ground Malabar pepper. The above compounds were identified by the classical methods of derivatization and degradation. They also reported the presence of epoxy dihydrocaryophyllene, cryptone and possibly citronellol and an azulene (Hasselstrom *et al.*, 1957). The presence of α- and β-pinenes, limonene and caryophellene in the hydrocarbon portion of black pepper oil has been confirmed by the use of infrared spectroscopy

Major pepper oil constituents identified by various researchers are listed below.

MONOTERPENE HYDROCARBONS AND OXYGENATED COMPOUNDS There are 15 monoterpene hydrocarbons identified so far and they are camphene, δ^3-carene, *p*-cymene, limonene, myrcene, *cis*-ocimene, α-phellandrene, β-phellandrene and α- and β-pinenes, sabinene, α- and γ-terpinenes, terpinolene and α-thujene.

About 43 oxygenated compounds of a monoterpenoid nature have been characterized. Popular oxygenated monoterpenes are borneol, camphor, carvacrol, *cis*-carveol, *trans*-carveol, carvone, carvetanacetone, 1,8-cineole, cryptone, *p*-cymene-8-ol, *p*-cymene-8-methyl ether, dihydrocarveol, dihydrocarvone, linalool, *cis*-2-menthadien-2-ol, 3,8(9)-*p*-menthadien-1-ol, 1(7)-*p*-menthadien-6-ol, 1(7)-*p*-menthadien-4-ol, 1,8(9)-*p*-menthadien-5-ol, 1,8(9)-*p*-menthadien-4-ol, *cis*-*p*-2-menthen-1-ol, myrtenal, myrtenol, methyl carvacrol, *trans*-pinocarveol, pinocamphone, *cis*-sabinene hydrate, *trans*-sabinene hydrate, 1-terpinen-4-ol, 1-terpinen-5-ol, α-terpeneol, 1,1,4,trimethylc yclohepta-2,4-dien-6-ol, phellandral, piperitone, citronellal, nerol, geraniol, isopinocamphone, methyl citronellate, methyl geranate, α-terpenyl acetate, terpenolene epoxide and *trans*-limonene epoxide (Pino and Borges, 1999).

Dilution and concentration experiments on samples of dried black pepper berries from India and Malaysia, as well as enantioselective analysis of optically active monoterpenes, indicated (±)-linalool, (+)-α-phellandrene, (–)-limonene, myrcene, (–)-α-pinene, 3-methylbutanal and methylpropanal as the most potent odorants of black pepper. Additionally, 2-isopropyl-3-methoxypyrazine and 2,3-diethyl-5-methylpyrazine were detected as important odorants of the black pepper sample from Malaysia, which had a mouldy, musty off-flavour (Jagella and Grosch, 1999a). Gamma irradiation was an effective means of decontamination, especially at 10 kGy, but caused losses in the major flavour components such as β-pinene and cineole in black pepper. Irradiation also induced the conversion of monoterpene hydrocarbons to alcohol terpenes in black pepper essential oils. Washing the spices slightly reduced microbial counts but generally had no effect on flavour constituents (Farag Zaied *et al.*, 1996).

SESQUITERPENE HYDROCARBONS AND OXYGENATED COMPOUNDS β-Caryophyllene is the major sesquiterpene hydrocarbon present in pepper oil. Other sesquiterpene hydrocarbons are also reported from black pepper oil. They are α-*cis*-bergamotene, α-*trans*-bergamotene, β-bisabolene, δ-and γ-cadinenes, calamenene, α-copaene, α- and β-cubebenes, *ar*-curcumene, β- and δ-elemenes, β-farnesene, α-guaiene, α- and γ-humulenes, isocaryophyllene, γ-muurolene, α-santalene, α- and β-selinenes, ledene, sesquisabinene and zingiberene.

About 20 oxygenated sesquiterpenes have been identified from pepper oil. They are 5,10(15)-cadinen-4-ol, caryophylla-3(12), 7(15)-dien-4-β-ol, caryophylla-2,7(15)-dien-4-β-ol, caryophylla-2,7(15)-dien-4-ol, β-caryophellene alcohol, caryophyllene ketone, caryophellene oxide, epoxy-dihydrocaryophellene, *cis*-nerolidol, 4,10,10-trimethyl-7-methylene bicycle-(6.2.0) decane-4-carboxaldehyde, cubenol, epi-cubenol, viridiflorol, α- and β-bisabolols, cubebol, elemol and γ-eudesmol.

MISCELLANEOUS COMPOUNDS Eugenol, methyl eugenol, myristicin, safrole, benzaldehyde, *trans*-anethole, piperonal, *m*-methyl acetophenone, *p*-methyl acetophenone, *n*-butyrophenone, benzoic acid, phenyl acetic acid, cinnamic acid and piperonic acid are some of the aromatic compounds characterized in pepper oil. Methyl heptenone, pinol, butyric acid, 3-methyl butyric acid, hexanoic acid, 2-methyl pentanoic acid, methyl heptanoate, methyl octanoate, 2-undecanone, *n*-nonane, *n*-tridecane, *n*-nonadecane and piperidine are the other compounds identified (Narayanan, 2000).

Wide variation in the chemical composition of pepper oil was observed by different research groups. This can be attributed to the effects of cultivar, agroclimatic variation,

variation in the maturity of raw material, differences in the method of obtaining the oil, non-resolution of constituents in early gas chromatographic analyses using packed columns, etc. Steam-distilled pepper oils usually contain about 70–80% monoterpene hydrocarbons, 20–30% sesquiterpene hydrocarbons and less than 4% oxygenated constituents. Oils prepared by vacuum distillation of oleoresin extracts differ in having less monoterpene hydrocarbons and more sesquiterpene hydrocarbons and oxygenated constituents.

The major monoterpene hydrocarbons present in pepper oil are α- and β-pinenes, sabinene and limonene. Chemical structures of major aroma compounds are illustrated in Fig. 2.1.

ANGULAR ROTATION OF OIL The optical rotation of black pepper oil is levorotatory (Shankaracharya *et al.*, 1997). George *et al.* (1988) observed the angular rotation pattern of pepper oil.

Variability in essential oil constituents

Lewis *et al.* (1969) reported the composition of the essential oil of 17 cultivars of Kerala, India. In the oils, monoterpene hydrocarbons ranged from 69.4 to 85%, sesquiterpene hydrocarbons from 15 to 27.6% and the rest was oxygenated constituents. The major monoterpene hydrocarbons, e.g. α-pinene, ranged from 5.9 to 12.8%, β-pinene from 10.6 to 35.5% and limonene from 22 to 31.1%. The major sesquiterpene hydrocarbon, β-caryophyllene, ranged from 10.3 to 22.4%. A Sri Lankan variety was also analysed by Lewis *et al.* (1969) and they found that the oil contained α-pinene to the extent of 22.1%, β-pinene 11.1%, sabinene 21.3%, limonene 11.1% and β-caryophyllene 16.6%. Zachariah (1995) evaluated 42 accessions of black pepper (*P. nigrum* L.) germplasm for essential oil and chemical constituents. Good variability was observed between the accessions for flavour and quality. Pinene content varied from 3.8 to 16.6%, sabinene from 2.2 to 33%, limonene from 3.6 to 21.2% and caryophyllene from 11.8 to 41.8%.

The state of Kerala in India is known for the many popular cultivars of black pepper. These cultivars exhibit wide variation in the percentage composition of major volatiles. Table 2.6 illustrates the list of compounds identified in the black pepper cultivars, e.g. Panniyur-1,

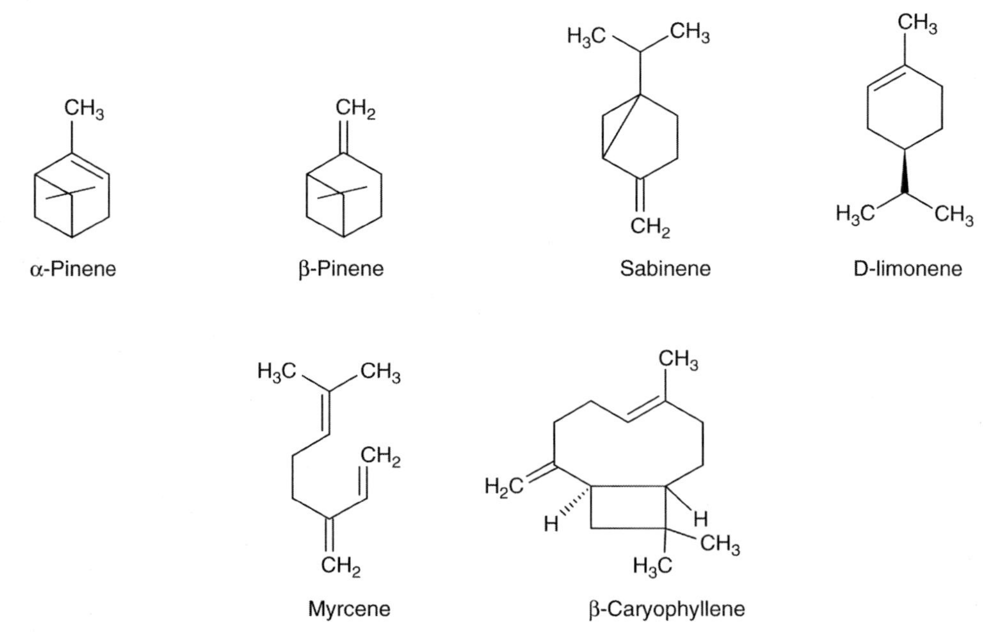

Fig. 2.1. Aromatic compounds of black pepper.

Table 2.6. Volatile compounds of four black pepper varieties.

No.	Compound	% Composition 1	2	3	4
1	α-Thujene	0.73	1.26	1.59	0.91
2	α-Pinene	5.28	6.18	5.07	5.32
3	Camphene	0.14	0.18	0.14	0.13
4	Sabinene	8.50	13.54	17.16	1.94
5	β-Pinene	11.08	10.88	9.16	6.40
6	Myrcene	2.23	2.30	2.20	8.40
7	α-Phellendrene	0.68	0.20	–	2.32
8	δ-3-Carene	2.82	0.18	–	1.03
9	α-Terpinene	–	–	0.39	1.13
10	*p*-Cymene	–	0.18	0.07	9.70
11	(*Z*)-β-Ocimene + β-phellendrene	–	0.15	0.23	0.37
12	Limonene	21.06	21.26	22.71	16.74
13	(*E*)-β-Ocimene	0.18	2.84	0.30	0.17
14	γ-Terpinene	0.01	0.49	–	0.03
15	*trans*-Sabinene hydrate	0.14	–	0.30	0.19
16	Terpinolene	0.10	0.20	0.22	0.08
17	*trans*-Linalool oxide (furanoid)[ti]	0.03	0.18	–	0.08
18	Unidentified	0.24	0.22	0.26	0.60
19	Linalool	0.22	0.22	0.46	0.28
20	*cis-p*-Menth-2-en-1-ol + *cis-p*-menth-2,8-diene-1-ol	0.04	0.04	0.05	0.02
21	*trans-p*-Menth-2-en-1-ol	0.01	0.01	0.01	0.01
22	Citronellal	0.02	0.03	0.03	0.01
23	*p*-Menth-8-en-1-ol	0.03	t	–	t
24	Boroneol	t	t	t	t
25	Terpinen-4-ol	0.19	0.32	0.52	0.18
26	α-Terpineol	0.10	0.17	0.12	0.07
27	Dihydrocarveol	0.01	–	0.02	0.02
28	*p*-Menth-8-en-2-ol	–	0.01	0.02	0.02
29	*trans*-Carveol	0.01	0.01	–	0.02
30	*cis*-Carveol + carvone	0.01	0.03	0.03	0.03
31	Piperitone	0.04	t	0.03	t
32	Carvone oxide	0.01	0.01	–	0.01
33	Myrtenol	0.20	0.04	0.11	0.04
	Unidentified	0.02	–	–	–
	Unidentified	0.02	t	t	–
34	α-Terpinyl acetate	0.86	1.22	1.33	1.05
35	Neryl acetate	0.20	0.07	0.05	0.13
36	Geranyl acetate	0.12	0.01	0.07	0.11
37	α-Cubebene/-δ-elemene	3.25	0.26	0.16	2.56
38	α-Copaene	0.82	0.49	0.44	0.71
39	β-Elemene	0.09	0.09	0.06	0.05
40	β-Caryophyllene	21.59	27.70	23.2	21.19
41	*trans*-α-Bergamotene	0.31	–	9.00	0.28
42	α-Humulene	0.21	0.20	–	0.29
43	(*E*)-β-Farnesene	0.08	0.22	0.11	0.13
44	α-Amorphene	1.51	1.53	0.03	1.28
45	α-Guaiene	0.11	0.07	1.54	0.10
46	Clovene[ti]	0.14	0.07	0.07	0.13
47	Germacrene-D[ti]	0.04	0.03	0.04	0.26
48	*ar*-Curcumene	0.26	0.12	0.04	0.29

Continued

Table 2.6. *Continued*

No.	Compound	% Composition 1	2	3	4
49	β-Selinene	0.64	0.87	1.37	0.63
50	α-Selinene	0.07	0.12	0.48	0.14
51	γ-Muurolene	0.73	0.93	0.16	0.58
52	(*E,E*)-α-Farnesene	0.72	–	0.47	0.72
53	β-Bisabolene[ti] + β-bisabolene[ti]	4.25	2.15	3.10	0.49
54	δ-Guaiene[ti]	0.82	0.17	0.09	1.85
55	Cuparene[ti]	1.38	0.09	0.14	0.04
56	δ-Cadinene	0.12	–	0.07	0.13
57	(*Z*)-Nerolidol	0.20	0.05	0.11	0.05
58	Elemol	0.11	0.06	0.07	0.08
59	Unidentified	0.04	0.02	0.07	0.03
60	(*E*)-Nerolidol	0.12	0.04	0.07	0.03
61	Caryophellene alcohol	0.07	0.02	0.04	0.02
62	Unidentified	0.03	0.11	0.07	0.07
63	Caryophellene oxide	0.90	0.35	0.38	0.25
64	Cedrol[ti]	0.07	–	0.05	0.05
65	α-Cadinol[ti]	1.51	0.29	0.12	1.27
66	α-Cadinol[ti]	0.26	0.12	0.15	0.25
67	β-Bisabolol	0.20	0.09	0.17	0.14

Notes: t = trace (< 0.01%), ti = tentative identification; 1 = Panniyur-1; 2 = Panniyur-2; 3 = Panniyur-3; 4 = Panniyur-4.
Source: Narayanan (2000).

Panniyur-2, Panniyur-3 and Panniyur-4. Other popular cultivars are Aimpiriyan, Narayakodi, Neelamundi, Uthirankotta, Karimunda, Kalluvally, Arakulammunda, Thommankodi, Kottanadan, Ottaplackal, Kuthiravally, Thevanmundi, Poonjaranmunda, Valiakaniakkadan and Subhakara. Some of the newly developed cultivars are Panniyur-1, Panniyur-2, Panniyur-3 and Panniyur-4. By adopting GC and GC-MS techniques, researchers have identified over 55 compounds from the volatile oil of these pepper cultivars. The major compounds identified were α- and β-pinene, sabinene, limonene, β-caryophyllene, myrcene, *p*-cymene and caryophellene oxide (Gopalakrishnan *et al.*, 1993; Menon *et al.*, 2000, 2002, 2003; Menon and Padmakumari, 2005). Table 2.7 illustrates this variability among different popular cultivars.

Gopalakrishnan *et al.* (1993) analysed four new genotypes of pepper (Panniyur-1, Panniyur-2, Panniyur-3 and Panniyur-4) by a combination of GC-MS and Kovats indices on a methyl silicone capillary column. The oils from the first three Panniyur genotypes contained α-pinene in the range of 5.07–6.18%, β-pinene 9.16–11.08%, sabinene 8.50–17.16%, limonene 21.06–22.71% and β-caryophyllene 21.57–27.70%. The oil from Panniyur-4 (culture 239) contained 5.32% α-pinene, 6.40% β-pinene, 1.94% sabinene, 8.40% myrcene, 9.70% *p*-cymene, 16.74% limonene and 21.19% caryophyllene.

Zachariah *et al.* (2005) conducted a study on the effect of grafting *P. nigrum* on *P. colubrinum* as rootstock. The cultivars used for grafting were Panniyur-1, -2, -3, -4 and -5, Malligesara, Pournami, Sreekara, Poonjaranmunda, Kuthiravally and Balankotta. The major essential oil constituents in grafts and non-grafts of pepper cultivars were pinene, sabinene and β-caryophyllene. Caryophyllene content varied from 12 to 27% in graft and from 7 to 29% in non-graft. Limonene content varied from 13 to 24% in graft and 13 to 22% in non-graft.

Essential oils obtained by hydrodistillation of green and black berries of Indian origin (cv. Thevanmundi) were analysed by GC and GC-MS methods and compared with the reported constituents of Sri Lankan green and

Table 2.7. Percentage composition of major volatiles in common black pepper varieties.

Variety/ compound	Limonene	β-Pinene	β-Caryophyllene	Sabinene	Caryophyllene oxide	δ-3-Carene	α-Pinene	Myrcene	*p*-Cymene	Elemol
Aimpiriyan	19.8–22.5	9.3–23.9	20.3–34.7	–	–	–	–	–	–	–
Narayakodi	9.5–19.5	4.8–15.6	29.8–52.9	4.4–24.6	2.3–3.9	–	–	–	–	–
Neelamundi	12.9–18.6	7.8–11.3	17–31.0	23.2–27.3	–	–	4.7–6.5	–	–	–
Uthirankotta	13.3–19.5	9.3–12.5	25.1–37.8	–	0.6–2.7	6.7–8.5	9.1–14.6	–	–	–
Panniyur-1,2,3	21.0–22.7	9.16–11.0	21.57–27.7	8.5–17.16	–	–	5–6.18	–	–	–
Panniyur-4	16.7	6.4	21.19	1.94	–	–	5.32	8.4	9.7	–
Karimunda	9.4–21.9	2.0–15.2	19.8–45.3	–	–	0.1–21.0	2.4–11.4	–	–	–
Kalluvally	9.4–21.9	2.0–15.2	19.8–45.3	–	–	0.1–21.0	2.4–11.4	–	–	–
Arakulammunda	9.4–21.9	2.0–15.2	19.8–45.3	–	–	0.1–21.0	2.4–11.4	–	–	–
Thommankodi	9.4–21.9	2.0–15.2	19.8–45.3	–	–	0.1–21.0	2.4–11.4	–	–	–
Kottandan	12.7–23.8	7.5–15.4	8.9–24.1	11.2–22.6	–	–	–	–	–	–
Ottaplackal	15.5–21.7	3.8–11.7	15.5–21.7	0.1–26.8	–	–	–	0–18.6	–	–
Kuthiravally	9.0–16.9	3.8–10.9	29.0–46	–	–	–	–	–	–	–
Cheriya kaniakadan	14.7–17.8	7.7–11.2	17.4–23.1	9.7–22.3	–	–	–	–	–	–
Thevanmundi	8.3–18.0	3.7–8.7	20.3–34.7	4.5–16.2	–	–	–	–	–	–
Poonjaranmunda	14.9–15.8	6.0–11.7	24.4–30.8	–	–	–	–	–	–	1.2–6.8
Valiakaniakkadan	12.9–18.6	–	23.0–38.4	12.9–17.3	–	0–10.5	2.9–6.3	–	–	–
KS-27	18.3–22.7	7.6–9.6	7.6–21.3	–	0.4–6	19.0–23.4	3.2–7.0	–	–	–

Source: Gopalakrishnan *et al.* (1993); Menon *et al.* (2000, 2002, 2003); Menon and Padmakumari (2005).

black pepper oils. The monoterpene hydrocarbons of Indian oils were similar to those of corresponding Sri Lankan oils, but the oils differed with regard to their sesquiterpene and oxygenated components. *β*-Pinene and caryophyllene occurred in all oils, sabinene in Sri Lankan oils only and car-3-ene in none of the oils (McCarron *et al.*, 1995).

AROMA COMPOUNDS IN PEPPER OIL Jirovetz *et al.* (2002) investigated the aroma compounds of the essential oils of dried fruits of black pepper (*P. nigrum*) and black and white 'Ashanti pepper' (*P. guineense*) from Cameroon by means of solid-phase microextraction (SPME) to identify the odorous target components responsible for the characteristic odour of these valuable spices and food flavouring products. By means of GC-flame ionization detection (FID) and GC-MS (using different polar columns), the main compounds of the SPME headspace samples of *P. nigrum* (black) and *P. guineense* (black and white) were found to be: *P. nigrum* (black) – germacrene D (11.01%), limonene (10.26%), β-pinene (10.02%), α-phellandrene (8.56%), β-caryophyllene (7.29%), α-pinene (6.40%) and *cis*-β-ocimene (3.19%); *P. guineense* (black) – β-caryophyllene (57.59%), β-elemene (5.10%), bicyclogermacrene (5.05%) and α-humulene (4.86%); and *P. guineense* (white) – β-caryophyllene (51.75%), *cis*-β-ocimene (6.61%), limonene (5.88%), β-pinene (4.56%), linalool (3.97%) and α-humulene (3.29%).

Sensory evaluation of pepper essential oil

The odour of pepper oil is described as fresh, dry-woody, warm-spicy and similar to that of the black peppercorn (Purseglove *et al.*, 1981). The flavour is rather dry-wood and lacks the pungency of the spice since the alkaloids are not extracted by steam distillation. Very few studies are reported in the literature on correlation of oil composition to odour characteristics. Hasselstrom *et al.* (1957) attribute the characteristic odour of pepper oil to the small amounts of oxygenated constituents present. Lewis *et al.* (1969) consider that a number of monoterpenes present in the oil are necessary for strong peppery top notes. Caryophellene-rich oils possess sweet floral odours, whereas oils with high pinene content give turpentine-like off-odours. By direct sniffing at the eluting port of the gas chromatographic column, the distinct odour of black pepper was ascribed to three areas of the late fractions. In earlier studies, these were identified as *trans*- and *cis*-bergamotenes and santalene. Information on descriptive odour analysis was used by Govindarajan (1977), who developed an odour description of pepper to evaluate horticultural varieties and trade types of pepper. Gopalakrishnan *et al.* (1993) have described the odour evaluation of four new genotypes of Indian pepper by descriptive odour profile based on a four-point category scale by subjecting the oils to a ranking test. The major odours are refreshing pinene-like, fresh green, camphoraceous, citrus/lemon-like, warm and spicy, peppery, sharp pungent, woody resinous, turmeric-like and musty/mouldy.

The most intense odour impressions of the essential oils of the various dried pepper fruits were given by professional perfumers as follows: *P. nigrum* (black) – fine, pleasant black pepper note; *P. guineense* (black) – black pepper top note; and *P. guineense* (white) – pleasant white pepper note (Jirovetz *et al.*, 2002) A GC-sniffing technique was used to correlate the single-odour impression of the identified pepper samples (*P. guineense* (black and white), *P. nigrum* (black)) and the following profile was reported. The main compounds, such as β-caryophyllene, germacrene D, limonene, β-pinene, α-phellandrene and α-humulene, as well as minor constituents, such as δ-carene, β-phellandrene, isoborneol, α-guaiene, sarisan, elemicin, calamenene, caryophyllene alcohol, isoelemicin, T-muurolol, cubenol and bulnesol, are of greatest importance for the characteristic pepper odour notes of these three Piper samples. Further aroma impressions can be attributed to mono- and sesquiterpenes, hexane, octane and nonane derivatives.

Fresh pepper was found to taste and smell differently when compared with dry pepper and pepper oil. The fresh pepper aroma compounds were isolated by Amberlite column chromatography and analysed by GC and

GC-MS. The major compounds were found to be *trans*-linalool oxide and α-terpineol, whereas the dry black pepper oil contained α- and β-pinenes, *d*-limonene and β-caryophyllene as major components. When fresh pepper oil was isolated by distillation and analysed by GC and GC-MS, the compounds were found to be of a different nature to that of fresh pepper aromatic compounds (Menon, 2000).

The sensory properties of black pepper oil obtained by steam distillation of ground peppercorns were analysed with the aid of column chromatography, high-resolution GC and GC-MS. A total of 46 compounds were identified, including (*E*)-β-ocimene, δ-guaiene, (*Z*)(*E*)-farnesol, δ-cadinol and guaiol, which are reported for the first time as volatile compounds of the essential oil. Sabinene and terpinen-4-ol appeared to be the most important contributors among the volatile compounds to the characteristic odour of black pepper oil (Pino *et al.*, 1990).

Flavour and off-flavour compounds of black and white pepper (*P. nigrum* L.) were evaluated by Jagella and Grosch (1999a,b). Enantioselective analysis of optically active monoterpenes indicated (±)-linalool, (+)-α-phellandrene, (−)-limonene, myrcene, (−)-α-pinene, 3-methylbutanal and methylpropanal as the most potent odorants of black pepper. Additionally, 2-isopropyl-3-methoxypyrazine and 2,3-diethyl-5-methylpyrazine were detected as important odorants of the black pepper sample from Malaysia, which had a mouldy, musty off-flavour. Omission tests indicated α- and β-pinene, myrcene, α-phellandrene, limonene, linalool, methylpropanal, 2- and 3-methylbutanal, butyric acid and 3-methylbutyric acid as key odorants. A storage experiment revealed that for ground black pepper, losses of α-pinene, limonene and 3-methylbutanal were mainly responsible for deficits in the pepper-like, citrus-like, terpene-like and malty notes after 30 days at room temperature. The musty/mouldy off-flavour of a sample of black pepper was caused by a mixture consisting of 2,3-diethyl-5-methylpyrazine (2.9 μg/kg) and 2-isopropyl-3-methoxypyrazine (0.2 μg/kg).

Non-volatiles

Pungency of black pepper

Piperine is the major constituent of pepper oleoresin (Borges and Pino, 1993). The isolation of piperanine, a new pungent component of black pepper oleoresin, is described and its structure is shown by synthesis to be *trans*-5-(3,4-methylenedioxyphenyl)-2-pentenoic acid piperidide. The pungency of black pepper (*P. nigrum* L.) was attributed to the presence of piperine, the structure of which was later proven to be *trans,trans*-5-(3,4-methylenedi-oxyphenyl)-2, 4-pentadienoic acid piperidide. Further investigations into the pungency of this spice revealed that unidentified materials other than piperine also contributed to its pungency.

The pungency of black pepper has been the subject of chemical investigations since the early 19th century. In 1819, Oersted isolated piperine, the most abundant alkaloid in pepper, as a yellow crystalline substance and its structure was later identified as the trans form of piperoyl piperidine (Narayanan, 2000). The pungent dark oily resin obtained after removal of piperine from the oleoresin was named as chavicine (Govindarajan, 1977). Chavicine was claimed to possess a far greater bite on the tongue than crystalline piperine, but later workers demonstrated that piperine in solution was very pungent. The controversy over which compound, e.g. piperine, its *cis-cis* isomer chavicine or other possible isomers-isopiperine (*cis-trans*) and isochavicine (*trans-cis*), was more pungent lasted almost a century. However, later investigations demonstrated that piperine was the major pungent principle and chavicine was a mixture of piperine and several minor alkaloids. The presence of chavicine and isopiperine has not been confirmed in pepper extracts, while isochavicine was shown to occur as an artefact of photolytic transformation of piperine. Five new minor alkaloids possessing a degree of pungency have been identified in pepper extracts. They are piperettine, piperylin, piperolein A and B and piperanine. Three trace constituents, e.g. peepuloidin, guineesine and pipericide, showing insecticidal properties have been identified recently.

The acetone extract of pepper showed the presence of 18 components, accounting for 75.59% of the total quantity. Piperine (33.53%), piperolein B (13.73%), piperamide (3.43%) and guineensine (3.23%) were the major components (Singh *et al.*, 2004).

Variation of piperine in relation to cultivars

Seven black pepper cultivars, namely Panniyur-2, Panniyur-3, Panniyur-4, Sreekara, Subhakara, KS-88 and Neelamundi (localcultivar), were evaluated for piperine, oleoresin and essential oil contents. Panniyur-4 recorded the lowest oleoresin (9.2%) and essential oil contents (2.1%) and relatively medium piperine content (4.4%). Panniyur-2 had poor yield but recorded the highest piperine content (6.6%). Neelamundi, KS-88 and Sreekara gave the highest oleoresin contents (13.9, 13.1 and 13.0%, respectively), while Sreekara and Subhakara gave the highest essential oil content (7.0 and 6.0%, respectively) (Radhakrishnan *et al.*, 2004). HP-813 ('IISR Malabar Excel') had reportedly high oleoresin and piperine content compared with other recently released black pepper cultivars (Sasikumar *et al.*, 2004). Varietal variation of oleoresin and piperine was reported by Kurian *et al.* (2002) in black pepper (*P. nigrum* L.) grown at Idukki District, Kerala. Mathew and Bhattacharyya (1990) showed that a slightly immature grade of 'half pepper' was economically advantageous and contained the highest levels of piperine (6.8%). Five different grades of four (*P. nigrum* L.) cultivars were analysed for their piperine percentage, oleoresin and volatile oil content. The garbled light special (GL special) grade of each cultivar had the highest piperine percentages. The Tellicherry garbled (TG) grade of Kalluvally had the highest volatile oil content at 3.4% (Sumathykutty *et al.*, 1989). Panniyur-6, released in 2000 for cultivation in all black pepper-growing tracts of Kerala, India, is a high-yielding clone with 8.27% oleoresin, 4.94% piperine and 1.33% volatile oil content. Panniyur-7, a seedling progeny of the Kalluvally type, a superior open-pollinated progeny, has 10.61% oleoresin, 5.57% piperine and 1.50% volatile oil content (Arya *et al.*, 2003). Structures of prominent pungent compounds are illustrated in Fig. 2.2.

Properties, synthesis and estimation of piperine

The alkaloid piperine generally is accepted as the active 'bite' component in black pepper. The homologues and analogues of piperine are minor or trace compounds and their contribution to pungency is small. Despite the controversy over the nature of pungent compounds in pepper, piperine content has been taken as a measure of the total pungency.

Piperine is a yellow crystalline substance having a melting point of 128–130°C. Piperine, $C_{17}H_{19}O_3N$, was shown to be a weak base which, on hydrolysis with aqueous alkali or nitric acid, yielded a volatile base $C_5H_{11}N$, later identified as piperidine. The acidic product of hydrolysis, piperine acid ($C_{12}H_{19}O_4$, m.p. 216–217°C), was shown to be 5-(3,4-methylene dioxy phenyl)-2,4, pentadienoic acid.

Due to controversy over the pungency of piperine and chavicine, a great amount of work

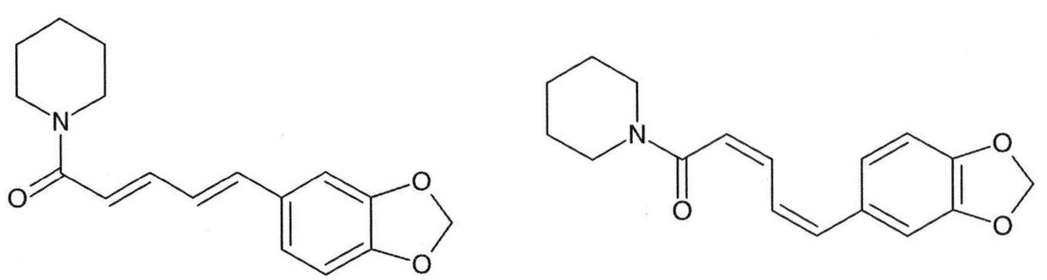

Piperine Chavicine

Fig. 2.2. Pungent compounds of black pepper.

was carried out to synthesize all the four isomers. Isopiperinic acid was synthesized from piperonylidene malonic acid, the structure of which was known, by elimination of carbon dioxide, and the *cis–trans* configuration was assigned to this acid (Narayanan, 2000).

Geisler and Gross (1990) isolated an acyltransferase from shoots of black pepper which catalysed the synthesis of piperine in the presence of piperoyl-coenzyme A and piperidine. The enzyme is classified as 'piperoyl-CoA: piperidine *N*-piperoyltransferase' (piperidine piperoyltransferase; EC 2.3.1).

Wood *et al.* (1988) developed the reversed-phase high-performance liquid chromatographic (HPLC) method for piperine determination in black pepper and its oleoresins. It employs bonded C18 stationary phase (ODS-2) and acetonitrile-aqueous acetic acid mobile phase with UV detection. As the spectrophotometric method which invariably yields higher results because of the contributions from other alkaloids such as piperyline and piperettine, the HPLC method relates more to piperine. Utilizing the UV absorption property of piperine, spectrophotometric estimation methods were developed by different groups using solvents such as benzene, ethanol, ethylene dichloride, acetone, ethyl acetate, chloroform and cyclohexane. Sowbhagya *et al.* (1990) proved that extractability with acetone was very efficient.

2.5. Medicinal and Pharmacological Properties

The therapeutic properties of black pepper oil include analgesic, antiseptic, antispasmodic, antitoxic, aphrodisiac, diaphoretic, digestive, diuretic, febrifuge, laxative, rubefacient and tonic (especially of the spleen).

Antioxidant activity

As a natural medicinal agent, black pepper in tea form has been credited for relieving arthritis, nausea, fever, migraine headaches, poor digestion, strep throat and even coma. It has also been used for non-medical applications, as an insecticide. Of course, black pepper is a favourite spice of cooks because of its dark colour and pungent aroma and flavour. Vijayakumar *et al.* (2004) studied the antioxidant efficacy of black pepper and piperine in rats with high-fat diet-induced oxidative stress. Thirty male rats (95–115 g) were divided into five groups. They were fed standard pellet diet, high-fat diet (20% coconut oil, 2% cholesterol and 0.125% bile salts), high-fat diet plus black pepper (0.25 g or 0.5 g/kg body weight) or high-fat diet plus piperine (0.02 g/kg body weight) for a period of 10 weeks. Significantly elevated levels of thiobarbituric acid reactive substances (TBARS), conjugated dienes (CD) and significantly lowered activities of superoxide dismutase (SOD), catalase (CAT), glutathione peroxidase (GPx), glutathione-*S*-transferase (GST) and reduced glutathione (GSH) in the liver, heart, kidney, intestine and aorta were observed in rats fed the high-fat diet as compared with the control rats. Simultaneous supplementation with black pepper or piperine lowered TBARS and CD levels and maintained SOD, CAT, GPx, GST and GSH levels to near those of the control rats. The data indicate that supplementation with black pepper or the active principle of black pepper, piperine, can reduce high-fat diet-induced oxidative stress. Antioxidant activity of essential oils was described by Dorman *et al.* (2000). Pepper and pepper-containing preparations are used for the treatment of intermittent fever, neuritis, cold, pains and diseases of the throat. In Chinese medicine, pepper is used for the treatment of malaria. Tipsrisukond *et al.* (1998) reported the antioxidant potential of black pepper essential oil and oleoresin extracted by supercritical carbon dioxide extraction.

Bio-enhancing ability

Piperine (1-piperoyl piperidine) is shown to possess bioavailability-enhancing activity with various structurally and therapeutically diverse drugs. Piperine's bioavailability-enhancing property may be attributed to increased absorption, which may be due to alteration in membrane lipid dynamics and change in the conformation of enzymes in the intestine. Piperine also stimulates leucine amino peptidase and glycyl–glycine dipeptidase activity, due to the alteration in enzyme kinetics. This suggests that piperine could modulate membrane dynamics due to its apolar nature by interacting with the surrounding lipids and hydrophobic portions in the protein vicinity, which may modify enzyme conformation. Ultrastructural studies with piperine showed an increase in microvilli length, with a prominent increase in free ribosomes and ribosomes on the endoplasmic reticulum in enterocytes, suggesting that synthesis or turnover of cytoskeletal components or membrane proteins may be involved in the observed effect. In conclusion, it is suggested that piperine may induce alterations in membrane dynamics and permeation characteristics, along with induction in the synthesis of proteins associated with cytoskeletal function, resulting in an increase in the small intestine absorptive surface, thus assisting efficient permeation through the epithelial barrier (Khajuria *et al.*, 2002).

The effect of piperine on the bioavailability and pharmacokinetics of propranolol and theophylline has been examined clinically by Bano *et al.* (1991). They established that the enhanced systemic availability of oral propranolol and theophylline could be exploited to achieve better therapeutic control using piperine.

Lee *et al.* (1984) projected the analgesic and antipyretic action of piperine. Gupta *et al.* (1998) studied the influence of piperine on nimesulide-induced antinociception. Piperine at a dose of 10 mg/kg significantly ($p < 0.001$) increased the analgesic activity of nimesulide administered at a submaximal dose of 6.5 mg/kg. This report emphasized the potential of piperine in enhancing the bioavailability of analgesics. Sunil *et al.* (2001) found the potential of piperine in inhibiting gastric emptying and gastrointestinal transit in rats and mice. Anticonvulsant activity of piperine on seizures induced by excitatory amino acid receptor agonists was studied by D'Hooge *et al.* (1996). In traditional Chinese medicine, a dried powder consisting of 1 radish:99 peppercorns is used to treat epilepsy. The effectiveness of the prescription may be due to the anticonvulsant actions of the principal component of pepper, the alkaloid piperine.

Rajinder *et al.* (2002) studied the mode of action of piperine. Through human liver microsomal studies, they established that piperine, a major constituent of black pepper, inhibits human *P*-glycoprotein and CYP3A4.

The effects of rifampicin alone or as a 24:1 (w/w) mixture of rifampicin and piperine (extracted from *P. nigrum*) against transcription of *Mycobacterium smegmatis* RNA polymerase, from rifampicin-resistant or -susceptible strains, was studied. The mixture of piperine and rifampicin showed remarkable growth inhibitory effects on *M. smegmatis*, and this inhibition was higher than that of rifampicin alone. Interestingly, piperine alone, even at higher concentration, did not inhibit the growth of *M. smegmatis*. The mixture of rifampicin and piperine abrogated nonspecific transcription catalysed by the microorganism's RNA polymerase. Here, too, the effect was higher than rifampicin alone, and piperine showed no effect when used independently. When RNA polymerase was purified from a rifampicin-resistant *M. smegmatis* strain, the enzymatic activity, otherwise resistant to rifampicin, decreased significantly in the presence of piperine along with rifampicin (Veena *et al.*, 2001).

2.6. International Specifications and Desirable Limits

The major requirements for international acceptance of black pepper and its products were listed by Tainter and Grenis (1993). Specifications for whole black pepper, white pepper, and powders are listed in Tables 2.8–2.11.

2.7. Conclusion

Black pepper, *P. nigrum*, also known as the king of spices, belongs to the family

Table 2.8. Whole black pepper: chemical and physical specifications.

Specification	Suggested limits
ASTA cleanliness specifications:	
Whole dead insects, by count	2
Mammalian excreta, mg/lb	1
Other excreta, mg/lb	5.0
Mould, % by weight	1
Insect-defiled/infested, % by weight	1
Extraneous, % by weight	1.00
General ISO specifications:	
Insect-infested and/or mouldy pieces by weight	Average of 1%
Mammalian excreta	Average of 1%
Foreign matter pickings and siftings by weight	
Volatile oil (% min.)	2.0
Moisture (% max.)	12.0
Ash (% max.)	5.0
Acid-insoluble ash (% max.)	0.5
Average bulk index (mg/100 g)	165

Table 2.9. Ground black pepper: chemical and physical specifications.

Specification	Suggested limits
Insect fragments	Average of 475 or more/50 g
Rodent hair fragments	Average of 2 or more/50 g
Volatile oil (% min.)	1.5
Moisture (% max.)	12
Total ash (% max.)	5
Acid-insoluble ash (% max.)	0.5
Crude fibre (% max.)	12.5
Non-volatile methylene chloride extract (% min.)	7.5
Starch (% max.)	30.0
Granulation (%)	70

Piperaceae, is cultivated for its fruit, which is usually dried, and used as a spice and seasoning. Black pepper has multiple uses in the processed food industry, in the kitchen, in perfumery, traditional medicine and even in beauty care. Pepper is valued for its pungency and flavour, which is attributed by the alkaloid piperine and the volatile oil.

Table 2.10. Whole white pepper: chemical and physical specifications.

Specification	Suggested limits
Cleanliness: whole dead insects by count	2
Mammalian excreta, mg/lb	1
Other excreta, mg/lb	1
Mould, % by weight	1
Insect-defiled/infested, % by weight	1
Extraneous, % by weight	0.50
Insect-infested and/or mouldy pieces by weight	Average of 1%
Mammalian excreta	Average of 1 mg/lb
Foreign matter pickings and siftings by weight	
Volatile oil (% min.)	1.5
Moisture (% max.)	14.0
Ash (% max.)	1.5
Acid-insoluble ash (% max.)	0.3
Average bulk index (mg/100 g)	150

Table 2.11. Ground white pepper: chemical and physical specifications.

Specification	Suggested limits
Insect fragments	Average of 475 or more/50 g
Rodent hair fragments	Average of 2 or more/50 g
Volatile oil (% min.)	1.5
Moisture (% max.)	14.0
Total ash (% max.)	1.5
Acid-insoluble ash (% max.)	0.3
Military specifications	
Volatile oil (ml/100 g min.)	1.0
Moisture (% max.)	15.0
Total ash (% max.)	3.0
Acid-insoluble ash (% max.)	0.3
Crude fibre (% max.)	5.0
Starch (% max.)	52.0
Non-volatile methylene chloride extract (% min.)	7.5
Granulation (% min. through a USS No. 30)	95
Bulk index (ml/100 g)	180

Note: USS = United States standard sieve size.

Piperine is the major constituent of pepper oleoresin. Black pepper oil contributes towards the aroma, oleoresin contributes towards the overall taste and the alkaloid piperine imparts pungency. β-Carophyllene is the major sesquiterpene hydrocarbon present in pepper oil. Other important sesquiterpene hydrocarbons are β-bisabolene, δ- and γ-cadinenes, calamenene, α-copaene, α- and β-cubebenes, *ar*-curcumene, β- and δ-elemenes, β-farnesene, α-guaiene, α- and γ-humulenes, isocaryophyllene, γ-muurolene, α-santalene, α- and β-selinenes, ledene, sesquisabinene and zingiberene. Piperine is shown to possess bioavailability-enhancing activity with various structurally and therapeutically diverse drugs.

The aroma and pungent constituents of pepper offer great future in pharmacology and industrial application.

References

Ahlert, B., Kjer, I., Liebl, B., Vries, T. de., Wedekind, R. and Vries, T. de (1998) Improvement of quality of black pepper at the farm level. *Zeitschrift fur Arznei and Gewurzpflanzen* 3(1), 15–20.

Arya, K., Rajagopalan, A., Satheesan, K.N., Nair, P.K.U., Mammotty, K.P., Zacharia, G. and Vanaja, T. (2003) Panniyur-6 and Panniyur-7 high yielding black pepper elections for Kerala. *Indian Journal of Genetics and Plant Breeding* 63(4), 363–364.

Bandyopadhyay, C., Narayan, V.S. and Variyar, P.S. (1990) Phenolics of green pepper berries (*Piper nigrum* L.). *Journal of Agricultural and Food Chemistry* 38(8), 1696–1699.

Bano, G., Raina, R.K., Zutshi, U., Bedi, K.L., Johri, R.K. and Sharma, S.C. (1991) Effect of piperine on bioavailability and pharmacokinetics of propranolol and theophylline in healthy volunteers. *European Journal of Clinical Pharmacology* 41(6), 615–617.

Borges, P. and Pino, J. (1993) Preparation of black pepper oleoresin by alcohol extraction. *Nahrung* 37(2), 127–130.

Borges, P., Fernandez, N. and Roncal, E. (2003) Isolation and characterization of essential oils of spice plants using steam distillation. *Alimentaria* 40(340), 125–127.

Chun, H., Shin, D.H., Hong, B.S., Cho, W.D., Cho, H.Y. and Yang, H.C. (2002) Biochemical properties of polysaccharides from black pepper. *Biological and Pharmaceutical Bulletin* 25(9), 1203–1208.

Dhas, P.H.A. and Korikanthimath, V.S. (2003) Processing and quality of black pepper: a review. *Journal of Spices and Aromatic Crops* 12(1), 1–14.

D'Hooge, R., Pei, Y.Q., Raes, A., Lebrun, P., vanBogaert, P.P. and Deyn, P.P. de (1996) Anticonvulsant activity of piperine on seizures induced by excitatory amino acid receptor agonists. *Arzneimittel Forschung* 46(6), 557–560.

Dorman, H.J.D., Surai, P. and Deans, S.G. (2000) *In vitro* antioxidant activity of a number of plant essential oils and phytoconstituents. *Journal of Essential Oil Research* 12(2), 241–248.

Farag Zaied, S.E.A., Aziz, N.H. and Ali, A.M. (1996) Comparing effects of washing, thermal treatments and gamma irradiation on quality of spices. *Nahrung* 40(1), 32–36.

Ferreira, S.R.S. and Meireles, M.A.A. (2002) Modeling the supercritical fluid extraction of black pepper (*Piper nigrum* L.) essential oil. *Journal of Food Engineering* 54(4), 263–269.

Ferreira, S.R.S., Meireles, M.A.A. and Cabral, F.A. (1993) Extraction of essential oil of black pepper with liquid carbon dioxide. *Journal of Food Engineering* 20, 121–133.

Ferreira, S.R.S., Nikolov, Z.L., Doraiswamy, L.K., Meireles, M.A.A. and Petenate, A.J. (1999) Supercritical fluid extraction of black pepper (*Piper nigrun* L.) essential oil. *The Journal of Supercritical Fluids* 14(3), 235–245.

Geisler, J.G. and Gross, G.G. (1990) The biosynthesis of piperine in *Piper nigrum*. *Phytochemistry* 29(2), 489–492.

George, K.M., Joy, M.T., Chandran, C.V. and Verghese, J. (1988) The angular rotation of black pepper oil. *Indian Perfumer* 32(1), 51–54.

Gopalakrishnan, M., Menon, N., Padmakumari, K.P., Jayalekshmy, A. and Narayanan, C.S. (1993) GC analysis and odor profiles of four new Indian genotypes of *Piper nigrum* L. *Journal of Essential Oil Research* 5(3), 247–253.

Govindarajan, V.S. (1977) Pepper chemistry, technology and quality evaluation. *Critical Reviews in Food Science and Nutrition* 9, 115–225.

Gupta, S.K., Velapandi, T., Sengupta, S., Mathur, P. and Sapra, P. (1998) Influence of piperine on nimesulide induced antinociception. *Phytotherapy Research* 12(4), 266–269.

Hasselstrom, T., Hewit, E.J., Konigsbacher, K.S. and Ritter, J.J. (1957) Composition of oil of black pepper. *Journal of Agricultural and Food Chemistry* 5, 53–55.

Jagella, T. and Grosch, W. (1999a) Flavour and off flavour compounds of black and white pepper (*Piper nigrum* L.). I. Evaluation of potent odorants of black pepper by dilution and concentration techniques. *European Food Research and Technology* 209(1), 16–21.

Jagella, T. and Grosch, W. (1999b) Flavour and off flavour compounds of black and white pepper (*Piper nigrum* L.). II. Odour activity values of desirable and undesirable odorants of black pepper. *European Food Research and Technology* 209(1), 22–26.

Jirovetz, L., Buchbauer, G., Ngassoum, M.B. and Geissler, M. (2002) Aroma compound analysis of *Piper nigrum* and *Piper guineense* essential oils from Cameroon using solid phase microextraction gas chromatography, solid phase microextraction gas chromatography mass spectrometry and olfactometry. *Journal of Chromatography* 976(1/2), 265–275.

Khajuria, A., Thusu, N. and Zutshi, U. (2002) Piperine modulates permeability characteristics of intestine by inducing alterations in membrane dynamics: influence on brush border membrane fluidity, ultrastructure and enzyme kinetics. *Phytomedicine* 9(3), 224–231.

Korány, K. and Amtmann, M. (1998) Gas chromatography/mass spectrometry measurements in the investigation of pepper aroma structures. *Rapid Communications in Mass Spectrometry* 11(6), 686–690.

Kurian, P.S., Backiyarani, S., Josephrajkumar, A. and Murugan, M. (2002) Varietal evaluation of black pepper (*Piper nigrum* L.) for yield, quality and anthracnose disease resistance in Idukki District, Kerala. *Journal of Spices and Aromatic Crops* 11(2), 122–124.

Lee, E.B., Shin, K.H. and Woo, W.S. (1984) Pharmacological study of piperine. *Archives of Pharmacal Research* 7, 127–132.

Lewis, Y.S., Nambudiri, E.S. and Krishnamurthy, N. (1969) Composition of pepper oil. *Perfumery and Essential Oil Records* 60, 259–262.

Mathew, A.G. and Bhattacharyya, S.C. (1990) Oil and oleoresin from Indian spices. In: Sen, N. and Sethi, K.L. (eds) *Proceedings of the 11th International Congress of Essential Oils, Fragrances and Flavours*, New Delhi, India, 12–16 November 1989, pp.189–195.

McCarron, M., Mills, A., Whittaker, D., Kurian, T. and Verghese, J. (1995) Comparison between green and black pepper oils from *Piper nigrum* L. berries of Indian and Sri Lankan origin. *Flavour and Fragrance Journal* 10(1), 47–50.

Menon, A.N. (2000) The aromatic compounds of pepper. *Journal of Medicinal and Aromatic Plant Sciences* 22(2/3), 185–190.

Menon, A.N. and Padmakumari, K.P. (2005) Essential oil composition of four major cultivars of black pepper (*Piper nigrum* L.) IV. *Journal of Essential Oil Research* 17, 206–208.

Menon, A.N., Padmakumari, K.P., Jayalekshmy, A., Gopalakrishnan, M. and Narayanan, C.S. (2000) Essential oil composition of four popular Indian cultivars of black pepper (*Piper nigrum* L.). *Journal of Essential Oil Research* 12(4), 431–434.

Menon, A.N., Padmakumari, K.P. and Jayalekshmy, A.J. (2002) Essential oil composition of four major cultivars of black pepper (*Piper nigrum* L.). *Journal of Essential Oil Research* 14(2), 84–86.

Menon, A.N., Padmakumari, K.P. and Jayalekshmy, A. (2003) Essential oil composition of four major cultivars of black pepper (*Piper nigrum* L.). III. *Journal of Essential Oil Research* 15(3), 155–157.

Narayanan, C.S. (2000) Chemistry of black pepper. In: Ravindran, P.N. (ed.) *Black Pepper* (*Piper nigrum* L.). Harwood Academic, Amsterdam, The Netherlands, pp. 143–162.

Peter, K.V. (2000) Futurology of black pepper In: Ravindran, P.N. (ed.) *Black Pepper* (*Piper nigrum* L.). Harwood Academic, Amsterdam, The Netherlands, pp. 481–487.

Pino, J.A. and Borges, P. (1999) Volatile components of spices. II. Black pepper. *Alimentaria* 301, 47–53.

Pino, J., Rodriguez Feo, G., Borges, P. and Rosado, A. (1990). Chemical and sensory properties of black pepper oil (*Piper nigrum* L.). *Nahrung* 34(6), 555–560.

Pruthi, J.S. (1993) *Major Spices of India: Crop Management and Post Harvest Technology*. Indian Council of Agricultural Research, New Delhi.

Purseglove, J.W., Brown, E.G., Green, C.L. and Robins, S.R.J. (1981) *Spices*, Volume I. Longman, New York, pp. 10–99.

Radhakrishnan, V.V., Madhusoodanan, K.J., Menon, P.P. and Thomas, J. (2004) Performance evaluation of selected varieties of pepper in the high ranges of Kerala. *Indian Journal of Arecanut, Spices and Medicinal Plants* 6(3), 87–88.

Rajinder, K.B., Glaeser, H., Becquemont, L., Klotz, U., Gupta, K.S. and Fromm, F.M. (2002) Piperine, a major constituent of black pepper, inhibits human P-glycoprotein and CYP3A4. *The Journal of Pharmacology and Experimental Therapeutics* 302(2), 645–650.

Ravindran, P.N. (2000) Introduction on black pepper. In: Ravindran, P.N. (ed.), *Black Pepper* (*Piper nigrum* L.). Harwood Academic, Amsterdam, The Netherlands, pp. 1–22.

Ravindran, P.N., Babu, K.N., Sasikumar, B. and Krishnamurthy, K.S. (2000a) Botany and crop improvement of black pepper. In: Ravindran, P.N. (ed.), *Black Pepper* (*Piper nigrum* L.). Harwood Academic, Amsterdam, The Netherlands, pp. 23–142.

Ravindran, P.N., Balachandran, I. and Chempakam, B. (2000b) Botany and crop improvement of black pepper. In: Ravindran, P.N. (ed.), *Black Pepper* (*Piper nigrum* L.). Harwood Academic, Amsterdam, The Netherlands, pp. 467–469.

Roessner, U., Wagner, C., Kopka, J., Tretheway, R.N. and Willmitzer, L. (2000) Simultaneous analysis of metabolites in potato tuber by gas chromatography mass spectrometry. *The Plant Journal* 23(1), 131–142.

Sasikumar, B., Haridas, P., George, K.J., Saji, K.V., Zachariah, T.J., Ravindran, P.N., Babu, K.N., Krishnamoorthy, B., Mathew, P.A. and Parthasarathy, V.A. (2004) 'IISR Thevam', 'IISR Malabar Excel' and 'IISR Girimunda' three new black pepper (*Piper nigrum* L.) clones. *Journal of Spices and Aromatic Crops* 13(1), 1–5.

Shankaracharya, N.B., Rao, L.J., Naik, J.P. and Nagalakshmi, S. (1997) Characterisation of chemical constituents of Indian Long Pepper (*Piper longum* L.). *Journal of Food Science and Technology, Mysore* 34(1), 73–75.

Shiratsuchi, H., Shimoda, M., Minegishi, Y. and Osajima, Y. (1993) Isolation and Identification of volatile flavour compounds in nonfermented coarse cut sausage. Flavour as a quality factor of non-fermented sausage. *Journal of Agricultural and Food Chemistry* 47, 647–652.

Singh, G., Marimuthu, P., Catalan, C. and Lampasona, M.P. de (2004) Chemical, antioxidant and antifungal activities of volatile oil of black pepper and its acetone extract. *Journal of the Science of Food and Agriculture* 84(14), 1878–1884.

Sowbhagya, H.B., Sampathu, S.R., Krishnamurthy, N. and Shankaranarayana, M.L. (1990) Stability of piperine in different solvents and its spectrophotometric estimation. *Indian Spices* 27, pp. 21–23.

Sumathykutty, M.A., Sankarikutty, B., Rajaraman, K., Narayanan, C.S. and Mathew, A.G. (1989) Quality criteria of pepper grades from pure cultivars. *Indian Perfumer* 33(2), 147–150.

Sunil, B., Bedi, K.L., Singh, A.K. and Johri, R.K. (2001) Piperine inhibits gastric emptying and gastrointestinal transit in rats and mice. *Planta Medica* 67(2), 176–179.

Tainter, R.D. and Grenis, A.T. (1993) *Spices and Seasonings: a Food Technology Handbook*. VCH Publishers, Inc., New York, pp.104–112.

Tipsrisukond, N., Fernando, L.N. and. Clarke, A.D. (1998) Antioxidant effects of essential oil and oleoresin of black pepper from supercritical carbon dioxide extractions in ground pork. *Journal of Agricultural and Food Chemistry* 46(10), 4329–4333.

Traxler, J.T. (1971) Piperanine, a pungent component of black pepper. *Journal of Agricultural and Food Chemistry* 19(6), 1135.

Veena, B., Saaket, V. and Dipankar, C. (2001) Piperine augments transcription inhibitory activity of rifampicin by severalfold in *Mycobacterium smegmatis*. *Current Science* 80(10), 1302–1305.

Vijayakumar, R.S., Surya, D. and Nalini, N. (2004) Antioxidant efficacy of black pepper (*Piper nigrum* L.) and piperine in rats with high fat diet induced oxidative stress. *Redox Reporter* 9(2), 105–110.

Wood, A.B., Barrow, M.L. and James, D.J. (1988) Piperine determination in pepper (*Piper nigrum* L.) and its oleoresins: a reversed phase high performance liquid chromatographic method. *Flavour and Fragrance Journal* 3(2), 55–64.

Zachariah, T.J. (1995) Essential oil and its major constituents in selected black pepper accessions. *Plant Physiology and Biochemistry New Delhi* 22(2), 151–153.

Zachariah, T.J., Mathew, P.A. and Gobinath, P. (2005) Chemical quality of berries from black pepper varieties grafted on *Piper colubrinum*. *Journal of Medicinal and Aromatic Plant Sciences* 27(1), 39–42.

Zhao, J. and Cranston, P.M. (1995) Microbial decontamination of black pepper by ozone and the effect of the treatment on volatile oil constituents of the spice. *Journal of the Science of Food and Agriculture* 68(1), 11–18.

3 Small Cardamom

B. Chempakam and S. Sindhu

3.1. Introduction

Small cardamom, known as the 'queen of spices', which belongs to the family of Zingiberaceae, is a rich spice obtained from the seeds of a perennial plant, *Elettaria cardamomum* Maton. It is one of the highly prized spices of the world and is the third most expensive spice after saffron and vanilla. Cardamom is one of those spices that cross the sweet/savoury boundary between desserts and main dishes. The original home of this precious spice is the mountains of the south-western parts of the Indian Peninsula. As early as the 4th century BC, cardamom was used in India as a medicinal herb and was an article of Greek and Roman trade. India had a virtual monopoly of cardamom until recently. Cardamom cultivation in India is confined to three states: Kerala, Karnataka and Tamil Nadu (Korikanthimath *et al.*, 2002) However, now it is cultivated in Guatemala, Sri Lanka, Thailand, Laos, Nepal, Vietnam, Costa Rica, El Salvador, Mexico and Tanzania (Mehra, 2001).

Cardamom shared about 60% of the total import value of US$204 million in 2004. World prices fell by 10% on average from US$6.64/kg in 2000 to US$4.30/kg in 2004. Oversupply of cardamom resulting from bumper crops and declining demand in the Arab countries resulted in lower unit values. Saudi Arabia is the major importer of cardamom, followed by Kuwait. With an export value of US$80 million, Guatemala was the world's main exporter of cardamom in 2004, followed by Nepal, which exports black cardamom. India is the world's main producer of cardamom, but the major quantity is used for internal consumption, with limited exports (International Trade Centre, 2006).

The major exporters of this famous spice are Costa Rica, Guatemala, Indonesia, Brazil, Nigeria, India, Thailand, Nicaragua and South Africa. On the other hand, Saudi Arabia claims to be its single largest importer. Kuwait follows Saudi Arabia in the importing list, but is nowhere near the leader. The major importers are Saudi Arabia, Kuwait, United Arab Emirates, China, Japan, Hong Kong, the Netherlands, Singapore and the USA.

World production of cardamom is estimated at 30,000 MT. Currently, the major producer is Guatemala, recording an average annual production of 18,000–20,000 MT. India is the second largest producer, with an average production of 11,000–12,000 MT. Indian cardamom is considered a superior quality in the international markets.

The countries in the western Asian region, such as Saudi Arabia, United Arab

Emirates and India, have maximum consumption and these countries share around 60% of the world's consumption. The Scandinavian countries like Denmark, Finland, Sweden, Norway and Iceland have around 16% share in the world consumption. The rest of the European countries have 14% share, Japan has 3% share and the USA has 2.5% share in world consumption. Global consumption of cardamom is estimated as 15,000–24,000 t. On the other hand, the current domestic demand for small cardamom has been estimated as 11,000 MT (http://www.nmce.com/commodityStudy/Cardamom_study.jsp).

Cardamom has a history as old as the human race. It is mentioned in ancient Vedic texts, by Theophrastus in the fourth century BC and later by Dioscorides in the 5th century BC. Cardamom grew in the gardens of the King of Babylon in 720 BC. It was used as early as the 4th century BC by Indian Ayurvedic experts and by ancient Greeks and Romans. Cardamom is believed to be a native of India. It was probably imported into Europe in 1214 AD. By 1000 AD, it was a good traded from India westwards (http://www.primaryinfo.com/cardamom.htm).

Data on the export of small cardamom from India for the period 1995/96 to 2005/06 are given in Table 3.1 (Anon., 2007).

Table 3.1. Export of small cardamom from India.

Year	Quantity (t)	Value (US$)
1995/96	527	308.81
1996/97	226	207.14
1997/98	370	301.67
1998/99	476	601.19
1999/2000	676	778.81
2000/01	1545	2016.19
2001/02	1031	1468.57
2002/03	682	1120.71
2003/04	757	879.05
2004/05	650	569.05
2005/06	875	643.09

Source: http://www.indianspices.com/.

3.2. Botany and Uses

Botany

Elettaria cardamomum var. Minor comprises all the cultivated types, while *E. cardamomum* var. Major Thw. denotes wild types native to southern India and Sri Lanka. Var. major is the more primitive variety from which the cultivated var. Minor is derived. All varieties and races are interfertile and the observed variations are due to natural crossing.

Cardamom grows abundantly in forests at 760–1500 m (2500–5000 ft) above sea level. It is widely cultivated in India, southern Asia, Indonesia and Guatemala, preferring shady locations and rich, moist, well-drained soil. Cardamom is a perennial bushy herb, growing to about 4.5 m (15 ft) with mauve-marked, orchid-like white flowers and very long, lance-shaped leaves. The tuberous underground rhizome is its real stem and the aerial shoot is a pseudostem formed by the encircling leaf sheaths. The leaves are distichous, long, alternate and lanceolate acuminate in shape. The flowers are borne on panicles and they emerge directly from the underground stem on long floral stalks. The flower-stalk proceeds from the base of the stem and lies on the ground, with the flowers arranged in a panicle.

The fruit is an ovoid, three-celled, loculicidally dehiscent capsule containing many seeds, which are covered by an aril. During drying, it is said to lose three-quarters of its weight. They are hermaphrodite and zygomorphic. The corolla is tubular, 3-lobed, pale green, androecium with petalloid labellum, white in colour with pink or purplish veins, composed of three modified stamens with an undulated edge. There are two further rudimentary staminodes and one functional stamen. The fruits are trilobular, ovoid or oblong, greenish-brown capsules containing about 15–20 seeds attached to an axile placenta. The light reddish- or dark reddish-brown seeds are irregularly 3-sided, transversely wrinkled or furrowed and are covered by a membranous aril. Each pod contains up to 20 aromatic, dark red-brown seeds that have a mild ginger flavour.

The seedpods are harvested by hand in dry weather during the autumn, just before they start to open. Then they are dried whole in the sun. The main harvest is in October and November of the third year after planting, after which the seeds are sorted according to size, form, colour, etc.

The basic chromosome number of *Elettaria* is x = 12 and the somatic number of *E. cardamomum* is 2n = 48 or 52.

Uses

Cardamom is used as an aromatic, carminative and stimulant. The seeds have a warm, slightly pungent aromatic flavour. It is used mainly as a flavouring agent in tea and food preparations. Cardamom oil is a precious ingredient in food preparations, perfumery, health foods, medicine and beverages.

Cardamom is also used internally for indigestion, nausea, vomiting and pulmonary disease with copious phlegm and also as a laxative to prevent stomach pain and griping, as well as flatulence. The seeds are also chewed to sweeten the breath and taken to detoxify caffeine in people drinking excessive amounts of coffee.

In India, it is used for many conditions, including asthma, bronchitis, kidney stones, anorexia and general debility, as well as for disorders of the urinary tract. It is also used for digestive upsets, soothing a spastic colon and relieving flatulence and constipation.

3.3. General Composition

The chemical composition of cardamom differs considerably with variety, region and age of the product. The general chemical composition is given in Table 3.2. The content of volatile oil in the seeds is strongly dependent on storage conditions, with an average yield from 2 to 5%. The oil is described as sweet, spicy, warm, lightly camphorated and citrusy (Robert, 1986; Boiswert and Hubert, 1998). The volatile oil contains about 1.5% α-pinene, 0.2% β-pinene, 2.8% sabinene, 1.6% myrcene, 0.2% α-phellandrene, 11.6% limonene, 36.3% 1,8-cineole, 0.7% γ-terpinene, 0.5% terpinolene, 3% linalool, 2.5% linalyl acetate, 0.9% terpinen 4-ol, 2.6% α-terpineol, 31.3% α-terpinyl acetate, 0.3% citronellol, 0.5% nerd, 0.5% geraniol, 0.2% methyl eugenol and 2.7% *trans*-nerolidol (Korikanthimath *et al.*, 1999). The basic cardamom aroma is produced by a combination of the major components, 1,8-cineole and α-terpinyl acetate (Lawrence, 1978).

Among the cultivated types, 'Malabar' and 'Mysore' are the major international trade groups. A third intermediate, termed 'Vazhukka', is also treated as international and is cultivated mainly in Kerala, India.

Table 3.2. Composition of cardamom.

Region of growth	Weight of 100 capsules (g)	Husk (%)	Seed (%)	Volatile oil (%)	Non-volatile ether extract (%)	Starch (%)	Crude fibre (5)	Protein ($N \times 6.25$) (%)
Karnataka								
Mudigere	23–24	25.5–28.0	72.0–74.5	8.6–8.9	2.0–3.6	47.0–48.0	6.9–6.8	8.8–11.3
Coorg	23–25	26.0–27.0	73.0–74.0	9.1–9.4	2.2–3.1	47.7–48.0	6.7–7.2	10.5
Kerala								
Wayanad	20–22	28.0–38.0	62.0–72.0	7.5–10.0	2.2–3.4	39.1–43.7	8.4–9.3	9.7–14.0
Alleppey Green	23	27.7	72.3	9.4–9.6	2.2.0	37.8	–	–
Tamil Nadu								
Yercaud	23–26	27.0	73.0	9.4–9.6	2.4	45.5	7.0	9.8
Nelliampathy	12–18	26.0–31.0	74.0	8.5–10.5	2.5–3.5	43.0–46.0	9.5–10.8	10.7–11.5

The Mysore cultivar comprises large plants with leafy stems up to 5 m, while the Malabar group comprises plants less than 3 m.

The Malabar and Mysore types differ in the composition of their volatile oils. The oil from var. Malabar is more camphory in aroma, due to the higher content of 1,8-cineole. Var. Mysore, or the commercial grade, known as 'Alleppey green', contains more α-terpinyl acetate, which contributes to the mild spicy flavour.

3.4. Chemistry

Volatiles

The volatile oil components in cardamom are summarized by Guenther (1975). The first detailed analysis of the oil was reported by Nigam *et al.* (1965). The oil has little mono- or sesquiterpenic hydrocarbons and is dominated by oxygenated compounds, all of which are potential aroma compounds. While many of the identified compounds (alcohols, esters and aldehydes) are commonly found in many spice oils (or even volatiles of many different foods), the dominance of the ether, 1,8-cineole and the esters, α-terpinyl and linalyl acetates in the composition make the cardamom volatiles a unique combination (Lewis *et al.*, 1966; Salzer, 1975; Korikanthimath *et al.*, 1997).

Table 3.3. Main components of volatile oil present in small cardamom.

Component	Total oil (%)
α-Pinene	1.5
β-Pinene	0.2
Sabinene	2.8
Myrcene	1.6
α-Phellandrene	0.2
Limonene	11.6
1,8-Cineole	36.3
γ-Terpinene	0.7
ρ-Cymene	0.1
Terpinolene	0.5
Linalool	3.0
Linalyl acetate	2.5
Terpinen-4-ol	0.9
α-Terpineol	2.6
α-Terpinyl acetate	31.3
Citronellol	0.3
Nerol	0.5
Geraniol	0.5
Methyl eugenol	0.2
trans-Nerolidol	2.7

Source: Lawrence (1978); Govindarajan *et al.* (1982).

The major components in cardamom oil are given in Table 3.3, while the trace components are grouped in Table 3.4.

The volatile oil, the most functionally important constituent of cardamom, varies from 6.6–10.6% in seeds for cv. Mysore and Malabar grown in India (Krishnamurthy, 1964; Krishnamurthy *et al.*, 1967; Korikanthimath

Table 3.4. Trace components in cardamom volatile oil.

Hydrocarbon	Acid	Carbonyl
α-Thujene	Acetic	3-Methyl butanal
Camphene	Propionic	2-Methyl butanal
α-Terpinene	Butyric	Pentanal
cis-Ocimene	2-Methyl butyric	Furfural
trans-Ocimene	3-Methyl butyric	8-Acetoxy carvotanacetone
Toluene	Alcohols and phenols	Cuminaldehyde
ρ-Dimethylstyrene	3-Methyl butanol	Carvone
Cyclosativene	ρ-Methyl-3-en-l-ol	Pinole
α-Copaene	Perillyl alcohol	Terpinene-4-yl-acetate
α-Ylangene	Cuminyl alcohol	α-Terpinyl propionate
γ-Cadinene	ρ-Cresol	Dihydro-α-terpinyl acetate
Δ-Cadinene	Thymol	

Sources: Lawrence (1978); Govindarajan *et al.* (1982).

et al., 1999). The oil content is low in the immature capsules in the order of 4–5%, while the husk oil is reported as 0.2%, having similar properties to seed oil (Rao *et al.*, 1925). Large differences are shown in the concentration of 1,8-cineole in the oils of var. Malabar and var. Mysore. In var. Mysore, linalool and linalyl acetate are markedly higher. This, along with a low content of 1,8-cineole, makes var. Mysore the largest selling Indian cardamom grade, Alleppey Green (Table 3.5). Differences in the oil content and the composition of the two main components (1,8-cineole and α-terpinyl acetate) in the germplasm collections at the Indian Institute of Spices Research Regional Station in Karnataka show variations, as indicated in Table 3.6.

Table 3.5. Concentration of 1,8-cineole and α-terpinyl acetate in cardamom oils from different origins.

		Percentage	
No.	Oil type	1,8-Cineole	α-Terpinyl acetate
1.	Guatemala I	36.40	31.80
2.	Guatemala II	38.00	38.40
3.	Guatemalayan Malabar type	23.40	50.70
4.	Guatemalayan I	39.08	40.26
5.	Guatemalayan II	35.36	41.03
6.	Synthite (commercial grade)	46.91	36.79
7.	Mysore type (Ceylon)	44.00	37.00
8.	Malabar type (Ceylon)	31.00	52.50
9.	Mysore I	49.50	30.60
10.	Mysore II	41.70	45.90
11.	Mysore	41.00	30.00
12.	Malabar I	28.00	45.50
13.	Malabar II	43.50	45.10
14.	Ceylon type	36.00	30.00
15.	Alleppey I	38.80	33.30
16.	Alleppey green	26.50	34.50
17.	Coorg green	41.00	30.00
18.	Mangalore I	56.10	23.20
19.	Mangalore II	51.20	35.60
20.	Papua New Guinea	63.03	29.09
21.	Cardamom oil (Indian origin)	36.30	31.30

Source: Govindarajan *et al.* (1982).

Table 3.6. Chemical quality profile of some cardamom accessions from the germplasm assemblage at the Indian Institute of Spices Research, Calicut, Kerala.

Acc. No.	Essential oil (%)	1,8-Cineole (% of oil)	α-Terpinyl acetate (% of oil)
APG7	5.5	51	29
APG12	8.6	37	43
APG 20	9.4	48	34
APG 23	9.4	39	39
APG 25	7.5	38	43
APG 27	6.9	34	33
APG 32	5.6	38	44
APG 44	7.1	34	38
APG 48	6.3	41	33
APG 54	6.8	44	34
APG 65	7.5	32	45
APG 69	8.3	49	28
APG 71	7.3	36	31
APG 87	5.7	42	32
APG 98	6.3	33	30
APG 106	10.0	43	39
APG 112	6.6	45	34
APG 117	6.3	38	43
APG 134	6.0	40	38
APG 135	9.9	45	40
APG 158	6.6	40	38
APG 175	9.8	43	40
APG 178	5.6	22	39
APG 180	6.8	31	32
APG 183	7.8	24	39
APG 187	8.0	22	48
APG 193	8.3	28	47
CCS-1	8.6	42	36
APG 215	6.0	23	55
"217	6.0	25	51
"218	7.8	24	52
"221	7.8	37	40

Source: Zachariah and Lukose (1992).

The aroma differences in various sources of cardamom are attributed to the proportion of esters and 1,8-cineole (Wijesekera and Jayawardena, 1973; Korikanthimath *et al.*, 1999). The flavour characteristics of some important volatile components of cardamom are given in Table 3.7 and the chemical structures of major aroma compounds are given in Fig. 3.1.

Table 3.7. Flavour characteristics of some important volatile components of cardamom.

Component	Flavour description	Use level (ppm)	Range of concentration in cardamom oil (%)
Esters			
α-Terpinyl acetate	Mildly herbaceous, sweet, spicy, variation in odour, mild spicy taste.	1–15	34.6–52.5
Linalyl acetate	Sweet floral fruity odour and taste, poor tenacity.	2–15	0.7–6.3
Ethers			
1,8-Cineole	Fresh, camphoraceous, cool odour and taste.	1–15	23–51
Alcohols			
Linalool	Floral woody with citrusy note, creamy floral taste at low levels.	2–10	1.4–4.5
α-Terpineol	Delicately floral-like, lilac-like.	5–40	1.4–3.3
Others			
Methyl eugenol	Musty, tea-like,mildly spicy.	5–15	1.3

Source: Bernhard *et al.* (1971); ASTA (American Spice Trade Association).

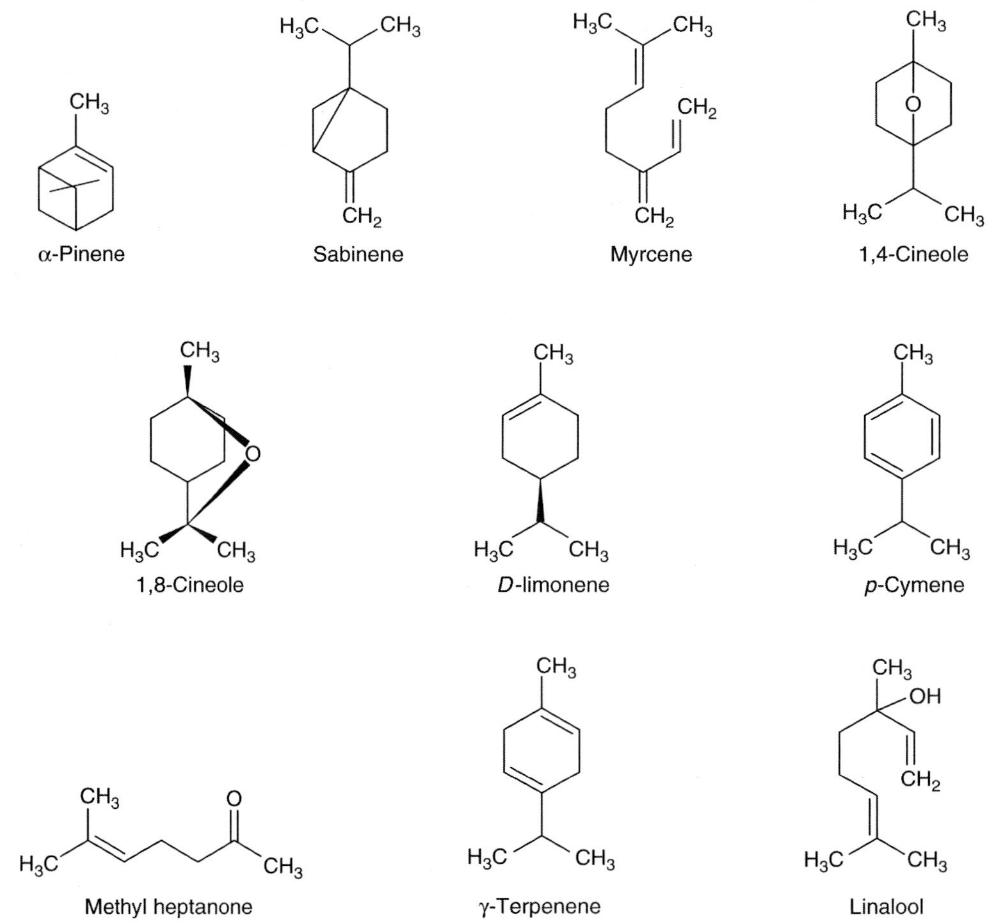

Fig. 3.1. Major essential oil components in small cardamom.

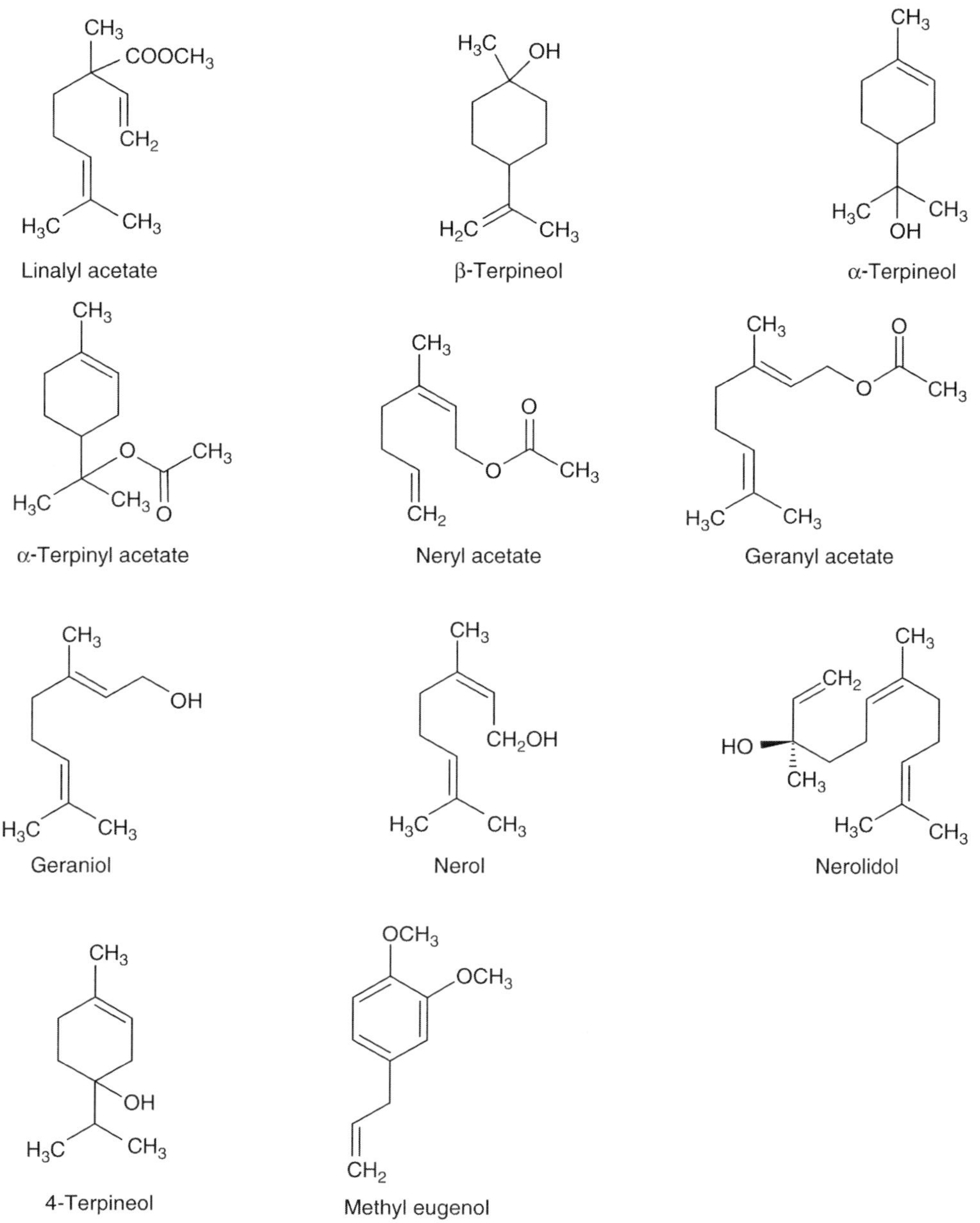

Fig. 3.1. *Continued*

Variability in oil composition

The value of cardamom as a food and beverage additive depends on the aroma components in the volatile oil. These are subject to differences according to varieties, maturity, processing, extraction techniques and storage.

VARIETY AND LOCATION Extracts of cardamom cultivated in Costa Rica have been analysed by GC-MS on various columns. Among the 122 compounds found, 56 represent over 99% of the total volatile fraction and 74 are reported for the first time (Table 3.8). Their distribution is similar to that described in other samples of *E. cardamomum* var.

Table 3.8. Major compounds identified in *Elettaria cardamomum* from Costa Rica.

Compound	Occurrence	Compound	Occurrence
Hydrocarbons		Farnesol	0.01
trans-4-*trans*-8,12-*trimethyl*-1,3, 7,11-Tridecatetraene	0.15	*trans*-Nerolidol	0.68
		α-Terpineol	2.97
cis-β-Ocimene	0.06	δ-Terpineol	0.07
trans-β-Ocimene	0.13	Terpinen-4-ol	1.76
Myrcene	2.21	*cis*-Carveol	1.5
α-Phellandrene	0.01	*trans*-Carveol	0.01
α-Terpinene	0.17	*trans*-ρ-Mentha-2,8-dien-1-ol	0.01
γ-Terpinene	0.55	Total	14.46
Terpinolene	0.36	Aldehydes	
Limonene	0.20	*trans*-2-Butenal	0.01
β-Elemene	0.03	2-Methylbutanal	0.01
Germacrene D	0.03	3-Methylbutanal	0.01
Humulene	0.01	Octanal	0.02
α-Pinene	1.51	*trans*-Dodec-5-enal	0.10
β-Pinene	0.15	Geranial	0.43
Camphene	0.02	Niral	0.33
Sabinene	3.78	Farnesal isomer	0.02
γ-Cadinene	0.12	Farnesal isomer	0.03
δ-Cadinene	0.03	Total	0.96
γ-Murolene	0.07	Esters	
α-Selinene	0.08	Octyl acetate	0.01
β-Selinene	0.23	Decyl acetate	0.01
β-Caryophyllene	0.01	Geranyl acetate	0.68
Total	9.91	Linalyl acetate	1.96
Alcohols		α-Terpinayl acetate	39.31
2-Methylbutan-1-ol	0.01	4-Terpinyl acetate	0.01
1-Octanol	0.21	Bornyl acetate	0.03
Geraniol	2.66	Geranyl propion	0.01
Nerol	0.11	Menthyl geraniate	0.03
Linalol	5.91	Total	42.05

Source: Noleau *et al.* (1987).

α-minor, grown in other countries (Noleau *et al.*, 1987).

Based on the physical and biochemical parameters and molecular techniques, traded (exported) cardamoms from India, Sri Lanka and Guatemala had been characterized (Thomas *et al.*, 2006). Indian cardamom was found to be superior for most of the physical quality parameters and for the biochemical traits. The GC profile of the oil of Indian cardamom also indicates a high quantity of α-terpinyl acetate and 1,8-cineole, which imparts aroma and flavour to the cardamom, thus reinforcing the legendary belief of the high intrinsic quality of the Indian cardamom. The GC-MS pattern of the volatile components in the cardamom oil extracted from the three traded cardamoms is given in Table 3.9.

EXTRACTION TECHNIQUES The oil is isolated either by traditional hydrodistillation or steam distillation. The major disadvantages of these methods are the loss of volatile components, longer periods of extraction and thermal or hydrolytic degradation of unsaturated or ester compounds (Tuan and Hangantileke, 1997; Khajeh *et al.*, 2004).

The volatile oil composition of cardamom seeds using supercritical CO_2 extraction shows the main components as follows: α-terpinyl acetate, 42.3%; 1,8-cineole, 21.4%; linalyl acetate, 8.2%; limonene, 5.6%; and linalool, 5.4%. The extract obtained using

Table 3.9. GC-MS-based comparison of essential oil constituents of Indian, Guatemalan and Sri Lankan cardamom.

Compound	Indian	Guatemalan	Sri Lankan
α-Phellandrene	0.37	ND	0.33
α-Pinene	2.37	2.30	2.50
Sabinene	4.93	4.61	4.32
Myrcene	2.62	2.56	2.34
Octanal	0.14	0.14	0.17
α-Terpinene 1,8-cineole	0.28	0.44	0.22
1,8-Cineole	27.59	26.99	26.85
trans-β-Ocimene	0.09	0.17	0.19
γ-Terpinene	0.62	0.51	0.19
cis-Sabinene hydrate	0.29	0.10	0.18
Tri cyclo heptane	ND	ND	0.17
α-Terpinolene	0.36	0.49	0.37
Linalool	1.23	6.99	5.44
4,8-Dimethyl-1,3,7-nona triene	0.20	ND	0.15
2-Cyclohexen-1-ol	0.20	0.20	0.19
δ-Terpineol	0.13	0.19	0.17
Terpineol-4	2.78	2.98	2.70
β-Fenchyl alcohol	2.97	6.61	4.23
Octyl acetate	ND	ND	0.11
Z-citral	0.36	ND	ND
Neral	ND	0.33	0.36
Nerol	2.56	ND	ND
β-Ocimene	ND	4.44	ND
Linalyl acetate	ND	ND	5.60
Geranial	0.56	0.51	0.69
δ-Terpinyl acetate	0.44	0.25	0.36
2,6-Octadienoic acid	0.37	ND	0.28
α-Terpinyl acetate	41.65	35.18	35.27
Neryl acetate	ND	0.30	0.45
2-Decenoic acid	0.17	ND	ND
Methyl cinnamate	0.14	ND	ND
Geranyl acetate	0.86	1.50	1.42
trans-β-Caryophyllene	ND	ND	0.13
Camphene	0.30	0.23	0.17
β-Selinene	1.55	0.30	1.29
α-Selinene	0.51	ND	0.37
α-Amorphene	0.32	ND	0.27
Germacrene	0.14	ND	0.12
Nerolidol	1.78	1.88	1.97

ND, not detected.
Source: Thomas *et al.* (2006).

hexane shows strong compositional differences, mainly of the following: limonene, 36.4%; 1,8-cineole, 23.5%; terpinolene, 8.6%; and myrcene, 6.6% (Marongiu *et al.*, 2004). A comparison with the hydrodistilled oil, obtained at a yield of 5.0%, does not reveal any consistent difference.

Recent methods for extraction of natural products using microwave energy, i.e. solvent-free microwave extraction (SFME), and also using supercritical CO_2 extraction, have been employed successfully in the case of cardamom (Lucchesi *et al.*, 2007).

SFME is based on the composition of microwave heating and distillation and is performed at atmospheric pressure. The composition of extracted oil varies with time, moisture content and irradiation power. The extraction time must be optimized to maximize the yield of the extraction without affecting the quality of the oil. Moisture content under the microwave treatment is critical, since water is an excellent absorber of microwave energy, which will provide a rise in temperature, and ruptures the essential oil cells by the *in situ* water, followed by the evaporation of water vapour. Irradiation power is directly related to the sample size. The power must be sufficient to reach the boiling point of water (100°C).

Six major compounds of the cardamom essential oil have been identified, namely: 1,8-cineole, α-terpinyl acetate, linalool, linalyl acetate, α-terpineol and terpin-4-ol, in order of importance. These six compounds represent almost 90% of the aromatic compounds of the essential oil from cardamom and all of them are oxygenated compounds.

A comparison of both the SFME method and hydrodistillation (HD), indicating the difference in the yields of the two major aromatic components, e.g. 1,8-cineole and α-terpinyl acetate, is shown in Table 3.10. Experiment 1 corresponds to the shortest extraction time, 23 min, the lowest irradiation power, 190 W and the lowest moisture content, while Experiment 8 consists of the longest extraction time, the highest irradiation power (340 W) and the highest moisture content (62%). In comparison, HD is characterized by a long extraction time (6 h) and high humidity level (~99%). Overall, the 1,8-cineole fraction seems to decrease with time, power and moisture, whereas α-terpinyl acetate seems to increase.

Monoterpene hydrocarbons are less valuable than oxygenated compounds in terms of their contribution to the fragrance of the essential oil. In the case of SFME, substantially higher amounts of oxygenated compounds are seen, as compared with HD. This is probably due to the diminution of thermal and hydrolytic effects during SFME. The more polar the compounds, the more readily microwave irradiation is absorbed, with better interaction between wave and matter, resulting in higher aromatic components. This corresponds with the higher levels of 1,8-cineole, which is more polar than α-terpinyl acetate.

STORAGE Quality parameters of cardamom oil obtained by supercritical carbon dioxide extraction and stored at 0°C or at ambient temperature (28 ± 3°) are compared with the quality of commercially steam-distilled oils at ambient temperature (Gopalakrishnan, 1994).

α-Pinene, sabinene and limonene are the major terpene hydrocarbons which undergo remarkable changes during storage. These hydrocarbons show 35–50% reduction during

Table 3.10. Comparison of oil composition: experimental data and the observed response value with different combinations of time, power and moisture.

Conditions of extraction (time (min)/power (W)/ humidity (%))	Experiment SFME R.R.I. (Relative Retention Indices)	Compounds					
		1,8-Cineole	Linalool	Terpene-4-ol	α-Terpineol	Linalyl acetate	α-Terpinyl acetate
23/190/38	Exp. 1	51.79	7.69	3.57	4.16	4.90	19.40
62/190/38	Exp. 2	42.51	7.84	3.95	4.59	6.18	24.78
23/340/38	Exp. 3	36.37	8.90	4.04	6.04	7.62	26.32
62/340/38	Exp. 4	44.32	9.64	4.05	4.30	5.57	22.59
23/190/62	Exp. 5	37.87	7.92	3.81	4.85	6.86	27.27
62/190/62	Exp. 6	40.86	7.73	3.71	4.54	6.26	27.33
23/340/62	Exp. 7	39.87	8.86	3.88	5.46	6.64	24.66
62/340/62	Exp. 8	35.47	7.93	3.83	5.02	7.39	29.22
	Hydrodistillation (HD)	26.23	5.29	2.60	3.88	3.63	45.45

Source: Lucchesi *et al.* (2007).

90 days of storage at 0°C in CO_2-extracted oil. α-Pinene and sabinene together are reduced from 7.1 to 0.4% at ambient conditions. Similarly, α-limonene is reduced from 2.3 to 0.5% during the same period of storage. The above hydrocarbons are reduced from 14.4 to 6.8% in the commercial oil stored at ambient conditions. Cineole contents decrease from 27.0 to 21.8% in the 0°C-stored samples, from 27.0 to 14.7% in the ambient temperature-stored samples and from 38.8 to 27.8% in commercial oil. Reductions in percentage proportion of these values are, respectively, 19, 45 and 28%. Changes also take place in the terpene alcohols but are not prominent. In the CO_2 extract stored at 0°C and in distilled oil, α-terpinyl acetate content increases during 90 days of storage. In other samples, a remarkable increase of this ester content is noted by 45 days. The two minor esters, geranyl acetate and linalyl acetate, also undergo minor changes in their contents in all of the samples during storage.

ENCAPSULATION OF CARDAMOM SEEDS The cardamom flavour is incorporated into processed foods, mainly by using the hydrodistilled cardamom oil or the solvent-extracted cardamom oleoresin (Govindarajan *et al.*, 1982). Solvent extraction of ground seeds of cardamom gives a greenish oleoresin containing about 70% volatile oil. It has the full flavour of the spice. At an elevated temperature, changes may occur in the volatile constituents. Cardamom volatile oils consist of terpenoids, which are generally unstable under detrimental conditions like acid, light, oxygen or heat. There is an increase in *p*-cymene, a terpene with a petroleum-like aroma, at the expense of the major constituent, α-terpinyl acetate, which contributes to the desirable flavour of this spice (Brennand and Heinz, 1970). These problems are overcome by microencapsulation, which is defined as the technique of packing minute particles of a core material within a continuous polymer film that is designed to release its contents in a predictable manner under a predetermined set of conditions (Beristain *et al.*, 2001). Microencapsulation of cardamom oleoresin using Gum Arabic could entrap the aroma for 6 weeks (Krishnan *et al.*, 2005). This technique will be highly useful in the preparation of many pastry products that use cardamom at higher oven temperatures (from 149 to 205°C) and beverages at lower ranges of pH.

EFFECT OF γ-IRRADIATION IN VOLATILE OIL COMPOSITION Currently, γ-irradiation is used for the decontamination of spices but its effect on essential oil composition is controversial and contradictory in cardamom (Ljubica, 1983; Klaus and Wilhelm, 1990; Maija *et al.*, 1990).

The parent yield of volatile essential oil isolated from non-irradiated (NI) and irradiated (I) samples of cardamom does not reveal any significant difference. GLC profiles of the two samples of NI and I oil do not show any variation in the retention time, but the relative percentage distribution of the major constituents in the oil exhibits clear-cut quantitative differences (Variyar *et al.*, 1998).

Fixed oils

Cardamom seeds also contain fixed oils, which are constituted mainly by oleic and palmitic acids (Table 3.11). The non-saponifiable fraction consists mainly of waxes (*n*-alkanes and

Table 3.11. The composition of the fixed (fatty) oil of cardamom seed.

Fatty acid	Total fixed oil (%)
Oleic	42.5–44.2
Palmitic	28.4–38.0
Linoleic	2.2–15.3
Linolenic	5.8
Caproic	5.3
Stearic	3.2
Hexadecenoic	1.9
Caprylic	5.3
Capric	< 0.1–3.8
Myristic	1.3–1.4
Arachidic	0.2–2.1
Hexadecanoic	1.9
Pentadecanoic	0.4
Lauric	0.2

Source: Verghese (1996).

n-alkenes) and sterols (β-sitosterol and γ-sitosterol) (Kasturi and Iyer, 1955; Gopalakrishnan *et al.*, 1990).

3.6. Medicinal and Pharmacological Properties

Cardamom essential oil traditionally has been used as a tonic to the digestive system, as well as a component of many sensual aphrodisiac blends. The oil has the aroma of freshly dried cardamom pods, far superior to the comparatively flat steam-distilled variety of this oil. Cardamom oil may relieve spasm, possibly making it beneficial for colitis, irritable bowel syndrome, indigestion and cramps. It may be of benefit where the digestive system is affected by nervous tension. In addition, cardamom oil can relieve nausea and may be useful for morning sickness in pregnancy.

Cardamom is a strong tonic and stimulant, is stomachic and carminative and, to a lesser degree, is listed as a neuromuscular antispasmodic. It is also reported as anti-inflammatory and analgesic (Al-Zuhair *et al.*, 1996) and is effective against post-operative nausea and vomiting (de Pradier, 2006).

The major medicinal properties of cardamom essential oil include the following:

- Anti-inflammatory
- Antimicrobial
- Antiseptic
- Carminative
- Digestive
- Diuretic
- Stimulant
- Stomachic
- Tonic and antispasmodic

Pharmacological Properties

Antimicrobial activity

Extract of cardamom seed displays a variable degree of antimicrobial activity on different microorganisms. Assays indicate that cardamom seed has inhibitory activity on *Mycobacterium smegmatis, Klebsiella pneumoniae, Staphylococcus aureus, Enterococcus faecalis, Micrococcus luteus* and *Candida albicans* (Agaoglu *et al.*, 2005). However no inhibitory activity was observed against *Pseudomonas aeruginosa*. The antimicrobial effect of the oil was tested against nine bacterial strains, one fungus and one yeast. The oil was 28.9% as effective as phenol, with minimal inhibitory concentration of 0.7 mg/ml (Badei *et al.*, 1991a,b; Kubo *et al.*, 1991).

Antioxidant effect

Cardamom oil is effective as an antioxidant for cottonseed oil. The effect is increased by increasing the content of the oil from 100 to 5000 ppm. Addition up to 100 ppm will not affect the characteristic odour of the cottonseed oil (Badei *et al.*, 1991b).

The essential oil of cardamom is used for its uplifting and invigorating properties and helps digestion and nausea. It is used as an aphrodisiac, is helpful in countering the irritation experienced during premenstrual tension (PMS) and works well on the respiratory system to ease coughs and warming the body.

Anti-inflammatory activity

A comparative study of the anti-inflammatory activity of the oil extracted from commercial *E. cardamomum* seeds, in doses of 175 and 280 μl/kg, and indomethacin, in a dose of 30 mg/kg, against acute carrageenan-induced planter oedema in male albino rats was performed, which proved to be marked (Al-Zuhair *et al.*, 1996).

The volatile oil of cardamom seeds is more effective than fixed oil in inhibiting the growth of the microbial species examined. The inhibitory effect of volatile oil against some pathogenic fungi was increased as volatile oil concentration increased and had a highly inhibitory effect on the selected pathogenic bacteria. Moreover, it exhibits highly cytotoxic and anticarcinogenic activities against human tumour cell lines. The cardamom volatile oil has a noticeable effect as an anti-inflammatory agent (El Bastawesy and Mohamed, 2005).

Insecticidal activity

The characteristic flavours and odours emanating from the volatile oils of spices are known to have various effects on insect pests, including stored insects (Jacobson, 1989; Shayya *et al.*, 1991). The volatile oil from cardamom acts as a potential grain protectant by killing various life stages of the stored product insects attacking wheat, e.g. *Tribolium castaneum* and *Sitophilus zeamais*, via contact and fumigant action (Huang *et al.*, 2000).

Culinary uses

Cardamom seeds are used widely for flavouring purposes in food and as a carminative. Despite their numerous applications in the cooking styles of Sri Lanka, India and Iran, 60% of the world production is exported to Arab countries (South-west Asia, North Africa), where the largest percentage is used to prepare coffee.

Cardamom is often employed for Oriental rice-and-meat dishes. To prepare these, meats (more rarely vegetables) are braised in a thick, aromatic sauce, then uncooked rice is added and cooked slowly so that it absorbs the sauce and all its flavours.

Sometimes, curry powders contain small amounts of cardamom; cardamom is also frequently added to the Northern Indian garam masala.

3.7. International Specifications and Desirable Limits

Trade varieties

The official cardamoms (from *E. cardamomum* Maton var. *minuscule* Burkhill) are now classified as 'shorts' and 'short-longs'. The former are usually broad and plump, the latter fine-ribbed and lighter than the shorts. The Malabar cardamoms have the highest value; they consist of both types. The Mysore cardamoms are considered the next best grade; they consist mostly of 'shorts' but are less pungent in flavour than the Malabar. Both the Malabar and the Mysore cardamoms are also grown on the island formerly known as Ceylon (Sri Lanka); hence the terms 'Malabar', 'Malabar–Mysore' and 'Malabar–Ceylon' cardamoms. Apart from this, there are the Mangalore and the Alleppy cardamoms, grown near the ports of Mangalore and Cochin, respectively.

Varieties and grades of cardamom (small)

1. Bold: is a popular export grade; 90% and above of the capsules have a diameter of 6.5 mm and above, matured and greenish in colour. Lt. Wt. is 415 g.
2. Super Bold: is a very special variety. All capsules are matured greenish and have a diameter of above 8 mm. Lt. Wt. is more than 450 g.
3. Extra Bold: the best in the export market. All capsules are matured, greenish and have a diameter of 7 mm and above. Lt. Wt. is 435 g.
4. Bulk: this grade of cardamom is produced as it is. It contains matured and immature capsules of all sizes from black, yellow and split cardamom.
5. Small: small-size cardamom between 5.5 mm and 6.5 mm in size. Cleaned and dust removed, husk and black capsules. Lt. Wt. is around 385 g.
6. Open/Splits: more than 60% of capsules are open and the colour is partly greenish/pale yellow. All capsules are matured and are 6.5 mm and above in size.
7. Seeds: black/brown colour seeds are the original content in every cardamom capsule. The husks are fully removed. Lt. Wt. is around 550–600 g.
8. Fruit: fruits are generally over-matured capsules, slightly yellowish in colour and Lt. Wt. over 425 g.

Commercial cardamom grades in Sri Lanka

As regards exports from Colombo on the island of Sri Lanka, cardamom is shipped under the following designations:

GREEN CARDAMOMS Kandy type – relatively large and of a dark greenish colour.
Copernicus type – slightly smaller than the 'Kandy', colour is generally green but some capsules have some off-colour.

Green Faq type – small cardamoms with a grey-green colour.

Green cardamoms are most suited for distillation. They are firm to the touch and tightly closed, thereby protecting the aromatic oil-containing seeds inside the pods. Green cardamoms are shipped chiefly from Alleppey and Mangalore and, to a lesser extent, from Sri Lanka.

BLEACHED CARDAMOMS Malabar half bleached – fair-sized, average quality of the season, rather small capsules.
Curtius – fair-sized, rather long capsules.
Cleophas – fair-sized, roundish capsules.
Bleached cardamoms are prepared by bleaching the green fruits. They are classified according to their size.

- Bold
- Medium bold
- Medium
- Small.

The bolder the cardamoms, the more expensive they are. In the USA, bleached cardamoms are used mostly in pickling spices and in packaged goods.

SEEDS Crispus type – freshly removed seeds, obtained by the husking of either green or bleached capsules.

Grading, Packing

Indian cardamom is offered to the international market in different grades, such as: 'Alleppey Green Extra Bold' (AGEB), 'Alleppey Green Bold' (AGB) and 'Alleppey Green Superior' (AGSO) (Tables 3.12–3.15).

After grading, the cardamom needs to be stored over a period of time and is normally kept in double-lined polythene bags. During storage, some storage pests do impair

Table 3.12. Specifications for Indian cardamom, physical characteristics.

Grade	Description	Size (mm)	Weight (g/l) min.	Colour	General characteristics
AG Alleppey green					
AGB	Extra bold	7.0	435 }	Green	Kiln-dried, 3-cornered and with ribbed appearance
AGS	Superior	5.0	385 }		
AG S1	Shipment	4.0	320–350	Light	
AGL	Light	3.5	260	Green	
CG Coorg green					
CGEB	Extra bold	8.0	450	Golden to light green	
CGB	Bold	7.5	435		Round, ribbed or smooth skin
CG-1	Superior	6.5	415	Light green	
CG-2	Mota, green	6.0	385	Green	
CG-3	Shipment	5.0	350	Cream	
CG-4	Light	3.5	280	Brown	
Bleached (half-bleached)					
BL-1		8.5	340	Pale	Fully developed round/ 3-cornered ribbed or smooth skin
BL-2		7.0	340	Creamy	
BL-3		5.0	300	Dull white	

Source: Indian standard specification for cardamom. IS: 1907–1966. Indian Standards Institution, New Delhi-1.

Table 3.13. Coorg clipped and bleachable white cardamoms: Agmark specifications.

Variety	Grade designation	Empty and malformed capsules by count (max.) (%)	Unclipped capsules by count (max.) (%)	Immature and shrivelled capsules (%) by weight	Size (mm)	Weight g/l (min.)
Coorg clipped cardamoms	Bold	5.0	0.0	0.0	8.5	435
	Coorg green or Mota green	5.0	3.0	4.0	6.0	385
	Shipment Light	3.0	5.0	7.0	4.0	350
					3.5	260
Bleachable white cardamoms	Mysore/Mangalore bleachable cardamom clipped	1.0		0.0	7.0	460
	Mysore/Mangalore bleachable cardamom unclipped	1.0		0.0	7.0	460
	Bleachable bulk cardamom clipped	2.0		0.0	4.3	435
	Bleachable bulk cardamom unclipped	2.0		0.0	4.3	435

the quality of the produce. Hence, there is a need to evolve a storage system that minimizes such infestation.

Table 3.14. Whole cardamom: chemical and physical specification.

Specification	Suggested limits
ASTA cleanliness specifications	
Whole dead insects, by count	4
Mammalian excreta, (mg/lb)	3.0
Other excreta, (mg/lb)	1.0
Mould, % by weight	1.0
Insect defiled, infested, % by weight	1.0
Extraneous, % by weight	0.5
FDA DALs (Food and Drug Administration Defect Action Levels)	None
Volatile oil (% min.)	3
Moisture (% max.)	12
Ash (% max.)	10
Acid-insoluble ash (% max.)	2
Average bulk index (mg/100g)	
Bleached	320
Green	250

Note: ASTA = American Spice Trade Association.

Equilibrium relative humidity studies have shown that cardamom dried and maintained at or below 10% moisture retains its original colour and avoids mould growth (Govindarajan *et al.*, 1982). If black polyethylene is used, the effect of light is further minimized and safe storage is possible for

Table 3.15. Ground cardamom: chemical and physical specification.

Specification	Suggested limits
FDA DALs	None
Volatile oil (% min.)	3
Moisture (% max.)	12
Total ash (% max.)	10
Acid-insoluble ash (% max.)	2
Military specification (EE-S-631J, 1981) (decorticated cardamom)	
Volatile oil (ml/100g) (% min.)	3
Moisture (% max.)	12
Total ash (% max.)	7
Acid-insoluble ash (% max.)	3
Granulation (% min. through a USS No. 40)	95
Bulk index (ml/100 g)	190

Table 3.16. Alleppey and Mangalore cardamom seeds: Agmark specifications.

Variety	Grade designation	Trade name	Extraneous matter (%) by wt	Light seeds (%) by wt	Weight min. (g/l)
Alleppey	AS1	Prime	1.0	3.0	675
cardamom	AS2	Shipment	2.0	5.0	460
seeds	AS3	Brokens	5.0	–	–
Mangalore	MS1	Prime	1.0	3.0	675
cardamom	MS2	Shipment	2.0	5.0	460
seeds	MS3	Brokens	5.0	–	–

4 months, which is required for port storage and trans-shipment.

It is advisable to make use of the dried cardamom capsules preferably within 12–15 months of harvest, after which time the pleasant flavour and aroma are likely to be affected. Stored samples should be tested frequently for storage pests.

3.8. Conclusion

Small cardamom (*E. cardamomum*), known for its sweet and highly aromatic characteristics, enjoys an enviable commercial value all over the world. It is an export-oriented crop and its export value lies mainly in the flavour composition. The major flavourant – α-terpinyl acetate – belongs to the class of isoprenoids, which constitute the largest and most widely distributed class of secondary metabolites. However, the natural quality degrades during the extraction process, storage and postharvest handling. Hence, research needs to be intensified to improve the flavour quality by modern biotechnological approches and also to maintain quality by adopting methods such as cryogrinding, supercritical carbon dioxide extraction, etc.

References

Agaoglu, S., Dostbil, N. and Alemdar, S. (2005) Antimicrobial effect of seed extract of cardamom (*Elettaria cardamomum* Maton). *YYU Veteriner Fakültesi Dergisi* 16(2), 99–101.

Al-Zuhair, H., El-Sayeh, B., Ameen, H.A. and Al-Shoora, H. (1996) Pharmacological studies of cardamom oil in animals. *Pharmacological Research* 34, 79–82.

Anon. (2007) *Arecanut and Spices Database*. Directorate of Arecanut and Spices Development, Calicut, India.

Badei, A.Z.M., El-Akel, A.T.M and Morsi, H.H.H. (1991a) Evaluation of chemical, physical and antimicrobial properties of cardamom essential oil. *Bulletin of the. Faculty of Agriculture, University of Cairo* 42(1), 183–197.

Badei, A.Z.M., Morsi, H.H.H. and El-Akel, A.T.M. (1991b) Chemical composition and antioxidant properties of cardamom essential oil. *Bulletin of the Faculty of Agriculture University of Cairo* 42(1), 199–215.

Beristain, C.I., Garcia, H.S. and Vernon-Carter, E.J. (2001) Spray-dried encapsulation of cardamom (*Elletaria cardamomum*) essential oil with mesquite (*Prosopis juliflora*) gum. *Zeitschrift für Lebensmittel-Technologie und -Verfahrenstechnik* 34, 398–401.

Bernhard, R.A., Wijeskere, R. and Chichedter, C.O. (1971) Terpinoids of cardamom oil and their comparative distribution among varieties. *Phytochemistry* 10, 171.

Boiswert, C. and Hubert, A. (eds) (1998) *L'ABCdaire des épices*. Flammarion, Paris.

Brennand, C.P. and Heinz, D.E. (1970) Effects of pH and temperature on the volatile constituents of cardamom. *Journal of Food Science* 35, 533–537.

de Pradier, E. (2006) A trial of a mixture of three essential oils in the treatment of postoperative nausea and vomiting. *International Journal of Aromatherapy* 16(1), 15–20.

El Bastawesy, A.M. and Mohamed, R.H. (2005) Evaluation of cardamom oil role as antimicrobial, anticarcenogenic and anti-inflammatory agents. *Egyptian Journal of Agricultural Research* 83(2), 789–809.

Gopalakrishnan, M., Narayanan, C.S. and Grenz, M. (1990) Non-saponifiable lipid constituents of cardamom. *Journal of Agriculture and Food Chemistry* 38(12), 2133–2166.
Gopalakrishnan, N. (1994) Studies on the storage quality of CO_2-extracted cardamom and clove bud oils. *Journal of Agriculture and Food Chemistry* 42, 796–798.
Govindarajan, V.S., Shanti Narasimhan, Raghuveer, K.G. and Lewis, Y.S. (1982) Cardamom production, technology chemistry and quality. CRC *Critical Review of Food Science and Nutrition* 16, 326.
Guenther, E. (1975) Cardamom. In: *The Essential Oils*. Vol 5. Robert E. Krieger Publishing, New York, p. 85.
http://www.indianspices.com/
http://www.nmce.com/commodityStudy/Cardamom_study.jsp
http://www.primaryinfo.com/cardamom.htm
Huang, Y., Lam, S.L. and Ho, S.H. (2000) Bioactivities of essential oil from *Elletaria cardamomum* (L.) Maton to *Sitophilus zeamais* Motschulsky and *Tribolium castaneum* (Herbst). *Journal of Stored Products Research* 36, 107–117.
International Trade Centre (2006) *World Markets in the Spice Trade – 2000–2004*. UNCTAD/WTO, Geneva, Switzerland.
Jacobson, M. (1989) Botanical pesticides: past, present, future. In: Arnason, J.T., Philogene, B.J. and Morland, P. (eds) *Insecticides of Plant Origin*. American Chemical Society, Washington, DC, pp. 1–10.
Kasturi, T.R. and Iyer, B.H. (1955) Fixed oil from *Elletaria cardamomum* seeds. *Journal of Indian Institute of Science* 37A, 106.
Khajeh, M., Yamini, Y., Sefidkon, F. and Bahramifar, N. (2004) Comparison of essential oil composition of *Carum copticum* obtained by supercritical carbon dioxide extraction and hydrodistillation methods. *Food Chemistry* 86, 587–591.
Klaus, W. and Wilhelm, G. (1990) Chemo metric evaluation of GC/MS profiles for the detection of γ-irradiation of spices exemplified with nutmeg. *Deutsch Lebensmittel-Rundsuchung* 86, 344–348.
Korikanthimath, V.S., Mulge, R. and Zachariah, T.J. (1997) Variation in yield and quality characters of cardamom clones. *Journal of Medicinal and Aromatic Plant Sciences* 19(4), 1024–1027.
Korikanthimath, V.S., Mulge, R. and Zachariah, T.J. (1999) Variations in essential oil constituents in high yielding selections of cardamom. *Journal of Plantation Crops* 27, 230–232.
Korikanthimath, V.S., Govardhan Rao and Hiremath, G.M. (2002) Cultivation of cardamom (*Eletteria cardamomum)* in valley bottoms under evergreen forest shade. *Journal of Medicinal and Aromatic Plant Sciences* 24, 53–59.
Krishnamurthy, M.N. (1964) Studies on curing aspects and utilization of cardamom. Assoc. thesis, CFTRI, Mysore, India.
Krishnamurthy, M.N., Padma Bhai, R. and Natarajan, C.P. (1967) Chemical composition of cardamom. *Journal of Food Science and Technology* 4, 170.
Krishnan, S., Kshirsagar, A.C. and Singhal, R.S. (2005) The use of gum arabic and modified starch in the microencapsulation of a food flavouring agent. *Carbohydrate Polymers* 62, 309–315.
Kubo, I., Himejima, M. and Muroi, H. (1991) Antimicrobial activity of flavor components of cardamom *Elattaria cardamomum (Zingiberaceae*) seed. *Journal of Agricultural and Food Chemistry* 39 (11, 1984–1986).
Lawrence, B.M. (1978) Major tropical spices – cardamom (*Elettaria cardamomum*). In: *Hagers Handbuch der pharmazeutischen Praxis* (Ed. Franz von Bruchhausen) Allured Publishers, Wheaton, Illinois, pp. 105–155.
Lewis, Y.S., Nambudiri, E.S. and Phillip, T. (1966) Composition of cardamom oils perfume. *Journal of Essential Oil Research* 57, 623–626.
Ljubica, J. (1983) Study on some chemical changes in irradiated pepper and parsley. *International Journal of Applied Radiation Isotopes* 34, 787–791.
Lucchesi, M.E., Smadja, J., Bradshaw, S., Louw, W. and Chemat, F. (2007) Solvent free microwave extraction of *Elletaria cardamomum* L.: a multivariate study of new technique for the extraction of essential oil. *Journal of Food Engineering* 79, 1079–1086.
Maija, S.A., Merja, M., Pia, M. and Sinikka, P. (1990) Methods for detection of irradiated spices. *Zeitschrift für Lebensmitteluntersuchung und -Forschung A* 190, 99–103.
Marongiu, B., Piras, A. and Porcedda, S. (2004) Cyclic monoterpene extract from cardamom oil as a skin permeation enhancer for indomethacin: *in vitro* and *in vivo* studies. 1. *Journal of Agriculture and Food Chemistry* 52(20), 6278–6282.
Mehra, K. (2001) Indian system of innovation in biotechnology – a case study of cardamom. *Technovation* 21, 15–23.

Nigam, M.C., Handa, K.L., Nigam, I.C. and Leo Levi (1965) Essential oils and their constituents: xxix. The essential oil of marihuana: composition of genuine Indian *Cannabis sativa L. Canadian Journal of Chemistry* 43 (12): 3372–3376.

Noleau, I., Toulemonde, B. and Richard, H. (1987) Volatile constituents of cardamom (*Elettaria cardamomum*) cultivated in Costa Rica. *Flavour and Fragrance Journal* 2, 123–127.

Rao, B.S., Subrorough, J.J. and Watson, H.E. (1925) Notes on some Indian essential oils. *Journal of the Indian Institute of Science* 8A, 143

Robert, J.L. (ed.) (1986) *Le livre des épices et herbes*. Flammarion, Paris.

Salzer, U.J. (1975) Analytical evaluation of seasoning extracts (oleoresins) and essential oils from seasonings. I. *International Flavours and Food Additives* 6(3), 151–157.

Shayya, E., Ravid, U., Paster, N., Juven, B., Zisman, U. and Pissarev, V. (1991) Fumigant toxicity of essential oils against four major stored-product insects. *Journal of Chemical Ecology* 17, 499–504.

Thomas, E., Jaleel, K., Zachariah, T.J., Syamkumar, S. and Sasikumar, B. (2006) Comparative quality characterization and molecular profiling of Indian, Sri Lankan and Guatemalan cardamoms. *Journal of Food, Agriculture and Environment* 4(2), 129–133.

Tuan, D.Q. and Hangantileke, S.G. (1997) Liquid CO_2 extraction of essential oil from star anise fruits (*Illicium verum* H.). *Journal of Food Engineering* 31, 47–57.

Variyar, P.S., Bandyopadhyay, C. and Thomas, P. (1998) Effect of γ-irradiation on the volatile oil constituents of some Indian spices. *Food Research International* 31(2), 105–109.

Verghese, J. (1996) The world of spices and herbs. *Spice India* 12–20.

Wijesekera, R.O.B. and Jayawardena, A.L. (1973) Recent developments in the production of spices and their essential oils in Ceylon. In: *Proceedings of the Conference on Spices*. Tropical Product Institute, London, pp. 159–167.

Zachariah, T.J. and Lukose, R. (1992) Aroma profile of selected germplasm accessions in cardamom (*Elletaria cardamomum* Maton). *Journal of Plantation Crops* 20 (Supplement), 310–312.

4 Large Cardamom

B. Chempakam and S. Sindhu

4.1. Introduction

India is the largest producer of large cardamom (*Amomum subulatum* Roxburgh), with an annual production of 4000 MT, followed by Nepal (2500 MT) and Bhutan (1000 MT) (Berrig *et al.*, 1993). More than 85% of the production within India is from Sikkim. An estimated 4000 t of large cardamom, valued at about Rs. 1.60 billion, is produced annually in Sikkim alone, which constitutes nearly 80% of total production from India. It is also called greater Indian or Nepal cardamom, which is a native of the Eastern Himalayan region. Large cardamom is the most important perennial cash crop of the region and is widely cultivated with Himalayan alder (*Alnus nepalensis*) as a shade tree (Sharma *et al.*, 2002).

Large cardamom is a shade-loving crop. It grows under dense (60–70% of full daylight interception) to light shade (26% full daylight interception) conditions. The daylight intensity required for optimum growth of cardamom is 5000–20,000 lux. Therefore, it is necessary to clear the undergrowth in virgin forest and regulation of overhead shade is essential, in such a way that at least 50% shade is maintained in the area. About 30 important tree species are used to provide shade to the cardamom plants. *A. nepalensis*, a deciduous, nitrogen-fixing and fast-growing tree, is the species most commonly underplanted with cardamom.

Being the largest producer and exporter of large cardamom, India enjoys near monopoly of this spice. The main production centres are the sub-Himalayan ranges spread across the Sikkim and Darjeeling districts of West Bengal. Several species of the genus *Amomum* are distributed all over the mountainous area from the Himalayas to southern China. Other species include *A. subulatum*, or large cardamom, grown in northern India and Nepal; *A. aromaticum*, or Bengal cardamom, grown in south-eastern India; *A. krervanh*, or Siam or Cambodian cardamom, growing wild under forest cover in Thailand, Cambodia, Lao PDR and Vietnam; *A. globosum*, or Chinese cardamom, grown in southern China; and *A. xanthiodes*, or bastard cardamom, growing wild under forest cover in Thailand (Subba, 1984; Rao *et al.*, 1993b; Singh and Singh, 1996). Figures for the export of large cardamom from India for the past 10 years are given in Table 4.1.

Large cardamom is also known as 'black cardamom'. The pods are used as a spice, in a manner similar to the green Indian cardamom pods, but it has a drastically different flavour, so it cannot be substituted in the same recipes, unless a different flavour

(eds V.A. Parthasarathy, B. Chempakam and T.J. Zachariah)

Table 4.1. Export of large cardamom from India.

Year	Quantity (t)	Value (US$)
1995/96	1677	291.42
1996/97	1628	288.09
1997/98	1648	300.95
1998/99	1288	302.14
1999/2000	1185	419.52
2000/01	1506	583.57
2001/02	1577	569.52
2002/03	1450	489.76
2003/04	924	293.81
2004/05	950	270.00
2005/06	1025	252.38

Source: Spices Board, Cochin, India.

is acceptable. Unlike green cardamom, this spice is used rarely in sweet dishes. Its strong, smoky flavour and aroma are derived from the traditional drying procedure, which involves drying over open flames. North-east Indian and South-east Asian countries are dominated by the *Amomum* species, while *Aframomum* species are prevalent in the African regions of Sierra Leone, Guinea Coast, Madagascar and Tanzania. The fruits are much larger in size in comparison with *Elettaria cardamomum* (small cardamom), but the seed size and anatomy are similar in all three genera.

Large cardamom, a perennial cash crop grown beneath the forest cover on the marginal lands of Sikkim Himalaya, has been a boon to the mountain people of the area (Sharma *et al.*, 2002). It is cultivated in an area of about 23,500 ha in Sikkim. Being a shade-loving plant, the hills of Sikkim provide an ideal environment. The plant grows at altitudes between 600 and 2000 m, where rainfall is between 1500 and 3500 mm and the temperature varies from 6°C (min) to 33°C (max) (Anon., 1991). Frost and hailstorms are injurious to the plants during flowering (Biswas *et al.*, 1988).

There are three popular varieties (cultivars) of large cardamom in Sikkim, e.g. Ramsey, Golsey and Sawney. The varietal differences are described by Gyatso *et al.* (1980), Subba (1984) and Rao *et al.* (1993a) (Table 4.2). In addition to these popular varieties, there are several other varieties, such as Ramla, Chivey Ramsey, Garday Seto Ramsey, Ramnag, Madhusay, Seto Golsey, Slant Golsey, Red Sawney, Green Sawney and

Table 4.2. Characteristics of different varieties of large cardamom.

Character/variety	Ramsey	Golsey	Sawney
Altitude	High	Low to middle	Middle
Extent of cultivation (%)	60	30	7
Status	Tall, vigorous wide clump growth	Less vigorous with erect leafy stem bearing stout upright leaves, clumps medium	Tall, vigorous, bent downwards
Stem colour	Maroonish with dense foliage	Greenish to maroonish	Pinkish with dark green foliage
Flowers	Yellowish and small, corolla tip with pink tinge at base	Yellowish-orange	Yellowish with pink tinge at base of corolla
Capsules	Smaller (16–30 seeds)	Bold to round (40–50 seeds)	Medium bold (30–40 seeds)
Essential oil (%)	1–8	2.3–5.0	1.8–2.5
Shade requirement	Deep shade	Less shade	Moderate to deep shade
Susceptibility to diseases	Susceptible to Chirkey and Foorkey at lower altitudes	Tolerant to Chirkey and Foorkey but susceptible to leaf spots	Susceptible to viral diseases

Source: Rao *et al.* (1993a).

Mingney. Rao *et al.* (1993b) reported a promising variety, Barlanga, from higher altitudes with desirable high-yielding characteristics, such as maximum ratio of mature tillers to productive spikes (1:3.6) and bold-size capsules (with 50–80 seeds). Surveys carried out by Biswas *et al.* (1986) revealed that Ramsey and Ramla were well suited to higher altitudes, Golsey to lower altitudes and Sawney was widely adaptable to different elevations.

Since *A. subulatum* Roxburgh is cultivated to a larger extent, and also has significant potential for trade, this chapter, unless otherwise specified, deals mostly with this variety.

4.2. Botany and Uses

Botany

Cultivation is carried out mainly in swampy places along the sides of mountain streams in Nepal, Bengal, Sikkim and Assam (eastern Himalayas). Usually, the plants are grown at an elevation of 765–1675 m above mean sea level, along small springs, on the moist and shady sides of mountain streams and along hilly slopes. The plant is a perennial herb having subterranean rhizomes, which give rise to leafy shoots and spikes. It matures during the third year of growth and its height ranges from 1.5 to 3.0 m. Leafy shoots are formed by long, sheath-like stalks encircling one another. The leaves are green or dark green, glabrous on both surfaces, with acuminate apex. Inflorescence is a dense spike on a short peduncle bearing 40–50 flower buds in an acropetal sequence. Flowering of cardamom commences in the third year after planting. Flowers appear during April and May and the capsules mature in September and October.

The fruit is a trilocular many-seeded capsule. The capsule wall is echinated and is reddish-brown to dark pink (Rao *et al.*, 1993a). The capsule morphology has been studied in detail by Gupta (1986). Harvesting is usually carried out during August to October.

Dried large cardamom capsules are, on average, 25 mm long, oval to blobose, greyish-brown to dark reddish-brown. The fruit contains 40–50 seeds, held together by a viscous sugary pulp. Though the fruits are clearly identifiable by their larger size and differences in shapes compared with small cardamom, the seeds are of nearly the same size as those of true cardamom. Histological features, sizes and orientation of cells in different layers of husk and seed have been described by Berger (1965).

Flowering

Large cardamom is essentially a cross-pollinated crop due to the heterocyclic nature of its flowers, though they are self-fertile. Each spike bears 40–50 flowers, which open in an acropetal sequence, but only 10–15 capsules are formed per spike. The flowers remain viable for 14 h after opening (Rao *et al.*, 1993b). They are borne on shortly peduncled spikes of about 5–6 cm in diameter. The number of inflorescences produced on each clump ranges from 20 to 45, depending on the age of the clump. Each inflorescence produces 30–50 flowers. The flowers are yellowish and measure 7.03 cm in length. The most conspicuous part of the flower is the yellowish labellum/lip, which provides a platform for visiting insects. The basal parts of the petals and the labellum are fused to form a corolla tube/nectar tube (3.07 cm long). The terminally expanded part of the labellum is 3.52 cm long and 1.4 cm wide. The midrib region of the labellum is a deeper yellow and the veins are translucent. The anther is solitary, borne on a filament about 1 cm long, originating from the tip of the corolla tube, and measures 10.6 mm in length. The stamen extends beyond the anther in the form of a rolled-up leafy hood/crest. The pistil is solitary; the ovary is 5.25 mm long and contains an average of 106.8 ovules. The style is long (5.03 cm) and delicate; it passes through the groove present between the two pollen sacs. The stigma is cup-shaped (1 mm deep and 1.5 mm wide) and slightly flattened, with a row of unicellular, non-receptive hairs on its margin. Only the inner surface of the cup is receptive. The stigma cup is pointed distally and the inner wall is lined with a viscous exudate. The stigma extends 1.5–2.0 mm beyond the level of the anther and is covered with the rolled-up extension of the stamen (crest) in the form of a hood. Two yellowish nectarines (3.93 mm long) are located at the base of the style and fill the entire space in the lower part of the corolla tube (Sinu and Shivanna, 2007).

Cytology of *Ammomum* indicates that the diploid chromosome number of *A. subulatum* is 48. However, variability is also reported with 2n = 26, 34, 42 and 44 (Sharma and Bhattacharya, 1959).

Uses

Large cardamom is the dried fruit of a perennial herbaceous plant. Its quality characteristics are different from those of small cardamom. It is valued for its acceptable taste, flavour and aroma. The spice is used in rice preparations and meat dishes, besides a wide range of beverages and sweets.

Large cardamom has a fresh and spicy aroma. By virtue of the traditional drying procedure over open flames, the spice also acquires a smoky flavour. The ground seeds are an optional ingredient in mixed preparations and spice masala mixtures, and are also used as a flavouring agent in confectionary, hot or sweet pickles and in beverages.

Large cardamom also possesses curative properties in the Ayurvedic and Unani systems of medicine (Mukherjee, 1972; Singh, 1978; Nambiar *et al.*, 1994). It is also used to flavour cardamom cola, prepared by blending caramer acid and carbonating mixture.

4.3. General Composition

The chemical composition of large cardamom (Table 4.3) differs with variety, region and age. The seeds contain 3% essential oil, which is dominated by 1,8-cineole (more than 70%). Smaller and variable amounts of limonene, terpinene, terpineol, terpinyl acetate and sabinene have also been reported. Table 4.4 compares the composition of large cardamom seeds with small cardamom seeds (Singh, 1978).

Table 4.3. General composition of large cardamom.

Component	Value (%)
Moisture	8.49
Protein	6.00
Total ash	4.01
Starch	43.21
Crude fibre	22.00
Non-volatile ether extract	2.31
Volatile ether extract	3.00
Alcohol extract	7.02
Volatile extract	2.80
Water-soluble ash	2.15
Alkalinity of water-soluble ash	0.90
Ash insoluble in acid	0.42
Volatile oil	2.80

Source: Pruthi (1993).

Table 4.4. Comparison of chemical analysis of large and small cardamom seeds.

Character	Large cardamom (average %)	Small cardamom (average %)
Moisture	8.49	8.30
Volatile oil	2.80	8.30
Protein	8.00	10.30
Crude fibre	22.00	9.20
Total ether extract	43.21	45.40
Alcohol extract	7.02	–
Total ash	4.01	5.00

Source: Singh (1978).

4.4. Chemistry

Chemistry of volatiles

Volatile oil is the principal aroma-giving compound in the large cardamom. Steam distillation of the crushed seeds gives a dark brown oil (2.5%) with a cineol-like aroma. The highest volatile oil content was recorded as 3.32% in the Golsey Dwarf variety, whereas the lowest was 1.95% in the White Ramna variety (Gupta, 1986). Cineole contributes to pungency, while terpinyl acetate contributes towards the pleasant aroma (Karibasappa, 1987). Karibasappa also reported that the cultivar Ramnag, followed by Golsey, had uniform-sized capsules with maximum values for capsule weight, capsule size, seeds per capsules, oleoresin content and volatile oil content.

Quantitative chromatographic analysis of the composition of distilled essential oil was reported previously by Nigam and Purohit (1960) and by Lawrence (1970). The major constituent of large cardamom essential oil is 1,8-cineole (65–80%), while the content of α-terpenyl acetate is low (traces to 5%). The monoterpene hydrocarbon content is in the range of 5–7%, of which limonene, sabinene, terpinene and pinene are significant components. The terpinols comprise approximately 5–7% of the oil. The high cineole and low terpenyl acetate probably account for the very harsh aroma of this spice in comparison with that of true cardamom (Pruthi, 1993).

Seed oil

The seed oil in *A. subulatum* has been the subject of several investigations. Nigam and Purohit (1960) obtained 2.5% oil from the seeds and fractionated the oil into different cineole-rich fractions. Lawrence (1970) separated the components of the oil by preparative gas chromatography, identified them by their IR spectra and retention data and found the major component, 1,8-cineole, in 74%. Patra *et al.* (1982) studied the oil by packed column GC and reported that its major components were sabinene (9.1%), γ-terpinene (16.2%) and 1,8-cineole (63.3%). In another study, Gupta *et al.* (1984) analysed oils derived from different strains of *A. subulatum* growing wild in Sikkim and found the 1,8-cineole content varied from 77 to 89%. The oil and volatile concentrate produced by liquid carbon dioxide extraction of *A. subulatum* were compared by Kaur *et al.* (1993).

Analysis of the steam-distilled volatile oil of the seeds of the large cardamom grown in Sikkim, India, using GC-MS, identified 25 components, of which 16.3% was monoterpene hydrocarbons and 75.3% was oxygenated monoterpenes, with 1,8-cineole [eucalyptol] (61.3%) being the major component. α-Terpineol, α- and β-pinene and allo-aromadendrene were also detected (Gurudutt *et al.*, 1996).

The large cardamom pericarp (husk) yielded 0.18% volatile oil by the Clevenger hydrodistillation method. This oil was analysed for physical parameters, e.g. specific gravity (0.9148), refractive index (1.4733) and optical rotation (–7.700). The volatile oil was subjected to GC-MS analysis and 37 compounds were identified, constituting > 98% of the total oil. The major compounds characterized were 1,8-cineole (38.7%), β-pinene (13.6%), α-terpineol (12.6%), spathulenol (8.3%), 4-terpineol (4.5%), germacrene D (3.0%), α-pinene (2.8%) and β-selinene (2.7%). GC and GC-MS data revealed that 1,8-cineole content was less than 50% when compared with the seed oil. Table 4.5 shows the major constituents separated by GC-MS (Rout *et al.*, 2003). Figure 4.1 gives the structures of the major chemical components in the volatile oil from seeds.

Steam distillation of the crushed seeds of large cardamom yielded 2.5% dark brown-coloured liquid with the following physical constants: specific gravity (29°C), 0.9142; refractive index (29°C), 1.460; optical rotation in chloroform, 18°C.

The physical and chemical quality attributes of large cardamom (*A. subulatum*) cultivars obtained from the north (Zongu Golsai, Ramsey and Golsai), south (Sawney and Golsai), east (Barlangey) and west (Zongu Golsai, Ramsey and Barlangey) regions of Sikkim, India, Bhutan (Bhutan large cardamom) and Nepal (Tedey K. cut) were evaluated (Naik *et al.*, 2006). GC analysis of the volatile oils showed that there was considerable variation among the cultivars with respect to α-pinene (3.2–4.5%), β-pinene (6.7–8.5%), 1,8-cineol (80.4–84.6%), 4-terpineol (0.60–1.30%) and α-terpineol (3.3–4.3%). Analysis for metals by AAS showed that the seeds contained cadmium (0.06, 0.07 and 0.07 ppm, respectively), lead (0.12, 0.37 and 0.24 ppm, respectively), copper (5.14, 9.68 and 6.33 ppm, respectively) and iron (28.51, 111.19 and 55.28 ppm, respectively). Analysis of the seed and capsules for heavy metals showed that there was considerable variation among regional cultivars with respect to iron content, whereas the cadmium,

Table 4.5. Percentage composition of oils from the seeds of *Amomum subulatum* growing in Sikkim.

Compound	Retention index calculated	RI literature	Fresh seed oil	Laboratory-dried seed oil
α-Thujene	934	931	0.1	0.1
α-Pinene	941	939	1.7	2.3
Camphene	953	953	0.1	0.1
β-Pinene	976	980	3.2	3.4
Myrcene	991	991	0.7	1.4
α-Phellandrene	1002	1005	t	t
δ-3-Carone	1014	1011	0.1	0.1
p-Cymene	1023	1026	t	t
1,8-Cineole	1033	1033	84.5	86.0
(*Z*)-β-Ocimene	1040	1040	t	t
(*E*)-β-Ocimene	1050	1050	t	0.6
γ-Terpinene	1057	1062	0.2	0.5
cis-Sabinene hydrate	1067	1067	t	t
Terpinolene	1084	1088	0.1	0.2
trans-Sabinene hydrate	1096	1097	t	t
Linalool	1101	1098	0.1	0.1
α-Fenchol	1110	1112	t	t
cis-*p*-Menth-2-en-1-ol	1118	1121	0.1	0.1
cis-Pincarveol	1133	1139	t	t
trans-*p*-Menth-2-en-1-ol	1137	1140	t	t
Isobomeol	1159	1156	t	t
cis-β-Terpineol	1163	1163	0.6	0.4
Terpinen-4-ol	1172	1177	1.4	1.1
α-Terpineol	1188	1189	4.6	2.9
trans-Piperitol	1205	1205	t	t
α-Terpinyl acetate	1350	1350	0.3	t
Germacrene D	1472	1480	0.1	0.1
γ-Cadinene	1508	1513	t	t
δ-Cadinene	1524	1524	t	t
(*E*)-Nerolidol	1570	1564	1.0	0.2
Spathulenol	1581	1576	0.1	t
T-Cadinol	1644	1640	0.1	t
α-Muurolol	1656	1670	t	t

t = trace.
Source: Rout *et al.* (2003).

lead and copper levels were of close range (Table 4.6).

Husk oil

The pericarp (husk) of large cardamom yielded 0.18% volatile oil by the Clevenger hydrodistillation method. This oil was analysed for physical parameters, *e.g.* specific gravity (0.9148), refractive index (1.4733) and optical rotation (–7.700).

The volatile oil was subjected to GC-MS analysis and 37 compounds were identified, constituting > 98% of the total oil. The major compounds characterized were 1,8-cineole (38.7%), β-pinene (13.6%), α-terpineol (12.6%), spathulenol (8.3%), 4-terpineol (4.5%), germacrene D (3.0%), α-pinene (2.8%) and β-selinene (2.7%). GC and GC-MS data revealed that the 1,8-cineole content was less than 50% when compared with the seed oil (Naik *et al.*, 2004).

Non-volatiles

Pigments

Extraction of fresh large cardamom pod husks with methanolic HCl yielded a

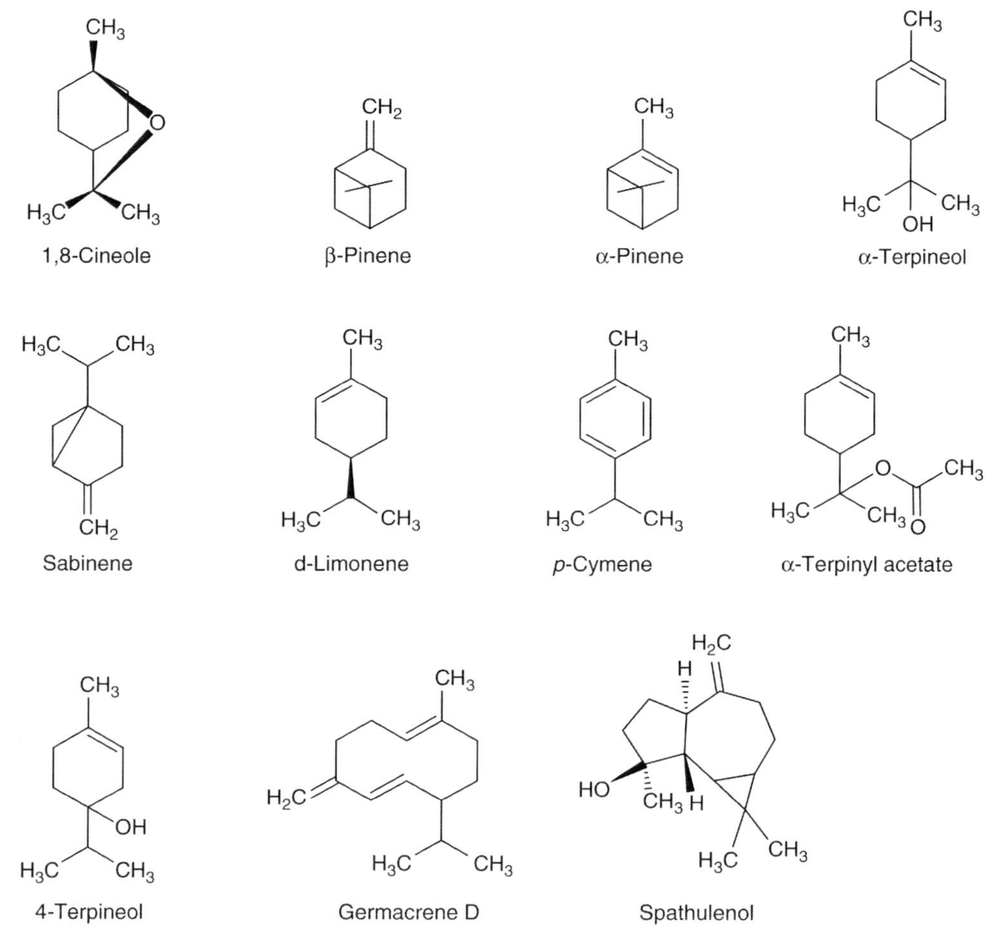

Fig. 4.1. Major essential oil components in large cardamom.

Table 4.6. Range of variation in the physical and chemical quality attributes of *Amomum subulatum* from the western regions of Sikkim, India.

Character	Minimum	Maximum	Mean
Moisture content (%)	7.2	14.9	10.5
Bulk density (g/l)	302.0	374.8	344.8
Husk to seed ratio	1:1.7	1:1.25	1:2.2
Anthocyanins (mg/100 g)	46.2	222.3	98.4
Volatile oil (%)	2.6	4.2	3.4
Total ash (%)	3.6	4.3	3.9
NVEE (%)	2.8	3.0	2.9
Cadmium (ppm)	0.06	0.07	0.07
Lead (ppm)	0.12	0.37	0.24
Copper (ppm)	5.14	9.68	6.33
Iron (ppm)	28.5	111.2	55.3

Note: NVEE - non-volatile ether extract.
Source: Naik *et al.* (2006).

mixture of two (deep pinkish-red) pigments. The pigments, present in the ratio of 1:2, were separated by paper chromatography, characterized as cyanidin 3-glucoside and cyanidin 3.5-diglucoside by chemical and spectroscopic analysis and confirmed by comparison with authentic samples (Naik *et al.*, 2004).

4.5. Medicinal and Pharmacological Uses

Large cardamom possesses the following medicinal properties: antiseptic (pulmonary), antispasmodic (neuromuscular), aphrodisiac, expectorant, anthelminthic, antibacterial (variable), cephalic, cardiotonic, diuretic, emmenagogue, sialogogue and stomachic.

Table 4.7. Distribution and diversified uses of different species of *Amomum*.

Species	Common name	Country	Use
A. aromaticum Roxb.	Bengal cardamom or Nepal cardamom	Eastern India, Pakistan	Rhizomes are used as condiment and flowering shoots are used in curries
A. A. compactum Soland	Round cardamom	Malaysia, Java	Fruits are used as condiment and spices
A. A. globosum Cour Cour	Round Chinese cardamom	China	Seeds are used as cardamom
A. Krervanw Pierre	–	Cambodia, Indo-China	Fruits are used as condiment and to flavour curries, sausages and cordials
A. maximum Roxb.	Java cardamom	Malaysia	Condiment
A. xanthioides Wall	Wild bastard Siamese cardamom	Burma, India	Condiment

Source: Arora (1985).

Table 4.7 summarizes the distribution and uses of the different species of *Amomum*.

Anti-inflammatory

In India, the spice is used broadly to treat infections in teeth and gums, to prevent and treat throat troubles, congestion of the lungs and pulmonary tuberculosis, inflammation of eyelids and also digestive disorders.

Species in the genus *Amomum* are also used in traditional Indian medicine. Among other species, varieties and cultivars, *A. villosum* is used in traditional Chinese medicine to treat stomach aches, constipation, dysentery and other digestive problems.

Antidote to snake venom

Reportedly, the spice is also used as an antidote for both snake and scorpion venom.

Hepatoprotective

The components in the volatile oil, e.g. 1,8-cineole, terpinene, terpiniol, sabinine, α-pinene and limonene, act as a tonic for the heart and liver, an appetizer, promote the elimination of bile and help reduce congestion of the liver. The oil is also useful in treating gonorrhoea.

Anti-ulcerogenic

Large cardamom fruit, commonly known as 'Heel kalan' or 'Bari Ilaichi', is used in the Unani system of medicine to treat gastrointestinal disorders. A crude methanolic extract and its different fractions, e.g. essential oil, petroleum ether (60–80°C), ethyl acetate and methanol fractions, were studied in rats for their ability to inhibit gastric lesions induced by aspirin, ethanol and pylorus ligature. A direct protective effect of ethyl acetate fraction on the gastric mucosal barrier was seen. The decrease observed in gastric motility brought about by essential oil and petroleum ether fractions suggests the gastroprotective action of the spice. These investigations validate the use of large cardamom in gastrointestinal disorders by Unani physicians (Jafri *et al.*, 2001).

Other uses

In medicine, cardamoms are fragrant adjuncts to other stimulants, bitters and purgatives. They are used in conditions like indigestion, vomiting, enlarged spleen, abdominal pains, rectal disease and mouth infections. The seed extract acts as a tonic for the heart and liver, is a bowel astringent and has hypnotic and appetizing properties. Cardamom skin is used for headaches, tooth ailments and stomatitis and its oil, applied to the eyelids, allays inflammation.

Large cardamom can also be put to a variety of industrial uses (Gupta *et al.*, 1984). The globous fruit stalks, usually discarded by farmers, can be used as a base of agarbathis (Pruthi, 1977; Chandrasekhar, 1987).

4.6. Specifications

The quality of large cardamom is based mainly on:

- External appearance, which provides the visual perception of quality; in particular, colour, uniformity of size, shape, consistency and texture.
- Flavour, influenced by the aromatic compounds.

The Spices Board of India has prepared a draft ISO with CFTRI, Mysore, which has been submitted to the Bureau of Indian Standards. The details are shown below:

Capsules	
1. Extraneous matter	Not more than 5% by weight.
2. Insect-damaged capsules	Not more than 5% by weight.
3. Moisture	Not more than 14% by weight.
4. Volatile oil % (ml/100 g)	Not less than 1.5% by weight.
5. Colour should be natural and capsules free from added colours.	
Seeds	
1. Moisture	Not more than 13% by weight.
2. Volatile oil	Not more than 2% by weight.
3. Total ash	Not more than 5% by weight.
4. Acid-insoluble ash	Not more than 2% by weight.
5. Extraneous matter	Not more than 2% by weight.
6. The seeds should be free from moulds and insects.	
7. Insect-damaged seeds	Not more than 2% by weight.
8. Colour and flavour	Should be natural and characteristic.

4.7. Conclusion

The demand for large cardamom in the export market is increasing steadily. The internal consumption of large cardamom is also increasing day by day, which is resulting in an exportable surplus. Used as a flavouring agent in various cuisines and also used in Ayurvedic medicines, India exports large cardamom to the Middle East, Japan and Russia.

Valued for its medicinal properties, this crop does not require much external input and is less labour-intensive, low-volume and non-perishable, with high economic returns. As the large cardamom grows well under shade in a humid environment, it can be cultivated under nitrogen-fixing tree species in moist wasteland along water channels, field bunds and terraces.

Large cardamom (*A. subulatum*) is the most important perennial cash crop of Sikkim, from where cultivation is spreading to the north-eastern states of India, Nepal and Bhutan. More than 85% of Indian production is from Sikkim. The spice has diversified uses in the fields of medicine and industry. Development of high-yielding superior varieties, combined with sustainable production, will definitely enhance the export value of the spice.

References

Anon. (1991) *Wealth of India*, Volume 1. Publications and Information Directorate, CSIR, New Delhi, pp. 227–229.

Arora, R.K. (1985) Genetic resources of less known cultivated food plants. *NBPGR Science Monograph* 9,102–103.

Berger, F. (1965) Neue Erkenninisse auf dem Gebiet der kardamonen forschung Tiel 1–5. *Gardian* 64, 836–839.

Berrig, Ch., Koechilin, R. and Tarnutzer, A. (1993) A preliminary profile with particular emphasis on horticulture and animal husbandry. Study group on institutions. Human Actions and Resource Management, Institute of Geography, University of Zurich, Switzerland, pp. 47–51.

Biswas, A.K., Gupta, R.K. and Bhutia, D.T. (1986) Characteristics of different plant parts of large cardamom. *Cardamom* 19(2), 7–11.

Biswas, A.K., Bhutia, D.T. and Gupta, R.K. (1988) Large cardamom: reasons for decline and ways for improving production. *ICRI Technical Bulletin*, 105 pp.

Chandrasekhar, R. (1987) Agro-based industrial development in the hills of Sikkim and Darjeeling districts. *Cardamom* 20(3), 5–7.

Gupta, P.N. (1986) Studies on capsule morphology of large cardamom cultivars (*Amomum subulatum* Roxb.). *Journal of Plantation Crops* 16, 371–375.

Gupta, P.N., Naqvi, A., Misra, L.N., Sen, T. and Nigam, M.C. (1984) Gas chromatographic evaluation of the essential oils of different strains of *Amomum subulatum* growing wild in Sikkim. *Perfume Kosmet* 65, 528–529.

Gurudutt, K.N., Naik, J.P., Srinivas, P. and Ravindranath, B. (1996) Volatile constituents of large cardamom (*Amomum subulatum* Roxb.). *Flavour and Fragrance Journal* 11, 7–9.

Gyatso, K., Tshering, P. and Basnet, B.S. (1980) *Large Cardamom of Sikkim*. Department of Agriculture, Government of Sikkim, India, p. 8.

Jafri, M.A., Farah, J.K. and Singh, S. (2001) Evaluation of the gastric antiulcerogenic effect of large cardamom (fruits of *Amomum subulatum* Roxb.). *Journal of Ethnopharmacology* 75(2–3), 89–94.

Karibasappa, G.S. (1987) Post harvest studies in large cardamom (*Amomum sublatum* Roxb.). *Sikkim Science Society Newsletter* 6(3), 2–10.

Kaur, S., Naik, S.N. and Maheswari, R.C. (1993) Composition of the essential oil of *Amomum subulatum* (Roxb.) (cardamom large) isolated by liquid CO_2 extraction and steam distillation. *Indian Perfumer* 37(3), 249–252.

Lawrence, B.M. (1970) Terpenes in two *Amomum* species. *Phytochemistry* 9, 665.

Mukherjee, D.K. (1972) Large cardamom. *World Crops* 25(1), 31–33.

Naik, P.J., Jagan Mohan Rao, L., Mohan Kumar, T.M. and Sampathu, S.R. (2004) Chemical composition of the volatile oil from the pericarp (husk) of large cardamom (*Amomum subulatum* Roxb.). *Flavour and Fragrance Journal* 19(5), 441–444.

Naik, P.J., Mohan, K.T.M., Sulochanamma, G., Ramesh, B.S. and Sampathu, S.R. (2006) Studies on quality attributes of cultivars of large cardamom (*Amomum subulatum* Roxb.). *Journal of Food Science and Technology* 43(3), 308–311.

Nambiar, V.P.K., Warrier, P.K. and Ramankutty, C. (eds) (1994) *Indian Medicinal Plants – A Compendium of 500 Species*, Volume I. Orient Longman Publishers, India, Coll No. AVS 2409, pp. 128–129.

Nigam, S.S. and Purohit, R.M. (1960) Chemical examination of the essential oil derived from the seeds of *Amomum subulatum* Roxb. *Perfumery and Essential Oil Record* 51(3), 121–122.

Patra, N.K., Siddiqui, M.S., Akhila, A., Nigam, M.C. and Naqvi, A.A. (1982) Chemical composition of the volatile oil from the pericarp (husk) of large cardamom (*Amomum subulatum* Roxb.). *Pafai Journal* 4(4), 29–31.

Pruthi, J.S. (1977) *Spices and Condiments*. National Book Trust of India, New Delhi.

Pruthi, J.S. (1993) *Major Spices of India – Crop Management and Post Harvest Technology*. ICAR Publications, New Delhi, pp.114–179.

Rao, Y.S., Anand, K., Sujatha, C., Naidu, R. and George, C.K. (1993a) Large cardamom (*Amomum subulatum* Roxb.) – a review. *Journal of Spices and Aromatic Crops* 2(1&2), 1–15.

Rao, Y.S., Gupta, U., Anand, K. and Naidu, R. (1993b) A note on large cardamom (*Amomum subulatum* Roxb.) germplasm collection. *Journal of Spices and Aromatic Crops* 2(1&2), 77–80.

Rout, P.K., Sahoo, D., Jena, K.S. and Rao, Y.R. (2003) Analysis of the oil of large cardamom (*Amomum subulatum* Roxb.) growing in Sikkim. *Journal of Essential Oil Research* 15, 265–266.

Sharma, A.K. and Bhattacharya, N.K. (1959) Cytology of several members of Zingiberaceae. *Cellule* 59, 229–349.
Sharma, R., Sharma, G. and Sharma, E. (2002) Energy efficiency of large cardamom grown under Himalayan alder and natural forest. *Agroforestry Systems* 56(3), 233–239.
Singh, D.B. (1978) Large cardamom. *Cardamom* 10(5), 3–15.
Singh, V.B. and Singh, K. (1996) *Large Cardamom. Spices*. Indian Institute of Plantation Management, Bangalore/New Age International Publishers, New Delhi, pp. 52–57.
Sinu, A.P. and Shivanna, K.R. (2007) Pollination biology of large cardamom (*Amomum subulatum*). *Current Science* 93(4), 25.
Subba, J.R. (1984) *Agriculture in the Hills of Sikkim*. Sikkim Science Society, Gangtok, India, p. 286.

5 Ginger

T. John Zachariah

5.1. Introduction

Ginger, the rhizome of *Zingiber officinale* Roscoe, one of the most widely used species of the family Zingiberaceae, is a common condiment for various foods and beverages. Ginger has been used traditionally for varied human ailments, to aid digestion and to treat stomach upset, diarrhoea and nausea. The ginger plant has a perennial, tuberous root or rhizome: the stems are erect, oblique, round, annual and invested by smooth sheaths of leaves, approximately 1 m in height. Ginger rhizome is generally consumed as a fresh paste, dried powder, slices preserved in syrup, candy (crystallized ginger) or for flavouring tea. In many countries, especially in India and China, fresh ginger is used to prepare vegetable and meat dishes and as a flavouring agent in beverages and many other food preparations (Shukla and Singh, 2006). Ginger is a natural dietary component which has antioxidant and anticarcinogenic properties (Manju and Nalini, 2005).

5.2. Botany

Ginger, *Z. officinale* Roscoe, is a monocotyledon, belonging to the family Zingiberaceae in the order Zingiberales.

The family Zingiberaceae of the suborder Scitaminae is comprised of some 40 genera and hundreds of species. The genus *Z. Boehmer*, the most important of which is represented by *Z. officinale* Roscoe, yields ginger and is grown extensively in many tropical countries. Fourteen species of this genus are reported to occur in India. In addition to *Z. officinale* Roscoe, other minor species which are used locally or which have some trade value include *Z. cassumunar* Roxb. (grown in Thailand), *Z. mioga* (Thunb) Roscoe (grown in Japan), *Z. zerumbet* Roscoe ex Smith (grown in the Asian tropics) and the rhizomes from the genera *Alpinia* and *Kaempfera*, which also belong to this family.

Z. officinale Roscoe is a slender perennial which reaches 60–100 cm in height. The stems (bearing flowers having narrow leaves and distichous, subsessile, linear lanceolate) are approximately 17.0 × 1.8 cm and are dark green with a prominent midrib sheathed at the base. The second and later stems are shorter and terminate in flat bracts with few ancillary flowers. The flowers are greenish-yellow streaked with purple and are sparse. The rare fruiting body is a triangular-oval capsule containing numerous irregular and blackish seeds.

The varieties under cultivation have evolved by unconscious selection and natural hybridization. Many named varieties grown in India show significant variation in yield of rhizome, dry weight, volatile oil, extractives and fibre content. These variations may be inherited and/or influenced by agroclimatic conditions. The cultivated varieties are generally known by the name of the region or area where they are grown regularly (Govindarajan,1982).

5.3. Products and End Uses

There are three primary products of the ginger rhizome: fresh ('green') ginger, preserved ginger in syrup or brine and dried ginger spice. Preserved ginger is prepared from the immature rhizome, while the more pungent and aromatic spice is prepared from harvesting and drying the mature rhizome. Fresh ginger, consumed as a vegetable, is harvested both when immature and mature. The preserved and dried products are the major forms in which ginger is traded internationally. Fresh ginger is of lesser importance in international trade but this is the major form in which ginger is consumed in the producing areas. Dried ginger is consumed in the preparation of its extractives, ginger oleoresin and ginger oil.

Dried ginger traditionally has been traded internationally in the whole or split forms and is ground in the consuming centres. The major use of ground dried ginger on a worldwide basis is for domestic culinary purposes, while in the industrialized Western countries it also finds extensive use in the flavouring of processed foods. Ground dried ginger is employed in a wide range of foodstuffs, especially in bakery products and desserts (Young *et al.*, 2002).

Preserved ginger is prepared in many ginger-growing countries, notably China, Hong Kong, Australia and India, but smaller quantities of fresh ginger are processed in some importing countries. It is used both for domestic culinary purposes and in the manufacture of processed foods such as jams, marmalades, cakes and confectionery.

Ginger oleoresin is obtained by solvent extraction of dried ginger and is prepared both in certain industrialized Western countries and in some of the spice-producing countries, most notably in Australia. This product possesses the full organoleptic properties of the spice, i.e. aroma, flavour and pungency, and finds similar applications to the ground spice in the flavouring of processed foods. The oleoresin is also used in certain beverages and, to a limited extent, in pharmaceutical preparations.

Ginger oil, obtained by steam distillation of the rhizome of *Z. officinale* Roscoe, is used in the beverage and fragrance industries (Wohlmuth *et al.*, 2006). This product possesses the aroma and flavour of the spice but lacks the pungency. It finds its main application in the flavouring of beverages and it is also used in confectionery and perfumery. Zhang *et al.* (2004) found the efficacy of ginger oil as a repellent to *Bemisia argentifolii* (Homoptera: Aleyrodidae) on tomato.

5.4. General Composition of the Ginger Rhizome

The ginger rhizome contains a little steam-volatile oil, fixed (fatty) oil, pungent compounds, resin, proteins, cellulose, pentosans, starch and mineral elements. Of these, starch is the most abundant and comprises 40–60% of the rhizome on a dry weight basis. The relative abundance of certain constituents can vary considerably between samples of ginger in both the fresh ('green') and the dried forms. The composition of the fresh rhizome is determined by the cultivar grown, the environmental conditions of growth and the stage of maturity at harvest. Further changes in the relative abundance of some constituents can also occur postharvest during the preparation and subsequent storage of dried ginger.

A typical analysis of a market sample of green ginger gave the following values (as percentages): moisture, 80.9; protein, 2.3; fat, 0.9; carbohydrates, 12.3; fibre, 2.4; and minerals, 1.2. The principal minerals and vitamins in mg/100 g are Ca, 20; P, 60; and Fe, 2.6; the vitamins, thiamine, 0.06; riboflavin, 0.03; niacin, 0.6; and ascorbic acid, 6.0. In addition to starch, the dominant carbohydrate, the rhizome contains 7.6% pentoses on a dry weight basis and small quantities of the free sugars, glucose, fructose and sucrose. Ginger contains 1.6–2.4% nitrogen on a dry weight basis, of which non-protein nitrogen is roughly one third. About 18.6% of the protein remains unextracted; the extracted proteins contain 35.6% albumin, 16.9% globulin, 11.0% prolamine and 17.9% glutelin, on total proteins (Govindarajan, 1982).

The fibre and volatile oil contents and the pungency level are the most important criteria in assessing the suitability of ginger rhizomes for particular processing purposes. The relative abundance of these three components in the fresh rhizome is governed by its state of maturity at harvest. Young, tender rhizomes lifted at the beginning of the harvesting season, about 5–7 months after planting, are preferred for the manufacture of preserved ginger since the fibre content is negligible and the pungency is mild. As the season progresses, the relative abundance (on a dry weight basis) of the volatile oil, the pungent constituents and the fibre increases. At about 9 months after planting, the volatile oil and pungent principle contents reach a maximum and thereafter their relative abundance falls as the fibre content continues to increase. The volatile oil content of Australian ginger has been reported to increase from about 1.8 to 4.4% (on a dry weight basis) through the harvesting season, but remains fairly constant at about 0.4% on a green weight basis during this period. The mid-season crop is therefore preferred in Australia for the distillation of the essential oil or the extraction of oleoresin, while the late-season crop is used for the preparation of the dried spice. In India, the volatile oil content of ginger has been reported to be at maximum between 215 and 260 days after planting (Purseglove *et al.*, 1981).

The drying of ginger usually leads to the loss by evaporation of some of the volatile oil and it is reported that this loss may be as high as 20% during sun drying. The extent of cleaning the rhizome prior to drying has a considerable influence on the volatile oil and fibre content of the end product. Removal of the outer cork skin not only reduces the fibre content but also enhances volatile oil loss through rupture of the oil cells which are near the skin. For this reason, the cleanly peeled Jamaican product tends to have somewhat lower volatile oil and fibre content than other commercial dried gingers which are only partially peeled or unpeeled.

The crude fibre content of unpeeled ginger may be as high as 10% (on a dry weight basis), but in commercial dried gingers it is usually in the range of 1.5–6%. The volatile oil content of commercial dried gingers has been reported to be 0.5–4.4% but, for the major types, the range is usually 1–3%.

Ginger oleoresin is prepared from dried ginger by extraction using a number of organic solvents. The oleoresin contains the organoleptically important volatile oil and pungent principles, together with fatty oil, palmitic and some other free fatty acids, resin and carbohydrates. The yield and the relative abundance of the components of the oleoresin are dependent, however, on the raw material and the solvent used and on the extraction conditions. Commercial dried gingers have been reported to provide oleoresins in yields of 3.5–10% and to contain 15–30% of volatile oil (Govindarajan, 1982). Table 5.1 gives the composition of ginger, spent ginger and by-products in commercial ginger samples. The proximate composition of cultivated varieties of ginger is illustrated in Table 5.2.

The care taken during preparation and subsequent storage of the dried spice and its oleoresin has an important influence on the organoleptic properties, and hence the quality, of these products. The major storage change with the dried spice, especially when in the ground state, is the evaporation of some volatile oil, which results in a flat odour and flavour, while the oleoresin is particularly prone to loss of pungency during storage by degradation of the gingerols. Heat treatment of the spice and its oleoresin can lead to degradation of both the volatile oil and the pungent principles, and this factor is of importance when either material is used in the flavouring of processed foods. The fatty oil of ginger is present at an abundance of 2–12% in dried gingers. Ginger fatty oil contains saturated and unsaturated fatty acids in a ratio of 46:53, and the major

Table 5.1. Composition of ginger, spent ginger, and by-products (commercial samples).

					Ash				Protein				Cold
								Crude	(N ×			Alcohol	water
	Moisture	NVEE	VEE	Fibre	Total	Sand	Lime	starch	6.25)	NVEE	VEE	extract	extract
Jamaican													
Natural	11.20	3.91	1.79	3.72	4.17	0.22	0.26	57.59	7.85	7.30	3.23	4.95	15.54
Limed	10.56	3.12	1.27	2.37	8.31	0.02	2.72	57.31	9.34	–	–	–	–
Cochin													
Rough	10.43	3.70	2.09	3.62	3.86	0.10	0.48	59.08	8.15	6.68	7.03	6.32	14.30
Scraped, limed	9.97	2.95	1.49	2.60	5.36	0.08	1.29	62.42	7.50	–	–	–	–
Calicut	–	–	–	–	–	–	–	–	–	6.42	4.62	7.64	13.08
African	9.97	5.35	2.73	4.66	4.00	0.11	0.25	56.74	7.92	8.49	7.17	6.36	12.62
Japanese	10.39	3.94	0.96	2.73	6.19	0.71	1.66	60.55	5.40	7.01	7.39	8.37	14.40
Wastes													
Scraggy	4.99	9.55	6.05	13.18	8.05	0.89	0.61	31.38	7.00	–	–	–	–
Cuttings	3.19	2.76	7.06	8.69	9.20	1.81	1.06	40.23	8.69	–	–	–	–
Residue (ginger after manufacture)	10.61	3.86	1.61	5.17	2.12	0.18	–	59.86	6.94	–	–	–	–
Residue (extract)	8.02	0.54	0.13	–	5.05	1.50	–	–	–	–	–	–	–

Note: NVEE – non-volatile ether extract; VEE – volatile ether extract.

Table 5.2. Proximate composition of cultivated varieties of ginger.

Variety	Moisture (%)	Starch (by acid hydrolysis) (%)	Crude protein (N × 6.25) (%)	Crude fibre (%)	Ash		Water extract (%)	Acetone extract (%)	Volatile oil (%)
					Total (%)	Acid insol. (%)			
Assam	10.00	40.4	10.3	9.70	7.50	0.03	22.4	9.3	2.4
Bajpai	12.50	55.4	12.8	5.94	5.84	0.07	18.5	4.9	1.7
Burdwan	9.00	42.4	13.9	6.37	5.90	0.23	17.8	6.5	1.8
Cheranad	16.50	51.7	13.5	6.68	7.67	0.59	16.0	5.6	1.4
China	9.60	52.8	11.8	8.00	9.28	0.28	15.5	5.8	1.4
Himachal Pradesh	10.00	55.9	11.5	9.40	7.03	0.01	14.4	5.0	1.6
Jorhat	11.00	55.9	11.7	9.80	6.12	0.23	25.8	8.3	2.0
Juggijan	10.00	56.9	13.1	5.94	6.81	–	17.9	5.8	1.7
Karkal	12.00	52.9	12.8	5.93	7.02	0.11	22.7	6.6	2.0
Kunnamangalam	10.00	55.6	12.9	4.79	5.46	0.13	17.0	6.4	1.5
Manantody	9.00	49.4	11.5	6.86	8.81	0.11	22.3	9.2	2.7
Manjeri	9.00	52.0	15.0	6.93	5.12	–	20.2	6.9	2.5
Maran	8.50	45.4	10.9	6.16	8.30	0.22	20.2	6.7	2.2
Mysore	10.00	50.9	12.0	7.76	6.05	0.02	17.8	7.8	2.7
Nadia	11.50	59.0	10.5	5.67	8.23	0.11	20.1	3.9	1.0
Narasapattom	10.50	56.7	11.1	6.22	5.89	–	17.8	6.3	1.6
Poona	11.00	54.0	12.7	6.28	5.89	0.13	22.8	5.4	1.4
Rio de Janeiro	11.00	52.9	12.6	7.14	6.23	0.05	19.6	7.2	1.7
Thingpuri	10.00	55.3	10.4	7.20	6.46	0.36	18.8	5.1	1.4
Thinladium	13.00	56.8	10.5	9.60	5.99	0.08	19.3	5.9	1.7
Uttar Pradesh	11.00	52.5	13.5	7.60	6.39	0.08	21.3	8.8	2.0
Valluvanad	12.50	47.6	13.7	7.50	6.18	0.24	20.5	7.9	2.0
Vengara	15.00	59.0	14.2	6.68	6.34	0.34	21.3	6.6	1.8
Wynad, local	10.00	50.7	13.7	8.20	6.33	0.02	19.7	6.8	1.9
Maximum	16.50	59.0	15.0	9.80	9.28	0.59	25.8	9.3	2.7
Minimum	8.50	40.4	10.3	4.79	5.12	–	14.4	3.9	1.0

Source: Govindarajan (1982).

Table 5.3. Fatty acid composition of ginger lipids.

Fatty acid	Carbon no.	Total fatty acids (%)
Caprylic acid	8	1.4
Capric acid	10	4.1
Lauric acid	12	7.6
Myristic acid	14	3.5
Pentadecanoic acid	15	0.4
Palmitic acid	16	23.2
Heptadecanoic acid	17	1.3
Stearic acid	18	3.3
Oleic acid	18:1	22.9
Linoleic acid	18:2	23.2
Linolenic acid	18:3	6.6
Arachidic acid	20	1.1
Saturated acids		45.9
Unsaturated acids		52.7
Saponification index		204

component acids are found to be palmitic, oleic and linoleic acids, each having a relative abundance of about 23%. Govindarajan (1982) reported a range of 5.8–15% lipid content among ginger varieties. The fatty acid composition of ginger lipids is given in Table 5.3.

John and Ferreira (1997), in their study in South Africa, found a ginger selection G9 with high crude fibre content of 6.8% on a dry basis, along with 3.06% oleoresin and 0.52% oil and selection G10, with dry ginger recovery of 27.5%.

5.5. Chemistry

The chemistry of *Z. officinale* has been the subject of sporadic study since the early 19th century. In common with some other pungent spices, considerable advances were made in the early part of the 20th century, but it has only been in recent years that a fairly clear understanding of the relationship of its chemical composition to its organoleptic properties has emerged.

Ginger, like pepper (*Piper nigrum*) and the fruits of the *Capsicum* species, owes its characteristic organoleptic properties to two classes of constituents: the odour and much of the flavour of ginger is determined by the constituents of its steam-volatile oil, while the pungency is produced by non-steam-volatile components, known as the gingerols, which possess a 1-(4′-hydroxy-3′-methoxyphenyl)-5-hydroxyzlkan-3-one structure.

Essential oil of ginger

The aroma and flavour of ginger are determined by the composition of its steam-volatile oil, which is comprised mainly of sesquiterpene hydrocarbons, monoterpene hydrocarbons and oxygenated monoterpenes. The monoterpene constituents are believed to be the most important contributors to the aroma of ginger and they tend to be relatively more abundant in the natural oil of the fresh ('green') rhizome than in the essential oil distilled from dried ginger. Oxygenated sesquiterpenes are relatively minor constituents of the volatile oil but appear to be significant contributors to its flavour properties.

Investigations of the aroma and flavour of ginger have been carried out almost exclusively on the steam-distilled essential oil obtained from dried ginger. However, it should be appreciated that this oil differs somewhat in its composition and organoleptic properties from the natural volatile oil present in dried ginger prior to distillation through the formation of artefacts during the distillation process and subsequent storage.

The steam-volatile oil content of some types of fresh ginger can be well over 4% on a dry weight basis. However, distillation of the more important dried gingers of commerce usually provides oil in the yields ranging from about 1 to 2.5%. As has been mentioned previously, oil distillation yield is influenced by a number of factors, which include the ginger cultivar, the state of maturity at harvest, the method of preparation and drying of the spice, its age and, to some extent, the distillation method. The best oil yields generally are obtained from partially scraped ginger from Nigeria. High yields (over 4%) may also be obtained from distillation of fresh skin scrapings discarded during the preparation of dried ginger.

Ginger oil prepared by steam distillation of dried ginger is obtained as a pale yellow to light amber mobile liquid whose viscosity increases on ageing or exposure to the air. The odour of the oil is described as warm, but fresh woody and spicy. The initial fresh top note has a peculiar resemblance to orange, lemongrass and coriander weed oil, while the sweet and heavy undertone is tenacious and rich. The organoleptic properties of ginger oils vary somewhat according to the geographical source of the dried ginger. African ginger oil tends to be darker in colour and exhibits a more fatty sweetness, while the Jamaican oil is usually very pale in colour and has pronounced odour freshness. The initial notes of freshly distilled Jamaican oil have a peculiar 'rubber-like' note, similar to that of nutmeg, which is hardly ever present in African oil. The citrus or lemon-like top note is a characteristic of Indian ginger oil, and this is even more pronounced in Australian oil (Purseglove *et al.*, 1981).

The physico-chemical properties of ginger oils can also vary considerably between individual samples and these differences are influenced by those same factors listed above which affect oil yields (Guenther, 1975). The optical rotation value is a notable variable and this tends to be abnormally low in oils which have been distilled from old material or in oils which have been stored exposed to air and light.

Compositional diversity of ginger oil

Ginger oil displays considerable compositional diversity but is typically characterized by a high content of sesquiterpene hydrocarbons, including zingiberene, *ar*-curcumene, β-bisabolene and β-sesquiphellandrene.

Yu *et al.* (1998) reported the presence of monoterpenes and sesquiterpenes as the main components in three samples of steam-distilled oil. They could not find pungent components in oil. Besides sesquiterpenes, the supercritical CO_2-extracted ginger oils contained 18.61–23.09% pungent components. These oils preserve the typical spicy odour and pungency of ginger.

The major sesquiterpene hydrocarbon constituent of ginger oil, (–)-α-zingiberene, was first isolated in 1900 and the elucidation of its structure was the subject of numerous studies during subsequent years (Purseglove *et al.*, 1981).

Australian ginger oil has a reputation for possessing a particular 'lemony' aroma due to its high content of the isomers, neral and geranial, often referred to collectively as citral. Fresh rhizomes of 17 clones of Australian ginger, including commercial cultivars and experimental tetraploid clones, were steam distilled and the resulting oils were analysed by GC-MS. The essential oils of 16 of the 17 clones, including the tetraploid clones and their parent cultivar, were found to be of substantially similar composition. These oils were characterized by very high citral levels (51–71%) and relatively low levels of the sesquiterpene hydrocarbons typical of ginger oil. The citral levels of most of these oils exceeded those reported previously for ginger oils. The neral–geranial ratio was shown to be remarkably constant (0.61 + or – 0.01) across all 17 clones. One clone, the cultivar 'Jamaican', yielded oil with a substantially different composition, lower citral content and higher levels of sesquiterpene hydrocarbons. Because this cultivar also contains significantly higher concentrations of pungent gingerols, it possesses unique aroma and flavour characteristics, which should be of commercial interest (Wohlmuth *et al.*, 2006).

The composition of the essential oil hydrodistilled from dried Nigerian ginger was determined by GC and GC-MS techniques. The oil yield was 2.4% and the oil consisted of 64.4% sesquiterpene hydrocarbons, 6.6% carbonyl compounds, 5.6% alcohols, 2.4% monoterpene hydrocarbons and 1.6% esters. The main compounds were zingiberene (29.5%) and sesquiphellandrene (18.4%). A number of constituents not previously reported in ginger oil were identified. These included 2,6-dimethyl hepten-l-ol, α-gurjunene, linalool oxide, isovaleraldehyde, 2-pentanone, cadinol, α- and γ-calacorene, eremophyllene, T-muurolol, α-himachallene, α-cubebene acetic acid, pinanol, α-santalene, geranyl propionate, geranoic acid, (*E,E*)-α-farnesene, *n*-methyl pyrrole and geranic acid (Onyenekwe and Hashimoto, 1999).

Miyazawa and Kameoka (1988) identified 72 components in the volatile oil extracted from the air-dried rhizomes. The main components were α-zingiberene (21.8%), geranial (9.9%), geraniol (9.4%), β-bisabolene (7.9%), nerol (7.1%), 1,8-cineol (6.2%), α-terpineol (5.6%), borneol (5.4%), β-phellandrene (3.1%), linalool (1.7%), methyl nonyl ketone (1.6%) and camphene (1.4%); the other components accounted for ~ 1% each of the volatile oil. Table 5.4 illustrates the different groups of volatile compounds detected in ginger oil.

Table 5.4. Constituents identified in ginger oils.

Sesquiterpene hydrocarbons	Oxygenated monoterpenes
(–)-α-Zingiberene	*d*-Borneol
β-Zingiberene	Bornyl acetate
(+)-*ar*-Curcumene	1:8 Cineol
(–)-β-Bisabolene	Citrals a & b
β-Elemene	Citronellyl acetate
β-Farnesene	Geraniol
γ-Selinene	Linalool
(–)-β-Sesquiphellandrene	α-Terpineol
Sesquithujene	
Sesquiterpene alcohols	**Miscellaneous compounds**
cis-β-Eudesmol	*n*-Heptane
trans-β-Eudesmol	*n*-Octane
Nerolidol	*n*-Nonane
cis-β-Sesquiphellandrol	*n*-Propanol
trans-β-Sesquiphellandrol	2-Heptanol
cis-Sabinene hydrate	*n*-Nonanol
Zingiberenol	2-Nonanol
	Acetaldehyde
Monoterpene hydrocarbons	Propionaldehyde
	n-Butyraldehyde
d-Camphene	Isovaleraldehyde
Δ-3-Carene	*n*-Nonanal
p-Cymene	*n*-Decanal
Cumene	Acetone
d-Limonene	Methyl heptanone
Myrcene	Methyl acetate
d-β-Phellandrene	Ethyl acetate
α-Pinene	Methyl caprylate
β-Pinene	Diethyl sulphide
Sabinene	Ethyl isopropyl sulphide
	Methyl allyl sulphide
	Chavicol

Essential oils from ginger rhizomes from India and Australia differed markedly in their terpenoid compositions. The main components of Indian ginger oil were the sesquiterpenoid hydrocarbons, *ar*-curcumene, zingiberene, α-farnesene, β-bisabolene and β-sesquiphellandrene, while the essential oil from the Australian ginger consisted mainly of the monoterpenoid hydrocarbons, camphene and phellandrene, and their oxygen-containing derivatives, neral, geranial and 1,8-cineol (Erler *et al.*, 1988). Ekundayo *et al.* (1988) could identify 54 constituents and among them (*E*)(*E*)-α-farnesene, viridiflorol and (*E*)(*E*)-farnesal had not been found previously in ginger.

The composition ranges for oils prepared from dried ginger are as follows: the sesquiterpene hydrocarbons are the most abundant component group (50–66%), the oxygenated sesquiterpene content is modest (up to 17%), while the rest consists substantially of monoterpene hydrocarbons and oxygenated monoterpenes. Among the sesquiterpene hydrocarbons, (–)-α-zingiberene predominates (20–30%) and is accompanied by lesser quantities of (–)-β-bisabolene (up to 12%), (+)-*ar*-curcumene (up to 19%) and farnesene (probably the β-isomer; up to 10%). γ-Selinene and β-elemene occur in relatively minor quantities. The quantitative balance in the oxygenated sesquiterpene group is less certain, but the abundance of zingiberol has been reported to range from zero to 0.1%. With the significant exception of the citrals, the relative abundance of the low-boiling monoterpene constituents generally is low and of a similar order (up to about 2%). The Australian oils were notable in exhibiting high citral contents in the range of 8–27% (averaging 19.3%) compared with 0.5–4.0% for oils from other sources. The ratio of citral a (geranial) to citral b (neral) in most samples was about 2:1. Many authors have suggested that the pronounced ‘citrus’ or lemon-like note of Australian ginger oils is related to their high citral content (Purseglove *et al.*, 1981). Figure 5.1 illustrates the major aroma compounds of ginger.

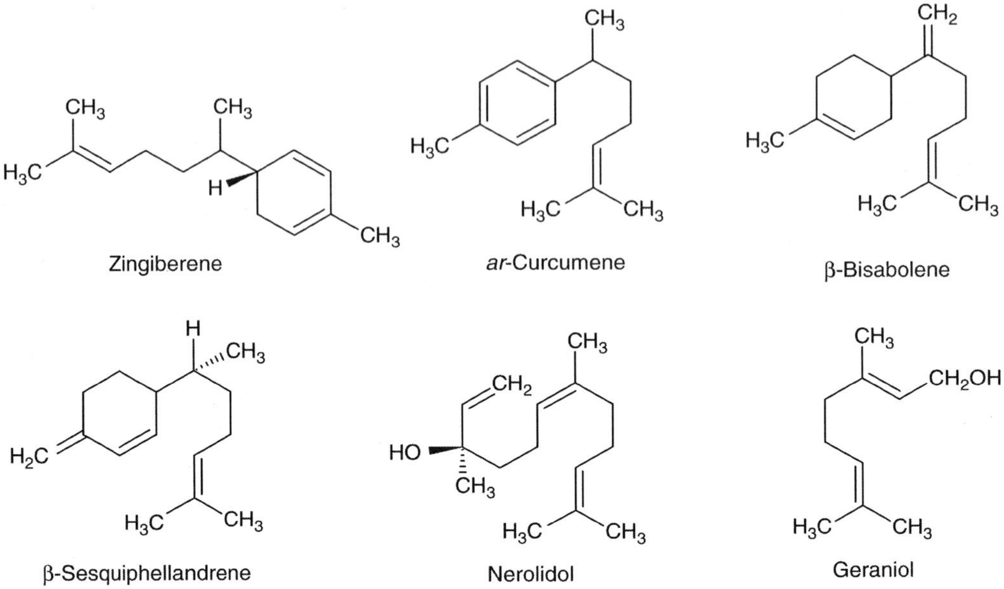

Fig. 5.1. Volatile compounds from ginger.

Biosynthesis of terpenes

The atomic ratios ($^{14}C/^{3}H$) of *ar*-curcumene, α-zingiberene and β-bisabolene obtained from the essential oil of ginger rhizomes fed with labelled mevalonic acid and farnesylpyrophosphate (FPP) reveal that the (2*E*,6*E*)-isomer of FPP is isomerized to the (2*Z*,6*F*)-isomer without loss of epimeric hydrogen (that means without a redox process); (2*Z*,6*E*)-FPP is cyclized to bisabolyl cation, which is the penultimate precursor of α-zingiberene, *ar*-curcumene and β-bisabolene; and two 1,2-hydrogen shifts take place during the formation of α-zingiberene, whereas one 1,2 shift has been observed during the formation of *ar*-curcumene (Rani, 1999).

Effect of shade and maturity on oil content

Sreekala and Jayachandran (2004) conducted studies on different levels of shade and maturity stage and its effect on the volatile oil and non-volatile ether extract. In general, volatile oil was highest under heavier shade levels (60 and 80%) and non-volatile ether extract was highest under 20% shade. They concluded that shade had a positive influence on the volatile oil, as well as on non-volatile ether extract of ginger. The study also indicated a general decrease of volatile oil content, as well as non-volatile ether extract with increasing periods of maturity.

Ratnambal *et al.* (1987) carried out a quality evaluation of ginger in relation to maturity. Quality parameters such as essential oil, oleoresin, gingerol, starch, protein and fibre contents were determined for 14 cultivars at 150, 180, 210 and 240 (full maturity) days after planting. Although the percentage contents of essential oils and oleoresin decreased with increasing maturity, the final yields/ha of these two quality components were at maximum at full maturity.

Variation in relation to genotypes and cultural practices

Twenty-four ginger genotypes were evaluated for yield and quality under rain-fed and irrigated conditions in Solan, Himachal Pradesh, India, during 1997/98. Significant variation among the cultivars was observed for oil, oleoresin, crude fibre and dry matter contents, irrespective of growing conditions. The effects of irrigation conditions were significant on crude fibre content only. Under rain-fed conditions, the highest ginger oil contents were recorded for SG 61 (2.28%), SG 702 (2.09%) and SG 699 (2.05%). Oleoresin content was highest in BDJR 1054 (5.31%). Himgiri (control) had the lowest crude fibre content (5.13%). The highest dry matter contents were observed in SG 692 (18.06%) and BDJR 1113 (17.93%). Yield per plant was highest in SG 646 (214 g). Under irrigated conditions, ginger oil contents were highest in SG 687 (2.18%) and SG 61 (2.15%). BDJR 1054 (5.27%) and SG 61 (5.27%) were superior in terms of oleoresin content. Himgiri also had the lowest crude fibre content (4.76%). Dry matter content was highest in SG 692 (18.07%). The highest yields per plant were recorded for Himgiri (232 g) and SG 646 (216 g). The results indicated that SG 61, SG 62, BDJR 1054 and SG 687 were suitable for ginger oil and oleoresin extraction, whereas SG 692 was the most suitable for dry ginger production and processing (Tiwari, 2003). Korla *et al.* (1999) reported variability in oil content in different ginger accessions cultivated in Himachal Pradesh, India.

The essential oils of the fresh rhizomes of *Z. officinale* and its three variants obtained by hydrodistillation were analysed by a combination of capillary GC and GC-MS. The variants investigated were *Z. officinale* Rosc. var. *officinale* (young common ginger), *Z. officinale* var. *rubrum*, *Z. officinale* Rosc. var. rubrum Theilade (halia bara) and *Z. officinale* Rosc. var. rubrum Theilade (halia padi). The analysis led to identification of 22, 40, 19 and 17 components comprising 71.8, 89.7, 74.4 and 84.1% of the total oils, respectively. The major components of the rhizome oils were found to be: *Z. officinale* Rosc. var. officinale (common ginger): zingiberene (16.70%), (*E,E*)-α-farnesene (13.10%), geranial (7.60%); *Z. officinale* var. rubrum (jahe merah): camphene (15.76%), geranial (12.66%) and *ar*-curcumene (9.71%); *Z. officinale* Rosc. var. rubrum Theilade (halia bara): geranial (28.43%), neral (14.20%) and geranyl acetate (8.77%); *Z.officinale* Rosc. var. rubrum Theilade (halia padi): geranial (28.62%), neral (15.58%) and β-sesquiphellandrene (6.44%). The investigation shows that closely related variants of ginger have one or two similar major components. Zingiberene, a characteristic component in the ginger variants, is always present in the range of 3–17% (Abd Malek *et al.*, 2005).

Sultan *et al.* (2005) evaluated the quality of ginger rhizomes imported from China and Thailand on the basis of their essential oil content and composition. Hydrodistillation yielded 0.98 and 1.58% essential oil (on a dry basis) in rhizomes from China and Thailand, respectively. Chemical analysis of the essential oil was carried out by GC-FID. The essential oil of the Thailand ginger sample contained α-pinene (3.59%), α-phellandrene (2.84%), myrcene (4.58%), β-pinene (0.74%), γ-terpinene (2.49%), 1,8-cineol (3.87%), citral (5.39%) and zingiberene (30.81%). The essential oil of the China ginger sample contained α-pinene (0.305%), α-phellandrene (1.02%), myrcene (4.82%), γ-terpinene (2.88%), 1,8-cineol (2.4%), α-terpinene (6.5%), citral (4.5%) and zingiberene (8.0%). The ginger sample from Thailand was found to be better in quality due to the higher percentage of essential oil.

The essential oil obtained from the rhizomes of *Z. officinale* from Cuba contained *ar*-curcumene (22.1%), zingiberene (11.7%), β-bisabolene (11.2%) and cadina-1,4-diene (12.5%) (Pino *et al.*, 2004).

The chemical composition of the essential oils obtained from the hydrodistillation of the rhizomes of the common ginger (*Z. officinale*) grown in Mauritius was analysed by Fakim *et al.* (2002). The oil was characterized by the presence of geranial (16.3%), neral (10.3%), zingiberene (9.5%), β-sesquiphellandrene (6.3%) and *ar*-curcumene (5.1%).

Gopalam and Ratnambal (1989) evaluated the essential oils of nine *Z. officinale* cultivars by gas chromatography. The total essential oil yield ranged from 1.5 to 2.2%.

Fresh ginger oil

Composition of ginger oil prepared from fresh ginger rhizomes was determined by gas chromatography (GC) and GC-MS techniques. The main sesquiterpene hydrocarbons identified were α-zingiberene (27–30%), α-curcumene

(8–9%), β-sesquiphellandrene (4.8%) and bisabolene (3.2%) (Antonious and Kochhar, 2003).

The principal composition difference between the oils distilled from dried and from fresh (green) ginger is that the latter usually contains a greater proportion of the lower-boiling components. Reports indicate that up to 20% of the volatile oil can be lost during sun drying of Indian ginger and that the lemon-like aroma becomes weaker in the process. The major oil loss expected during the drying of ginger is of the lower-boiling components, which include the citrals. The fresh (green) ginger of Australia, Cochin and Calicut is characterized by a pronounced fresh, lemon-like aroma and it is possible that retention of this characteristic in Australian ginger oils arises as much from more careful drying methods, in which volatile-oil losses are minimized, as from intrinsic composition differences between the gingers of Australia and India.

Although several workers have monitored the changes in the volatile oil content of fresh ginger as the rhizome matures, there are not many reports of similar studies concerned with the possible changes in oil composition during maturation.

Storage losses

Post-distillation changes in the properties of ginger oils also can occur during either storage or use. Exposure of ginger oils to light and air results in an increase in viscosity, the formation of non-volatile (polymeric) residues and a decrease in the optical-rotation value. This change is associated with a decrease in the relative abundance of (–)-α-zingiberene and (–)-β-sesquiphellandrene and a concomitant increase in the relative abundance of (+)-*ar*-curcumene. Reports suggest that (+)-*ar*-curcumene is not, in fact, a true natural component of ginger oil but is an artefact which can be produced during the distillation stage by transformation of the other more labile sesquiterpenes. Ginger oil is heat-sensitive and detrimental changes in its composition and its aroma and flavour can occur on heating above 90°C.

Flavour of ginger oil

Govindarajan (1982), in his review on ginger, suggested the desirable flavour requirements for good quality ginger oil.

1. Citral and citronellyl acetate, being powerful co-determinants of the odour.
2. Zingiberene and β-sesquiphellandrene as the main components of the freshly prepared oil.
3. *ar*-Curcumene, increasing with storage, being indicative of the age of the oil or the process condition.
4. The ratio of zingiberene + β-sesquiphellandrene to *ar*-curcumene = 2:3 as a characteristic of the oil.

The lemony note is attributed to citrals together with α-terpineol, while β-sesquiphellandrene and *ar*-curcumene are regarded as partly responsible for the characteristic ginger flavour. Nerolidol was considered to contribute to the woody note; and *cis*- and *trans*-β-sesquiphellandrol were suspected as significant contributors to the ginger flavour. In combination, these compounds accounted for 85% of the taste panel's response.

Various groups of workers have made certain suggestions on the odour quality of ginger and its oil from different sources. Thus, the mild and delicate odour of Jamaican ginger, the characteristic 'lemony note' of Cochin ginger and the 'pungent and camphoraceous' odour of African ginger have been recorded from early days (Govindarajan, 1982). The dominance of sesquiterpenes, particularly zingiberene, is considered characteristic of ginger. However, sensory panel tests attribute the flavour of ginger oil to low-boiling monoterpene hydrocarbons and oxygenated compounds. High citral (geranial and neral) content in Australian ginger (about 19%) is reported to give the characteristic citrus-like odour. Based on the suggestion of various sensory panels, the odour and flavour profile of three different ginger samples is listed in Table 5.5 (Govindarajan, 1982).

Bartley and Jacobs (2000) found that the drying of peeled rhizomes in mechanical dryers led to a reduction in gingerol content, an increase in terpene hydrocar-

Table 5.5. Odour and flavour profile of ginger.

Origin		Jamaican	African	Sierra Leone (commercial)
Impact Odour	Initial	SWEET; mildly spicy	HARSH; strongly spicy	HEAVY; full body
		WARM; smooth	WARM; irritating	WARM; spicy
		LEMONY	LEMONY; terpeny	LEMONY; fruity
				TERPENEY: turpentine-like, pine-like
		Slightly EARTHY; woody	EARTHY; dirty; woody	Slightly MUSTY
				Slightly HARSH;
		Slightly irritating	Strongly MUSTY	PENETRATING; IRRITATING after short time
	Persistence	Persistent; characteristic	Persistent; characteristic	Hardly changes for several days on smelling strip
	Dry-out	Characteristic over long period	Characteristic over long period	Characteristic over long period with slight increase in EARTHY notes
Flavour		LEMONY; terpeney	FLAT; dirty	MILD; FLAT; LEMONY; fruity citrus-like Slightly PINE-like; terpeney
		WARM; pleasant	BITTER; harsh	
		SMOOTH; mildly spicy	WOODY EARTHY; green mustiness	WARMING
		Slightly WOODY	DRYING; astringent	Slightly DRYING; astringent
		Slightly BITTER		
		Slightly IRRITATING		Slightly BITTER
	Aftertaste	FRESH; pleasant	HARSH; unpleasant	Initially pleasing SPICY; later BITTER

bons and the conversion of some monoterpene alcohols to their corresponding acetates.

Non-volatiles

Ginger oleoresin

Ginger oleoresin was extracted from rhizomes with ethanol, isopropanol or liquid carbon dioxide. All oleoresin samples had monoterpenes and sesquiterpenes. Carboxylic acids were found in organic solvent extracts for an extraction time of 2 h. The components responsible for the pungent characteristic of the oleoresin gingerols were detected in samples obtained with organic solvent for an extraction time of 6 h and in samples obtained with CO_2 liquid for an extraction time of 2 h (Nobrega *et al.*, 1997).

Oleoresin is the total soluble extractive in a specified solvent. From the functional point, the best oleoresin is one which contains all the flavour components of the material contributing to aroma, taste, pungency and related sensory factors which, when diluted to the original concentration in the original material, truly recreates the sensory quality of the original spice.

Ginger oleoresin should contain predominantly the aroma and pungency contributed mainly by the volatile oils, gingerols and related compounds.

Extraction is achieved by percolation of the solvents acetone, alcohol or ethylene dichloride at room temperature, through a column of coarsely ground ginger. Oleoresin ginger is a dark golden brown viscous oil. The principal components are the volatile oil dominated by high-boiling sesquiterpenes and the pungency-stimulating gingerols, with a minor amount of shogaols. Good ginger oleoresin contains 20–25% volatile oil, 25–30% pungency stimuli and the rest are non-flavour compounds such as fats, waxes and carbohydrates.

Pungent principles

The non-volatile pungent oil obtained by solvent extraction of the spice was named gingerol (Purseglove *et al.*, 1981). Spiro and Kandiah (1989) studied the extraction kinetics of [6]-gingerol, 1-(4′-hydroxy-3′-methoxyphenyl)-5-hydroxydecan-3-one (Kandiah and Spiro, 1990).

In 1917, a crystalline, optically inactive, pungent keto-phenol was isolated from the alkali-soluble fraction of an ethereal extract of ginger. This compound was named zingerone and its structure was proposed as 4-hydroxy-3-methoxyphenylethyl methyl ketone. On the basis of a number of other experiments, investigators concluded that gingerol was a mixture of at least two compounds in which zingerone was condensed with homologous straight-chain aldehydes. Gingerol was isolated from both ginger and grains of paradise and studies were carried out on its derivative, gingeryl methyl ether (Purseglove *et al.*, 1981).

Column chromatography of ginger oleoresin furnished a fraction from which [6]-gingerol was obtained by preparative TLC. Naturally occurring [6]-dehydro shogaol was synthesized following condensation of dehydrozingerone with hexanal, whereas zingerone and 3-hydroxy-1-(4-hydroxy-3-methoxyphenyl) butane were obtained by hydrogenation of dehydro zingerone with 10% Pd/C. The structures of the compounds were established by ^{1}H NMR, ^{13}C NMR and mass (EI-MS and ES-MS) spectral analysis (Agarwal *et al.*, 2001).

The pungent group includes gingerols, shogaols, paradols and zingerone that produce a 'hot' sensation in the mouth. The gingerols, a series of chemical homologues differentiated by the length of their unbranched alkyl chains, were identified as the major active components in the fresh rhizome (Govindarajan, 1982). In addition, the shogaols, another homologous series and the dehydrated form of gingerols, are the predominant pungent constituents in dried ginger. Paradol is similar to gingerol and is formed on hydrogenation of shogaol. Shukla and Singh (2006) reported the effect of [6]-gingerol and [6]-paradol in suppressing the induction of apoptosis. Figure 5.2 illustrates the gingerols, paradols and shogaols of ginger oleoresin.

Nakasone *et al.* (1999) reported that gingerol content, antioxidant activity and pungency intensity were higher in tetraploid types of the *Z. officinale* cultivars Kintoki, Oshoga and Sanshu than in diploid types. Diploid and tetraploid types of cv. Kintoki (Wohlmuth *et al.*, 2005) had the highest gingerol content among the diploids and tetraploids, respectively. Pungency intensity decreased in the order Kintoki > Oshoga > Sanshu. Pungency intensity was correlated with gingerol content in all except the diploid type of cv. Kintoki.

Another pungent compound from the alkali soluble portion of an ethereal extract of ginger was isolated in 1918. This compound was named shogaol, from the Japanese name for ginger, and its structure was proposed as 1-(4′-hydroxy-3′-methoxyphenyl)-dec-4-en-3-one. Subsequent investigations (Bhattarai *et al.*, 2001) showed that the gingerol derivative, gingeryl methyl ether, yielded methyl zingerone and *n*-hexanal on decomposition and no evidence for the formation of *n*-heptanal was obtained.

No further investigations on the composition of the pungent principles of ginger were reported in the following 30 years, although considerable effort was devoted to the synthesis of zingerone analogues (Purseglove *et al.*, 1981). The pungency of ginger was attributed in the first half of the 20th century to two pungent substances,

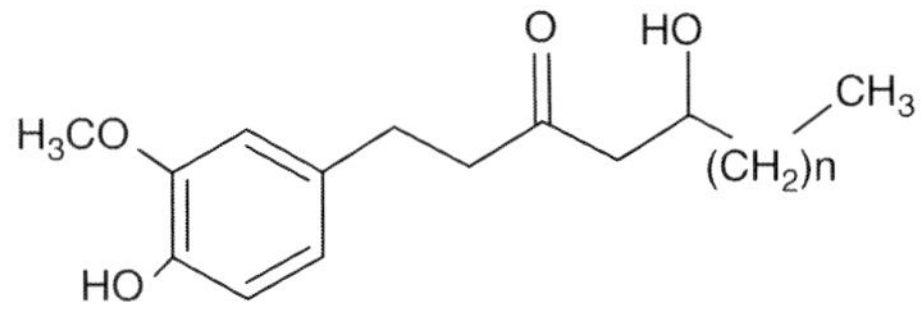

Gingerols
$n = 4$: [6]-gingerol
$n = 6$: [8]-gingerol
$n = 8$: [10]-gingerol

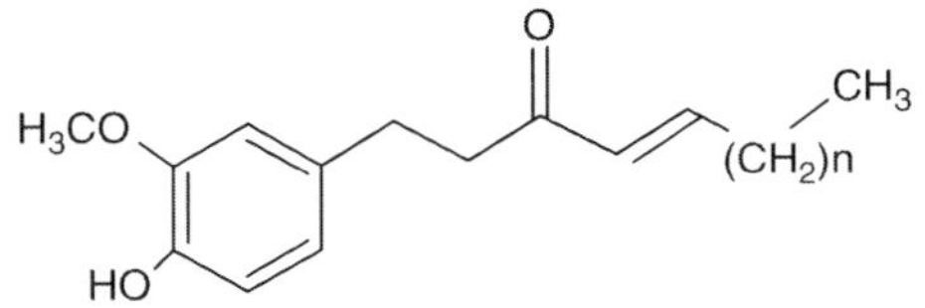

Shogaols
$n = 4$: [6]-Shogaol
$n = 6$: [8]-Shogaol
$n = 8$: [10]-Shogaol

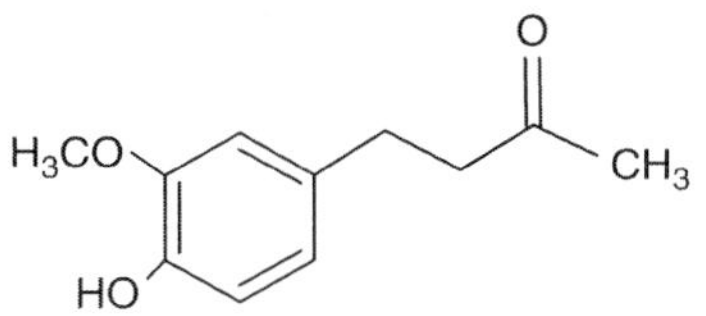

Zingerone

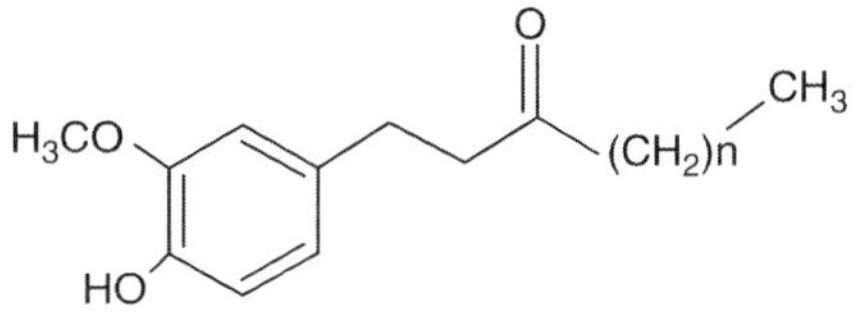

Paradols
$n = 6$: [6]-Paradol
$n = 8$: [8]-Paradol
$n = 10$: [10]-Paradol

Fig. 5.2. Non-volatile pungent constituents of ginger.

gingerol and zingerone, whose relative abundance in ginger was uncertain.

The composition and the relative status of the pungent principles of ginger were clarified in 1969. Gingerol was isolated in a crystalline state from Australian (Buderim) ginger by a solvent extraction procedure, which confirmed that it was composed of a series of homologous compounds. The major constituents were found to be condensation products of zingerone with saturated straight-chain aldehydes of chain lengths 6, 8 and 10. These compounds were named [6]-, [8]- and [10]-gingerols according to the length of the aldehyde unit, and the relative abundance of these compounds in the sample was estimated as 53:17:30, respectively (Vernin and Parkanyl, 2005).

The pungent principles of commercial Japanese ginger have been studied by two groups of workers who found [6]-gingerol to be the major constituent, together with smaller quantities of [8]- and [10]-gingerols (Chen *et al.*, 1986).

Kim *et al.* (2002) conducted quantitative analysis of 6-gingerol in 14 samples collected throughout the region of the Republic of Korea; these contained an average of 0.359%. The content of 6-gingerol decreased during processing (0.306%).

Evidence obtained from the investigations of Australian and Japanese gingers suggests that [6]-gingerol is the most abundant principle of ginger and is accompanied by several other gingerol homologues and analogues, the [8]- and [10]-gingerols being prominent. Knowledge of the relative pungency values of the individual compounds is, at present, rather limited, as is the case with the pungent principles of pepper (*P. nigrum*), but they are expected to differ somewhat according to the length of the side-chain. Reports by various groups indicate that gingerols have the greatest relative pungency values, followed by shogaols and then zingerone (Purseglove *et al.*, 1981).

Solvent extracts

Analysis of unmodified, partially purified fractions from the dichloromethane extracts of organically grown fresh Chinese white and Japanese yellow varieties of ginger resulted in the detection of 20 hitherto unknown natural products (Jolad *et al.*, 2004). These compounds include paradols, dihydroparadols, gingerols, acetyl derivatives of gingerols, shogaols, 3-di-hydro shogaols, ginger diols, mono- and di-acetyl derivatives of ginger diols, 1-dehydro ginger diones, diarylheptanoids and methyl ether derivatives of some of these compounds. Thermal degradation of gingerols to gingerone, shogaols and related compounds has also been demonstrated (Jolad *et al.*, 2004). The study resulted in the identification of 115 compounds, including 31 new compounds. These also include iso-gingerol, shogaols, gingerdiols and other derivatives.

Stability of the pungent principles

Govindarajan (1982) demonstrates that the gingerols are susceptible to chemical transformation to less pungent degradation products and that these reactions can occur by poor handling during the preparation, storage and use of dried ginger and its oleoresin, with consequent deterioration of quality.

The gingerols can undergo a retro aldol reaction at the β-hydroxy ketone group to yield zingerone and aliphatic aldehydes, such as hexanal. This reaction can occur by base catalysis or by the action of heat, and with oleoresins it proceeds rapidly at temperatures above 200°C. The process is detrimental not only because of reducing the pungency level, but also from the production of off-flavours by the liberated aldehydes.

The second, and more important, transformation to which the gingerols are prone is a dehydration at the β-hydroxy ketone group to form the corresponding, less pungent shogaols. This reaction is influenced markedly by pH and temperature. Under alkaline conditions, the dehydration occurs readily at room temperature, but higher temperatures are required under acid conditions. With oleoresins, the reaction proceeds at five times the rate under acid conditions than at pH7 (Surh and Lee, 1994).

The pH of ginger oleoresins is normally in the range of 3–5 and thus dehydration of the gingerols tends to proceed by the acid-catalysed mechanism during extraction and subsequent storage.

Shogaol formation can occur even during the drying of the ginger rhizome and it is more extensive with whole, dried ginger than with sliced, dried ginger. This is attributed to the longer drying time required for whole ginger.

The shogaols have also been found to be susceptible to acid pH and heat treatment and they probably transform to non-pungent polymers. Thus, the pungency of oleoresins decreases steadily on storage as the gingerols are first transformed to the shogaols, which are in turn degraded (Purseglove *et al.*, 1981).

Pungency losses can also occur during use of the spice or its oleoresin in food, either by excessive heat treatment or, for example, by using potassium carbonate in ginger-flavoured baked goods.

Biosynthesis

Biosynthesis of these active metabolites takes place through the phenylpropanoid pathway. Assays for enzymes in the phenylpropanoid pathway identified the corresponding enzyme activities in crude extracts from leaf, shoot and rhizome tissues from ginger. These enzymes included phenylalanine ammonia lyase, polyketide synthases, *p*-coumaroyl shikimate transferase, *p*-coumaroyl quinate transferase, caffeic acid *O*-methyltransferase and caffeoyl-CoA *O*-methyltransferase, which were evaluated because of their potential roles in controlling production of gingerols. All crude extracts possessed activity for all of these enzymes, with the exception of polyketide synthases. Thioesterase activities that cleaved phenylpropanoid pathway CoA esters were found to be present at high levels in all tissues, especially in ginger. These activities may shunt phenylpropanoid pathway intermediates away from the production of gingerols, thereby potentially playing a regulatory role in the biosynthesis of these compounds (Ramirez Ahumada *et al.*, 2006). Phenylalanine ammonia lyase (PAL) is the first enzyme of the phenyl propanoid pathway and, as such, is the entry point into potential pathways leading to the formation of gingerols (Ramirez Ahumada *et al.*, 2006). This enzyme catalyses the non-oxidative deamination of L-phenylalanine to afford *trans*-cinnamic acid ammonium ion. Other enzymes involved in the gingerol biosynthesis are *p*-coumaroyl shikimate transferase, *p*-coumaroyl quinate transferase, caffeic acid *O*-methyltransferase and caffeoyl-CoA *O*-methyltransferase.

Pungency evaluation

The pungency of ginger is mild compared with capsicum. Gingerols and shogaols have now been considered essential components of ginger extracts. Zingerone, which was once considered as the constituent responsible for pungency, is now known to be present only in long-stored extracts or poorly processed extracts. Studies conducted on the extent of the pungency of ginger oleoresin indicate that pungency decreases in the order of gingerol > shogaol > zingerone. The proportion of gingerols and shogaols varies widely, 98:2 to 50:50, or even reversed to 40:60 or 20:80, depending on the source, such as fresh commercial ginger, and stored and poorly processed oleoresin. Table 5.6 indicates the pungency stimuli and related components in some of the well-traded ginger oleoresin samples.

Irradiation on oleoresin

Oleoresin and gingerol contents in γ-irradiated dried ginger rhizomes were evaluated to determine the effect of radiation and storage on these constituents. Both whole and ground samples of dried rhizomes were irradiated (with 0, 5 or 10 kGy doses of γ-rays from a 60 Co source) in sealed polyethylene pouches, which were then transferred into screw-capped plastic containers and stored at 23–26°C. The oleoresin and gingerol contents were monitored for 9 months. Radiation treatment (10 kGy) reduced the decrease of the oleoresin content of ginger during the storage period by 14% in unground samples and by 11% in ground samples. There was a dose-dependent decrease in the 6-gingerol content of the ginger during the storage period. It decreased by 65.6, 67.4 and 70.4% for the 0, 5 and 10 kGy irradiated ground ginger samples, respectively, while the corresponding values for the unground ginger samples were 37.8, 40.0 and 44.3%, respectively (Onyenekwe, 2000).

Estimation of pungent compounds

He *et al.* (1998) developed a gradient elution reversed-phase HPLC technique for separation of pungent principles. HPLC-UV-electrospray MS was used successfully to identify the individual pungent constituents in the chromatogram of ginger extract. Seven compounds were identified positively as the major pungent constituents of

Table 5.6. Pungency stimuli and related components and pungency of fresh, commercial and stored oleoresins.

Sample	Gingerols (% in oleoresin)		Shogaols (% in oleoresin)		Scoville heat units (× 10^3)	
	Pungent	Non-pungent	Pungent	Non-pungent	Estimated	Calculated
Fresh oleoresins						
Wynad Manantody	32.62	14.54	1.82	3.21	28.16	27.20
Kurupumpadi	26.33	9.87	1.26	2.59	24.64	21.64
Rio de Janeiro	29.64	11.92	1.39	2.74	27.98	24.31
Sierra Leone	30.97	14.61	2.45	4.20	29.97	26.90
China	22.16	7.76	1.83	2.00	17.09	19.36
Commercial and stored oleoresins						
Dry ginger, coated	20.26	6.22	4.09	1.24	21.36	21.37
Dry ginger, 1 years' storage	13.23	3.33	10.24	3.10	25.30	25.65
Jamaican, > 3 years' storage	8.77	3.88	12.91	3.80	28.73	25.94
China (alkali treated)	2.45	1.23	15.55	3.80	24.61	25.09
Dry ginger, > 3 years' storage	4.55	2.30	9.60	4.20	42.10	17.70

ginger based on their UV spectra, $[M + H]^+$, $[M + Na]^+$ and characteristic sodiated dimer $[2M + Na]^+$ ions, and comparison to data for the purified standards: [6]-gingerol and [6]-shogaol. The pungent compounds were assigned as [6]-gingerol, [8]-gingerol, [10]-gingerol, [6]-shogaol, [8]-shogaol, [10]-shogaol and [6]-gingediol. Another eight minor compounds were identified tentatively as gingerol analogues. Isolation of 6-, 8- and 10-gingerol from ginger rhizome by HPLC was reported by Gorecki *et al.* (1997) and Hiserodt *et al.* (1998).

Bartley (1995) developed a new method for the determination of pungent compounds in ginger in which the analysis of gingerols and shogaols was by negative ion electrospray-MS of crude extracts, without previous chromatographic separation. Extracts prepared by supercritical fluid extraction contained 6-gingerol (70 g/kg) and only small amounts of 6-shogaol (< 2 g/kg). The low concentration of shogaols, which are degradation products of gingerols, indicates the mild nature of the extraction and analytical procedures.

Yoshikawa *et al.* (1993) conducted studies on qualitative and quantitative analysis of bioactive principles and chemical change of constituents during processing in ginger rhizome by means of high-performance liquid chromatography and gas liquid chromatography. It was found that all Japanese samples of dry ginger and fresh ginger contained 6-gingerol, 6-dehydrogingerdione and galanolactone as major constituents, whereas galanolactone was not detected in any imported rhizome; 6-dehydrogingerdione was detected in Vietnamese but not in Chinese or Taiwanese rhizomes. Processing of fresh ginger rhizome was accompanied by large decreases in the contents of 6-gingerol and galanolactone.

An optimized high-speed countercurrent chromatographic (CCC) technique was described for the isolation of [6]-, [8]- and [10]-gingerols, in quantities between 40 and 500 mg, from powdered root ginger. Minor modifications to the procedure allowed the separation of [4]-gingerol. The purity of the isolated gingerols was assessed by proton NMR. Of the four gingerols isolated, only [8]-gingerol was shown to have a significant impurity (about 10%), with the purest being [6]-gingerol at > 97% (Farthing and O'Neill, 1990). Spiro *et al.* (1990) studied the kinetics of extraction of the oleoresin constituents [6]-gingerol and hexahydrocurcumin from ground ginger rhizome.

A sensitive and accurate high-performance TLC (HPTLC) method has been developed by Rai *et al.* (2006) to determine the quantity of 6-gingerol in rhizomes of *Z. officinale*. Methanol extracts of rhizomes from three different sources were used for HPTLC with the mobile phase consisting of *n*-hexane and diethyl ether (40:60 v/v). The Rf of 6-gingerol was found to be 0.40. The calibration plot was linear in the range of 250–1200 ng of 6-gingerol and the correlation coefficient of 0.9997 was indicative of good linear dependence of peak area on concentration. The mean quantity of 6-gingerol was found to be 60.44 ± 2.53 mg/g of ginger extract. The method permits reliable quantification of 6-gingerol and good resolution and separation of 6-gingerol from other constituents of ginger.

Cellular localization of pungent principles

Zarate and Yeoman (1994) conducted studies of the cellular localization of the phenolic pungent principle of ginger, *Z. officinale* Roscoe. Examination of cryosections from mature rhizomes, immature rhizomes and adventitious roots of ginger showed the presence of yellow-pigmented cells. There was a correlation between the number of pigmented cells and the amount of [6]-gingerol, the main pungent principle in ginger, in these organs. It was also shown histochemically and micro-spectrophotometrically that the yellow cells contain, in addition to flavonoid-like compounds, possibly curcumin derivatives, phenolics and large amounts of lipid material. Low-temperature scanning electron microscopy demonstrated the oil content of these cells. Therefore, it would appear that the site of accumulation of flavonoids, phenolics including gingerol and the essential oils of ginger lies within the same cell type.

Variation in culture systems

The accumulation of [6]-gingerol and [6]-shogaol (phenolic pungent principles of ginger) was much higher in culture systems of ginger where morphological differentiation was apparent. Cultures grown on a callus-inducing medium also accumulated these metabolites, but to a lesser extent. There is a positive relationship between product accumulation and morphological differentiation, although unorganized callus tissue also seems to possess the necessary biochemical machinery to produce and accumulate some phenolic pungent principles. In contrast to earlier studies with the intact plant, there was no positive correlation between the amount of [6]-gingerol accumulated and the number of pigmented cells in either of the culture systems investigated (Zarate and Yeoman, 1996).

Metabolic profiling, using GC-MS and LC-ESI-MS, was used to determine whether chemical differences existed between greenhouse-grown or *in vitro* micropropagation-derived plants. Three different ginger lines were analysed. The constituent gingerols and gingerol-related compounds, other diarylheptanoids and methyl ether derivatives of these compounds, as well as major mono- and sesquiterpenoids, were identified (Ping *et al.*, 2004). Principal component analysis and hierarchical cluster analysis revealed chemical differences between lines (yellow ginger versus white ginger and blue ring ginger) and tissues (rhizome, root, leaf and shoot). However, this analysis indicated that no significant differences existed between growth treatments (conventional greenhouse-grown versus *in vitro* propagation-derived plants). These findings suggest that the biochemical mechanisms used to produce the large array of compounds found in ginger are not affected by *in vitro* propagation (Ma and Gang, 2006).

5.6. Medicinal and Pharmacological Properties

Anticancer properties

Ginger, a natural dietary component, has been known to have antioxidant and anticarcinogenic properties. Manju and Nalini (2005) demonstrated the chemopreventive efficacy of ginger in colon cancer. They had investigated the effect of ginger on the initiation and post-initiation stages of 1,2-dimethyl hydrazine (DMH)-induced colon carcinogenesis in male Wistar rats. The number of tumours, as well as the incidence of cancer, was decreased significantly on treatment with ginger. Shukla and Singh (2006) attributed the anticancer properties to the presence of pungent vallinoids, e.g. [6]-gingerol and [6]-paradol, shogaols, zingerone, etc.

Anti-inflammatory effect

Ginger contains pungent phenolic substances with pronounced antioxidative and anti-inflammatory activities. The antitumour-promoting activity of [6]-gingerol, a major pungent principle, was investigated using a two-stage mouse skin carcinogenesis model. Topical application of [6]-gingerol on to shaven backs of female ICR mice prior to each topical dose of 12-*O*-tetradecanoylphorbol-13-acetate (TPA) inhibited 7,12-dimethylb enz[a]anthracene-induced skin papillomagenesis significantly. The compound also suppressed TPA-induced epidermal ornithine decarboxylase activity and inflammation (Park *et al.*, 1998).

[6]-Gingerol, a pungent ingredient of ginger, has antibacterial-, anti-inflammatory and antitumour-promoting activities. Kim *et al.* (2005) found its antiangiogenic activity *in vitro* and *in vivo*. Angiogenesis, the formation of new blood vessels from pre-existing endothelium, is fundamental in a variety of physiological and pathological processes, including wound healing, embryonic development, chronic inflammation and tumour progression and metastasis. *In vitro*, [6]-gingerol inhibited both the VEGF- and bFGF-induced proliferation of human endothelial cells and caused cell cycle arrest. Vascular endothelial growth factor (VEGF) is the most important angiogenic factor associated with induction and maintenance of neovasculature in human tumours. Studies on colorectal cancer showed that bevacizumab,

a VEGF monoclonal antibody in combination with cytotoxic therapy, has positive effects on patient survival. [6]-Gingerol also blocked capillary tube-like formation by endothelial cells in response to VEGF and strongly inhibited sprouting of endothelial cells in rat aorta and formation of new blood vessels in the mouse cornea in response to VEGF (Kim *et al.*, 2005).

The mechanism of potentiation of prostaglandin (PG) F2α-induced contraction of mouse mesenteric veins by (±)-[6]-gingerol was investigated *in vitro* by Hata *et al.* (1998). (±)-[6]-Gingerol (0.3 mM) potentiated the maximum contraction response elicited by 0.28 mM PGF2α in the presence of intact vascular endothelium, but not in its absence (de-endothelialized preparations). The potentiating effect was inhibited completely by cyclooxygenase inhibitors (0.2 mM aspirin and 0.2 mM indomethacin [indometacin]) and partly by calcium antagonists (2 μM verapamil, 8 nM nitrendipine and 1 μM ryanodine), but was not inhibited by nordihydroguaiaretic acid (NDGA, a lipoxygenase inhibitor) or ONO-3708, a thromboxane (TX) A2 antagonist. The potentiation by (±)-[6]-gingerol was also observed in mesenteric veins of streptozotocin-diabetic mice where the enhancement of PGF2α-induced contraction was caused mainly by activation of lipoxygenase. The potentiation of PGF2α-induced contraction by (±)-[6]-gingerol may be caused by a cyclooxygenase-dependent release of vasoconstrictors other than PGF2α and TXA2, or by inhibiting vasorelaxants released from endothelial cells of mouse mesenteric veins (Jolad *et al.*, 2005).

The effects of gingerols ([6]-gingerol, [8]-gingerol and [10]-gingerol) isolated from rhizomes of ginger on Ca^{2+}-ATPase activity of cardiac sarcoplasmic reticulum (SR) were investigated. Ca^{2+}-ATPase activity and Ca^{2+}-pumping activity were increased by these compounds in a concentration-dependent manner. It was suggested that both the *o*-methoxyphenol and hydrocarbon chain of compounds were necessary for the activation of the Ca^{2+}-pumping ATPase activity of SR (Ohizumi *et al.*, 1996). Jiang *et al.* (2006) could establish the anti-inflammatory property of [6]-, [8]- and [10]-gingerols.

Antiplatelet effect

Guh *et al.* (1995) studied the antiplatelet effect of gingerol isolated from *Z. officinale*. Gingerol (0.5–20 μM) concentration dependently inhibited the aggregation and release reaction of arachidonic acid and collagen-induced rabbit platelets, but not those induced by platelet-activating factor U46619 and thrombin. Gingerol (0.5–10 μM) also concentration-dependently inhibited thromboxane B2 and prostaglandin D2 formation caused by arachidonic acid and completely abolished phosphoinositide breakdown induced by arachidonic acid, but had no effect on that of collagen, PAF or thrombin, even at concentrations as high as 300 μM. In human platelet-rich plasma, gingerol and indomethacin [indometacin] prevented the secondary aggregation and blocked ATP release from platelets induced by ADP (5 μM) and adrenaline [epinephrine] (5 μM), but had no influence on primary aggregation. The highest antiplatelet effect was obtained when platelets were incubated with gingerol for 30 min, and this inhibition was reversible. It is concluded that the antiplatelet action of gingerol is due mainly to the inhibition of thromboxane formation.

Antioxidant effect

Kikuzaki and Nakatani (1993) evaluated the antioxidant effects of some ginger constituents. The non-volatile fraction of the dichloromethane extract of ginger rhizomes exhibited a strong antioxidative activity using linoleic acid as the substrate in ethanol-phosphate buffer solution. The fraction was purified by chromatographic techniques to provide five gingerol-related compounds and eight diarylheptanoids. Among them, 12 compounds exhibited higher activity than α-tocopherol. The activity was probably dependent on side-chain structures and substitution patterns on the benzene ring. Aeschbach *et al.* (1994) also studied the antioxidant effect of zingerone from ginger. Manju and Nalini (2005) demonstrated the chemopreventive efficacy of ginger in colon cancer. They investigated the effect of ginger on the initiation and post-initiation stages of 1,2-dimethyl hydrazine (DMH)-induced colon carcinogenesis in male Wistar rats. The number of tumours, as well as the incidence of cancer, was decreased significantly on treatment with ginger. Shukla and Singh (2006) attributed the anticancer properties to the presence of pungent vallinoids, e.g. [6]-gingerol and [6]-paradol, shogaols, zingerone, etc.

Studies by Stoilova *et al.* (2007) established the antioxidant activity of ginger extract. Total phenols of the alcoholic ginger extract are about 870.1 mg/g dry extract. 2,2-Diphenyl-1-picril hydrazyl radical (DPPH) scavenging reached 90.1% and exceeded that of butylated hydroxyl toluene (BHT). The antioxidant activity in a linoleic acid/water emulsion system determined by means of thiobarbituric acid-reactive substances (TBARS) was highest at 37°C – 73.2 and 71.6% when the formation of conjugated dienes was inhibited. Ginger extract inhibited hydroxyl radicals by 79.6% at 37°C and 74.8% at 80°C, which showed a higher antioxidant activity than quercetin (Stoilova *et al.*, 2007). Nakatani (2003) also demonstrated the antioxidant property of gingerol-related compounds and diarylheptanoids from common ginger.

Anti-ulcer principles

Yoshikawa *et al.* (1994) investigated the stomachic principles in ginger. They detected an anti-ulcer principle, 6-gingesulphonic acid, and three monoacyl digalactosyl glycerols, ginger glycolipids A, B and C, from ginger rhizome from Taiwan. Dried rhizome of ginger is used in Chinese and Japanese traditional medicines to treat headaches, nausea, stomach ache and colds. The structures of 6-gingesulphonic acid and gingerglycolipids A–C were elucidated from chemical and physico-chemical analyses. The determination of the absolute stereostructure of (+)-angelicoidenol-2-*O*-β-D-glucopyranoside was based on its synthesis from *d*-borneol. In an experiment involving a sensory panel of six volunteers, 6-gingesulphonic acid was considered less pungent than the previously isolated phenolic compounds, 6-gingerol and 6-shogaol. Oral administration of 6-gingesulphonic acid, 6-shogaol and 6-gingerol at 150 mg/kg reduced HCL/ethanol-induced gastric lesions in male Wistar rats by 92.7, 70.2 and 57.5%, respectively. Oral administration of acetone extract of ginger at 1000 mg/kg and zingiberene (the main terpenoid from the acetone extract) at 100 mg/kg inhibited gastric lesions by 97.5 and 53.6%, respectively. 6-Gingerol (the pungent principle) at 100 mg/kg inhibited gastric lesions by 54.5%. The results suggest that zingiberene and 6-gingerol are the constituents which act as protectants against gastric lesions in medications containing ginger (Yamahara *et al.*, 1988).

Yamahara *et al.* (1990) studied the gastrointestinal motility-enhancing effect of ginger and its active constituents. Powdered rhizome of ginger was extracted with acetone and the extract was evaporated to dryness at < 40°C to give a residue (yield 3.4%), which included volatile oils and bitter substances. This was fractionated on silica gel columns and the compounds isolated were identified from spectral data. The transport of a charcoal meal in mice was dose-dependently accelerated by oral administration of the whole dry extract or its constituents, [6]-shogaol and [6]-, [8]- and [10]-gingerol. Endo *et al.* (1990) elucidated structures of antifungal diarylheptenones, gingerenones A, B, C and isogingerenone B, which were isolated from the rhizomes.

Anticonvulsive and analgesic effect

Naora *et al.* (1992) evaluated the potential of ginger as a cardiotonic and analgesic. Preparations of the rhizomes are used in traditional Chinese medicine as an antipyretic, cardiotonic, anticonvulsive or analgesic. The pharmacokinetics of [6]-gingerol, the main active component of *Z. officinale*, were investigated in 12-week-old male Wistar rats suffering from acute renal failure induced by bilateral nephrectomy, or showing acute hepatic failure induced by a single oral administration of carbon tetrachloride (CCl_4). The objective was to clarify the contribution of the kidney and liver to the elimination process of [6]-gingerol. There were no significant differences between control and nephrectomized rats in their plasma concentration–time curves and other pharmacokinetic parameters, suggesting that renal excretion was not involved in the disappearance of [6]-gingerol from rat plasma. In contrast, hepatic intoxication with CCl_4 elevated the plasma concentration of [6]-gingerol at the terminal phase; the elimination half-life of [6]-gingerol increased significantly from 8.5 (control) to 11.0 min (CCl_4-treated). Most of the [6]-gingerol (> 90%) was bound to serum protein and binding was hardly affected by CCl_4-intoxication. It was concluded that [6]-gingerol was partly eliminated by the liver.

Ginger is known to warm the body, curing chills caused by the common cold. An acetone extract of ginger rhizomes (administered orally) inhibited serotonin-induced hypothermia and serotonin-induced diarrhoea significantly. When the extract was fractionated on silica gel, the main active constituent against both disorders was found to be [6]-shogaol. Other anticathartic components were [6]-dehydrogingerdione, [8]-gingerol and [10]-gingerol (Huang *et al.*, 1990).

Cardiovascular effect

Ghayur *et al.* (2005) reported the hypotensive, endothelium-dependent and -independent vasodilator and cardio-suppressant and stimulant effects of aqueous extract (Zo.Cr) of ginger. Zo.Cr, which tested positive for saponins, flavonoids, amines, alkaloids and terpenoids, induced a dose-dependent (3.0–10.0 mg/kg) fall in the arterial blood pressure (BP) of anaesthetized rats, which was partially blocked by atropine (1 mg/kg). In isolated endothelium-intact rat aorta, Zo.Cr (0.01–5.0 mg/ml) relaxed the phenylephrine (1 μM)-induced contractions effect partially blocked by atropine (1 μM). An atropine-resistant vasodilator activity was also noted from ginger phenolic constituents 6-, 8- and 10-gingerol, while 6-shogaol showed a mild vasodilator effect. The data indicate that the aqueous ginger extract lowers BP through a dual inhibitory effect mediated via stimulation of muscarinic receptors and blockade of Ca^{2+} channels, and this study provides a sound mechanistic basis for the use of ginger in hypertension and palpitations.

Abdel Aziz *et al.* (2005) reported the potential of different extracts (ethanolic, hexane and aqueous) of ginger and the essential oil in 5-HT3 receptor antagonistic effects. [6]-Gingerol showed maximum potential.

Based on clinical trials, Geiger (2005) postulated that a 5% solution of essential oil of ginger is an effective post-operative nausea and vomiting (PONV) prevention when administered pre-operatively, nasocutaneously, concurrently with conventional therapies, to general anaesthesia patients at high risk of PONV.

Gingerols, the pungent constituents of ginger, were assessed as agonists of the capsaicin-activated vanilloid receptor (VR1). [6]-Gingerol and [8]-gingerol evoked capsaicin-like intracellular Ca^{2+} transients and ion currents in cultured dorsal root ganglion neurons. These effects of gingerols were blocked by capsazepine, the VR1 receptor antagonist. The potency of gingerols increased with the increasing size of the side-chain and with the overall hydrophobicity in the series. It is concluded that gingerols represent a novel class of naturally occurring VR1 receptor agonists that may contribute to the medicinal properties of ginger, which have been known for centuries. The gingerol structure may be used as a template for the development of drugs acting as moderately potent activators of the VR1 receptor (Dedov *et al.*, 2002).

5.7. International Specifications

Table 5.7 illustrates the specifications for whole Indian Calicut/Cochin ginger based on size and other limits. The general specifications by various importing countries are given in Table 5.8. Table 5.9 depicts the specification for ginger oleoresin given by the Essential Oil Association and the Indian Standards. The specification for ginger oil is given in Table 5.10.

Table 5.7. Ginger whole, Indian grade designation and requirements.

			Requirements				
Source	Type	Grade	Size (length in mm)	Size tolerance (max. %)	Extraneous matter (max. %)	Light pieces (max. %)	Lime as CaO (max. %)
Cochin/Calicut	Non-bleached	Garbled	15	3.0	2.0	0.0	–
		Ungarbled special	15	7.0	3.0	5.0	–
		Good	15	15.0	4.0	10.0	–
		Non-specified			As in contract		
		Garbled	15	3.0	2.0	0.0	2.5
		Ungarbled special	15	7.0	3.0	5.0	4.0
		Good	15	15.0	4.0	10.0	6.0
		Non-specified			As in contract		

Table 5.8. Specifications for ginger, whole and pieces in importing countries.

	USA	UK	Germany	India	ISO
Moisture (% max.)	–	12.0	12.0	10.0	12.0
Total ash (% max.)					
Unbleached	7.0	8.0	7.0	7.0	8.0
Bleached	–	12.0	–	–	12.0
Calcium, as CaO (% max.)					
Unbleached	1.0	1.1	1.1	1.0	1.1
Bleached	–	2.5	–	–	2.5
Volatile oil (ml/100 g; min.)	–	1.0	1.5	–	1.5
Starch (% min.)	42.0	–	–	–	–
Crude fibre (% max.)	8.0	–	–	–	–
Lead (ppm max.)	–	–	–	10.0	–
Cold water soluble extract (% min.)	12.0 15.0 (Jamaican ginger)	11.4	11.4	10.0	11.4
Alcohol soluble extract (% min.)	–	5.1	5.4	4.5	5.1
Water soluble ash (% min.)	2.0	1.9	–	1.7	1.9
Acid insoluble ash (% max.)	2.0	2.3	2.3	1.0	2.3

Table 5.9. Specification for ginger oleoresin.

	EOA-243	Indian Standard: 7826–1975
Raw material	Rhizomes of *Z. officinale* Roscoe, source specified as Cochin, African, Jamaican	Rhizomes of *Z. officinale* Roscoe
Solvents	–	Passes relevant IS specification
Preparation	By solvent extraction and subsequent solvent removal	By solvent extraction of ground rhizomes
Description	A dark brown viscous to highly viscous liquid with characteristic odour and flavour of ginger	Viscous dark brown or reddish brown liquid with characteristic odour and flavour
Appearance and odour		
Volatile oil content	18–35 ml/100 g	16–35% (v/w)
Specific gravity		0.8640–0.8758 at 30°C
Refractive index	1.4880–1.4970 at 20°C	1.4880–1.4979 at 20°C
Optical rotation	–30° to –60°	–30° to –60°
Gingerol content (% by mass, min.)	–	18
Residual solvent in oleoresin (ppm, max.)	Meets Federal Food, Drug and Cosmetic Act regulations	Acetone, ethylene dichloride, trichloroethylene: 30 isopropanol, methanol: 50 hexane: 25
Solubility	Alcohol and benzyl benzoate: soluble with sediment in all proportions	–
	Fixed oil: slightly soluble	–
	Glycerine: insoluble	–
	Mineral oil: insoluble	–
	Propylene glycol: insoluble	–
Odour testing	–	Warm, spicy, sweet and aromatic; compare with soxhlet extract of genuine dried ginger and for absence of off-odour or by smelling strips
Taste testing	–	Warm, pungent, biting and cool 20 mg stirred into 100 ml boiling neutral soup; compare with standard oleoresin and passed if there is no off-taste
Containers	Glass or suitably lined containers, filled full and closed tightly	Glass, aluminium, or food-grade high-density polyethylene containers filled full and tightly closed

5.8. Conclusion

Ginger is one of the most extensively used spices because of its wide range of application. It is used fresh and in the preserved or dried form. The potential of ginger in the culinary, non-culinary and medicinal fields is based on the chemistry of volatile oil and non-volatile pungent principles. The oil yield is about 2–3% and the oil consists of 64% sesquiterpene hydrocarbons, 6% carbonyl compounds, 5% alcohols, 2% monoterpene hydrocarbons and 1% esters. The main compounds are zingiberene (29.5%) and sesquiphellandrene (18.4%). The pungent compounds of ginger include gingerols, shogaols, paradols and zingerone, which produce a 'hot' sensation in the mouth. The composition of these constituents varies, based on maturity, genotype and agroclimatic conditions. Ginger has proven anti-inflammatory, antioxidant and anti-ulcer principles.

Table 5.10. Specification for ginger oil.

	EOA-13	Indian Standard: 761–1955
Source	Rhizomes of *Z. officinale* Roscoe	Rhizomes of *Z. officinale* Roscoe
Preparation	Direct steam distillation of coarse, ground, dried rhizome	Water or steam distillation of coarse ground rhizome
Colour and appearance	Light yellow to yellow	Clear liquid, free from sediment; light yellow or greenish yellow
Specific gravity	0.870–0.888 at 25°C	0.868–0.880 at 30°C
Optical rotation	–28° to –45°	–28° to –45°
Refractive index	1.4880–1.4940 at 20°C	1.4840–1.4894 at 30°C
Saponification number/value	Not more than 20	20 (max)
Stability	Alkali: relatively stable to weak alkali; unstable to strong alkali	–
	Acid: unstable in presence of strong acids	–
Solubility	Alcohol: soluble, with turbidity	Alcohol, absolute: soluble, with slight turbidity
	Benzyl benzoate, diethyl phthalate, fixed oils, mineral	–
	Oil: soluble in all portions	
	Glycerin, propylene glycol: practically insoluble	
Odour and taste	–	Characteristic and persistent odour of ginger; lacking pungent taste
Containers and storage	Glass or tin-lined containers, filled full; cool storage protected from light	Closed containers, as agreed in contract

References

Abd Malek, S.N., Ibrahim, H., Hong, S.L., Lee, G.S., Chan, K.S. and Mohd Ali, N.A. (2005) The essential oils of *Zingiber officinale* variants. *Malaysian Journal of Science* 24(2), 37–43.

Abdel Aziz, H., Nahrstedt, A., Petereit, F., Windeck, T., Ploch, M. and Verspohl, E.J. (2005) 5-HT3 receptor blocking activity of arylalkanes isolated from the rhizome of *Zingiber officinale*. *Planta Medica* 71(7), 609–616.

Aeschbach, R., Loliger, J., Scott, B.C., Murcia, A., Butler, J., Halliwell, B. and Aruoma, O.I. (1994) Antioxidant actions of thymol, carvacrol, 6-gingerol, zingerone and hydroxytyrosol. *Food and Chemical Toxicology* 32(1), 31–36.

Agarwal, M., Walia, S., Dhingra, S. and Khambay, B.P.S. (2001) Insect growth inhibition, antifeedant and antifungal activity of compounds isolated/derived from *Zingiber officinale* Roscoe (ginger) rhizomes. *Pest Management Science* 57(3), 289–300.

Antonious, G.F. and Kochhar, T.S. (2003) Zingiberene and curcumene in wild tomato. *Journal of Environmental Science and Health Part B, Pesticides, Food Contaminants, and Agricultural Wastes* 38(4), 489–500.

Bartley, J.P. (1995) A new method for the determination of pungent compounds in ginger (*Zingiber officinale*). *Journal of the Science of Food and Agriculture* 68(2), 215–222.

Bartley, J.P. and Jacobs, A.L. (2000) Effects of drying on flavour compounds in Australian-grown ginger (*Zingiber officinale*). *Journal of the Science of Food and Agriculture* 80(2), 209–215.

Bhattarai, S., Tran, V.H. and Duke, C.C. (2001) The stability of gingerol and shogaol in aqueous solutions. *Journal of Pharmaceutical Sciences* 90(10), 1658–1664.

Chen, C., Kuo, M. and Ho, C. (1986) High performance liquid chromatographic determination of pungent gingerol compounds of ginger (*Zingiber officinale* Roscoe). *Journal of Food Science* 51(5), 1364–1365.

Dedov, V.N., Tran, V.H., Duke, C.C., Connor, M., Christie, M.J., Mandadi, S. and Roufogalis, B.D. (2002) Gingerols: a novel class of vanilloid receptor (VR1) agonists. *British Journal of Pharmacology* 137(6), 793–798.

Ekundayo, O., Laakso, I. and Hiltunen, R. (1988) Composition of ginger (*Zingiber officinale* Roscoe) volatile oils from Nigeria. *Flavour and Fragrance Journal* 3(2), 85–90.

Endo, K., Kanno, E. and Oshima, Y. (1990) Structures of antifungal diarylheptenones, gingerenones A, B, C and isogingerenone B, isolated from the rhizomes of *Zingiber officinale. Phytochemistry* 29(3), 797–799.

Erler, J., Vostrowsky, O., Strobel, H. and Knobloch, K. (1988) Essential oils of ginger, *Zingiber officinalis* Roscoe. *Zeitschrift fur Lebensmittel Untersuchung und Forschung* 186(3), 231–234.

Fakim, A.G., Maudarbaccus, N., Leach, D., Doimo, L. and Wohlmuth, H. (2002) Essential oil composition of Zingiberaceae species from Mauritius. *Journal of Essential Oil Research* 14(4), 271–273.

Farthing, J.E. and O' Neill, M.J. (1990) Isolation of gingerols from powdered root ginger by countercurrent chromatography. *Journal of Liquid Chromatography* 13(5), 941–950.

Geiger, J.L. (2005) The essential oil of ginger, *Zingiber officinale*, and anaesthesia. *The International Journal of Aromatherapy* 15, 7–14.

Ghayur, M.N., Gilani, A.H., Afridi, M.B. and Houghton, P.J. (2005) Cardiovascular effects of ginger aqueous extract and its phenolic constituents are mediated through multiple pathways. *Vascular Pharmacology* 43(4), 234–241.

Gopalam, A. and Ratnambal, M.J. (1989) Essential oils of ginger. *Indian Perfumer* 33(1), 63–69.

Gorecki, P., Segiet Kujawa, E., Gaedcke, F., Kurt, H. and Feistel, B. (1997) Isolation of [6]-gingerol from *Zingiber officinale* Roscoe by chromatographic methods. *Herba Polonica* 43(3), 202–207.

Govindarajan, V.S. (1982) Ginger – chemistry, technology and quality evaluation: part I. *CRC Critical Reviews* 17(1), 1–96.

Guenther, E. (1975) *The Essential Oils*, Volume V. Robert E. Kreiger Publishing Co., Inc., New York, pp. 85–131.

Guh, J.H., Ko, F.N., Jong, T.T. and Teng, C.M. (1995) Antiplatelet effect of gingerol isolated from *Zingiber officinale. Journal of Pharmacy and Pharmacology* 47(4), 329–332.

Hata, Y., Pancho, L.R., Nojima, H. and Kimura, I. (1998) Endothelium dependent potentiation of prostaglandin F2α-induced contractions by (±)-[6]-gingerol is inhibited by cyclooxygenase but not lipoxygenase inhibitors in mouse mesenteric veins. *Biological and Pharmaceutical Bulletin* 21(8), 792–794.

He, X.G., Lian, L.Z. and Lin, L.Z. (1998) High performance liquid chromatography electrospray mass spectrometric analysis of pungent constituents of ginger. *Journal of Chromatography, A* 796(2), 327–334.

Hiserodt, R.D., Franzblau, S.G. and Rosen, R.T. (1998) Isolation of 6-, 8-, and 10-gingerol from ginger rhizome by HPLC and preliminary evaluation of inhibition of Mycobacterium avium and Mycobacterium tuberculosis. *Journal of Agricultural and Food Chemistry* 46(7), 2504–2508.

Huang, Q.R., Matsuda, H., Sakai, K., Yamahara, J. and Tamai, Y. (1990) The effect of ginger on serotonin induced hypothermia and diarrhoea Yakugaku Zasshi. *Journal of the Pharmaceutical Society of Japan* 110(12), 936–942.

Jiang, H.L., Xie, Z.Z., Koo, H.J., McLaughlin, S.P., Timmermann, B.N. and Gang, D.R. (2006) Metabolic profiling and phylogenetic analysis of medicinal Zingiber species: tools for authentication of ginger (*Zingiber officinale* Rosc.). *Phytochemistry* 67(15), 1673–1685.

John, R.P. and Ferreira, D.I. (1997) Yield and quality evaluation of five ginger (*Zingiber officinale* Roscoe) selection in the Northern Province of South Africa. *Journal of the Southern African Society for Horticultural Sciences* 7(1), 26–29.

Jolad, S.D., Lantz, R.C., Solyom, A.M., Chen, G.J., Bates, R.B. and Timmermann, B.N. (2004) Fresh organically grown ginger (*Zingiber officinale*): composition and effects on LPS-induced PGE2 production. *Phytochemistry* 65(13), 1937–1954.

Jolad, S.D., Lantz, R.C., Chen, G.J., Bates, R.B. and Timmermann, B.N. (2005) Commercially processed dry ginger (*Zingiber officinale*): composition and effects on LPS-stimulated PGE2 production. *Phytochemistry* 66(13), 1614–1635.

Kandiah, M. and Spiro, M. (1990) Extraction of ginger rhizome: kinetic studies with supercritical carbondioxide. *International Journal of Food Science and Technology* 25(3), 328–338.

Kikuzaki, H. and Nakatani, N. (1993) Antioxidant effects of some ginger constituents. *Journal of Food Science* 58(6), 1407–1410.

Kim, E.C., Min, J.K., Kim, T.Y., Lee, S.J., Yang, H.O., Han, S., Kim, Y.M. and Kwon, Y.G. (2005) [6]-Gingerol, a pungent ingredient of ginger inhibits angio genesis *in vitro* and *in vivo. Biochemical and Biophysical Research Communications* 335, 300–308.

Kim, H.K., Kim, Y.A., Hwang, S.W. and Ko, B.S. (2002) Quantitative analysis of 6-gingerol in the Zingiberis Rhizoma by processing methods. *Korean Journal of Pharmacognosy* 33(4), 291–295.

Korla, B.N., Tiwari, S.S. and Goyal, R.K. (1999) Evaluation of ginger clones for quality attributes under rain-fed and irrigated conditions. *Horticultural Journal* 12(1), 39–44.

Ma, X.Q. and Gang, D.R. (2006) Metabolic profiling of *in vitro* micro-propagated and conventionally greenhouse grown ginger (*Zingiber officinale*). *Phytochemistry* 67(20), 2239–2255.

Manju, V. and Nalini, N. (2005) Chemopreventive efficacy of ginger, a naturally occurring anticarcinogen during the initiation, post-initiation stages of 1,2-dimethylhydrazine induced colon cancer. *Clinica Chimica Acta* 358(1–2), 60–67.

Miyazawa, M. and Kameoka, H. (1988) Volatile flavour components of Zingiberis Rhizoma (*Zingiber officinale* Roscoe). *Agricultural and Biological Chemistry* 52(11), 2961–2963.

Nakasone, Y., Yonaha, M., Wada, K., Adaniya, S., Kikuzaki, H. and Nakatani, N. (1999) Evaluation of pungency in the diploid and tetraploid types of ginger (*Zingiber officinale* Roscoe) and antioxidative activity of their methanol extracts. *Japanese Journal of Tropical Agriculture* 43(2), 71–75.

Nakatani, N. (2003) Biologically functional constituents of spices and herbs (2002's JSNFS award for excellence in research). *Journal of Japanese Society of Nutrition and Food Science* 56(6), 389–395.

Naora, K., Ding, G., Hayashibara, M., Katagiri, Y., Kano, Y. and Iwamoto, K. (1992) Pharmacokinetics of [6]-gingerol after intravenous administration in rats with acute renal or hepatic failure. *Chemical and Pharmaceutical Bulletin* 40(5), 1295–1298.

Nobrega, L.P., Monteiro, A.R., Meireles, M.A.A. and Marques, M.O.M. (1997) Comparison of ginger (*Zingiber officinale* Roscoe) oleoresin obtained with ethanol and isopropanol with that obtained with pressurized CO_2. *Cienciae Tecnologia de Alimentos* 17(4), 408–412.

Ohizumi, Y., Sasaki, S., Shibusawa, K., Ishikawa, K. and Ikemoto, F. (1996) Stimulation of sarcoplasmic reticulum Ca^{2+}-ATPase by gingerol analogues. *Biological and Pharmaceutical Bulletin* 19(10), 1377–1379.

Onyenekwe, P.C. (2000) Assessment of oleoresin and gingerol contents in γ irradiated ginger rhizomes. *Nahrung* 44(2), 130–132.

Onyenekwe, P.C. and Hashimoto, S. (1999) The composition of the essential oil of dried Nigerian ginger (*Zingiber officinale* Roscoe). *European Food Research and Technology* 209(6), 407–410.

Park, K.K., Chun, K.S., Lee, J.M., Lee, S.S. and Surh, Y.J. (1998) Inhibitory effects of [6]-gingerol, a major pungent principle of ginger, on phorbol ester-induced inflammation, epidermal ornithine decarboxylase activity and skin tumor promotion in ICR mice. *Cancer Letters* 129(2), 139–144.

Ping, M.J., Xiao, L.J., Yang, L. and Zhong, L.L. (2004) Diarylheptanoids from the rhizomes of *Zingiber officinale*. *Phytochemistry* 65(8), 1137–1143.

Pino, J.A., Marbot, R., Rosado, A. and Batista, A. (2004) Chemical composition of the essential oil of *Zingiber officinale* Roscoe L. from Cuba. *Journal of Essential Oil Research* 16(3), 186–188.

Purseglove, J.W., Brown, E.G., Green, C.L. and Robbins, S.R.J. (1981) *Spices*, Volume 2. Longman, New York, pp. 447–531.

Rai, S., Mukherjee, K., Mal, M., Wahile, A., Saha, B.P. and Mukherjee, P.K. (2006) Determination of 6-gingerol in ginger (*Zingiber officinale*) using high performance thin layer chromatography. *Journal of Separation Science* 29(15), 2292–2295.

Ramirez Ahumada, M. del C., Timmermann, B.N. and Gang, D.R. (2006) Biosynthesis of curcuminoids and gingerols in turmeric (*Curcuma longa*) and ginger (*Zingiber officinale*): identification of curcuminoid synthase and hydroxycinnamoyl-CoA thioesterases. *Phytochemistry* 67(18), 2017–2029.

Rani, K. (1999) Cyclisation of farnesyl pyrophosphate into sesquiterpenoids in ginger rhizomes (*Zingiber officinale*). *Fitoterapia* 70(6), 568–574.

Ratnambal, M.J., Gopalam, A. and Nair, M.K. (1987) Quality evaluation in ginger (*Zingiber officinale* Rosc.) in relation to maturity. *Journal of Plantation Crops* 15(2), 108–117.

Shukla, Y. and Singh, M. (2006) Cancer preventive properties of ginger: a brief review. *Food and Chemical Toxicology* 45(5), 683–690.

Spiro, M. and Kandiah, M. (1989) Extraction of ginger rhizome: kinetic studies with acetone. *International Journal of Food Science and Technology* 24(6), 589–600.

Spiro, M., Kandiah, M. and Price, W. (1990) Extraction of ginger rhizome: kinetic studies with dichloromethane, ethanol, 2-propanol and an acetone–water mixture. *International Journal of Food Science and Technology* 25(2), 157–167.

Sreekala, G.S. and Jayachandran, B.K. (2004) Ginger oil and non-volatile ether extract as influenced by shade levels. *South Indian Horticulture* 52(1/6), 244–247.

Stoilova, I., Krastanv, A., Stoyanova, A., Denev, P. and Gargova, S. (2007) Antioxidant activity of a ginger extract (*Zingiber officinale*). *Food Chemistry* 102, 764–770.

Sultan, M., Bhatti, H.N. and Iqbal, Z. (2005) Chemical analysis of essential oil of ginger (*Zingiber officinale*). *Pakistan Journal of Biological Sciences* 8(11), 1576–1578.

Surh, Y.J. and Lee, S.S. (1994) Enzymic reduction of [6]-gingerol, a major pungent principle of ginger, in the cell-free preparation of rat liver. *Life Sciences* 54(19), PL321–PL326.

Tiwari, S.K. (2003) Evaluation of ginger genotypes for yield and quality attributes under rain-fed and irrigated conditions. *Annals of Agricultural Research* 24(3), 512–515.

Vernin, G. and Parkanyl, C. (2005) Chemistry of ginger. In: Ravindran, P.N. and Nirmal Babu, K. (eds) *Ginger, The Genus Zingiber*. CRC Press, New York, pp. 87–180.

Wohlmuth, H., Leach, D.N., Smith, M.K. and Myers, S.P. (2005) Gingerol content of diploid and tetraploid clones of ginger (*Zingiber officinale* Roscoe). *Journal of Agricultural and Food Chemistry* 53(14), 5772–5778.

Wohlmuth, H., Smith, M.K., Brooks, L.O., Myers, S.P. and Leach, D.N. (2006) Essential oil composition of diploid and tetraploid clones of ginger (*Zingiber officinale* Roscoe) grown in Australia. *Journal of Agricultural and Food Chemistry* 54(4), 1414–1419.

Yamahara, J., Mochizuki, M., Huang, Q.R., Matsuda, H. and Fujimura, H. (1988) The anti-ulcer effect in rats of ginger constituents. *Journal of Ethnopharmacology* 23(2–3), 299–304.

Yamahara, J., Huang, Q.R., Li, Y.H., Xu, L. and Fujimura, H. (1990) Gastrointestinal motility enhancing effect of ginger and its active constituents. *Chemical and Pharmaceutical Bulletin* 38(2), 430–431.

Yoshikawa, M., Hatakeyama, S., Chatani, N., Nishino, Y. and Yamahara, J. (1993) Qualitative and quantitative analysis of bioactive principles in Zingiberis Rhizoma by means of high performance liquid chromatography and gas liquid chromatography. On the evaluation of Zingiberis Rhizoma and chemical change of constituents during Zingiberis Rhizoma processing Yakugaku Zasshi. *Journal of the Pharmaceutical Society of Japan* 113(4), 307–315.

Yoshikawa, M., Yamaguchi, S., Kunimi, K., Matsuda, H., Okuno, Y., Yamahara, J. and Murakami, N. (1994) Stomachic principles in ginger. III. An anti-ulcer principle, 6-gingesulfonic acid, and three monoacyldigalactosylglycerols, gingerglycolipids A,B, and C, from Zingiberis Rhizoma originating in Taiwan. *Chemical and Pharmaceutical Bulletin* 42(6), 1226–1230.

Young, H.Y., Chiang, C.T., Huang, Y.L., Pan, F.P. and Chen, G.L. (2002) Analytical and stability studies of ginger preparations. *Journal of Food and Drug Analysis* 10(3), 149–153.

Yu, Z., Wu, H.M. and Ding, J.K. (1998) The volatile chemical components of fresh *Zingiber officinale. Acta Botanica Yunnanica* 20(1), 113–118.

Zarate, R. and Yeoman, M.M. (1994) Studies of the cellular localization of the phenolic pungent principle of ginger, *Zingiber officinale* Roscoe. *New Phytologist* 126(2), 295–300.

Zarate, R. and Yeoman, M.M. (1996) Changes in the amounts of [6]-gingerol and derivatives during a culture cycle of ginger, *Zingiber officinale. Plant Science Limerick* 121(1), 115–122.

Zhang, W., McAuslane, H.J. and Schuster, D.J. (2004) Repellency of ginger oil to *Bemisia argentifolii* (Homoptera: Aleyrodidae) on tomato. *Journal of Economic Entomology* 97(4), 1310–1318.

6 Turmeric

B. Chempakam and V.A. Parthasarathy

6.1. Introduction

Turmeric, *Curcuma longa* L. (Zingiberaceae), has been attributed a number of medicinal properties in the traditional system of medicine for treating several common ailments (Sivananda, 1958; Nadkarni and Nadkarni, 1976; Raghunath and Mitra, 1982; Dash, 1987). It belongs to the genus *Curcuma*, which consists of several plant species with underground rhizomes and roots. About 40 species of the genus are indigenous to India, indicating the Indian origin (Velayudhan *et al.*, 1999). Originally, it had been used as a food additive to improve the palatability, storage and preservation of food. The spice is cultivated in warm, rainy regions like India, China, Indonesia, Jamaica and Peru (Govindarajan, 1980).

India is the major producer and exporter of turmeric at present, even though the crop is grown in several countries, e.g. Pakistan, Malaysia, Myanmar, Vietnam, Thailand, the Philippines, Japan, China, Korea, Sri Lanka, the Caribbean Islands and Central America. It is estimated officially that about 80% of the world production of turmeric is from India alone.

Important states in India that grow turmeric are Andhra Pradesh (56,822 ha), Orissa (23,640 ha), Tamil Nadu (1728 ha), West Bengal (11,731 ha), Assam (12,066 ha), Maharashtra (6644 ha) and Karnataka (6153 ha). Andhra Pradesh stands first in terms of production (283,541 t) and productivity (4989 kg/ha), followed by Tamil Nadu with production at 64,536 t and productivity of 1268 kg/ha. India is the largest producer and exporter of turmeric in the world. Among the spices grown in India, turmeric ranked fifth in area (161,300 ha), second in production (653,600 t) and third in export (both in quantity (35,556 t) and value (Rs.121.7 crores)) during 1999–2000 (Spices Board, 2000). In India, turmeric is produced from 230 districts in 22 states (Table 6.1).

6.2. Botany and Uses

The use of turmeric dates back nearly 4000 years to the Vedic culture in India, where it was used as a culinary spice and had some religious significance. The name derives from the Latin term *terra merita*, meaning 'meritorious earth', referring to the colour of ground turmeric, which resembles a mineral pigment. In many languages, turmeric is named simply as 'yellow root'. The botanical name is *Curcuma longa* L. Val. syn. *Curcuma domestica* Val., belonging to the family Zingiberaceae.

Table 6.1. Turmeric area and production in India (2004–2005).

State	Area (thousand ha)	Production (thousand t)
Andaman & Nicobar Islands	0.1	0.2
Andhra Pradesh	61.4	419.0
Arunachal Pradesh	0.5	1.8
Assam	11.7	8.4
Bihar	2.8	2.8
Chhattisgarh	0.6	0.6
Gujarat	1.0	14.1
Haryana	0.6	7.0
Himachal Pradesh	0.1	0.1
Karnataka	5.4	26.4
Kerala	2.8	5.7
Madhya Pradesh	0.6	0.5
Maharashtra	9.0	9.0
Manipur	0.6	0.4
Meghalaya	1.6	8.8
Mizoram	0.3	2.3
Nagaland	0.6	3.1
Orissa	23.9	56.8
Rajasthan	0.1	0.2
Sikkim	0.5	1.7
Tamil Nadu	21.6	118.5
Tripura	1.5	4.3
Uttar Pradesh	1.4	0.7
West Bengal	12.6	24.5
All India	161.3	716.9

Source: Spices Board. Quoted by Global AgriSystem Pvt. Ltd., New Delhi.

Botany

Turmeric is an erect perennial herb with thick and fleshy rhizomes and leaves in sheaths, characteristic of the family Zingiberaceae. The plant reaches a height of about 1 m. Leaves are alternate, obliquely erect or sub-sessile. The leaf number ranges from 7–12. The leaf length ranges from 30–45 cm with a breadth of between 14 and 16 cm, with the petiole equalling the blade. The surrounding leaf sheaths taper near the leaf and broaden near the base, forming the pseudostem of the plant. The pseudostem is tall and robust, with oblong/elliptic leaves narrowed at the base. The inflorescence is a cylindrical, fleshy, central spike of 10–15 cm, arising through the pseudostem.

Flowering

Flowers are seen occasionally on cylindric spikes, bearing numerous greenish-white bracts. The upper bracts are white in colour, while the lower bracts are green (Parry, 1969; Nadkarni and Nadkarni, 1976). About 30 flowers are produced in a spike (Nazeem and Rema Menon, 1994). Flowering is acropetal within a spike. The first flush of flowers open completely in 5–12 days. The time of anthesis is between 6.00 and 6.30 a.m. Anther dehiscence takes place at the time the flowers open. The mode of pollination is by insects. The flowering period is between June and October. Cultivars of *C. longa* flower rarely and the setting of seeds is hardly seen. Studies on the morphology and anthesis of turmeric did not result in any suitable technique for controlled pollination. The number of days taken for flowering varied from 118 to 143 in *C. longa* and from 95 to 104 in *C. aromatica*. Flowering to seed maturity took 23–29 days in *C. aromatica*. Pollen stainability in aceto-carmine varied from 45–48% in *C. longa* and from 68–74% in *C. aromatica*.

C. longa is a triploid with a somatic chromosome number of 63 ($2n = 3 \times 63$). Rhizomes are bigger, with a stout mother rhizome, with branching primary and secondary fingers exhibiting a yellow to bright orange-yellow core. Most of the high-yielding varieties are long-duration types which take 8–9 months to mature.

The underground rhizome consists of two distinct parts: the egg-shaped primary or mother rhizome, which is an extension of the stem, and the long cylindrical, multibranched secondary rhizomes, growing downward from the primary rhizomes. Rhizomes are of orange-brown, pale yellow or reddish-yellow in colour.

Uses

In India and Nepal, turmeric rhizomes, popularly known as 'Haldi' rhizomes, are used as a household remedy (Eigner and Scholz, 1999). Turmeric has been used internally as a stomachic, tonic and blood purifier and externally in the prevention and treatment of skin diseases (Anon., 2001). It is also effective against biliary disorders, cough, anorexia, diabetic wounds, rheumatism and sinusitis (Araujo and Leon, 2002).

Over the past few years, there has been increasing interest in turmeric due to its medicinal properties. This is evidenced by the large number of scientific studies published on this topic. Turmeric is used widely as a food colourant and is one of the principal ingredients in curry powder. It has long been used in both Ayurvedic and Chinese medicine as an anti-inflammatory, to treat digestive disorders and liver problems, and for the treatment of skin diseases and wound healing. The active ingredient in turmeric is curcumin. It is still used in rituals of the Hindu religion and as a dye for holy robes, being natural, unsynthesized and cheap. Although as a dye it is used similarly to saffron, the culinary uses of the two spices should not be confused. Turmeric cannot be used to replace saffron in food dishes.

Epidemiological observations suggest that turmeric consumption may reduce certain forms of cancer and render other protective biological effects in humans, which is attributed to its constituent – curcumin (Radha *et al.*, 2006). Thus, it is used effectively as anti-inflammatory, antiangiogermic, antioxidant, anticancerous, etc. The curcuminoids, which are administered orally, enter the blood circulation and are present as glucuronides and glucuronide sulphate conjugate forms (Asai and Miyazawa, 2000) and hence the physiological effects expressed are due to these conjugates.

In the food industry, turmeric powder is used in mustard paste, curry powder, etc. In Asian countries, dry or fresh turmeric, as well as ground turmeric, are used for vegetable and meat dishes and soup-like dishes (Govindarajan, 1980). Oleoresin extracted from turmeric is used in brine pickle (Eiserlie, 1966; Cripps, 1967) and also in mayonnaise and relish formulations. It is also used in non-alcoholic beverages, for garnishing and in some ice creams (Perotti, 1975). In all cases, it is used mainly as a colouring agent, replacing synthetic colours, e.g. tartrazine, which were used formerly (Govindarajan, 1980).

Value-added products

India is the global leader in value-added products of turmeric and exports. Value-added products from turmeric include curcuminoids, dehydrated turmeric powder, oils and oleoresin. Turmeric, like other spices, is available whole, as ground and as oleoresin. The institutional sector in the West buys ground turmeric and oleoresins, while in the industrial sector, whole dry turmeric is preferred.

Ground turmeric

Dried turmeric is powdered by disc-type attrition mills to obtain 60–80 mesh powder for use in various end products. The rhizomes contain 4–6% of volatile oil and there is a great chance of losing the oil when powdered. Since curcuminoids, the colour constituents of turmeric, deteriorate on exposure to light and, to a lesser extent, under heat and oxidative conditions, it is important that ground turmeric is packed in a UV protective packaging and stored appropriately (Buescher and Yang, 2000). Powdered turmeric is packed in bulk, in a variety of containers, fibreboard drums, multiwalled bags and tin containers. The colour of turmeric will not be affected for up to 6 months in any of the packaging or storage conditions (Balasubramanian *et al.*, 1979). Turmeric powder is a major ingredient in curry powders and pastes. In the food industry, it is used mostly to colour and flavour mustard (ASTA, 2002). It is also used in chicken bouillon and soups, sauces, gravies and dry seasonings, and also as a colourant in cereals (Tainter and Grenis, 2001).

Turmeric oil

Dried rhizomes and leaves are used industrially to extract the volatile oil. Dried rhizomes contain 5–6% and leaves contain about 1.0–1.5% oil. Generally, the oil is extracted by steam distillation. Supercritical extraction using liquid carbon dioxide is a relatively new technique for extracting volatile oil and oleoresin. The peculiar turmeric aroma is imparted by *ar*-turmerone, the major aroma principle in the oil.

Turmeric oleoresin

Turmeric oleoresin is the organic extract of turmeric and is added to food items as a spice and colouring agent. Supercritical CO_2 extraction and molecular distillation to extract and separate turmeric oleoresin is the latest high-technology process followed to obtain quality product. Turmeric oleoresin essentially is used in institutional cooking in meat and fish products and certain products such as mustard, pickles and relish formulae, butter and cheese. It is obtained by the solvent extraction of the ground spice with organic solvents like acetone, ethylene dichloride and ethanol for 4–5 h (Krishnamurthy *et al.*, 1976). It is orange-red in colour. Oleoresin yield ranges from 7.9 to 10.4%. Curcumin, the principal colouring matter, forms one-third of a good-quality oleoresin.

Curcumin

Curcumin, or curcuminoids concentrate, for use as a food colour, is not a regular article of commerce because the cheaper turmeric oleoresin has been found suitable for most current uses. Curcumin is included in the list of colours with a restricted use because it has been allotted a low ADI (acceptable daily intake) of 0–1 mg/kg body weight/day. Curcumin gives a bright yellow colour even at doses of 5–200 ppm. A variety of blends are available to suit the colour of the product (Henry, 1998).

6.3. General Composition

The general composition of turmeric is given in Table 6.2. Turmeric contains protein (6.3%), fat (5.1%), minerals (3.5%), carbohydrates (69.4%) and moisture (13.1%). The rhizomes contain curcuminoids (2.5–6%) and are responsible for the yellow colour. Curcumin (diferuloylmethane) comprises Curcumin I (curcumin), Curcumin II (demethoxycurcumin) and Curcumin III (bisdemethoxycurcumin), which are found to be natural antioxidants (Ruby *et al.*, 1995). A new curcuminoid, cyclocurcumin, has been isolated from the nematocidally active fraction of turmeric. The fresh rhizomes also contain two new natural phenolics, which possess antioxidant and anti-inflammatory activities, and also two new pigments. The essential oil (5.8%) obtained by steam distillation of rhizomes has α-phellandrene (1%), sabinene (0.6%), cineol (1%), borneol (0.5%), zingiberene (25%) and sesquiterpines (53%) (Kapoor, 1990).

6.4. Chemistry

Volatiles

Turmeric, dried and cured, generally yields from 1.5 to 5.0% volatile oil. However, *C. aromatica* Salisb. is generally high in volatile oil (4–8%) and low in curcuminoids

Table 6.2. Chemical composition of turmeric trade samples.

Source	Moisture	Starch	Protein	Fibre	Ash	Fixed oil	Volatile oil	Alcohol extractives
China	9.0	48.7	10.8	4.4	6.7	8.8	2.0	9.2
Pulena	9.1	50.1	6.1	5.8	8.5	7.6	4.4	7.3
Alleppey	8.1	50.4	9.7	5.8	6.0	7.5	3.2	4.4
Indian	13.1	69.4	6.3	2.6	3.5	5.1	5.8	–
Alleppey (Finger)	11.0	30.8	–	4.0	–	–	3.4	24.2
Alleppey (Bulbs)	12.0	26.3	–	4.6	–	–	3.4	16.2
Kadur	19.0	32.1	–	3.7	–	–	4.5	16.3
Duggirala	11.0	32.8	–	1.8	–	–	2.9	13.9

Source: Govindarajan (1980).

(1.5%). Turmeric owes its aromatic taste and smell to the oil present in the rhizome. Analysis of the oil, obtained by steam distillation of the powdered rhizome, followed by fractional distillation and derivatization, shows that the components are a mixture of predominantly sesquiterpene ketones and alcohols (Kelkar and Rao, 1933). The residue on steam distillation yields mainly sesquiterpene alcohols. Besides these major components, they have also identified a mixture of low-boiling terpenes, *d*-sabinene, α-phellandrene, cineole, borneol and the higher-boiling sesquiterpene, zingiberene, in substantial amounts (25%).

Turmerones

Rupe *et al.* (1934) could identify two major ketonic sesquiterpenes – *ar*-turmerone and turmerone ($C_{15}H_{20}O$, $C_{15}H_{22}O$) – responsible for the aroma of turmeric. In addition, *p*-cymene, β-sesquiphellandrene and sesquiterpene alcohols have also been reported (Malingre, 1975).

The ratio of turmerone to *ar*-turmerone as reported by Rupe *et al.* (1934) is 60:40, while analysis of the volatile oils from commercial oleoresin shows a ratio of 80:20 (Salzer, 1977). The physico-chemical properties of turmerone and *ar*-turmerone are given in Table 6.3. The effect of maturity on the major components of rhizome oil from turmeric grown in Sri Lanka also indicated *ar*-turmerone (24.7–48.9%) and turmerone (20–39%) as the major compounds (Cooray *et al.*, 1988; Nigam and Ahmed, 1991). Golding *et al.* (1982) demonstrated the structure of the two turmerones and defined them as 2-methyl-6-(4-methylcyclohexa-2,4-dien-1-yl)hept-2-en-4-one (5, 'α-turmerone') and 2-methyl-6-(4-methylenecyclohex-2-en-1-yl)hept-2-en-4-one (2, 'β-turmerone'). *ar*-Turmerone has been proven as a potential antivenom agent (Ferreira *et al.*, 1992).

GC-MS analysis of rhizome oils from five different *Curcuma* species shows variations in the major components, e.g. *ar*-turmerone (2.6–70.3%), α-turmerone (trace–46.2%) and zingiberene (trace–36.8%). Li *et al.* (1997) analysed a series of oils produced from several Zingiberaceae plants, including the rhizome oil of turmeric, using GC-MS. Thirty-five components were identified and turmerone (49%), *ar*-curcumene (15%) and *ar*-turmerone (6.4%) were the major compounds.

Separation and purification of three turmerones, e.g. *ar*-turmerone, α- and β-turmerone, from turmeric oil extracted by supercritical carbon dioxide gave 71% purity by weight. Subsequently, purification using a normal-phase silica gel 60 column could separate and purify three major turmerones with 86% purity by weight of *ar*-turmerone and 81% purity by weight of α- and β-turmerone. These were identified by liquid–solid chromatography, NMR qualification and HPLC quantification, respectively (Li-Hsun Chang *et al.*, 2006).

Two new sesquiterpene ketoalcohols – turmeronol A and turmeronol B – have also been reported from the dried rhizome (Imai *et al.*, 1990). Five sesquiterpenes, e.g. germacrone-13-al, 4-hydroxybisabola-2, 10-diene-9-one, 4-methoxy-5-hydroxybisabola-2,10-diene-9-one, 2,5-dihydroxybisabola-3,10-diene and procurcumadiol, have been isolated and identified

Table 6.3. Physico-chemical properties of turmerone and *ar*-turmerone.

	$C_{15}H_{22}O$	$C_{15}H_{20}O$
Molecular weight	218	216
Boiling point	125–126°C/10 mm	159–160°C/10 mm
Refractive index	n^{25}_{D}, 1.5960	n^{20}_{D}, 1.5219
Optical rotation	α^{25}_{D}, −69.70	α^{20}_{D}, +84°
Absorption	λ_{max}, 234–235 nm	λ^{EtOH}_{max}, 234–235 nm λε 11,750
Derivatives	Semicarbazide, 104–106°C	2,4-DNP-hydrazone, 134°C
Melting point	Semicarbazone, 110–120°C	Semicarbazone, 109°C

by NMR (^{1}H and ^{13}C) spectroscopy. The study indicates more of bisabolene-type sesquiterpenes in turmeric (Ohshiro *et al.*, 1990).

Liquid chromatography–electrospray ionization tandem mass spectrometry (LC–ESI-MS/MS) coupled to DAD analysis is used as an online tool for identification of diarylheptanoids in fresh turmeric rhizome extracts. Based on their mass spectra, from both negative and positive mode LC–ESI-MS/MS analysis, and supported by their DAD spectra, 19 diarylheptanoids were identified. Among these 19 compounds, curcumin, demethoxycurcumin and bisdemethoxycurcumin were identified by comparing their chromatographic and spectral data with those of authentic standard compounds (Hongliang Jiang *et al.*, 2006).

Variability in volatile oil constituents

GEOGRAPHICAL LOCATION Rhizome oil of fresh turmeric from French Polynesia shows 20 components, with zingiberene (16.7%), *ar*-turmerone (15.5%) and α-phellandrene (10.6%) being the major ones (Vabirua-Lechat *et al.*, 1996). GC and GC-MS of the rhizome oils from turmeric grown in Bhutan shows *ar*-turmerone (16.7–25.7%), α-turmerone (30.1–32.0%) and β-turmerone (14.7–18.4%), as reported by Sharma *et al.* (1997). Leaf oil from turmeric grown on the plains of North India shows 20 compounds by GC-MS, dominated by monoterpenes, sesquiterpene hydrocarbons and oxygenated sesquiterpenes (57.1, 3.3 and 2.1%, respectively). The major compounds identified are: *p*-cymene (25.4%), 1,8-cineole (18%), *cis*-sabinol (7.4%) and β-pinene (6.3%) (Garg *et al.*, 2002). In another study, the rhizome oil of Chinese origin, when analysed through GC-MS, gave 17 chemical constituents, of which turmerone (24%), *ar*-turmerone (18%) and germacrone (11%) were the major compounds (Zhu *et al.*, 1995).

The monoterpenes from green leaves and fresh rhizomes of *C. longa* L. grown in India have been analysed by GC and GC-MS. With the exception of the absence of myrcene, the leaf oil largely resembled the Nigerian counterpart. The rhizome oil did not contain β-pinene, but did contain all the other components of the leaf oil in different proportions. Of interest is the fact that the rhizome volatiles included car-3-ene, α-terpinene, γ-terpinene and terpinolene (McCarron *et al.*, 1995). Variations were also observed in the volatle oil composition of rhizomes collected from the sub-Himalayan region of the Tarai in India (Garg *et al.*, 1999).

EXTRACTION TECHNIQUES Extraction techniques to maximize the yield of essential oil and pigments, especially *ar*-turmerone (α- and β-), turmerone and the curcuminoids, have been reported (Manzan *et al.*, 2003). By varying the distillation time and autoclave pressure, higher yields can be obtained. Thus, with extraction by volatile solvents, the best yield (5.49 wt%) is obtained while using 0.175, 0.124 and 0.088 mm particles at 40°C and 6 h extraction.

Supercritical fluid extraction (SFE), mainly by supercritical carbon dioxide (SC-CO_2), can be used to extract volatile oils from natural products and does not produce substantial thermal degradation or organic solvent contamination. Sanagi *et al.* (1993) have reported a process for the direct analysis of turmeric using online coupling of supercritical fluid extraction (SFE) with supercritical fluid chromatography (SFC). The extraction rate of turmeric oil in SC-CO_2 was measured as a function of pressure, temperature and flow rate at constant extraction time. The total oil yield decreased with temperature at constant flow rate, but increased with flow rate at constant pressure and temperature. The optimum pressure for the extraction yield was found to be 22.5 MPa (Gopalan *et al.*, 2000a).

GC-MS analysis of the hydrodistilled oil indicates β-turmerone (11–36%), α-turmerone (19–24%) and *ar*-turmerone (4–14%) (Kojima *et al.*, 1998).

The composition of essential oil extracted using SC-CO_2 has been compared with that of steam-distilled oil by GC-MS (Gopalan *et al.*, 2000b). Out of the 21 components identified, *ar*-turmerone and turmerone constituted about 60% of the total oil. Analysis of the cyclohexane extract of turmeric by GC-MS coupled with Pseudo Sadtler retention indices reveals a series of saturated and unsaturated fatty acids, along with sesquiterpenes. The fatty acids reported are: tetradecanoic acid, *cis*-9-hexadecenoic acid, hexadecanoic acid, *cis-cis*-9,12-octadecenoic acid, *cis-trans*-9-octadecenoic acid, octadecanoic acid and eicosa decanoic acid.

A precise comparative study on the components of the oil from leaves, flowers, rhizomes and roots of turmeric showed that the oils from rhizomes and roots were more similar, compared with the oil from leaves and flowers, indicating the presence of biogenetically linked characters (Leela *et al.*, 2002). The volatile oils from flowers and leaves were dominated by monoterpenes, while the major part of the oil from roots and rhizomes contained sesquiterpenes (Table 6.4). Another study (Oguntimein *et al.*, 1990) on the leaf essential oil of turmeric also showed α-phellandrene (47.7%) and terpenoline (28.9%) as the major constituents (Table 6.5).

Curcuminoids

Chemistry

Curcumin, $C_{21}H_{20}O_6$, m.p. 184–185°C was isolated as early as 1815 (Vogel and Pelletier,

Table 6.4. Composition of essential oils from various vegetative parts of *Curcuma longa* L.

	Concentration (%)			
Component	Leaf	Flower	Root	Rhizome
α-Pinene	2.1	0.4	0.1	0.1
β-Pinene	2.8	0.1	0.1	t
Myrcene	2.3	0.2	t	0.1
α-Phellandrene	32.6	–	0.1	0.1
δ-3-Carene	1.1	0.6	–	–
α-Terpinene	1.3	0.1	–	–
p-Cymene	5.9	1.1	3.3	3.0
β-Phellandrene	3.2	t	–	t
1,8-Cineole	6.5	4.1	0.7	2.4
z-β-Ocimene				
E-β-Ocimene	0.2	–	–	–
γ-Terpinene	0.4	–	–	–
Terpinolene	1.5	–	–	–
Linalool	26.0	7.4	0.1	0.3
1,3,8-Paramenthatriene	0.7	1.1	0.1	–
p-Methyl acetophenone	0.2	0.3	–	–
p-Cymen-8-ol	0.1	0.3	t	t
α-Terpineol	0.8	26.0	1.5	0.3
Thymol	0.4	1.1	0.1	0.2
Carvacrol	0.3	–	0.1	–
γ-Curcumene	0.1	–	0.3	0.1
ar-Curcumene	0.1	t	0.4	0.1
α-Zingiberene	0.2	1.9	7.0	6.3
β-Bisabolene	0.5	0.8	t	t
β-Sesquiphellandrene	–	0.9	2.3	1.3
E-Nerolidal	0.3	1.1	t	2.6
Dehydrocurcumene	0.1	1.1	–	–
ar-Tumerone	–	–	4.3	2.2
Turmerone	0.1	1.2	46.8	31.1
Curlone	0.9	1.0	–	10.0
Curcuphenol	0.2	0.3	0.6	10.6
6*S*-7*R*-Bisabolene	t	t	0.6	0.5
Others	0.1	0.4	1.2	0.9

t = trace.

Table 6.5. Composition of the essential oil of *Curcuma longa* L. leaves.

Components	Composition (%)	Identification
α-Pinene	2.2	GC-MS, ^{1}H-NMR
β-Pinene + myrcene	6.3	^{1}H-NMR
α-Phellandrene	47.7	GC-MS, ^{1}H-NMR and ^{13}C-NMR
δ-3-Carene	1.2	GC-MS, ^{1}H-NMR
α-Terpinene	1.8	GC-MS, ^{1}H-NMR
p-Cymene	1.2	GC-MS, ^{1}H-NMR
Limonene + 1,8-cineol	6.0	GC-MS, ^{1}H-NMR
γ-Terpinene	2.0	GC-MS, ^{1}H-NMR and ^{13}C-NMR
Terpenoline	28.9	GC-MS, ^{1}H-NMR
α-Terpineol	0.8	GC-MS
4-Terpineol	< 0.2	GC-MS
Sabinol	< 0.2	GC-MS
	< 0.2	GC-MS

Source: Oguntimein *et al.* (1990).

1818). It is insoluble in water but soluble in ethanol and acetone. The structure and synthesis of curcumin as a diferuloylmethane was confirmed by the work of Lampe (1910), and also by Majeed *et al.* (1995).

The main coloured substances in the rhizomes are curcumin (1,7-*bis* (4-hydroxy-3-methoxy prenyl)-1, 6-heptadiene-3, 5-dione) and two related demethoxy compounds, demethoxy curcumin and *bis*-demethoxy curcumin, which belong to the group of diarylheptanoids (see Fig. 6.1). Besides these three forms of curcuminoids, three minor constituents have also been isolated (Srinivasan, 1952) that are supposed to be geometrical isomers of curcumin. One of these is assumed to be a *cis-trans* geometrical isomer of curcumin based on its UV spectrum, lower m.p. and lower stability when compared with curcumin, which has a *trans–trans* configuration. Heller (1914) isolated an isomer of curcumin with a diketone structure.

A new curcuminoid, cyclocurcumin, possessing nematocidal activity, was isolated from the mother liquor by repeated purification as a yellow gum (Kiuchi *et al.*, 1993). It had the same molecular formula as curcumin, but differed in structure by an intramolecular Michael addition of the enol-oxygen to the enone group. The chemical structures of these components are given in Fig. 6.2.

Properties

The physico-chemical properties of the three curcuminoids are given in Table 6.6 (Khalique and Amin, 1967; Roughley and Whiting, 1973). The absorption spectra of these three components vary slightly, with their maxima at: 429 nm, curcumin; 424 nm, demethoxycurcumin; and 419 nm, bisdemethoxycurcumin. The three curcuminoids also exhibit fluorescence under ultraviolet and, after separation on thin-layer plates, can be estimated directly by fluorescence-densitometer when irradiated at 350 nm (Jentzsch *et al.*, 1959). The fluorescence spectra of curcuminoids showed distinct excitation at 435 nm and emission at 520 nm (Maheswari and Singh, 1965). The three components, curcumin, demethoxycurcumin and bisdemethoxycurcumin, have been estimated variously by thin-layer chromatographic separation to be present in the ratio 60:30:10 (Perotti, 1975), 47:24:29 (Jentzsch *et al.*, 1959) and 49:39:22 (Roughley and Whiting, 1973).

Curcumin gives vanillic acid and ferulic acid on boiling with alkali. Oxidation with potassium permanganate yields vanillin, while hydrogenation gives a mixture of tetrahydro- and hexahydro-derivatives.

Curcumin is reported to prevent DNA damage, even in individuals who may be genetically susceptible to toxic effects of

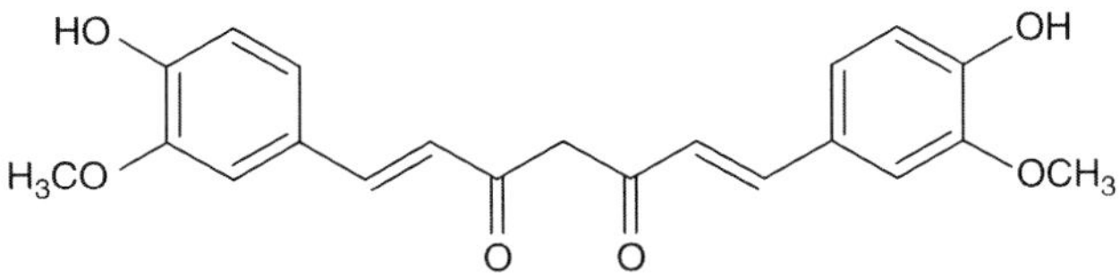

Curcumin

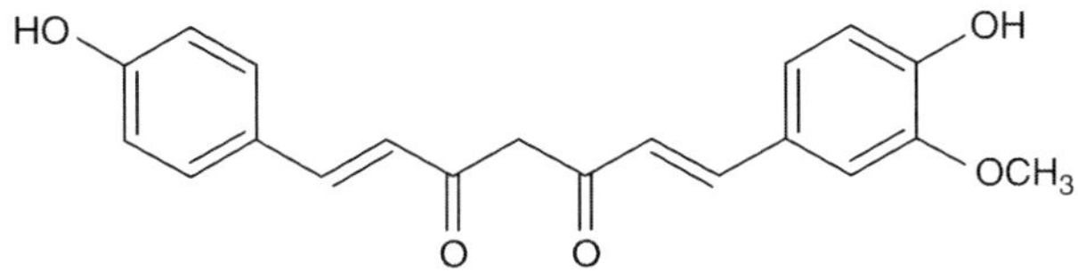

Demethoxycurcumin

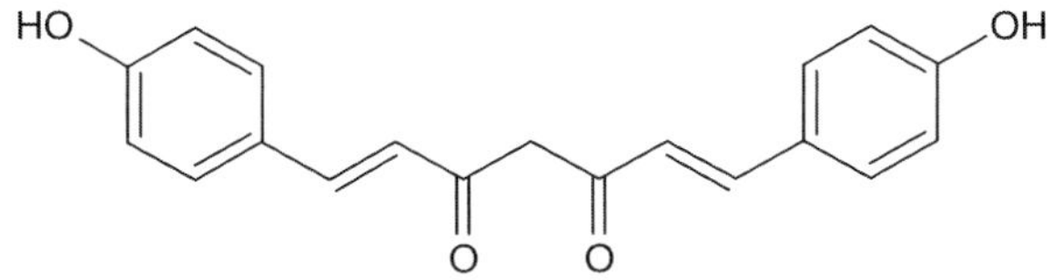

Bisdemethoxycurcumin

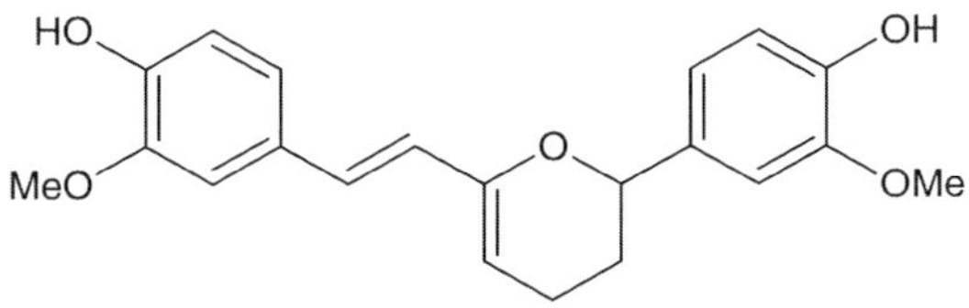

Cyclocurcumin

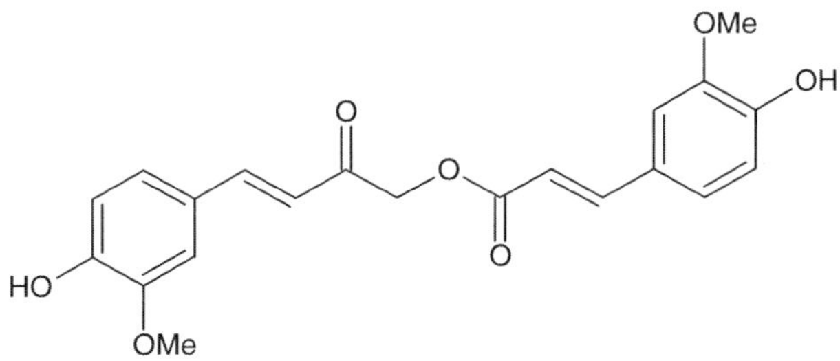

Calebin

Fig. 6.1. Chemical structures of curcuminoids in turmeric.

xenobiotic exposures, and is also able to exert antimutagenic/anticarcinogenic properties at levels as low as 0.1–0.5% in the diet (Polasa *et al.*, 2004).

Isolation

Isolation of curcuminoids from turmeric has been reported by numerous workers. However, extraction of pure curcumin from the plant

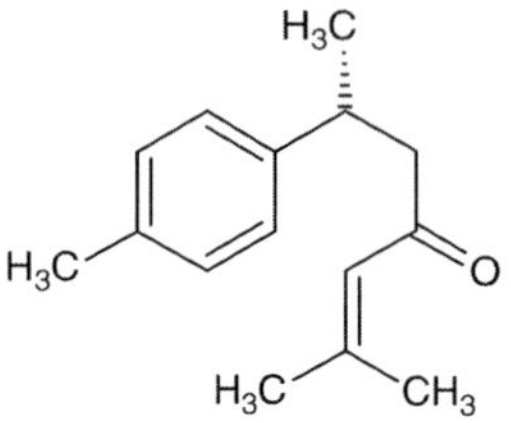

ar-Turmerone

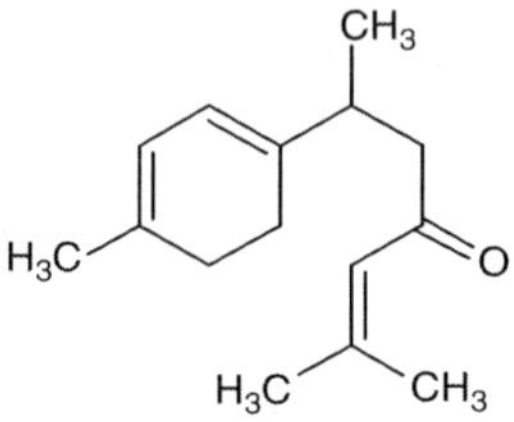

Turmerone

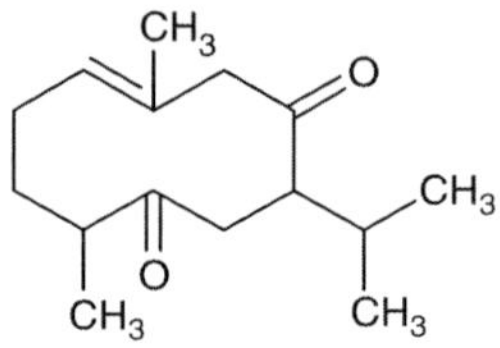

Curdione

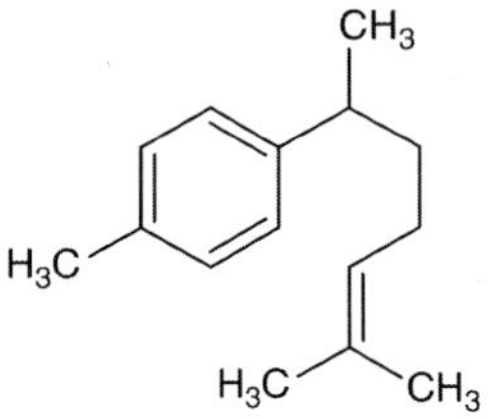

ar-Curcumene

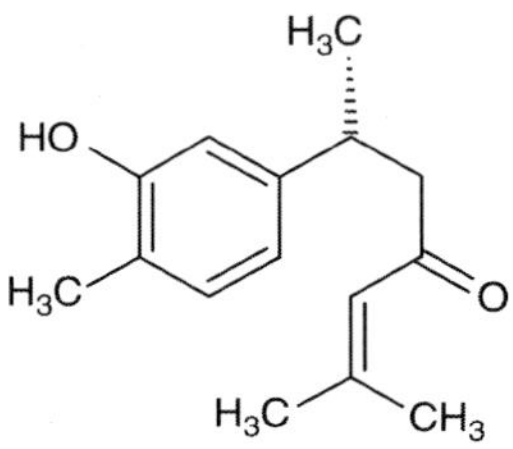

Turmeronol A

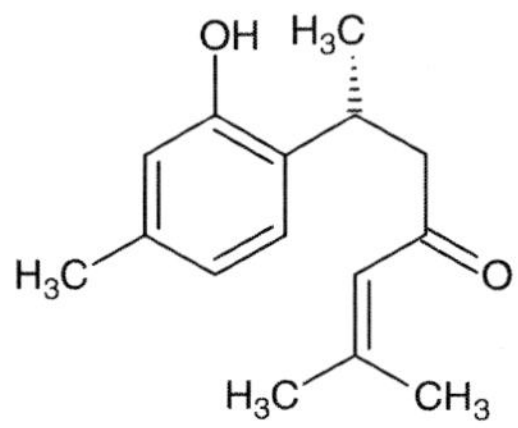

Turmeronol B

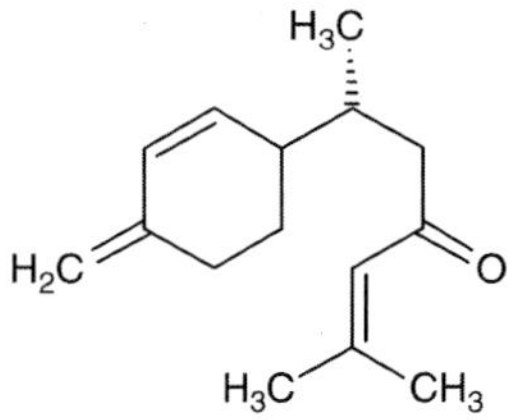

Curlone

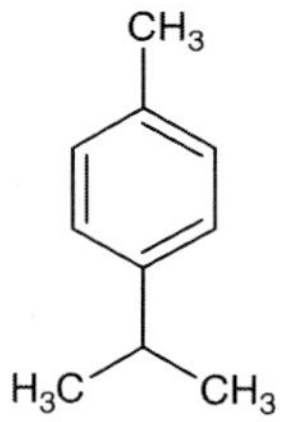

para-Cymene

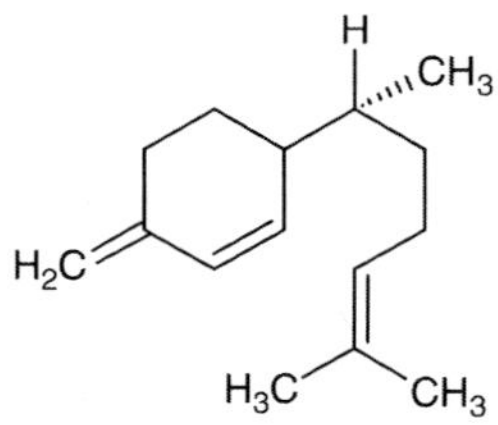

β-Sesquiphellandrene

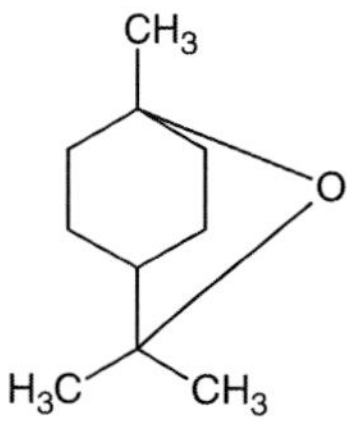

1,8-Cineole

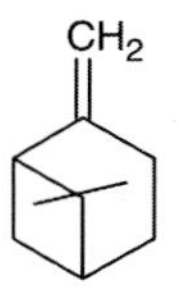

β-Pinene

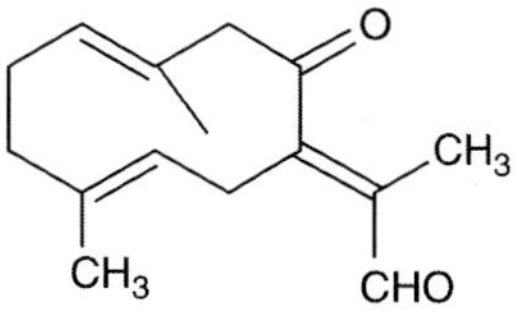

Germacrone-13-al

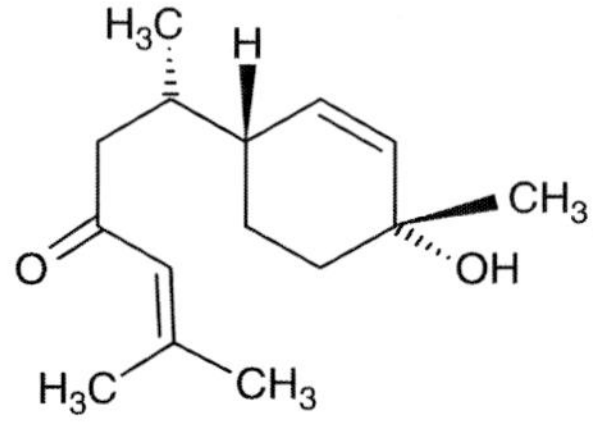

4-Hydroxybisabola-2, 10-diene-9-one

Fig. 6.2. Chemical structures of constituents in turmeric oil.

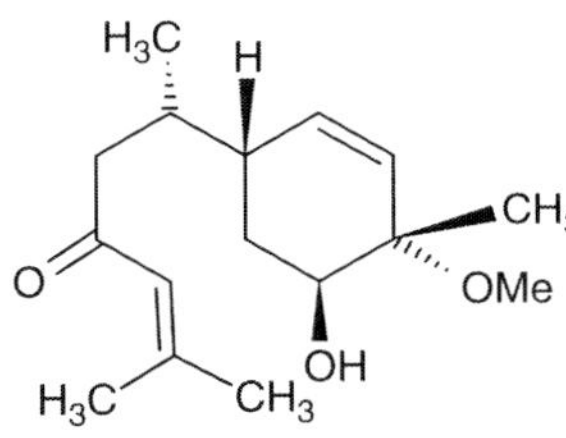

4-Methoxy-5-hydroxybisabola-2, 10-diene-9-one

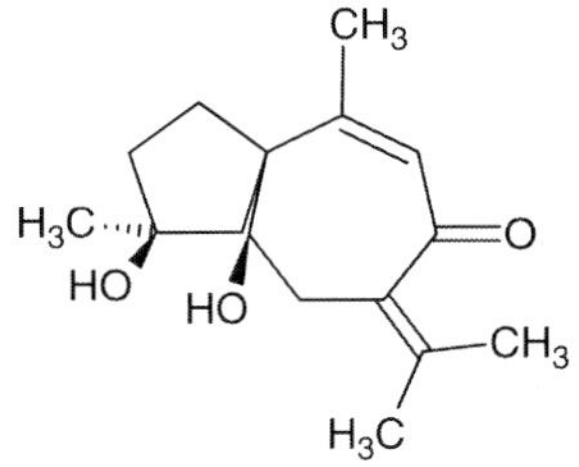

Procurcumadiol

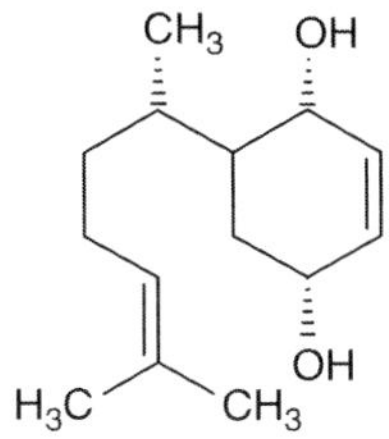

2,5-Dihydroxybisabola-3, 10-diene

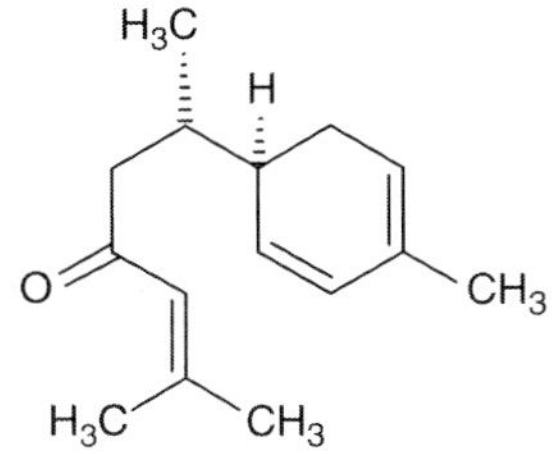

α-Turmerone

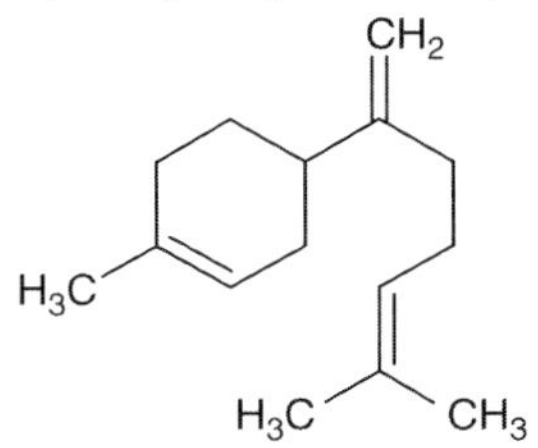

β-Bisabolene

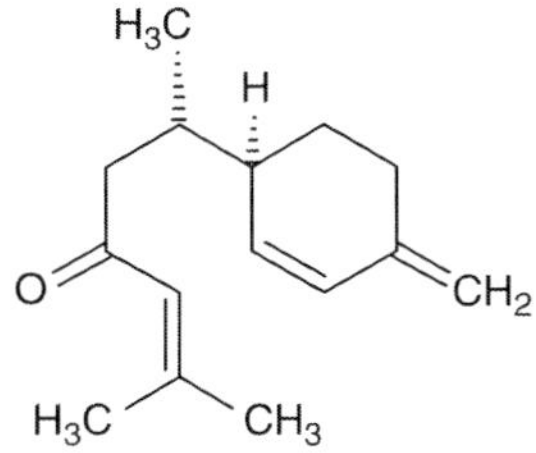

β-Turmerone

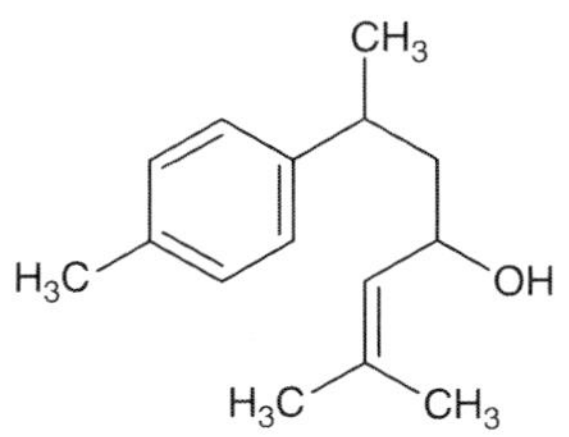

α-Turmerol

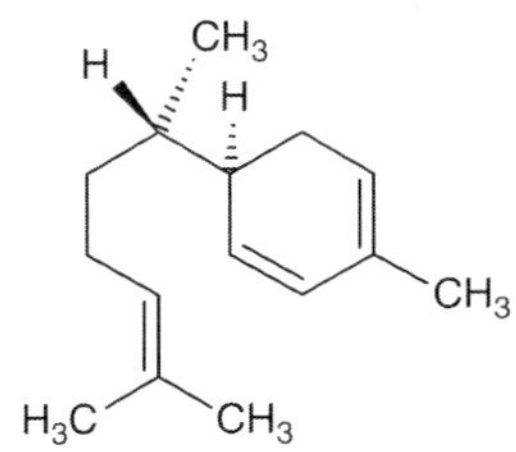

α-Zingiberene

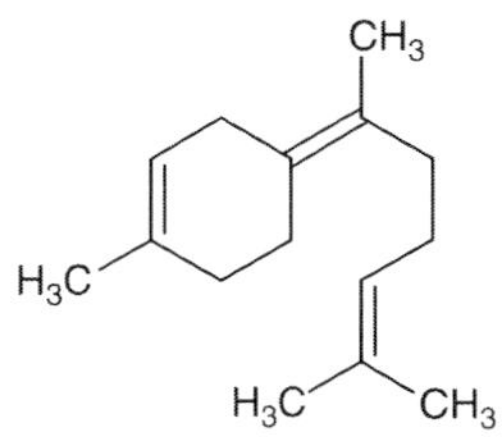

CIS-γ-Bisabolene

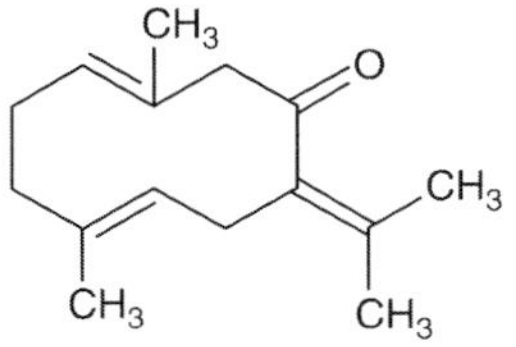

Germacrone

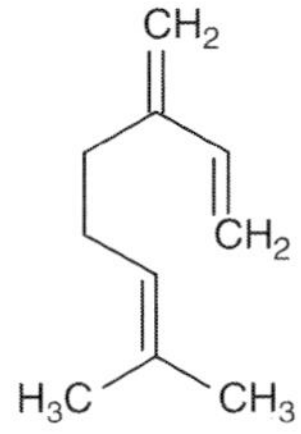

Myrcene

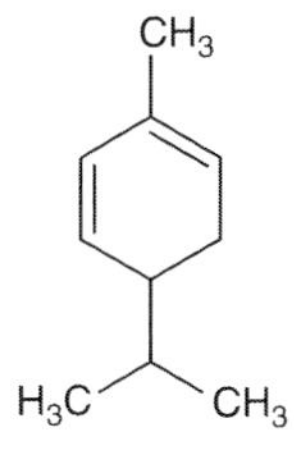

α-Phellandrene

Fig. 6.2. *Continued*

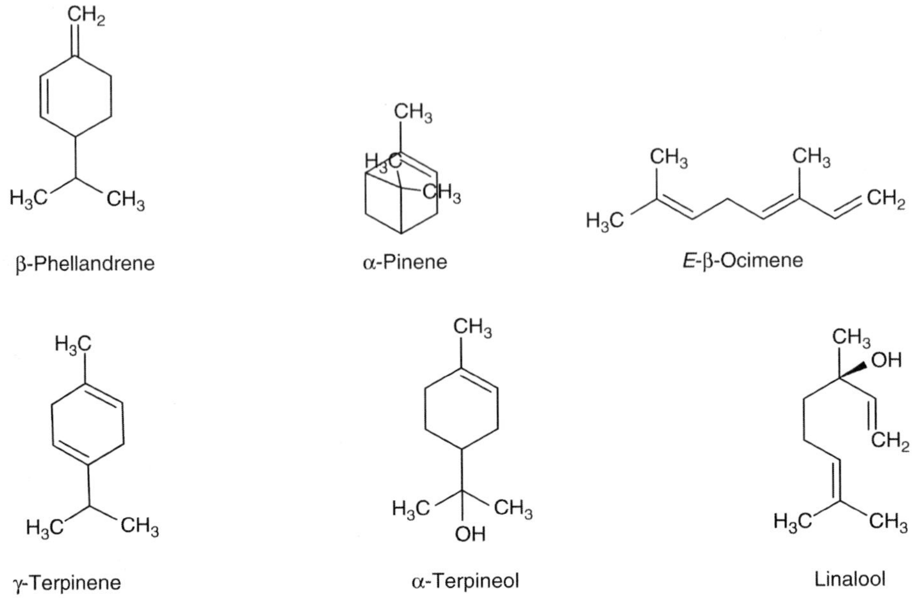

Fig. 6.2. *Continued*

Table 6.6. Physico-chemical properties of curcuminoids.

Chemical name	Di-cinnamoyl methane	4-Hydroxy cinnamoyl (feruloyl) methane	*bis*-4-Hydroxy cinnamoyl methane
Common name	Curcumin	Demethoxycurcumin	Bisdemethoxycurcumin
Molecular weight	368.4	338.0	308.1
Melting point (°C)	184–186	172.5–174.5	224
Solubility:			
Hexane or ether	Insoluble	Insoluble	Insoluble
Alcohol or acetone	Soluble	Soluble	Soluble
Relative R*f*:			
Chloroform: ethanol (25:1)	1.00	0.66	0.41

is tedious and a purified extract of the three forms is available on the market with curcumin (75–81%), demethoxycurcumin (15–19%) and bisdemethoxycurcumin (2.2–6.6%).

Various isolation methods are summarized below:

1. Rhizomes are extracted with hexane to remove the fatty components and then extracted with benzene, which gives a yield of 1.1%. This is further purified by crystallization from ethanol to yield orange-yellow needles (Janaki and Bose, 1967).
2. Sastry (1970) also reports isolation of curcumin and related demethoxy compounds by extraction with organic solvents, but the yield is very poor (1.5–2.0%), while hot and cold percolation methods give higher yield (Krishnamurthy *et al.*, 1976).
3. Numerous similar methods are available for isolation by solvent extraction using organic solvents (Zhang and Yang, 1988; Verghese and Joy, 1989; Verghese, 1993; Xianchun *et al.*, 1993).
4. SC-CO_2 extraction modified by 10% ethanol is more efficient (Baumann *et al.*, 2000), but the high operating pressure creates problems. Microwave-assisted extraction (MAE) technique for selective and rapid extraction of curcuminoids is also possible.

Turmeric powder irradiated for 2 and 4 min with microwave showed marginally higher extraction of curcuminoids in 60 min by acetone (Dandekar and Gaikar, 2002).

The composition and yield of curcuminoids extracted using the existing techniques, e.g. hydrodistillation, low-pressure solvent extraction, soxhlet and SC-CO_2 extraction, were compared (Braga *et al.*, 2003). The highest yield was obtained using soxhlet extraction with ethanol and the lowest (2.1%) with hydrodistillation. The best solvent for SC-CO_2 extraction is a mixture of ethanol and isopropanol. *ar*-Curcumene, *z*-(γ)-atlantone and (*E*)-(γ)-atlantone constituted about 60% of the light fraction.

Analysis

Direct spectrophotometric absorption between 420 and 430 nm is the most common method used (BSI, 1983), even though other compounds absorbing in this region will influence the values. Hence, only pure standards are to be used (Karasz *et al.*, 1973).

Direct flourimetric assay for curcumin also poses problems due to the relative flourescence intensities of the three curcuminoids, which are in the ratio of 1:2.2:10.4 at equimolar concentrations (Tonnesen and Karlsen, 1983). Other methods, based on the molar absorptivity by developing coloured complexes with alkalis, strong acids and boric acid, are not consistent (Karasz *et al.*, 1973; Krishnamurthy *et al.*, 1976; Janssen and Gole, 1984) due to the unstable nature of the complex, except with boric acid (Dyrssen *et al.*, 1972; Janssen and Gole, 1984).

Other spectroscopic methods, e.g. IR, NMR and MS, can be used for the identification and characterization of curcuminoids (Roughley and Whiting, 1973; Govindarajan, 1980; Unterhalt, 1980; Tonnesen and Karlsen, 1986).

Commercial produce of curcumin contains the three curcuminoids that can be estimated using HPLC (Tonnesen and Karlsen, 1983), but separation is strongly dependent on the chromatographic conditions. Due to the labile nature of the curcuminoids, C_{18} columns are preferred for HPLC analysis (Khurana and Ho, 1988). Using a ternary mobile phase, e.g. methanol:2% acetic acid:acetonitrile, the three curcuminoids can be separated and quantified. The advantage of this method lies in the estimation of individual curcuminoids from varieties of turmeric rhizomes (Jayaprakasha *et al.*, 2002).

Structure–activity relationships

The anti-inflammatory activity of curcumin and its derivatives is associated with the hydroxyl and phenol groups in the molecule, which are also essential for the inhibition of prostaglandins, PG synthetase and leucotriene synthesis (LT) (Kiuchi *et al.*, 1982, 1992; Iwakami, *et al.*, 1986). Claeson *et al.* (1993, 1996) suggested that the anti-inflammatory action and the antiparasitic activity were associated with the β-dicarbonylic system with conjugated double bonds (dienes) (Araujo *et al.*, 1998, 1999). The better skin penetration and lipophylicity is attributed to the presence of a diene ketone system. Calebin-A, a novel curcuminoid isolated from turmeric, protects neuronal cells from β-amyloid insult. The hydroxy group at para-position of this compound is most critical for the expression of biological activity (Kim *et al.*, 2001).

Chemical synthesis

Curcumin was synthesized starting from carbomethoxy feruloyl chloride which, on condensation with ethyl acetoacetate, gave the ester (Lampe, 1910). The ester on hydrolysis and loss of carbomethoxy feruloyl chloride gave the diferuloyl compound. The hydrolysis of this compound released the carboxymethyl and acetyl groups and gave curcumin identical to the natural curcumin (Mayer, 1943).

Another synthesis of curcumin from acetyl acetone and vanillin was reported earlier by Pabon (1964), which gave an 80% yield of curcumin. Starting with other related aldehydes, curcumin derivatives and related compounds are also obtained. The synthesis by this method has been used recently in biosynthetic studies (Roughley and Whiting, 1973).

Biosynthesis

Initial investigations into the biosynthesis of the curcuminoids and gingerols were performed many years ago (Denniff and Whiting, 1976; Macleod and Whiting, 1979; Denniff *et al.*, 1980). These initial radiotracer feeding studies suggested that these compounds were derived from intermediates in the phenylpropanoid pathways that were condensed with other molecules, derived in turn from the acetate and short- and medium-chain fatty acid pathways.

The structure established for curcumin (Pabon, 1964) suggested the reasonable biosynthetic mechanism involving two cinnamyl units combining with the central carbon atom of malonate (Geissman and Crout, 1969). The proposed scheme was tested by Roughley and Whiting in 1973, by incorporation of labelled precursors. They concluded that an alternate scheme might exist for curcumin biosynthesis, which involved a cinnamate starter extending by five acetate or malonate units; cyclization of the chain would give the second aromatic ring. Biosynthesis would be completed by hydroxylation and methylation. They also found that none of the cinnamic acids was quite as well incorporated into curcumin as phenyl alanine, and caffeic acid was relatively unacceptable as a precursor.

In the turmeric plant, synthesis of curcumin starts 120 days after planting and reaches optimum at 180–190 days after flowering. Curcumin synthesis is highly dependent on location, agroclimatic conditions, genotypes and cultural practices. Thus, multilocational trials carried out at different districts in India (Kerala, Tamil Nadu, Andhra Pradesh and Himachal Pradesh) indicated significant variations in curcumin content (Zachariah *et al.*, 1998).

Phenylalanine ammonia lyase (PAL), the first enzyme of the biosynthetic sequence, and a flavonoid glucosyltransferase, the last enzyme, appear to be located in the lumen of the membranes. Cinnamate 4-hydroxylase is membrane embedded, while other enzyme activities appear to be weakly associated with the cytoplasmic phase of endoplasmic reticulum membranes (Hrazdina and Wagner, 1985). PAL, which initiates the series of reactions leading to curcumin synthesis, was studied during the early germination phase. The activity was at a maximum in the leaves as compared with the roots, rhizomes and pseudostem, indicating that the conversion of phenylalanine to cinnamic acid takes place mostly in the leaves. Studies on the localization of PAL activity in various cell fractions showed maximum activity in the microsomal fraction. Using tracer studies with labelled phenyl alanine and malonate, it can be seen that the cinnamate pathway might lead to curcumin synthesis, ruling out the alternate pathway, involving acetate (Neema, 2005).

Recently, curcuminoid synthase has been identified as being capable of forming the curcuminoids in turmeric (Maria *et al.*, 2006). This activity required malonyl-CoA and phenylpropanoid pathway-derived hydroxycinnamoyl-CoA esters as substrates, suggesting that the corresponding protein was a polyketide synthase or an enzyme that was closely related. It is postulated that this activity could be the result of a single enzyme, or of multiple enzymes in sequence.

Polysaccharides

Besides curcuminoids and oils, *C. longa* also contains some polysaccharides. Three acidic polysaccharides were isolated from turmeric ('Ukon') by hot water extraction, followed by precipitation with ethanol, with remarkable activity on the reticuloendothelial system (RES) (Gonda *et al.*, 1990). The components were purified on a column of DEAE Sephadex A-25 and named as Ukon A, Ukon B and Ukon C.

Another neutral polysaccharide, Ukon D, was isolated by the same group as having activity on the RES, being composed of L-arabinose, D-galactose, D-glucose and D-mannose in the molar ratio 1:1:12:2 (Gonda *et al.*, 1992).

A novel water-soluble peptide, 'turmerin', was isolated from *C. longa* with antioxidant activity (Srinivas *et al.*, 1992). It has been found to be an efficient antioxidant/DNA-protectant/antimutagen. Turmerin forms 0.1% of the dry weight of turmeric and is obtained in a crystalline form. It is a heat-stable, non-cyclic peptide containing 40 amino acid residues, with a blocked N-terminal and leucine at the C-terminal. The peptide is insensitive to trypsin and pepsin, heat and UV radiation. Turmerin contains three residues of methionine that are partly responsible for the antioxidant activity.

Oleoresin

Turmeric oleoresin functions chiefly as a food colour, and secondarily, in some of the products, to impart a characteristic mild spicy aroma compatible with mustard, pickles, relish formulae, etc. The product from industrial practice using good, clean turmeric, with a curcuminoid content of 4.5–5.0%, is a highly viscous, deep brownish-orange product, obtained in a yield of about 12%. This analyses 30–40% as curcumin, 15–20% volatile oil and has the characteristic fresh, clean, mildly pungent, woody-pungent, woody-spicy aroma of turmeric. Alcohol and acetone are good extractants and (as with ginger) the yields can also be expected to be high because of extraction of non-flavour components. The analytical characteristics of turmeric oleoresin are given in Table 6.7.

The use of ethylene dichloride has the advantage of having relatively selective extraction of the flavour constituents, water immiscible, non-flammable and of sufficiently low boiling point, but requiring no refrigeration. However, impurities in the solvent, such as traces of high-boiling fractions accumulating in the product and creating off-flavour problems, and possible corrosivity to the equipment in the presence of water, are serious problems to contend with. In the view of many manufacturers, acetone appears to be the choice solvent for extraction of good-quality turmeric oleoresin (Govindarajan, 1980).

Soxhlet extraction of turmeric powder with acetone gives in 4–5 h a yield of about 5.0%, containing 42% curcuminoids; prolonged extraction up to 24 h gave only a fractionally higher yield of curcuminoids. Extraction by batch and counter-current cold percolation in columns gave a slightly higher yield of 5.7–5.9% with 42.2–46.0% curcuminoids and extraction efficiencies varying from 72.9 to 81.7%. Acetone as solvent was slightly superior to alcohol and ethylene dichloride (Krishnamurthy *et al.*, 1976).

Upon appropriate dilution with a vegetable oil, propylene glycol or polysorbates,

Table 6.7. Analytical characteristics of turmeric oleoresin.

	Oleoresin analysis (%)											
	Soxhlet extraction						Cold percolation extraction					
Raw material	Yield		Curcuminoids in extractives		Recovery of curcuminoids		Yield		Curcuminoids in extractives		Recovery of curcuminoids	
Cultivar	F	C	F	C	F	C	F	C	F	C	F	C
Tekurpeta	5.5	3.2	38.7	49.1	98.4	86.8	5.1	3.6	35.0	36.8	98.1	73.6
Miraj-26	8.6	4.7	29.0	50.8	86.8	82.5	8.9	7.9	32.4	32.7	82.9	78.7
Erode	7.9	5.8	38.6	56.9	92.7	100.0	7.9	5.8	40.0	49.5	95.5	87.3
Rajpuri	8.5	6.0	37.1	57.5	91.6	100.0	8.2	6.7	41.5	43.9	98.3	85.2
Gadhvi	9.9	5.9	29.8	50.9	85.1	86.0	10.0	6.8	31.4	34.8	90.4	68.2
Waigon	9.9	5.8	34.2	56.3	96.9	92.9	9.0	9.5	38.7	33.2	100.0	90.0
Sugandham	9.7	6.1	32.4	53.1	87.0	90.0	9.9	6.9	36.0	38.9	98.6	74.0
Kuchupudi	9.5	6.7	40.1	53.1	94.8	88.3	10.3	8.8	41.8	38.8	100.0	84.4
Alleppey	8.1	8.2	57.7	57.6	86.0	86.6	9.6	8.5	49.4	49.7	90.1	80.0

Note: F = fine powder, 60 mesh; C = coarse powder, 30 mesh; extracting solvent: acetone (Krishnamurthy *et al.*, 1976).

the oleoresin gives a bright yellow liquid with the characteristic turmeric aroma and a slightly bitter and pungent taste (Govindarajan, 1980; Weiss, 2002). The oleoresin may also be spray-dried to a powder on a sugar matrix, such as maltodextrin, and can be used as a colourant in dry cereals or beverages. The advantage of spray-dried turmeric oleoresin over ground turmeric powder is that it is devoid of starch, the predominant component in dried rhizome, proteins and other fibres (Buescher and Yang, 2000).

Turmeric oleoresin is used essentially in institutional cooking in meat and certain processed products, such as prepared mustard, pickles and relish formulae, for frozen fish fillets, frozen potato croquettes, butter and cheese. The aroma of turmeric is, however, due to high-boiling components and is rather difficult to remove unless curcuminoids are isolated by crystallization (Perotti, 1975).

6.5. Medicinal and Pharmacological Uses

Several pharmacological and medicinal properties of turmeric are widely known (Ammon and Wahl, 1991; Eigner and Scholz, 1999; Araujo and Leon, 2001). The rhizome extracts of turmeric, apart from the ethanolic and methanolic ones, have been examined for their biological activities and have been in use for centuries.

Anti-inflammatory activity

Curcuminoids and other constituents of turmeric are well known for their anti-inflammatory activity. Turmeric extract, volatile oils from turmeric and curcuminoids were reported to possess this property in different experimental models of inflammation in mice, rats, rabbits and pigeons (Arora *et al.*, 1971; Ghatak and Basu, 1972; Chandra and Gupta, 1972). Thus, curcuminoids are effective against carrageenan-induced oedema in rats (Srivastava *et al.*, 1985) and mice (Srimal and Dhawan, 1985). Administration of curcuminoids to patients who have undergone surgery or suffered from trauma reduced inflammation to a comparable level with phenylbutazone (Satoskar *et al.*, 1986). Oral administration of curcumin at a dose of 3 mg/kg was also found to be effective in reducing inflammation associated with various forms of arthritis (Srimal and Dhawan, 1973; Chandra and Gupta, 1972). The antirheumatic properties of curcuminoids were also tested successfully in patients with diagnosed rheumatoid arthritis (Deodhar *et al.*, 1980).

Curcumin also enhances wound healing in diabetic rats and mice (Sidhu *et al.*, 1999) and in H_2O_2-induced damage in human keratinocytes and fibroplasts (Phan *et al.*, 2001). The wound healing effect is expressed by inhibiting the activation of NF-kappaB transcription factor through the prevention of the p65 unit of the factor.

Antioxidant effect

Curcuminoids are natural phenolic compounds with potent antioxidant properties, which were reported as early as 1975 (Sharma, 1976). Both turmeric and curcumin inhibit generation of superoxide and hydroxyl free radicals (Reddy and Lokesh, 1992; Subramonian *et al.*, 1994; Ruby *et al.*, 1995). The antioxidant properties of curcumin in the prevention of lipid peroxidation are also well recognized (Sreejayan and Rao, 1994). The three forms of the pigment have dual prolonged antioxidant activity, e.g. preventing the formation of free radicals, as well as intervening in their propagation. In fact, the antioxidant activity has been attributed to its unique conjugated structure, which includes two methoxy phenols and an enol form of β-diketone, with the typical radical trapping ability as a chain-breaking antioxidant (Masuda *et al.*, 2002).

Curcumin also has the potential to prevent oxidative damage to the arterial wall. Thus, administration of 500 mg of curcuminoids daily to healthy humans for 7 days reduced lipid peroxides by 33% and blood cholesterol by 29%, indicating a possible role of curcumin in reducing cardiovascular diseases (Soni and Kuttan, 1992). An *in vitro* study on the comparison of the antioxidant activity of curcuminoids and tetrahydrocurcumin using rabbit erythrocyte membrane showed highest activity with tetrahydrocurcumin (Osawa, 1995).

Antimutagenic and anticancerous property

The anticancer action of curcumin has been studied in a standard model of radiation-induced tumour in rat mammary gland (Inano *et al.*, 2000). In animal studies, curcuminoids inhibited capsaicin-induced mutagenic changes in mouse bone marrow. Additionally, mice maintained on turmeric or curcuminoid-enriched diets, when challenged with carcinogens, excreted low levels of mutagenic metabolites, as well as carcinogens (Polasa *et al.*, 1992; Usha *et al.*, 1994).

Curcumin is reported to prevent DNA damage, even in individuals who may be genetically susceptible to the toxic effects of xenobiotic exposures, and is also able to exert antimutagenic/anticarcinogenic properties at levels as low as 0.1–0.5% in the diet (Polasa *et al.*, 2004).

Chemopreventive and bioprotectant property

Numerous studies have been published on the positive effects of turmeric, both in the prevention of cancer and in the recovery from chemotherapy and radiation treatment (Stoner and Mukhtar, 1995; Khafif, *et al.*, 1998; Kawamori *et al.*, 1999; Bush *et al.*, 2001; Jung *et al.*, 2005). In addition to its capacity to intervene in the initiation and growth of cancer cells and tumours, and to prevent their subsequent spread throughout the body by metastasis, curcumin increases cancer cells' sensitivity to certain drugs commonly used to combat cancer, rendering chemotherapy more effective.

Curcuminoids can also act as photochemoprotective agents that provide protection against UVB radiation-induced oxidative stress. This inhibition of UVB radiation-induced damage can reduce the incidence of skin cancer (Afaq *et al.*, 2002).

Antiviral, antimicrobial and antiparasitic activity

Antiviral

An interesting property of curcuminoids is their anti-HIV effect, which has been demonstrated during *in vitro* and *in vivo* experiments, including a limited number of human studies (Lin *et al.*, 1994). HIV infection is characterized by a complex command system, the structural part of which is called 'long terminal repeat' (LTR), which results in virus activation or inactivation. Drugs that interfere with LTR may be of potential therapeutic value in delaying active HIV infection and the progression of AIDS. Curcumin has been found to inhibit activation of the LTR and to decrease HIV replication effectively (Li *et al.*, 1993).

Antimicrobial

Curcuminoids have also been shown to exhibit antimicrobial properties. The antibacterial effects of alcoholic extract of turmeric, curcumin and oil from turmeric have been studied by Banerjee and Nigam (1978) and Bhavanishankar and Srinivasamurthy (1979). Extracts from turmeric, as well as the active principles, the curcuminoids, were found to inhibit the growth of numerous Gram-positive and Gram-negative bacteria, fungi and the intestinal parasite, *Entamoeba histolytica*. The ethanol extract of turmeric has been reported to have anti-amoebic activity against *E. histolytica in vitro* (Dhar *et al.*, 1968). Curcumin at concentrations of 2.5–50.0 mg/100 ml inhibited *in vitro* growth of *Staphylococcus aureus* (Bhavani Shankar and Srinivasamurthy, 1979). Interestingly, the antibacterial and antiviral activities of curcumin were enhanced significantly by illumination with visible light (Tonnesen *et al.*, 1987; Dahl *et al.*, 1989; Pervaiz, 1990).

Curcumin also inhibits *in vitro* production of aflatoxins – toxins produced by the mould *Aspergillus parasiticus*, which may grow and contaminate poorly preserved foods and is a potent biological agent causing injury to the liver, often resulting in liver cancer (Madhyastha and Bhat, 1985; Polasa *et al.*, 1992; Jayaprakasha *et al.*, 2001).

Antiparasitic

Curcumin possesses leishmanicidal effects *in vitro* and is more potent than the standard leishmaniasis drug, pentamidine. LD50 for leishmanicidal activity *in vitro* is found to be 37.6 ± 3.5 μM (Koide *et al.*, 2002).

Antidiabetic property

The efficacy of turmeric and curcumin on the blood sugar and polyol pathway in diabetic albino rats showed significant reduction in blood sugar and glycosylated haemoglobin levels. This could be due to a decreased influx of glucose into the polyol pathway, leading to an increased NADPH/NADP ratio and elevated activity of the antioxidant enzyme, glutathione peroxidase. The activity of sorbitol dehydrogenase, an enzyme that catalyses the conversion of sorbitol to fructose, is also lowered significantly on treatment with turmeric or curcumin (Arun and Nalini, 2002).

Dietary curcumin can alleviate dangerous secondary complications induced by diabetes. The beneficial effects of dietary curcumin on diabetic nephropathy are probably mediated through the hypolipidaemic effects of curcumin (Babu and Srinivasan, 1997).

Curcumin has been proved to be an effective hypolipidaemic agent (Babu and Srinivasan, 1997). One study validated the role of dietary curcumin in maintaining healthy serum cholesterol levels in diabetic rats. Employing a high-cholesterol diet for the diabetic rats, curcumin exhibited a lowering of cholesterol and phospholipid in treated animals as compared with curcumin-free controls. Liver cholesterol, triglycerides and phospholipid elevated under diabetic conditions were lowered by dietary curcumin. Curcumin induces a higher rate of cholesterol catabolism, which is evidenced by the higher activity of liver cholesterol-7α-hydroxylase.

Other pharmacological properties

Antiangiogenic effect

Studies on the effect of curcumin on the growth of Ehrlich ascites tumour cells and endothelial cells *in vitro* prove curcumin to be a potent angio-inhibitory compound, as demonstrated by inhibition of angiogenesis in two angiogenesis assay systems *in vivo*, e.g. peritoneal angiogenesis and chorioallantoic membrane assay. The angio-inhibitory effect of curcumin *in vivo* is corroborated by the results on down-regulation of the expression of proangiogenic genes by curcumin (Gururaj *et al.*, 2002).

Antithrombotic effect

In a comparative study on the protective effects of curcumin and aspirin in mice subjected to induced thrombotic challenge, curcumin exhibited a dose-related antithrombotic effect (Srivastava *et al.*, 1985).

Hepatoprotective effect

Curcumin and turmeric protect the liver against several toxicants both *in vitro* and *in vivo*. Reddy and Lokesh (1992) found that oral administration of curcumin (30 mg/kg body weight) for 10 days lowered the liver and serum lipid peroxide levels, serum alanine aminotransferase (ALAT), aspartate aminotransferase (ASAT) and lactate dehydrogenase (LDH), enhanced by i.p. injection of iron in rats.

Curcumin and Alzheimer's disease

Accumulation of β-amyloid proteins in the brain is one of the hallmarks of Alzheimer's disease. Dietary curcumin at low dose (160 ppm) and high dose (5000 ppm) significantly lowered oxidized proteins and interleukin 1-β, a pro-inflammatory cytokine elevated in the brains of mice. With low dose, the astrocytic marker, GFAP, was reduced and insoluble β-amyloid (Abeta), soluble Abeta and plaque burden were reduced significantly by 43–50% (Lim *et al.*, 2001).

6.6. Standards and Grade Specifications

Standards generally emphasize the attributes that guarantee genuineness and purity, while grade specifications additionally emphasize quality. The Indian standard specification for turmeric, whole and powder, incorporating the country's Agmark grade specifications, is summarized in Table 6.8. Grading is undertaken at the major trade centres in each state and is based on recognized commercial-type fingers/rounds, external appearance, maturity and hardness, colour of core, pungency and aroma, moisture and freedom from extraneous matter. The range of values generally accepted in the UK and the USA is given in Table 6.9. The WHO/FAO specifications (FAO, 1976; WHO, 1976) on turmeric and curcumin as colour additives to food, besides specifying limits for chemical components and heavy metals, also define methods for testing the absence of artificial colouring matter and a method of assay of curcuminoids, similar to the method of the American Spice Trade Association (Englewood and Cliffs, 1958).

ASTA sampling guidelines are as follows: precisely weighed samples are passed through a sieve (US Standard No. 8 or standard pepper sieve No. 9) with a white paper underneath to observe foreign matter, insects and mammalian excreta (Table 6.10). Rhizomes are examined for mould and defiling insects. Foreign matter is reported by count (insects) or by weight (ASTA, 2001).

EU member countries, such as the UK, Germany and the Netherlands, have their own specifications. The European Spice Association (ESA) has a set of 'quality minima for herbs and spices', but has yet to finalize the cleanliness specification standards for spices and spice products. Extraneous matter and foreign matter should not exceed 1 and 2%, respectively, and should be free from live and/or dead insects, insect fragments and rodent contamination visible to the naked eye (corrected if necessary for abnormal vision). Salmonella must be absent in (at least) 25 g of material; yeast and mould, 10^5/g (target), absolute maximum 10^6/g; *E. coli*, 10^2/g (target), absolute maximum 10^3/g (Table 6.11).

Specifications for turmeric oleoresin

The US Code of Federal Regulations 21 CFR 73.615 defines turmeric oleoresin as the 'combination of flavour and color principles obtained from turmeric (*Curcuma longa* L.)

Table 6.8. Analytical specifications for turmeric (whole and powder).

			Ash							
	Sample	Moisture (max.)	Total (max.)	Acid-insol. (max.)	Starch (max.)	Crude fibre (max.)	Volatile oil (max.)	Colour as curcumin	Lead (max. ppm)	Chromate test
Whole	BP	8–10	6–9	–	–	4–6	2–5	–	–	–
	US	9	7	0.5	–	6	4	5	–	–
	DDR	–	7	–	–	–	2.6	3–4	–	–
Powder	Indian	10	7	1.5	60.0	–	–	–	1.5	Negative
	WHO	10	7	1.5	–	–	–	–	3	Negative

All values except lead and chromate as percentage of weight.
BP: British Pharmacopoiea; US: United States; DDR: Deutsche Demokratische Repulik; WHO: World Health Organisation.
Source: Specification for turmeric powder, IS.2446, Indian Standards Institution, Delhi, India (1961).

Table 6.9. Indian specification for turmeric grades.

Grade designation	Pieces max.: wt%	Foreign matter max.: wt%	Defectives max.: wt%	Bulbs max.: wt%	Characteristics
Fingers (general)					Finger-like shape; breaks with a metallic twang; free from damage from weevils, over-boiling, etc.
Special	2.0	1.0	0.5	2.0	
Good	3.0	1.5	1.0	3.0	
Fair	5.0	2.0	1.5	5.0	
Fingers, Alleppey					As above
Good	5.0	1.0	3.0	4.0	As above, admixture of other curcuma varieties allowed at a maximum 2, 5 and 10% in the three grades, respectively.
Fair	7.0	1.5	5.0	5.0	
Fingers, Rajapore	7.0	1.0	3.0	2.0	
Special		1.5	5.0	3.0	
Good		2.0	7.0	5.0	
Fair					
Bulbs (round)					
Special	–	1.0	1.0	–	Be well-developed, smooth, sound, free from rootlets.
Good	–	1.5	3.0	–	
Fair	–	2.0	5.0	–	The Rajapore variety has a higher allowance of 3, 5 and 7% defectives in the three grades, respectively.

Source: *Marketing of Turmeric in India*, Agric. Market, Ser. No. 148, Directorate of Marketing and Inspection, Govt. of India, Nagpur (1965).

Table 6.10. ASTA cleanliness specifications for turmeric.

Whole insects, dead (number)	Excreta, mammalian (mg/kg)	Excreta, other (mg/kg)	Mould (% by weight)	Insect-defiled/ infested (% by weight)	Extraneous foreign matter[1] (% by weight)
3	11.1	11.1	3	2.5	0.5

Note: [1]Extraneous matter includes, but is not restricted to: stones, dirt, wire, string, stems, sticks, non-toxic foreign seeds, excreta, manure and animal contamination.

by extraction using any one or the combination of the following solvents: acetone, ethyl alcohol, ethylene dichloride, hexane, isopropyl alcohol, methyl alcohol, methylene chloride, or trichloroethylene'. The residue allowed in the product should not exceed the residue for that specific solvent: 30 ppm for acetone and chlorinated solvents; 50 ppm for methanol, ethanol and isopropanol. Turmeric oleoresin is described as

Table 6.11. ESA quality minima for turmeric.

Turmeric product	Total ash (% w/w) max. (ISO 928)	Acid-insoluble ash (% w/w) max. (ISO 930)	Moisture (% w/w) max. (ISO 939)	Volatile oil (v/w) min. (ISO 6571)
Whole	8	2	12	2.5
Ground	9	10	10	1.5

a 'deep red or orange red, somewhat viscid liquid, with characteristic odor'. Turmeric oleoresin is valued for its curcumin content, but there is no standard specification for a minimum amount of curcumin.

6.6. Conclusion

Turmeric, which belongs to a group of aromatic spices, had been used originally as a food additive in curries to improve the storage condition, palatability and preservation of food. Turmeric is a key component of curries, curcumin being the principal ingredient. The medicinal properties of curcuminoids have been well researched. In particular, these compounds block several enzymes required for the growth of tumours and may therefore have a role to play in future cancer treatments. As a treatment, it also has some enticing attributes. Curcumin is a non-toxic, highly promising natural antioxidant compound having a wide spectrum of biological functions. It is expected that, in the near future, curcumin may find application as a novel drug to control various diseases, including inflammatory disorders, carcinogenesis and oxidative stress-induced pathogenesis. New research also suggests that turmeric may play a vital role in fighting HIV/AIDS, particularly HIV Type 1. With the support of nanotechnology, the therapeutic effect of turmeric can be enhanced. In future, the therapy can be used for curing various diseases. Not only does turmeric slow cancer growth, it has also been found to correct the cystic fibrosis defect in mice, to help prevent the onset of alcoholic liver disease and may slow down other serious brain diseases like multiple sclerosis and Alzheimer's disease. Future research programmes need to focus on these aspects.

References

Afaq, F., Adhami, V.M. and Ahmad, N. (2002) Botanical antioxidants for chemoprevention of photocarcinogenesis. *Plant Biosciences* 7(7), 84–92.

Ammon, H.P.T. and Wahl, M.A. (1991) Pharmacology of *Curcuma longa*. *Planta Medica* 57, 1–7.

Ammon, H.P.T., Safayhi, H., Mack, T. and Sabieraj, J. (1993) Mechanism of anti-inflammatory actions of curcumine and boswellic acids. *Journal of Ethnopharmacology* 38(2–3), 113–119.

Anon. (2001) *Wealth of India – Raw Materials*, Volume II. Council of Scientific and Industrial Research, New Dehli, India.

Araujo, C.A.C. and Leon, L.L. (2002) Biological activities of *Curcuma longa* L. *Memórias do Instituto Oswaldo Cruz* 96, 723–728.

Araujo, C.A.C., Alegrio, L.V., Castro, D., Lima, M.E.F. and Leon, L.L. (1998) *Leishmania amazonensis*: *in vivo* experiments with diarylheptanoids from Leguminosae and Zingiberaceae plants. *Memórias do Instituto Oswaldo Cruz* 93, 306–310.

Araujo, C.A.C., Alegrio, L.V., Castro, D., Lima, M.E.F., Gomescardoso, L. and Leon, L.L. (1999) Studies on the effectiveness of diarylheptanoid derivatives against *Leishmania amazonensis*. *Memórias do Instituto Oswaldo Cruz* 94, 791–794.

Arora, R.B., Basu, N., Kapoor, V. and Jain, A. (1971) Anti-inflammatory studies on *Curcuma longa*. *The Indian Journal of Medical Research* 59, 1289–1295.

Arun, N. and Nalini, N. (2002) Efficacy of turmeric on blood sugar and polyol pathway in diabetic albino rats. *Plant Foods for Human Nutrition* 57(1), 41–52.

Asai, A. and Miyazawa, T. (2000) Occurrence of orally administered curcuminoid as glucuronide and glucuronide/sulfate conjugates in rat plasma. *Life Sciences* 67, 2785–2793.

ASTA (2001) *Cleanliness Specifications Manual, Method 14.1*. American Spice Trade Association, Washington DC, pp. 13–1713.

ASTA (2002) *A Concise Guide to Spices, Herbs, Seeds and Extractives*. American Spice Trade Association, Washington DC, pp. 48–50.

Babu, P.S. and Srinivasan, P. (1997) Hypolipidemic action of curcumin, the active principle of turmeric *Curcuma longa* in streptozotocin induced diabetic rats. *Molecular and Cellular Biochemistry* 166(1–2), 169–175.

Balasubramanian, N., Kumar, K.R. and Anandaswamy, B. (1979) Packing and storage studies on ground turmeric (*Curcuma longa* L.) in flexible consumer packages. *Indian Spices* 16(12), 10–13.

Banerjee, A. and Nigam, S.S. (1978) Antimicrobial efficacy of the essential oil of *Curcuma longa*. *Indian Journal of Medical Research* 68, 864.

Baumann, W., Rodrigues, S.V. and Viana, L.M. (2000) Pigments and their solubility in and extractability by supercritical CO_2. Part 1. The case of curcumin. *Brazilian Journal of Chemical Engineering* 17, 323–328.

Bhavani Shankar, T.N. and Srinivasamurthy, V. (1979) Effect of turmeric fractions on the growth of some intestinal and pathogenic bacteria *in vitro*. *Indian Journal of Experimental Biology* 17, 1363–1366.

Braga, M.E.M., Leal, P.F., Carvalho, J.O.E. and Meireles, M.A.A. (2003) Comparison of yield, composition, and antioxidant activity of turmeric (*Curcuma longa* L.) extracts obtained using various techniques. *Journal of Agricultural and Food Chemistry* 51, 6604–6611.

BSI (1983) *British Standard Methods of Test For Spices and Condiments*, Part 13. BSI, London.

Buescher, R. and Yang, L. (2000) Turmeric. In: Lauro, G.L. and Fancis, F.J. (eds) *Natural Food Colorants. Science and Technology*. Marcel Dekker, New York, pp. 205–226.

Bush, J.A., Cheung, K.J.J. and Li, G. (2001) Curcumin induces apoptosis in human melanoma cells through a Fas receptor/caspase-8 pathway independent of p53. *Experimental Cell Research* 271(2), 305–311.

Chandra, D. and Gupta, S.S. (1972) Anti-inflammatory and antiarthritic activity of volatile oil of *C. longa*. *Indian Journal of Medical Research* 60, 138–142.

Claeson, P., Panthong, A., Tuchinda, P., Reutrakul, V., Kanjanapothi, D. and Taylor, W.C. (1993) Three non-phenolic diarylheptanoids with anti-inflammatory activity from *Curcuma xanthorrhiza*. *Planta Medica* 59, 451–454.

Claeson, P., Pongprayoon, U., Sematong, T., Tuchinda, P., Reutrakul, V. and Soontornsaratune, P. (1996) Non-phenolic linear diarylheptanoids from *Curcuma xanthorrhiza:* a novel type of topical anti-inflammatory agents: structure activity relationship. *Planta Medica* 62, 236–240.

Cooray, N.F., Jansz, E.R., Ranatunga, J. and Wimalasena, S. (1988) Effect of maturity on some chemical constituents of turmeric (*Curcuma longa* L.). *Journal of the National Science Council of Sri Lanka* 16, 39–51.

Cripps, H.P. (1967) Oleoresin turmeric, application in pickle production. *Glass Packer Process* 46, 24.

Dahl, T.A., Bilski, P., Reszka, K.J. and Chignell, C.F. (1989) Photokilling of bacteria by the natural dye, curcumin. *Archives in Microbiology* 151, 183.

Dandekar, D.V. and Gaikar, V.G. (2002) Microwave assisted extraction of curcuminoids from *Curcuma longa*. *Separation Science and Technology* 37, 2669–2690.

Dash, V.B. (1987) *Illustrated Materia Medica of Indotibetan Medicine*. Classic India Publication, Delhi, pp. 334–335.

Denniff, P. and Whiting, D.A. (1976) Biosynthesis of (6)-gingerol – the pungent principle of *Zingiber officinale*. *Journal of the Chemical Society for Chemical Communication* 18, 711.

Denniff, P., Macleod, I. and Whiting, D.A. (1980) Studies in the biosynthesis of [6]-gingerol, pungent principle of ginger (Zingiber officinale). *Journal of the Chemical Society – Perkin Transactions I*, 2637–2644.

Deodhar, S.D., Sethi, R. and Srimal, R.C. (1980) Preliminary studies on antirheumatic activity of curcumin (diferuloyl methane). *Indian Journal of Medical Research* 71, 632–634.

Dhar, M.L., Dhar, M.M., Dhawan, B.M., Mehrotra, B.N. and Ray, C. (1968) Screening of Indian plants for biological activity. *Indian Journal of Experimental Biology* 6, 232.

Dyrssen, D.W., Novikov, Y.P. and Uppstrdm, L.R. (1972) Studies on the chemistry of the determination of boron with curcumin. *Analytica Chimica Acta* 60, 139–151.

Eigner, D. and Scholz, D. (1999) *Ferula asafoetida* and *Curcuma longa* in traditional medicinal treatment and diet in Nepal. *Journal of Ethnopharmacology* 67, 1–6.

Eiserlie, R.J. (1966) The role of oleoresin turmeric in the pickling process. *Glass Packer Process* 45, 48–49.

Englewood Cliffs (1958) *Official Analytical Methods, Color Power of Turmeric*, Method 18.0. American Spice Trade Association, Washington, DC.

Ferreira, L.A.F., Henriques, O.B., Andreoni, A.A.S., Vital, G.R.F., Campos, M.M.C., Habermehl, G.G. and Moraes, V.L.G. de (1992) Antivenom and biological effects of *ar*-turmerone isolated from *Curcuma longa* (Zingiberaceae). *Toxicon* 30, 1211–1218.

Food and Agricultural Organization (1976) *Turmeric in Specifications for the Identity of Some Food Colors, Flavor Enhancers*, WHO Food Additives Series No.13, Rome, 1976.

Garg, S.N., Bansal, R.P., Gupta, M.M. and Kumar, S. (1999) Variation in the rhizome essential oil and curcumin contents and oil quality in the land races of turmeric *Curcuma longa* of North Indian plains. *Flavour Fragance Journal* 14, 315–318.

Garg, S.N., Mengi, N., Patra, N.K., Charles, R. and Kumar, S. (2002) Chemical examination of the leaf essential oil of *Curcuma longa* L. from the North Indian plains. *Flavour and Fragrance Journal* 17, 103–104.

Geissman, T.A. and Crout, D.H.G. (1969) *Organic Chemistry of Secondary Plant Products*. Freeman Cooper, San Francisco, CA.

Ghatak, N. and Basu, N. (1972) Sodium curcuminate as an effective anti-inflammatory agent. *Indian Journal of Experimental Biology* 10, 235–236.

Golding, B.T., Pombo, E. and Christopher, J.S. (1982) Turmerones: isolation from turmeric and their structure determination. *Journal of the Chemical Society for Chemical Communications* 6, 363–364.

Gonda, R., Tomoda, M., Shimizu, N. and Kanari, M. (1990) Characterization of polysaccharides having activity on the reticuloendothelial system from the rhizome of *C. longa*. *Chemical and Pharmaceutical Bulletin* 38(2), 482–486.

Gonda, R., Takeda, K., Shimizu, N. and Tomoda, M. (1992) Characterization of a neutral polysaccharide having activity on the reticuloendothelial system from the rhizomes of *Curcuma longa*. *Chemical and Pharmaceutical Bulletin* 40(1), 185–188.

Gopalan, B., Goto, M., Kodama, A. and Hirose, T. (2000a) Response surfaces of total oil yield of turmeric (*Curcuma longa*) in supercritical carbon dioxide. *Food Research International* 33, 341–345.

Gopalan, B., Goto, M., Kodama, A. and Hirose, T. (2000b) Supercritical carbon dioxide extraction of turmeric (*Curcuma longa*). *Journal of Agricultural and Food Chemistry* 48, 2189–2192.

Govindarajan, V.S. (1980) Turmeric – chemistry, technology and quality. *CRC Critical Reviews in Food Science and Nutrition* 12(3), 200–213.

Gururaj, A., Belakavadi, M., Venkatesh, D., Marme, D. and Salimath, B. (2002) Molecular mechanisms of antiangiogenic effect of curcumin. *Biochemical and Biophysical Research Communications* 297(4), 934–942.

Heller, G. (1914) Berichte der Deutschen Chemischen Gesellschaft, 47, 2998. Cited by Govindarajan, V.S. (1980).

Henry, B. (1998) Use of capsicum and turmeric and natural colors. *Indian Spices* 35(3), 7–14.

Hongliang Jiang, Barbara, N.T. and Gang, R.D. (2006) Use of liquid chromatography-electrospray ionization tandem, mass spectrometry to identify diarylheptanoids in turmeric rhizome. *Journal of Chromatography* A 1111(1), 21–31.

Hrazdina, G. and Wagner, G.J. (1985) Metabolic pathways as enzyme complexes: evidence for the synthesis of phenylpropanoids and flavonoids on membrane associated enzyme complexes. *Archives of Biochemistry and Biophysics* 237(1), 88–100.

Imai, S., Morikiyo, M., Furihata, K., Hayakawa, Y. and Seto, H. (1990) Turmeronol A and turmeronol B, new inhibitors of soybean lipoxygenase. *Agricultural and Biological Chemistry* 54, 2367–2371.

Inano, H., Onoda, M., Inafuku, N., Kubota, M., Kamada, Y., Osawa, T., Kobayashi, H. and Wakabayashi, K. (2000) Potent preventive action of curcumin on radiation-induced initiation of mammary tumorigenesis in rats. *Carcinogenesis* 12, 1835–1841.

Iwakami, S., Shibuya, M., Tseng, C.F., Hanaoka, F. and Sankawa, U. (1986) Inhibition of arachidonate 5-lipoxygenase by phenolic compounds. *Chemical and Pharmaceutical Bulletin* 34, 3960–3963.

Janaki, N. and Bose, J.L. (1967) An improved method for the isolation of curcumin from turmeric, *Curcuma longa* L. *Journal of the Indian Chemical Society* 44, 985–989.

Janssen, A. and Gole, T. (1984) Thin-layer chromatographic determination of curcumine (turmeric) in spices. *Chromatographia* 18, 546–549.

Jayaprakasha, G.K., Negi, P.S., Anandha Ramakrishanan, C. and Sakariah, K.K. (2001) Chemical composition of turmeric oil – a by-product from turmeric oleoresin industry and its inhibitory activity against different fungi. *Zeitschrift fur Naturforschung* 56c, 40–44.

Jayaprakasha, G.K., Jaganmohan Rao, L. and Sakariah, K.K. (2002) An improved HPLC method for the determination of curcumin, demethoxycurcumin and bisdemethoxycurcumin. *Journal of Agricultural and Food Chemistry* 50, 3668–3672.
Jentzsch, K., Gonda, T. and Holler, H. (1959) Paper chromatographic and pharmacological investigation on curcuma pigments. *Pharmaceutica Acta Helvitiae* 34, 181–188.
Jung, E.M., Lim, J.H., Lee, T.J., Park, J., Choi, K.S. and Kwon, T.K. (2005) Curcumin sensitizes tumor necrosis factor-related apoptosis-inducing ligand (TRAIL)-induced apoptosis through reactive oxygen species-mediated up-regulation of death receptor 5 (DR5). *Cancer Biology* 26(11), 1905–1913.
Kapoor, L.D. (1990) *Handbook of Ayurvedic Medicinal Plants*. CRC Press, Boca Raton, Florida, p. 185.
Karasz, A.B., DeCocca, F. and Bokus, L. (1973) Detection of turmeric in foods by rapid fluorimetric method and by improved spot test. *Journal of the Association of Official Analytical Chemists* 56, 626–628.
Kawamori, T., Lubet, R., Vernon, E.S., Kelloff, G.J., Kaskey, R.B., Rao, V.R. and Reddy, B.S. (1999) Chemopreventive effect of curcumin, a naturally occurring anti-inflammatory agent, during the promotion/progression stages of colon cancer. *Cancer Research* 59, 597–601.
Kelkar, N.C. and Rao, B.S. (1933) Studies in Indian essential oils: essential oils from *C. longa*. *Journal of Indian Institute of Science* 17A, 7–15.
Khafif, A., Schantz, S.P., Chou, T.C., Edelstein, D. and Sacks, P.G. (1998) Quantitation of chemopreventive synergism between (–)-epigallocatechin-3-gallate and curcumin in normal, premalignant and malignant human oral epithelial cells. *Carcinogenesis* 19, 419–424.
Khalique, A. and Amin, M.N. (1967) Examination of *Curcuma longa* L. Part I. Constituents of the rhizome. *Scientific Research in East Regional Lab (Pakistan)* 4(4), 193–197.
Khurana, A. and Ho, C.T. (1988) High-performance liquid chromatographic analysis of curcuminoids and their photooxidative decomposition compounds in *Curcuma longa* L. *Journal of Liquid Chromatography* 11, 2295–2304.
Kim, D.S.H.L., Park, S.Y. and Kim, J.Y. (2001) Curcuminoids from *Curcuma longa* L. (Zingiberaceae) that protect PC12 rat pheochromocytoma and normal human umbilical vein endothelial cells from ba(1–42) insult. *Neuroscience Letters* 303, 57–61.
Kiuchi, F., Shibuya, M. and Sankawa, U. (1982) Inhibitors of prostaglandin biosynthesis from *Alpinia officinarum*. *Chemical and Pharmaceutical Bulletin* 30, 2279–2282.
Kiuchi, F., Iwakami, S., Shibuya, M., Hanaoka, F. and Sankawa, U. (1992) Inhibition of prostaglandin and leukotriene biosynthesis by gingerols and diarylheptanoids. *Chemical and Pharmaceutical Bulletin* 40, 387–391.
Kiuchi, F., Goto, Y., Sugimoto, N., Akao, N., Kondo, K. and Tsuda, Y. (1993) Nematocidal activity of turmeric: synergistic action of curcuminoids. *Chemical and Pharmaceutical Bulletin* 41, 1640–1643.
Koide, T., Nose, M., Ogihara, Y., Yabu, Y. and Ohta, N. (2002) Leishmanicidal effect of curcumin *in vitro*. *Biological and Pharmaceutical Bulletin* 25, 131–133.
Kojima, H., Yanai, T., Toyota, A., Hanani, E. and Saiki, Y. (1998) Essential oil constituents from *Curcuma aromatica*, *C. longa* and *C. xanthorrhiza* rhizomes. In: Ageta, H., Aimi, N., Ebizaka, Y., Fujita, T. and Honda, G. (eds) *Towards Natural Medicine Research in the 21st Century*. Elsevier, Amsterdam, pp. 531–539.
Krishnamurthy, N., Mathew, A.G., Nambudiri, E.S., Shivasankar, S., Lewis, Y.S. and Natarajan, C.P. (1976) Oil and oleoresin of turmeric. *Tropical Science* 18(1), 37–45.
Lampe, V., Milobedeska, J. and Kostanecki, V. (1910) Synthese von p, p′-dioxy – and p-oxy dicinnamoylmethane. *Berichte der Deutschen Chemischen Gesellschaft* 43, 2163.
Leela, N.K., Aldo, T., Shaji, P.M., Sinu, P.J. and Chempakam, B. (2002) Composition of essential oils from turmeric (*Curcuma longa* L.). *Acta Pharmaceutica* 52, 137–141.
Li, C.J., Zhang, L.J., Dezube, B.J. and Crumpacker, C.S. (1993) Three inhibitors of human type 1 immunodeficiency virus long terminal repeat directed gene expression and virus replication. *Proceedings of the National Academy of Science* (USA) 90, 1839–1842.
Li, Z.O.Q., Luo, L., Ma, G.Y., Zhang, Y.Q. and Zhao, G.J. (1997) GC/MS study on chemical constituents from Zingiberaceae plants. *Guangpu Shiyanshi* 14, 1–5.
Li-Hsun Chang, Jong, T.T., Huang, H., Nien, Y. and Chang, J.C. (2006) Supercritical carbon dioxide extraction of turmeric oil from *Curcuma longa* Linn and purification of turmerones. *Separation and Purification Technology* 47(3), 119–125.
Lim, G.P., Chu, T., Yang, F., Beech, W., Frantschy, S.A. and Cole, G.M. (2001) The curry spice curcumin reduces oxidative damage of amyloid pathology in an Alzheimer transgenic mouse. *Journal of Neurosciences* 21, 8370–8377.
Lin, J.K., Huang, T.S., Shih, C.A. and Lin, J.Y. (1994) Molecular mechanism of action of curcumin, in food phytochemicals II: teas, spices, and herbs. *American Chemical Society* 20, 196–203.

Macleod, I. and Whiting, D.A. (1979) Stages in the biosynthesis of [6]-gingerol in *Zingiber officinale*. *Journal of the Chemical Society for Chemical Communications* 8, 1152–1153.

Madhyastha, M.S. and Bhat, R.V. (1985) Evaluation of substrate potentiality and inhibitory effects to identify high risk spices for aflatoxin contamination. *Journal of Food Science* 50(2), 376–378.

Maheswari, P. and Singh, U. (1965) *Dictionary of Economic Plants in India*. Indian Council of Agricultural Research, New Delhi.

Majeed, M., Badmaev, V., Shivakumar, U. and Rajendran, R. (1995) *Curcuminoids Antioxidant Phytonutrients*. Nutriscience Publishers, Inc., Piscataway, New Jersey.

Malingre, T.R. (1975) *Curcuma xanthorrhiza* Roxb., temoe lawak, als plant met galdrijrende werking. *Pharmaceutisch Weekblad* 110, 601–606.

Manzan, A.C.C.M., Toniolo, F.S., Bredow, E. and Povh, N.P. (2003) Extraction of essential oil and pigments from *Curcuma longa* (L.) by steam distillation and extraction with volatile solvents. *Journal of Agricultural and Food Chemistry* 51, 6802–6807.

Maria, C.R. del, Barbara, N.T. and David, R.G. (2006) Biosynthesis of curcuminoids and gingerols in turmeric (*Curcuma longa*) and ginger (*Zingiber officinale*): identification of curcuminoid synthase and hydroxycinnamoyl-CoA thioesterases. *Phytochemistry* 67, 2017–2029.

Masuda, T., Toi, Y., Bando, H., Maekawa, T., Takeda, Y. and Yamaguchi, H. (2002) Structural identification of new curcumin dimers and their contribution to the antioxidant mechanism of curcumin. *Journal of Agricultural and Food Chemistry* 50, 2524–2530.

Mayer, F. (1943) *The Chemistry of Natural Coloring Matters*. Translated and revised by Cook, A.H. Reinhold, New York, p. 93.

McCarron, M., Mills, A.J., Whittaker, D., Sunny, T.P. and Verghese, J. (1995) Comparison of the monoterpenes derived from green leaves and fresh rhizomes of *Curcuma longa* L. from India. *Flavour and Fragrance Journal* 10, 355–357.

Nadkarni, K.M. (ed.) and Nadkarni, A.K. (1976) *Indian Materia Medica*, Revised Edition. Popular Prakashan, Bombay, India.

Nazeem, P.A. and Rema Menon (1994) Blossom biological and hybridisaton studies in turmeric (*Curcuma longa*). *South Indian Horticulture* 42(3), 161–167.

Neema, A. (2005) Investigations on the biosynthesis of curcumin in turmeric (*Curcuma longa* L.). PhD thesis, Calicut University, Kerala, India, p. 171.

Nigam, M.C. and Ahmed, A. (1991) *Curcuma longa*: terpenoid composition of its essential oil. *Indian Perfumer* 3, 201–205.

Oguntimein, B.O., Weyerstahl, P. and Marshall, H. (1990) Essential oil of *Curcuma longa* L. leaves. *Flavour and Fragrance Journal* 5, 89–90.

Ohshiro, M., Kuroyanagi, M. and Ueno, A. (1990) Structures of sesquiterpenes from *Curcuma longa*. *Phytochemistry* 29, 2202–2205.

Osawa, T., Sugiyama. Y., Inayoshi, M. and Kawakishi. S. (1995) Antioxidative activity of tetrahydro curcuminoids. *Bioscience, Biotechnology and Biochemistry* 59(9), 1609–1612.

Pabon, H.J.J. (1964) Synthesis of curcumin and related compounds. *Recueil des Travaux Chimiques des Pays-Bas* 83, 379–386.

Parry, J.W. (1969) *Spices. Volume 1 – The Story of Spices and Spices Described. Volume II – Morphology, Histology and Chemistry*. Chemical Publishing, New York.

Perotti, A.G. (1975) Curcumin – a little known but useful vegetable colour. *Industrial Aliment for Products and Vegetables* 14(6), 66–68.

Pervaiz, S. (1990) Antitumor and antiviral activity of curcumin and light. *Proceedings of the Annual Meeting of the American Association of Cancer Research Abstract* A2325.

Phan, T.T., See, P., Lee, S.T. and Chan, S.Y. (2001) Protective effects of curcumin against oxidative damage on skin cell *in vitro*: its implication for wound healing. *Journal of Trauma* 51, 927–931.

Polasa, K., Sesikaran, B. and Krishnaswamy, K. (1992) Antimutagenicity of curcumin and turmeric. *Proceedings of the Nutritional Society of India* 36, 107–109.

Polasa, K., Naidu, A.N., Ravindranath, I. and Krishnaswamy, K. (2004) Inhibition of B(a)P induced strand breaks in presence of curcumin. *Mutation Research* 557, 203–213.

Radha, K.M., Anoop, K.S., Gaddipati, J.P. and Richkab, C.S. (2006) Multiple biological activities of curcumin: a short review. *Life Sciences* 78(18), 2081–2087.

Raghunath, K. and Mitra, R. (1982) *Pharmacognosy of Indigenous Drugs, 1*. Central Council for Ayurveda and Siddha, New Delhi, India, pp. 376–392.

Reddy, A.C.P. and Lokesh, B.R. (1992) Studies on spice principles as antioxidants in the inhibition of lipid peroxidation of rat liver microsomes. *Molecular and Cellular Biochemistry* III, 117–124.
Roughley, P.J. and Whiting, D.A. (1973) Experiments in the biosynthesis of curcumin. *Journal of the Chemical Society – Perkin Transactions-I* 20, 2379–2388.
Ruby, A.J., Kuttan, G., Dinesh Babu, K., Rajasekharan, K.N. and Kuttan, R. (1995) Antitumor and antioxidant activity of natural curcuminoids (*Curcuma longa*) on iron-induced lipid peroxidation in the rat liver. *Food and Chemical Toxicology* 32, 279–283.
Ruby, A.J., Kuttan, G., Dinesh Babu, K., Rajasekharan, K.N. and Kuttan, R. (1995) Antitumor and antioxidant activity of natural curcuminoids. *Cancer Letters* 94, 79–83.
Rupe, H., Clar, G., Pfau, A.St and Plattner, P. (1934) Volatile plant constituents. II. Turmerone, the aromatic principle of turmeric oils (in German). *Helvetica Chimica Acta* 17, 372.
Salzer, U.J. (1977) The analysis of essential oils and extracts (oleoresins) from seasonings – a critical review. *CRC Critical Review for Food Science and Nutrition* 9(4), 345.
Sanagi, M.M., Ahmad, U.K. and Smith, R.M. (1993) Application of supercritical fluid extraction and chromatography to the analysis of tumeric. *Journal of Chromatographic Science* 31, 20–25.
Sastry, B.S. (1970) Curcumin content of turmeric. *Research and Industry* 15, 258–260.
Satoskar, R.R., Shah, S.J. and Shenoy, S.G. (1986) Evaluation of anti-inflammatory property of curcumin (diferuloyl methane) in patients with postoperative inflammation. *International Journal of Clinical Pharmacology and Theoretical Toxiclogy* 24, 651–654.
Sharma, O.P. (1976) Antioxidant activity of curcumin and related compounds. *Biochemical Pharmacology* 25, 1811–1812.
Sharma, R.K., Misra, B.P., Sarma, T.C., Bordoloi, A.K., Pathak, M.G. and Leclercq, P.A. (1997) Essential oils of *Curcuma longa* L. from *Bhutan. Journal of Essential Oil Research* 9, 589–592.
Sidhu, G.S. *et al.* (1999) Curcumin enhances wound healing in streptozotocin induced diabetic rats and genetically diabetic mice. *Wound Repair and Regeneration* 7, 362–374.
Sivananda, S. (1958) *Home Remedies*. The Yoga Vedanta University, Sivanad Nagar, India.
Soni, K.B. and Kuttan, R. (1992) Effect of oral curcumin administration on serum peroxides and cholesterol levels in human volunteers. *Indian Journal of Physiology and Pharmacology* 36(4), 273–275 and 239–243.
Spices Board (2000) Quoted by Global Agri System Pvt. Ltd, New Delhi.
Sreejayan, N. and Rao, M.N.A. (1994) Curcuminoids as potent inhibitors of lipid peroxidation. *The Journal of Pharmacy and Pharmacology* 46, 1013–1016.
Srimal, R.C. and Dhawan, N. (1983) Pharmacology of diferuloyl methane (curcumin), a non-steroidal anti-inflammatory agent. *The Journal of Pharmacy and Pharmacology* 25, 447–452.
Srimal, R.C. and Dhawan B.N. (1985) Pharmacological and clinical studies on *Curcuma longa*. Monograph, Hamdard National Foundation, New Delhi, section 313(ii), 131–138.
Srinivas, L., Shalini, V.K. and Shylaja, M. (1992) Turmerin: a water soluble antioxidant peptide from turmeric. *Archives of Biochemistry ad Biophysics* 292, 617–623.
Srinivasan, K.R. (1952) The coloring matter in turmeric. *Current Science* 311–313.
Srivastava, R. and Srimal, R.C. (1985) Modification of certain inflammation-induced biochemical changes by curcumin. *Indian Journal of Medical Research*, 81, 215–223.
Srivastava, R., Dikshit, M., Srimal, R.C. and Dhawan, B.N. (1985) Antithrombotic effect of curcumin. *Thrombosis Research* 40, 413–417.
Stoner, G.D. and Mukhtar, H. (1995) Polyphenols as cancer chemopreventive agents. *Cell Biochemical Supplement* 22, 169–180.
Subramonian, M., Sreejayan Rao, M.N.A., Devasagayam, T.P.A. and Singh, B.B. (1994) Diminuition of singlet oxygen induced D.N.A. damage by curcumin and related antioxidants. *Mutation Research* 311, 249–255.
Tainter, D.R. and Grenis, A.T. (2001) *Spices and Seasonings. A Food Technology Handbook*, 2nd edn. Wiley-VCH, New York.
Tonnesen, H.H. and Karlsen, J. (1983) High-performance liquid chromatography of curcumin and related compounds. *Journal of Chromatography* 259, 367–371.
Tonnesen, H.H. and Karlsen, J. (1986) Studies on curcumin and curcuminoids. VII. Chromatographic separation and quantitative analysis of curcumin and related compounds. *Zeitschrift fur Lebensmittel-Untersuchung und Forschung* 182, 215–218.
Tonnesen, H.H., DeVries, H., Karlsen, J. and Van Henegouwen, G.B. Studies on curcumin and curcuminoids. IX. Investigation of the photobiological activity of curcumin using bacterial indicator systems. *Journal of Pharmaceutical Science* 76, 371–373.

Unterhalt, B. (1980) Curcuma and seine verwendubg in spicessent. *Zeitschrift fur Lebensmittel-Untersuchung und Forschung* 170, 425–428.

Usha, K., Nagabhushan, M., Nair S.C., Rao P.U. and Rahiman M.A. (1994) The possible mode of action of cancer chemopreventive spice, turmeric. *Journal of the American College of Nutrition* 13, 519.

Vabirua-Lechat, I., Menut, C., Lamaty, G. and Bessiere, J.M. (1996) Huiles essentielles de Polynesia Français. *Rivista Italiana Eppos (Numero Speciale)* 1996, 627–638.

Velayudhan, K.C., Muralidharan, V.K., Amalraj, V.A., Gautam, P.L., Mandal, S. and Dinesh Kumar (1999) *Curcuma Genetic Resources. Scientific Monograph No 4.* National Bureau of Plant Genetic Resources, New Delhi.

Verghese, J. (1993) Isolation of curcumin from *Curcuma longa* L. rhizome. *Flavour and Fragrance Journal* 8(6), 315–319.

Verghese, J. and Joy, M.T. (1989) Isolation of the colouring matter from dried turmeric (*Curcuma longa* L.) with ethyl acetate. *Flavour and Fragrance Journal* 4, 31–32.

Vogel and Pelletier (1818) *Journal of Pharmaceutical Sciences* 2, 50. Cited in: Govindarajan, V.S. (1980).

Weiss, E.A. (2002) *Spice Crops*. CAB International, Wallingford, UK.

WHO (1976) WHO Food Additives, series no.7. Specifications for the identity and purity same food colours, enhancers, thickening agents and certain food additives. FAO Nutrition Meetings Report Series, Rome, no. 546, 1975.

Xianchun, W., Xie, J., Shan, C. and Qi, Z. (1993) Extraction of curcumin from Jianghuang (*Curcuma longa*) by ethanol reflux. *Shipin Kexue (Beijing)* 165, 45–47.

Zachariah, T.J., Sasikumar, B. and Nirmal Babu (1998) Variation for quality components in ginger and turmeric and their interaction with environments. In: *Biodiversity, Conservation and Utilization of Spices, Medicinal and Aromatic Plants*. IISR, Calicut, India, pp. 116–120.

Zhang, L. and Yang, Z. (1988) Extraction of curcumine from rhozoma *Curcumas longae*. *Faming Zhuanli Shenqing Gongkai Shuongshu CN* 87,101,355 (Cl. C09B61100), Oct 12, 10.

Zhu, L.F., Li, Y.H., Li, B.L., Ju, B.Y. and Zhang, W.L. (1995) *Aromatic Plants and Essential Constituents (Supplement 1)*. South China Institute of Botany, Chinese Academy of Sciences, Hai Feng Publ. Co. Distributed by Peace Book Co. Ltd., Hong Kong.

7 Cinnamon and Cassia

N.K. Leela

7.1. Introduction

Cinnamon and its close relative, cassia, are among the earliest, most popular spices used by mankind. The genus *Cinnamomum* (family: Lauraceae) consists of 250 species of trees and shrubs distributed in South-east Asia, China and Australia. In India, it is represented by 26 species, of which 12 each are reported from North-east and South India. The true cinnamon, *Cinnamomum verum* syn. *C. zeylanicum*, is a native of Sri Lanka and South India. Cassia cinnamon is derived from different sources, such as Chinese cassia (*C. cassia* syn. *C. aromatica*) from China and Vietnam, Indonesian cassia (*C. burmannii*) from Sumatra and the Java region and Indian cassia (*C. tamala*) from the north-eastern region of India and Myanmar (Burma) (Baruah and Nath, 2004).

Sri Lanka is the major cinnamon-producing country in the world and it controls 60% of the world cinnamon trade. About 24,000 ha are under cinnamon cultivation in Sri Lanka, producing 12,000 t 'quills' (long, compound rolls of cinnamon bark measuring up to 1m in length) per year. Sri Lanka produces the best quality of cinnamon bark, mainly as quills. It also produces annually around 120 t leaf oil and 4–5 t bark oil. Cinnamon leaf oil is produced in Sri Lanka and the Seychelles, though the bark oil is distilled in importing countries. The UAE is the major buyer of Chinese cassia (*C. cassia*) bark and leaf oil (from China). Vietnam cassia (*C. cassia*) is exported to the USA. Indonesian cassia (*C. burmannii*) is also an important export product to the USA. *C. tamala* (*tejpat*) is grown for its leaves, which are used extensively in flavouring. It is inferior to other cassias because of the lower oil content in its bark and leaves. The Seychelles, Madagascar and India also produce cinnamon in small quantities. The Seychelles produces around 600 t quills yearly from about 34,000 ha of cinnamon. India exported 18.12 t of oleoresin and 17.39 t of cassia oil in 2002/03 (Rema *et al*., 2006). The area and production of cinnamon in India for 1994 to 2005 are indicated in Table 7.1.

7.2. Botany and Uses

Cinnamon is a small evergreen tree, 10–15 m tall. The bark is used widely as a spice. The leaves are ovate-oblong in shape, 7–18 cm long. The flowers, which are arranged in panicles, have a greenish colour and have a rather disagreeable odour. The fruit is a purple 1-cm berry containing a single seed (http://en.wikipedia.org/wiki/Cinnamon).

Table 7.1. Area and production of cinnamon in India.

Year	Area (ha)	Production (t)
1994/95	714	300
1995/96	655	364
1996/97	833	371
1997/98	745	1176
1998/99	720	1661
1999/2000	718	1658
2000/01	701	1658
2001/02	727	1659
2002/03	739	1659
2003/04	757	1659
2004/05	774	1659

Source: DASD (2007).

The cinnamon of commerce is the dried inner bark of the tree, *C. verum*. It is an essential item in curry powders and *masalas*. The bark oil, bark oleoresin and leaf oil are important value-added products from cinnamon. Bark oil is used in the food and pharmaceutical industries. Cinnamon leaf oil is cheaper than bark oil and is used in the flavour industry. Cinnamon oleoresin, obtained by solvent extraction of the bark, is used mainly for flavouring food products such as cakes and confectionary. As in the case of cinnamon, the volatile oil and oleoresin from cassia are also used extensively in flavouring, especially soft drinks and other beverages.

The leaf of the Indian cassia, known as *tejpat* in Hindi, is a spice that has a clove-like odour and a faint pepper-like aroma. It is a popular flavouring agent in north Indian vegetarian and non-vegetarian preparations. Apart from *C. tamala*, *C. sulphuratum*, *C. bejolghota* and *C. impressinervium* are also traded as *tejpat* in North-east India (Baruah *et al.*, 2000). The essential oil is used in the flavouring and formulation of liquors and confections.

7.3. General Composition

The dried bark of *C. zeylanicum* contains 59.5% carbohydrates, 20.3% fibre, 9.9% moisture, 4.6% protein, 2.2% fat and 3.5% total ash. It also contains 1.6% calcium, 0.05% phosphorus, vitamin A (175 IU), vitamin B_1 (0.14 mg/100 g), vitamin B_2 (0.21 mg/100 g), vitamin C (39.8 mg/100 g) and niacin (1.9 mg/100 g). The composition varies depending on the geographical origin of the spice and the processing conditions. According to different authors, the following range of variation was observed: carbohydrates, 16.6–22.6%; fibre, 25.6–30.5%; moisture, 5.4–11.4%; protein, 3.0–4.5%; volatile oil, 0.3–2.8%; and fixed oil, 0.3–1.9% (Pruthi, 1976).

The variation in the composition of various cassia bark is as follows: moisture (6.5–11.9%); crude fibre (12.0–28.8%); carbohydrate (6.9–32.0%); protein (3.1–3.4%); fixed oil (0–2.1%); volatile oil (0.5–5.1%); and cold alcohol extract (4.6–16.7%) (Pruthi, 1976). The nutritional composition of cinnamon is given in Table 7.2. Of these, the constituent of commercial importance is the volatile oil.

7.4. Chemistry

The dried inner bark of cinnamon and cassia contains volatile oil, fixed oil, tannin, resin,

Table 7.2. Nutritional composition of cinnamon per 100 gm.

Composition	[1]USDA	[2]ASTA
Water (g)	9.52	10.00
Food energy (Kcal)	261	355
Proteins (g)	3.89	4.50
Fat (g)	3.18	2.20
Carbohydrates (g)	79.85	79.8
Ash (g)	3.55	3.50
Calcium (g)	1.23	1.60
Phosphorus (mg)	61	50
Sodium (mg)	26	10
Potassium (mg)	500	400
Iron (mg)	38.07	4.10
Thiamine (mg)	0.077	0.140
Riboflavin (mg)	0.14	0.21
Niacin (mg)	1.30	1.90
Ascorbic acid (mg)	28.46	40.00
Vitamin A activity (RE)	26	26

Source: Tainter and Grenis (1993); [1]Handbook 8–2, data for *C. verum* and *C. cassia* combined. [2]American Spice Trade Association.

proteins, cellulose, pentosans, mucilage, starch, calcium oxalate and mineral elements. The relative abundance of these components varies considerably according to location, age of the tree, climatic condition, season, time of harvest and duration of storage.

Volatiles

Cinnamon possesses a delicate, spicy aroma, which is attributed to its volatile oil. Volatile components are present in all parts of cinnamon and cassia. They can be classified broadly into monoterpenes, sesquiterpenes and phenylpropenes.

Cinnamon (C. verum)

Cinnamon yields mainly leaf and bark oils, which are used in perfumery and flavouring. The major component of leaf oil is eugenol, while that of bark oil is cinnamaldehyde. Volatile components do occur in other parts, including root bark, fruits, flowers, twigs and branches.

BARK OIL The volatile oil content in cinnamon bark varies from 0.4 to 2.8% (Angmor *et al.*, 1972; Wijesekera, 1978; Krishnamoorthy *et al.*, 1996). The physico-chemical properties of cinnamon oil are indicated in Table 7.3. Senanayake *et al.* (1978) reported that the oil from the stem bark of a commercial sample contained 75% cinnamaldehdyde, 5% cinnamyl acetate, 3.3% caryophyllene, 2.4% linalool and 2.2% eugenol. Krishnamoorthy *et al.* (1988) observed variation in the bark oil content in plants with purple leaf flushes (1.84%) and those with green flushes (1.43%). Bernard *et al.* (1989) studied the composition of volatiles from *C. zeylanicum* bark by two methods, namely, direct distillation and extraction using TTE (1,1,2-trichloro-1,2,2-trifluoroethane) followed by hydrodistillation. Both methods were comparable, yielding 0.98–1.1% volatile oil. However, compositional differences were observed in both the oils. The TTE extract had a higher cinnamaldehyde content (84.1%) compared with the direct hydrodistilled oil (75%). α-Pinene, 1,8-cineole and *p*-cymene, which were present in minor amounts in the hydrodistilled oil, were absent in the TTE product. There was less linalool, β-caryophyllene and cinnamyl acetate in the oil obtained by the TTE method compared with the direct distillation method.

The volatile oil from the stem bark of Madagascan origin was rich in eugenol (Medici *et al.*, 1992). Krishnamoorthy *et al.* (1996) reported 2.7–2.8% volatile oil in the bark of the cinnamon varieties *Navashree* and *Nithyasree*, with 58–68% cinnamaldehyde content. Nath *et al.* (1996) recorded a chemotype of *C. verum* with 84.7% benzyl benzoate in bark oil from the Brahmaputra Valley, India.

Kaul *et al.* (2003) analysed essential oil profiles of various parts of cinnamon. The oil yields of different plant parts were: 0.40% in tender twigs; 0.36% in the pedicels of buds and flowers; 0.04% in buds and flowers; 0.33% in the pedicels of fruits; and 0.32% in fruits. The tender twig oil was richer in *α*-phellandrene (3.4%), limonene (1.6%) and (*E*)-cinnamaldehyde (4%). The volatile oils from pedicels were richer in (*E*)-cinnamyl acetate (58.1–64.5%), *β*-caryophyllene (9.6–11.1%) and neryl acetate (1.4–2.0%). Higher amounts of (*Z*)-cinnamyl acetate (6.1%), *α*-humulene (2.2%),

Table 7.3. Physico-chemical properties of *Cinnamomum verum* oil.

	Bark oil	Leaf oil
Specific gravity	1.021–1.070 (at 20°C)	1.044–1.062 (at 30°C)
Refractive index	1.567–1.614 (at 20°C)	1.522–1.530 (at 30°C)
Optical rotation (°)	−1°–0° (at 20°C)	3.60° (at 30°C)
Eugenol content (%)	–	65–87.2

Source: Baslas (1967); Bernard *et al.* (1989).

δ-cadinene (2.2%), humulene epoxide I (5%), α-muurolol (4.9%) and α-cadinol (2.4%) were observed in the oil of buds and flowers. However, all the oils contained linalool (3.6–27.4%), (*E*)-cinnamyl acetate (22.0–64.5%) and β-caryophyllene (6.9–11.1%) as their major compounds.

LEAF OIL *Cinnamomum verum* leaves contained 0.24–3.0% volatile oil, depending on the location and method of distillation (Angmor *et al.*, 1972; Wijesekera, 1978; Rao *et al.*, 1988; Krishnamoorthy *et al.*, 1996; Raina *et al.*, 2001). The principal component of leaf oil, namely, eugenol, varied from 65 to 92% (Senanayake *et al.*, 1978). Krishnamoorthy *et al.* (1996) recorded 75–78% eugenol in the leaves of the cinnamon varieties *Navashree* and *Nithyasree*. Several chemotypes of *C. zeylanicum* have been reported, based on the chemical composition of leaf oil. Guenther (1953) and Rao *et al.* (1988) reported two chemical races of *C. zeylanicum* from Bubhaneshwar, India, one rich in eugenol (83.1–88.6%) and the other dominated by benzyl benzoate (63.6–66.0%). Another chemotype with 85.7% linalool in leaf oil was reported by Jirovetz *et al.* (2001). Nath *et al.* (1996) reported a chemotype of *C. verum* growing in the Brahmaputra Valley containing benzyl benzoate as its major component (65.4%) in leaf oil. Two chemotypes of *C. verum* from Brazil were reported by Koketsu *et al.* (1997); one rich in eugenol (94.14–95.09%) and the other predominated by eugenol and safrole (with 55.08–58.66% eugenol and 29.57–39.52% safrole, respectively). According to Variyar and Bandyopadhyay (1989), eugenol-type is the most commonly occurring chemical race of *C. verum*.

Higher oil content was reported in cinnamon leaf from Hyderabad (4.7%) compared with that from Bangalore (1.8%) (Mallavarapu *et al.*, 1995). The two oils were of eugenol type and differed with respect to the relative amounts of linalool, cinnamaldehyde, cinnamyl alcohol, cinnamyl acetate and benzyl benzoate. The essential oil of the leaves of *C. zeylanicum* from Cameroon contained eugenol (85.2%), (*E*)-cinnamaldehyde (4.9%), linalool (2.8%) and β-caryophyllene (1.8%) (Jirovetz *et al.*, 1998).

The oils from the leaves and bark of *C. zeylanicum* from Madagascar contained cinnamaldehyde and camphor as the major components (Chalchat and Valade, 2000). The leaf oil from Little Andaman Island contained 47 constituents, representing 99.96% of the oil. The main constituents were eugenol (76.60%), linalool (8.5%), piperitone (3.31%), eugenyl acetate (2.74%) and cinnamyl acetate (2.59%) (Raina *et al.*, 2001).

The leaves harvested in summer gave the highest oil recovery (1.84%) and eugenol content (83%), whereas in the rainy season, the concentration of esters, namely, eugenyl acetate and benzyl benzoate, were comparatively higher (Kaul *et al.*, 1996). Cinnamon leaves affected by leaf spot disease yielded less oil (1.2%), but the eugenol content was unaffected (Kaul *et al.*, 1998). Rao *et al.* (2006) reported that the essential oil content (1.9–2.2%) and the chemical composition of *C. verum* leaves were not affected by storage up to a period of 15 months.

ROOT BARK OIL The root bark of *C. zeylanicum* yielded 2–3% volatile oil. The major component of the volatile oil of the root bark of *C. zeylanicum* from Ghana was camphor (Angmor *et al.*, 1972). Senanayake *et al.* (1978) reported that oil from the root bark of cinnamon contained camphor (56.2%) and 1,8-cineole (11.7%) as chief components.

FLOWER OIL The buds and flowers of cinnamon contained 0.04% volatile oil. The oil contained (*E*)-cinnamyl acetate (22%) as the chief constituent; β-caryophyllene (9.8%), humulene epoxide-1 (5%), α-muurolol (4.9%), linalool (3.6%), α-cadinol (2.4%), α-humulene (2.2%) and δ-cadinene (2.2%) were the minor components (Kaul *et al.*, 2003). Cinnamon flowers from Karnataka, India, recorded 0.5% (v/w) volatile oil (Jayaprakasha *et al.*, 2000). The oil consisted of 23% hydrocarbons and 74% oxygenated compounds and was dominated by (*E*)-cinnamyl acetate (41.98%), *trans*-α-bergamotene (7.97%), caryophyllene oxide (7.29%), α-cadinol (6.35%), tetradecanal (5.05%), globulol (3.8%), α-copaene (3.3%), benzyl benzoate (3.19%), δ-cadinene (2.97%), *α*-humulene (2.4%) and *n*-heptadecane (2.14%) (Jayaprakasha *et al.*, 2000).

FRUIT OIL The fruits and pedicels of fruits of cinnamon yielded 0.32 and 0.33% volatile oil, respectively (Kaul *et al.*, 2003). The fruit oil was dominated by 43.4% (*E*)-cinnamyl acetate and 27.4% linalool (Table 7.5). The chief constituents of oil from the pedicels of fruits were (*E*)-cinnamyl acetate (58.1%), linalool (13.1%) and *β*-caryophyllene (11.1%). The oil from the pedicels of fruits had a higher level of (*E*)-cinnamyl acetate and a lower level of linalool compared with that of fruits (Kaul *et al.*, 2003). The fruit oil of cinnamon grown in South India consisted of 20.8–32.8% hydrocarbons and 63.7–73.4% oxygenated compounds. *trans*-Cinnamyl acetate (42–54%) and *β*-caryophyllene (9–14%) were the major compounds in the oil (Jayaprakasha *et al.*, 1997). Syamasundar *et al.* (2000) studied the variation in the composition of unripe and ripe fruits of cinnamon. The oil from unripe fruits was dominated by δ-cadinene (19.15%), *α*-pinene (11.47%), *β*-pinene (10.51%), (*E*)-cinnamyl acetate (7.11%) and γ-cadinene (8.05%), whereas the ripe fruits contained γ-cadinene (23.48%), *α*-pinene (11.52%), (*E*)-cinnamyl acetate (8.62%) and *α*-muurolene (8.22%) as chief components.

In the fruit oil of *C. zeylanicum* from Bangalore (South India), *α*-pinene (11.2%), *β*-pinene (9.2%), *β*-caryophyllene (11.0%), *α*-muurolene (6.1%), γ-cadinene (7.1%), δ-cadinene (13.6%) and *α*-muurolol (9.8%) predominated (Mallavarapu and Ramesh, 2000). The constituents of essential oil from various parts of *C. verum* are indicated in Tables 7.4 and 7.5.

Chinese cassia (C. cassia)

Bark and leaf oils were extracted from *C. cassia*. Both the oils contained mainly cinnamaldehyde. The physico-chemical properties of cassia oil are indicated in Table 7.6.

BARK OIL The bark oil content in *cassia* ranged from 1.2 to 4.9%. The principal component in the oil, cinnamaldehyde, varied between 61.5 and 91% (Krishnamoorthy *et al.*, 1999). The bark oil from Nigeria contained a high level of benzyl benzoate (Lockwood, 1979). Headspace composition of cinnamon and cassia quills of different origins showed that the cinnamaldehyde and benzaldehyde contents were in the range 2.3–86.2 and 0.5–40.5%, respectively (Vernin *et al.*, 1994).

Jayatilake *et al.* (1995) examined the composition of bark oil from 25 samples of *C. cassia* and the major components identified were (*E*)-cinnamaldehyde (92–98%), (*Z*)-cinnamaldehyde (0.8–2.7%), *β*-caryophyllene (0.4–3.6%), coumarin (0.1–1.6%) and *α*-ylangene (0.1–2.7%). Analysis of Chinese cassia oil by the HPLC method and supercritical CO_2 extraction indicated cinnamaldehyde content as 68.2–71.9 and 73.9–74.4%, respectively (Ehlers *et al.*, 1995; Lawrence, 2001).

Evaluation for chemical constituents in open-pollinated seedling progenies of *C. cassia* accessions from Calicut (India) showed that these contained 1.20–4.95% bark oil, 6.0–10.5% bark oleoresin and 0.40–1.65% leaf oil. The principal component of both the oils, namely, cinnamaldehyde, varied from 40.7–86.0 and 61.9–91.5%, respectively, in leaf and bark oils (Krishnamoorthy *et al.*, 1999). The bark oil of *C. cassia* from the Yunnan province was dominated by cinnamaldehyde (80.40–88.50%) (Li *et al.*, 1998). The bark oil from China recorded 65.5% *E*-cinnamaldehyde, 8.7% coumarin, 3.6% cinnamyl acetate and 2.7% 2-methoxy cinnamaldehyde as chief components, whereas in the Australian oil, cinnamaldehyde (87%), benzaldehyde (4.7%), 2-phenyl ethanol (2.5%) and 3-phenyl propanal (2%) predominated (Vernin *et al.*, 1990). Li and Yuan (1999) reported cassia oil from China containing 67.12% *E*-cinnamaldehyde, 6.17% methyl salicylate, *E*-2-methoxy cinnamaldehyde (7.40%) and (*E*)-cinnamyl acetate (3.47%) as major components.

LEAF OIL Cassia leaves yielded 0.4–1.6% volatile oil and the major constituent in the oil, namely, cinnamaldehyde, ranged from 40.7 to 86% (Krishnamoorthy *et al.*, 1999). The leaf oil from Nigeria was rich in benzyl benzoate (Lockwood, 1979). The leaf oil of Australian cassia recorded 77.2% cinnamaldehyde,

Table 7.4. Volatiles from *Cinnamomum verum.*

Sl no.	Compound	Leaf[1]	Stem bark[2]	Root bark[2]	Flower[3]	Fruit[4]
1	*n*-Hexanol	–	t	–	–	–
2	*n*-Hexane-2-ol	+	–	–	–	–
3	(*Z*)-Hex-3-en-1-ol	–	–	–	+	–
4	Heptan-2-one	+	–	–	–	–
5	α-Thujene	+	–	–	–	–
6	α-Pinene	+	+	+	–	+
7	Camphene	+	+	+	–	+
8	Sabinene	+	+	+	–	–
9	β-Pinene	+	+	+	–	+
10	Myrcene	+	+	+	–	+
11	α-Phellandrene	+	+	+	–	+
12	Δ-3-Carene	+	+	t	–	–
13	α-Terpinene	+	+	+	–	+
14	γ-Terpinene	+	+	+	–	+
15	*p*-Cymene	+	+	+	–	+
16	1,8-Cineole	+	+	+	–	+
17	Limonene	–	+	+	–	+
18	(*E*)-β-Ocimene	+	+	+	–	+
19	(*Z*)-β-Ocimene	–	–	–	–	+
20	*trans*-Ocimene	–	t	+	–	–
21	*Cis*-Ocimene	–	t	+	–	–
22	2-Phenyl ethyl benzoate	–	t	–	+	–
23	Methyl cinnamate	–	t	t	–	–
24	Fenchone	–	t	+	–	–
25	Furfural	–	+	+	–	–
26	*cis*-Linalool oxide (furanoid)	+	–	–	–	–
27	Terpinolene	+	+	+	–	+
28	*t*-Linalool oxide (furanoid)	+	+	+	–	–
29	Linalool	+	+	+	–	+
30	*α*-Ylangene	–	+	–	–	–
31	Linalool acetate	–	t	+	–	–
32	Bornyl acetate	–	+	+	–	–
33	2-Phenyl ethyl alcohol	+	+	t	–	–
34	Camphor	+	t	+	–	–
35	Citronellal	+	–	–	–	–
36	2-Phenyl acetaldehyde	–	t	+	–	–
37	Borneol	+	+	+	+	+
38	Methyl chavicol	+	t	+	–	–
39	Methyl eugenol	–	t	–	–	–
40	Methyl isoeugenol	–	–	t	–	–
41	Terpinen-4-ol	+	+	+	–	+
42	*β*- Caryophyllene	+	+	+	–	–
43	Ethyl cinnamate	–	–	t	–	–
44	Methyl cinnamate	–	t	t	–	–
45	*Z*-Methyl cinnamate	+	–	–	–	–
46	(*Z*)-Cinnamaldehyde	+	–	–	–	–
47	Cinnamaldehyde	–	+	+	–	–
48	(*E*)-Cinnamaldehyde	–	+	+	+	–
49	Nerol	+	t	–	–	+
50	Geraniol	–	+	+	–	+
51	Piperitone	+	+	+	–	–
52	Safrole	+	t	+	–	–
53	Benzyl alcohol	–	t	t	–	–
54	Eugenol	+	+	+	–	–

Continued

Table 7.4. *Continued*

Sl no.	Compound	Leaf[1]	Stem bark[2]	Root bark[2]	Flower[3]	Fruit[4]
55	Isoeugenol	–	+	–	–	–
56	Acetyl eugenol	–	+	+	–	–
57	(*Z*)-Cinnamyl acetate	+	–	–	–	+
58	(*E*)-Cinnamyl acetate	–	–	–	+	+
59	Cinnamyl acetate	–	+	t	–	–
60	Cinnamyl alcohol	–	+	+	–	–
61	Farnesol	–	+	–	–	–
62	(*E*)-β-Farnesene	+	–	–	–	+
63	Eugenyl acetate	+	–	–	–	–
64	α-Humulene	–	+	+	–	+
65	Cadalene	–	–	–	+	–
66	*epi*-x-Bisabolol	–	–	–	+	–
67	*n*-Heptadecane	–	–	–	+	–
68	2-Heptadecanone	–	–	–	+	–
69	α-Selinene	+	–	–	–	–
70	*β*-Selinene	–	t	–	–	–
71	δ-Cadinene	+	–	–	+	+
72	α-Cadinene	–	–	–	–	+
73	γ-Cadinene	–	–	+	–	–
74	Geranial	–	t	–	–	–
75	(*E*)-Nerolidol	+	–	–	–	+
76	Nerolidol	–	–	–	+	–
77	Spathulenol	+	–	–	–	+
78	Caryophyllene oxide	-	t	+	+	+
79	Isocaryophyllene oxide	–	–	–	–	+
80	β-Caryophyllene oxide	+	t	+	–	–
81	*β*-Caryophyllene	–	–	–	–	+
82	Humulene epoxide	+	–	–	–	+
83	*t*-Cadinol	+	–	–	–	–
84	α-Cadinol	+	–	–	+	+
85	α-*n*-Hexyl cinnamaldehyde	+	–	–	–	–
86	Geranyl acetate	–	t	–	–	–
87	Geranyl benzoate	+	–	–	–	–
88	Phenylethyl-*n*-decanoate	+	–	–	–	–
89	Phenylethylanthranilate	+	–	–	–	–
90	Benzaldehyde	–	+	–	+	–
91	Cuminaldehyde	–	+	+	–	–
92	Hydrocinnamaldehyde	–	+	+	+	–
93	Hydrocinnamic acid	–	–	t	–	–
94	α-Terpineol	–	+	+	+	+
95	Phenyl ethyl acetate	–	+	t	–	–
96	3-Phenyl propyl acetate	–	+	t	+	–
97	α-Copaene	–	–	–	+	+
98	*E*-Cinnamyl alcohol	–	–	–	+	–
99	*t*-Bergamotene	–	–	–	+	–
100	α-Humulene	–	+	+	+	–
101	Germacrene D	–	–	–	+	+
102	Globulol	–	–	–	+	+
103	Tetradecanal	–	–	–	+	–
104	Phenol	–	t	t	–	–
105	2-Vinyl phenol	–	+	–	–	–
106	Coumarin	–	+	–	–	–

Continued

Table 7.4. *Continued*

Sl no.	Compound	Leaf[1]	Stem bark[2]	Root bark[2]	Flower[3]	Fruit[4]
107	Vanillin	–	t	–	–	–
108	Benzyl benzoate	–	+	+	+	–
109	(*E*)-2-Hexenal	–	–	–	–	+
110	Tricyclene	–	–	–	–	+
111	α-Muurolene	–	–	–	–	+
112	γ-Muurolene	–	–	–	–	+
113	*cis*-Calaminene	–	–	–	–	+
114	Elemol	–	–	–	–	+
115	*B*-Elemene	–	–	–	–	+
116	α-Fenchyl alcohol	–	–	–	–	+
117	Isoborneol	–	–	–	–	+
118	Isobornyl acetate	–	–	–	–	+
119	1-*epi*-Cubebenol	–	–	–	–	+
120	*t*-Cadinol	–	–	–	–	+
121	Cubenol	–	–	–	–	+
122	α-Muurolol	–	–	–	–	+
123	Selin-11-en-4a-ol	–	–	–	–	+
124	4-Hydroxy-3, 4-dihydrocalacorene	–	–	–	–	+

t = trace.
[1]Raina *et al.* (2001); [2]Senanayake (1977); [3]Jayaprakasha *et al.* (2000); [4]Jayaprakasha *et al.* (1997).

Table 7.5. Volatile constituents of essential oils from different parts of cinnamon.

Compound	Area (%)				
	Tender twigs	Pedicels of buds + flowers	Buds + flowers	Pedicels of fruits	Fruits
α-Thujene	0.2	0.1	–	0.1	0.2
α-Pinene	2.3	0.7	0.1	0.8	4.2
Camphene	1.0	0.2	–	0.4	1.1
Sabinene	0.1	–	–	–	0.1
β-Pinene	0.9	0.4	–	0.5	1.9
Myrcene	0.4	0.3	–	0.3	0.5
α-Phellandrene	3.4	2.2	0.2	1.7	2.3
δ-3-Carene	0.1	0.1	–	–	0.1
α-Terpinene	0.2	0.1	–	0.1	0.2
p-Cymene	0.6	0.4	–	0.2	0.2
Limonene	1.6	0.9	–	0.8	1.5
1,8-Cineole	0.1	0.1	0.1	0.2	0.2
(*Z*)-β-Ocimene	0.1	0.1	0.2	0.1	0.1
(*E*)-β-Ocimene	0.1	0.1	0.1	0.1	0.2
γ-Terpinene	0.1	0.1	–	0.1	0.1
Terpinolene	0.3	0.3	–	0.2	0.3
Linalool	15.2	11.3	3.6	13.1	27.4
Borneol	0.3	0.3	0.2	0.4	0.5
Terpinen-4-ol	0.1	0.1	0.2	0.1	0.1
α-Terpineol	0.8	0.6	0.5	0.8	1.0
(*E*)-Cinnamaldehyde	4.0	0.8	0.2	0.9	–
Eugenol	0.8	0.1	1.0	0.3	0.1
Neryl acetate	1.2	2.0	0.6	1.4	0.9

Continued

Table 7.5. *Continued*

Compound	Tender twigs	Pedicels of buds + flowers	Buds + flowers	Pedicels of fruits	Fruits
	Area (%)				
(*Z*)-Cinnamyl acetate	2.0	0.7	6.1	1.5	1.1
β-Elemene	0.1	0.1	0.8	0.1	0.1
(*E*)-Cinnamyl acetate	49.4	64.5	22.0	58.1	43.4
β-Caryophyllene	8.3	9.6	9.8	11.1	6.9
α-Humulene	0.8	0.8	2.2	1.1	0.9
Germacrene D	0.1	0.1	0.2	–	0.1
α-Muurolene	0.1	0.2	0.9	0.4	0.5
δ-Cadinene	0.4	0.3	2.2	0.4	0.3
Caryophyllene oxide	0.3	0.2	1.4	0.4	0.2
Globulol	–	–	0.4	0.1	0.1
Humulene epoxide I	0.1	0.3	5.0	0.3	0.3
α-Muurolol	0.2	0.2	4.9	0.4	0.7
α-Cadinol	0.2	0.2	2.4	0.3	0.3

Source: Kaul *et al.* (2003).

Table 7.6. Physico-chemical properties of *Cinnamomum cassia* oil.

Specification	*C. cassia* (at 15°C)	Indonesian cassia (at 30°C)	Indian cassia (at 30°C) HP	Indian cassia (at 30°C) UP
Yield of oil (%)		1.32	–	–
Specific gravity	1.055–1.070	0.9593	0.9349	0.9563
Refractive index	1.6000–1.6060	1.5215	1.4090	1.5038
Optical rotation (°)	1–6	+19	+1.30	+0.12
Solubility in ethanol	Readily soluble in 80% alcohol	Soluble in 1:1 alcohol	–	–
Aldehyde (total %)	75–90	27.15	–	–

15.3% coumarin and 3.6% cinnamyl acetate as chief constituents (Senanayake, 1977). Zhu *et al.* (1993) determined the leaf oil composition of *C. cassia* from China. Fifteen components contributing to 96.5% of oil were identified, among which cinnamaldehyde (74.1%), 2-methoxycinnamaldehyde (10.5%) and cinnamyl acetate (6.6%) were the major ones. The constituents from the cassia oil are listed in Table 7.7.

The chief constituents of leaf oil of cinnamon cassia from Yunnan province were cinnamaldehyde (64.1–68.3%), (*E*)-2-methoxy cinnamaldehyde (8.4–10.5%) and (*E*)-cinnamyl acetate (4.5–12.5%) (Li *et al.*, 1998). The structures of the chief volatile components of cinnamon oil are shown in Fig. 7.1.

Indonesian cassia (C. burmannii)

Indonesian cassia of commerce is the dried bark of *C. burmannii*. Most of the bark produced is exported and domestic consumption is very small. The main importing countries are the USA, Germany and the Netherlands. The chemical composition of Indonesian cassia was thought to be similar to that of Chinese cassia, but later studies showed the existence of various chemotypes.

BARK OIL Bark yields 0.5–2.0% volatile oil. It is a colourless to brownish-yellow liquid having an odour similar to that of Ceylon cinnamon bark oil (Purseglove *et al.*, 1981). The oil contained 80–95% cinnamaldehyde (Guenther,

Table 7.7. Volatiles from *Cinnamomum cassia*.

Sl no.	Compound	Leaf oil	Bark oil
1	α-Pinene	+	+
2	Camphene	+	+
3	β-Pinene	+	+
4	Myrcene	+	t
5	α-Phellandrene	+	t
6	Limonene	+	+
7	1,8-Cineole	+	+
8	δ-3-Carene	+	t
9	*p*-Cymene	+	+
10	Camphor	+	+
11	Benzaldehyde	+	+
12	Linalool	+	+
13	Terpinolene	t	+
14	β-Caryophyllene	+	t
15	α-Humulene	t	+
16	β-Elemene	–	t
17	Isoborneol	+	+
18	Borneol	+	+
19	α-Terpineol	t	+
20	Geraniol	t	+
21	Carvone	+	+
22	2-Methoxy benzaldehyde	+	+
23	Safrole	–	t
24	γ-Elemene	t	+
25	δ-Cadinene	t	t
26	β-Cadinene	–	t
27	Hydro cinnamaldehyde	+	+
28	Phenyl acetaldehyde	+	t
29	Methyl eugenol	+	t
30	(*E*)-Cinnamaldehyde	+	+
31	α-Copaene	+	+
32	Vanillin	t	t
33	Salicylaldehyde	+	+
34	2-Phenenethyl alcohol	+	t
35	Benzyl alcohol	t	–
36	Acetophenone	t	+
37	Eugenol	+	+
38	(*Z*)-Isoeugenol	+	+
39	(*E*)-Cinnamyl acetate	+	+
40	γ-Muurolene	t	t
41	Anisaldehyde	+	t
42	2-Phenethyl acetate	t	–
43	β-Bisabolene	t	t
44	β-Bisabolol	t	t
45	α-Muurolol	+	+
46	Coumarin	+	t
47	(*E*)-Cinnamic acid	+	+
48	(*E*)-2-Methoxycinnaldehyde	+	t
49	Hydrocinnamic acid	+	+
50	4-Hydroxy-2-phenethyl alcohol	+	+
51	Caryophyllene oxide	+	+
52	Patchoulene	+	+
53	Octanoic acid	t	t

Continued

Table 7.7. *Continued*

Sl no.	Compound	Leaf oil	Bark oil
54	3-Phenylpropyl acetate	+	+
55	Nonanoic acid	t	t
56	Guaicol	t	+
57	(*E*)-Cinnamyl alcohol	+	+
58	(*E*)-Ethyl cinnamate	+	t
59	Benzyl benzoate	+	t
60	Methyl alaninate	t	–
61	Guaicyl cinnamate	t	t
62	Decanoic acid	t	t
63	Undecanoic acid	+	+
64	Dodecanoic acid	t	t
65	Benzoic acid	+	+
66	Salicylic acid	t	+

t = trace.
Source: Li *et al.* (1998).

1950). Lawrence (1967) also reported the presence of α-terpineol, coumarin and benzaldehyde in bark oil. Xiao-duo *et al.* (1991) identified a chemotype having 1,8-cineole (51.4%) as the major constituent in bark oil. α-Terpineol, camphor and terpenen-4-ol were other prominent components of this oil.

LEAF OIL The leaves yield 0.4–0.9% oil and the oil composition shows wide variation. Yu-Jing *et al.* (1987) studied the composition of the leaf oil of *C. burmannii* and found that borneol (70.8%) was the chief component. 1,8-Cineole, bornyl acetate and 4-carene were the other dominant compounds. Later, Xiao-duo *et al.* (1991) identified a chemotype having 1,8-cineole (28.5%) as the major constituent. Borneol (16.5%), α-terpineol (6.4%), *p*-cymene (6.1%), spathulenol, terpenen-4-ol, bornyl acetate and β–caryophyllene were other prominent components of this leaf oil. Chen *et al.* (1997) reported *C. burmannii* leaf oil containing 96.3–99.7% safrole.

Indian cassia

Indian cassia is known as *tejpat* in Hindi. The leaves of *C. tamala*, *C. sulphuratum*, *C. bejolghota* and *C. impressinervium* are also traded as 'tejpat' in North-east India. The leaf and bark of Indian cassia yield 0.03–2.50% oil (Lewis *et al.*, 1977; Bradu and Sobti, 1988; Hussain *et al.*, 1988; Nath *et al.*, 1994b; Ahmed *et al.*, 2000). The physico-chemical properties of *C. cassia* leaf oil are indicated in Table 7.6. The chemical composition of tejpat oil was studied by several workers and the results exhibited variations. Several chemotypes of tejpat oil have been identified based on the leaf oil composition.

C. TAMALA The leaves of *C. tamala* yield 0.1–0.2% essential oil. Several chemotypes have been identified in this species. The essential oil of *C. tamala* fruit from the Brahmaputra Valley was dominated by eugenol (73.6%) (Baruah *et al.*, 2004). Bradu and Sobti (1988) reported *C. tamala* leaf oil with cinnamaldehyde and linalool as major constituents. Nath *et al.* (1994b) recorded oil rich in linalool (60.7%). Another chemotype of *C. tamala* leaf oil rich in eugenol was reported from North-east India (Nath *et al.*, 1999). Oil from Pakistan contained β-caryophyllene (25.3%), linalool (13.4%) and caryophyllene oxide (10.3%) as major constituents (Ahmed *et al.*, 2000). Two chemotypes of *C. tamala*, namely, cinnamaldehyde type and eugenol type, were reported by Hussain *et al.* (1988). *C. tamala* from Nepal contained linalool 54.7%, α-pinene (9.7%), *p*-cymene (6.4%), β-pinene (4.5%) and limonene (2.6%) as major components.

C. SULPHURATUM *C. sulphuratum* is another *Cinnamomum* species that is traded as 'tejpat'. Nath *et al.* (1994a) reported 92.66% linalool

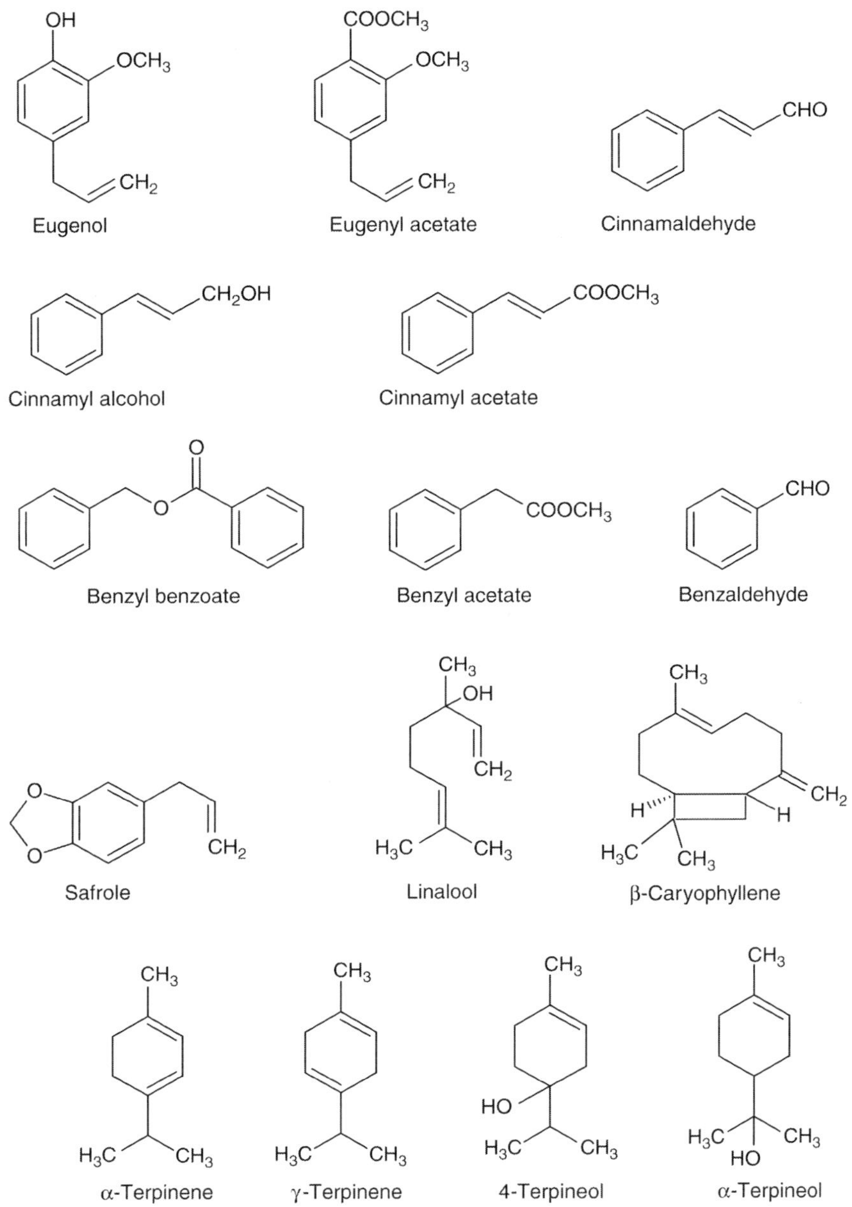

Fig 7.1. Major volatile components of cinnamon.

followed by geranial (2.2%) and citrotronellol (1.47%) in leaf oil from North-east India. Baruah *et al.* (1999) identified a chemotype rich in (*E*)-cinnamaldehyde (65.6%) in bark oil and neral (17.6%), geranial (27.8%) and genaniol (23.2%) as the major components in leaf oil. Another report by Baruah *et al.* (2001) indicated the existence of a variant with methyl cinnamate as the chief constituent in leaf (72.4%) and stem bark oils (83.9%). A chemotype predominated by cinnamaldehyde in the oil of both leaf and stem bark was reported by Baruah *et al.* (2002). This variant from North-east India contained cinnamaldehyde (50%),

1,8-cineole (14.6%), camphor (11%), linalool (7.04%), camphene (1.70%), α-pinene (0.83%), β-pinene (1.06%), myrcene (0.15%), γ-terpinene (0.80%), borneol (0.41%) and terpinene-4-ol (0.47%) in its leaf oil. Its stem bark contained cinnamaldehyde (64.3%), cinnamyl alcohol (1.93%), borneol (1.6%), linalool (1.3%) and 1,8-cineole (1.23%), besides other minor constituents.

C. BEJOLGHOTA Baruah *et al.* (1997) investigated the composition of essential oils of *C. bejolghota* from North-east India. They identified 27 components in the leaf and panicle essential oils, representing 92.8% of total oils. Linalool was the major component of leaf essential oil (57.4%) and panicle oil (68.2%). The stem bark oil contained α-terpineol (18.2%) and (*E*)-nerolidol (15.3%) as main components.

Essential oils from the bark and flowers of *C. bejolghota* from two locations in Assam were analysed by Choudhury *et al.* (1998). Bark from both areas yielded 0.08% essential oil. The essential oil yield in flowers was 0.13% from Jorhat and 0.6% from Sibsagar. The major constituents of the bark oils from Jorhat were 1,8-cineole (31.3%), α-terpineol (21.3%) and linalool (20%), but Sibsagar oil was predominated by 19.9% linalool, 12.7% α-terpineol and 7.2% 1,8-cineole. The flower oil from Jorhat and Sibsagar constituted mainly α-pinene and β-pinene.

C. IMPRESSINERVIUM The leaf oil of *C. impressinervium* cultivated in North-east India constituted 8.3% eugenol (Nath and Baruah, 1994). Cinnamaldehyde and eugenol type oil have been reported from *C. impressinervium* (Kya and Min, 1970; Nath and Baruah, 1994). Table 7.8 reveals the constituents of Indian cassia oil.

Table 7.8. Chemical constituents of Indian cassia oil.

Sl no.	Compound	Leaf oil[1]	Fruit oil[2]
1	α-Pinene	+	–
2	β-Pinene	+	–
	Car-3-ene	–	–
3	Benzaldehyde	+	–
4	Linalool	+	t
5	Benzyl acetate	+	–
6	Linalool acetate	+	–
7	α-Terpineol	+	+
8	Terpenene-4-ol	–	+
9	Geraniol	+	–
10	Limonene	+	–
11	1,8-Cineole	–	+
12	Cinnamic aldehyde	+	t
13	Eugenol	+	+
14	Benzyl cinnamate	+	–
15	α-Phellandrene	+	–
16	*p*-Cymene	+	–
17	Ocimene	+	–
18	α-Terpinene	+	–
19	Camphor	+	+
20	Borneol	+	t
21	β-Caryophyllene	+	+
20	Cadinene	+	–
22	Eugenol acetate	+	–
23	Safrole	+	–
24	Caryophellene oxide	–	+
25	Methyl cinnamate	–	+
26	Ethyl cinnamate	–	+
27	Eugenylacetate	+	+
28	Methyl eugenol	–	t
29	Isoeugenol	–	+
30	α-Humulene	–	+
31	α-Farnesene	–	+

t = trace.
Source: [1]Bradu and Sobti (1988); [1]Nath *et al.* (1994b); [1]Nath *et al.* (1999) [2]Baruah *et al.* (2004).

Non-volatiles

Phytochemistry of cinnamon and cassia deals mainly with the volatile components. Besides essential oil constituents, several diterpenes have been isolated from the genus. These include cinncassiols A, B, C_1 and their glucosides, cinncassiols C_2 and C_3, cinncassiols D_1, D_2 and D_3 and their glucosides, cinncassiol E, cinnzeylanol, cinnzeylanin, anhydrocinnzeylanol and anhydrocinnzeylanin. In addition to these, several benzyl isoquinoline alkaloids, flavanol glucosides, coumarin, β-sitosterol, cinnamic acid, protocatechuic acid, vanillic acid and syringic acid have been isolated from the aqueous extract of *C. cassia*. Some important secondary metabolites isolated from *Cinnamomum* and cassia are listed in Table 7.9. Figure 7.2

Table 7.9. Non-volatile constituents from Cinnamon and cassia.

Sl no.	Compound	Plant part	References
1	Lyoniresinol 3α-*O*-β-D-glucopyranoside	*C. cassia* stem bark	Miyamura *et al.*, 1983
2	3,4,5-Trimethoxy phenol β-D apiofuranosyl (1–6)-β-D-glucopyranoside	"	"
3	Syringaresinol	"	"
4	5,7,3′-Trimethyl (–)-epicatchin	"	"
5	5,7-Dimethyl-3′, 4′-di-*O*-methylene (±) epicatchin	"	"
6	Cinnamic aldehyde cyclic glycerol-1,3-acetol (9,2′-trans)	"	"
7	Cinnamic aldehyde cyclic glycerol-1,3-acetol (9,2′-cis)	"	"
8	Cinncassiol D_4	"	Nohara *et al.*, 1982
9	Cinncassiol-D_4-glucoside	"	"
10	2'-Hydroxy cinnamaldehyde	"	Kwon *et al.*, 1996
11	3-(2-Hydroxy phenyl)-propanoic acid	"	Tanaka *et al.*, 1989
12	Cinncassiol E	*C. cassia*	Nohara *et al.*, 1985

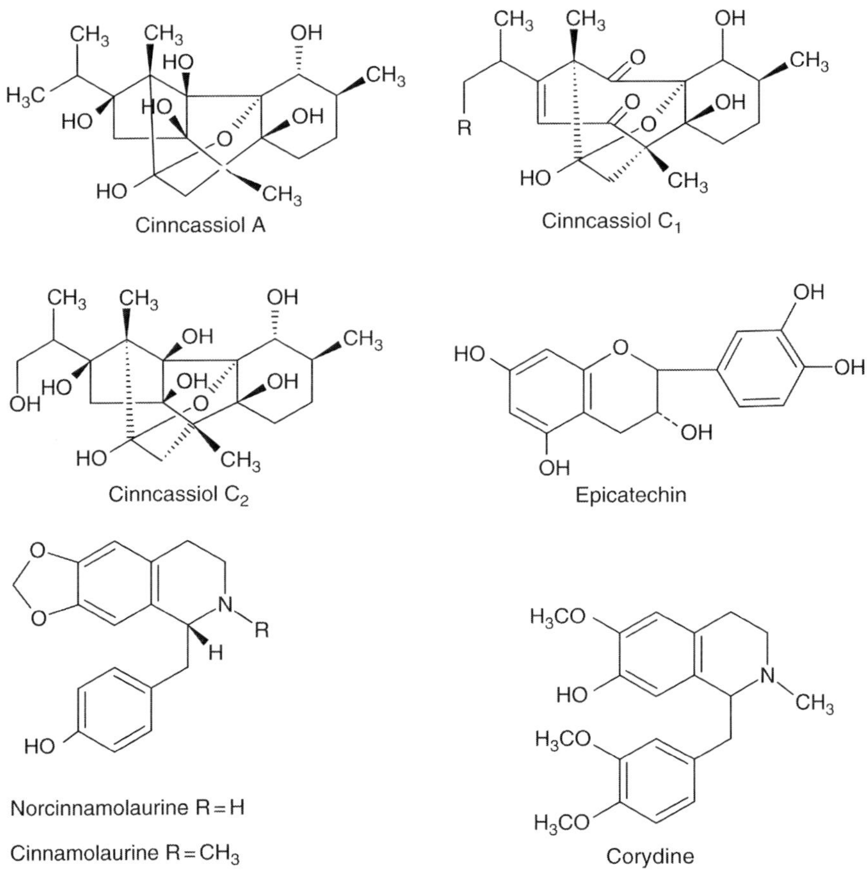

Fig. 7.2. Non-volatiles from cinnamon.

shows the structures of a few representative non-volatiles of cinnamon and cassia.

7.5. Medicinal and Pharmacological Uses

Cinnamon bark is used widely as a spice. It has been used as a flavouring agent since ancient times. The ground spice is used for flavouring baked products such as cakes, biscuits, puddings, chewing gum and desserts. The bark oil and leaf oil are used in the manufacture of perfumes, soaps and toothpastes, and also as a flavouring agent for liquors and in dentifrices. The leaf oil is also used as a source of eugenol for the production of synthetic vanillin and isoeugenol. Besides these, cinnamon and cassia have a broad spectrum of medicinal and pharmacological application.

Cinnamon

Cinnamon is used as an ingredient in many 'Ayurvedic' and 'Unani' medicinal preparations. The bark of *C. zeylanicum* is an aphrodisiac, anthelmintic and tonic. It is useful in the treatment of *vata*, biliousness, parched mouth, bronchitis, diarrhoea, itching, heart disease and urinary disease. The bark is a carminative and expectorant; it is useful in hydrocoele, flatulence, headache, piles, etc. (Kirtikar and Basu, 1984). Cinnamon possesses various biological activities, such as antioxidant, antimicrobial, antidiabetic and antiallergic.

Antioxidant activity

For many centuries, cinnamon and its essential oil have been used as preservatives in food, due to the antioxidant property of cinnamon. Deterioration of food is due to lipid peroxidation. *In vivo* lipid peroxidation causes tissue damage, which can lead to inflammatory diseases. Phenolic compounds, such as hydroxy cinnamaldehyde and hydroxycinnamic acid, present in the cinnamon extract, act as scavengers of peroxide radicals and prevent oxidative damages (Wu *et al.*, 1994).

Anti-inflammatory activity

Cinnamon is reported to possess anti-inflammatory activity (Kirtikar and Basu, 1975). The ethanolic extract (70%) of cinnamon was effective on acute inflammation in mice (Kubo *et al.*, 1996). A herbal ophthalmic preparation, called 'Opthacare', containing 0.5% cinnamon was found to be effective as an anti-inflammatory agent on ocular inflammation in rabbits (Mitra *et al.*, 2000).

Antidiabetic activity

In Ayurveda and folklore medicines, cinnamon is used in the treatment of diabetes. Cinnamon is reported to reduce the blood glucose level in non-insulin-dependent diabetics. Therapeutic studies have proved the potential of cinnamaldehyde as an antidiabetic agent. Cinnamaldehyde inhibits aldose reductase, a key enzyme involved in the 'polyol' pathway. This enzyme catalyses the conversion of glucose to sorbitol in insulin-insensitive tissues in diabetic patients. This leads to accumulation of sorbitol in chronic complications of diabetes, such as cataract, neuropathy and retinopathy. Aldose-reductase inhibitors prevent conversion of glucose to sorbitol, thereby preventing several diabetic complications (Lee, 2002).

Antipyretic and analgesic effects

A decoction of dried twigs of cinnamon can produce an antipyretic effect in mice. Studies conducted in anaesthetized dogs and guinea pigs indicated that cinnamaldehyde, or sodium cinnamate, also produced the hypothermic and antipyretic effects (*Chinese Materia Medica*, 1996). It also causes a hypotensive effect, which is due mainly to vasodilation of peripheral vessels. Cinnamaldehyde produced an analgesic effect in mice (Wang, 1985).

Immunological effects

Nephritis is an autoimmune disease caused by activation of the complement system. Cinnamon cortex and cinnamon oil inhibited complement formation *in vitro*. Cinncassiol C_1 and its glucoside, the cinncassiols C_2 and C_3 and cinncassiol D_1 and its glucoside were reported to possess anticomplementary activity. A water-soluble polysaccharide isolated from the cinnamon extract showed complement system activity (Tang and Eisenbrand, 1992). 2-Hydroxycinnamaldehyde and 2-benzyloxy cinnamaldehyde isolated from the stem bark of cinnamon possessed immunomodulatory effects (Koh *et al.*, 1999).

Antibacterial activity

Cinnamon bark (*C. zeylanicum*) oil showed an inhibitory effect against the Gram-positive bacteria *Bacillus cereus*, *Micrococcus luteus*, *Staphylococcus aureus* and *Enterococcus faecalis*; the Gram-negative bacteria *Alcaligenes faecalis*, *Enterobacter cloacae*, *Escherichia coli* and *Pseudomonas aeruginosa*; the fungi *Aspergillus niger* and *Rhizopus oligosporus*; and the yeast *Candida albicans* (Chao *et al.*, 2000). Cinnamaldehyde possessed strong antibacterial activity against nine strains of bacteria, including *E. coli*, *P. aeruginosa*, *E. faecalis*, *S. aureus*, *Staphylococcus epidermidis*, methicillin-resistant *S. aureus* (MRSA), *Klebsiella pneumoniae*, *Salmonella* sp. and *Vibrio parahaemolyticus*. Cinnamaldehyde is beneficial to human health, having the potential to be used for medical purposes and to be utilized as an antibacterial additive in making paper products (Chang *et al.*, 2001).

Antimicrobial activities

Cinnamon oil and extracts possess various antimicrobial activities against several bacteria, fungi, etc. Aqueous extract from cinnamon (*C. zeylanicum*, Blume) inhibited the replication of the influenza virus (Mancini *et al.*, 1999). Oral administration of *t*-cinnamaldehyde and *O*-methoxycinnamaldehyde inhibited candidiasis at a MIC (minimum inhibitory concentration) value of 0.03–0.05 mg/ml. Cinnamaldehyde was an effective antifungal agent against *Alternaria alternata*, a plant pathogen. It exhibited strong antifungal activity against the fungus *Malassexia furfur* (Ferhout *et al.*, 1999). It also exhibited strong nematicidal activity against the root-knot nematode, *Meloidogyne javanica*; EC_{50} values for juvenile immobilization and hatching inhibition *in vitro* were 15 and 11.3 μl/l, respectively (Oka, 2001). Cinnamaldehyde, cinnamic acid, cinnamyl alcohol and eugenol possessed antibacterial, astringent, carminative and stomachic effects (Lee and Ahn, 1998).

Insecticidal activity

Cinnamon oil exhibited fumigant toxicity to adults of *Acanthoscelides oblectus* and inhibited its reproduction through ovicidal and larvicidal action. Both cinnamaldehyde and cinnamyl alcohol showed ovicidal and larvicidal activity (Roger and Hamraoui, 1994). Cinnamaldehyde possessed antifeedant activity against *Ceratitis capitata*, a pest causing damage to fruit crops (Moretti *et al.*, 1998).

Nematicidal activity

Cinnamon oil possessed strong nematicidal activity against the male, female and juveniles of pinewood nematode *Bursaphelenchus xylophilus* (Park *et al.*, 2005). Cinnamyl acetate, the active ingredient in the oil, at a concentration of 32.81 μg/l resulted in 50% mortality of nematodes.

Cassia

The bark of *C. cassia* is a tonic, stomachic and carminative. The bark is useful in inflammation, headache and piles. *C. cassia* bark and its essential oil are used in various medicinal preparations. The essential oil from the stem bark and cinnamaldehyde regulates the triggering of hepatic drug-metabolizing enzymes by the formation of a glutathione-conjugate (Choi *et al.*, 2001). The bark exhibits promising anticancerous and antitumour activities.

Anti-ulcerogenic activity

Aqueous extract of *C. cassia* bark prevented stress-induced ulcers in rats (Akira *et al.*, 1986). Tanaka *et al.* (1989) isolated 3-(2-hydroxy phenyl)-propanoic acid and its *O*-glucoside from *C. cassia*, which prevented ulcerogenesis. The anti-ulcerogenic effect was attributed to the improvement in gastric blood flow and gastric cytoprotection. Other compounds associated with the anti-ulcerogenic property of cinnamon are cassioside, cinnamoside and 3,4,5-trimethoxyphenol-β-D-apiofuranosyl-(1→6)-β-D-glucopyranoside (Shiraga *et al.*, 1988).

The extract from *C. cassia* bark inhibited skin and pulmonary tumours. The extract also inhibited the growth of gastric and colon cancer lines (Jin Wong *et al.*, 1994). 2'-Hydroxycinnamaldehyde isolated from the stem bark of *C. cassia* inhibited the activity of farnesyl protein transferase, an enzyme involved in the initiation of tumour formation (Kwon *et al.*, 1996). 2'-Hydroxycinnamaldehyde and 2'-benzyloxycinnamaldehyde from *C. cassia* bark inhibited *in vitro* growth of 29 human cancer cell lines (Lee *et al.*, 1999; Kim *et al.*, 2004). 3-(2-Hydroxyphenyl)-propanoic acid and its *O*-glucoside prevented serotonin-induced ulcerogenesis in rats. 3-(2-Hydroxyphenyl)-propanoic acid also inhibited gastric ulcers induced by phenyl butazone, ethanol and water immersion stress, although it failed to prevent indomethacin-induced ulcers. Pharmacological studies showed that 3-(2-hydroxyphenyl)-propanoic acid hardly inhibited the secretion of gastric acid but promoted gastric blood flow (Tanaka *et al.*, 1989).

Pesticidal activity

The oil from *C. cassia* exhibited high toxicity to the adult beetles of the storage pest, *Lasioderma serricorne* (Kim-Soon *et al.*, 2003).

C. tamala

The leaf of *C. tamala* is a brain tonic, anthelmintic, diuretic, is good for the liver and spleen and is useful in inflammation. The leaves are a stimulant and are used in rheumatism and diarrhoea. Its bark is useful for the treatment of gonorrhoea (Kirtikar and Basu, 1984). The essential oil from *C. tamala* exhibits antifungal, antidermatophytic, hypoglycaemic and hypolipidaemic effects.

Antifungal activity

The essential oil of *C. tamala* was effective in inhibiting the growth of *Fusarium moniliforne*, a postharvest fungal pathogen of cereal crops (Baruah *et al.*, 1996). The leaf oil exhibited fungitoxicity against *Aspergillus flavus* and *A. parasiticus* at 3000 and 1000 ppm, respectively. The active compound was identified as eugenol.

Hypoglycaemic and hypolipidaemic effects

Oral administration of 50% alcoholic extract of leaves lowered plasma glucose levels in streptozotocin-induced hyperglycaemic rats. The extract also exhibited antihypercholesterolaemic and antihyperglycaemic effects in rats (Sharma *et al.*, 1996).

Antidermatophytic activity

C. tamala oil possessed antidermatophytic activity against the ringworm fungi, *Microsporium audouini* and *Trichophyton metagrophytes* at 500 ppm. The ointment containing essential oil showed promising efficacy as a herbal antifungal agent in treating dermatomycosis of guinea pigs (Yadav *et al.*, 1999).

7.6. ISO Specifications

Cinnamon

Sri Lanka has been the traditional producer and exporter of cinnamon and its value-added products, such as bark oil and leaf oil. Cinnamon bark oil is a high-value oil and Sri Lanka is the only supplier of this commodity, with an annual production of around 2.8–3.0 t only. Western Europe, especially France, is the major importer, followed in recent times by the USA.

The world demand for cinnamon leaf oil is around 120–150 t per annum, a demand met mainly by Sri Lanka. The USA and Western Europe are the largest consumers of leaf oil.

Though it is a costly oil, there is no international standard for cinnamon bark oil. The higher the cinnamaldehyde content, the higher the price. In the USA, the Essential Oil Association (EOA) standard specifies an aldehyde content of 55–78% (EOA, 1975).

However, in the case of leaf oil, international standards do exist. In this case, a phenol content of 75–85% has been specified for oil of Sri Lankan origin (ISO, 1977). Cinnamaldehyde is another constituent of leaf essential oil, contributing to the total flavour, and the specification limits its content to 5%. In the USA, the FMA (Fragrance Materials Association) specifies the eugenol content (80–88%) in cinnamon leaf oil in terms of its solubility in KOH (FMA, 1992).

Cassia

Cassia oil is distilled from a mixture of leaves, twigs and fragments of bark and there is only one type of cassia oil. It is used mainly for flavouring soft drinks, confectionary and liquors; its use in perfumery is limited due to its skin-sensitizing properties (Coppen, 1995). The world trade in cassia oil is controlled by export from China. Imports to the USA are rising, mainly because of the boon in the soft drinks industry. Cassia oil has specific ISO standards (ISO, 1974). The total annual production of cassia oil is estimated to be more than 500 t (Coppen, 1995). The oils distilled in Indonesia (from *C. burmannii*) and Vietnam (from *C. cassia*) are also being marketed as cassia oil, but these are less valued and are much less widely traded. Indonesian cassia has a good market in the USA.

7.7. Conclusion

Phytochemical studies of cinnamon and related species are restricted mainly to the volatile oil and its constituents. Recently, the chemical composition of the essential oils of a few rare species has been researched and new aroma sources have been identified. The chemistry of the genus *Cinnamomum* is interesting, as there exist several chemotypes within a species. So far, except for *C. cassia*, very little attention has been paid to the non-volatiles of the genus. This is an area worth exploring.

References

Ahmed, A., Choudhary, M.I., Farooq, A., Demirci, B., Demirci, F. and Baser, K.H.C. (2000) Essential oil constituents of the spice *Cinnamomum tamala* (Ham.) Nees and Eberm. *Flavour and Fragrance Journal* 15, 388–390.

Akira, T., Tanaka, S. and Tabata, M. (1986) Pharmacological studies on the antiulcerogenic activity of Chinese cinnamon. *Planta Medica* 6, 440–443.

Angmor, J.E., Dicks, D.M., Evans, W.C. and Sandra, D.K. (1972) Studies on *Cinnamomum zeylanicum*. Part 1. The essential oil components of *C. zeylanicum* Nees grown in Ghana. *Planta Medica* 21, 416–420.

Baruah, A. and Nath, S.C. (2004) *Indian Cassia*. In: Ravindran, P.N., Babu, K.N. and Shylaja, M. (eds) *Cinnamon and Cassia: The Genus Cinnamomum*. CRC Press, London, pp. 199–210.

Baruah, P., Sharma, R.K., Singh, R.S. and Ghosh, A.C. (1996) Fungicidal activity of some naturally occurring essential oils against *Fusarium moniliforme*. *Journal of Essential Oil Research* 8, 411–412.

Baruah, A., Nath, S.C., Hazarika, A.K. and Sarma, T.C. (1997) Essential oils of leaf, stem bark and panicle of *Cinnamomum bejolghota* (Buch-Ham) Sweet. *Journal of Essential Oil Research* 9, 293–295.

Baruah, A., Nath, S.C. and Leclerck, P.A. (1999) Leaf and stem bark oils of *Cinnamomum sulphuratum* Nees. from Northeast India. *Journal of Essential Oil Research* 11, 194–196.

Baruah, A., Nath, S.C. and Boissya, C.L. (2000) Systematics and diversities of *Cinnamomum* species used as 'tejpat' in Northeast India. *Journal of Economic and Taxonomic Botany* 41, 361–374.

Baruah, A., Nath, S.C. and Hazarika, A.K. (2001) Methyl cinnamate, the major component of the leaf and stem bark oils of *Cinnamomum sulphuratum* Nees. *Indian Perfumer* 45, 39–41.

Baruah, A., Nath, S.C. and Hazarika, A.K. (2002) Essential oils of *Cinnamomum sulphuratum* Nees. A new chemotype source of spice value from Northeast India. *Indian Perfumer* 46, 89–92.

Baruah, A., Nath, S.C. and Hazarika, A.K. (2004) Fruit essential oil of a variant of *Cinnamomum tamala* Nees. *Indian Perfumer* 48, 437–438.

Baslas, K.K. (1967) *Chemistry of Indian Oils* Part II. *Perfumer and Essential Oil Review* 622–628.

Bernard, T., Perineau, F., Delmas, M. and Gaset, A. (1989) Extraction of essential oils by refining of plant materials. II. Processing of products in the dry state: *Illicium verum* Hooker (fruit) and *Cinnamomum zeylanicum* (bark). *Flavour and Fragrance Journal* 4, 85–90.

Bradu, B.L. and Sobti, S.N. (1988) *Cinnamomum tamala* in North West Himalayas; evaluation of various chemical types for perfumery value. *Indian Perfumer* 32, 334–340.

Chalchat, J.C. and Valade, I. (2000) Chemical composition of leaf oils of *Cinnamomum* from Madagascar: *C. zeylanicum* Blume, *C. camphora* L., *C. fragrans* Baillon and *C. angustifolium*. *Journal of Essential Oil Research* 12, 537–540.

Chang, S.T., Chen, P.F. and Chang, S.C. (2001) Antibacterial activity of leaf essential oils and their constituents from *Cinnamomum osmophloem*. *Journal of Ethnopharmacology* 77(1), 123–127.

Chao, S.C., Young, D.G. and Oberg, C.J. (2000) Screening for inhibitory activity of essential oils on selected bacteria, fungi and viruses. *Journal of Essential Oil Research* 12(5), 639–649.

Chen, B.Q., Xu, Y., Zeng, F.X., Yu, X.J., Ding, J.K.D. and Wu, Y. (1997) Studies on the introduction and chemical constituents of essential oil of *Cinnamomum burmannii f. heyneanum*. *Acta Botanica Yunnanica* 14, 105–110.

Chinese Materia Medica (1996) (English translation) p. 51.

Choi, J., Lee, K.T., Ka, H., Jung, W.T., Jung, H.J. and Park, H.J. (2001) Constituents of the essential oil of the *Cinnamomum cassia* stem bark and the biological properties. *Archives of Pharmaceutical Research* 24(5), 418–423.

Choudhury, S., Ahamed, R., Barthel, A. and Leclercq, P.A. (1998) Composition of the bark and flower oils of *Cinnamomum bejolghota* (Buch.-Ham.) sweet from two locations of Assam, India. *Journal of Essential Oil Research* 10, 245–250.

Coppen, J.J.W. (1995) *Flavours and Fragrances of Plant Origin, Non-wood Forest Products 1*. FAO, Rome.

DASD (Directorate of Arecanut and Spices Development) (2007) Arecanut and spices database. Premaja, P. and Manoj Kumar, K. (eds). Directorate of Arecanut and Spices Development, India, pp. 110.

Ehlers, D., Hilmer, S. and Bartholomae, S. (1995) Quoted from Lawrence (2001) Progress in essential oils. *Perfumer Flavorist* 26, 44–57.

EOA (1975) *Oil of Cinnamon Bark*. Ceylon EOA No. 87. Essential Oil Association of USA, Washington, DC.

Ferhout, H., Bohatier, J., Guillot, J. and Chalchat, J.C. (1999) Antifungal activity of selected essential oil of cinnamaldehyde and carvacrol against *Mallssezia furfur* and *Candida albicans*. *Journal of Essential Oil Research* 11, 119–129.

FMA (1992) Cinnamon leaf oil. 5pp. FMA Monographs Volume 1. Fragrance Materials Association of the United States, Washington, DC.

Guenther, E. (1950) *The Essential Oils*. Volume II. Van Nostrand Co., Inc., New York, p. 516.

Guenther, E. (1953). *The Essential Oils*. Volume IV. Van Nostrand Co., Inc., Princeton, New Jersey.

Hussain, A., Viramani, O.P., Sharma, A., Kumar, A. and Misra, L.N. (1988) *Major Essential Oil-bearing Plants of India*. CIMAP, Lucknow, India, pp. 213–214.

ISO (1974) Oil of cassia. International Standard ISO 3216–1974 (E.). International Organization for Standardization, Paris, 2 pp.

ISO (1977) Oil of cinnamon leaf. International Standard ISO 3524–1977 (E.). International Organization for Standardization, Paris, 2 pp.

Jayaprakasha, G.K., Rao, L.J. and Sakariah, K.K. (1997) Chemical composition of the volatile oil from the fruits of *Cinnamomum zeylanicum* Blume. *Flavour and Fragrance Journal* 12(5), 331–333.

Jayaprakasha, G.K., Rao, L.J.M. and Sakariah, K.K. (2000) Chemical composition of the flower oil of *Cinnamomum zeylanicum* Blume. *Journal of Agriculture and Food Chemistry* 48, 4294–4295.

Jayatilake, A., Poole, S.K., Poole, C.F. and Chichila, T.M.P. (1995) Simultaneous microsteam distillation – solvent-extraction for the isolation of semivolatile flavour compounds from *Cinnamomum* and their separation by series coupled-column gas chromatography. *Analytica Chimica Acta* 30, 147–162.

Jin Wong, H., Kyoungttwa, L. and Jim, E.S. (1994) Anti neoplastic effects of extracts from tradition medicinal plants. *Korean Journal of Pharmacognosy*, 25, 171–177.

Jirovetz, L., Buchbauer, G., Ngassoum, M.B. and Eberhardt, R. (1998) Analysis and quality control of the essential oil of the leaves of *Cinnamomum zeylanicum* from Cameroon. *Ernahrung* 22(10), 443–445.

Jirovetz, L., Buchbauer, G., Ruzika, J., Shafi, M.P. and Rosamma, M.K. (2001) Analysis of *Cinnamomum zeylanicum* Blume leaf oil from South India. *Journal of Essential Oil Research* 13, 442–443.

Kaul, P.N., Bhattacharjee, A.K. and Rao, B.R R. (1996) Seasonal variation in the composition of the essential oil of cinnamon (*Cinnamomum zeylanicum* Blume) leaves. *Indian Perfumer* 40(2), 36–38.

Kaul, P.N., Bhattacharjee, A.K. and Rao, B.R.R. (1998) Chemical composition of the essential oils of cinnamon (*Cinnamomum zeylanicum* Blume) leaves affected by leaf spot disease. *Indian Perfumer* 42(1), 33–35.

Kaul, P.N., Bhattacharya, A.K., Rao, B.R.R., Syamasundar, K.V. and Ramesh, S. (2003) Volatile constituents of essential oils isolated from different parts of cinnamon (*Cinnamomum zeylanicum* Blume). *Journal of Science of Food and Agriculture* 83(1), 53–55.

Kim, S.Y., Lee, K.J., Shin, Y.H. and Lee, C.H. (2004) Physico-chemical properties of 2'-benzoyloxycinnamaldehyde. *International Journal of Pharmaceutics* 287(1–2), 21–26.

Kim-Soon, I.L., Park Chan, Ohh Myung Hee, Cho Hyung Chan and Ahn Young Joon (2003) Contact and fumigant activities of aromatic plant extracts and essential oils against *Lasioderma serricorne* (Coleoptera: Anobiidae). *Journal of Stored Products Research* 39(1), 11–19.

Kirtikar, K.R. and Basu, B.D. (1975) Indian medicinal plants, vol Iii. Bischen Singh, Mahendrapal Singh, Dharadun, pp 2145–2150.

Kirtikar, K.R. and Basu, B.D. (1984) *Indian Medical Plants*. Volume 3. Goyal Offset Printers, New Delhi, pp. 2143–2155.

Koh, W.S., Yoon, S.Y., Kwon, B.M., Jeong, T.C., Nam, K.S. and Han, A.U. (1999) Cinnamaldehyde inhibits lymphocyte proliferation and modulates T-cell differentiation. *International Journal of Immunopharmacology* 20, 643–660.

Koketsu, M., Gonclaves, S.L., Godoi, R.L., Lopez, D. and Morsbach, N. (1997) The bark and leaf essential oils of cinnamon (*Cinnamomum verum* Presl.) grown at Parana, Brazil. *Ciência Tecnologia Alimentos* 17(3), 281–285.

Krishnamoorthy, B., Gopalam, A. and Abraham, J. (1988) Quality parameters of cinnamon (*Cinnamomum verum*) in relation to flush colour. *Indian Cocoa Arecanut and Spices Journal* XII (2), 38.

Krishnamoorthy, B., Rema, J., Zachariah, T.J., Jose, A. and Gopalam, A. (1996) Navasree and Nithyasreee – two high yielding and high quality cinnamon (*Cinnamomum verum* – Bercht & Presl.). *Journal of Spices and Aromatic Crops* 5(1), 28–33.

Krishnamoorthy, B., Zachariah, T.J., Rema, J. and Mathew, P.A. (1999). Evaluation of selected Chinese cassia (*Cinnamomum cassia* Blume) accessions for chemical quality. *Journal of Spices and Aromatic Crops* 8(2), 193–195.

Kubo, M., Ma, S., Wu, J. and Matsuda, H. (1996) Antiinflammatory activities of 70% methanolic extract of *Cinnamomi cortex. Biological and Pharmaceutical Bulletin* 19, 1041–1045.

Kwon, B.M., Cho, Y.K., Lee, S.H., Nam, J.Y., Bok, S.H., Chun, S.K., Kim, J.A. and Lee, I.R. (1996) 2-Hydroxycinnamaldehyde from stem bark of *Cinnamomum cassia. Planta Medica* 62,183–184.

Kya, P. and Min, N.C. (1970) Studies on some local *Cinnamomum* species. *Union Burma Journal of Life Sciences*, 3: 197–204.

Lawrence, B.M. (1967) A review of some commercial aspects of cinnamon. *Perfumer and Essential Oil Review* April, 236–241.

Lawrence, B.M. (2001) Progress in essential oils. *Perfumer Flavorist* 26, 44–57.

Lee, C.W., Hong, D.H., Han, S.B., Park, S.H., Kim, H.K., Kwon, B.M. and Kim, H.M. (1999) Inhibition of human tumour growth by 2'-hydroxy and 2'-benzyloxy cinnamaldehydes. *Planta Medica* 65, 263–266.

Lee, H.S. (2002) Inhibitory activity of *Cinnamomum cassia* bark derived component against rat lens aldose reductase. *Journal of Pharmacy and Pharmaceutical Sciences* 5, 226–230.

Lee, H.S. and Ahn, Y.J. (1998) Growth inhibiting effects of *Cinnamomum cassia* bark derived materials on human intestinal bacteria. *Journal of Agricultural and Food Chemistry* 46, 8–12.

Lewis, Y.S., Krishnamurthy, N., Shivasankar, S. and Natarajan, C.P. (1977) Quality of cinnamon marketed in India. *Current Science* 46, 832–833.

Li, L.L. and Yuan, W.J. (1999) Cinnamon oil analysis by GC and GC/MS. *Fujian Fenxi Ceshi* 8, 1121–1125.

Li, Z.Q., Luo, L., Huang, R. and Xia, Y.Q. (1998) Cinnamon oil analysis by GC and GC/MS. *Yunnan Daxue Xuebao, Ziran Kexueban* 20 (supplement), 337–339.

Lockwood, G.B. (1979) The major constituents of the essential oils of *Cinnamomum cassia* Blume growing in Nigeria. *Planta Medica* 36(4), 380–381.
Mallavarapu, G.R. and Ramesh, S. (2000) Essential oil of the fruits of *Cinnamomum zeylanicum Blume*. *Journal of Essential Oil Research* 12, 628–630.
Mallavarapu, G.R., Ramesh S., Chandrasekhara, R.S., Rao, B.R.R., Kaul, P.N. and Bhattacharya, A.K. (1995) Investigation of the essential oil of cinnamon leaf grown at Bangalore and Hyderabad. *Flavour and Fragrance Journal* 10, 239–242.
Mancini, D.A.P., Dias, A.L.F., Pinto, J.R. and Mancini, F.J. (1999) Antioxidant aqueous extract from cinnamon (*Cinnamomum zeylanicum*, Blume) as inhibitors of influenza virus. *Revista Brasileira de Ciencias Farmaceuticas* 35(1), 155–160.
Medici, D.D., Pieretti, S., Salvatore, G., Nicoletti, M. and Rasoanaivo, P. (1992) Chemical analysis of essential oils of Malagasy medicinal plants by gas chromatography and NMR spectroscopy. *Flavour and Fragrance Journal* 7, 275–281.
Mitra, S.K., Sundaram, R. and Venkataranganna, M.V. (2000) Antiinflammatory, antioxidant and antimicrobial activity of '*Opthacare*' brand, a herbal eye drop. *Phytomedicine* 7, 123–127.
Miyamura, M., Nohara, T., Tomimatsu, T. and Nishioka, I. (1983) Seven aromatic compounds from the bark of *Cinnamomum cassia*. *Phytochemistry* 22, 215–218.
Moretti, M.D.L., Bazzoni, E., Passino, G.S. and Prota, R. (1998) Antifeedant effects of some essential oils on *Ceratitis capitata*. Wied. (Diptera, Tephritidae). *Journal of Essential Oil Research* 10, 405–412.
Nath, S.C. and Baruah, A.K.S. (1994) Eugenol as the major component of the leaf oil of *Cinnamomum impressinervium* Meissn. *Journal of Essential Oil Research* 6, 211–212.
Nath, S.C., Hazarika, A.K., Baruah, R.N., Singh, R.S. and Ghosh, A.C. (1994a) Major components of the leaf oil of *Cinnamomum sulphuratum* Nees. *Journal of Essential Oil Research* 6, 77–78.
Nath, S.C., Hazarika, A.K. and Singh, R.S. (1994b) Essential oil of leaves of *Cinnamomum tamala* Nees and Ebarm. from North East India. *Journal of Spices and Aromatic Crops* 3, 33–35.
Nath, S.C., Modon, G. and Baruah, A. (1996) Benzyl benzoate the major component of the leaf and stem bark oil of *Cinnamomum zeylanicum* Blume. *Journal of Essential Oil Research* 8, 327–328.
Nath, S.C., Baruah, A. and Hazarika, A.K. (1999) Essential oils of the leaves of *Cinnamomum* Schaefer members. *Indian Perfumer* 43, 473–478.
Nohara, T., Kashiwada, Y., Tomimatsu, T. and Nishioka, I. (1982) Studies on the constituents of *Cinnamomum* cortex VII. Two novel diterpenes from bark of *Cinnamomum* cassia. *Phytochemistry* 21, 2130–2132.
Nohara, T., Kashiwada, Y. and Nishioka, I. (1985) Cinncassiol E, a diterpene from the bark of *Cinnamomum cassia*. *Phytochemistry* 24, 1849–1850.
Oka, Y. (2001) Nematicidal activity of essential oil components against the root-knot nematode *Meloidogyne javanica*. *Nematology* 3, 159–164.
Park, K.I., Park, J.Y., Kim, K.H., Choi, K.S., Choi, I.H., Kim, C.S. and Shin, S.C. (2005) Nematicidal activity of plant essential oils and components from garlic (*Allium sativum*) and cinnamon (*Cinnamomum verum*) oils against the pine wood nematode (*Bursaphelenchus xylophilus*). *Nematology* 7, 767–774.
Pruthi, J.S. (1976) *Spices and Condiments*. National Book Trust, New Delhi, pp. 86–90.
Purseglove, J.W., Brown, I.G., Green, C.L. and Robbins, S.R.J. (1981) *Spices*. Volume 1. Longman, London, pp. 100–173.
Raina, V.K., Srivastava, S.K., Aggarwal, K.K., Ramesh, S. and Kumar, S. (2001) Essential oil composition of *Cinnamomum zeylanicum* Blume leaves from Little Andaman, India. *Flavour and Fragrance Journal* 16(5), 374–376.
Rao, B.R.R., Rajput, D.K., Kaul, P.N., Bhattacharya, A.K. and Naqvi, A.A. (2006) Effect of short and long-term storage on essential oil content and composition of cinnamon (*Cinnamomum verum* Bercht & Presl.) leaves. *Journal of Spices and Aromatic Crops* 15(1), 19–24.
Rao, Y.R., Paul, S.C. and Dutta, P.K. (1988) Major constituents of essential oils of *Cinnamum zeylanicum*. *Indian Perfumer* 32, 86–89.
Rema, J., Leela, N.K., Jacob, T.K. and Krishnamoorthy, B. (2006) *Cinnamon and Indian Cassia*. In: Parthasarathy, V.A. and Rajeev, P. (eds) *Major Spices – Production and Processing*. Indian Institute of Spices Research, Calicut, India, pp. 289–306.
Roger, R.C. and Hamraoui, A. (1994) Inhibition of reproduction of *Acanthoscilides oblectus* Say (Coleoptera), a kidney bean bruchid by aromatic essential oils. *Crop Protection* 13, 624–628.
Senanayake, U.M. (1977) The nature, description and biosynthesis of volatiles of *Cinnamomum* spp. PhD Thesis, University of New South Wales, Kensington, Australia (quoted from Senanyake and Wijesekera, 2004).

Senanayake, U.M., Lee, T.H. and Wills, R.B.H. (1978) Volatile constituents of Cinnamon (*Cinnamomum zeylanicum*) oils. *Journal of Agriculture and Food Chemistry* 26, 822–824.

Sharma, S.R., Dwivedi, S.K. and Swarup, D. (1996) Hypoglycaemic and hypolipidaemic effects of *Cinnamomum tamala* Nees leaves. *Indian Journal of Experimental Biology* 34, 372–374.

Shiraga, Y., Okano, K., Akira, T., Fukaya, C., Yokoyama, K., Tanaka, S., Fukui, H. and Tabata, M. (1988) Structures of potent antiulcerogenic compounds from *Cinnamomum cassia. Tetrahedron* 44, 4703–4711.

Syamasundar, K.V., Ramesh, S. and Chandrasekhara, R.S. (2000) Volatile constituents of *Cinnamomum zeylanicum* Blume fruit oil. In: *Spices and Aromatic Plants: Challenges and Opportunities in the New Century*. Indian Society for Spices, IISR, Calicut, India, pp. 284–286.

Tainter, D.R. and Grenis, A.T. (1993) *Spices and Seasonings – A Food Technology Handbook*. VCH Publishers Inc., New York.

Tanaka, S., Yoon, Y.H., Fukui, H., Tabata, M., Akira, T. and Okan, K. (1989) Antiulcerogenic compounds isolated from Chinese cinnamon. *Planta Medica* 55, 245–248.

Tang, W. and Eisenbrand, G. (1992) *Chinese Drugs of Plant Origin – Chemistry, Pharmacology and Use in Traditional Medicine*. Springer, Berlin, pp. 319–330.

Variyar, P.S. and Bandyopadhyay, C. (1989) On some chemical aspects of *Cinnamomum zeylanicum. PAFAI Journal* Oct–Dec, 35–38.

Vernin, C., Vernin, G., Metzger, J. and Puigol, I. (1990) La canelle, Premire partie. Analyse CPG/Sm Banuqe SPECMA d'huile essentielle de canelle de Ceylan et de Chine. *Parfumes, Cosmetiques, Aromes* 93, 85–90.

Vernin, G., Vernin, C., Metzger, J., Pujol, L. and Parkanyi, C. (1994) GC/MS analysis of cinnamon and cassia essential oils: a comparative study. In: Charalambous, G. (ed.) *Spices, Herbs and Edible Fungi*. Elsevier Science, B.V., Amsterdam, pp. 411–425.

Wang, J.H. (1985) Chinese herbal pharmacology. In: *Chinese Materia Medica*. Shanghai Science and Technology Press, Shanghai, China, pp. 27–28.

Wijesekera, R.O.B. (1978) The chemistry and technology of cinnamon. *CRC Critical Reviews in Food Science and Nutrition* 10, 1–30.

Wu, T.S., Leu, Y.L., Chan, Y.Y., Yu, S.M., Teng, C.M. and Su, J.D. (1994) Lignans and an aromatic acid from *Cinnamomum philippinense. Phytochemistry* 36, 785–788.

Xiao-duo, J., Quan-Long, P., Garafoo, H.M. and Pannel, L.K. (1991) Essential oil of the leaf, bark and branch of *Cinnamomum burmannii* Blume. *Journal of Essential Oil Research*, 3: 373–375.

Yadav, P., Dubey, N.K., Joshi, V.K., Chansouria, J.P.N. and Yadav, P. (1999) Antidermatophytic activity of essential oil of *Cinnamomum. Journal of Medicinal and Aromatic Plant Sciences* 21, 347–351.

Yu-Jing, L., Liang-Jeng, Z., Bi-Yao, L., Langtian, M., Zhao-Lun, L. and Liang-Shi, J. (1987) Studies on meipian tree (*Cinnamomum burmannii* physiological type) as a new source of natural d-borneol. *Acta Botanica Sinica* 29, 373–375.

Zhu, I.F., Li, Y.H., Li, B.L., Lu, B.Y. and Xia, N.H. (1993) *Aromatic Plants and their Essential Constituents*. South China Institute of Botany. Chinese Academy of Sciences, Pace Book Co, Hong Kong.

8 Clove

N.K. Leela and V.P. Sapna

8.1. Introduction

Clove (*Syzygium aromaticum* (L.) Merril. & Perry, syn. *Eugenia aromaticum* or *E. caryophyllata*) is one of the most ancient and valuable spices of the Orient. It is a member of the family *Myrtaceae*. The clove of commerce is its dried unopened flower buds. The word 'clove' was derived either from the Latin word *clavus*, or the French form *clou*, meaning 'nail'. The buds resemble irregular nails.

Area and production of clove in India for the period 1995 to 2005 are indicated in Table 8.1. There was a marginal increase in the area under cultivation from 2270 to 2528 ha over these years, whereas production decreased from 2455 to 1815 t. Tanzania, Indonesia, Madagascar, Cameroon and Sri Lanka are the major clove-exporting countries. In recent years, world production of clove has averaged around 80,000 t a year. Indonesia is the world's largest producer at 50,000–60,000 t per annum. It is used mainly in the preparation of *kretek* cigarettes. Singapore is the entrepôt for the clove trade. Saudi Arabia, the USA, France and India are the major importing countries.

8.2. Botany and Uses

Clove is a medium-sized tree, which grows to a height of 10–20 m, that can live up to 100 years or more. The bark is grey, the leaves are elliptical in shape and fragrant with crimson flowers. The flowers are hermaphrodite with a fleshy hypanthium surrounded by sepals. The fruit is a purple drupe, about 2.5 cm long.

Cloves are best used whole. The flavour deteriorates quickly once it is powdered. Whole and ground cloves are used to enhance the flavour of meat and rice dishes. They are used widely in curry powders and masalas. In North Indian cuisine, cloves are used in almost every sauce or side dish made, mostly mixed with other spices. In South India, they find extensive use in 'biriyanis' to enhance the flavour of the rice. They are highly valued in medicine as a carminative and stimulant. Cloves are said to be a natural anthelmintic.

The spice is used throughout Europe and Asia and is smoked in a type of cigarette, known locally as *kretek*, in Indonesia and in occasional coffee bars in the West, mixed with marijuana to create marijuana spliffs. Clove cigarettes (Indonesian *kretek*) are cigarettes made with a complex blend of tobacco, cloves and a flavouring 'sauce'. Cloves are also an important incense material in Chinese and Japanese culture.

Oil of clove is used extensively for flavouring all kinds of food products, such as meats, sausages, baked goods, confectionery, candies, table sauces, pickles, etc. Clove oil

Table 8.1. Area and production of clove in India.

Year	Area (ha)	Production (t)
1994/95	2270	2455
1995/96	2300	2439
1996/97	2222	1836
1997/98	2273	1698
1998/99	2308	1056
1999/2000	2795	1633
2000/01	1881	979
2001/02	1891	1048
2002/03	2127	1374
2003/04	2431	1811
2004/05	2528	1815

Source: DASD (2007).

Table 8.2. Nutrient composition of 100 g of clove.

Composition	USDA (ground)
Water (g)	5.40–6.86
Food energy (Kcal)	323
Protein (g)	5.98
Fat (g)	20.06
Carbohydrate (g)	61.22
Ash (g)	5.88
Ca (g)	0.646
P (mg)	105
Na (mg)	243
K (mg)	1102
Fe (mg)	8.68
Thiamin (mg)	0.115
Riboflavin (mg)	0.267
Niacin (mg)	1.458
Ascorbic acid (mg)	80.81
Vitamin A (RE)	53

Source: Tainter and Grenis (1993).

is used in aromatherapy and oil of cloves is used widely to treat toothache. It is used in medicine for its antibacterial, antiseptic and antibiotic properties. The oil has many industrial applications and is used extensively in perfumes, soaps and as a clearing agent in histological work. It is an ingredient in many toothpastes and mouthwashes. It is also used for flavouring oral preparations and chewing gums. The chief constituent of the oil, eugenol, is used in the preparation of synthetic vanillin and isoeugenol (Pruthi, 1976).

8.3. General Composition

The composition of the clove varies according to the agroclimatic conditions under which it is grown, processed and stored. The dried clove bud contains carbohydrates, fixed oil, steam-volatile oil, resins, tannins, proteins, cellulose, pentosans and mineral elements. Carbohydrates comprise about two-thirds of the weight of the spice (Purseglove *et al.*, 1981). The dried dark and flower buds also contain nutrients like proteins, minerals, vitamins, etc. Nutrient composition of 100 g of clove is indicated in Tables 8.2 and 8.3. It is evident from Table 8.2 that 61% of clove is carbohydrates, 20% is fat and the rest is contributed by secondary metabolites, vitamins and minerals. Cloves are an excellent source of manganese, a very good source of dietary fibre, vitamin C, vitamin K and ω-3 fatty acids and a good source of calcium and magnesium. Volatile oil can be extracted from the leaf, stem and buds of clove. Volatile oil is present in oval cavities, two or three rows below the epidermis. The major component of the volatile oil is a phenol, namely eugenol. Phenolic activity is greater at the outer glandular regions of the hypanthium than in the inner aerenchymatous spongy tissue.

8.4. Chemistry

Volatiles

Clove yields three types of volatile oil – oil extracted from the leaves, the stem and the buds. These oils differ considerably in yield and quality. The yield and composition of the oil obtained are influenced by its origin, season, variety and quality of raw material, maturity at harvest, pre- and post-distillation treatments and method of distillation. The chief component of the oil is eugenol.

Bud oil

Good-quality clove buds contain 15–20% essential oil (Gopalakrishnan *et al.*, 1988; Pino *et al.*, 2001; Raina *et al.*, 2001; Zachariah *et al.*, 2005). The oil is dominated by eugenol (70–85%), eugenyl acetate (15%) and

Table 8.3. Nutrient values and weights for edible portion of clove.

Nutrient	Units	Value per 100 g	Number of data points	Std. error	1.00 × 1 tsp 2.1 g
Proximates:					
Water	g	6.86	343	0.241	0.14
Energy	kcal	323	0	0	7
Energy	kj	1350	0	0	28
Protein	g	5.98	73	0.106	0.13
Total lipid (fat)	g	20.07	299	0.273	0.42
Ash	g	5.88	384	0.049	0.12
Carbohydrate, by difference	g	61.21	0	0	1.29
Fibre, total dietary	g	34.2	0	0	0.7
Sugars, total	g	2.38	1	0	0.05
Sucrose	g	0.02	1	0	0.00
Glucose (dextrose)	g	1.14	1	0	0.02
Fructose	g	1.07	1	0	0.02
Maltose	g	0.00	1	0	0.00
Galactose	g	0.15	1	0	0.00
Minerals:					
Calcium	mg	646	6	64.353	14
Iron	mg	8.68	7	1.500	0.18
Magnesium	mg	264	5	4.867	6
Phosphorus	mg	105	4	2.887	2
Potassium	mg	1102	6	68.431	23
Sodium	mg	243	6	12.368	5
Zinc	mg	1.09	5	0.278	0.02
Copper	mg	0.347	0	0	0.007
Manganese	mg	30.033	0	0	0.631
Selenium	mcg	5.9	1	0	0.1
Vitamins:					
Vitamin C, total ascorbic acid	mg	80.8	1	0	1.7
Thiamin	mg	0.115	2	0	0.002
Riboflavin	mg	0.267	4	0.012	0.006
Niacin	mg	1.458	1	0	0.031
Vitamin B_6	mg	0.590	2	0	0.012
Folate, total	mcg	93	0	0	2
Folate, food	mcg	93	0	0	2
Folate, DFE	mcg_DFE	93	0	0	2
Vitamin B_{12}	mcg	0.00	0	0	0.00
Vitamin B_{12}, added	mcg	0.00	0	0	0.00
Vitamin A	IU	530	1	0	11
Vitamin A	mcg_RAE	27	0	0	1
Retinol	mcg	0	0	0	0
Vitamin E (α-tocopherol)	mg	8.52	1	0	0.18
Vitamin E, added	mg	0.00	0	0	0.00
Vitamin K (phylloquinone)	mcg	141.8	1	0	3.0
Lipids:					
Fatty acids, total saturated	g	5.438	0	0	0.114
14:0	g	0.022	0	0	0.000
16:0	g	3.967	0	0	0.083
18:0	g	0.847	0	0	0.018
Fatty acids, total monounsaturated	g	1.471	0	0	0.031
16:1 undifferentiated	g	0.089	0	0	0.002
18:1 undifferentiated	g	1.337	0	0	0.028

Continued

Table 8.3. *Continued*

Nutrient	Units	Value per 100 g	Number of data points	Std. error	1.00 × 1 tsp 2.1 g
20:1	g	0.022	0	0	0.000
Fatty acids, total polyunsaturated	g	7.088	0	0	0.149
18:2 undifferentiated	g	2.586	0	0	0.054
18:3 undifferentiated	g	4.257	0	0	0.089
20:4 undifferentiated	g	0.045	0	0	0.001
22:5 n-3	g	0.022	0	0	0.000
Phytosterols	mg	256	0	0	5
Others:					
β-Carotene	mcg	84	0	0	2
α-Carotene	mcg	0	1	0	0
β-Cryptoxanthin	mcg	468	0	0	10
Lycopene	mcg	0	1	0	0
Lutein + zeaxanthin	mcg	0	1	0	0

Source: USDA (2005).

β-caryophyllene (5–12%), which together make up 99% of the oil. *β*-Caryophyllene, which was earlier thought to be an artefact of distillation, was first reported as a constituent of bud oil by Walter (1972). The constituents of the oil also include methylamylketone, methylsalicylate, *α*- and *β*-humulene, benzaldehyde, *β*-ylangene and chavicol. The minor constituents like methylamylketone, methylsalicylate, etc., are responsible for the characteristic pleasant odour of cloves. The physico-chemical properties of clove oils are shown in Table 8.4.

Gopalakrishnan *et al.* (1984) characterized six sesquiterpenes, namely: α-cubebene (1.3%), α-copaene (0.4%), *β*-humulene (9.1%), β-caryophyllene (64.5%), γ-cadinene (2.6%) and δ-cadinene (2.6%), in the hydrocarbon fraction of the freshly distilled Indian clove bud oil. The oil from the Malagasy Republic (Madagascar) was dominated by eugenol (72–73%), eugenyl acetate (6.3–7.8%) and caryophellene (15.7%) (Lawrence and Reynolds, 1989).

The clove bud and stem oils from Madagascar were also dominated by eugenol, eugenyl acetate and β-caryophyllene. The stem oil contained a higher level of eugenol, whereas the eugenyl acetate content was higher in the bud oil. The oil from clove bud contained 73.5–79.7% eugenol and 4.5–10.7% eugenyl acetate, while the stem oil contained 76.4–84.8% eugenol and 1.5–8.0%

Table 8.4. Physico-chemical properties of clove oil.

Characteristic	Bud oil	Stem oil	Leaf oil
Colour	Colourless to pale yellow	Yellow to dark brown	Straw coloured or very pale
Specific gravity (25°C)	1.051–1.054	1.050–1.055	1.040–1.054
Optical rotation	−1°35′ to −0°25′	−1°30′ to −0°32′	−1°40′ to −0°40′
Refractive index (20°C)	1.531–1.537	1.531–1.539	1.531–1.538
Solubility	Soluble in 1 vol. of 70% ethanol	Soluble in 1–2 vol. of 70% ethanol	Soluble in 1.0–1.5 vol. of 70% ethanol
Total phenols (%)	91–93	88–93	78–93

Source: Guenther (1950).

eugenyl acetate. Both contained 7.3–12.4% β-caryophyllene and 1.0–1.4% α-humulene (Gaydou and Randriamiharisoa, 1987).

The essential oils of clove buds of Indian and Madagascan origins were analysed by Srivastava *et al.* (2005). The oil from Madagascar was richer in eugenol (82.6%) and eugenyl acetate (6%) compared with that of India (70 and 2.1%, respectively), whereas the Indian oil contained a higher level of *β*-caryophyllene (19.5 against 7.2% in Madagascan oil) (Srivastava *et al.*, 2005).

The neutral fraction of the bud oil from Madagascar contained β-caryophyllene (75.64%), *α*-humulene (14.12%) and δ-cadinene (2.34%) as the major components (Muchalal and Crouzet, 1985).

Pino *et al.* (2001) identified 36 compounds of the volatile oil of clove buds. The major components of the bud oil were eugenol (69.8%), β-caryophyllene (13%) and eugenyl acetate (16.1%) (Pino *et al.*, 2001). The chief components of clove oil from various regions are listed in Table 8.5, which indicates quantitative variations of the individual components of the oil from different regions. Zachariah *et al.* (2005) reported that clove buds from India contained 12.9–18.5% oil, of which 44–55% was eugenol, whereas the pedicels contained 3.0–7.7% oil with 60.0–72.4% eugenol.

Wild uncultivated trees in Molucca yielded 3.0–7.7% bud oil. The oil contained no eugenol and was quite different from the bud oil from cultivated trees (Guenther, 1950).

Analysis of clove bud oil extracted with liquid and supercritical carbon dioxide showed significant qualitative and quantitative compositional differences compared to oil obtained by the conventional hydrodistillation process. The parameters such as pressure, temperature, contact time, etc., affect the extraction of the bud flavour from the spice (Gopalakrishnan *et al.*, 1990). Guan *et al.* (2007) compared the essential oil obtained by four different extraction techniques; namely, hydrodistillation, steam distillation, solvent extraction and supercritical carbon dioxide extraction (SFE). The study showed that temperature had the largest effect on the eugenol content of the extracts and particle size had the maximum effect on oil yield. Among these techniques, the oil obtained by SFE and steam distillation had a desirable, pale yellow colour. Hydrodistilled oil had the lowest content of eugenol and eugenyl acetate. Extraction yield of SFE was twice as high as that obtained by steam and hydrodistillation. The SFE method yielded the highest content of eugenol + eugenol acetate in the oil. Clove oil obtained by steam distillation yielded the highest eugenol content, followed by SFE. Hydrodistillation yielded oil with a high *β*-caryophyllene content, whereas the SFE-extracted oil had the lowest *β*-caryophyllene content. GC-MS analysis of the clove oils obtained by different methods showed that the composition of the clove oil was almost similar, but the relative concentration of the identified compounds was apparently different. Among the four methods evaluated, SFE was the optimum method for obtaining high quality oil (Guan *et al.*, 2007). The oil yield was influenced largely by particle size and the eugenol content by temperature.

Leaf oil

Clove leaves yield 3.0–4.8% essential oil (Raina *et al.*, 2001). In Zanzibar, oil is distilled from dried fallen leaf or fresh leaf

Table 8.5. Major constituents of clove oil from different locations.

Constituent	A	B	C	D
β-Caryophyllene	4.35	5.13	9.86	12.50
α-Humulene	0.54	0.60	1.10	1.36
Eugenol	88.95	92.35	86.89	84.77
Eugenol acetate	5.54	1.25	1.59	ng

Note: A: Madagascan clove bud oil; B: Zanzibar clove stem oil; C: Madagascan clove leaf oil; D: Indonesian clove leaf oil; ng = not given.

after trimming the upper part of the tree. Crude leaf oil is harsh and woody, with a phenolic, sweet aroma quite different from bud oil. Rectified oil is clear pale yellow in colour with a sweeter, less harsh, dry woody odour close to that of eugenol. The oil contained 94.4% eugenol followed by β-caryophyllene (2.9%), nerol (0.79%) and β-caryophyllene oxide (0.67%) (Raina *et al.*, 2001). The leaf oil from Cuba contained 31 volatile compounds. Eugenol (78.1%) and β-caryophyllene (20.5%) were the main constituents in the oil (Pino *et al.*, 2001). Cuban leaf oil contained a higher amount of β-caryophyllene compared with that from Little Andaman.

The leaf oil from Madagascar contained 22 constituents, the chief constituents being eugenol (82.0%) and β-caryophyllene (13.0%). It contained a higher level of β-caryophyllene compared with bud oil (7.2%) (Srivastava *et al.*, 2005; Table 8.7). A commercial sample of leaf oil obtained in Germany contained 76.8% eugenol and 17.4% β-caryophyllene as the chief components (Jirovetz *et al.*, 2006).The constituents of various clove oils are indicated in Tables 8.6 and 8.7.

The essential oil content during the different stages of leaf growth revealed that the eugenol content in the leaves increased from 38.3 to 95.2% with maturity, while the contents of eugenyl acetate (51.2 to 1.5%) and caryophyllene (6.3 to 0.2%) decreased (Gopalakrishnan and Narayanan, 1988). Clove bud and leaf oil contain various classes of compounds, e.g. monoterpenes, sesquiterpenes, aldehydes and ketones (Vernin *et al.*, 1994), which are indicated in Table 8.8.

Clove stem oil

Clove stem yields 6% volatile oil (Gopalakrishnan *et al.*, 1988). The oil is a pale to light yellow liquid containing 80.2% eugenol and 6.6% β-caryophyllene, besides several minor components. Stem oil is used mainly in flavouring and perfumery and also to adulterate bud oil. Stem oil from Madagascar contains 77.10% eugenol and 11.20% β-caryophyllene as the major compounds (Gaydou and Randriamiharisoa, 1987). The chief volatile compounds from clove are indicated in Fig.8.1.

Fruit oil

Ripe fruits yield 2% of oil, which is comprise of 50–55% eugenol.

Clove bud concrete

Clove bud concrete is another important value-added product from buds, extracted using petroleum ether and benzene. It is olive to pale brown, having a sweet, rich spicy aroma similar to that of dried buds.

Clove concrete, on treatment with benzene/petroleum ether, produces a viscous, olive-green semi-solid (at low temperature), namely bud *absolute*, soluble in alcohols of different proportions. *Absolute* lacks caryophyllene and contains the same constituents as those present in unprocessed bud. Clove oleoresin is an extremely concentrated product, containing all the flavouring ingredients soluble in the solvent used, and is much closer to the original clove odour and flavour. Menon and Narayanan (1992) studied the glycosidically bound volatiles in clove leaves and buds using hydrolysing enzymes. When β-glucosidase was used for hydrolysis, eugenol was the major compound liberated from both buds and leaves, with *cis*- and *trans*-isoeugenol, nerolidol and farnesol in minor amounts, whereas hydrolysis using α–amyloglucosidase yielded farnesol as the major compound (Table 8.9).

Non-volatiles

So far, a few non-volatiles have been isolated from clove, which include tannins, sterols, triterpenes and flavonoids. These are listed below. Wild uncultivated trees of the Moluccas contained the crystalline compounds eugenone, eugenine, eugenitin and isoeugenitol (Guenther, 1950).

Tannins

Cloves contain 10–13% tannin, which has the same chemical composition as

Table 8.6. Volatiles of clove oils.

Component	% Composition			
	Leaf oil[1]	Bud oil[2]	Stem oil[2]	Leaf oil[2]
α-Pinene	–	0.42	–	–
β-Pinene	–	0.44	0.16	0.09
2-Hexanone	–	0.48	0.13	0.09
2-Heptanone + 1,8-cineol	–	0.50	0.11	0.10
α-Terpinene + limonene	–	0.53	0.14	–
p-Cymene	–	0.56	0.09	0.07
2-Heptanol	–	0.60	–	–
2-Nonanol	–	0.64	–	0.02
Benzaldehyde	–	0.69	–	0.01
β-Terpineol (t)	–	0.81	0.01	0.02
α-Cubebene	–	0.90	0.66	0.70
α-Terpineol	–	0.92	0.93	0.96
β-Caryophyllene	2.91	1.00	7.22	7.59
Benzyl alcohol	–	1.10	0.59	0.57
δ-Cadinene	–	1.18	0.31	0.44
α-Caryophyllene	–	1.34	0.07	0.22
Isoeugenol	–	1.51	–	1.00
Eugenol acetate	–	1.54	24.59	16.71
Farnesol (c,t)	–	1.59	–	–
Farnesol (t,t)	–	1.66	–	0.93
Vanillin	–	1.82	0.89	1.15
Asarone (t)	–	2.06	1.17	1.47
(*E*)-β-Ocimene	0.03	–	–	–
Linalool	0.08	–	–	–
Terpinen-4-ol	0.03	0.87	–	0.01
Nerol	0.79	–	–	–
Eugenol	94.41	1.41	59.14	60.82
α-Copaene	0.04	–	–	–
α-Humulene	0.36	10.60	1.24	1.44
(*E,E*)-α-Farnesene	0.06	–	–	–
γ-Cadinene	0.18	1.14	0.45	0.45
(*E*)-Nerolidol	0.03	–	–	–
β-Caryophyllene oxide	0.67	–	–	–
Humulene oxide II	0.07	–	–	–
t-Cadinol	0.07	–	–	–
Cadalene	0.18	–	–	–
Hexadecyl acetate	0.09	–	–	–

Source: [1]Raina *et al.* (2001); [2]Gopalakrishnan *et al.* (1988).

gallotannic acid. Eugeniin and ellagitannin were isolated from cloves by Nonaka *et al.* (1980). Tanaka *et al.* (1993) isolated eugenol glucoside gallate, a chromone *C*-glycoside, galloyl and hexahydroxy diphenyl esters of 2,4,6-trihydroxy acetophenone-3-glucopyranoside from clove leaves. Further, two ellagitannins, namely, syzyginin A (1,2,3-tri-*O*-galloyl–4,6-(*S*)-tergalloyl-β-D-glucoside) and syzyginin B, were also isolated from the leaves by Tanaka *et al.* (1996).

Triterpenes

Cloves contain about 2% of the triterpene, oleanolic acid. Narayanan and Natu (1974) isolated maslinic acid from clove buds. From clove, 2α-hydroxyoleanolic acid was isolated by Brieskorn *et al.* (1975).

Table 8.7. Essential oil composition of clove bud and leaf from India and Madagascar.

Compound	Bud oil (India)	Bud oil (Madagascar)	Leaf (Madagascar)
n-Octane	–	0.1	–
α-Pinene	–	0.1	–
(*E*)-β-Ocimene	–	t	–
Methyl benzoate	–	t	–
Linalool	0.1	t	–
m-Methyl acetophenone	t	0.1	–
Methyl salicylate	0.3	0.1	0.1
Nerol	t	0.1	–
Carvone	0.1	0.2	0.1
Chavicol	t	0.1	0.1
Linalyl acetate	t	0.1	0.1
Anethole	t	–	–
Eugenol	70.0	82.6	82.0
n-Butyl benzoate	1.3	–	–
α-Cubebene	–	t	–
Methyl eugenol	–	t	–
α-Ylangene	–	t	–
iso-Eugenol-1	0.8	0.1	0.1
Vanillin	t	–	–
α-Copaene	0.1	0.1	–
β-Caryophyllene	19.5	7.2	13.0
(*E*)-α-Bergamotene	1.3	0.2	0.4
α-Humulene	1.9	0.8	1.5
allo-Aromadendrene	0.3	0.1	–
Germacrene D	0.1	–	–
Eugenyl acetate	2.1	6.0	0.4
α-Selinene	0.1	0.3	–
Calamenene	0.1	0.1	–
γ-Cadinene	0.8	0.2	0.3
δ-Cadinene	0.2	–	–
(*E*)-Nerolidol	0.1	0.4	0.2
Caryophyllene oxide	0.4	0.3	0.5
Humulene epoxide I	–	0.1	–
Humulene epoxide II	0.1	t	0.1
Cubenol	–	0.1	–
t-Cadinol	0.1	0.1	0.2
t-Muurolol	–	t	–
epi-α-Cadinol	–	0.1	0.1
α-Cadinol	0.1	0.1	0.1
n-Heptadecane	–	0.1	0.2
Benzyl *n*-octanate	–	–	0.1
Myristic acid	–	–	0.1
iso-Propyl myristate	–	–	0.1
Oleic acid	–	–	0.1

t = trace.
Source: Srivastava *et al.* (2005).

Sterols

Sterols isolated from clove include sitosterol, stigmasterol and campesterol (Brieskorn *et al.*, 1975).

Flavonoids

Achromone *C*-glucoside, isobiflorin (5,7-dihydroxy–2-methoxychromone-8-*C*-β-D-glucopyranoside) and biflorin were isolated from

Table 8.8. Classification of clove essential oil constituents.

Structural type	Compounds
Monoterpenes	α-Thujene, α-pinene, myrcene, limonene, *p*-cymene
Sesquiterpenes and their derivatives	*t*-α-Bergamotene, α-cubebene, α-copaene, β-caryophyllene, germacrene D, α-amorphene, β-selinene, α-farnesene, viridiflorene; β-himalachene, valencene, γ-cadinene, zonarene, calamenene, calacorene, caryophyllene and humulene oxides, palustrol, α-cadinol, 4,4-dimethyl-8-bicyclo(6.2.3.0) tridecan-1-ol, 4(12),8(13)-caryophylladien-5β-ol,11,11-dimethyl-8-methylene bicyclo (1.2.0)-3-undecen-5β-ol
Aledehydes and ketones	Valeraldehyde, caproaldehyde, benzaldehyde 2 (or) 3-methoxybezaldehyde, coniferaldehyde, cuminaldehyde, geranial. 2-Hexanone, 2-heptanone, 2-octanone, 2-nonanone, 6-methyl-5-hepten-2-one, fenchone, carvone, acetophenone, 2-hydroxy-4,6-dimethoxy-5-methylacetophenone
Esters	Ethyl caproate, methyl caprylate, ethyl caprylate, methyl palmitate, methyl stearate, methyl linoleate, benzyl tiglate.
Acetates	*Sec*-heptyl, *sec*-nonyl, α-terpenyl, benzyl, phenyl, β-phenethyl, eugenyl, sterallyl
Benzoates	Methyl, ethyl, *n*-propyl, *sec*-heptyl, benzyl
Cinnamate	Ethyl
Salicylates	Methyl, benzyl
Alcohols	Methanol, 2-heptanol, 2-nonalol, linalool, benzyl alcohol
Phenols and their derivatives	Methyl chavicol, *trans*-anethol, methyleugenol, eugenol, chavicol, vanillin
Heterocycles	Furfural, 5-methylfur fural, dimethyl furfural, furfuryl alcohol, 5-methylfurfuryl alcohol, γ-decalactone
Other compounds	Naphthalene

the ethanolic extract of cloves (Zhang and Chen, 1997). From the ethanol extract of the seeds, apigenin 6-*C*-[β-D-xylopyranosyl-(1→2″)-β-D-galactopyranoside]-7-*O*-β-D-glucopyranoside and apigenin-6-*C*-[β-D-xylopyranosyl- (1→2″)-β-D-galactopyranoside]-7-*O*-β-D-(6-*O*-*p*-coumarylglucopyranoside) were isolated (Nassar, 2006). The flavonoids, kaempferol and rhamnetin, isolated from clove are antioxidants. Chemical structures of a few non-volatile constituents are indicated in Fig.8.2.

8.5. Medicinal and Pharmacological Uses

India's traditional Ayurveda healers have used cloves since ancient times to treat respiratory and digestive ailments. Like many culinary spices, cloves help relax the smooth muscle lining of the digestive tract and eating cloves is said to be aphrodisiac. Aqueous extract of clove flower bud inhibits immediate hypersensitivity in rats by inhibition of histamine release from mast cells *in vivo* and *in vitro* (Kim *et al.*, 1998).

Cloves are more often used to assist the action of other herbal remedies rather than alone. When not available, allspice is substituted. It is spicy, warming, stimulant, anodyne, anaesthetic (topical), antiemetic, antigriping (added to other herbs), vermifuge, uterine stimulant, stomachic, aromatic, carminative, antiseptic, antiviral, antibacterial, antifungal, antispasmodic, expectorant, aphrodisiac and promotes salivation and digestive juices. The oil is expectorant, anaesthetic, emmenogogue; it affects the kidney, spleen and stomach and has preservative properties. Tea made from clove bud (other herbs/spices can be used or added to cloves, such as allspice, bay, cinnamon and marjoram) has been used to relieve bronchitis, asthma, coughs, a tendency to infection, tuberculosis, altitude

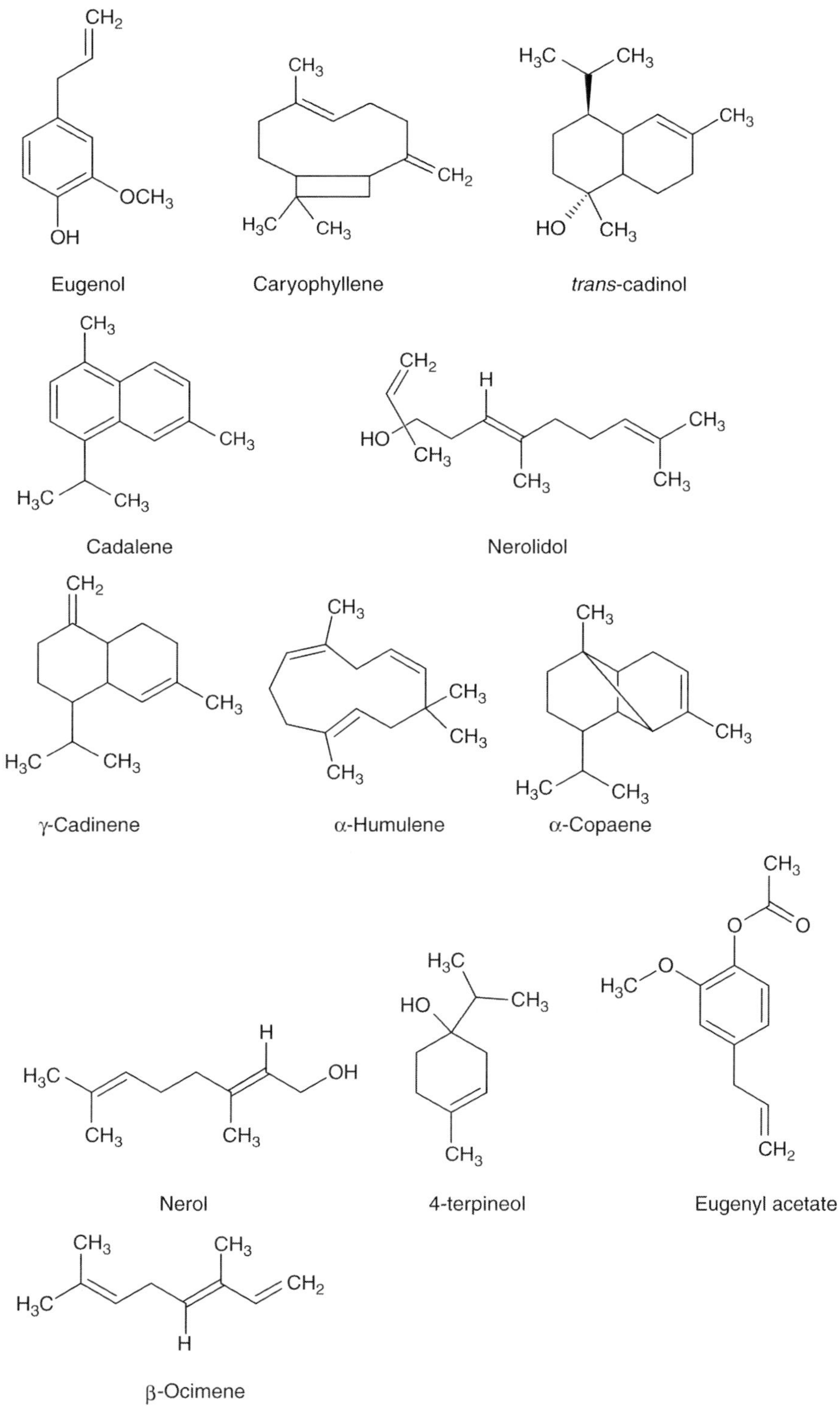

Fig. 8.1. Volatiles from clove.

Table 8.9. Aglycones in clove buds and leaves.

Component	Clove buds		Clove leaves	
	α-Amyloglucosidase	β-Glucosidase	α-Amyloglucosidase	β-Glucosidase
Pentan-3-ol	t	t	t	t
Hexen-3-ol	t	t	t	t
Heptan-3-ol	t	0.5	1.5	t
Octanol	t	t	t	t
Nonan-2-ol	t	t	t	t
Linalool oxide	t	0.5	t	0.4
Linalool	t	0.9	t	0.9
Benzyl alcohol	t	t	t	t
β-Phenylethyl alcohol	t	t	t	t
α-Terpineol	1.1	t	t	t
Nerol	0.5	t	t	0.5
Geraniol	0.3	2.6	t	0.8
Eugenol	3.4	62.6	12.0	76.1
cis-Isoeugenol	2.5	1.6	2.3	1.1
trans-Isoeugenol	2.0	3.7	2.6	3.6
cis-Nerolidol	4.1	1.8	0.8	0.3
Farnesol	54.3	1.9	59.8	2.0

t = trace.
Source: Menon and Narayanan (1992).

sickness, nervous stomach, nausea, diarrhoea, flatulence, indigestion, dyspepsia and gastroenteritis.

In Chinese medicine cloves are used as a kidney tonic (especially for impotence associated with deficient yang), to warm the body, increase circulation and as a digestive aid. They are also used for nausea, vomiting, flatulence, hiccups, stomach chills, fever, caries, toothache, cholera, colic, cracked nipples, diarrhoea, dyspepsia, halitosis (chewing on the whole clove), unusual uterine bleeding, nasal polyps and impotence. The root is used for a weaker effect. The oil is employed for diarrhoea, halitosis, hernia, nausea and toothache.

Ethanolic extract (50%) of clove produced a significant and sustained increase in the sexual activity of normal male rats, without any conspicuous gastric ulceration or adverse effects. Thus, the resultant aphrodisiac activity of the extract lends support to claims for its traditional usage in sexual disorders. In traditional Chinese medicine it is used to treat indigestion, diarrhoea, hernia, ringworm and other fungal infections. In Ayurveda, cloves are used to treat respiratory and digestive ailments, flatulence, nausea and vomiting. The medieval German herbalists used cloves as part of an antigout mixture. Clove is believed to have a cooling effect on the stomach. A paste of clove was applied to the forehead for relief from colds. It has powerful local antiseptic and mild anaesthetic actions.

Clove bud oil has various biological activities such as antibacterial, antifungal, antioxidant and insecticidal properties. The high level of eugenol present in the essential oil imparts strong biological and antimicrobial activity.

Clove oil is an active ingredient in several mouthwash products and a number of over-the-counter toothache pain-relief preparations. It is also used to disinfect root canals. For toothache, clove tea has been used in combination with chamomile or sage.

Eugenol is shown to alleviate neuropathic pain (Guénette *et al.*, 2007). Eugenol inhibits 5-lipoxygenase activity and leukotriene-C4 in human PMNL cells (Raghavenra *et al.*, 2006).

Clove oil is used to prepare microscopic slides for viewing. It is used to treat

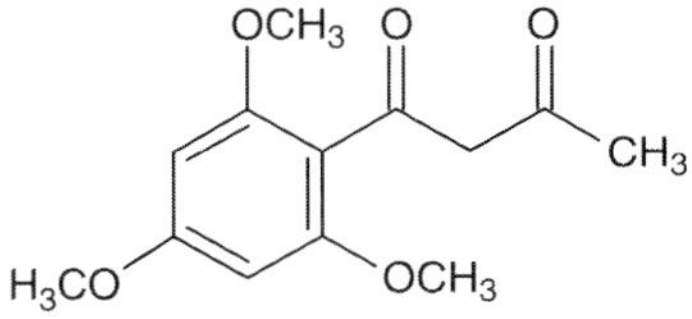

Eugenone

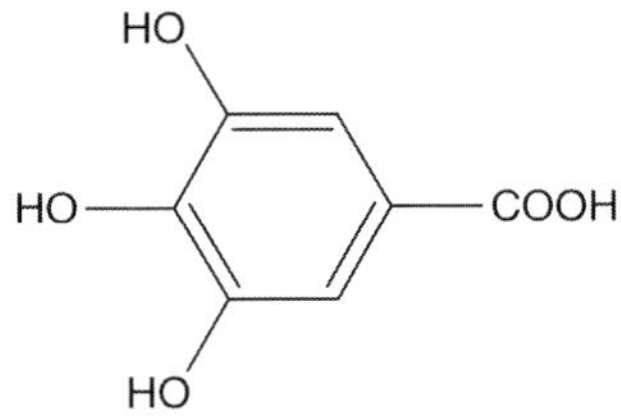

Gallic acid

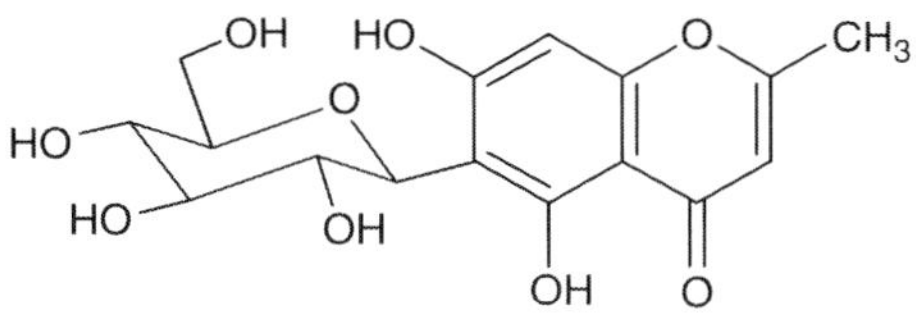

Biflorin

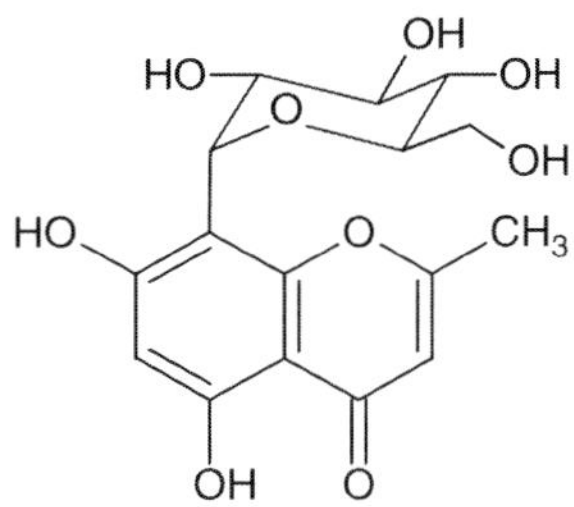

Isobiflorin

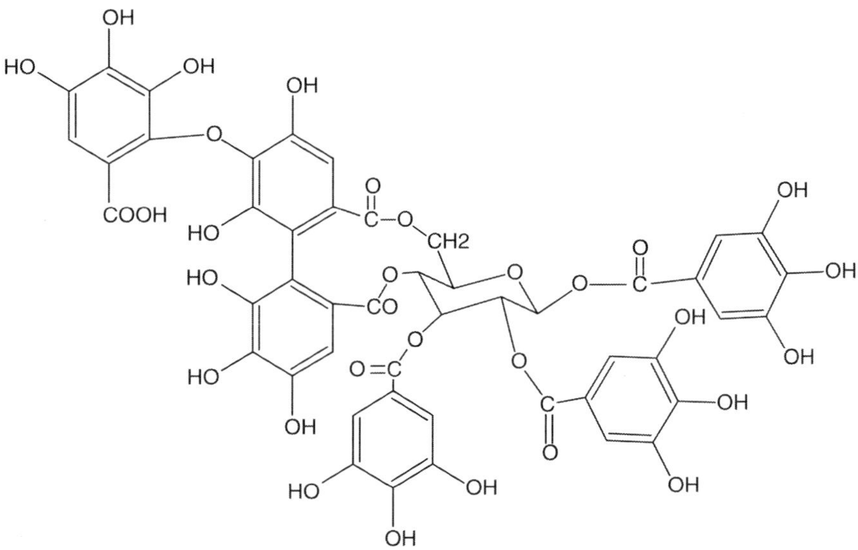

Syzyginin A

Continued

Fig. 8.2. Non-volatiles from clove.

flatulence, colic, indigestion and nausea. Eugenol is used in germicides and perfumes, in the synthesis of vanillin and as a sweetener or intensifier. A recent review by Chaieb *et al.* (2007) lists the chemical composition and biological activity of clove essential oil.

Antimicrobial activity

Clove exhibits potent antimicrobial activity against *Bacillus subtilis*, *Escherichia coli* and *Saccharomyces cerevisiae* (De *et al.*, 1999). Essential oils from clove and eugenol show various degrees of inhibition against

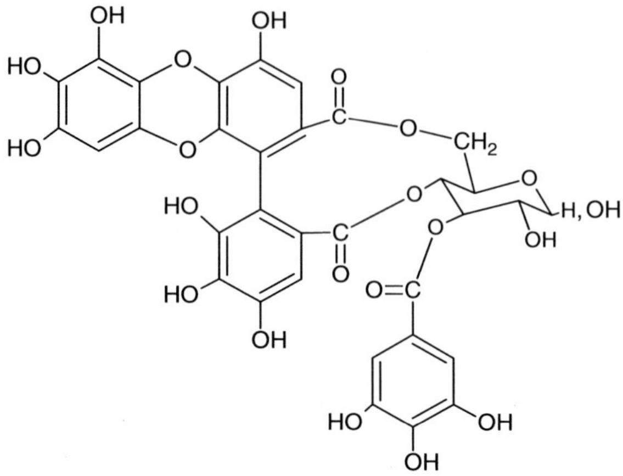

Syzyginin B

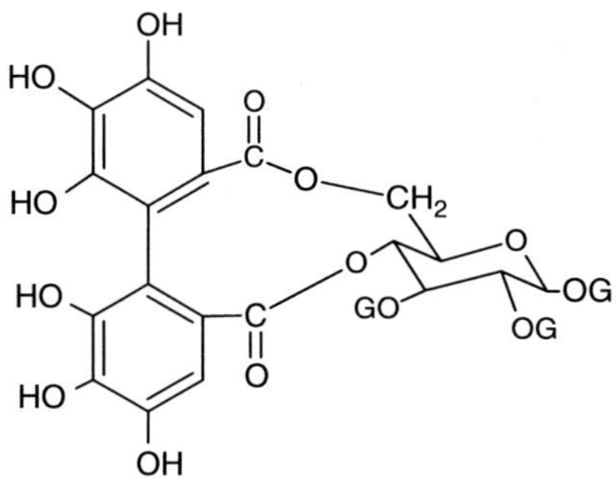

Eugenin
G = galloyl

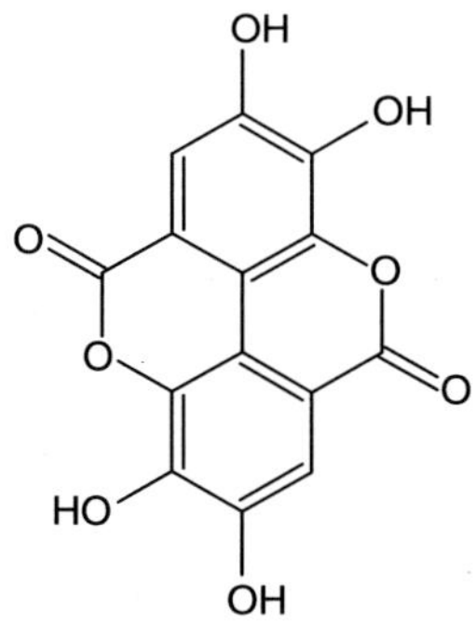

Ellagic acid

Fig. 8.2. *Continued*

Aspergillus niger, S. cerevisiae, Mycoderma sp., *Lactobacillus acidophilus* and *B. cereus*, as estimated by the paper disc agar diffusion method (Meena and Sethi, 1994). The oil also inhibits the growth of *Fusarium verticilloides* (Veluti *et al.*, 2004). Clove oil (1% v/w) inhibits *Listeria monocytogenes* in chicken frankfurters (Mytle *et al.*, 2006). It has excellent antimicrobial properties and is used in food preservation (Smith Palmer *et al.*, 1998, 2001).

Clove extracts show high antifungal activity against *Rhizoctonia solani* (Lee Sang *et al.*, 2003). Clove oil and eugenol are reported to possess significant antifungal activity against rye bread spoilage fungi (Suhr and Nielsen, 2003). Clove oil shows antifungal activity against the fungi belonging to *Eurotium, Aspergillus* and *Penicillium* species, commonly causing deterioration of bakery products (Guynot *et al.*, 2003). Eugenol possesses antifungal activity against *Cladosporium herbarum, Penicillium glabrum, P. expansum* and *A. niger* (Martini *et al.*, 1996; Kong Qiu *et al.*, 2004).

Clove bud oil causes inhibition of both mycelial growth and aflatoxin production of *A. parasiticus* (Farag *et al.*, 1989; Gowda *et al.*, 2004). Clove oil, at concentrations $> 100\,\mu g/ml$, results in reduction in the aflatoxin production in liquid cultures (Sinha *et al.*, 1993). Clove oil inhibits the growth and production of fumonisin B_1 by *F. proliferatum* (Veluti *et al.*, 2003).

Antibacterial activity

Cloves are one of Mother Nature's premium antiseptics. A few drops of the oil in water can stop vomiting and an infusion relieves nausea. Essential oil of clove is effective against *Streptococci*, *Staphylococci* and *Pneumomocci* bacteria.The volatile oils of clove exhibited considerable inhibitory effects and antibacterial activity against several genera of bacteria, including animal and plant pathogens and food poisoning and spoilage bacteria (Deans and Ritchie, 1987; Dorman and Deans, 2000).

Clove kills intestinal parasites and exhibits broad antimicrobial properties, thus supporting its traditional use as a treatment for diarrhoea, intestinal worms and other digestive ailments. Clove essential oil is strongly antimicrobial, antiseptic, haemostatic and anti-inflammatory. Because of its strong antiparasitic action, clove is also included in Dr Huda Clark's protocol for elimination of parasites from the digestive system. It has also been found that a 0.05% solution of eugenol is sufficient to kill *B. tuberculosis*. Clove oil showed antimicrobial activity against some human pathogenic bacteria resistant to certain antibiotics (Arora *et al.*, 1999; Lopez *et al.*, 2005).

Antioxidant activity

Clove essential oil has the highest antioxidant capability of any essential oil, perhaps one of the highest known for a food or supplement. It has been included in some 'longevity' formulae for this reason. Clove and eugenol possess strong antioxidant activity, which is comparable to the activities of the synthetic antioxidants, BHA and pyrogallol (Dorman *et al.*, 2000). Essential oil from clove leaf possesses scavenging activity against the 2,2-diphenyl-1-picryl hydrazyl (DPPH) radical at concentrations lower than the concentrations of eugenol, butylated hydroxytoluene (BHT) and butylated hydroxyanisole (BHA). It also shows a significant inhibitory effect against hydroxyl radicals and acts as an iron chelator (Gulcin *et al.*, 2004; Jirovetz *et al.*, 2006). Clove oil is also commonly used for numbing tooth pain and the healing of mouth and gum sores. The oil can also be used to assist in the breaking of tobacco addiction (Ananda Aromatherapy). The antioxidant activity of clove bud extract and its major aroma components, eugenol and eugenyl acetate, are comparable to that of the natural antioxidant, α-tocopherol (Lee and Shibamoto, 2001). Eugenol inhibits 5-lipoxygenase activity and leukotriene-C4 in human PMNL cells (Raghavenra *et al.*, 2006).

Anti-inflammatory activity

Eugenol, the primary component of clove's volatile oils, functions as an anti-inflammatory substance. In animal studies, the addition of clove extract to diets already high in anti-inflammatory components (like cod liver oil, with its high ω-3 fatty acid content) brings a synergistic effect. In some studies, it further reduces inflammatory symptoms by another 15–30%. Clove also contains a variety of flavonoids, including kaempferol and rhamnetin, which also contribute to clove's anti-inflammatory and antioxidant properties. Another constituent of clove oil, β-caryophyllene, also contributes to the anti-inflammatory activity (Ghelardini *et al.*, 2001).

Anaesthetic effect

Clove oil is used as a safe anaesthetic for aquatic research. Tricaine or MS-222, the only anaesthetic registered in North America, is a very effective anaesthetic for several fish species but its application in the field is limited because the US Food and Drug Administration guidelines demand a 21-day withdrawal period after exposure to MS-222 before the fish enters the food chain. In this context, clove oil is found to be an alternative to MS-222 for use as a fish anaesthetic. Exposure of channel catfish (*Ictalurus punctatus*) to clove oil at a concentration of 100 mg/l induced anaesthesia within 1 min. The fish recovered from a 10 min period of anaesthesia within 4 min after removal from the anaesthetic solution.

Clove oil is therefore used as a safe anaesthetic for channel catfish (Waterstrat, 1999). The anaesthetic effect of clove oil and eugenol for use in aquaculture and aquatic research was also reported by Soltani *et al.* (2001) and Jayathilake *et al.* (2003). Clove oil and eugenol were reported as an acceptable anaesthetic for rabbit fish (*Saiganus lineatus*), coral reef fish (*Pomacentrus amboinensis*) and rainbow trout (*Oncorhynchus mykiss*) for use in aquaculture and aquatic research (Soto and Burhanuddin, 1995; Munday and Wilson, 1997; Keene *et al.*, 1998). It was also found to be useful as a crab anaesthetic (Morgan *et al.*, 2001). β-Caryophellene is also reported to be an anaesthetic (Ghelardini *et al.*, 2001).

Mosquito-repellent activity

Clove oil exhibits repellent activity on *Anopheles albimanus*, *Aedes aegypti*, *A. dirus* and *Culex quinquefasciatus* (Barnard, 1999; Trongtokit *et al.*, 2005).

Insecticidal activity

Eugenol, isoeugenol and methyl eugenol cause contact toxicity to the storage pathogens, *Sitophilus zeamis* and *Tribolium costaneum*. These compounds have similar toxicity to *S. zeamis* at LD_{50} 30 µg/mg insect, while for *T. costaneum* the order of potency is isoeugenol > eugenol > methyl eugenol (Ho *et al.*, 1994; Huang *et al.*, 2002). The clove leaf and bud oils show potent insecticidal activity against the human headlouse (*Pediculus capitis*) (Yang *et al.*, 2003a,b).

Antithrombotic activity

Clove oil inhibits human platelet aggregation induced by arachidonic acid (AA), platelet-activating factor (PAF) or collagen. Clove oil is a more effective inhibitor for aggregation induced by AA and PAF (IC_{50}: 4 and 6 µM, respectively) than collagen (IC_{50}: 132 µM). It inhibits platelet aggregation and thromboxane synthesis and acts as an antithrombotic agent. Eugenol and acetyl eugenol are more potent than aspirin in inhibiting platelet aggregation induced by arachidonate, adrenaline and collagen. In arachidonate-induced aggregation eugenol is on par with indomethacin (Srivastava, 1990).

Anticancerous activity

Clove has strong anticancerous properties. The sesquiterpenes, β-cayophyllene, β-cayophyllene epoxide, α-humulene, α-humulene epoxide and eugenol present in clove oil showed potent anticarcinogenic activity by inducing the detoxifying enzyme, glutathione-*S*-transferase, in mouse liver and small intestine (Zheng *et al.*, 1992).

Antiviral activity

Clove is a potent antiviral agent and eugenin isolated from clove buds showed antiviral activity against *Herpes simplex* virus at a concentration of 10 µg/ml (Kim *et al.*, 2001; Chaieb *et al.*, 2007).

Antipyretic effect

Eugenol, the chief constituent of clove oil, has marked antipyretic activity when given intravenously, intragastrically and centrally to rabbits made febrile by interleukin-1. Eugenol was more effective in reducing fever than acetaminophen and it reduced fever primarily through a central action similar to that of common antipyretic drugs, such as acetaminophen (Feng and Lipton, 1987).

Toxicity studies

Cloves can cause local skin irritation, pulmonary oedema, mouth sensitivity and sudden lower airway closure. In addition, smoking clove cigarettes can damage soft tissues and injure the airway linings (Fetrow and Avila, 1999).

8.6. ISO Specifications

Quality requirements for various clove products is country specific. The American Spice Trade Association (ASTA) and Food and Drug Administration (FDA) recommendations for whole spice (clove) are illustrated in Table 8.10.

Table 8.10. ASTA cleanliness specifications.

Items	Suggested limit whole clove
Whole dead insect (mg/lb)	4.00
Mammalian excreta (mg/lb)	5.00
Other excreta (mg/lb)	8.00
Mould, % by weight	1.00
Insect-defiled/infested, % by weight	1.00
Extraneous, % by weight	1.00

The specific requirement for ground clove is for a minimum quercitannic content, maximum clove stem content and minimum 15% volatile ether extract. The ISO specifications of clove oil are illustrated in Table 8.11.

8.7. Conclusion

The spice clove and its value-added products are used extensively for flavouring food and confectionery. Clove oil has many industrial and pharmacological applications. Most of the studies conducted so far pertain to the clove volatiles and very little attention has been paid to the non-volatile constituents. Therefore, the phytochemical studies and biological activities of non-volatiles are worth examining. This may lead to identifying new properties and novel molecules.

Table 8.11. Specifications of clove products.

Oil of clove leaves	ISO 3141:1997	*Syzygium aromaticum*
Oil of clove buds	ISO 3142:1997	*S. aromaticum*
Oil of clove stems	ISO 3143:1997	*S. aromaticum*
Cloves, whole and ground (powdered)	ISO 2254:1980	*S. aromaticum*

Source: http://www.fao.org/inpho/content/documents/vlibrary/ad420e/AD420e37.htm.

References

Arora, D.S., Jasleen Kaur, J. and Kaur, J. (1999) Antimicrobial activity of spices. *International Journal of Antimicrobial Agents* 12(3), 257–262.

Barnard, D.R. (1999) Repellency of essential oils to mosquitoes (Diptera: Culicidae). *Journal of Medical Entomology* 36(5), 625–629.

Brieskorn, C.H., Münzhuber, K. and Unger, G. (1975) Crataegolsäure und steroidglukoside aus blütenknospen von *Syzygium aromaticum*. *Phytochemistry* 14, 2308–2309.

Chaieb, K., Hajlaoui, H., Zmantar, T., Kahla-Nakbi, A.B., Rouabhia, M., Mahdouani, K. and Bakhrouf, A. (2007) The chemical composition and biological activity of clove essential oil, *Eugenia caryophyllata* (*Syzigium aromaticum* L. Myrtaceae): a short review. *Phytotherapy Research* 21, 501–506.

DASD (Directorate of Arecanut and Spices Development) (2007) In: Premaja, P. and Manoj Kumar, K. (eds) *Arecanut and Spices Database*. Naseema Printers and Publishers, Kochi, India, p. 110.

De, M., De, A.K. and Banerjee, A.B. (1999) Antimicrobial screening of some Indian Spices. *Phytotherapy Research* 13(7), 616–618.

Deans, S.G. and Ritchie, G. (1987) Antibacterial properties of plant essential oils. *International Journal of Food Microbiology* 5, 165–180.

Dorman, H.J.D. and Deans, S.G. (2000) Antimicrobial agents from plants: antibacterial activity of plant volatile oils. *Journal of Applied Microbiology* 88(2), 308–316.

Dorman, H.J.D., Surai, D. and Deans, S.G. (2000) *In vitro* antioxidant activity of a number of plant essential oils and phyto constituents. *Journal of Essential Oil Research* 12, 241–248.

Farag, R.S., Daw, Z.Y. and Abo Raya, S.H. (1989) Influence of some spice essential oils on *Aspergillus parasiticus* growth and production of aflotoxins in a synthetic medium. *Journal of Food Science* 54(1), 74–76.

Feng, J. and Lipton, J.M. (1987) Eugenol: antipyretic activity in rabbits. *Neuropharmacology* 26, 1775–1778.

Fetrow, C.W. and Avila, J.R. (1999) *The Complete Guide to Herbal Medicines*. Springhouse Corporation, Springhouse, PA.

Gaydou, E.M. and Randriamiharisoa, R. (1987) Multidimensional analysis of gas chromatographic data, application to the differentiation of clove bud and clove stem essential oils from Madagascar. *Perfumer and Flavorist* 12, 45–51.

Ghelardini, C., Galeotti, N., Di Cesare Mannelli, L., Mazzanti, G. and Bartolini, A. (2001) Local anaesthetic activity of β-caryophyllene 11. *Farmaco* 56, 387–389.

Gopalakrishnan, M., Narayanan, C.S. and Mathew, A.G. (1984) Sequiterpene hydrocarbons from clove oil. *Lebensmittel-Wissenschaft Und-Technologie B* 17, 42–43.

Gopalakrishnan, N. and Narayanan, C.S. (1988) Composition of clove leaf oil during leaf growth. *Indian Perfumer* 32(2), 130–132.

Gopalakrishnan, N., Narayanan, C.S. and Mathew, A.G. (1988) Chemical composition of Indian clove bud, stem and leaf oils. *Indian Perfumer* 32, 229–235.

Gopalakrishnan, N., Shanti, P.P.V and Narayanan, C.S. (1990) Composition of clove bud oil extracted using carbon dioxide. *Journal of the Science of Food and Agriculture* 50, 111–117.

Gowda, N.K.S., Malathi, V. and Suganthi, R.U. (2004) Effect of some chemical and herbal compounds on growth of *Aspergillus parasiticus* and aflatoxin production. *Animal Food Science and Technology* 116, 281–291.

Guan, W., Li, S., Yan, R., Tang, S. and Quan, C. (2007) Comparison of essential oils of clove buds extracted with supercritical carbon dioxide and other three traditional extraction methods. *Food Chemistry* 101(4), 1558–1564.

Guénette, S.A., Ross, A., Marier, J., Beaudry, F. and Vachon, P. (2007) Pharmacokinetics of eugenol and its effects on thermal hypersensitivity in rats. *European Journal of Pharmacology* 562, Issues 1–2, 7 May 2007, pp 60–67.

Guenther, E. (1950) *The Essential Oils*, Volume IV. Van Nostrand Publishing Co., New York, pp. 396–437.

Gülçin, I., Sat, I.G., Beydemir, S., Elmastas, M. and Kufrevioglu, O.I. (2004) Comparison of antioxidant activity of clove (*Eugenia caryophylata* Thunb) buds and lavender (*Lavandula stoechas* L.). *Food Chemistry* 87, 393–400.

Guynot, M.E., Ramos, A.J., Setó, L., Purroy, P., Sanchis, V. and Marín, S. (2003) Antifungal activity of volatile compounds generated by essential oils against fungi commonly causing deterioration of bakery products. *Journal of Applied Microbiology* 95, 893.

Ho, S.H., Cheng, L.P.L., Sim, K.Y. and Tan, H.T.W. (1994) Potential of cloves (*Syzygium aromaticum* (L.) Merr. and Perry as a grain protectant against *Tribolium castaneum* (Herbst) and *Sitophilus zeamais* Motsch. *Postharvest Biology and Technology* 4, 179–183.

Huang, Y., Ho, S.H., Lee, H.C. and Yap, Y.L. (2002) Insecticidal properties of eugenol, isoeugenol and methyleugenol and their effects on nutrition of *Sitophilus zeamais* Motsch. (Coleoptera: Curculionidae) and *Tribolium castaneum* (Herbst) (Coleoptera: Tenebrionidae). *Journal of Stored Products Research* 38, 403–412.

Jayathilake, P.S., Edirisinghe, U. and Silva, D.D.N. de (2003) Use of clove oil to anesthetize female wild guppy (*Poecilia reticulata*) for a short duration. *Tropical Agricultural Research* 15, 235–241.

Jirovetz, L., Buchbauer, G., Stoilova, I., Stoyanova, A., Krastanov, A. and Schmidt, E. (2006) Chemical composition and antioxidant properties of clove leaf essential oil. *Journal of Agriculture and Food Chemistry* 54(17), 6303–6307.

Keene, J.L., Noakes, D.L.G., Moccia, R.D. and Soto, C.G. (1998) The efficacy of clove oil as an anaesthetic for rainbow trout, *Oncorhynchus mykiss* (Walbaum). *Aquaculture Research* 29(2), 89–101.

Kim, H.J., Lee, J.S., Woo, E.R., Kim, M.K., Yang, B.S., Yu, Y.G., Park, H. and Lee, Y.S. (2001) Isolation of virus cell fusion inhibitory components from *Eugenia caryophyllata*. *Planta Medica* 67, 277–279.

Kim, H.M., Lee, E.H. and Hong, S.H. (1998) Effect of *Syzygium aromaticum* extract on immediate hypersensitivity in rats. *Journal of Ethnopharmacology* 60, 125–131.

Kong Qiu, L., Ain, Song YizHong, Zhang Lili, Chen LiuYong and Qing Fang, L. (2004) Natural antifungal compounds from *Syzygium aromaticum*. *Acta Agriculturae Shanghai* 20(3), 68–72.

Lawrence, B.M. and Reynolds, R.J. (1989) Progress in essential oils. *Perfumer and Flavourist* 14(2), 45–55.

Lee, K.G. and Shibamoto, T. (2001) Antioxidant property of aroma extract isolated from clove buds [*Syzygium aromaticum* (L.) Merr. et Perry]. *Food Chemistry* 74(4), 443–448.

Lee Sang, Myeong, Jo HyunJin, Kim DongSoo, Kim DoWan, Lee DongWoon, Choo HoYul, and Lee Chongkyn (2003) Effect of clove extract on antifungal activity against *Rhizoctonia solani* and plant growth. *KFRI Journal of Forest Science Seoul* 66, 105–111.

Lopez, P., Sanchez, C., Batle, B. and Nerin, C. (2005) Solid and vapour phase antimicrobial activities of six essential oils: susceptibility of selected foodborne bacterial and fungal strains. *Journal of Agriculture and Food Chemistry* 53, 6338–6346.

Martini, H., Weidenborner, M., Adams, S. and Kunz, B. (1996) Eugenol and carvacrol: the main fungicidal compounds in clove and savory. *Italian Journal of Food Science* 8, 63–67.

Meena, M.R. and Sethi, V. (1994) Antimicrobial activity of essential oils from spices. *Journal of Food Science and Technology* 31(1), 68–70.

Menon, A.N. and Narayanan, C.S. (1992) Glycosidically bound volatiles of clove *Syzygium aromaticum* (L.) Merr. *et* Perry (*Myrtaceae*). *Flavour and Fragrance Journal* 7(3), 155–157.

Morgan, J., Cargill, C. and Groot, R. (2001) The efficacy of clove oils as anesthetic for decapod crustaceans. *Bulletin of the Aquaculture Association of Canada* 101, 27–31.

Muchalal, M. and Crouzet, J. (1985) Volatile components of clove essential oil (*Eugenia caryophyllus* Spreng.) neutral fraction. *Agricultural and Biological Chemistry* 49(6), 1583–1589.

Munday, P.L. and Wilson, S.K. (1997) Comparative efficacy of clove oil and other chemicals in anaesthetization of *Pomacentrus amboinensis*, a coral reef fish. *Journal of Fish Biology* 51(5), 931–938.

Mytle, N., Anderson, G.L., Doyle, M.P. and Smith, M.A. (2006) Antimicrobial activity of clove (*Syzgium aromaticum*) oil in inhibiting *Listeria monocytogenes* on chicken frankfurters. *Food Control* 17(1), 102–107.

Narayanan, C.R. and Natu, A.A. (1974) Triterpene acids of Indian clove buds. *Phytochemistry* 13(9), 1999–2000.

Nassar, M.I. (2006) Flavonoid triglycosides from the seeds of *Syzygium aromaticum. Carbohydrate Research* 341, 160–163.

Nonaka, G., Harada, M. and Nishioka, I. (1980) Eugeniin, a new ellagitannin from cloves. *Chemical and Pharmacological Bulletin* 28, 685–687.

Pino, J.A., Marbot, R., Aguero, J. and Fuentes, V. (2001) Essential oil from buds and leaves of clove (*Syzygium aromaticum* (L.) Merr. et Perry) grown in Cuba. *Journal of Essential Oil Research* 13(4), 278–279.

Pruthi, J.S. (1976) *Spices and Condiments*. National Book Trust, New Delhi, p. 91–98.

Purseglove, J.W., Brown, E.G, Green, C.L. and Robins, S.R.J. (1981) *Spices*, Volume 1. Longman, London, p. 229–287.

Raghavenra, H., Diwakr, B.T., Lokesh, B.R. and Naidu, K.A. (2006) Eugenol, the active principle from cloves inhibits 5-lipoxygenase activity and leukotriene-C4 in human PMNL cells. *Prostaglandins, Leukotrienes and Essential Fatty Acids* 74, 23–27.

Raina, V.K., Srivastava, S.K., Aggarwal, K.K., Syamasundar, K.V. and Kumar, S. (2001) Essential oil composition of *Syzygium aromaticum* leaf from Little Andaman, India. *Flavour and Fragrance Journal* 16(5), 334–336.

Sinha, K.K., Sinha, A.K. and Prasad, G. (1993) The effect of clove and cinnamon oils on growth and aflatoxin production by *Aspergillus flavus. Letters in Applied Microbiology* 16(3), 114–117.

Smith Palmer, A., Stewart, J. and Fyfe, L. (1998) Antimicrobial properties of plant essential oils and essence against five important food born pathogens. *Letters in Applied Microbiology* 26, 118–122.

Smith Palmer, A., Stewart, J. and Fyfe, L. (2001) The potential application of plant essential oils as natural foods preservatives in soft cheese. *Food Microbiology* 18, 463–470.

Soltani, M., Omidbeigi, R., Rezvani, S., Mehrabi, M.R. and Chitsaz, H. (2001) Study of anaesthetic effects induced by clove flower (*Eugina caryophillata*) on rainbow trout (*Oncorhynchus mykiss*) under various water quality conditions. *Journal of the Faculty of Veterinary Medicine, University of Tehran* 56(4), ar85–ar89.

Soto, C.G. and Burhanuddin (1995) Clove oil as a fish anaesthetic for measuring length and weight of rabbitfish (*Siganus lineatus*). *Aquaculture* 136(1–2), 149–152.

Srivastava, A.K., Srivastava, S.K. and Syamsundar, K.V. (2005) Bud and leaf essential oil composition of *Syzygium aromaticum* from India and Madagascar. *Flavour and Fragrance Journal* 20(1), 51–53.

Srivastava, K.C. (1990) Antiplatelet components from common food spice clove (*Eugenia caryophyllata*) and their effects on prostanoid metabolism. *Planta Medica* 56(6), 501–502.

Suhr, K.I. and Nielsen, P.V. (2003) Antifungal activity of essential oils evaluated by two different application techniques against rye bread spoilage fungi. *Journal of Applied Microbiology* 94(4), 665–674.

Tainter, R.D. and Grenis, T.A. (1993) *Spices and Seasonings Food Science and Technology*. VCH Publishers, New York.

Tanaka, T., Orii, Y., Nonaka, G. and Nishioka, I. (1993) Tannins and related compounds. CXXIII. Chromone, acetophenone and phenyl propanoid glycosides and their galloyl and/or hexahydroxy phenoyl esters from leaves of *Syzygium aromaticum* Merr and Perry. *Chemical Pharmaceutical Bulletin* 28, 685–687.

Tanaka, T., Orii, Y., Nonaka, G., Nishioka, I. and Kouno, I. (1996) Syzyginins, A and B, two ellagitannins from *Syzygium aromaticum*. *Phytochemistry* 43, 1345–1348.

Trongtokit, Y., Rongsriyum, Y., Komalamisra, N. and Apiwathnasorn, C. (2005) Comparative repellency of 38 essential oils against mosquito bites. *Phytotherapy Research* 19, 303–309.

USDA (2005) National Nutrient Database for Standard Reference, Release 18.

Veluti, A., Sanchis, V., Ramos, A.J. and Marin, S. (2003) Inhibitory effect of cinnamon, clove, lemongrass, oregano and palmarosa essential oils on growth and fumonisin B_1 production by *Fusarium proliferatum* in maize grain. *International Journal of Food Microbiology* 89, 145–154.

Veluti, A., Sanchis, V., Ramos, A.J. and Marin, S. (2004) Effect of essential oils of cinnamon, clove, lemongrass, oregano and palmarosa on growth and fumonisin B1 production by *Fusarium verticillioides* in maize. *Journal of the Science of Food and Agriculture* 84(10), 1141–1146.

Vernin, G., Vernin, E., Merzger, J., Piyol, L. and Parkanyi, C. (1994) Developments in food science 34. In: Charalambous, G. (ed.) *Spices, Herbs and Edible Fungi*. Elsevier, Amsterdam, pp. 483–499.

Walter, R.H. (1972) β-Caryophyllene in native clove bud oil. *Phytochemistry* 211(1), 405–406.

Waterstrat, P.R. (1999) Induction and recovery from anesthesia in channel catfish *Ictalurus punctatus* fingerlings exposed to clove oil. *Journal of the World Aquaculture Society* 30, 250–255.

www.anandaapothecary.com/aromatherapy-essential-oils/clove-essentioal-oil.html

Yang, Y.C., Lee, S.H., Clark, J.M. and Ahn, Y.J. (2003a) Insecticidal activity of plant essential oils against *Pediculus humanus capitis* (Anoplura: Pediculidae). *Journal of Medical Entomology* 41, 699–704.

Yang, Y.C., Lee, S.H., Clark, J.M. and Ahn, Y.J. (2003b) Ovicidal and adulticidal effects of of *Eugenia caryophyllata* bud and leaf oil compounds on *Pediculus capitis*. *Journal of Agricultural and Food Chemistry* 54, 4884–4888.

Zachariah, T.J., Krishnamoorthy, B., Rema, J. and Mathew, P.A. (2005) Oil constituents in bud and pedicel of clove (*Syzygium aromaticum*). *Indian Perfumer* 49, 313–316.

Zhang, Y.W. and Chen, Y. (1997) Isobiflorin, a chromone-C-glucoside from cloves (*Eugenia caryophyllata*). *Phytochemistry* 45, 401–403.

Zheng, G.Q., Kenney, P.M. and Lam, L.K.T. (1992) Sesquiterpenes from clove (*Eugenia caryophyllata*) as potential anticarcinogenic agents. *Journal of Natural Products* 55(7), 999–1003.

9 Nutmeg and Mace

N.K. Leela

9.1. Introduction

Nutmeg (*Myristica fragrans* Houtt.) belongs to the family Myristicaceae, with about 18 genera and 300 species. The genus *Myristica* is distributed from India and South-east Asia to North Australia and the Pacific Islands. Five species occur in India, including *M. fragrans*, *M. beddomeii* and *M. malabarica* (Anon., 1962). Nutmeg has its origins in the Spice Islands of Indonesia. It was widely popular in Europe and India for its flavouring and medicinal properties. The name nutmeg is derived from the Latin word *nux muscatus*, meaning 'musky nut'. Two important spices are derived from the fruit – nutmeg and mace. The spice nutmeg is the dried kernel of the seed and mace is the dried aril surrounding it. Both the spices have similar flavour. However, nutmeg is reported to be slightly sweeter than mace and is more preferred in food. Besides nutmeg and mace, a number of other products, namely oleoresin, nutmeg butter and essential oils, are also derived from *M. fragrans*. These value-added products find varied use in the food, medicine and perfume industries. The pericarp of the fruit is used in Grenada to make a jam called *morne delice*. In Indonesia, the fruit is sliced finely, cooked and crystallized to make a fragrant candy called *manisan pala* (nutmeg sweets) (http://en.wikipedia.org/wiki/Nutmeg; May 2007).

Nutmeg is produced in the tropical areas of Indonesia and the West Indies. World production of nutmeg is estimated to average between 10,000 and 12,000 t per year, with annual world demand estimated at 9000 t; production of mace is estimated at 1500–2000 t. Indonesia and Grenada dominate production and exports of both products, with a world market share of 75 and 20%, respectively. Other producers include India, Malaysia, Papua New Guinea, Sri Lanka and the Caribbean Islands. The principal import markets are the European Community, the USA, Japan and India. Singapore and the Netherlands are major re-exporters (http://en.wikipedia.org/wiki/Nutmeg; May 2007).

The East Indian Islands of Siauw, Sangihe, Ternate, Ambon, Banda and Papua produce highly aromatic nutmeg, traded as East Indian nutmeg. Grenada produces the West Indian nutmeg, which is milder in flavour and lighter in colour. International trade in nutmeg is either of the East or West Indian nutmeg, with a negligible quantity of wild 'Bombay' nutmeg imported by the USA. The USA is the largest individual market for whole nutmegs. US importers prefer the East Indian type of deep brown, aromatic nutmeg and orange-red nutmeg

(eds V.A. Parthasarathy, B. Chempakam and T.J. Zachariah)

Table 9.1. Area and production of nutmeg in India.

Year	Area (ha)	Production (t)
1994–95	4,756	2,895
1995–96	5,345	1,471
1996–97	6,419	2,044
1997–98	6,592	2,102
1998–99	6,401	1,603
1999–00	7,110	1,773
2000–01	7,517	1,919
2001–02	7,849	1,985
2002–03	8,704	2,184
2003–04	10,010	2,525
2004–05	10,010	2,530

Source: DASD (2007) Directorate of Arecanut and Spices Development.

and mace in their whole form. Indonesia traditionally has been the principal supplier of nutmeg and mace to the US market, accounting for an average 65% of total US imports of nutmeg per year in terms of volume (Krishnamoorthy and Rema, 2000). The area and production of nutmeg in India for the period 1994–2005 are shown in Table 9.1. The figures indicate that, though the area under nutmeg cultivation has doubled, production has slightly decreased over the period.

9.2. Botany and Uses

Nutmeg is a medium-sized tree reaching a height of 4–10 m. It is dioecious, with male and female flowers occurring on different trees. The fruits are pendulous, broadly pyriform, yellow smooth, 7–10 cm long, fleshy, splitting open into two halves when ripe, showing the ovoid, 2–3 cm-long, dark brown shining seeds, with hard seedcoat surrounded by a lanciate red aril attached to the base of the seed. The seed of nutmeg is large, with ruminate endosperm and is considered as the most primitive among the flowering plants (Krishnamoorthy and Rema, 2000). The first harvest of nutmeg trees is carried out 7–9 years after planting and the trees reach their full potential after 20 years.

Mace is called 'flower of nutmeg' in France. Arab traders first introduced it into Europe during the 11th century to flavour beer. Its popularity as a spice reached its height during the 17th century when the Dutch monopolized the nutmeg trade in the Spice Islands. Mace and nutmeg have different colour and flavour profiles. Mace blades are smooth and shiny and are about 4 cm long. Mace can be pale orange, yellowish-brown, orange or reddish-brown in colour. The colour and flavour of mace varies depending on its origins – reddish-orange from Indonesia and brownish-yellow from the West Indies. It is spicy and bitter, with clove-like and piney overtones. Its aroma is terpeny. Mace is more aromatic than nutmeg, but has more bitter notes. Ground mace is lighter in colour than ground nutmeg. The rind of nutmeg constitutes about 80% fruit weight. It is used to produce nutmeg rind preserve, candy, pickle, chutney and powder (Joshi *et al.*, 1996).

Although whole nutmeg is available, ground nutmeg is more popular. The spice in the ground form is used mainly in the food processing industry. It loses its flavour easily when ground. Therefore, it is generally grated just before cooking or baking. It complements chocolate, fruits, custards, vanilla, coconut milk, lemongrass and *kari* leaves. Nutmeg provides an intense, sweet, spicy aroma to pastries, cakes, sweet rolls, banana bread, pumpkin pies, apple pies, ice cream, chocolate and lemon desserts. Nutmeg is also used in cheese fondues and it enhances savoury products such as vegetable stews, sauces, processed meat and pork patties.

Europeans use it in mashed potatoes, rice dishes, soups, rice puddings, pies, eggnog, biscuits and milk-based drinks. Italians flavour spinach with nutmeg for stuffed pastas. It is also a favourite spice of the Dutch, who use it in potatoes and other vegetables. Nutmeg is an important ingredient in the French spice blend that is used to flavour meats. Along with mace, it is an important ingredient in spice blends of India, the Middle East and North Africa. Nutmeg is one of the ingredients in the pungent *garam masalas* of North India. The aromatic spice blends of the Middle East and Asia also contain nutmeg. It is generally used sparingly during the cooling process. In Indonesia, nutmeg is used in sauces and curries, and nutmeg pulp is made into a local jam. Nutmeg has become a popular spice in Caribbean cooking and is added to jerk seasoning, pastries, ice creams, fruit punches, eggnogs, breads and cakes.

Traditionally, nutmeg has been used to treat digestive disorders, such as nausea and diarrhoea, and kidney ailments. South-east Asians also treat fevers, headaches and bronchial problems with nutmeg. The Chinese consider it to be an aphrodisiac.

Mace is sold as either whole or ground spice and is used in savoury dishes. It is used to flavour milk-based sauces and processed meats like sausages. Soups, pickles, ketchup and chutneys are also seasoned with mace. In Europe and the USA, mace is usually used in light-coloured products, such as cream soups, cream sauces, crackers, pie fillings and cakes. Mace is also used with fish and in vegetable purees, meat stews and meat pies. Commercially, mace flavours frankfurters, doughnuts, pickles, preserves, ice cream, confectionary, icings, sausages, knockwurst, ham, soup mixes and poultry.

Mace pairs well with fruit, sugar, chocolate and milk-based products, cakes, cookies and doughnuts. It provides an intense aroma to Middle Eastern and Asian foods. In Middle Eastern, Iranian and northern Indian recipes, mace is often combined with nutmeg. Arabs add mace to mutton and lamb dishes and many spice blends. It is ground and sprinkled over North Indian *pulaos*, lamb and other meat dishes to give aroma. Ground mace is used commonly in South-east Asian, Chinese and Indian food such as *garam masalas*, curries, sauces, puddings, cakes, pies and cookies. Asian Indians traditionally have treated stomach pains, dysentery, vomiting and the symptoms of malaria with mace. It is also chewed to prevent foul breath (Uhl, 2000).

Nutmeg oil and mace oil are used mainly in flavouring soft drinks, canned foods and meat products (Krishnamoorthy and Rema, 2000). Nutmeg oil is used in cosmetics, men's perfume and toiletries, due to its aromatic properties. Mace oil is also used to a limited extent in perfumes and soaps. The oleoresin of nutmeg finds use in the preparation of meat and vegetables, and to flavour milk dishes and punches. The fleshy outer cover of the fruit is crystallized, pickled or made into jellies.

Myristicin, which imparts hallucinogenic properties, is also reported to be an effective insecticide. The lignan types of the constituents in the nut are anticarcinogenic. Camphene present in the oil is used in the manufacture of camphor and related compounds and has strong antibacterial, antifungal and insecticidal properties. Pinene of the essential oil of nutmeg is used to make camphor, solvents, plasticizers, perfume bases and synthetic pine oil. Dipentine is used in the manufacture of resins and is used as a wetting and dispersing agent. Myristic acid is used in the food industry as a flavour ingredient and also finds use in the preparation of soap, liquid detergents, shampoos, shaving creams, perfumes, plastics, in compounding rubber, paints and greases, in the synthesis of esters for flavours and perfumes and as a component of good-grade additives.

9.3. General Composition

The principal constituents of nutmeg are fixed oil, volatile oil and starch. It also contains proteins, cellulose, pentosans, resin and mineral elements. The percentages of constituents differ between the spices and this is a consequence of geographical origin, quality and duration of storage and even growing locations. The flavour and therapeutic action are due to the volatile oil. Al-Bataina *et al.* (2003) analysed the element composition of nutmeg, which is indicated in Table 9.2. Nutmeg is a good source of potassium, magnesium and phosphorus.

Nutmeg is reported to contain moisture, 14.3%; protein, 7.5%; ether extract, 36.4%; carbohydrates, 28.5%; fibre, 11.6%; and mineral matter, 1.7%; calcium, 0.12%; phosphorus, 0.24%; and iron, 4.6 mg/100 g. It contains volatile oil (6–16%); starch

Table 9.2. Elemental contents of Nutmeg.

Element	Quantity
Mg (%)	0.45
Al (mg/kg)	210
Si (mg/kg)	167
P (%)	0.17
S (%)	0.14
Cl (mg/kg)	402
K (%)	0.63
Ca (%)	0.30
Ti (mg/kg)	13
Mn (mg/kg)	124
Fe (mg/kg)	151
Cu (mg/kg)	73
Zn (mg/kg)	44
Br (mg/kg)	95
Rb (mg/kg)	32
Sr (mg/kg)	11

Source: Al-Bataina *et al.* (2003).

(14.6–24.2%); pentosans (2.25%); furfural (1.5%); and pectin (0.5–0.6%). It is only a fair source of vitamins (Pruthi, 1976). Table 9.3 shows the nutritional composition of nutmeg (Tainter and Grenis, 1993).

Table 9.3. Nutritional composition of nutmeg per 100 g.

Composition	USDA Handbook 8-2 (ground)	ASTA
Water (gm)	6.23	4.00
Food energy (Kcal)	525	565
Protein (gm)	5.84	7.00
Fat (gm)	36.31	38.90
Carbohydrate (gm)	49.29	47.30
Ash (gm)	2.34	2.00
Calcium (gm)	0.184	0.20
Phosphorous (mg)	213	200
Sodium (mg)	16	10
Potassium (mg)	350	400
Iron (mg)	3.04	2.20
Thiamine (mg)	0.346	0.360
Riboflavin (mg)	0.057	0.250
Niacin (mg)	1.299	9.400
Vitamin A activity (RE)	10	10

Source: Tainter and Grenis (1993).

9.4. Chemistry

There are several reviews on the phytochemical studies of nutmeg (Satyavathy *et al.*, 1987; Thakur *et al.*, 1989; Ross, 2001; Latha *et al.*, 2005). The constituents of nutmeg can be classified broadly into terpenoids, fatty acids, phenolic acids, lignans, neolignans and miscellaneous compounds.

Fixed oil

Nutmeg contains 25–50% lipids as fixed oil comprising mainly of myristic, petroselinic and palmitic acids (Purseglove *et al.*, 1981). There are two general methods by which the fixed oil of nutmeg is extracted. In one method, ground nutmeg is subjected to intense hydraulic pressure and heat (heated plates in the presence of steam) while, in the other method, the ground nutmeg is extracted by refluxing with a solvent-like diethyl ether. Both processes will result in the crude fixed oil containing significant quantities of essential oil in the range 10–12%. Prior steam distillation will lead to a significant reduction of essential oil in the prepared fixed oil.

The extracted or expressed fixed oil is an orange-coloured aromatic semi-solid, also known as concrete or nutmeg butter (because it has the consistency of butter at room temperature). It melts at 45–51°C and has a specific gravity of 0.990. It is completely soluble in hot alcohol, but sparingly soluble in cold. However, it is freely soluble in ether and chloroform. The major component of fixed oil is trimyristin.

Nutmeg butter consists of mainly saturated fats (90%) with 10% unsaturated fats (Purseglove *et al.*, 1981). The fatty acid profile of nutmeg and mace is given in Table 9.4. Nutmeg contains 35.7% total lipids, 74.9% of which is myristic acid. Similarly, mace contains 30.4% total lipids, in which oleic acid (40.3%) and palmitic acid (31.3%) predominate (Chandrashekar *et al.*, 1995). Irradiation of nutmeg leads to the breakdown of triacyl glycerols, releasing free fatty acids (Niyas *et al.*, 2003). Sensory

Table 9.4. Fatty acid profile of nutmeg and mace.

Fatty acid	Nutmeg (%)	Mace (%)
C 14:0	74.9	0.3
C 14:1	–	–
C 16:0	9.7	31.3
C 16:1	–	–
C 18:0	6.9	2.9
C 18:1	3.8	40.3
C 18:2	1.8	19.8
C 18:3	1.0	2.9
C 20:0	2.0	1.8

Source: Chandrashekar *et al.* (1995).

evaluation of the irradiated spice is also very different from the non-irradiated spice.

Volatile oil

Volatile oil is extracted from both nutmeg and mace. Usually, it is obtained by hydro/steam distillation. Nutmeg oil is a colourless or yellow liquid with the characteristic odour and taste of nutmeg. The oil is insoluble in water, but soluble in alcohol. It keeps best in cool, tightly closed containers protected from light. Depending on the type, its flavour can vary from a sweetly spicy to a heavier taste. It has a clove-like, spicy, sweet, bitter taste with a terpeny, camphor-like aroma. It is sweeter in flavour than mace. The physico-chemical properties of East Indian and West Indian nutmeg oil from different regions, as described by Guenther (1952), are given below:

Physico-chemical properties	East Indian oil	West Indian oil
Specific gravity at 25°C	0.880–0.913	0.859–0.865
Optical rotation	+7°53′ to +22°10′	+25°45′ to +38°32′
Refractive index at 20°C	1.4776–1.4861	1.4729–1.4746

Extensive analyses have been carried out on the volatile oil of nutmeg and these have provided the major classes of compounds constituting the oil as: monoterpene hydrocarbons, 61–88%; oxygenated monoterpenes, aromatic ethers, sesquiterpenes, aromatic monoterpenes, alkenes, organic acids and miscellaneous compounds. The essential oil content in nutmeg from South India ranged from 3.9 to 16.5%, whereas in mace it varied from 6 to 26.1% (Maya *et al.*, 2004). The major constituents of the oil were sabinene, myristicin, elemicin and safrole. The accessions A9/71 and A9/95 contained a higher level of sabinene and a lower level of myristicin and elemicin. The oil composition of nutmeg and mace from Karnataka, South India is indicated in Table 9.5. Sabinene, pinene and terpinen-4-ol were the chief components of the nutmeg oil, whereas the mace oil contained sabinene, terpinen-4-ol, elemicin and γ-terpinene. It was interesting to note that both the oils contained low levels of myristicin. One notable difference between the oils was that mace contained a higher level of elemicin, whereas nutmeg contained a higher level of pinenes.

It must be noted that the composition of distilled volatile oil is not identical to the natural oil in the kernel or oleoresin extract. The kernel consists of 30–55% oil and 45–60% solid matter. The volatile oil accounts for 5–15% of the nutmeg kernel, while the fixed oil accounts for 24–40%.

Depending on the source, the essential oil of nutmeg contains mainly sabinene (15–50%), α-pinene (10–22%) and β-pinene (7–18%), with myrcene (0.7–3%), 1,8-cineole (1.5–3.5%), myristicin (0.5–13.5%), limonene (2.7–4.1%), safrole (0.1–3.2%) and terpinen-4-ol (0–11%). The contents depend on whether the oil is of West Indian, Indian or Sri Lankan origin (http://www.chem.uwimona.edu.jm: 1104/lectures/nutmeg.html).

On organoleptic grounds, West Indian oils are reported to be weaker in odour and less spicy than East Indian oils. Baldry *et al.* (1976) suggested that the difference in composition of the East and West Indian oils was more a reflection of the proportion of the constituting compounds than the absence of constituents. The major quantitative differences were lower proportions of α-pinene, safrole and myristicin and higher proportions of sabinene in the West Indian oils.

Table 9.5. Composition of the essential oils of nutmeg and mace.

Compound	RI BP-1	RI BP-20	Nutmeg (%)	Mace (%)	Method of identification
α-Thujene	925	1019	2.2	1.3	RI, MS, PE
α-Pinene	933	1019	13.6	1.3	RI, MS, PE
Camphene	945	1055	0.3	t	RI, PE
Sabinene	973	1114	32.1	23.5	RI, MS, PE
β-Pinene	975	1100	12.9	1.2	RI, MS, PE
Myrcene	984	1154	2.2	1.8	RI, MS, PE
α-Phellandrene	997	1168	0.7	1.6	RI, PE
δ-3-Carene	1006	1138	0.8	2.5	RI, PE
α-Terpinene	1012	1168	2.2	3.9	RI, PE
p-Cymene	1014	1261	0.7	0.2	RI, MS, PE
Limonene	1024	1186	4.0	3.9	RI, MS, PE
1,8 Cineol+ β-phellandrene	1024	1195	2.3	2.2	RI, MS, PE
(*E*) β-Ocimene	1040	1235	–	t	RI, PE
γ-Terpinene	1054	1235	3.9	6.6	RI, PE
trans-Sabinene hydrate	1058	1466	0.5	0.2	RI, MS
α,*p*-Dimethyl styrene	1076	1425	–	0.1	RI
Terpinolene	1082	1268	1.2	3.3	RI, PE
Linalool	1086	1550	0.8	0.7	RI, MS, PE
cis-Sabinene hydrate	1086	1537	–	t	RI
(*Z*)-*P*-menth-2-en-1-ol	1109	1564	0.4	0.9	RI, MS
(*E*)-*P*-menth-2-en-1-ol	1126	1632	0.3	0.6	RI, MS
Terpinen-4-ol	1167	1605	7.2	23.6	RI, MS, PE
α-Terpineol	1176	1694	0.6	1.3	RI, MS, PE
cis-Piperitol	1184	1683	–	0.2	RI, MS
trans-Piperitol	1193	1751	–	0.3	RI, MS
Nerol	1210	1780	–	t	RI, PE
Geraniol	1235	1860	–	t	RI, PE
Safrole	1268	1874	2.8	0.7	RI, MS
Bornyl acetate	1271	1605	–	0.7	RI, MS
Terpinen-4-yl acetate	1290	1656	–	t	RI, PE
α-Terpinyl acetate	1332	1734	–	0.1	RI, PE
Eugenol	1335	2199	0.4	0.1	RI, MS, PE
Geranyl acetate	1358	1750	–	0.1	RI, PE
Methyl eugenol	1374	2028	1.6	3.7	RI, PE
β-Cubebene	1384	1550	0.1	t	RI
β-Caryophellene	1422	1588	0.2	0.1	RI, PE
trans-α-Bergamotene	1433	1598	0.1	–	RI
trans-Methyl isoeugenol	1464	2126	0.2	0.2	RI, MS, PE
Germacrene D	1480	1712	0.1	t	RI
Myristicin	1494	2284	2.6	2.3	RI, MS, PE
Elemicin	1527	2254	2.4	10.5	RI, MS, PE
cis-Isoelemicin	1564		–	t	RI
trans-Isoelemicin	1614	2470	–	0.5	RI, MS

t = trace.
Source: Mallavarapu and Ramesh (1998).

Essential oil yield and composition of nutmeg and mace collected from Grenada revealed that the quality of the two oils was very similar, but they showed variation in the quantity of the components. The nutmeg oils showed 85–93% monoterpene hydrocarbons, 6.6–12% oxygenated monoterpenes and sesquiterpenes and 3.5% aromatic ethers, while the corresponding values for the mace oils were 75–94%, 4.7–17.6%

and 0–5.9%, respectively. Worm-eaten nutmegs had a higher content of volatile oils than sound nutmegs since, in the former, the starch and fixed oil had been eaten selectively by insects (http://www.fao.org/docrep/x5047E/x5047E0a.htm 2007).

Prolonged storage of nutmeg oil leads to the loss of few volatile components and thereby changes in the oil composition. Sanford and Heinz (1971) observed that the content of myristic acid might serve as an indicator of the age of ground nutmeg.

Nutmeg and mace oils from Indonesia, the West Indies and Papua New Guinea showed clear-cut differences in the composition. The oils differed in the composition of the aromatic ether fraction. The most abundant aromatic ether was myristicin in East Indian oils, elemicin in the West Indian oils and safrole in the Papuan oils (Ehlers *et al.*, 1998). Chang Yen *et al.* (1996) reported that Grenadian nutmeg oils contained more sabinene and less myristicin and safrole than oils from other geographic regions.

Moyler *et al.* (1994) reported that CO_2 extract of West Indian nutmeg contained sabinene (20.7%), α-pinene (6.4%), β-pinene (5.5%), limonene (2.8%) and elemicin (3.1%) as major components.

Essential oil recovered by traditional steam distillation contains 5–10% of the hallucinatory compound, myristicin. Nguyen *et al.* (1998) reported that extraction using supercritical carbon dioxide yielded a volatile fraction containing less myristicin.

Nutmeg seeds from Nigeria contained sabinene (49.09%), α-pinene (13.19%), α-phellandrene (6.72%) and terpinen-4-ol (6.43%) as major constituents (Ogunwande *et al.*, 2003). Table 9.6 gives the composition of volatile oil of nutmeg from four different sources, e.g. Grenada, Nigeria, the UK and Pakistan. Simpson and Jackson (2002) made a comparative study of the East Indian, Jamaican and other West Indian oils and observed that the Jamaican oil contained lower quantities of the phenyl propanoids, myristicin and safrole, than the East Indian oils. Jamaican oil contained three isomers of ocimene, namely (*E*)-β-ocimene, (*Z*)-β-ocimene and (*Z*)-α-ocimene, which was not recorded in other oils. They suggested that α-pinene, β-pinene and terpinen-4-ol could be used as markers to distinguish between West Indian Oils (Table 9.7).

Nutmeg oil from Papua New Guinea contained α-pinene (22.6%), sabinene (15.8%), β-pinene (15.2%) and myristicin (13.2%) as the chief constituents (Wossa *et al.*, 2004). Sait and Satyaputra (1995) reported that deterpenation of nutmeg oils resulted in lowering of the medicinal quality of the oil due to increase in myristicin concentration.

Schubert and Mosandl (1991) determined the enantiomeric distribution of linalool in nutmeg oil and found the distribution of the component as 3*R*-(–)-linalool (83%) and 3*S*-(+)-linalool (17%). Hener *et al.* (1991) examined the enantiomeric distribution of three monoterpene hydrocarbons in mace oil as: (1*R*,5*R*)-(+)-α-pinene (20%); (1*S*,5*S*)-(–)-α-pinene (80%); (1*R*,5*R*)-(+)-β-pinene (26%); (1*S*,5*S*)-(–)-β-pinene (74%); (4*R*)-(+)-limonene (78%); (4*S*)-(–)-limonene (22%). The enantiomeric proportions of the monoterpene hydrocarbon fraction of East Indian nutmeg oil are indicated in Table 9.8 (Konig *et al.*, 1992).

Essential oil distilled from the ethnolic extract of nutmeg contained a relatively higher content of terpinen-4-ol, elemicin and myristicin and a lower content of sabinene and pinene compared with the steam-distilled oil (Borges and Pino, 1993). Zhu *et al.* (1995) reported that nutmeg oil from Guangdong, China, contained 27.63% α-pinene, 26.84% sabinene, 15.52% β-pinene and 7.39% myristicine besides the minor constituents.

According to Ravid *et al.* (1996) the enantiomeric distribution of terpinen-4-ol in nutmeg oil was *S*(+)- 73% and *R*(–)- 27%. Liquid CO_2 extract of nutmeg was rich in sabinene (36.64%), α-pinene (11.05%), β-pinene (10.19%), myristicin (6.98%), elemicin (3.29%) and myrcene (3.15%) (Spricigo *et al.*, 1999).

African nutmeg (*Monodora myristica* Dund Gaerth) seeds yielded 4.6% essential oil. The oil contained 75% monoterpene hydrocarbons (Table 9.9). The major components were α-phellandrene (50.4%), α-pinene (5.5%) and myrcene (4.35%). The oil contained 3% sesquiterpene hydrocarbons and oxygenated compounds such

Table 9.6. Volatile oil composition of nutmeg from different sources.

Compound	Grenada[1] (%)	Nigeria[2] (%)	UK[3] (%)	Pakistan[4] (%)
α-Thujene	1.2	–	1.2	1.30
α-Pinene	10.2	13.19	22.0	4.90
Camphene	< 0.1	0.21	0.4	0.10
β-Pinene	8.0	2.42	21.5	4.60
Sabinene	57.0	49.09	15.4	1.90
Myrcene	2.2	3.09	1.9	1.60
δ-3-Carene	0.3	–	0.9	0.60
α-Phellandrene	0.3	6.72	0.7	0.60
α-Terpinene	0.9	–	1.2	3.50
Limonene	2.9	0.56	3.9	3.20
γ-Terpinene	1.7	–	1.8	7.80
p-Cymene	1.0	3.30	1.9	6.50
Terpinolene	0.4	0.49	0.9	2.40
p-Cymenene				0.10
α-Cubebene	< 0.1	0.18	–	–
α-Copaene	0.3	–	–	0.10
α-Bergamotene	< 0.1	0.17	–	–
β-Caryophyllene	< 0.1	0.83	–	–
β-Bisabolene	< 0.1	–	–	–
δ-Cadinene	< 0.1	–	–	–
1,8-Cineole	0.8	–	–	0.10
β- Phellandrene				2.70
trans-Sabinene hydrate	2.4	1.62	–	–
Citronellol	< 0.1	–	–	–
Decanal	< 0.1	–	–	–
Linalool	0.2	0.47	0.5	0.40
cis-Sabinene hydrate	0.8	–	–	–
Camphor	0.1	–	–	–
Bornyl acetate	6.4	1.10	0.1	–
Terpinen-4-ol	< 0.1	6.43	5.7	31.30
Citronellyl acetate	0.3	–	–	–
α-Terpineol	0.2	0.54	0.7	5.20
Geranyl acetate	< 0.1	0.29	–	0.20
Geraniol	0.1	–	0.1	–
Methyl eugenol	< 0.1	0.23	0.6	0.80
Eugenol	0.7	0.81	0.4	0.20
Elemicin	0.9	–	–	4.80
Myristicin	–	1.85	9.4	7.10
cis-p-Menth-2-en-1-ol	–	0.43	–	0.10
t-p-Menth-2-en-1-ol	–	0.33	–	0.20
cis-piperitol	–	0.11	–	0.10
t-Piperitol	–	0.14	–	0.10
Safrole	–	1.34	–	2.00
Borneol	–	–	0.3	0.10
α-Terpinyl acetate	–	0.19	0.1	0.10
Nerylacetate		–	0.2	–
Methyl myristate		–	0.3	–
Germacrene-D	–	0.33	–	–
Bicyclogermacrene	–	0.13	–	–
α-Asarone	–	1.10	–	–
p-Cymene-8-ol	–	–	–	0.30
tetra-Decanoic acid	–	–	–	2.90

Continued

Table 9.6. *Continued*

Compound	Grenada[1] (%)	Nigeria[2] (%)	UK[3] (%)	Pakistan[4] (%)
Cis-1,2, epoxy-terpinen-4-ol	–	–	–	1.10
Cis-p-menth-3-ene-1,2-diol	–	–	–	0.20
(*E*)-methyl isoeugenol	–	–	–	0.01

Source: [1]Pino *et al.* (1991); [2]Ogunwande *et al.* (2003); [3]Dorman and Deans (2004), [4]Ur-Rahman (1999).

Table 9.7. Comparison of nutmeg oils from Grenada, Indonesia and Jamaica.

Compound	Grenada (%)	Indonesia (%)	Jamaica (%)
α-Pinene	13.2	26.5	19.9
β-Pinene	8.0	15.0	18.8
Myrcene	3.4	3.7	4.7
α-Phellandrene	0.7	0.9	1.6
α-Terpinene	4.2	2.0	2.1
Limonene	4.4	3.6	4.8
p-Cymene	0.8	0.6	< 0.1
Linalool	0.3	0.2	0.3
Terpinen-4-ol	4.7	3.0	17.8
α-Terpineol	0.3	0.6	0.4

Source: Simpson and Jackson (2002).

Table 9.8. Enantiomeric proportion of monoterpene hydrocarbons of East Indian nutmeg oil.

Compound	Content (%)
(+)-α-Thujene	0.19
(–)-α-Thujene	1.66
(–)-α-Pinene	21.74
(+)-α-Pinene	5.69
(–)-Camphene	0.36
(–)-β-Pinene	10.60
(+)-β-Pinene	7.65
(+)-Sabinene	26.84
(–)-Sabinene	0.51
Myrcene	2.88
(+)-α-Phellandrene	0.82
(–)-δ-3-Carene	1.43
α-Terpinene	3.69
P-Cymene	1.49
(–)-β-Phellandrene	0.20
(+)-β-Phellandrene	2.39
(–)-Limonene	3.47
(+)-Limonene	2.23
γ-Terpinene	4.94
Terpinolene	2.28

Source: Konig *et al.* (1992).

as germacrene-D-4-ol (Onyenekwe *et al.*, 1993). Wang *et al.* (2004) reported 39.63% myristicin in nutmeg seeds in market samples from China.

Mace oil

Depending on its origin, mace has 7–14% essential oil and about 30% fixed oil. The physico-chemical properties of mace oil are given below (Guenther, 1952):

Physico-chemical properties	East Indian oil	West Indian oil
Specific gravity at 25°C	0.923–0.947	0.860–0.892
Optical rotation	+2°42′ to +11°48′	+21°18′ to +41°30′
Refractive index at 20°C	1.486–1.494	1.472–1.470

It contains the same aroma compounds as nutmeg but in smaller amounts, mainly monoterpenes (87.5%), monoterpene alcohols (5.5%) and other aromatics (7%). Like nutmeg essential oil, the main constituents of mace essential oil are sabinene, α-pinene, β-pinene, myrcene, limonene, 1,8-cineole, terpinen-4-o1, myristicin, γ-terpinene and safrole. Mace oil is more expensive than nutmeg oil.

The mace oleoresin is amber to dark red in colour; 7 kg of mace oleoresin are equivalent to 100 kg of freshly ground spice. Mace butter, which has fixed oils and volatiles, has 60% unsaturated fats and 40% saturated fats. Ground mace contains vitamin A, phosphorus, potassium, magnesium, sodium and calcium. East Indian mace oil contained α-pinene (16.3%), β-pinene (10.6%),

Table 9.9. Composition of essential oil of African nutmeg.

Compound	DB-5 column (%)	Supelcowax 10M (%)
α-Thujene	1.99	2.01
α-Pinene	5.52	5.44
β-Pinene	0.35	0.24
Myrcene	4.85	4.35
Δ^2-Carene	0.64	0.66
α-Phellandrene	51.81	49.18
α-Terpinene	0.17	0.14
P-Cymene	8.63	8.42
Limonene + β-phellandrene	3.42	2.84
(*Z*)-β-Ocimene	0.37	0.54
(*E*)-β-Ocimene	0.17	0.39
Linalool	1.40	0.16
(*Z*)-Sabinene hydrate	0.19	1.44
(*E*)-Sabinene hydrate	< 0.10	–
MW 152	0.15	–
α-Terpineol	0.35	0.51
MW 152 (m/z: 91, 92, 81, 79)[a]	1.53	1.65
Carvacrol	0.50	–
Carvacryl acetate	0.18	–
β-Caryophyllene	0.37	0.34
Unidentified	0.22	–
α-Caryophyllene	0.27	0.33
MW 204	0.12	–
Valencene + MW 222	0.10	0.35[a]
α-Muurolene	0.47	0.40
γ-Cadinene	1.62	–
δ-Cadinene	3.02	3.93
Germacrene-D-4-ol + spathulenol	9.16[a]	9.79
ι-Cadinol + γ-muurolol	1.13[a]	0.27
α-Cadinol	1.38	0.70
(m/z: 43, 153, 135, 71)[a]	0.23	0.69
(m/z: 41–43, 69, 93, 109–119)[a]	0.28	1.57[a]

[a]Mixed peak,
Source: Onyenekwe *et al.* (1993).

sabinene (12.5%), α-terpinene (7.5%), γ-terpinene (11.6%) and terpinen-4-ol (14.2%) as major constituents (Lawrence, 1979). The constituents of the oil were compiled by Lawrence (1997, 2000, 2005). A commercial sample of mace from a Karachi market was predominated by terpinen-4-ol (20%) and methyl eugenol (13.3%) (Ur-Rahman, 1999). Spricigo *et al.* (1999) reported that supercritical extraction of nutmeg seeds at 90 bars and 23°C resulted in the co-extraction of fatty oil and essential oil, with myristic acid and myristicin predominating in each group. The extraction yield was improved by reduction in particle size.

The constituents of nutmeg and mace oils of Indian origin were determined by Mallavarapu and Ramesh (1998). They found that both nutmeg and mace oils were dominated by monoterpenes, but the mace oil contained higher quantities of oxygenated monoterpenes and phenyl propanoid ethers (Table 9.5). Gopalakrishnan (1992) reported that mace oil contained higher myristicin and elemecin than nutmeg oil.

Leaf oil

The leaves of *M. fragrans* yield 0.5–2.0% volatile oil. Zachariah *et al.* (2000) reported that myristicin and elemicin were present in the oil of the nut, mace and leaves. The leaf oil of nutmeg was predominated by monoterpenes (Madhavan *et al.*, 1991). The major constituents identified were sabinene + β-pinene (25.95–38.90%), α-pinene (8.62–34.64%) and limonene (4.17–8.96%) (Table 9.10). The phenolic profile and essential oil content of leaves could be used to determine the sex of the nutmeg (Packiyasothy *et al.*, 1991). The adult male plants showed additional phenolic compounds compared with that of the female adult plants. The leaf oil content of the female adult plants was higher than that of the male adult plants.

Flower oil

The composition of nutmeg flowers was determined by Madhavan *et al.* (1991). The chief components of the oil are sabinene, pinenes, α-terpineol, terpinen-4-ol and elemicin (Table 9.10).

Pericarp oil

Nutmeg pericarp oil contained 16 monoterpenes (60%), nine monoterpene alcohols (29%), eight aromatic ethers (7%), three sesquiterpenes (1%), six esters (1%) and eight other minor components. The components were similar to those in nutmeg and mace oils but differed substantially in concentration (Table 9.11). The chief constituents were α-pinene, α-terpineol and terpinen-4-ol. The sabinene, myristicin and safrole concentrations were much lower, while the terpinen-4-ol and α-terpineol contents were much higher than in nutmeg and mace oils (Choo *et al.*, 1999). The chief volatiles from nutmeg and mace are indicated in Fig. 9.1.

Non-volatiles

The non-volatiles from nutmeg include lignans, phenolic acids, glycosides, sterols and miscellaneous compounds.

Table 9.10. Chemical composition of nutmeg leaf and flower oil.

Compound	Leaf oil (%)	Flower oil (%)
α-Pinene	8.62–34.64	9.51
Camphene	0–0.57	0.49
Sabinene + β-pinene	25.95–38.90	14.05
Myrcene + α-phellandrene	2.92–7.55	4.42
δ-3-Carene	0.70–2.08	–
α-Terpinene	2.51–6.48	3.67
p-Cymene	2.49–4.78	4.55
Limonene	4.17–8.96	4.40
β-Phellandrene +1,8 cineol	1.79–4.59	1.46–4.41
γ-Terpinene	1.80–5.96	4.63
Linalool	0.95–2.17	3.95
Terpinolene	0–0.28	0.48
β-Terpineol	0–0.21	0.36
Borneol	0–0.09	0.14
Terpinen-4-ol	1.89–6.78	11.63
α-Terpineol	0.35–1.74	5.88
Geraniol	0.02–0.08	–
Bornyl acetate + α-fenchyl acetate	0–0.59	0.31
Terpinen-4-yl acetate	0.32–2.82	0.46
α-Terpinyl acetate	0.02–0.47	0.64
Geranyl acetate	0.51–1.42	0.31
β-Caryophellene	0–0.07	0.10
Methyl eugenol	0.17–0.48	0.58
Eugenol	0.10–2.16	0.71
(*Z*)-isoeugenol	0.14–1.00	0.63
Nerolidol	0–1.86	0.96
(*E*)-isoeugenol	0–0.40	0.56
Myristicin	0.44–1.85	3.79
Elemicin	0.19–3.51	8.76

Source: Madhavan *et al.* (1991).

Lignans

Several lignans and neolignans have been isolated from nutmeg. Hada *et al.* (1988) isolated eight neolignans and five lignans from mace. Kuo *et al.* (1989) isolated five lignans, namely otobain, malabaricone A, otobanone, cagayanin and cagayanone, from *M. cagayanesis*. From mace, Orabi *et al.* (1991) isolated two resorcinols, namely malabaricone-B and malabaricone-C. Four diarylnonanoids, malabaricones A–D, were isolated from the fruit rind of *M. malabarica* Lam by Purushothaman *et al.* (1977). The absolute configuration of

Table 9.11. Composition of nutmeg pericarp oil.

	Relative retention index (KI)		Relative peak area (%)	
Compound	SPB-1	Supelcowax-10	Fresh	Freeze-dried
α-Thujene	921	–	0.05	nd
α-Pinene	926	1026	15.2	14.1
α-Fenchene	–	1053	0.1	0.1
Camphene	943	1062	0.2	0.1
Sabinene	962	1124	0.4	1.2
β-Pinene	971	1112	6.8	9.0
β-Myrcene	982	1165	3.3	4.1
α-Phellandrene	997	–	1.9	1.7
Δ^3-Carene	1006	1147	3.7	1.7
α-Terpinene	1011	1181	3.4	6.1
β-cymene	1012	1274	0.7	0.4
Limonene	1021	1202	5.7	4.6
β-Phellandrene	1022	1210	5.4	2.8
β-Ocimene	1039	1255	0.1	0.2
γ-Terpinene	1051	1248	5.2	9.1
Unknown alcohol	1060	–	0.1	0.2
Unknown (alkyl aromatics)	1074	1421	0.4	0.2
α-Terpinolene	1080	1288	6.5	5.1
Linalool	1085	1539	1.8	0.7
D-Fenchyl alcohol	1098	1576	0.4	0.1
cis-p-Menth-2-en-1-ol	1106	1560	0.2	0.2
Unknown (alcohol)	1124	–	0.2	0.1
Unknown (ester)	–	1324	0.1	0.1
Terpinen-4-ol	1168	1603	13.5	19.1
1-Borneol	1170	1656	0.1	0.1
α-Terpineol	1180	1697	15.5	8.2
Nerol	1218	1783	0.4	0.5
Safrole	1264	1871	1.5	0.8
endo-Bornyl acetate	1270	1857	0.1	0.1
Eugenol	1329	2160	0.1	0.1
Citronellyl acetate	1333	1667	0.2	nd
Geranyl acetate	1359	1742	0.05	nd
β-Citronellol	–	1752	0.05	nd
p-Cymen-8-ol	–	1832	0.2	0.2
α-Copaene	1375	–	0.1	0.2
Methyl eugenol	1393	2004	0.1	0.1
trans-Caryophyllene	1417	1595	0.4	nd
Isoeugenol	1420	2338	4.5	3.1
β-Farnesene	1447	–	0.1	0.3
Myristicin	1491	2256	1.0	3.0
Elemicin	1512	2221	0.2	0.1
2,6-Dimethoxy-4- (2-propenyl) phenol	1546	2544	0.1	0.1
Methyl isoeugenol	–	2194	0.1	nd
Unknown	–	2185	0.2	0.1
t-Muurol	–	2190	0.1	0.2
Unknown	–	2424	0.2	0.3
Unknown	–	2688	0.4	0.5

nd = no data.
Source: Choo *et al.* (1999).

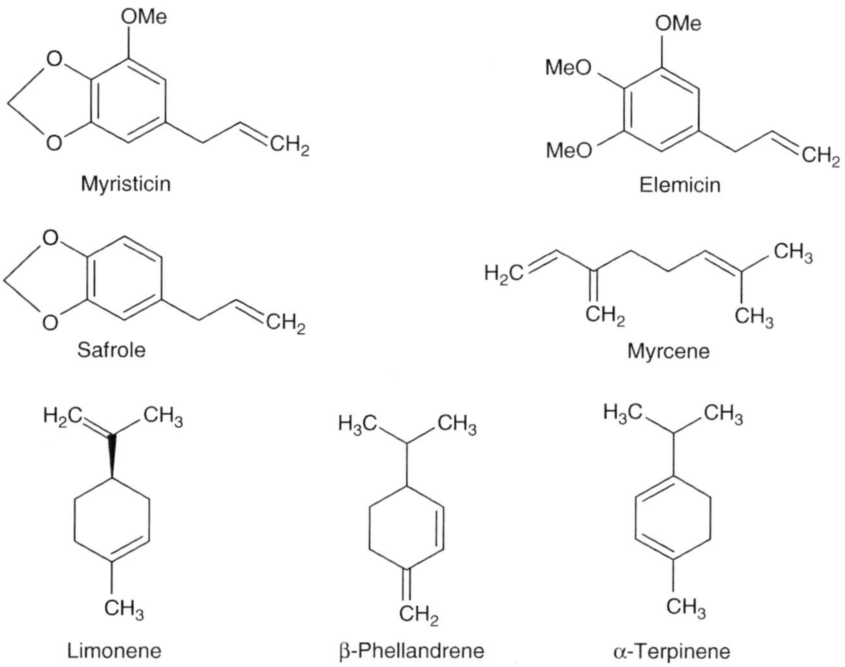

Fig. 9.1. Volatiles from nutmeg.

8-*O*-4′ neolignans, e.g. erythro-$\Delta^{8'}$-4,7-dihydroxy-3,3′,5′-trimethoxy-8-*O*-4′-neolignan and $\Delta^{8'}$-4-hydroxy-3,3′,5′-trimethoxy-8-*O*-4′-neolignan, was determined by Kasahara *et al.* (1995). Zacchino and Badano (1991) reported the absolute configuration of erythro-(3–4-methylene dioxy 7-hydroxy-1′-allyl-3′,5′-dimethoxy)-8-*O*-4′-neolignan by chemical synthesis. Gokhale *et al.* (2004) determined the chemical constituents of false nutmeg (*M. malabarica*). The neolignans, fragransol C and D and myristicanol A and B, were isolated from mace by Rastogi and Mehrotra (1995). Purushothaman and Sarada (1980) isolated a neolignan, dihydro-di-isoeugenol, from the hexane and chloroform extracts of *M. fragrans* arils. The chemical structure of mace lignan was determined by Woo *et al.* (1987). The lignans isolated from nutmeg are shown in Table 9.12. The structures of a few representatives of lignans are indicated in Fig. 9.2.

Sterols

Among various sterols, sitosterol was predominant in nutmeg. Other sterols reported were campesterol, lanosterol and desmosterol (Izaldin *et al.*, 1987).

Phenolic acids

Variyar and Bandyopadhyay (1995) determined the phenolic acid profile of nutmeg and mace. The chief phenolic acids in nutmeg were caffeic acid and vanillic acid, whereas ferulic acid and synapic acids predominated in mace.

Aromaglycosides

The principal glycosidically bound volatiles of nutmeg were *p*-cymene-7-ol rutinoside and glucosides of eugenol, methyl eugenol and α-terpineol. During radiation processing, the glycosides break down to their aglycones. Among these glycosides, glucoside of α-terpineol was found to be the most sensitive to radiation, while *p*-cymene-7-ol rutinoside was the least sensitive (Ananthakumar *et al.*, 2006). Jukic *et al.* (2006) isolated the glycosidically bound volatiles and identified the

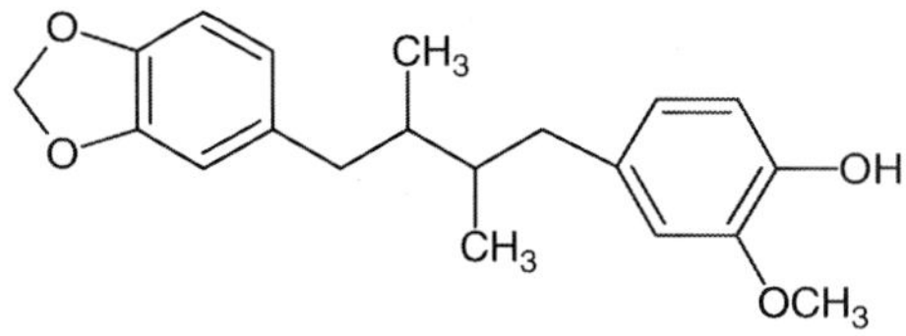

Macelignan

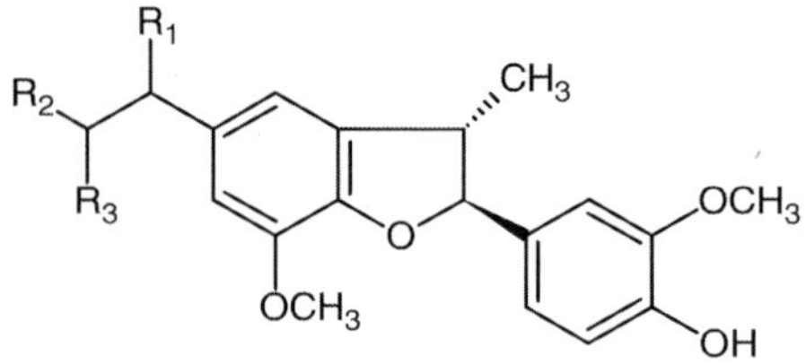

Fragransol A $R_1 = OCH_3$, $R_2 = OH$, $R_3 = CH_3$

Fragransol B $R_1 = R_2 = H$, $R_3 = OH$

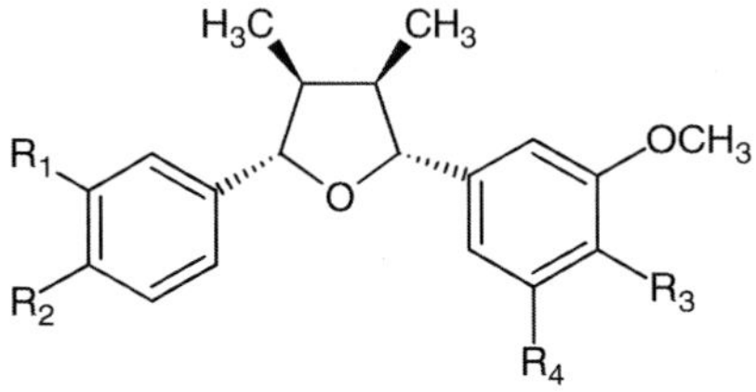

Fragransin D_1 $R_1 = R_3 = R_4 = OCH_3$, $R_2 = OH$

Fragransin E_1 $R_1 = R =$ O-CH_2-O, $R_3 = OH$, $R_4 = H$

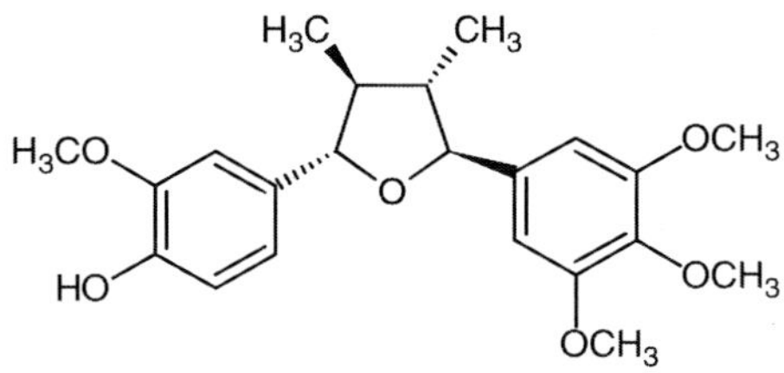

Fragransin D_2

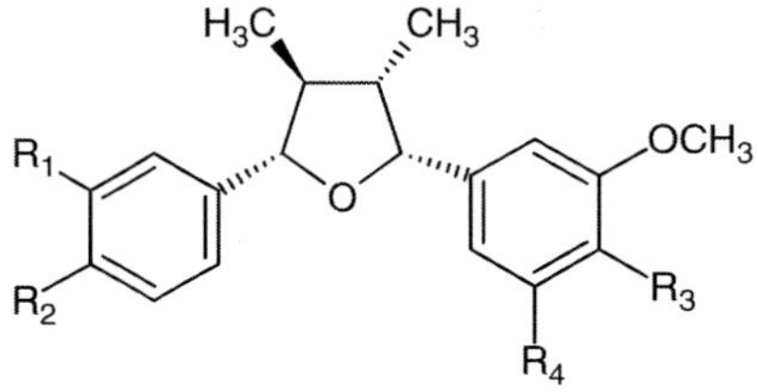

Fragransin D_3 $R_1 = R_3 = R_4 = OCH_3$, $R_2 = OH$

Austrobailignan-7 R_1, $R_2 =$ O-CH_2-O, $R_3 = OH$, $R_4 = H$

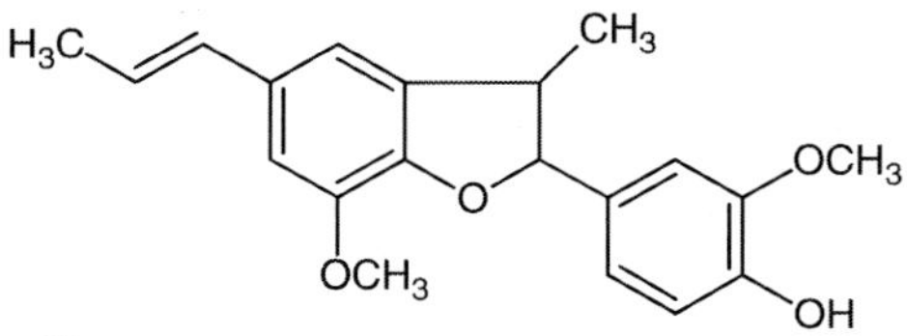

Dehydrodiisoeugenol

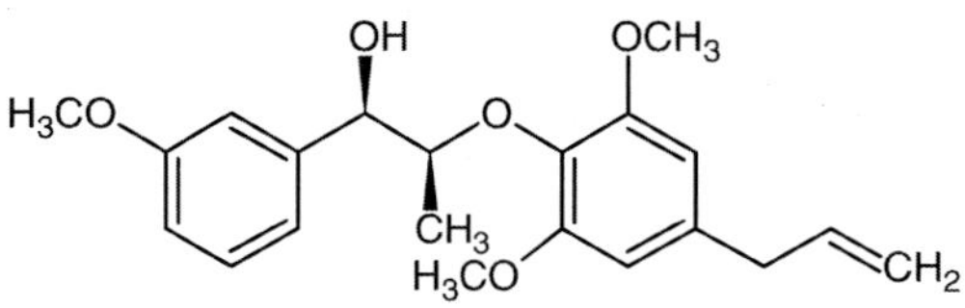

8-*O*-4′ neolignans

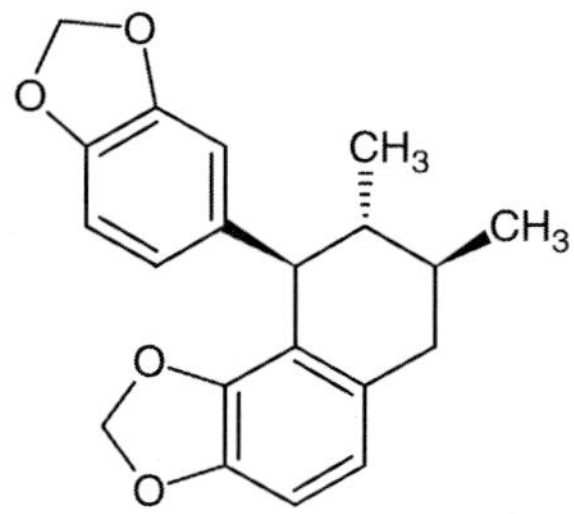

Otobain

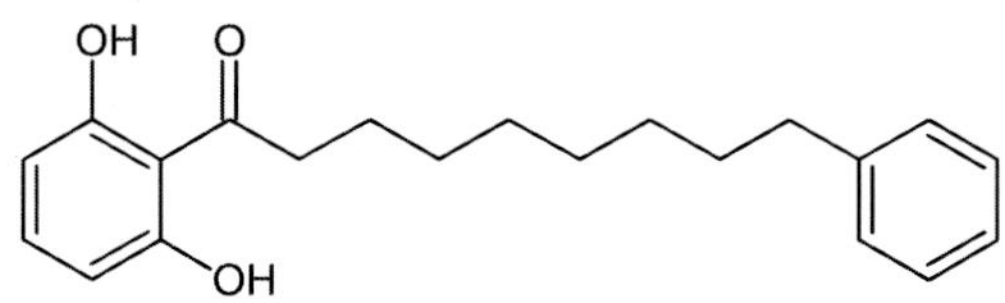

Malabaricone A

Fig. 9.2. Lignans from nutmeg.

Table 9.12. Lignans and neolignans from nutmeg.

Lignans and neolignans	References
Erythro-2-(4″-allyl-2″,6″-dimethoxy phenoxyl)-1-(3′,4′,5′-trimethoxy phenyl)-propan-1,3-diol	
Threo-2-(4″-allyl-2″,6″-dimethoxy phenoxyl)-1-(4′-hydroxy- 3′-methoxy phenyl) propan-1-ol	
Threo-1-(4′-hydroxy-3′–methoxyphenyl)-2-(2″;- methoxy-4″-(1‴-(E-)-propenyl) phenoxy-propan-1-ol	Hada *et al.* (1988)
Erythro-1-(4′-hydroxy-3′–methoxyphenyl)-2-(2″;-methoxy-4″-(1‴-(E-)-propenyl) phenoxy-propan-1-ol	
Threo-1-(4′-hydroxy-3′-methoxy phenyl)-1-methoxy 2-(2″-methoxy 4″-(1‴-(E)-propenyl) phenoxy)-propane	
Erythro-1-(4′-hydroxy-3′-methoxy phenyl)-1-methoxy 2-(2″-methoxy 4″-(1‴-(E)-propenyl) phenoxy)-propane	
Fragransol-A	Hada *et al.* (1988) Hattori *et al.* (1988)
Fragransol B	Hada *et al.* (1988)
Fragransol C	Hattori *et al.* (1988)
Fragransol D	Hattori *et al.* (1988)
Fragransol D1	Hada *et al.* (1988)
Fragransin A2	
Fragransin B1	
Fragransin B2	
Fragransin B3	
Fragransin C1	Hattori *et al.* (1987a)
Fragransin C2	
Fragransin C2A	
Fragransin C3A	
Fragransin C3B	
Fragransin D2	
Fragransin D3	
Fragransin E1	
Austrobilignan -7	Hada *et al.* (1988)
erythro-$\Delta^{8'}$-4, 7-dihydroxy-3,3′,5′-trimethoxy-8-O-4′-neolignan	Isogai *et al.* (1973)
$\Delta^{8'}$-4-hydroxy-3,3′,5′-trimethoxy-8-O-4′-neolignan	Hattori *et al.* (1987a)
Erythro-(3-4-methylene dioxy 7-hydroxy-1′-allyl-3′,5′-dimethoxy)-8-O-4′-neolignan	Forrest *et al.* (1974)
Propan-1-ol Erythro-2-(4-allyl-2,6-dimethoxy phenoxyl)-1-(3-hydroxy-4,5 -di methoxy phenyl)	
Propan-1-ol Erythro-2-(4-allyl-2,6-dimethoxy phenoxyl)-1-(4-hydroxy- 3,5-di methoxy phenyl)	
Propan-1-ol Erythro-2-(4-allyl-2-methoxy-phenoxy)-1-(4-hydroxy-3 methoxy phenyl)	Hattori *et al.* (1987b)
Propan-1-ol-Threo-1-(4-hydroxy-3,5-dimethoxy phenyl)-2-(2methoxy–4-(1-t-propenyl)-phenoxy	
Propan-1-ol-threo-2-(4-allyl-2,6-dimethoxy phenoxyl)-1-(4-hydroxy-3-methoxy phenyl)	
Propane-2-(4-allyl-2,6-dimethoxy phenoxy)-1-(4-hydroxy-3-methoxy phenyl)	
Propane-erythro-2-(4-allyl-2,6-dimethoxy phenoxy)-1-(4-hydroxy-3-methoxy phenyl)-1-(4-hydroxy-3-methoxy phenyl)-1-methoxy	

main aglycones as isoeugenol (46.1%) and methoxy eugenol (27.7%). The aglycones, namely pulegone (5.6%), *cis*-isoeugenol (3.7%), β-thujone (3.4%), cuminol (3.3%), isoelemicin (3%), eugenol (2.8%), isoeugenol (2.3%) and terpinen-4-ol (1%), were present in minor quantities (Jukic *et al.*, 2006).

Miscellaneous compounds

Isogai *et al.* (1973) isolated five phenyl propanoids from the seed kernel of nutmeg. Park *et al.* (1998) isolated dihydroguaiaretic acid from mace.

9.5. Medicinal and Pharmacological Uses

Nutmeg is used more commonly in Oriental medicine than in Western medicine. Medicinally, it is known for its stimulative and carminative properties. The seeds are carminative, stomachic, astringent, deodorant, narcotic and aphrodisiac, and useful in flatulence, nausea and vomiting. The antioxidant property of nutmeg was reviewed by Krishnamoorthy and Rema (2000). Both nutmeg and mace are used in the pharmaceutical industries. Powdered nutmeg is rarely administered alone, but it enters into the composition of numerous medicines as aromatic adjuncts.

Oil of nutmeg is useful in the treatment of inflammation of the bladder and urinary tract, halitosis, dyspepsia, flatulence, impotence, insomnia and skin diseases. It is also used externally as a stimulant and ointment as a counter-irritant. The essential oil contains several compounds, most of which are valuable in industry. Most of the pharmacological properties of nutmeg are attributed to the compounds present in the essential oil. Mace oil possesses almost identical physiological and organoleptic properties to nutmeg oil. Nutmeg butter is a mild external stimulant used in the form of ointments, hair lotions and plaster, and used against rheumatism, paralysis and sprains.

Another application of nutmeg essential oils is in aromatherapy, which is gaining importance these days. The main constituents of nutmeg and mace – myristicin, elemicin and isoelemicin – when presented in aroma form, act as stress relievers. In Japan, many companies diffuse such aromas through air ventilation systems to improve the work environment, as well as the quality of the air.

Both nutmeg and mace contain the active ingredient myristicin, which possesses narcotic properties. Nutmeg butter also contains elemicin and myristicin, which cause psychotropic effects. Ingestion in large quantities produces narcosis, delirium, drowsiness, epileptic convulsions and even death. It also causes temporary constipation and difficulty in urination and increased fat deposition in the liver. Powdered nutmeg is used occasionally as a hallucinogenic drug, but such use is dangerous as an excessive dose of mace has a narcotic effect and symptoms of delirium and epileptic convulsions appear 1–6 h after consumption.

Antimicrobial activity

Nutmeg oil showed strong antibacterial activity against 25 genera of bacteria (Dorman and Deans, 2000; 2004). It exhibited potent activities against *Bacillus subtilis*, *Escherichia coli* and *Saccharomyces cerevisiae* (De *et al.*, 1999). The resorcinols, malabaricone B and malabaricone C, isolated from mace, exhibited strong antimicrobial activities against *Staphylococcus aureus* and *Candida albicans*. Methylation of reduction of these resourcinols resulted in diminished activity (Orabi *et al.*, 1991). Dehydroxy-isoeugenol and 5′-methoxy dehydro-di-isoeugenol were the antibacterial principles isolated from mace against *Streptococcus mutants* (Hattori *et al.*, 1986).

Ur-Rahman *et al.* (1999) reported the antifungal properties of essential oil from nutmeg. Phenyl propanoids and 8-*O*-4′ neolignans isolated from mace exhibited antifungal properties (Zacchino *et al.*, 1997, 1999). Methanol extract of *M. fragrans* (seed), at MIC of 12.5 μg/ml, inhibited *Helicobacter pylori*, the Gram-negative bacterium associated with the development of gastritis and peptic ulcer diseases (Mahady *et al.*, 2005).

Insecticidal activity

Nutmeg oil has strong antifeedant activity, fumigant toxicity and contact toxicity against the stored product insects, *Tribolium castaneum* and *Sitophilus zeamis*. The

LC_{50} values of contact toxicity for adults of *S. zeamais* and *T. castaneum* were 1.7 and 18 mg/cm^2, respectively, whereas LC_{50} values for fumigant action were 4.5 and 7.7 gm/cm^2, respectively. The larvae (10–16 days old) of *T. castaneum* were more susceptible than the adults and the susceptibility of the larvae decreased with age. Nutmeg oil also affects the hatching of *T. castaneum* eggs and the subsequent survival of the larvae in the concentration range 1.4–3.2 mg/cm^2. The production of F_1 progeny of both *T. castaneum* and *S. zeamais* exposed to media treated with nutmeg oil was reduced at all concentrations tested. F_1 progeny production was totally suppressed at nutmeg oil concentrations of 1.05 g/100 g rice for *T. castaneum* and 0.35 g/100 g wheat for *S. zeamais*. At 20 g nutmeg oil/100 ml, the feeding deterrence index of *T. castaneum* was only about 7%, whereas that of *S. zeamais* was 33%. Nutmeg oil inhibits oviposition and adult emergence against cowpea storage bruchid, *Callosobruchus maculatus* Fabricius (Adedire, 2002).

Larvicidal properties have been reported in mace; the larvicidal principle in mace against second stage larvae of *Toxocara canis* was identified as diarylnonanoid, malabaricone C (Nakamura *et al.*, 1988).

Hypolipidaemic effect

The ethanolic extract of nutmeg kernels showed hypolipidaemic effect in albino rabbits. Administration of 500 mg/kg of the extract daily for a period of 60 days in the hyperlipidaemic rabbits resulted in significantly lower levels of lipoprotein lipids (total cholesterol: 574 ± 61 versus 210 ± 27 mg/dl; low-density lipoprotein (LDL) cholesterol: 493 ± 57 versus 131 ± 25 mg/dl; and triglycerides: 108 ± 14 versus 67 in control versus experimental) (Ram *et al.*, 1996).

Antioxidant activity

The antioxidant activity of the essential oil and petroleum ether extract of aril has been reported by Dorman *et al.* (1995, 2000a,b), Saito *et al.* (1976) and Kimura *et al.* (1979). Patro *et al.* (2005) determined 1,1-diphenyl-2-picrylhydrazyl (DPPH) scavenging activity of methanol extract of *M. malabarica* and found that it was mainly due to the diarylnonanoid, malabaricone C, which prevented both Fe (II)- and 2,2′-azobis-(2-amidinopropane)-dihydrochloride-induced lipid peroxidation (LPO) of rat liver mitochondria more efficiently than curcumin. Malabaricone C also prevents the γ-ray-induced damage of pBR322 plasmid DNA. The radioprotective activity is related to its hydroxyl radical scavenging property (Patro *et al.*, 2005). The aglycone fraction from glycosidically bound volatiles of nutmeg had a higher antioxidant activity compared with free volatiles from its essential oil (Jukic *et al.*, 2006). Tomaino *et al.* (2005) reported higher antioxidant activity of nutmeg oil at 180°C. This could be due to the volatilization of the hydrocarbons of the oil at higher temperature, resulting in the accumulation of phenolic constituents in the remaining oil.

Administration of eugenol (10.7 mg/kg of body weight/day) removes the oxidative stress imposed by CCl_4 in rats. Eugenol, an allyl benzene and ingredient of nutmeg, inhibits the accumulation of lipid peroxidation products in red blood cells and maintains the activities of the antioxidant enzymes such as superoxide dismutase, catalase, glutathione peroxidase, glutathione-*S*-transferase(s), glutathione reductase and glucose-6-phosphate dehydrogenase at normal levels (Kumaravelu *et al.*, 1996). Acetone extract of mace containing lignans inhibited lipid oxidation and prevented oxidative damage to cells (Chatterjee *et al.*, 2007).

Macelignan, isolated from nutmeg, showed a protective effect against *tert*-butylhydroperoxide (*t*-BHP)-induced cytotoxicity in the human hepatoma cell line, HepG2. The lignan reduced significantly the cell growth inhibition and necrosis caused by *t*-BHP. It also reduced intracellular reactive oxygen species (ROS) formation and DNA damaging effect caused by *t*-BHP (Sohn *et al.*, 2007).

Antiamoebic activity

The essential oil from nutmeg at a concentration of 0.5 µl/ml is active against *Entamoeba hystolytica* (De Blasi *et al.*, 1990).

Nematicidal activity

The essential oil of nutmeg exhibits strong nematicidal activity against the root-knot nematode, *Meloidogyne incognita* (Gotke *et al.*, 1990).

Antibacterial activity

Chloroform extract of seeds showed potent antibacterial activity against Gram-positive and Gram-negative bacteria. Trimyristin and myristic acid isolated from the extract also showed good antibacterial activity (Narasimhan and Dhake, 2006).

Anticancer activity

The essential oil of nutmeg possesses excellent anticarcinogenic activity, which has been well documented in studies involving animals. The essential oil interferes with the activities of the host enzymes associated with activation and detoxication of xenobiotic compounds, including chemical carcinogens and mutagens. Banerjee *et al.* (1994) studied the influence of essential oil from nutmeg on the activities of hepatic carcinogen-metabolizing enzymes, namely, cytochrome P-450, aryl hydrocarbon hydroxylase, and glutathione-*S*-transferase and acid-soluble sulfhydryl level in Swiss albino mice. Nutmeg oil induced the cytochrome P450 level significantly ($p < 0.05$). Aryl hydrocarbon hydroxylase activity was reduced significantly. Glutathione-*S*-transferase activity was elevated significantly ($p <$ 0.1–0.001) compared with controls. The acid-soluble sulfhydryl was elevated significantly by the essential oils of nutmeg ($p < 0.05$) (Banerjee *et al.*, 1994).

An aqueous suspension of mace at the dose levels of 0.025 or 0.1 g per animal per day was administered by oral gavage to dams from day 1 of lactation and continued daily for 14 or 21 days. Dams receiving mace treatment and their F_1 pups showed significantly elevated hepatic sulfhydryl content, glutathione-*S*-transferase and glutathione reductase activities and cytochrome b5 content. Hepatic cytochrome P450 content decreased in dams ($P < 0.05$) receiving the lower mace dose for 21 days and the F_1 pups ($P < 0.001$), but increased in dams receiving the higher dose for both time periods ($P < 0.001$) and the lower dose for 14 days ($P < 0.05$). Only the 14-day-old pups of dams receiving either mace dose showed significantly elevated ($P < 0.001$) levels of hepatic glutathione peroxidase (Chhabra and Rao, 1994).

Mace prevented 3-methylcholanthrene (MCA)-induced carcinogenesis in the uterine cervix of Swiss albino mice (Hussain and Rao, 1991). Oral administration of mace at a dose of 10 mg/mouse/day for 7 days before and 90 days following carcinogen thread insertion reduced cervical carcinoma incidence by 50%.

Essential oil from nutmeg suppressed the formation of DNA adducts by aflatoxin B1 *in vitro* in a microsomal enzyme-mediated reaction (Hashim *et al.*, 1994). A diet containing 1% mace inhibited the DMBA-induced papillomagenesis in the skin of male Swiss albino mice (Jannu *et al.*, 1991).

Supplementation of nutmeg in rats with liver damage caused by lipopolysaccharide (LPS) plus D-GalN plasma aminotranferase activities showed potent hepatoprotective activity. Myristicin, one of the essential oil constituents of nutmeg, was found to possess extraordinarily potent hepatoprotective activity. Myristicin markedly suppressed LPS/D-GalN-induced enhancement of serum TNF-α concentrations and hepatic DNA fragmentation in mice. These findings suggest that the hepatoprotective activity of myristicin might be due, at least in part, to the inhibition of TNF-α release from macrophages. However, further studies are needed to elucidate the hepatoprotective mechanism(s) of myristicine (Morita *et al.*, 2003).

Antifungal activity

Methanol extract of nutmeg seeds showed potent antifungal activity *in vitro* and *in vivo* against several plant pathogens. Three lignans isolated from the methanol extract, namely erythro-austrobailignan-6 (EA6), meso-dihydroguaiaretic acid (MDA) and nectandrin-B (NB), showed varied antimicrobial activity depending on the target species. *Alternaria alternata*, *Colletotrichum coccodes*, *C. gloeosporioides*, *Magnaporthe grisea*, *Agrobacterium tumefaciens*, *Acidovorax konjaci* and *Burkholderia glumae* were relatively sensitive to the three lignans. *In vivo*, all three compounds suppressed the development of rice blast and wheat leaf rust effectively. EA6 and NB were highly active against the development of barley powdery mildew and tomato late blight, respectively. Both MDA and NB also inhibited the development of rice sheath blight moderately (Cho *et al.*, 2007).

Anticariogenic activity

Methanol extract of nutmeg possessed strong inhibitory activity against *Streptococcus mutans*, a pathogen associated with the occurrence of dental caries. The minimum inhibitory concentration (MIC) of macelignan against *S. mutans* was 3.9 μg/ml, which was much lower than that of other natural anticariogenic agents such as 15.6 μg/ml of sanguinarine, 250 μg/ml of eucalyptol, 500 μg/ml of menthol and thymol, and 1000 μg/ml of methyl salicylate. In the bactericidal test, macelignan at a concentration of 20 μg/ml inactivated *S. mutans* completely in 1 min. The specific activity and fast effectiveness of macelignan against oral bacteria strongly suggest that it could be employed as a natural antibacterial agent in functional foods or oral care products (Chung *et al.*, 2006).

Antimicrobial activity

Nutmeg extract possessed strong antibacterial activity against non-pathogenic and pathogenic *E. coli*, but the strain O157 showed more sensitivity to β-pinene than non-pathogenic *E. coli* strains (Takikawa *et al.*, 2002).

Neurotoxicity

Myristicin, 1-allyl-3,4-methylenedioxy-5-methoxybenzene, a naturally occurring alkenylbenzene found in nutmeg, produced neurotoxic effects. Myristicin at ≥ 0.5 mM concentration showed cytotoxicity in human neuroblastoma cells (Lee *et al.*, 2005).

9.6. ISO Specifications

To export spices and spice products, exporting countries have to comply with the specifications laid down by regulatory agencies in the importing countries. Specifications for whole and ground nutmeg and mace are indicated in Tables 9.13 to 9.16.

ESA quality minima for nutmeg are as follows:

Product (whole form)	ASH% w/w max.	AIA% w/w max.	H_2O% w/w max.	V/O% v/w min.
Nutmeg	3 (ISO)	0.5 (ISO)	12 (ESA)	6.5 (ESA)

The following methods are used for quality evaluation:

Moisture	: ISO 939
Total ash	: ISO 928
Acid-insoluble ash	: ISO 930
Volatile oil	: ISO 6571

For whole-scale commercial purposes, ISO 6577 specifications are followed for nutmeg

Table 9.13. Whole nutmeg: chemical and physical specifications.

Specification	Suggested limits	
	Broken	Whole
ASTA cleaniless specifications		
Whole dead insects, by count	4	4
Mammalian excreta (mg/lb)	5	0
Other excreta (mg/lb)	1.0	0.0
Mold, % by weight	1	2
Insect-defiled/infested, % by weight	1	2
Extraneous, % by weight	0.50	0.00
FDA DALs		
Insect-infested and/or mouldy pieces by count	Av. of 10%	
Volatile oil (% min.)	7.0	
Moisture (% max.)	8.0	
Ash (% max.)	3.0	
Acid-insoluble ash (% max.)	0.5	
Average bulk index (mg/100 g)	N/A	

Source: Tainter and Grenis (1993).

Table 9.14. Ground nutmeg: chemical and physical specifications.

Specification	Suggested limits
FDA DALs (6 subsamples)	
Insect fragments/10 gm	Av. of 100 or more
Rodent hairs/10 gm	Av. of 1 or more
Volatile oil (% min.)	6.0
Moisture (% max.)	8.0
Total ash (% max.)	3.0
Acid-insoluble ash (% max.)	0.5
Military specifications (EE-S-63IJ, 1981)	
Volatile oil (ml/100 g, min.)	7.5
Moisture (% max.)	8.0
Total ash (% max.)	3.0
Acid-insoluble ash (% max.)	0.5
Non-volatile ether extract (% min.)	25.0
Granulation (% min. through USS #20)	95
Bulk index[2] (ml/100 g)	180

Source: Tainter and Grenis (1993).

Table 9.15. Whole mace: chemical and physical specifications.

Specification	Suggested limits
ASTA cleaniless specifications	
Whole dead insects, by count	4
Mammalian excreta (mg/lb)	3
Other excreta (mg/lb)	1.0
Mould, % by weight	2.00
Insect-defiled/infested, % by weight	1.00
Extraneous, % by weight	0.50
FDA DALs	
Insect infested and/or mouldy pieces by weight	Av. of 3%
Mammalian excreta	Av. of 3 mg per lb
Foreign matter through a 20-mesh sieve	Av. of 1.5%
Volatile oil (% min.)	15.0
Moisture (% max.)	8.0
Ash (% max.)	5.0
Acid-insoluble ash (% max.)	0.5
Average bulk index (mg/100 g)	N/A

Source: Tainter and Grenis (1993).

Table 9.16. Ground mace: chemical and physical specifications.

Specification	Suggested limits
FDA DALs	None
Volatile oil (% min.)	14.0
Moisture (% max.)	8.0
Total ash (% max.)	5.0
Acid-insoluble ash (% max.)	0.5
Military specifications (EE-S-63IJ, 1981)	
Volatile oil (ml/100 g, min.)	12.0
Moisture (% max.)	6.0
Total ash (% max.)	3.5
Acid-insoluble ash (% max.)	0.5
Non-volatile ether extract (% max.)	20.0–35.0
Granulation (% min. through USS #20)	95
Bulk index[2] (ml/100 g)	205

Source: Tainter and Grenis (1993).

and mace (http://www.indianspices.com/html/s1490qua.htm).

9.7. Conclusion

Nutmeg and mace are the two major primary products of *M. fragrans* and are considered commercially as spices. The constituents of nutmeg can be classified broadly into terpenoids, fatty acids, phenolic acids, lignans, neolignans and miscellaneous compounds. The oil content ranges from 3.9 to 16.5% in nutmeg, whereas in mace it varies from 6.0 to 26.1%. The major constituents of the essential oil in nutmeg and mace are sabinene and pinenes. The chief flavour-contributing components, namely myristicin and elemicin, are present in low concentrations. Myristicin is reported to be a potent hepatoprotective principle in nutmeg. However, further studies are needed to elucidate the hepatoprotective mechanism of myristicin. The lignan types of constituents possess anticarcinogenic property, which has been well documented. The use of essential oils in aromatherapy has become an exciting field in most countries. Apart from the hallucinogenic property of the spice, the other beneficial medicinal aspects will gain more importance in future research programmes, especially in the fields of cancer and aromatherapy.

References

Adedire, C.O. (2002) Use of nutmeg *Myristica fragrans* (Houtt.) powder and oil for the control of cowpea storage bruchid, *Callosobruchus maculates* Fabricius. *Zeitschrift- fur Pflanzenkrankheiten-und-Pflanzenschutz* 109(2), 193–199.

Al-Bataina, B.A., Maslat, A.O. and Al-Kofahi, M.M. (2003) Element analysis and biological studies on ten oriental spices using XRF and Ames test. *Journal of Trace Elements in Medicine and Biology* 17(2), 85–90.

Ananthakumar, A., Variyar, P.S. and Sharma, A. (2006) Estimation of aroma glycosides and their changes during radiation processing. *Journal of Chromatography A* 1108, 252–257.

Anon. (1962) *The Wealth of India Raw Materials*, Volume 6. CSIR, New Delhi, pp. 473–480.

Baldry, J., Dougan, J., Mathews, W.S., Nabney, J., Pickering, G.R. and Robinson, F.V. (1976) Composition and flavour of nutmeg oils. *International Flavours and Food Additives* 7, 28–30.

Banerjee, S., Sharma, R., Kale, R.K. and Rao, A.R. (1994) Influence of certain essential oils on carcinogen-metabolizing enzymes and acid-soluble sulfhydryls mouse liver. *Nutrition and Cancer* 21, 263–269.

Borges, P. and Pino, J. (1993) Nota. Obtencion de oleorresina de nuez moscada (*Myristica fragrans* Houtt.) *Revista Española de Ciencia y Tecnología de Alimentos* 33, 209–215.

Chandrashekar, K., Vijayalaxmi, V. and Deosthale, Y.G. (1995) Fatty acid profile of some Indian Spices. *Journal of Food Science and Technology* Mysore 32(5), 403–405.

Chang Yen, I., Sookram, R. and McGaw, D. (1996) Yield and chemical composition of essential oils of Grenadian nutmegs. *Tropical Agriculture* 73(4), 301.

Chatterjee, S., Niaz, Z., Gautam, S., Adhikari, S., Variyar, P.S. and Sharma, A. (2007) Antioxidant activity of some phenolic constituents from green pepper (*Piper nigrum* L.) and fresh nutmeg mace (*Myristica fragrans*). *Food Chemistry* 101, 515–523

Chhabra, S.K. and Rao, A.R. (1994) Trans mammary modulation of xenobiotic metabolizing enzymes in liver of mouse pups by mace (*Myristica fragrans* Houtt.) *Journal of Ethnopharmacology* 42, 169–177.

Cho, J.Y., Choi, G.J., Son, S.W., Jang, K.S., Lim, H.K., Lee, S.O., Sung, N.D., Cho, K.Y. and Kim, J.C. (2007) Isolation and antifungal activity of lignans from *Myristica fragrans* against various plant pathogenic fungi. 1. *Pest Management Science* 63(9), 935–940.

Choo, L.C, Wong, S.M. and Liew, K.Y. (1999) Essential oil of nutmeg pericarp. *Journal of the Science of Food and Agriculture* 79(13), 1954–1957.

Chung, J.Y., Choo, J.H., Lee, M.H. and Hwang, J.K. (2006) Anticariogenic activity of macelignan isolated from *Myristica fragrans* (nutmeg) against *Streptococcus mutans*. *Phytomedicine* 13(4), 261–266.

DASD (Directorate of Arecanut and Spices Development) (2007) *Arecanut and Spices Database*. Premaja, P. and Manoj kumar, K. (eds), Directorate of Arecanut and Spices Development, Calicut, India, pp. 110.

De, M., De, A.K. and Banerjee, A.B. (1999) Antimicrobial screening of some Indian spices. *Phytotherapy Research* 13(7), 616–618.

De Blasi, V., Debrot, S., Menoud, P.A., Gendre, L. and Schowing, J. (1990) Amoebicidal effect of essential oils. *Journal of Toxicology In Vitro* 10(6), 361–373.

Dorman, H.J.D. and Deans, S.J. (2000) Antimicrobial agents from plants:antibacterial activity of plant volatile oils. *Journal of Applied Microbiology* 88, 308–315.

Dorman, H.J.D. and Deans, S.J. (2004) Chemical composition, antimicrobial and *in vitro* antioxidant properties of *Monarda citriodora* var. *citriodora, Myristica fragrans, Origanum vulgare* ssp. hirsutum, *Pelargonium* species and *Thymus zygis* oils. *Journal of Essential Oil Research* 16, 145–150.

Dorman, H.J.D., Deans, S.G., Noble, R.C. and Surai, P. (1995) Evaluation of *in vitro* plant essential oil as natural antioxidants. *Journal of Essential Oil Research* 7, 645–651.

Dorman, H.J.D., Surai, P. and Deans, S.G. (2000a) *In vitro* antioxidant activity of a number of plant essential oils and phytoconstituents. *Journal of Essential Oil Research* 12, 241–248.

Dorman, H.J.D., Figueiredo, A.C., Barroso, J.G. and Deans, S.G. (2000b) *In vitro* evaluation of antioxidant activity of essential oils and their components. *Flavour and Fragrance Journal* 15(1), 12–16.

Ehlers, D., Kirchhoff, J., Gerard, D. and Quirin, K.W. (1998) High performance liquid chromatography analysis of nutmeg and mace oils produced by supercritical CO_2 extraction – comparison with steam distilled oils, comparison of East Indian, West Indian and Papuan oils. *International Journal of Food Science and Technology* 33(3), 215–223.

Forrest, J.E., Heacock, R.A. and Forrest, T.P. (1974) Diarylpropanoids from nutmeg and mace (*Myristica fragrans* Houtt.). *Journal of Chemical Society – Perkin Transactions* I 1974, 205–209.

Gokhale, S.B., Joshi, P.V., Tatiya, A.U., Bakliwal, S.R. and Fursule, R.A. (2004) Some neglected spices in India. *Natural Product Radiance* 3(5), 343–346.

Gopalakrishnan, M. (1992) Chemical composition of nutmeg and mace. *Journal of Spices and Aromatic Crops* 1(1), 49–54.

Gotke, N., Maheswari, M.L. and Mathur, V.K. (1990) Nematicidal activity of *Myristica fragrans* against *Meloidogyne incognita. Indian Perfumer* 34, 105–107.

Guenther, E. (1952) *The Essential Oils*, Volume 5. Robert E. Krieger Publishing Company, New York, pp. 59–81.

Hada, S., Hattori, M., Tezuka, Y., Kikuchi, T. and Namba, T. (1988) New neolignans and lignans from the aril of *Myristica fragrans. Phytochemistry* 27, 563–568.

Hashim, S., Aboobaker, V.S, Madhubala, R., Bhattacharya, R.K. and Rao, A.R. (1994) Modulatory effects of essential oils from spices on the formation of DNA adduct by aflatoxin B1 *in vitro. Nutrition and Cancer* 21, 169–175.

Hattori, M., Hada, S., Watahiki, A., Ihara, H., Shu, Y.Z., Kakiuchi, N., Miznno, T. and Namba, T. (1986) Studies on dental caries prevention by traditional medicines. X. Antibacterial action of phenolic components from mace against *Streptococcus mutans. Chemical and Pharmacological Bulletin* 34, 3885–3893.

Hattori, M., Hada, S., Kawata, Y., Tezuka, Y., Kikuchi, T. and Namba, T. (1987a) New 2,5-bis-aryl-3,4-dimethyltetrahydrofuran lignans from the aril of *Myristica fragrans. Chemical and Pharmacological Bulletin* 35(8), 3315–3322.

Hattori, M., Hada, S., Shu, Y.Z., Kakiuchi, N. and Namba, T. (1987b) New acyclic bis-phenyl propanoids from the aril of *Myristica fragrans. Chemical and Pharmacological Bulletin* 35, 668–674.

Hattori, M., Yang, X.W., Shu, Y.Z., Kakiuchi, N., Tezuka, Y., Kikuchi, T. and Namba, T. (1988) New constituents of the aril of *Myristica fragrans. Chemical and Pharmacological Bulletin* 36(2), 648–653.

Hener, U., Kreis, P. and Mosandl, A. (1991) Enantiomeric distribution of α-pinene, β-pinene and limonene in essential oils and extract. Part 3. Oils for alcoholic beverages and seasonings. *Flavour and Fragrance Journal* 6, 109–111.

http://en.wikipedia.org/wiki/Nutmeg; May 2007.

http://www.chem.uwimona.edu.jm: 1104/lectures/ nutmeg.html.

http://www.fao.org/docrep/x5047E/x5047E0a.htm 19-6–07; Section VI – Composition and constituents of nutmeg and mace.

http://www.indianspices.com/html/s1490qua.htm.

Hussain, S.P. and Rao, A.R. (1991) Chemopreventive action of mace (*Myristica fragrans* Houtt) on methylcholanthrene-induced carcinogenesis in the uterine cervix in mice. *Cancer Letters* 56(3), 231–234.

Isogai, A., Murakoshi, S., Suzuki, A. and Tamura, S. (1973) Structure of new dimeric phenyl propanoids from *Myristica fragrans* Houtt. *Agricultural and Biological Chemistry* 37, 1479–1481.

Izaldin, M.H., Al-Khatib, S.A., Hanifa, M., Mehdi, A.W.R. and Al-Shabibi, M.M. (1987) Gas-liquid chromatographic determination of fatty acids and sterols of selected Iraqi foods. *Journal of Food Composition and Analysis* 1, 59–64.

Jannu, L.N., Hussain, S.P. and Rao, A.R. (1991) Chemo preventive action of mace on DMBA-induced Papilloma gene is in the skin of mice. *Cancer Letters* 56, 59–63.
Joshi, G.D., Relekar, P.P., Pawar, C.D. and Magdum, M.B. (1996) Preliminary studies on utilization of nutmeg (*Myristica fragrans* Van Houllen) fruit spice rind. *Indian Cocoa Arecanut and Spices Journal* 20(1), 5–9.
Jukic, M., Politeo, O. and Milos, M. (2006) Chemical composition and antioxidant effect of free volatile aglicones from nutmeg compared to its essential oil. *Croatica Chemica Acta* 79, 209–214.
Kasahara, H., Miyazawa, M. and Kameoka, H. (1995) Absolute configuration 8-*0*-4 -neolignans from *Myristica fragrans*. *Phytochemistry* 40, 1515–1517.
Kimura, Y., Saito, Y., Sakamoto, T., Shinbo, M. and Kameyama, Y. (1979) Food antioxidant from mace. Patent-*Japan Kokai Tokkyo Koho* 79, 130, 486–489.
Konig, W.A., Kruger, A., Icheln, D. and Runge, T. (1992) Enantiomeric composition of the chiral constituents in essential oils. Part 1: monoterpene hydrocarbons. *Journal of High Resolution Chromatography* 15, 184–189.
Krishnamoorthy, B. and Rema, J. (2000) Nutmeg and mace. In: Peter, K.V. (ed.) *Handbook of Herbs and Spices*. CRC Press, New York, pp. 238–248.
Kumaravelu, P., Subramaniyam, S., Dakshinamoorthy, D.P. and Devaraj, N.S. (1996) The antioxidant effect of eugenol on CC14-induced erythrocyte damage in rats. *Journal of Nutritional Biochemistry* 7(1), 23–28.
Kuo, Y.I.I., Lin, S.T. and Wu, R.E. (1989) Three new lignans from the nutmeg of *Myristica cagayanesis*. *Chemical and Pharmaceutical Bulletin* 37(9), 2310–2312.
Latha, P.G., Sindhu, P.G., Suja, S.R., Geetha, B.S., Pushpangadan, P. and Rajasekharan, S. (2005) Pharmacology and chemistry of *Myristica fragrans* Houtt – a review. *Journal of Spices and Aromatic Crops* 14(2), 94–101.
Lawrence, B.M. (1979) Major tropical spices – nutmeg and mace (*M. fragrans* Houtt.). In: *Essential Oils 1978*. Allured Publishing Corp., Carol Stream, Illinois, pp. 50–103.
Lawrence, B.M. (1997) Progress in essential oils. *Perfumer and Flavorist* 22(4), 57–74.
Lawrence, B.M. (2000) Progress in essential oils. *Perfumer and Flavorist* 25(5), 52–71.
Lawrence, B.M. (2005) Progress in essential oils. *Perfumer and Flavorist* 30(5), 52–66.
Lee, B.K., Kim, J.H., Jung, J.W., Choi, J.W., Han, E.S., Lee, S.H., Ko, K.H. and Ryu, J.H. (2005) Myristicin-induced neurotoxicity in human neuroblastoma SK-N-SH cells. *Toxicology Letters* 157(1), 49–56.
Madhavan, L., Gopalakrishnan, M. and Narayanan, C.S. (1991) Chemical composition of nutmeg leaf and flower oils. *Indian Perfumer* 35(3), 161–163.
Mahady, G.B., Pendland, S.L., Stoia, A., Frank, A., Hamill, D., Fabricant, B.M., Dietz, L.R. and Chadwick, L.R. (2005) *In vitro* susceptibility of *Helicobacter pylori* to botanical extracts used traditionally for the treatment of gastrointestinal disorders. *Phytotherapy Research* 19, 988–991.
Mallavarapu, G.R. and Ramesh, S. (1998) Composition of essential oils of nutmeg and mace. *Journal of Medicinal and Aromatic Plant Sciences* 20(3), 746–748.
Maya, K.M., Zachariah, T.J. and Krishnamoorthy, B. (2004) Chemical composition of essential oils of nutmeg (*Myristica fragrans* Houtt.) accessions. *Journal of Spices and Aromatic Crops* 13(2), 135–139.
Morita, T., Jinno, K., Kawagishi, H., Arimoto, Y., Suganuma, H., Inakuma, T. and Sugiyama, K. (2003) Hepatoprotective effect of myristicin from nutmeg (*Myristica fragrans*) on lipopolysaccharide/D-galactosamine-induced liver injury. *Journal of Agriculture and Food Chemistry* 51(6), 1560–1565.
Moyler, D.A., Bowing, R.M. and Stevens, M.A. (1994) CO_2 extraction of essential oils. Part V. Nutmeg and mace oils. In: Charalambous, G. (ed.) *Spices, Herbs and Edible Fungi*. Elsevier Science, B.V., Amsterdam, pp. 145–170.
Nakamura, N., Kiuchi, F., Tsuda, Y. and Kondo, K. (1988) Studies on crude drugs effective on visceral larva migrans. V. The larvicidal principle in mace (aril of *Myristica fragrans*). *Chemical and Pharmacological Bulletin* 36(7), 2685–2688.
Narasimhan, B. and Dhake, A.S. (2006) Antibacterial principles from *Myristica fragrans* seeds. *Journal of Medicinal Food* 9(3), 395–399.
Nguyen, U., Anstee, M. and Evans, D.A. (1998) Extraction and fractionation of spices using supercritical fluid carbon dioxide. Presented at *The 5th International Symposium on Supercritical Fluids*, Nice, France.
Niyas, Z., Variyar, P.S., Gholap, A.S. and Sharma, A. (2003) Effect of gamma irradiation on the lipid profile of nutmeg (*Myristica fragrans* Houtt.). *Journal of Agricultural and Food Chemistry* 51, 6502–6504.
Ogunwande, I.A., Olawore, N.O., Adeleke, K.A. and Ekundayo, O. (2003) Chemical composition of essential oils of *Myristica fragrans* Houtt (nutmeg) from Nigeria. *Journal of Essential Oil Bearing Plants* 6(1), 21–26.
Onyenekwe, P.C., Ogbadu, G.H., Deslauriers, H., Gagnon, M. and Collin, G.J. (1993) Volatile constituents of the essential oil of *Monodora myristica* (Gaertn) Dunal. *Journal of the Science of Food and Agriculture* 61(3), 379–381.

Orabi, K.Y., Mossa, J.S. and El-Feraly, F.S. (1991) Isolation and characterization of two antimicrobial agents from mace (*Myristica fragrans*). *Journal of Natural Products* 54(3), 856–859.

Packiyasothy, E.V., Jansz, E.R. and Dharmadasa, H.M. (1991) Studies on some chemical components of nutmeg (*Myristica fragrans* Houtt) leaf directed at determination of sex of seedlings. *Journal of the National Science Council of Sri Lanka* 19(2), 91–97.

Park, S., Lee, D.K. and Yang, C.H. (1998) Inhibition of fos-jun-DNA complex formation by dihydroguaiaretic acid and *in vitro* cytotoxic effects on cancer cells. *Cancer Letters* 127, 23–28.

Patro, B.S., Bauri, A.K., Mishra, S. and Chattopadhya, S. (2005) Antioxidant activity of *Myristica malabarica* extracts and their constituents. *Journal of Agricultural and Food Chemistry* 53(17), 6912–6918.

Pino, J.A., Borges, P. and Mollinedo, B. (1991) Nota aceite esencial denuez moscada (*Myristica fragrans* H.): obtenction y caracterizacion de su composcion quimica. *Revista de Agroquímica y Tecnología de Alimentos* 31, 411–416 (Lawrence, B.M., 1997).

Pruthi, J.S. (1976) *Spices and Condiments*. National Book Trust, New Delhi, India, pp. 287.

Purseglove, J.W., Brown, E.G., Green, C.L. and Robbins, S.R.J. (1981) *Spices*. Volume 1. Longman, London, pp. 174–228.

Purushothaman, K. and Sarada, A. (1980) Chemical examination of the aril of *Myristica fragrans* (Jathipathri). *Indian Journal of Chemistry* 19B, 236–238.

Purushothaman, K., Sarada, A. and Connolly, J.D. (1977) Malabaricones A–D, novel diarylnonanoids from *Myristica malabarica* Lam. (*Myristicaceae*). *Journal of Chemical Society – Perkin Transactions* I, 587–588.

Ram, A., Lauria, P., Gupta, R. and Sharma, V.N. (1996) Hypolipidaemic effect of *Myristica fragrans* fruit extract in rabbits. *Journal of Ethnopharmacology* 55(1), 49–53.

Rastogi, R.P. and Mehrotra, B.N. (1995) *Compendium of Indian Medicinal Plants*. Volume IV (1985–1989). Central Drug Research Institute, Lucknow and Publication Information Directorate, Council of Scientific and Industrial Research, New Delhi, p. 493.

Ravid, U., Putievsky, E., Katzir, I. and Ikan, R. (1992) Determination of the enantiomeric composition of terpinen-4-ol in essential oils using a permethylated β-cyclodextrin coated chiral capillary column. *Flavour and Fragrance Journal* 7, 49–52.

Ravid, U., Putievsky, E. and Katzir, I. (1996) Sterochemical analysis of borneol in essential oils using permethylated β-cyclodextrin as a chiral stationary phase. *Flavour and Fragrance Journal* 11, 191–195.

Ross, I.A. (2001) *Medicinal Plants of the World. Volume 2: Chemical Constituents. Traditional and Modern Medicinal Uses*. Humana Press, Totowa, New Jersey, pp. 333–343.

Sait, S. and Satyaputra, I. (1995) The effect of deterpenation process on the medicinal quality of nutmeg oil. *Warta* IHP 12(1–2), 41–43.

Saito, Y., Kimura, Y. and Sakamoto, T. (1976) The antioxidant effects of petroleum ether soluble and insoluble fractions from spices. *Eito To Shokuryo* 29, 505–510.

Sanford, J.K. and Heinz, D.E. (1971) Effects of storage on the volatile composition of nutmeg. *Phytochemistry* 10(6), 1245–1250.

Satyavathy, G.V., Gupta, A.K. and Tandon, N. (1987) *Myristica* Boehmer (*Myristicaceae*). In: *Medicinal Plants of India*, Volume 2. Indian Council of Medical Research, New Delhi, pp.306–309.

Schubert, V. and Mosandl, A. (1991) Chiral compounds of essential oils. VIII. Stereo differentiation of linalool using multidimensional gas chromatography. *Phytochemical Analysis* 2, 171–174.

Simpson, G.I.C. and Jackson, Y.A. (2002) Comparison of the chemical composition of East Indian, Jamaican and other West Indian essential oils of *Myristica fragrans* Houtt. *Journal of Essential Oil Research* 14(1), 6–9 (http://www.chem.uwimona.edu.jm:1104/lectures/nutmeg.html).

Sohn, J.H., Han, K.L., Choo, J.H. and Hwang, J.K. (2007) Macelignan protects HepG2 cells against tert-butylhydroperoxide-induced oxidative damage. 1. *Biofactors* 29(1), 1–10.

Spricigo, C.B., Pinto, L.T., Bolzan, A. and Novais, A.F. (1999) Extraction of essential oil and lipids from nutmeg by liquid carbon dioxide. *Journal of Supercritical Fluids* 15, 253–259.

Tainter, R.D. and Grenis, T.A. (1993) *Spices and Seasonings Food Science and Technology*. VCH Publishers, New York, p. 226.

Takikawa, A., Abe, K., Yamamoto, M., Ishimaru, S., Yasui, M., Okubo, Y. and Yokoigawa, K. (2002) Antimicrobial activity of nutmeg against *Escherichia coli* O157. *Journal of Biosciences and Bioengineering* 94(4), 315–320.

Thakur, R.S., Puri, H.S. and Hussain, A. (1989) *Current Status of Chemistry and Pharmacology Major Medicinal Plants of India*. Vedams Books, New Delhi.

Tomaino, A., Cimino, F., Zimbalatti, V., Venuti, V., Sulfaro, V., De Pasquale, A. and Saija, A. (2005) Influence of heating on antioxidant activity and the chemical composition of some spice essential oils. *Food Chemistry* 89, 549–554.

Uhl, S.R. (2000) *Handbook of Spices Seasonings and Flavorings*. Technomic Publishing, Lancaster, Pennsylvania, pp.126–139.

Ur-Rahman, M.I., Choudhary, A., Farooq, A., Ahmed, M.Z., Iqbal, B. and Demirici, F. (1999) Antifungal activities and essential oil constituents of some spices from Pakistan. *Third International Electronic Conference on Synthetic Organic Chemistry (ECSOC-3)*, September 1–30, 1999, 10 pp.

Variyar, P.S. and Bandyopadhyay, C. (1995) Estimation of phenolic acids in cinnamon, clove, cardamom, nutmeg and mace by high performance liquid chromatography. *Journal of Spices and Aromatic Crops* 4(2), 129–134.

Wang, Y., Yang, X.W., Tao, H.Y. and Liu, H.X. (2004) GC-MS analysis of essential oils from seeds of *Myristica fragrans* in Chinese market. *Zhongguo Zhong Yao Za Zhi* [China Journal of Chinese Materia Medica] 29(4), 339–342.

Woo, W.S., Shin, K.H., Wagner, H. and Lotter, H. (1987) The structure of macelignan from *Myristica fragrans*. *Photochemistry* 26(5), 1542–1543.

Wossa, S.W., Rali, T. and Leach, D.N. (2004) Analysis of essential oil composition of some selected spices of Papua New Guinea. *Papua New Guinea Journal of Agriculture, Forestry and Fisheries* 47(1/2), 17–20.

Zacchino, S.A. and Badano, H. (1991) Enantioselective synthesis and absolute configuration assignment of erythro-(3,4-methylenedioxy-7-hydroxy-1′-allyl-3′,5′-dimethoxy)-8-*O*-4′-neolignan and its acetate, isolated from nutmeg (*Myristica fragrans*). *Journal of Natural Products* 54(1), 155–160.

Zacchino, S.A., Rodriguez, G., Pezzenati, G. and Orellana, G. (1997) *In vitro* evaluation of antifungal properties of 8-*O*-4′-neolignans. *Journal of Natural Products* 60, 659–662.

Zacchino, S.A., Lopez, S.N. and Pezzenati, G.D. (1999) *In vitro* evaluation of antifungal properties of phenylpropanoids and related compounds acting against dermatophytes. *Journal of Natural Products* 62, 1353–1357.

Zachariah, T.J., Shajiprabha, S.P. and Krishnamoorthy, B. (2000) Major chemical constituents of nut, mace and leaf of *Myristica fragrans*. Spices and aromatic plants: challenges and opportunities in the new century. Contributory papers, *Centennial Conference on Spices and Aromatic Plants*, Calicut, Kerala, India, pp. 297–300.

Zhu, L.F., Li, Y.H., Li, B.L., Lu, B.Y. and Zhang, W.L. (1995) *Aromatic Plants and Essential Constituents*. Supplement 1. South China Institute Botany, Chinese Academy of Sciences. Hai Feng Publishing Co., Peace Book Co., Hong Kong.

10 Coriander

V.A. Parthasarathy and T. John Zachariah

10.1. Introduction

Coriander (*Coriandrum sativum*) is an annual and herbaceous plant, belonging to the Apiaceae family. Coriander (*C. sativum* L.) is a culinary and medicinal plant. Native of southern Europe and the western Mediterranean region, this herb is cultivated worldwide (Weiss, 2002). This species, rich in linalool, has potential for use as an essential oil. It has been used as an analgesic, carminative, digestive, antirheumatic and antispasmodic agent. Its fruits (commonly called 'seeds') are used for flavouring candies, in cookery, perfumery, beverages and in the tobacco industry. It was one of the earliest spices used by mankind. Coriander was used in Egypt for medicinal and culinary purposes as early as 1550 BC and is mentioned in the Ebers Papyrus.

It has been used as a flavouring agent in food products, perfumes and cosmetics. Coriander has been credited with many medicinal properties. Powdered seeds or dry extract, tea, tincture, decoction or infusion have been recommended for dyspeptic complaints, loss of appetite, convulsions, insomnia and anxiety (Msaada *et al.*, 2007). The essential oils and other extracts from coriander have been shown to possess antibacterial, antioxidant, antidiabetic, anticancerous and antimutagenic properties.

10.2. Botany

The genus *Coriandrum* L. has two species, which are native to the eastern Mediterranean region. *C. sativum* L. is cultivated and is coriander. *C. sativum* is approximately 30–100 cm in height, with glabrous, greatly divided, strong-smelling leaves. The mature fruits have a fresh and pleasant flavour and are largely used all over the world in ground or volatile isolate form for flavouring sweets, beverages, tobacco products and baked goods and as a basic ingredient for curry powder. The essential oil obtained from its fruits at amounts ranging from approximately 0.5 to 2.5% is used both in flavours and in the manufacture of perfumes and soaps (Carrubba *et al.*, 2002).

A type with ovoid fruits was introduced into South-east Asia initially from India, a globoid-fruited type later from China and the typical European cultivars much later. However, it is cultivated in the region only as a domestic plant. It was also believed that an extract of the whole plant, or fruit, could preserve a person's spirit or essence after death and, for this reason, such compounds were an important accessory at funerals. In commerce, coriander exists in two categories: the small-fruited *C. sativum* L. var. *microcarpum* DC. (diameter 1.5–3.0 mm) and the larger-fruited *C. sativum* L. var. *vulgare* Alef. (diameter 3.0–5.0 mm). The former is exemplified by the volatile oil-rich Russian coriander, while the latter includes Moroccan, Indian and some other Asiatic types of coriander, all of which have very low volatile oil contents (Purseglove *et al.*, 1981).

10.3. Uses

The coriander plant yields two primary products which are employed for flavouring purposes: the fresh green herb and the spice. The latter is the dried form of the whole mature seed capsule (fruit) but it is frequently and incorrectly termed 'coriander seed' in commerce. The odour and flavour of these two products are markedly different, and the herb is generally disliked and little used in Europe and North America, except in certain specialist applications. However, the herb is much appreciated for culinary flavouring purposes in Asia, the Middle East and Central and South America, where it is produced in substantial quantities for domestic consumption. By contrast, the spice is of considerable importance as an item of international trade and it is used widely in the whole or ground form for flavouring applications. The spice is also employed for the preparation of its steam-distilled essential oil and of its solvent-extracted oleoresin, both these products being utilized by the aroma and flavour industries. The residues remaining after distillation of the spice may be employed as a cattle feed and interest has also been expressed in the erstwhile USSR in the recovery of the fatty oil for possible use as a lubricant in the metallurgical industries (Purseglove *et al.*, 1981).

The major use of the spice on a worldwide basis is in flavouring applications in the ground form and its main outlet is as an ingredient of curry powders, of which it comprises some 25–40% of the spice mixture. The ground spice is also extensively used domestically as a flavouring agent and by manufacturers of processed foods in baked goods, sauces and meat dishes. The whole spice is employed in pickling and in the flavouring of certain alcoholic beverages, particularly gin. The spice mainly enters trade in the whole form and is ground in the importing centres.

The oleoresin, prepared by solvent extraction of the spice, is prepared on a relatively small scale in Russia and in some Western countries.

10.4. General Composition

Dried, ripe coriander fruit contain steam-volatile oil, fixed (fatty) oil, proteins, cellulose, pentosans, tannins, calcium oxalate and minerals. The major constituents are fibre (23–36%), carbohydrates (about 20%), fatty oil (16–28%) and proteins (11–17%). The residues remaining after distillation of the essential oil contain high fat and protein, which is useful as animal feed.

The saponifiable portion of the fatty oil accounts for about 90% of the total fixed oil and is characterized by a very high content of octadecenoic acids. Petroselinic and oleic acid occur at similar levels and jointly comprise 74–85%, linoleic 7–16% and palmitic 4–8%, of the constituent fatty acids. During prolonged storage of the spice, the free fatty acid content gradually increases and this is a good indicator of the age of the material. The contents of fatty acids, sterols and total tocopherols in a deodorized oil derived from coriander seeds (yield up to 28%) are compared with those in sunflower oil and tests on the biological effects of coriander oil are reported by Mironova *et al.* (1991). Of the fatty acids present, total $C_{18:1}$ acids (petroselinic acid + oleinic acids) constituted 80–82% and petroselinic acid alone 50–60%, and the food value was lower than that of sunflower oil. Kim *et al.* (1996) found the production of petroselinic acid from cell suspension cultures of *C. Sativum*.

Ripe, dried fruit contains essential oil, fixed oil, proteins, cellulose, pentosans, tannins, calcium oxalate and minerals. The major constituents are fibre (23–36%), carbohydrates (13–20%), fatty oil (16–28%), proteins (11–17%) and essential oil (1–1.5%), with small-fruited cultivars (up to 2.0%). These amounts and ratios can vary with cultivar, season and maturity at harvest. Analysis of air-dried fruits gave the following average content: water 11%, crude protein 11%, fatty oil 19%, carbohydrate 23% (including starch 11%, pentosans 10%, sugar 2%), crude fibre 28% and minerals 5%. Fruit contains two small seeds, each enclosed in a mericarp, which is concave on the commissural, convex on the dorsal side; the testa is attached to the fruit wall. The seed is whitish to yellowish with two flat, thin, circular cotyledons and a conical radicle; there is copious grey–white endosperm and 1000 seed weight is 7–17 g (Uma Pradeep *et al.*, 1993).

Ross and Murphy (1992) found triacylglycerol rich in octadecenoic acid as the major lipid component found even in very young coriander seeds. The storage albumins accumulated at a slightly earlier stage of seed development. Data on subcellular localization of the two classes of seed protein showed that the storage albumins were confined to the soluble protein fraction of seed homogenates and were localized in the large, electron-dense, protein storage bodies' characteristic of multiple mature seeds. In contrast, the oleoresins were located in the floating oil-body fraction and were confined to the surface of oil storage bodies.

10.5. Chemistry

The composition of the volatile oil, which determines the odour and flavour character, has been of particular fascination to chemists. In the unripe fruits and the vegetative parts of the plant, aliphatic aldehydes predominate in the steam-volatile oil and are responsible for the peculiar, fetid-like aroma. On ripening, the fruits acquire a more pleasant and sweet odour and the major constituent of the volatile oil is the monoterpene alcohol, linalool.

The primary quality determinant of the spice is the content and composition of its steam-volatile oil. The volatile oil content of the spice can vary considerably according to the type and source and usually ranges from 0.1 to 1.7% and, in some cases, up to 2.7%. European coriander is mainly of the small-fruited type and usually has a volatile oil content greater than 0.4%, with the highest values exhibited by some Russian cultivars. Moroccan and Indian corianders are mainly large-fruited types, globular in shape in the former and egg-shaped in the latter, and their volatile oil contents are usually less than 0.4%. In addition to the variation in volatile oil content, distinctions are made on an organoleptic basis between various sources of coriander. Indian and Moroccan corianders generally are regarded as inferior to European types and the traditional Russian spice is regarded as the best (Purseglove *et al.*, 1981). During storage, some of the volatile oil can be lost by evaporation, but the rate of loss and the extent of organoleptic deterioration are dependent on the physical form of the spice and on the conditions and duration of storage. Whole coriander fruits, which have been carefully dried, normally can be stored with minimal deterioration over a considerable period. However, significant oil losses can occur with damaged or split fruit during storage. The most serious problem is encountered with ground coriander, which undergoes a rapid volatile oil loss and a marked organoleptic deterioration within a matter of weeks if left exposed to the atmosphere.

Coriander oleoresin is prepared by solvent extraction of the spice. The oleoresin contains the volatile oil, fatty oil and some other extractives, but their relative abundance is dependent on the raw material, the processing procedure and the particular solvent used. Coriander oleoresins commonly contain about 90% fatty oil and about 5% steam-volatile oil.

Fixed oil

The fruit contains a dark brownish-green fixed oil with a similar pleasant odour to the essential oil and often solidifies with storage. Fatty oil content and composition in ripe fruit endosperm ranges from 12 to 25%, mainly dependent on environmental conditions. The saponifiable portion of the fatty oil accounts for about 90% of the total and contains a very high content of octadecenole acids; petroselinic and oleic acid occur at similar levels and together comprise 75–85%, linolenic 7–16% and palmitic 4–8%, but the relative abundance of each is a cultivar characteristic of an Indian oil (Suh *et al.*, 1999).

Suh *et al.* (1999) studied the isoforms of acyl carrier protein involved in seed-specific fatty acid synthesis in coriander seed. It produces unusual monoenoic fatty acids which constitute over 80% of the total fatty acids of the seed oil. The initial step in the formation of these fatty acids is the desaturation of palmitoyl-ACP (acyl carrier protein) at the DELTA4 or DELTA6 positions to produce DELTA4-hexadecenoic acid (16:1DELTA4) or DELTA6-hexadecenoic acid (16:1DELTA6), respectively.

Mustafaev *et al.* (1989) found certain parameters useful in the determination of the quality of fatty oil in coriander fruits. Drying to ~16% moisture, grinding in a pirouette mill for 60 s, ageing in a thin layer for 2 h for evaporation of volatile oils, comminution in chloroform in a Foss–Lett apparatus, filtration and measurement of acid number and peroxide number are better techniques to be adopted.

Ramadan and Morsel (2002) extracted coriander seed oil with chloroform/methanol (2:1, v/v) and the amount of total lipids was 28.4% of seed weight. The major fatty acid was petroselinic acid (65.7% of the total fatty acid methyl esters), followed by linoleic acid (Mekhedov *et al.*, 2001). Chromatography on a silica column with solvent of increasing polarity yielded 93.0% neutral lipids, 4.14% glycolipids and 1.57% phospholipids. Fatty acid profiles of neutral lipid subclasses, triacylglycerols and sterol content were determined using gas-liquid chromatography. Six triacylglycerol molecular species were detected, but one component ($C_{54:3}$) corresponding to tripetroselinin and/or dipetroselinoyl oleoyl glycerol comprised more than 50% of the total triacylglycerols. Sterol content was estimated to be at a high level (5186 µg/g oil). Stigmasterol, beta-sitosterol, Delta 5-avenasterol and campesterol were found to be the sterol markers. The major individual phospholipid subclasses were phosphatidylcholine, followed by phosphatidylethanolamine, phosphatidylinositol and phosphatidylserine.

One group of components which play a major role in the medicinal potential are phenolic acids. In all, seven phenolic acids, e.g. tannic, gallic, caffeic, cinnamic, chlorogenic, ferulic and vanillic acids, could be identified. Several parts of the spices, for instance, seeds, leaves, barks, rhizomes, latex, stigmas, floral buds and modified stems, were used in the study (Singh *et al.*, 2004).

Volatile oil

Composition

Studies of the composition of the spice oil were initiated early in the 19th century and the major constituent was first isolated in 1852 by Kawalier (Purseglove *et al.*, 1981). This compound was characterized subsequently as an alcohol and was named coriandrol in a later investigation. The identity of coriandrol was established eventually as an optically active form (dextrorotary) of the monoterpene alcohol, linalool.

Detailed study of the coriander spice oil composition showed that the *d*-linalool content ranged from 60 to 70% and the hydrocarbon content was about 20%. α- and β-pinenes, dipentene (limonene), *p*-cymene, α- and γ-terpinenes, *n*-decanal, geraniol and l-borneol were also identified as constituents of the spice oil.

The linalool content of coriander spice oils has been reported in the literature to range between 25 and 80%, but values below 55% have been found mainly with coriander indigenous to Asia. With regard to the linalool content and some other composition characteristics, European coriander spice oils may be placed at one end of the spectrum and Indian oils at the other end. For most European oils, the monoterpene hydrocarbon content is between 16 and 30%, the linalool ranges between 60 and 75% and the remainder is comprised largely of other oxygenated monoterpenes. Russian oils traditionally have exhibited high linalool contents (69–75%) and this feature has been maintained with the new cultivars selected for a combination of desirable characteristics. In European spice oils, the major monoterpene hydrocarbon constituents are γ-terpene (up to 10%), geranyl acetate (up to 5%), geranyl and camphor (each up to 4%) and geraniol (up to 2%).

Ishikawa *et al.* (2003) could obtain 33 compounds, including two new monoterpenoids, four new monoterpenoid glycosides, two new monoterpenoid glucoside sulphates and two new aromatic compound glycosides, from the water-soluble portion of the methanol extract of coriander fruit. Their structures were clarified by spectral investigation. The major constituents of coriander essential oils (dried fruits, herb prior to flowering and flowering herb) are listed below.

Constituents identified in coriander essential oils

MONOTERPENE HYDROCARBONS *p*-Cymene, camphene, Δ-3-carene, limonene (dipentene), myrcene, *cis*- and *trans*-ocimene, α-phellandrene, β-phellandrene, α-pinene, β-pinene, sabinene, α-terpinene, γ-terpinene, terpinolene, α-thujene.

MONOTERPENE OXIDES AND CARBONYLS Camphor, 1,8-cineole, linalol oxide, carvone, geranial.

MONOTERPENE ALCOHOLS Borneol, citronellol, geraniol, linalool, nerol, α-terpineol, 4-terpinenol.

MONOTERPENE ESTERS Bornyl acetate, geranyl acetate, linalyl acetate, α-terpinyl acetate.

SESQUITERPENES β-Caryophyllene, caryophellene oxide, elemol, nerolidol.

PHENOLS Anethole, myristicin, thymol.

MISCELLANEOUS COMPOUNDS Acetic acid, α-*p*-dimethyl styrene.

ALIPHATIC HYDROCARBONS Heptadecane, octadecane.

ALIPHATIC ALCOHOLS Decanol, dodecanol.

ALIPHATIC ALDEHYDES Octanal, nonanal, decanal, undecanal, dodecanal, tridecanal, tetradecanal, 3-octenal, 2-decenal, 5-decenal, 8-methyl-2-nonenal, 8-methyl-5-nonenal, 6-undecenal, 2-dodecenal, 7-dodecenal, 2-tridecenal, 8-tridecenal, 9-tetradecenal, 10-pentadecenal, 3,6-undecadienal, 5,8-tridecadienal.

Figure 10.1 illustrates the major volatile compounds of coriander.

Properties of the oil

The essential oil obtained on steam distillation of the spice is a colourless or pale yellow liquid. Its aroma is pleasant, sweet, somewhat woody-spicy aromatic and the floral balsalmic undertone and peppery woody top note are characteristic features. The organoleptic properties of the distilled oil tend to deteriorate during prolonged storage, as is the case with the spice, especially if it is left exposed to light and air. The inclusion of unripe fruits or other overground parts of the plant during distillation of the spice imparts an obnoxious odour to the oil (Burt, 2004).

Turysheva *et al.* (1989) showed that the oil from material with a high proportion of split fruits was suitable for obtaining the perfumery fractions of (+)-linalool and (+)-linalyl acetate.

Physico-chemical properties

As in the case with the oil distillation yield, the physico-chemical constants of the distilled oil tend to vary somewhat according to the type of spice used, its age and, to a certain extent, the processing procedure followed.

The physico-chemical properties of European spice oils fall within fairly close limits, with the optical rotation value usually ranging between +9° and +13°, while

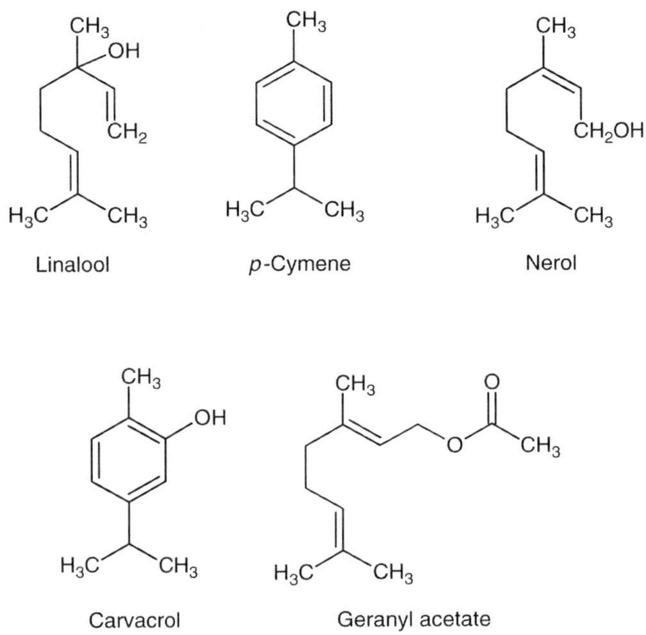

Fig. 10.1. Volatile compounds of coriander.

the ester number normally is well below the upper limit of the standard specifications. Data in the literature for Moroccan spice oils indicate an upper limit of +10° for the optical rotation value and ester numbers of up to 30. Traditional Indian spice oils are notable in exhibiting refractive index values at or below the lower limit of the standard specifications and in possessing very high ester numbers (Borges *et al.*, 1990).

The inclusion of unripe fruits or leaf and stalk material of the plant together with the spice during distillation results in a change in the physico-chemical constants. The specific gravity, optical rotation and alcohol (calculated as linalool) values are reduced somewhat, while the carbonyl value is enhanced (Purseglove *et al.*, 1981; Smallfield *et al.*, 1994).

Frank *et al.* (1995), by adopting capillary gas chromatography online coupled with isotope ratio mass spectrometry (cGC-IRMS), compared compounds of coriander with those of commercially available spices and essential oils. They found that isotopic effects among genuine monoterpenes are determined exclusively by the influence of enzymatic reactions during secondary biogenetic pathways.

Porter and Lammerink (1994) studied the density of coriander essential oil over the temperature range 20 to 60°C. The density of the oil decreased as temperature increased. There was some variation between oils in the temperature coefficients with the change in density. The density differentials between oil and water and their temperature coefficients varied markedly between different oils. A preliminary separation coefficient is used to indicate the effect of oil density and condensate water viscosity on oil separation at different temperatures. Separation of less dense oils with small differentials would benefit more from increased temperatures than less dense oils with large differentials (*C. sativum*). It was necessary to use both density and viscosity data to determine whether temperature control of a separator was required to obtain efficient separation of an oil from the condensate stream following steam distillation.

Shubert and Mosandl (1991) studied the chiral compounds of essential oils with reference to stereo differentiation of linalool using multidimensional gas chromatography. The enantiomeric distribution of linalool content is useful in quality control of essential oils. The main characteristics of coriander oil are: specific gravity (20°C) 0.863–0.875; refractive index (20°C) 1.4620–1.4720; optical rotation (25°C) +8° to +15°; almost insoluble in water, soluble in 3 vols 70% alcohol, very soluble in chloroform, ether and glacial acetic acid. A fairly wide variation occurs in these characteristics since cultivar, time of harvest, storage and method of distillation all affect the type of oil produced. Inclusion of unripe fruits or leaf and stalk material, for example, reduces the specific gravity, optical rotation and linalool values, while the carbonyl value is enhanced. Physico-chemical properties of European oils are normally within fairly close limits; the optical rotation value is usually in the range +9° to +13°, while the ester number is normally well below the upper limit of standard specifications. Data for Moroccan oils show an upper limit for the optical rotation value of +10° and ester numbers up to 30.

Localization of volatile oil

Volatile oil canals are present in all of the organs of the coriander plant, but the odour is remarkable in the fruit and the content of its volatile oil changes during the course of fruit ripening. In the unripe fruit, two types of volatile oil canals are present. One type is located on the periphery of the fruit and these contain a volatile oil which is similar in composition to that of the vegetative organs in being comprised predominantly of aldehydes. The second type of canals are buried in the mericarp of the fruit kernel and the composition of their volatile oil is very different, containing linalool as the major component, together with some other oxygenated monoterpenes and monoterpene hydrocarbons. As the fruit ripens on the plant, the peripheral canals become flattened, commence to lose their volatile oil and the odour of the fruit changes. On drying to around 7% moisture content, the outer canals lose their volatile oil completely but the inner canals remain intact and the characteristic odour and composition of the volatile oil of the spice are attained (Purseglove *et al.*, 1981).

Variation due to distillation techniques

Hydrodistillation of coriander seeds yields about 0.35% (w/w) oil based on dry weight (Msaada *et al.*, 2007). Extraction of ripe fruits of *C. sativum* by steam distillation and by supercritical fluid extraction (SFE), using CO_2, was compared. The percentage composition of the 40 identified compounds was compared with the composition of commercial coriander oil extracted by hydrodistillation. The oil obtained by SFE showed some quantitative and qualitative differences, giving a superior aroma compared with that of the oil obtained by hydrodistillation (Anitescu *et al.*, 1997).

Smallfield *et al.* (1994) found chopping and immediate distillation gave an average increase in oil yield of 10.4 l/ha, compared with unchopped herb. However, oil yields of chopped herb fell after storage for more than 4 h. The oil composition was also affected by chopping and storage.

Microwave-assisted hydrodistillation (MWHD) and hydrodistillation (HD) were carried out for the analysis of volatile components in whole and ground fruits of *C. sativum* L. (coriander seed). Fruits were distilled using a microwave oven modified to fit a Clevenger apparatus. The effect of microwave energy on the yield and composition of the essential oil was investigated against classical hydrodistillation. Essential oils of all samples were analysed by GC-FID and gas chromatography-mass spectrometry (GC-MS). A decrease in the linalool content of the coriander oil (from 80.0 to 75.5%) was observed in the microwave-assisted hydro distilled ground fruits. Microwave-assisted distillation appeared to increase the amounts of fatty acids, for example, tetradecanoic acid (from 2.8 to 8.8%) and hexadecanoic acid (from 1.9 to 6.0%) in coriander oil (Kosar *et al.*, 2005).

High-pressure CO_2 extracts are, in many cases, superior to solvent oleoresins and essential oils when used as food flavourings and have significantly lower spice equivalent levels. Anitescu *et al.* (1997) proved the advantages of SFE using CO_2 in a two-stage separation system.

Kerrola and Kallio (1993) isolated volatile components of coriander fruits by supercritical carbon dioxide (SC-CO_2) extraction and analysis using GC-MS. The composition of SC-CO_2 extract was compared with that of liquid CO_2 (LCO_2) extract. Oxygenated monoterpenes were found to comprise 80% of the SC-CO_2 and 82% of the LCO_2 extracts, linalool being the main compound (67%) in both extracts. The sensory characteristics of the aroma of CO_2 extracts were evaluated and compared with those of the freshly ground spice. The SC-CO_2 extract was more terpenous and less sweet than the reference spice. The aroma of LCO_2 extract was characterized as pomerance-like and sweet. All odour characteristics were evaluated to be less intense than those in freshly ground coriander fruits.

Analysis of the essential oil content of seeds of certain cultivars, harvested at various stages of maturity and extracted by hydro-distillation from seeds that had been ground either in a pestle and mortar or a blender, indicated that the maximum yield of essential oil was obtained from ripening seeds (50% turning yellow). The use of a blender also increased the essential oil yield (Sanjeev *et al.*, 1994).

Skapska *et al.* (2004) observed the changes in the volatiles of coriander induced by ultra-high pressure (UHP) heat treatment in helium. The treatments did not affect the total essential oil content. The most important sensory constituent (linalool) was not significantly affected by the treatments. Several monoterpenes (sabinene, myrcene, limonene, γ-terpinene and α-terpinolene) decreased in pressurized samples and increased in non-pressurized samples.

Carbon dioxide (CO_2) extraction of oils from dried whole plants produced linalool, camphor, γ-terpinene and α-terpinene (Reis Machado *et al.*, 1993).

Stoyanova *et al.* (2003) reported an oil yield of 0.8% by CO_2 extraction, which is lower than that of conventional extraction methods. The major volatile compounds were petroselinic + oleic acids ($C_{18:1}$, 66.9%) and palmitic acid ($C_{16:0}$, 8.6%).

Oliver (2003) reported pure (S)-(+)-linalool from oil of coriander by formation and recrystallization of its 3,5-dinitrobenzoate ester, followed by regeneration and distillation. Fathima *et al.* (2001) noted that microwave drying affected the colour, appearance and odour properties of coriander.

Crushing intensity and distillation time were evaluated for their effects on the oil yield and composition of steam-distilled essential oil from fruits of *C. sativum* var. *microcarpum* L. A comparison of oils produced by laboratory- and pilot-scale stills showed that the two still types gave comparable yields and oil composition. Maximum oil yield was less from the light-crushed fruits, but the rate of oil recovery was significantly ($P < 0.05$) higher. From the light-crushed fruits, 95% of the maximum yield was extracted in 22.5 min compared with 32 and 39 min for the standard and heavy-crushed fruits, respectively. The effect of crushing intensity on oil composition was most pronounced on the low-boiling-point α-pinene and on the higher-boiling point geranyl acetate. Crushing had little effect on linalool content, but distillation time could be manipulated to alter the linalool content of the oil (Smallfield *et al.*, 2001).

Boelens *et al.* (1990) compared the quality of essential oils from various plant materials for hydrodiffusion and hydrodistillation. Hydrodiffusion oils contained higher levels of oxygenated monoterpenes and benzenoids and possessed preferred olfactive and organoleptic properties.

During distillation of the spice, a number of transformations may occur in the composition of the volatile oil. Small quantities of *p*-cymene may be produced as an artefact, linalyl acetate and some other esters may be hydrolysed and *d*-linalool may rearrange to its optically inactive isomer geraniol. Potter *et al.* (1993) found structural similarity in the compounds of *Polygonum odoratum* to that of *C. sativum*.

Influences of conditions of growth

In addition to intrinsic differences between cultivars in their fruit yield and volatile oil content, these properties are also influenced strongly by the location and environmental conditions of growth.

One of the most important influences is the site location or, more precisely, the latitude of cultivation. It has been noted for some time that in the northern hemisphere there is a marked trend for volatile oil contents to increase as one progresses from south to north. Even in the extremely cold and short growing seasons in Norway and Siberia, exceptionally good volatile oil contents, superior to those of many areas in central and southern Europe, have been reported (Gil *et al.*, 2002).

The climatic and weather conditions during growth are regarded as more important than the nature of the soil with regard to the volatile oil content. Reports indicate that the best oil yields are obtainable in cool, rather wet summers (Purseglove *et al.*, 1981). The combined application of fertilizers N80, S12 and Zn2.5 yielded 1% essential oil content compared with 0.7% without nutrient application (Manure *et al.*, 2000).

Hussien (1995) studied the influence of different sources of N (ammonium nitrate, calcium nitrate or urea) on essential oil yield and composition of coriander in successive seasons. Ammonium nitrate resulted in the highest oil concentration and yield in *C. sativum*.

There was a significant interaction between N and P for essential oil yield but not for seed yield (Tiwari and Banafar, 1995). Irrigation experiments using diluted sea water in potted plants were conducted by Boselah (1995). Plants were irrigated from 75 days after sowing with diluted sea water containing salts at 2000, 4000 or 6000 ppm. Controls were irrigated with tap water (250 ppm salt). Plant growth parameters, seed yield, essential oil concentration and yield/plant decreased with increasing salinity in both seasons of the trial. GLC analysis of the essential oil from the seeds revealed the presence of α-pinene, β-pinene, cineole, linalool, limonene, terpineol, carvone, cymene, eugenol and caryophyllene. Linalool was the main component at most salinities (56.67% in controls, reaching 75.87% at 6000 ppm salt). Some components, including cineole, increased in concentration with increasing salinity, while others, such as α-pinene, decreased. Compounds such as limonene were detected in saline irrigation treatments only and not in controls.

Gil *et al.* (2002) made a study of the essential oil composition of coriander fruits in plants growing in environments differing in soil conditions and weediness level. The variation in the oil composition was related to the relative proportion of the constituents and not to the presence/absence of a particular component. Location, fertilization and weediness also affected the chemical profile. The European landrace showed a more stable concentration of the major components than the Argentinean landrace.

Garg *et al.* (2004) conducted a study on the yield, mineral composition and quality of coriander grown in sodic soil. In both crops, N, P, K, Ca and Mg contents were higher in seeds than in stover, whereas the total nutrient uptake was higher in stover than in seeds. The essential oil content of *C. sativum* seeds was 0.32%, and this could give a mean volatile oil yield of 4.54 l/ha, depending on the seed yield. The main constituents of the oil were *d*-linalool (81.97%), geraniol (1.67%) and geranyl acetate (2.37%) (Oliveira *et al.*, 2004). Maurya (1990) studied the effect of micronutrients on yield and essential oil content of coriander and the best result was obtained with $CuSO_4$ in cv. Rajendra Swati.

Kalra *et al.* (2003) observed a negative correlation between seed yield and essential oil content and susceptibility to stem gall disease, and a tight relationship between days to flowering and days to maturity, oil yield and late maturity. The accession CIMAP 2096 was early maturing with a high degree of tolerance to major diseases and seeds rich in essential oil. Venkateswarlu *et al.* (1992) showed the effect of sowing time on essential oil yield in different *C. sativum* L. cultivars.

Abou Aly and Gomaa (2002) evaluated the effect of inoculation with diazotrophs (*Azotobacter chroococcum* or *Azospirillum brasilense*) combined with either *Bacillus megatherium* var. *phosphaticum* or *Glomus mosseae*, in the presence of half the recommended dose of N, on the growth and yield of coriander. Inoculation with *A. chroococcum* or *A. brasilense*, combined with *G. mosseae*, gave significant increases in vegetative growth, total carbohydrates, photosynthetic characteristics, essential oils, seed parameters and N, P and K contents. GLC analysis of seed volatile oil composition showed that linalool was the major constituent.

Khattab and Omer (1999) carried out field experiments in Egypt to investigate the effect of excessive fertilization with microelements on growth, yield and chemical composition in coriander. Plant growth, yield and essential oil content were increased significantly following application of 100 ppm microelements, with highest values recorded in the 200 ppm treatment. Further increases in microelement concentration (300 and 400 ppm) decreased all parameters. Irrigation increased the seed yield and essential oil content, but decreased seed weights and percentage linalool in the oil. Ayanoglu *et al.* (2002) evaluated 43 coriander lines for 2 years in the winter season for their yield and oil content. Essential oil content ranged from 0.18 to 0.60%. (S)-(+)-linalool is referred to occasionally by the trivial name, coriandrol. However, the linalool from coriander is not 100% (S) and typically includes 12 to 15% of the (R)-(–)-isomer (Oliver, 2003).

Oil constituents and maturity

Essential oil recovery and composition showed variation with maturity. *C. sativum* L. fruits grown in Tunisia gave essential oils at the initial, middle and final stages of maturity, with yields of 0.01, 0.12 and 0.35%, respectively. Essential oil at the first stage of maturity consisted mainly of monoterpene alcohols (14.6%), especially linalool (10.96%). Other constituents were monoterpene aldehydes (2.07%), ethers, hydrocarbons and monoterpene ketones, as well as phenols and sesquiterpenes.

Essential oil at the middle stage of maturity constituted monoterpene alcohols (76.77%), ketones (3.43%), esters (2.85%) and ethers (1.87%). Major constituents at this stage were linalool (76.3%), *cis*-dihydro carvone (3.21%), geranyl acetate (2.85%) and anethole (1.41%). Essential oils of mature fruit (final stage) were predominated by monoterpene alcohols (88.5%) and ketones (2.61%). Linalool (87.54%) and *cis*-dihydrocarvone (2.36%) were the major constituents at this stage (Msaada *et al.*, 2007).

Table 10.1 illustrates the oil composition of coriander fruit at three stages of maturity. Table 10.2 gives the percentage distribution of the classes of volatiles at those three stages of maturity. Carrubba *et al.* (2002) found that the age of the fruits seemed to generate rather identifiable effects, such as a decreasing trend for α- and γ-terpinene, terpinolene and linalool and an increase in *p*-cymene. Cultivars which flower and mature earlier generally have larger fruits with a lower volatile oil content, but have a better appearance and are more suited for spice usage than for distillation (Kalra *et al.*, 2003).

Geographical variation

Indigeneous Indian coriander oils differ from European oils in terms of low linalool content, and another distinguishing observation is the comparatively high ester content. The large-fruited type of Indian coriander revealed the presence of 21% linalyl acetate and 42% linalool in the distilled oil. Small-fruited coriander indigenous to the Kulu valley area in India was found to contain 48% linalool and, interestingly, a substantial proportion of citronellol. The large-fruited Indian coriander also appears to be distinct from other sources of the spice in the unique occurrence of the phenol, thymol (up to 7%).

Pino *et al.* (1993) investigated the compositional difference of the volatile oil of coriander fruits of different geographical origins (fruits from Russia and commercial oil samples from Italy, Albania and India) by means of gas liquid chromatography, column chromatography and coupled gas chromatography-mass spectrometry. All oils had high linalool content (49–59%). The Russian and Albanian oil samples were more fruity and coriander-like than the others and were of a higher organoleptic quality, presumably due to lower *p*-cymene content.

Raal *et al.* (2004) evaluated oil isolated by hydrodistillation from different geographical sources of Europe. Among the 37 components isolated, the major constituent

Table 10.1. Essential oil composition (% w/w) of coriander fruit at three stages of maturity.

Sl no.	Compound	Immature	Intermediate	Mature
1	Heptanal	t	t	t
2	α-Thujene	t	t	t
3	α-Pinene	0.01	t	0.02
4	Sabinene	t	t	0.03
5	β-Pinene	t	0.20	0.05
6	δ^3-Carene	0.09	0.10	0.02
7	α-Terpinene	t	t	0.01
8	*p*-Cymene	t	t	t
9	Limonene	0.04	t	0.02
10	1,8-Cineole	0.23	0.14	0.20
11	(Z)-β-Ocimene	0.08	t	t
12	γ-Terpinene	t	t	t
13	*cis*-Linalool oxide (furanoid)	0.32	0.32	0.27
14	Terpinolene	0.02	0.18	0.15
15	Linalool	10.96	76.33	87.54
16	*trans*-Linalool oxide (furanoid)	0.27	t	t
17	Camphor	0.86	0.13	0.17
18	Borneol	0.08	0.28	0.34
19	Menthol	0.14	0.16	0.05
20	Terpinene-4-ol	t	t	t
21	*p*-Cymen-8-ol	1.36	t	t
22	*cis-hex*-3-Enyl butyrate	t	t	0.01
23	α-Terpineol	0.39	t	0.05
24	*cis*-Dihydrocarvone	0.01	3.21	2.36
25	Nerol	1.53	t	t
26	β-Citronellol	0.11	t	0.52
27	Neral	1.42	0.10	0.13
28	Carvone	0.10	0.09	0.08
29	Geraniol	t	t	t
30	Geranial	0.65	t	0.03
31	Anethole	0.05	1.41	0.01
32	Thymol	0.02	0.99	1.85
33	Carvacrol	1.04	0.11	0.46
34	δ-Elemene	t	0.05	0.01
35	Eugenol	0.09	t	0.01
36	Neryl acetate	t	t	t
37	Geranyl acetate	46.27	2.85	0.83
38	β-Caryophyllene	0.02	0.07	0.03
39	α-Humulene	0.09	t	0.02
40	Germacrene-D	0.04	0.19	0.05
41	Eugenyl acetate	t	t	0.07
	Total identified	66.29	86.91	95.39

t = trace (< 0.01%).

of the oil was linalool (58.0–80.3%). The other characteristic compounds present in the oils were γ-terpinene (0.3–11.2%), α-pinene (0.2–10.9%), *p*-cymene (0.1–8.1%), camphor (3.0–5.1%) and geranyl acetate (0.2–5.4%). The content of linalool as aliphatic terpene showed high correlations with the content of cyclic terpenes in essential oils.

Figueiredo *et al.* (2004) evaluated the composition of coriander essential oil from Brazil. Among the 18 compounds

Table 10.2. Percentage of major classes of volatile compounds at three stages of maturity.

Class	Immature	Intermediate	Mature
Monoterpene hydrocarbons	0.24	0.48	0.30
Aromatic hydrocarbons	t	t	t
Monoterpene alcohols	14.66	76.77	88.51
Phenols	1.06	1.10	2.31
Monoterpene esters	46.27	2.85	0.90
Monoterpene ketones	0.97	3.43	2.61
Monoterpene aldehydes	2.07	0.10	0.16
Monoterpene ethers	0.87	1.87	0.48
Sesquiterpenes	0.15	0.31	0.11
Non-terpenic	t	t	0.01

t = trace (< 0.01%).

quantified, the main components of the oil were linalool (77.48%), γ-terpinene (4.64%), α-pinene (3.97%), limonene (1.28%), geraniol (0.64%) and 2-decenal (0.16%).

Comparison of five commercial oils hydrodistilled from fruits produced in two areas of Buenos Aires Province with that of Russian essential oils by the Argentinean fragrance and flavour industry indicted the presence of 20 components. This accounted for 96.6–99.7% of total essential oil composition. The main constituents were linalool (68.9–83.7%), γ-terpinene (2.2–5.1%), camphor (3.2–4.8%), α-pinene (1.0–6.5%), geraniol (1.4–3.2%) and geranyl acetate (0.8–3.8%). The contents of *cis*- and *trans*-linalool oxide (0.1–0.4%) were low (Bandoni *et al.*, 1998). Chislova (1988) found, on the basis of a study of 100 *C. sativum* oils, that the essential oil content was highest (2.61%) in a Soviet form from Georgia (Vr. K350), which also had the highest fatty oil content (21.44%).

Dobos and Novak (2005), after 3-year trials in Austria with 36 different coriander, found the fatty oil content ranged between 20 and 29% and the essential oil content ranged between 0.2 and 1.3%. The content of some essential oil compounds differed significantly.

Low essential oil content cultivars from India contain little or no camphor, myrcene and limonene, but considerable linalool. Despite the relatively low essential oil content, this type may be preferred because of its specific flavour.

Harvesting and drying influences

The volatile oil content of the fruit reaches maximum while it is still unripe and during ripening; it diminishes owing to a collapse of the peripheral volatile oil canals. However, since the volatile oil present in the peripheral canals imparts a rather fetid, bedbug-like odour to the fruit, harvesting is delayed until the fruit has ripened and the characteristic, sweet odour of the spice has developed.

In fully ripe fruit, traces of the peripheral canal volatile oil may be present, but the bulk of the volatile oil is enclosed in four internal oil ducts which are located in pairs buried below the internal surface of each half-kernel and facing one another. These internal volatile oil reservoirs are dead-ended with dense epithelial cells through which the oil penetrates with difficulty. Under conditions of natural drying on the plant or carefully controlled artificial drying, very little volatile oil loss occurs owing to the structural characteristics of the fruit. Studies have shown that volatile oil losses of 4–5% can occur while fruits dehydrate under natural conditions from a moisture content of 35% down to 12%. The main losses of volatile oil during natural drying were found to occur during the initial stages of dehydration. Changes in volatile oil content during artificial drying were found to be dependent on the initial moisture content and the temperature and duration of drying. With fruits harvested when their moisture content was below 18% and then dried at temperatures below 90°C, losses of volatile oil were insignificant. Interestingly, the volatile oil content actually increased slightly under certain artificial drying conditions, reaching its maximum (2–5%) between 80 and 90°C, as a result of an apparent stimulation of the biosynthetic pathways responsible for volatile oil formation. However, fruits which were harvested with high initial moisture contents (greater than 18%) were found to lose 5–12% of their volatile oil during artificial drying, and the losses increased progressively as the drying temperature increased beyond 100°C (Purseglove *et al.*, 1981).

Cultivar variation

Arganosa *et al.* (1998) reported the variation in oil recovery and linalool content in relation to cultivars. They found that the six large seed selections of coriander gave 0.83–0.90% oil and 60.7–62.5% linalool in the oil. They also reported that early seeding resulted in the highest seed yields, seed weights, linalool content and essential oil yield for both small-seed and large-seed cultivars.

Kalra *et al.* (1999) evaluated 16 coriander genotypes to assess the oil yield potential of these diverse genotypes in the subtropical climate of northern India. C-1 had the highest seed yield and oil yield, as well as a high resistance against stem gall (*Protomyces macrosporus*), a major threat to coriander cultivation. Small-seeded and large-seeded cultivars of coriander grown at several locations in western Canada were evaluated for seed weight, essential oil content of seeds and linalool concentration of essential oils. Large-seeded cultivars had 1000-seed weights of 9.2–9.9 g, seed essential oil contents of 0.83–0.9% and linalool contents of 60.7–62.5%. Small-seeded cultivars had 1000-seed weights of 7.1–7.4 g, seed essential oil contents of 1.22–1.30% and linalool contents of 63.9–66.2%. Early sowing resulted in the highest seed yields, seed weights, linalool contents and essential oil yields for both small- and large-seeded cultivars. Splitting the fruits prior to sowing reduced costs and did not affect seed yield, seed size or chemical composition of the essential oil adversely (Arganosa *et al.*, 1998).

Rao *et al.* (2004) isolated essential oil from three local cultivars (Jaipur local, Surabhi and Bangalore local) and three advanced breeding lines (LC-12, LC-13 and LC-21) of coriander grown under identical conditions in Bangalore, Karnataka, India, by hydrodistillation of whole herbs and leaves at vegetative and flowering stages. The essential oil yields varied from 0.03 to 0.06 ml/100 g in the case of whole herbs and from 0.04 to 0.12 ml/100 g in leaf samples. Altogether, 26 compounds were identified, with decanol, (E)-2-decen-1-ol, decanal, (E)-2-undecen-1-ol, (E)-2-dodecenol and (E)-2-tetradecenal as the major compounds. The percentage of identified compounds ranged from 83.92 to 96.45%, except in one herb oil (Surabhi, vegetative stage), which was 75.44%. Significant compositional differences were observed between herb oils and leaf oils, at both vegetative and flowering stages.

Storage influences

Reports indicate weight losses in coriander of about 3–5% over a 2-year period for whole, dried fruit stored in heaps or cotton bags, and deterioration of the organoleptic quality was not great. Loss of up to 23% over 1 year for the whole spice stored in paper or cotton bags, along with some organoleptic deterioration, occurred, even with samples in which oil losses were slight. These differences can be attributed to the care taken during harvesting and drying and also to intrinsic differences between cultivars for resistance to oil permeation. Reports indicate that volatile oil diffusion through the whole spice can be extensive. Some authors reported 40% of the volatile oil in the husk and the remainder dissolved in the fatty oil of the kernel (Purseglove *et al.*, 1981).

Polshkov (2001) studied the effect of storage time and condition on the composition of coriander oil after 1, 2, 4, 6, 9 and 12 months by GC-MS analysis. In the fresh specimen, the monoterpene hydrocarbons, γ-terpinene and α-pinene, were detected in considerable amounts and camphor, linalyl acetate and geranyl acetate in smaller amounts. Aldehydes (neral, 2-decenal and undecenal) were also found. During 1 year of storage in the dark, the composition of the oil changed insignificantly. Storage of coriander seed oil showed a decrease in γ-terpinene, *p*-cymene and linalool contents by 12.2, 20.8 and 20.7%, respectively. In general, the composition of the stored oil was close to the average values for fresh specimens. During storage of the oil in light, the composition changed significantly. After 2 months, γ-terpinene was not detected, but new compounds were found. With further storage, the contents of α-thujene, sabinene, β-pinene, β-myrcene and ocimene decreased and had disappeared by 9 months. The contents of α-pinene, limonene and linalyl acetate after 12 months decreased by six fold, geranyl acetate by 2.5-fold and linalool by threefold. The content of *p*-cymene, α-terpineol and 4-terpineol increased, respectively, by 3.2-, eight- and 6.5-fold. The camphor content showed an insignificant increase of about 2%. The greatest increase was noted for linalool oxide, from 1.40 to 14.38%. The structure of the new compounds, which appeared after 2–4 months, is identified as an isomer of *p*-cymene-3-methylisopropylbenzene, geranial, carvacrol and eugenol (Polshkov, 2001). The relative concentrations of the aldehydes (in particular, the main one, (*E*)-2-decenal) decreased and the relative concentrations of alcohols rose with increased duration of storage (Smallfield *et al.*, 1994).

With fully ripe fruits which suffer only slight oil loss during drying and storage, the changes in oil composition appear to be minor and quantitative rather than qualitative. Studies indicate that, under conditions of natural drying or carefully controlled artificial drying, a slight enhancement of the linalool content occurs and proportionately greater losses of the monoterpene hydrocarbons rather than of the oxygenated constituents are encountered. However, it was found that artificial drying at temperatures over 100°C resulted in the formation of a 'burnt note' and a lowering of the linalool content. Substantial oil losses are observed during storage of damaged, crushed or ground spice, and the changes in oil composition appear to parallel those encountered in high-temperature drying. The maximum deterioration during storage of the distilled spice oil occurs when it is left exposed to light and air. Oil storage transformations include those mentioned for distillation but oxidative changes, including the conversion of linalool to geranial, are the most important (Purseglove *et al.*, 1981).

Changes in the volatile oil during plant development

At the seedling stage, the coriander plant acquires a distinctive aroma, reminiscent of the stink bug, and this is retained by the leaves and stalk throughout the period of plant development until the final phase of fruit ripening, during which the vegetative organs shrivel up. The immature fruits of the plant also possess the odour but, as they ripen, their odour acquires a greater similarity to that of the dried spice. On full ripening and drying of the fruit, the bug-like aroma disappears. In Western markets, the odour and flavour of the herb is generally disliked and for this reason only fully ripe fruits are employed for distillation of the spice oil. In India, however, the leaves are appreciated for their flavouring properties and are incorporated into many dishes. The fresh leaves possess a high content of vitamin C and carotene, but the vitamin C content diminishes rapidly on storage. The odour of the green plant and flowers acts as an excitant and attractant for bees, which play an important role in fertilizing the flowers (Kalra *et al.*, 2000; Kohara *et al.*, 2006).

Dehydration occurs as fruits mature owing to collapse of peripheral volatile oil canals, as noted, and approximately one-third of the essential oil present in the immature fruit is lost. Oil content of mature fruit depends basically on its position; in small-fruited varieties it is highest in the central umbels, lower in the second-order and lowest in the first-order umbels. The major oil component is *d*-linalool (coriandrol) between 55 and 75%, depending mainly on cultivar, location and ripeness at harvest. Other constituents are α-pinene and γ-terpinene up to 8%, camphor up to 6% and geraniol up to 2% (Anitescu *et al.*, 1997). Each of the vegetative organs of the plant (leaves, stalks and root) contains volatile oil, which is detectable at an early stage of seedling development. The content of the volatile oil in the vegetative organs increases progressively with plant development and reaches a maximum (about 0.1–0.2% on a fresh weight basis) in the overground parts at flowering stage.

Leaf oil

Volatiles were recovered from the fresh green leaves of *C. sativum* by steam codistillation with pentane and analysed by capillary GC-MS. A total of 41 compounds were detected. Constituents identified included alkenals in the C_9–C_{16} range, C_7–C_{17} alkanals, C_{10}–C_{12} primary alkenols and alkanols, and nonane. The estimated mass of volatiles recovered was 4 mg/g (wet weight of leaves), with aldehydes accounting for 82.6% and alcohols 16.6% of the compounds detected (Potter and Fagerson, 1990).

Freshly cut coriander leaves are one of the most widely used culinary herbs, the dried leaves being less popular. Leaves are often cut from plants subsequently harvested for seed and in India it was found that one cutting generally had no effect, but two reduced seed yield. Some cultivars can be cut more often than others and still produce a reasonable seed yield.

Fresh foliage contains 0.1–0.2% essential oil, which reaches its maximum at flowering. Some 40 compounds have been identified in the oil, with decyl and nonyl aldehydes accounting for about 80% and alcohols 16% (Potter and Fagerson, 1990). Herbage also contains (per 100 g fresh foliage) up to 12 mg pro-vitamin A, up to 60 mg vitamin B_2 and up to 250 mg vitamin C; the last diminishes rapidly on storage. The seedling acquires the distinctive coriander smell, which is retained by the leaves and stalk until the final phase of fruit ripening, when the vegetative parts wither.

10.6. Medicinal and Pharmacological Properties

Antioxidant activity

Leaf and seed extracts of coriander and coriander oil were tested for their antioxidant activity by Wangensteen *et al.* (2004) by using different bioassay techniques. Positive correlations were found between total phenolic content in the extracts and antioxidant activity. Coriander leaves showed stronger antioxidant activity than the seeds and, in both parts of coriander, the ethyl acetate extract contributed to the strongest activity. It was suggested that addition of coriander to food would increase the antioxidant content and may have potential as a natural antioxidant and thus inhibit unwanted oxidation processes (Wangensteen *et al.*, 2004).

The biochemical effects of coriander seeds (10% powdered seeds added to the diet) on tissue lipid parameters in 1,2-dimethyl hydrazine (DMH)-induced colon cancer in rats were studied after 15 and 30 weeks. The spice diet was given during the initial 15-week period of carcinogen administration only. The study shows that the concentrations of cholesterol and the cholesterol:phospholipids ratio decreased, while the level of phospholipids increased significantly in the DMH control group compared with the spice-administered group. Faecal dry weight, faecal neutral sterols and bile acids showed a sharp increase in the coriander-fed group compared with the DMH-administered group. Thus, coriander plays a protective role against the deleterious effects on lipid metabolism in experimental colon cancer (Chithra and Leelamma, 2000).

Essential oils from commercial samples of coriander were analysed by GC-MS and assayed for their antibacterial, antifungal and antioxidant activities. Twenty-five genera of bacteria and one fungal species (*Aspergillus niger*) were used as test organisms. The essential oils showed a high degree of inhibition against all the microorganisms tested (Baratta *et al.*, 1998).

Kaur and Kapoor (2002) found that antioxidant activity correlated significantly and positively with total phenolics.

Health benefits

Coriander essential oils showed inhibition against *Listeria monocytogenes*, *Staphylococcus aureus*, *Escherichia coli* O:157:H7, *Yersinia enterocolitica*, *Pseudomonas aeruginosa*, *Lactobacillus plantarum*, *A. niger*,

Geotrichum and *Rhodotorula*. Coriander and basil were also highly inhibitory (MLC, 25 to 50 ppm) to *E. coli* O:157:H7 and to the other bacteria and fungi tested (Elgayyar *et al.*, 2001).

Coriander incorporated into the diet (62.5 g/kg) and drinking water (2.5 g/l, prepared by 15 min decoction) reduced hyperglycaemia of streptozotocin-diabetic mice. Aqueous extract of coriander (1 mg/ml) showed increases of 1.6-fold in 2-deoxyglucose transport and 1.4-fold in glucose oxidation and incorporation of glucose into glycogen (1.7-fold) comparable with 10–8 M-insulin. In acute 20 min tests, 0.25–10.00 mg/ml aqueous extract of coriander evoked a stepwise 1.3–5.7-fold stimulation of insulin secretion from a clonal B-cell line. Sequential extraction with solvents revealed insulin-releasing activity in hexane and water fractions, indicating a possible cumulative effect of more than one extract constituent. These results demonstrate the presence of anti-hyperglycaemic, insulin-releasing and insulin-like activity in coriander (Gray and Flatt, 1999; Selvan, 2003).

Padmanabha and Rangaswamy (1996) studied the effects of phosphine on compounds in the essential oils having double bonds, using α-pinene and linalool as model compounds. α-Pinene and linalool interacted with phosphine to form monomers and dimers via a free radical pathway. Phosphine has potential as a fumigant for coriander seeds, which are susceptible to infestation by the spice beetle, *Stegobium paniceum*, during storage. Coriander essential oil evoked a marked analgesic activity in mice in a study conducted by Afifi *et al.* (1994).

Ertas *et al.* (2005) investigated the potential effects of dietary supplementation by coriander seed (considered as a lipolytic and antioxidant compound) on carcass lipid composition of quails. Their aim was to reduce saturated fatty acid consumption and to increase essential fatty acids (particularly n^3 unsaturated acids) in alimentation. Dietary supplementation by coriander seed affected the lipid composition of carcass greatly by decreasing saturated fatty acid (SFA) contents (palmitic and stearic acids) and by increasing monounsaturated and polyunsaturated fatty acid (MUFA and PUFA) proportions in comparison with the control group ($p < 0.01$). The highest dosage of coriander seed (4% added to the ration) systematically induced the greatest effects on fatty acid composition. Consequently, dietary supplementation by coriander seed would improve the quality of the lipid carcass of quails by lowering SFA proportions and by enhancing the contents of PUFA, particularly of n^3 PUFA.

Insecticidal effect

Pascual Villalobos (2003) found the potential of plant essential oils against stored-product beetle pests. Coriander oil (10 µl) showed activity against the bruchid *Callosobruchus maculates*, the cereal storage pest.

Aflatoxin control

The inhibitory effects of the essential oils of coriander on the mycelial growth and ochratoxinA production by *A. ochraceus* NRRL 3174 were studied by Basilico and Basilico (1999). Sage and coriander showed no important effect at any of the concentrations studied. Meena and Sethi (1994) also studied the potential of coriander oil in the control of *A. niger*, *Saccharomyces cerevisiae*, *Mycoderma* sp., *L. acidophilus* and *Bacillus cereus*.

Pradeep *et al.* (2003) reported the efficacy of coriander essential oil on seed mycoflora and seedling quality of some crop species.

Tolkunova (2002) investigated the influence of essential oils on microbiological indicators of meat products. The formulation of horsemint–fennel–coriander was found effective against Gram-positive microorganisms. Introducing protein to the incubation environment was found to have little effect on the antibacterial activity of the essential oil formulations. His study showed that the essential oil formulations exerted a bacteriostatic action on the development of mesophilic aerobic and facultative anaerobic bacteria in cooked sausages. Significant increases in the storage life of sausages in both natural and artificial casings were associated with the use of the essential oil formulations.

10.7. International Specification

The chemical and physical specification for whole coriander is given in Table 10.3 and for ground coriander in Table 10.4.

10.8. Conclusion

The coriander plant yields two primary products which are employed for flavouring purposes: the fresh green herb and the spice. The odour and flavour of these two products are markedly different, and the herb generally is disliked and little used in Europe and North America, except in certain specialist applications. The major use of the spice on a worldwide basis is in flavouring applications in the ground form and its main outlet is as an ingredient of curry powders, of which it comprises about 25–40% of the spice mixture. The ground spice is used extensively as a flavouring agent domestically and by manufacturers of processed foods in baked goods, sauces and meat dishes. The whole spice is employed in pickling and in the flavouring of certain alcoholic beverages. Detailed study of the oil composition of coriander spice showed that the *d*-linalool content ranged from 60 to 70% and the hydrocarbon content was about 20%. α- and β-pinenes, dipentene (limonene), *p*-cymene, α- and γ-terpinenes, *n*-decanal, geraniol and l-borneol were also identified as constituents of the spice oil. Leaf and seed extracts showed antioxidant, insulin-like and anti-aspergillus activity. Considering these potentials, coriander biomolecules possess a tremendous future in the health-related industry.

Table 10.3. Whole coriander: chemical and physical specifications.

Specification	Suggested limits
ASTA cleanliness specifications	
Whole dead insects, by count	4
Mammalian excreta (mg/lb)	3
Other excreta (mg/lb)	10.0
Mould, % by weight	1.00
Insect-defiled/infested, % by weight	1.00
Extraneous, % by weight	0.50
FDA Detect Action Levels (condimental seed)	
Adulteration with mammalian excreta	Av. of 3 mg/lb
Volatile oil (% min.)	0.3
Moisture (% max.)	9.0
Ash (% max.)	6.0
Acid-insoluble ash (% max.)	1.0
Average bulk index (mg/100 g)	285

Table 10.4. Ground coriander: chemical and physical specifications.

Specification	Suggested limits
Volatile oil (% min.)	0.2
Moisture (% max.)	9.0
Total ash (% max.)	6.0
Acid-insoluble ash (% max.)	1.0
Military specifications	
Volatile oil (ml/100 g)	trace
Moisture (% max.)	10.0
Total ash (% max.)	7.0
Acid-insoluble ash (% max.)	1.0
Granulation (% min. through USS No. 30)	95
Bulk index (ml/100 g)	200

References

Abou Aly, H.E. and Gomaa, A.O. (2002) Influence of combined inoculation with diazotrophs and phosphate solubilizers on growth, yield and volatile oil content of coriander plants (*Coriandrum sativum* L.). *Bulletin of Faculty of Agriculture, Cairo University* 53(1), 93–113.

Afifi, N.A., Ramadan, A., El Kashoury, E.A. and El Banna, H.A. (1994) Some pharmacological activities of essential oils of certain umbelliferous fruits. *Veterinary Medical Journal Giza* 42(3), 85–92.

Anitescu, G., Doneanu, C. and Radulescu, V. (1997) Isolation of coriander oil: comparison between steam distillation and supercritical CO_2 extraction. *Flavour and Fragrance Journal* 12(3), 173–176.

Arganosa, G.C., Sosulski, F.W. and Slikard, A.E. (1998) Seed yields and essential oil of northern grown coriander (*Coriandrum sativum* L.). *Journal of Herbs, Spices and Medicinal Plants* 6(2), 23–32.

Ayanoglu, F., Mert, A., Aslan, N. and Gurbuz, B. (2002) Seed yields, yield components and essential oil of selected coriander (*Coriandrum sativum* L.) lines. *Journal of Herbs, Spices and Medicinal Plants* 9(2/3), 23, 71–76.

Bandoni, A.L., Mizrahi, I. and Juarez, M.A. (1998) Composition and quality of the essential oil of coriander (*Coriandrum sativum* L.) from Argentina. *Journal of Essential Oil Research* 10(5), 581–584.

Baratta, M.T., Dorman, H.J.D., Deans, S.G., Biondi, D.M. and Ruberto, G. (1998) Chemical composition, antimicrobial and antioxidative activity of laurel, sage, rosemary, oregano and coriander essential oils. *Journal of Essential Oil Research* 10(6), 618–627.

Basilico, M.Z. and Basilico, J.C. (1999) Inhibitory effects of some spice essential oils on *Aspergillus ochraceus* NRRL 3174 growth and ochratoxin A production. *Letters in Applied Microbiology* 29(4), 238–241.

Boelens, M.H., Valverde, F., Sequeiros, L., Jimenez, R., Bhattacharyya, S.C. (ed.), Sen, N. (ed.) and Sethi, K.L. (1990) Ten years of hydrodiffusion oils. In: *Volume 4. Chemistry Analysis and Structure*. Proceedings of the 11th International Congress of Essential Oils, Fragrances and Flavours, New Delhi, India, 12–16 November, 1989. Aspect Publishing, London, pp. 121–126.

Borges, P., Pino, J. and Rosado, A. (1990) The isolation of volatile oil from coriander fruit by steam distillation. *Nahrung* 34(9), 831–834.

Boselah, N.A.E. (1995) Effect of different levels of salinity on the growth, yield and volatile oil constituents of coriander (*Coriandrum sativum* L.) plants. *Annals of Agricultural Science, Moshtohor* 33(1), 345–358.

Burt, S. (2004) Essential oils: their antibacterial properties and potential applications in foods – a review. *International Journal of Food Microbiology* 94(3), 223–253.

Carrubba, A., Torre, R., Prima, A.D., Saiano, F. and Alonzo, G. (2002) Statistical analysis on the essential oil of Italian coriander (*Coriandrum sativum* L.) fruits of different ages and origins. *Journal of Essential Oil Research* 14, 389–396.

Chislova, L.S. (1988) Useful forms of coriander. *Selektsiya I Semenovodstvo Moskva* 6, 33.

Chithra, V. and Leelamma, S. (2000) *Coriandrum sativum* effect on lipid metabolism in 1,2-dimethyl hydrazine induced colon cancer. *Journal of Ethnopharmacology* 71(3), 457–463.

Dobos, G. and Novak, J. (2005) Comparison of the composition of the essential oil of some winter annually cultivated coriander accessions (*Coriandrum sativum* L.). *Zeitschrift fur Arznei and Gewurzpflanzen* 10(3), 144–145.

Elgayyar, M., Draughon, F.A., Golden, D.A. and Mount, J.R. (2001) Antimicrobial activity of essential oils from plants against selected pathogenic and saprophytic microorganisms. *Journal of Food Protection* 64(7), 1019–1024.

Ertas, O.N., Guler, T., Cftc, M., Dalklc, B. and Ylmaz, O. (2005) The effect of a dietary supplement coriander seeds on the fatty acid composition of breast muscle in Japanese quail. *Revue de Médecine Vétérinaire* 156(10), 514–518.

Fathima, A., Begum, K. and Rajalakshmi, D. (2001) Microwave drying of selected greens and their sensory characteristics. *Plant Foods for Human Nutrition* 56(4), 303–311.

Figueiredo, R.O. de, Marques, M.O.M., Nakagawa, J. and Ming, L.C. (2004) Composition of coriander essential oil from Brazil. *Acta Horticulturae* 629, 135–137.

Frank, C., Dietrich, A., Kremer, U. and Mosandl, A. (1995) GC-IRMS in the authenticity control of the essential oil of *Coriandrum sativum* L. *Journal of Agricultural and Food Chemistry* 43(6), 1634–1637.

Garg, V.K., Singh, P.K. and Katiyar, R.S. (2004) Yield, mineral composition and quality of coriander (*Coriandrum sativum*) and fennel (*Foeniculum vulgare*) grown in sodic soil. *Indian Journal of Agricultural Sciences* 74(4), 221–223.

Gil, A., Fuente, E.B. de la, Lenardis, A.E., Lopez Pereira, M., Suarez, S.A., Bandoni, A., Baren, C. van, Leo Lira, P. di and Ghersa, C.M. (2002) Coriander essential oil composition from two genotypes grown in different environmental conditions. *Journal of Agricultural and Food Chemistry* 50(10), 2870–2877.

Gray, A.M. and Flatt, P.R. (1999) Insulin releasing and insulin like activity of the traditional anti diabetic plant *Coriandrum sativum* (coriander). *British Journal of Nutrition* 81(3), 203–209.

Hussien, M.S. (1995) Response of growth, yield and essential oil of coriander and dill to different nitrogen sources. *Egyptian Journal of Horticulture* (publ) 1996, 22(1), 1–10.

Ishikawa, T., Kondo, K. and Kitajima, J. (2003) Water soluble constituents of Coriander. *Chemical and Pharmaceutical Bulletin* 51(1), 32–39.

Kalra, A., Patra, N.K., Singh, H.P., Singh, H.B., Mengi, N., Naqvi, A.A., Sushil Kumar, and Kumar, S. (1999) Evaluation of coriander (*Coriandrum sativum*) collection for essential oil. *Indian Journal of Agricultural Sciences* 69(9), 657–659.

Kalra, A., Parameswaran, T.N., Ravindra, N.S., Rao, M.G. and Kumar, S. (2000) Effects of planting date and dinocap applications on the control of powdery mildew and yields of seed and seed oil in coriander. *Journal of Agricultural Science* 135(2), 193–197.

Kalra, A., Gupta, A.K., Katiyar, N., Srivastava, R.K. and Kumar, S. (2003) Screening of *Coriandrum sativum* accessions for seed and essential oil yield and early maturity. *Plant Genetic Resources Newsletter* 133, 19–21.

Kaur, C. and Kapoor, H.C. (2002) Antioxidant activity and total phenolic content of some Asian vegetables. *International Journal of Food Science and Technology* 37(2), 153–161.

Kerrola, K. and Kallio, H. (1993) Volatile compounds and odor characteristics of carbon dioxide extracts of coriander (*Coriandrum sativum* L.) fruits. *Journal of Agricultural and Food Chemistry* 41(5), 785–790.

Khattab, M.E. and Omer, E.A. (1999) Influence of excessive fertilization with micro nutrients on the growth, yield, essential oil and micro elements of some Apiaceae plants. *Egyptian Journal of Horticulture* 26(3), 249–266.

Kim, S.W., Park, M., Bae, K., Rhee, M. and Liu, J. (1996) Production of petroselinic acid from cell suspension cultures of *Coriandrum sativum. Phytochemistry* 42(6), 1581–1582.

Kohara, K., Sakamoto, Y., Hasegawa, H., Kozuka, H., Sakamoto, K. and Hayata, Y. (2006) Fluctuations in volatile compounds in leaves, stems, and fruits of growing coriander (*Coriandrum sativum* L.) plants. *Journal of the Japanese Society of Horticultural Science* 75(3), 267–269.

Kosar, M., Ozek, T., Goger, F., Kurkcuoglu, M. and Baser, K.H.C. (2005) Comparison of microwave assisted hydrodistillation and hydrodistillation methods for the analysis of volatile secondary metabolites. *Pharmaceutical Biology* 43(6), 491–495.

Manure, G.R., Shivaraj, B., Farooqui, A.A. and Surendra, H.S. (2000) Yield attributes, seed, essential oil yield and oil content of coriander (*Coriandrum sativum* L.) as influenced by the graded levels of nitrogen, sulphur and zinc nutrition in red sandy loam soils. Spices and aromatic plant challenges and opportunities in the new century. Contributory papers, *Centennial Conference on Spices and Aromatic Plants*, Calicut, Kerala, India, 20–23 September, 2000. Indian Society for Spices, Calicut, India, pp. 139–144.

Maurya, K.R. (1990) Effect of micronutrients on yield and essential oil content of coriander (*Coriandrum sativum* Linn.). *Indian Perfumer* 34(4), 263–265.

Meena, M.R. and Sethi, V. (1994) Antimicrobial activity of essential oils from spices. *Journal of Food Science and Technology* Mysore 31(1), 68–70.

Mekhedov, S., Cahoon, E.B. and Ohlrogge, J. (2001) An unusual seed-specific 3-ketoacyl ACP synthase associated with the biosynthesis of petroselinic acid in coriander. *Plant Molecular Biology* 47(4), 507–518.

Mironova, A.N., Filippova, G.I., Fedina, N.I., Volkova, Z.D., Kozlova, V.L., Alymova, T.B., Gorshkova, E.I. and Bykova, S.F. (1991) Chemical and biological properties of coriander fatty oil. *Voprosy-Pitaniya* 1, 59–62.

Msaada, K., Hosni, K., Taarit, M.B., Chahed, T., Kchouk, M.E. and Marzouk, B. (2007) Changes on essential oil composition of coriander (*Coriandrum sativum* L.) fruits during three stages of maturity. *Food Chemistry* 102, 1131–1134.

Mustafaev, S.K., Melekhina, O.V., Ksandopulo, S.Yu and Stadnik, N.G. (1989) Determination of the quality of fatty oil in coriander fruits. *Izvestiya Vysshikh Uchebnykh Zavedenii, Pishchevaya Tekhnologiya* 4, 87.

Oliveira, A.P. de, Araujo, L.R. de, Mendes, J.E.M.F., Dantas, O.R. Jr. and Silva, M.S. da (2004) Effect of phosphorus fertilization on the yield of coriander in soil with low levels of phosphorus. *Horticultura Brasileira* 22(1), 87–89.

Oliver, J.E. (2003) (S)(+)-Linalool from oil of coriander. *Journal of Essential Oil Research* 15(1), 31–33.

Padmanabha, K.K. and Rangaswamy, J.R. (1996) Characterisation of interaction products of alpha-pinene and linalool with phosphine. *Indian Journal of Chemistry Section B, Organic including Medicinal* 35(6), 611–614.

Pascual Villalobos, M.J. (2003) Volatile activity of plant essential oils against stored product beetle pests. Advances in stored product protection. *Proceedings of the 8th International Working Conference on Stored Product Protection*, York, UK, 22–26 July 2002. CAB International, Wallingford, UK, pp. 648–650.

Pino, J., Borges, P. and Roncal, E. (1993) Compositional differences of coriander fruit oils from various origins. *Nahrung* 37(2), 119–122.

Polshkov, A.N. (2001) Composition of coriander oil and its change during storage. *Russian Agricultural Sciences* C/C of Doklady – Rossiiskaia Akademiia Sel'skokhoziaistvennykh Nauk, Part 12. Allerton Press, New York, pp. 33–35.

Porter, N.G. and Lammerink, J.P. (1994) Effect of temperature on the relative densities of essential oils and water. *Journal of Essential Oil Research* 6(3), 269–277.

Potter, T.L. and Fagerson, I.S. (1990) Composition of coriander leaf volatiles. *Journal of Agricultural and Food Chemistry* 38(11), 2054–2056.

Potter, T.L., Fagerson, I.S., Craker, L.E., Palevitch, D. and Putievsky, E. (eds) (1993) Composition of Vietnamese coriander leaf oil. *International Symposium on Medicinal and Aromatic Plants,* Tiberias on the Sea of Galilee, Israel, 22–25 March 1993. *Acta Horticulturae* 344, 305–311.

Pradeep, A.G., Lokesh, S. and Rai, V.R. (2003) Efficacy of some essential oils on seed mycoflora and seedling quality of some crop species saved by farmers. *Advances in Plant Sciences* 16(1), 53–58.

Purseglove, J.W, Brown, E.G., Green, C.L. and Robbins, S.R. J. (1981) *Spices.* Volume 2. Longman, New York, pp. 736–788.

Raal, A., Arak, E. and Orav, A. (2004) Chemical composition of coriander seed essential oil and their conformity with EP standards. *Agraarteadus* 15(4), 234–239.

Ramadan, M.F. and Morsel, J.T. (2002) Oil composition of coriander (*Coriandrum sativum* L.) fruit-seeds. *European Food Research and Technology* 215(3), 204–209.

Rao, V.K., Roy, T.K., Shivashankara, K.S. and Varalakshmi, B.A. (2004) A comparative study of whole herb and leaf essential oils of coriander. *Journal of Essential Oil Bearing Plants* 7(1), 49–55.

Reis Machado, A.S., Gomes de Azevedo, E. and Nunes da Ponte, M. (1993) High pressure carbon dioxide extraction from coriander plants. Headspace analysis. *Journal of Essential Oil Research* 5(6), 645–649.

Ross, J.H.E. and Murphy, D.J. (1992) Biosynthesis and localisation of storage proteins, oleosins and lipids during seed development in *Coriandrum sativum* and other Umbelliferae. *Plant Science Limerick* 86(1), 59–70.

Sanjeev, A., Choudhary, G.R., Sharma, R.K. and Agrawal, S. (1994) Improving the yield of essential oil in coriander (*Coriandrum sativum*). *Indian Cocoa, Arecanut and Spices Journal* 18(2), 45–47.

Selvan, M.T. (2003) Role of spices in medicine. *Indian Journal of Arecanut, Spices and Medicinal Plants* 5(4), 129–133.

Shubert, V. and Mosandl, A. (1991) Chiral compounds of essential oils. VIII: Stereodifferentiation of linalool using multidimensional gas chromatography. *Phytochemical Analysis* 2(4), 171–174.

Singh, U.P., Singh, D.P., Maurya, S., Maheshwari, R., Singh, M., Dubey, R.S. and Singh, R.B. (2004) Investigation on the phenolics of some spices having pharmacotherapeutic properties. *Journal of Herbal Pharmacotherapy* 4(4), 27–42.

Skapska, S., Bal, K., Jendrzejczak, Z., Morawski, A., Fonberg Broczek, M., Windyga, B. and Karowski, K. (2004) Changes in the volatiles of coriander and caraway induced by UHP heat treatment in helium. *Herba Polonica* 50(3/4), 152–160.

Smallfield, B.M., Perry, N.B., Beauregard, D.A., Foster, L.M. and Dodds, K.G. (1994) Effects of post harvest treatments on yield and composition of coriander herb oil. *Journal of Agricultural and Food Chemistry* 42(2), 354–359.

Smallfield, B.M., Klink, J.W. van, Perry, N.B. and Dodds, K.G. (2001) Coriander spice oil: effects of fruit crushing and distillation time on yield and composition. *Journal of Agricultural and Food Chemistry* 49(1), 118–123.

Stoyanova, A., Damianova, S., Konakchiev, A. and Djurdjev, I. (2003) Super carbon dioxide extracts of spices. Part I. Coriander (*Coriandrum sativum* L.). *Herba Polonica* 49(3/4), 235–239.

Suh, M.C., Schultz, D.J. and Ohlrogge, J.B. (1999) Isoforms of acyl carrier protein involved in seed specific fatty acid synthesis. *Plant Journal* 17(6), 679–688.

Tiwari, R.J. and Banafar, R.N.S. (1995) Application of nitrogen and phosphorus increases seed yield and essential oil of coriander. *Indian Cocoa, Arecanut and Spices Journal* 19(2), 51–55.

Tolkunova, N.N. (2002) The influence of essential oils on microbiological indicators of meat products. *Pishchevaya Promyshlennost* 12, 56–58.

Turysheva, N.A., Pelipenko, T.V. and Kaneeva, E.F. (1989) Qualitative changes in the essential oil when coriander fruits split. *Izvestiya Vysshikh Uchebnykh Zavedenii, Pishchevaya Tekhnologiya* 2, 123–124.

Uma Pradeep, K., Geervani, P. and. Eggum, B.O. (1993) Common Indian spices: nutrient composition, consumption and contribution to dietary value. *Plant Foods for Human Nutrition* (formerly Qualitas Plantarum) 44(2),137–148.

Venkateswarlu, K., Reddy, S.E. and Reddy, Y.N. (1992) Investigations on the effect of sowing time on essential oil yield of *Coriandrum sativum* L. cultivars. *South Indian Horticulture* 40(3), 155–158.

Wangensteen, H., Samuelsen, A.B. and Malterud, K.E. (2004) Antioxidant activity in extracts from coriander. *Food Chemistry* 88(2), 293–297.

Weiss, E.A. (2002) *Spice Crops*. CAB International, Wallingford, UK, p. 411.

11 Cumin

Shamina Azeez

11.1. Introduction

Cumin (*Cuminum cyminum*) (or *Jeerah*) is of the family Apiaceae, has been used as a spice since ancient times and is native to the eastern Mediterranean, extending to East India. Cumin seeds are used for their unique aroma and are popular in North African, Middle Eastern, Western, Chinese, Indian and Mexican cuisine. Cumin is a very popular spice in Western to Central Asia (Near and Middle East), in Central and South America, Burma, India and Indonesia. Indian cumin finds extensive use in foods, beverages, liquors, medicines, toiletries and perfumery, and is grown in the mild climes of Gujarat, Rajasthan and Uttar Pradesh. The best seeds of black cumin (*Nigella sativa*) come from Egypt, where they grow under almost perfect conditions in oases and where they are irrigated until the seed pods form (Weiss, 2002.) Black cumin is the fruit of a related plant that grows wild in Iran and the northern Indian region of Kashmir.

Indian cumin is exported in different forms – natural seed, powdered and essential oil – to the USA, Singapore, Japan, the UK and North Africa. India is the world's largest producer and consumer of cumin, with annual production ranging between 0.1 and 0.2 million t. India exports cumin to Bangladesh, Brazil, Japan, Malaysia, Nepal, Singapore, the UAE, the UK, the USA and many other countries, and cumin seed powder to the UK, the USA, etc. (Peter, 2000.) Of the 80,000–170,000 t of cumin grown in India each year, only about 10% is exported. Consumption by the rest of the world is between 25,000 and 30,000 t. In India, it is cultivated almost exclusively in Rajasthan and Gujarat and is harvested at the beginning of the year, while in Turkey, Syria and Iran it is harvested from May to August. Iran, Turkey and Syria also cultivate cumin mainly for export (Weiss, 2002.) Historically, Iran has been the main supplier of cumin, but today the major sources are India, Syria, Pakistan and Turkey (http://www.theepicentre.com/Spices/cumin.html).

11.2. Botany and Uses

The basic chromosome number of cumin is $x = 7$, with cumin being a diploid, $2n = 14$. Cumin is a small, slender, glabrous herbaceous annual of the parsley family, usually reaching 25 cm (some varieties being double this height) and tends to droop under its own weight. The blue-green linear leaves are 5–10 cm long, pinnate or bipinnate, thread-like leaflets. The white or pink flowers bloom in small compound umbels. The fruit is a lateral fusiform or ovoid achene 4–5 mm long, containing a single seed. The plants bloom in June and July. The seeds normally are ready 4 months after planting. The plants are cut when the seeds turn brown, threshed and dried like the other Umbelliferae (Vedamuthu *et al.*, 1994).

The seeds come as paired or separate carpels and are 3–6 mm long. They have a striped pattern of nine ridges and oil canals and are hairy, brownish in colour, boat-shaped; tapering at each extremity, with tiny stalks attached. They resemble caraway seeds, being oblong in shape and longitudinally ridged, but are lighter in colour and, unlike caraway, have minute bristles hardly visible to the naked eye. They are available dried, or ground to a brownish-green powder. The seeds are strongly aromatic. The aroma is characteristic and is modified by frying or dry toasting. Cumin looks deceptively simple, for its nutty peppery flavour, which is penetrating and peppery with slight citrus overtones (http://www.theepicentre.com/Spices/cumin.html).

Cumin is often confused with a few other spices. In Indian recipes, cumin is frequently confused with caraway, which it resembles in appearance, though not in taste, cumin being far more powerful. This is due to a misunderstanding of the Indian word *jeera*. The term usually means cumin, but occasionally can mean caraway. The use of the terms 'black cumin' for nigella and 'sweet cumin' for aniseed or fennel further confounds this confusion. As a general rule, interpret *jeera* or *zeera* (jira, zira) as cumin and *kalonji* as nigella. Cumin is distinguished easily from the other Umbelliferae by its flavour, and its shape and colour are quite different from nigella. Cumin is hotter to the taste, lighter in colour and larger than caraway (*Carum carvi*). The distantly related *Bunium persicum* and the unrelated *Nigella sativa* are both sometimes called black cumin.

Cumin fell out of favour in Europe, except in Spain and Malta, during the Middle Ages but is used more widely today, mainly as a carminative; it was introduced to the Americas by Spanish colonists. It is now grown mostly in Iran, Uzbekistan, Tajikistan, Turkey, Morocco, Egypt, India, Syria, Mexico and Chile. In most countries of Northern and Eastern Europe, cumin is of little importance as a traditional flavouring and is considered an alien spice, an oriental variety of caraway, comparable to, but distinct from, the native spice caraway ('foreign caraway'). Today, cumin usage in Europe is restricted to flavouring cheese in the Netherlands and France (Farrell, 1985), but it is experiencing a revival due to new-found appreciation of its culinary and therapeutic properties. Cumin essential oil is also used in cosmetics and toiletries to scent creams and lotions and in perfumes, with a reported maximum use of about 0.4% (Weiss, 2002).

11.3. General Composition

Bouquet: Strong, heavy and warm with a spicy-sweet aroma.
Flavour: Pungent, powerful, sharp and slightly bitter.
Hotness scale: 3.

Nutritional profile

The nutrient content of cumin is detailed in Table 11.1. One tsp of cumin is equivalent to 2 g. As the chart reveals, cumin is a very good source of iron and manganese as per the ratings of the World's Healthiest Foods. The amount of iron is 1.32 mg and of manganese 0.06 mg; the daily value being 7.3 and 3%, respectively. The nutrient density of iron is 17.6 and that of manganese is 7.2 (ESHA Research, Salem, Oregon). Reports by Barakat *et al.* (2003) on the elemental composition of spices using X-ray fluorescence (XRF) are revealed in Table 11.2, along with the report by Christensen *et al.* (1968) on the elemental composition of spices and herbs employing direct-reading emission spectroscopy.

Amino acid composition

Badr and Georgiev (1990) determined the crude protein, true protein, non-protein nitrogen and amino acid composition in cumin seeds supplied by Bulgaria, Egypt and Turkey. Bulgarian cumin had the highest content of crude protein (23%), whereas the Egyptian seeds contained the lowest percentage (18%). Eighteen amino acids were identified in all cumin seeds, of which eight were essential amino acids. The first limiting amino acid was tryptophan.

Table 11.1. Nutrient profile of cumin (in 2g of seeds).

Nutrient	Amount	% Daily value
Calories	7.50	
Calories from fat	4.00	
Calories from saturated fat	0.28	
Protein (g)	0.36	
Carbohydrates (g)	0.88	
Dietary fibre (g)	0.22	0.88
Total fat (g)	0.44	
Saturated fat (g)	0.04	
Monounsaturated fat (g)	0.28	
Polyunsaturated fat (g)	0.06	
Water (g)	0.16	
Ash (g)	0.16	
Vitamins		
Vitamin A (IU)	25.40	0.51
Vitamin A (RE)	2.54	
A – carotenoid (RE)	2.54	0.03
A – beta carotene (μg)	15.24	
Thiamin – B_1 (mg)	0.02	1.33
Niacin – B_3 (mg)	0.10	0.50
Niacin equiv	0.10	
Vitamin C	0.16	0.27
Vitamin E alpha equiv	0.02	0.10
Vitamin E (IU)	0.04	
Vitamin E (mg)	0.02	
Folate (μg)	0.20	0.05
Vitamin K (μg)	0.11	0.14
Minerals		
Calcium (mg)	18.62	1.86
Copper (mg)	0.02	1.00
Iron (mg)	1.32	7.33
Magnesium (mg)	7.32	1.83
Manganese (mg)	0.06	3.00
Phosphorus (mg)	9.98	1.00
Potassium (mg)	35.76	
Selenium (μg)	0.10	0.14
Sodium (mg)	3.36	
Zinc (mg)	0.10	0.67
Saturated fats		
16:0 Palmitic acid (g)	0.02	
Mono fats		
18:1 Oleic (g)	0.28	
Poly fats		
18:2 Linoleic acid (g)	0.06	
Other fats		
Omega 6 fatty acids (g)	0.06	

Source: ESHA Research, Salem, Oregon.

11.4. Chemistry

Volatiles

Cumin oil

The average characteristics of commercial oil are: specific gravity (25°C), 0.900–0.935; optical rotation (20°C), +4° to +8°; refractive index (20°C), 1.495–1.509. The oil is almost insoluble in water, but soluble in ten volumes of 80% alcohol, ether and chloroform. The essential oil of cumin consists of hydrocarbons (30–50%), aldehydes and ketones (50–70%), alcohols (2–5%) and ethers less than 1%, and their relative abundance depends mainly on cultivar and maturity at harvest (Weiss, 2002).

The essential oil content of cumin seed ranges from 2.3 to 5.0%. Cumin fruits have a distinctive bitter flavour and a strong, warm aroma due to their abundant essential oil content. Of this, 40–65% is cuminaldehyde (4-isopropylbenzaldehyde), the major constituent and important aroma compound, and also the bitterness compound reported in cumin (Hirasa and Takemasa, 1998). The odour is best described as penetrating, irritating, fatty, overpowering, curry-like, heavy, spicy, warm and persistent, even after drying out (Farrell, 1985.) The characteristic flavour of cumin is probably due to dihydrocuminaldehyde and monoterpenes (Weiss, 2002).

In the essential oil, apart from cuminaldehyde, perilla aldehyde (4-(1-methylethenyl)-1-cyclohexene-1-carboxaldehyde), cumin alcohol or 4-isopropylbenzyl alcohol, α-pinene and β-pinene (21%), dipentene, *p*-cymene, β-phellandrene and limonene (Fig. 11.1) have been reported by Baser *et al.* (1992).

The composition of Indian cumin oil, which contains less cuminaldehyde, is:

Table 11.2. Elemental composition of cumin.

X-ray fluorescence assay[1]		Emission spectroscopy[2]	
Element	Composition	Element	Composition
Aluminum (mg/kg)	105	Calcium (%)	1.0
Silica (mg/kg)	396	Phosphorus (%)	0.49
Phosphorus (mg/kg)	384	Potassium (%)	2.2
Sulphur (mg/kg)	700	Sodium (%)	0.22
Chloride (%)	0.14	Magnesium (%)	0.45
Potassium (%)	0.66	Aluminium (ppm)	570
Calcium (%)	0.37	Barium (ppm)	< 10
Manganese (mg/kg)	15	Iron (ppm)	> 500
Iron (mg/kg)	210	Strontium (ppm)	190
Copper (mg/kg)	56	Boron (ppm)	50
Zinc (mg/kg)	34	Copper (ppm)	9.1
Strontium (mg/kg)	7	Zinc (ppm)	56
		Manganese (ppm)	40
		Chromate (ppm)	5.5

Source: [1]Barakat *et al.* (2003); [2]Christensen *et al.* (1968).

cuminaldehyde (15–40%); terpinenes (18–29%); *α*-pinene (1.3%); *β*-pinene (2.0%); cineole, *p*-cymine and limonene (variable); and menthadienal (30%). According to Gachkar *et al.* (2007), the major compounds in the essential oils from cumin, extracted by hydrodistillation and characterized by GC and GC-MS, were *α*-pinene (29.1%), 1,8-cineole (17.9%) and linalool (10.4%) (Table 11.3).

The important aroma compounds of roasted cumin are the pyrazines, their various alkyl derivatives – particularly, 2,5- and 2,6-dimethyl pyrazine – and also substituted pyrazines 2-alkoxy-3-alkylpyrazines: 2-ethoxy-3-isopropyl pyrazine, 2-methoxy-3-*sec*-butyl pyrazine, 2-methoxy-3-methyl pyrazine, in addition to a sulphur compound, 2-methylthio-3-isopropyl pyrazine.

EFFECT OF MICRONUTRIENTS ON OIL El-Sawi and Mohamed (2002) reported that the application of micronutrients (50 mg/l of Zn and Mn), as single and combined treatments, had significant positive effects on the growth measurements and chemical composition of cumin plants. Combined treatment of the two micronutrients gave the highest values. In the herb and seed oils, 21 constituents were identified. Cumin aldehyde was found to be the main component at concentrations of 53.6% for seed oil and 40.5% for herb oil. Perilla aldehyde, *α-cis*-bergamotene, acoradiene and 4-(1-methylethyl) benzoic acid are the newly identified components in the seed and herb oils, which also contain considerable amounts of oxygenated monoterpenes and small amounts of monoterpenoid and sesquiterpene hydrocarbons (Table 11.4).

However, the composition of the volatile oil obtained from the herb also differs markedly from that of the seeds. Eleven components out of 21 are similar in both herb and seed oils, while some differences have been observed between the relative amounts of *β*-pinene, *α*-terpinene, *p*-cymene, *α*-terpineol, perilla aldehyde, thymol, *α-cis*-bergamotene, acoradiene and 4-(1-methylethyl) benzoic acid in the herb and seed oils.

Application of microelements spray increased the main constituents, such as cumin aldehyde, *p*-cymene, *α*-terpineol, thymol and acoradiene. On the other hand, spraying the cumin plant with microelements decreased other constituents, especially *β*-pinene. No marked differences between the relative percentages of the minor constituents of cumin herb oil, due to application of trace element treatments, were observed. However, for seed oil, an increase in cumin aldehyde, acoradiene and

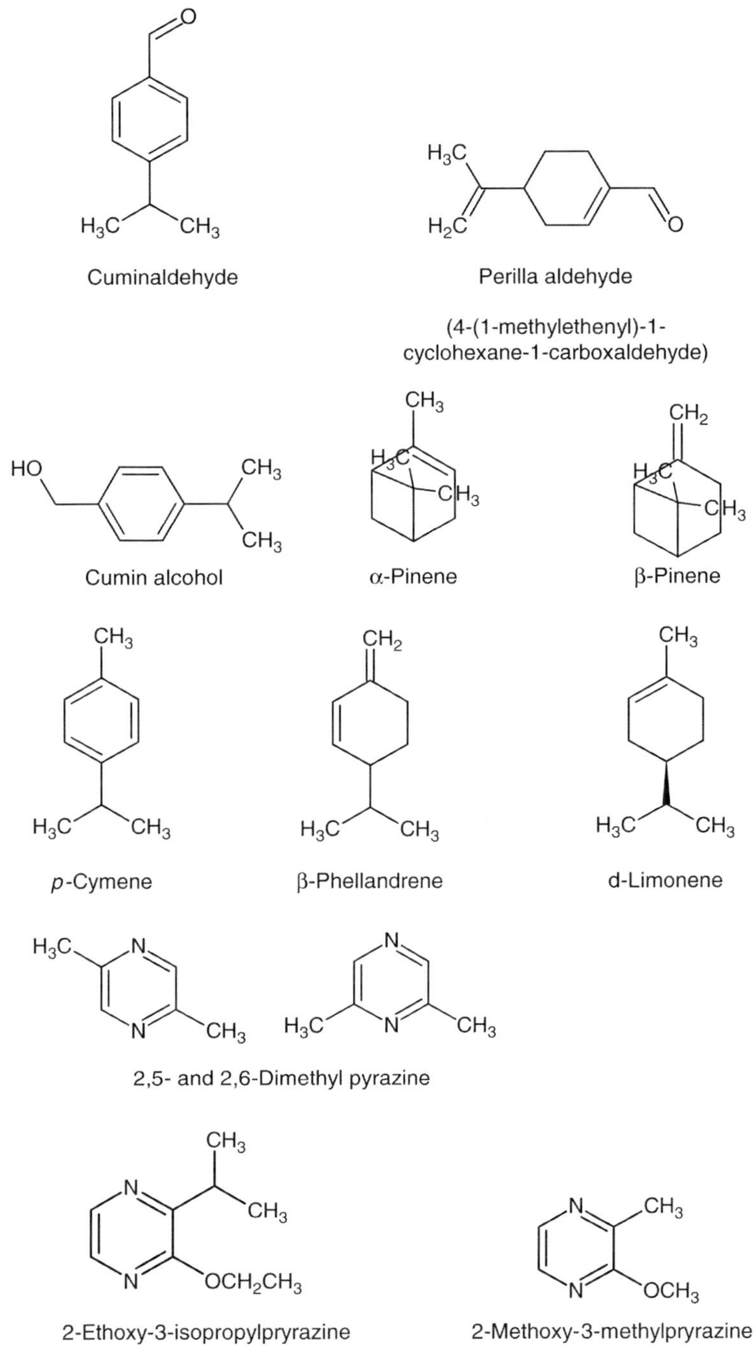

Fig. 11.1. Major chemical constituents in cumin.

Table 11.3. Chemical composition of *Cuminum cyminum* essential oil.

Compound	%	Compound	%
Isobutyl isobutyrate	0.8	Terpinene-4-ol	0.5
α-Thujene	0.3	α-Terpineole	3.17
α-Pinene	29.1	*trans*-Carveole	0.4
Sabinene	0.6	*cis*-Carveole	0.07
Myrcene	0.2	Geraniol	1.1
δ-3-Carene	0.2	Linalyl acetate	4.8
p-Cymene	0.3	Methyl geranate	0.2
Limonene	21.5	α-Terpinyl acetate	1.3
1,8-Cineole	17.9	Neryl acetate	0.09
(*E*)-Ocimene	0.1	Methyl eugenol	1.6
γ-Terpinene	0.6	β-Caryophyllene	0.2
Terpinolene	0.3	α-Humulene	0.2
Linalool	10.4	Spathulenol	0.07
α-Campholenal	0.03	Caryophylleneb epoxide	0.1
trans-Pinocarveole	0.07	Humulene epoxide II	0.08
δ-Terpineole	0.09	Acetocyclohexane dione (2)	0.4

Source: Gachkar *et al.* (2007).

propyl tiglate and a decrease in β-pinene was obvious (El-Sawi and Mohamed, 2002).

Extraction techniques

Cumin oil is usually obtained by steam distillation of the milled spice; hydrodiffusion gives a higher yield and, more recently, supercritical gaseous extraction is claimed to give oil closer to the aroma and taste of the spice (Eikani *et al.*, 1999). The yields of cumin seed oil with steam distillation are 2.3–3.6%, with liquid carbon dioxide it is 4.5% and with ethanol it is 12%. The major components are cuminaldehyde, cuminyl alcohol, *p*-mentha and 1.3-dien-7-al, the minimum perceptible levels being at 0.2 ppm. Naik *et al.* (1989) reported that liquid CO_2 extraction was quicker than steam distillation for the quantitative extraction of cumin oil without loss of active flavour components, at 58 bar and 20°C.

Solvent-free microwave extraction (SFME) is a recently developed 'green' technique, performed in atmospheric conditions without adding any solvent or water and being applied to the extraction of essential oil from fresh plant or dried materials. The essential oil is evaporated by the *in situ* water in the plant materials. Wang *et al.* (2006) observed that an improved SFME, in which a kind of microwave absorption solid medium such as carbonyl iron powders (CIP) was added and mixed with the sample, could be applied to the extraction of essential oil from the dried cumin without any pretreatment. GC-MS analysis of the compositions of essential oil extracted by four kinds of extraction methods – improved SFME, conventional SFME, microwave-assisted hydrodistillation and conventional hydrodistillation – revealed no obvious difference in the quality of essential oils.

Behera *et al.* (2004) concluded that the optimum conditions for the conventional roasting method were 125°C for 10 min and, in the microwave processing method, the best conditions were 730 W for 10 min. The yields and physico-chemical properties of the volatile oils were similar in both cases. Changes were observed in the optical rotation values, which indicated differences in the chemical compositions. GC and GC-MS analysis of optimized condition samples showed that microwave-heated samples could better retain the characteristic flavour compounds of cumin (i.e. total aldehydes) than conventionally roasted samples (Table 11.5). Earlier GC reports showed cuminaldehyde as the only major aldehyde present in Indian cumin oil, but this study revealed the

Table 11.4. Effect of micronutrients (Zn and Mn at 50 mg/l) on the herb and seed oil composition of cumin.

Compound	Herb oil (%)	Seed oil (%)
Monoterpene hydrocarbons		
α-Pinene	–	1.27
Sabinene	–	0.26
β-Pinene	2.31	6.26
Myrcene	–	0.72
α-Phellandrene	–	0.75
3-Carene	–	0.81
α-Terpinene	2.70	0.95
ρ-Cymene	3.51	1.54
γ-Terpinene	1.70	1.06
Oxygenated monoterpenes		
Terpinene-4-ol	1.89	–
α-terpineol	2.22	0.84
Cuminaldehyde	40.54	53.55
Perilla aldehyde	3.14	1.13
Thymol	2.38	1.27
Cumin alcohol	–	2.10
Sesquiterpene hydrocarbons		
α-*cis*-Bergamolene	2.46	1.21
β-Caryophyllene	–	3.14
cis-β-Farnesene	–	1.72
Acoradiene	7.64	11.46
Cuparene	2.09	–
Oxygenated sesquiterpenes		
Caryophyllene oxide	3.42	–
Carotol	0.67	–
Daucol	0.77	–
Acids and esters		
Propyl tiglate	0.34	–
Hydrocinnamyl acetate	–	2.34
Benzoic acid 4-(1 methylethyl)-	5.36	1.09
ρ-Anisyl acetate	–	2.11
Menth-8-ene-3-ol, acetate	3.36	–
Hexadecanoic acid	0.23	–
Others		
Octanal	0.23	–
Estragole	3.27	–
Total		
No. of compounds	21	21
Monoterpene hydrocarbons	10.22	13.62
Oxygenated monoterpenes	50.17	58.89
Sesquiterpene hydrocarbons	12.19	17.53
Oxygenated sesquiterpene	4.68	–
Acids and esters	9.29	5.54
Others	3.5	–

Source: El-Sawi and Mohamed, 2002.

presence of two more aldehydes, *p*-mentha-1,3-dien-7-al and *p*-mentha-1,4-dien-7-al. Thus, the microwave treatment, in spite of losing terpene hydrocarbons, retained aldehydes in the volatile oil, making it the best choice as an alternative heating medium for processing (Table 11.6).

Pruthi and Misra (1963) reported some important physico-chemical changes that occurred during drying, milling and mechanical mixing operations in the manufacture of curry powder. Losses in weight, moisture, volatile oil and volatile-reducing substances (VRS) as a measure of aroma, as a consequence of the increase in temperature, are given in Table 11.7.

Non-volatiles

Oleoresin

The oleoresin of cumin is brownish to yellowish-green in colour, which tends to darken on ageing, and 100 g contains 60 ml of volatile oil. One kg of the oleoresin is equivalent to 20 kg of freshly ground cumin in aroma and flavour characteristics (Farrell, 1985.) Cumin oleoresin or absolute is produced in very small quantities, either by the end-user or made to order.

Other non-volatile components

The tissue of the fruits contains fatty oil with resin, mucilage and gum, malates and albuminous matter and, in the outer seed-coat, there are significant amounts of tannin. The yield of ash is about 8%. Dried cumin fruits contain essential oil with over 100 different chemical constituents, including abundant sources of the essential fatty acids, oleic acid (3%), linoleic acid (34%), flavonoid glycosides, tannins, resins and gum (Singh *et al.*, 2006).

Ceska *et al.* (1986) reported the presence of photoactive furocoumarins not previously detected from the fruits of cumin, using sensitive methods like HPLC and photobiological bioassay. Harborne and Williams (1972) surveyed the fruit flavonoids in some 100 species representing all

Table 11.5. Comparison of chemical compositions of essential oils of fresh and optimally processed cumin seeds.

Compound	Fresh (μl/100 g)	Conventional (μl/100 g) (125°C; 10 min)	Microwave (μl/100 g) (730 W; 10 min)
α-Pinene	18.7	3.94	2.58
β-Pinene	544	286	8.05
β-Myrcene	44.6	18.9	174
p-Cymene	943	992	845
γ-Terpinene	1565	425	115
Terpinene-4-ol	12.4	21.7	83.2
trans-Verbenol	4.28	–	45.1
Myrtenal	6.88	6.81	–
Cuminaldehyde	1038	819	920
p-Mentha-1,3-dien-7-al	316	100	349
p-Mentha-1,4-dien-7-al	1034	297	2.24
cis-Farnesene	5.99	–	2.18

Source: Behera *et al.* (2004).

Table 11.6. Comparison of terpene hydrocarbons and aldehydes from the essential oils of fresh and optimally processed (conventional and microwave) cumin seeds.

Compound	Fresh sample (%)	Conventional roasting (125°C; 10 min) (%)	Microwave heating (730 W; 10 min) (%)
Monoterpenes	56.4	58.1	45.0
Sesquiterpenes	0.108	–	0.085
Aldehydes	43.2	41.0	50.0
Alcohols	0.3	0.73	7.68
Ratio of aldehyde/hydrocarbons	0.765	0.705	1.11

Source: Behera *et al.* (2004).

the major tribes of the Umbelliferae. Of the 25 flavone and flavonol glycosides detected, by far the most common were luteolin 7-glucoside and quercetin 3-rutinoside. The discovery of apigenin and luteolin 7-glucuronosylglucosides in cumin supports its removal from the Apieae and transfer to the Caucalideae.

The polar portion of the methanolic extract of cumin fruit contains two sesquiterpenoid glucosides, cuminoside A and B, and two alkyl glycosides isolated together with five known compounds (Takayanagi *et al.*, 2003). Their structures were established as (1*S*,5*S*,6*S*,10*S*)-10-hydroxyguaia-3,7(11)-dien-12,6-olide β-D-glucopyranoside(1*R*,5*R*,6*S*,7*S*,9*S*,10*R*,11*R*)-1,9-dihydroxyeudesm-3-en-12,6-olide 9-*O*-β-D-glucopyranoside, methyl β-D-apiofuranosyl-(1→6)-β-D-glucopyranoside and ethane-1,2-diol 1-*O*-β-D-apiofuranosyl-(1→6)-β-D-glucopyranoside.

Table 11.7. Changes in physico-chemical properties during drying and milling of cumin, in the process of making curry powders.

Parameter	Value initial	After milling
Moisture (%)	6.00	4.00
Volatile oil (%)	5.75	4.60
Volatile-reducing substances (meq $KmnO_4$/g)	270	260

Note: Temperature during milling and mixing: 85°C. Percentage loss in weight: during drying, 8.65; during milling, 2.00.
Source: Pruthi and Misra, (1963).

In addition, three glycosides of 2-*C*-methyl-D-erythritol: 1-*O*-β-D-glucopyranoside, 3-*O*-β-D-glucopyranoside and 4-*O*-β-D-glucopyranoside, were identified from cumin fruit (Kitajima *et al.*, 2003). Though the phosphate of 2-*C*-methyl-D-erythritol was known to be one of the first precursors of isoprenoids in the non-mevalonate pathway, and was considered to be a common constituent in Umbelliferous plants, its glycosides were found for the first time.

11.5. Culinary Uses and Medicinal Properties

The oil of cumin is an essential part of kummel liqueur and German baked goods; it is also used in perfumery. In medicine, it is used as a stimulant, an antispasmodic and a carminative. It is used mainly as a seasoning in curry powders, soups, stews, sausages, cheeses, pickles, meats and chutneys (Farrell, 1985).

Culinary uses

Cumin is available both in its whole seed form and as a powder. To make the best of their aroma and flavour, whole cumin seeds are lightly roasted before use. It is preferable to buy whole cumin seeds instead of cumin powder since the latter loses its flavour more quickly. Cumin seeds and cumin powder should be kept in an airtight glass container in a cool, dark and dry place. Ground cumin will keep for about 6 months, while the whole seeds will stay fresh for about a year. This spice should be used with restraint, as it can surpass all the other flavours in a dish (http://www.royalthai-cuisine.com/). The oil is a partial substitute for the powdered spice in similar food products and the highest average maximum use level is about 0.025% (247 ppm) in relishes and condiments (Weiss, 2002).

Cumin is used mainly where spicy foods are prepared. It is used in Indian, Eastern, Middle Eastern, Mexican, Portuguese and Spanish cookery. Cumin also forms an essential part of curry powder, chilli powder, *sambar* powder, garam masala and of the Bengali spice mixture, *panch phoron*, besides being used in Northern Indian tandoori dishes. In imperial North Indian cuisine (Mughal or Mughlai), the mixture of cumin is prepared to relish sweet and aromatic flavours. This spice mixture is sometimes used for cooking, but more frequently sprinkled over the dishes before serving. Legumes, especially lentils, are normally seasoned with cumin.

The flavour of cumin also plays a major role in Mexican, Thai and Vietnamese cuisines. It can be found in some Dutch cheeses, like Leyden cheese, and in some traditional bread from France. Cumin is also very popular in Western to Central Asia. In South-eastern and Eastern Asia, cumin is less valued, but used occasionally. Cumin is very important in Burmese cooking and it plays a role in the cooking styles of Thailand and Indonesia. In China proper, cumin is a rare spice used only for a small number of recipes. The patterning theory of spice use reveals that cumin is most suitable for Eastern cooking (e.g. Indian and South-east Asian) and does not show suitability for any Western cooking (except American) (Hirasa and Takemasa, 1998).

Medicinal properties

In traditional medicine, cumin has varied uses; it is used to treat hoarseness, jaundice, dyspepsia and diarrhoea. Its seeds have stomachic, diuretic, carminative, stimulant, astringent and abortifacient properties. It has myriad physiological effects, such as:

- Gastrointestinal, reproductive, nervous and immune systems.
- Hypoglycaemic.
- Hypolipidaemic.
- A very good source of iron.
- Chemoprotective.
- Antimicrobial.
- Antioxidant.
- Tyrosinase inhibitor activity.

Gastrointestinal system

Cumin can be used to stimulate the appetite and relieve dyspepsia and diarrhoea. It may stimulate the secretion of pancreatic enzymes, which could explain its effect on the digestive system. In the West, though now used mainly in veterinary medicine as a carminative, it is making a comeback as its medicinal properties are being recognized. It remains a traditional herbal remedy in the East (Amin, 2000).

Reproductive system

It is emmenagogic and antispasmodic. It is believed to increase lactation and reduce nausea in pregnancy (Weiss, 2002).

Nervous system

Mahyar *et al.* (2006) report the effect of the fruit essential oil of cumin on the epileptiform activity induced by pentylenetetrazol (PTZ), using the intracellular technique. The results demonstrate that extracellular application of the essential oil of cumin (1 and 3%) dramatically decreases the frequency of spontaneous activity induced by PTZ in a time- and concentration-dependent manner. In addition, it showed protection against PTZ-induced epileptic activity by increasing the duration and decreasing the amplitude of after-hyperpolarization potential (AHP) following the action potential, the peak of action potential and inhibition of the firing rate.

Hypoglycaemic property

The orally administered seed powder (2 g/kg) lowered the blood glucose levels in hyperglycaemic rabbits (Jain *et al.*, 1992). Cumin also decreased the glucose tolerance curve and hyperglycaemic peak (Aslam *et al.*, 2003). Dhandapani *et al.* (2002) reported that the oral administration of 0.25 g/kg body weight of cumin for 6 weeks to diabetic rats resulted in significant reduction in blood glucose and an increase in total haemoglobin and glycosylated haemoglobin, with a decrease in body weight and reduction in plasma and tissue cholesterol, phospholipids, free fatty acids and triglycerides. Histological observations demonstrate that fatty changes and inflammatory cell infiltrates in diabetic rat pancreas are reduced significantly by supplementation with cumin. Moreover, cumin supplementation is found to be more effective than glibenclamide in the treatment of diabetes mellitus.

Hypolipidaemic property

Cumin decreased significantly the plasma levels of cholesterol, triglycerides and phospholipids and activity of the enzymes, aspartate transaminase, alkaline phosphatase and gamma glutamyl transferase (enzymes that are non-specific indicators of tissue damage such as liver disease (alcoholic liver disease, chronic hepatitis, cirrhosis, obstructive jaundice, hepatic cancer), myocardial infarction, pancreatitis and muscle-wasting diseases) when compared with the normal control group (Aruna *et al.*, 2005). The activity of phospholipases A and C (enzymes that catalyse the splitting of phospholipids into fatty acids and other lipophilic substances by the addition of water) also decreased significantly in the liver of treated rats. The results obtained indicated that cumin could decrease the lipid levels in alcohol and thermally oxidized oil-induced hepatotoxicity.

Source of iron

Cumin seeds are a very good source of iron, an integral component of haemoglobin (see the section on general composition) and also part of enzyme systems for energy production and metabolism. Also, iron is instrumental in keeping the human immune system healthy. Iron is particularly important for women, growing children and adolescents.

Antioxidant activity

Essential oil from spice materials including cumin was investigated on sunflower oil, stored at 70°C and was found to possess excellent antioxidant effects, better than those of the synthetic antioxidant, butylated hydroxytoluene (Singh *et al.*, 1998).

In a study by Chipault *et al.* (1952) to test the stabilizing effect of 36 different spices on lard by the active oxygen method at 98.6°C, the antioxidant index of ground cumin was found to be 1.3, while that of the petroleum ether-soluble fraction was 1.1 and of the alcohol-soluble fraction 1.2. The study of the antioxidant property of spices in oil-in-water emulsion by Chipault *et al.* (1955) revealed that cumin could increase the mean stability of these emulsions to 33.5 h, against 15.5 h for control, the antioxidant index (ratio of mean stability of sample to mean stability of control) being 2.6.

Chemoprotective property

Cumin seeds may also have anticarcinogenic properties, as they protect laboratory animals from developing stomach or liver tumours. This effect may stem from cumin's free radical scavenging abilities and its ability to enhance the liver's detoxification enzymes. Gagandeep *et al.* (2003) found that tumour burden reduced significantly with different doses of cumin seeds in mice with benzo(a)pyrene-induced forestomach tumorigenesis and 3-methylcholanthrene (MCA)-induced uterine cervix tumorigenesis. In the latter case, tumour burden was reduced to 27.3% on a diet of 5% cumin seeds and to 12.5% on a diet of 7.5% cumin seeds, compared with the MCA-treated control group (66.67%). Cumin augments the levels of carcinogen/xenobiotic metabolizing phase I enzymes, cytochrome P-450 (cyt P-450) and cytochrome b_5 (cyt b_5), the levels of cyt P-450 reductase and cyt b_5 reductase, and the phase II enzymes, e.g. glutathione-*S*-transferase and DT-diaphorase. In the antioxidant system, significant elevations of superoxide dismutase and catalase activities are also observed. Elevation of reduced glutathione levels and inhibition of lipid peroxidation are also noticed. These results strongly suggest the cancer chemopreventive potential of cumin seed, which could be attributed to its ability to modulate carcinogen metabolism. Cumin seeds could also decrease significantly the incidence of both B[*a*]P-induced neoplasia and 3'MeDAB-induced hepatomas in Wistar rats (Aruna and Sivaramakrishnan, 1992).

Contrary to the above reports, Barakat *et al.* (2003) reported a very weak oxidative mutagenicity in cumin, in the strain TA102 of *Salmonella typhimurium*, an oxidative mutation detector (Levin *et al.*, 1982), using the Ames test. Spices like cumin, aniseed, black pepper and ginger contain safrole, a natural mutagenic compound, which is degraded by cooking and/or irradiation (Frage and Abozeid, 1997.) It has been reported that cumin and black pepper may also have a protecting effect on the colon by decreasing the activity of bacterial β-glucuronidase and mucinase (Nalini *et al.*, 1998).

Inhibition of platelet aggregation and alteration of eicosanoid biosynthesis

The eicosanoids – prostaglandins, prostacyclins, thromboxanes and leukotrienes – are derived from omega-3 or omega-6 fats and are signalling molecules which exert complex control over diverse bodily functions such as inflammation and immunity, and are messengers in the central nervous system. Srivastava (1989) reported that the ethereal extract of both cumin and turmeric inhibited arachidonate-induced platelet aggregation. Extracts from these spices inhibited thromboxane B_2 production from exogenous (^{14}C) arachidonic acid (AA) in washed platelets; a simultaneous increase in the formation of lipoxygenase-derived products was also observed.

Antimicrobial activity

The essential oil of cumin exhibits strong antimicrobial activity against *Escherichia coli*, *Staphylococcus aureus* and *Listeria monocytogenes*. Complete death time on exposure to cumin oil was 20, 180 and 90 min for *E. coli*, *S. aureus* and *L. monocytogenes*, respectively (Gachkar *et al.*, 2007).

Lawrence (1992) reported that cumin oil showed fungitoxic, fungicidal, antibacterial and larvicidal activity due to the cuminaldehyde content. The undiluted oil also has a distinct phytotoxic effect on mammals, but not due to the cuminaldehyde content. Marked antifungal activity is seen against the following fungi: *Penicillium notatum*, *Aspergillus niger*, *A. fumigatus*, *Microsporum canis* (Afifi *et al.*, 1994), *Pseudallescheria boydii* and *A. flavus* (Atta-ur-Rahman *et al.*, 1999). Cumin seed and/or callus extracts and essential oils inhibit bacteria (particularly *S. aureus*) and fungi (*Fusarium moniliforme*), as well as polio and Coxsackie viruses (Jain *et al.*, 1992).

The volatile oil of cumin, which is a mixture of about 32 components, principally cuminaldehyde (40.7%), terpinene (16.7%) and *p*-cymene (14.5%), inhibits *Curvularia lunata* and *F. moniliforme* by 100%, while the acetone extract is 85% effective in inhibiting the mycelial growth of *A. ochraceus, A. flavus* and *P. citrinum* (Singh *et al.*, 2006).

Singh and Upadhyay (1991) found that the essential oil of cumin seeds inhibited mycelial growth of *A. flavus* and *A. niger* completely at 3000 ppm, inhibition at 1000 ppm being 85–89%. The aldehyde fraction, separated using $NaHSO_3$ and HCl, contained only cuminaldehyde. This gave 100% inhibition of both fungi at 1000 ppm. Farag *et al.* (1989) found that the essential oils of cumin and other spices inhibited the total aflatoxin production of *A. parasiticus* at relatively low concentrations, although not as effectively as thyme oil.

Among the 60 constituents of the cumin oil identified by GC, GC-MS and olfactometry as essential volatiles, cuminaldehyde (36%), *β*-pinene (19.3%), *p*-cymene (18.4%) and γ-terpinene (15.3%) are the principal components showing high antimicrobial activity against the mould *A. niger*, the Gram-positive bacteria, *Bacillus subtilis* and *S. epidermidis*, as well as the yeasts, *Saccharomyces cerevisiae* and *Candida albicans* (Jirovetz *et al.*, 2005).

Ovicidal property

Fumigant activity of essential oil vapours distilled from cumin against the eggs of two stored-product insects, the confused flour beetle, *Tribolium confusum*, and the Mediterranean flour moth, *Ephestia kuehniella*, has been reported by Tunç *et al.* (2000). The exposure to vapours of essential oils resulted in 100% mortality of the eggs. At a concentration of 98.5 μl cumin essential oil/l air, the LT_{99} value for *E. kuehniella* was 127.0 h. A peculiar use of the oil is as a bird repellent to reduce the nuisance of roosting birds on buildings (Clark, 1998).

Tyrosinase inhibitor activity

Tyrosinase inhibitors prevent browning in food because they inhibit the oxidation caused by the enzyme tyrosinase. Cuminaldehyde is identified as a potent mushroom tyrosinase monophenol monooxygenase inhibitor from cumin seeds. It inhibits the oxidation of L-3,4-dihydroxyphenylalanine (L-DOPA) by mushroom tyrosinase with an ID_{50} of 7.7 g/ml (0.05 mM). Its oxidized analogue, cumic acid (*p*-isopropylbenzoic acid), also inhibits this oxidation with an ID_{50} of 43 g/ml (0.26 mM). These two inhibitors affect mushroom tyrosinase activity in different ways (Kubo and Kinst-Hori, 1998).

11.6. Quality Specifications

Amin (2000) gives the quality specifications expected of cumin. The spice is traded as whole or ground seed, or as essential oil. Cumin seeds are not a commonly allergenic food and are not known to contain measurable amounts of goitrogens, oxalates or purines. The regulatory status of cumin and cumin oil in the USA is regarded generally as safe, GRAS 2340 and GRAS 2343.

The specific quality indices are: seed moisture, < 6%; total ash, 7%; acid-insoluble ash, 1.5%; volatile oil, minimum 2%; foreign organic matter, 2%.

Powdered seed specifications

The powdered seeds are yellowish-brown with an aromatic, slightly camphoraceous odour and taste. The characteristics are:

- The epicarp, composed of a layer of colourless cells, polygonal in surface view with thin sinuous walls and a faintly and irregularly striated cuticule; stomata are fairly frequent and, very occasionally, cicatrices may be present. Underlying the epicarps, the thin-walled cells of the palisade are sometimes visible.

- The covering trichomes, which are usually found attached to small fragments of the epicarp, are pluricellular, multiseriate and rounded at the apex, vary in length and are composed of fairly thick-walled cells.
- The sclereids from the mesocarp are of two main types: single layer and elongated cells. They are found frequently associated with the vascular tissue.
- The fairly numerous pale yellowish-brown fragments of the vittae are composed of fairly large, thin-walled cells, polygonal in surface view.

Volatile oil specifications

Colourless or pale yellow; specific gravity (25°C), 0.905–0.925; optical rotation (20°C), +3 to +8; refractive index, 1.501–1.506; solubility (80% ethanol), 8 vol; aldehydes (as cumin aldehyde), 40–52%.

Cumin essential oil can be adulterated with synthetic cumin aldehyde, which is difficult to detect, though methods such as stable isotope ratio analysis and selective ion monitoring help in detecting adulteration of this kind (Amin, 2000).

Cleanliness specifications and defect action levels

The ASTA cleanliness specifications (Anon., 1991), the Defect Action Levels (DAL), as set by the Food and Drug Administration (FDA) and the DAL as prescribed by the USFDA for spices (http://www.indianspices.com) for cumin, are presented in Table 11.8.

Table 11.8. ASTA cleanliness specifications and Defect Action Levels (DAL) prescribed by FDA and USFDA for cumin.

Specification	Suggested limit
ASTA cleanliness specifications	
Whole insects, dead (No.)	4
Mammalian excreta (mg/lb)	3
Other excreta (mg/lb)	5
Mould (% by weight)	1
Insect-defiled/infested (% by weight)	1
Extraneous foreign matter (% by weight)	0.5
Ash (% max)	9.5
Acid-insoluble ash (% max)	1.5
FDA DAL	
Volatile oil (% min)	2.5
Moisture (% max)	9.0
Ash (% max)	8.0
Acid-insoluble ash (% max)	1.0
Average bulk index (mg/100 g)	240
USFDA DAL	
Sand and grit (AOAC 975.48)	Average of 9.5% or more ash and/or 1.5% or more acid-insoluble ash

11.7. Conclusion

In summary, the seed of the plant *C. cyminum* of the family Apiaceae and a native from the eastern Mediterranean to East India has been used as a spice since Biblical times. India is the largest producer and consumer of the spice, which is most suited for Eastern cooking (e.g. Indian and South-east Asian) and does not show suitability for Western cooking (except American).

Cumin is a very good source of iron and manganese. Eight of the 18 amino acids identified in cumin seeds are essential amino acids, the limiting amino acid being tryptophan. Cumin oil is obtained usually by steam distillation of the milled spice; hydrodiffusion gives a higher yield. Solvent-free microwave extraction (SFME) is the most efficient extraction system reported to date, followed by supercritical gaseous extraction. The essential oil of cumin consists of hydrocarbons, aldehydes and ketones, alcohols and ethers. The essential oil content of the cumin seed ranges from 2.3 to 5%, of which 40–65% is cuminaldehyde. The other chief constituents reported are perilla aldehyde, cumin alcohol, α-pinene and β-pinene, dipentene, *p*-cymene, β-phellandrene, 1,8-cineole, linalool and limonene. Usually, cumin is roasted before its use in cooking; the aroma compounds of toasted cumin are the pyrazines, their various alkyl derivatives (particularly, 2,5- and 2,6-dimethyl pyrazine, and also substituted pyrazines 2-alkoxy-3-alkylpyrazines: 2-ethoxy-3-isopropyl pyrazine, 2-methoxy-3-*sec*-butyl pyrazine, 2-methoxy-3-methyl pyrazine; a sulphur compound, 2-methylthio-3-isopropyl pyrazine) were also found. Dried cumin fruits contain essential oil, with 22% fatty oil, 18% protein, 14 free amino acids, flavonoid glycosides, tannins, resins and gum.

Apart from its use in cooking and cosmetics, it has a number of documented medicinal uses: it has effects on the gastrointestinal system, reproductive system, nervous system and immune system; and hypoglycaemic, hypolipidaemic, antimicrobial, antioxidant and chemoprotective activity.

The regulatory status of cumin and cumin oil in the USA is regarded generally as safe, GRAS 2340 and GRAS 2343. The ASTA cleanliness specifications and the DAL as set by the FDA and the DAL as prescribed by the USFDA for spices are described by Anon. (1991) and Potty and Krishnakumar (2001).

References

Afifi, N.A., Ramadan, A., El-Kashoury, E.A. and El-Banna, H.A. (1994) Some pharmacological activities of essential oils of certain umbelliferous fruits. *Veterinary Medical Journal Giza* 42(3), 85–92.

Amin, Gh. (2000) Cumin. In: Peter, K.V. (ed.), *Handbook of Herbs and Spices*. Woodhead Publishing, Cambridge, UK, pp. 164–167.

Anon. (1991) *Clean Spices: A Guidebook for Shippers of Products to the US Spice Trade*. ASTA, Washington, DC, pp. 4.

Aruna, K. and Sivaramakrishnan, V.M. (1992) Anticarcinogenic effects of some Indian plant products. *Food and Chemical Toxicology* 30(11), 953–956.

Aruna, K., Rukkumani, R., Suresh Varma, P. and Venugopal, M.P. (2005) Therapeutic role of *Cuminum cyminum* on ethanol and thermally oxidized sunflower oil induced toxicity. *Phytotherapy Research* 19(5), 416–421.

Aslam, M., Jafri, M.A., Javed, K. and Singh, S. (2003) Evaluation of antidiabetic drugs from plant sources. In: Singh, S., Govil, J.N. and Singh, V.K. (eds) *Recent Progress in Medicinal Plants. Volume 2, Phytochemistry and Pharmacology*. Studium Press, Houston, Texas, pp. 105.

Atta-ur-Rahman, Choudhary, M.I., Farooq, A., Ahmed, A., Iqbal, M.Z., Demirci, B., Demirci, F. and Baser, K.H.C. (1999) Antifungal activities and essential oil constituents of some spices from Pakistan. Proceedings of the *Third International Electronic Conference on Synthetic Organic Chemistry (ECSOC-3)*, 1–30 September, www.reprints.net/ecsoc-3.htm.

Badr, F.H. and Georgiev, E.V. (1990) Amino acid composition of cumin seed (*Cuminum cyminum* L.). *Food Chemistry* 38(4), 273–278.

Barakat, A.A.B., Ahmed, O.M. and Mahmoud, M.A.K. (2003) Element analysis and biological studies on ten oriental spices using XRF and Ames test. *Journal of Trace Elements in Medicine and Biology* 17(2), 85–90.

Baser, K.H.C., Kurkcuoglu, M. and Ozek, T. (1992) Composition of Turkish cumin seed oil. *Journal of Essential Oil Research* 4, 133–138.

Behera, S., Nagarajan, S. and Jagan Mohan Rao, L. (2004) Microwave heating and conventional roasting of cumin seeds (*Cuminum cyminum* L.) and effect on chemical composition of volatiles. *Food Chemistry* 87(1), 25–29.

Ceska, O., Chaudhary, S.K., Warrington, P.J. and Ashwood-Smith, M.J. (1986) Photoactive furocoumarins in fruits of some umbellifers. *Phytochemistry* 26(1), 165–169.

Chipault, J.R., Mizuna, G.R., Hawkins, J.M. and Lundberg, W.O. (1952) The antioxidant property of natural spices. *Food Research* 17, 46.

Chipault, J.R., Mizuna, G.R., Hawkins, J.M. and Lundberg, W.O. (1955) The antioxidant property of spices in oil-in-water emulsions. *Food Research* 20, 443.

Christensen, R.E., Beckman, R.M. and Birdsall, J.J. (1968) Some mineral elements of commercial spices and herbs as determined by direct reading emission spectroscopy. *Journal of the Association of Official Analytical Chemists* 51, 1003.

Clark, L. (1998) Dermal contact repellants for starlings: foot exposure to natural plant products. *Journal of Wilderness Management* 61(4), 1352–1358.

Dhandapani, S., Subramanian, V.R., Rajagopal, S. and Namasivayam, N. (2002) Hypolipidemic effect of *Cuminum cyminum* L. on alloxan-induced diabetic rats. *Pharmacological Research* 46(3), 251–255.

Eikani, M.H., Goodarzani, I. and Mirza, M. (1999) Supercritical carbon dioxide extraction of cumin seeds. *Flavour and Fragrance Journal* 14(1), 29–31.

El-Sawi, S.A. and Mohamed, M.A. (2002) Cumin herb as a new source of essential oils and its response to foliar spray with some micro-elements. *Food Chemistry* 77(1), 75–80.

Farag, R.S., Daw, Z.Y. and Abo-Raya, S.H. (1989) Influence of some spice essential oils on *Aspergillus parasiticus* growth and production of aflatoxins in a synthetic medium. *Journal of Food Science* 54, 74–76.

Farrell, K.T. (1985) In: *Spices, Condiments and Seasonings*. The AVI Publishing Co., Inc., Westport, Connecticut, pp. 98–100.

Frage, S.E.A. and Abozeid, M. (1997) Degradation of the natural mutagenic compound safrole in spices by cooking and irradiation. *Nahrung* 41, 359–361.

Gachkar, L., Davood, Y., Mohammad, B.R., Masood, T., Shakiba, A.A. and Iraj, R. (2007) Chemical and biological characteristics of *Cuminum cyminum* and *Rosmarinus officinalis* essential oils. *Food Chemistry* 102(3), 898–904.

Gagandeep, S., Dhanalakshmi, E., Mendiz, A., Rao, R. and Kale, R.K. (2003) Chemopreventive effects of *Cuminum cyminum* in chemically induced forestomach and uterine cervix tumors in murine model systems. *Nutrition and Cancer* 47(2), 171–180.

Harborne, J.B. and Williams, C.A. (1972) Flavonoid patterns in the fruits of the umbelliferae. *Phytochemistry* 11(5), 1741–1750.

Hirasa, K. and Takemasa, M. (1998) Cooking with spices. In: *Spice Science and Technology*. Marcel Dekker, New York, pp. 70.

http://www.royalthai-cuisine.com.

http://www.theepicentre.com/Spices/cumin.html.

Jain, S.C., Purohit, M. and Jain, R. (1992) Pharmacological evaluation of *Cuminum cyminum*. *Fitoterapia* 63(4), 291–294.

Jirovetz, L., Buchbauer, G., Stoyanova, A.S., Georgiev, E.V. and Damianova, S.T. (2005) Composition, quality control and antimicrobial activity of the essential oil of cumin (*Cuminum cyminum* L.) seeds from Bulgaria that had been stored for up to 36 years. *Flavour and Fragrance Journal* 21(1), 186–191.

Kitajima, J., Ishikawa, T., Fujimatu, E., Kondho, K. and Takayanagi, T. (2003) Glycosides of 2-C-methyl-D-erythritol from the fruits of anise, coriander and cumin. *Phytochemistry* 62(1), 115–120.

Kubo, I. and Kinst-Hori, I. (1998) Tyrosinase inhibitors from cumin. *Journal of Agricultural and Food Chemistry* 46(12), 5338–5341.

Lawrence, B.M. (1992) Progress in essential oils. *Perfumer and Flavourist* 17(4), 42–44.

Levin, D.F.E., Hollstein, M., Christman, M.F., Schwiers, E.A. and Ames, B.N. (1982) A new Salmonella tester strain (TA 102) with AT base pairs at the site of mutation detects oxidative mutagens. *Proceedings of the National Academy of Sciences, USA* 79, 7445–7449.

Mahyar, J., Farshad, N., Samira, D. and Mohammad, K. (2006) Effects of the fruit essential oil of *Cuminum cyminum* Linn. (Apiaceae) on pentylenetetrazol-induced epileptiform activity in F1 neurones of *Helix aspersa*. *Journal of Ethnopharmacology* 104(1–2), 278–282.

Naik, S.N., Lentz, H. and Maheshwari, R.C. (1989) Extraction of perfumes and flavors from plant materials with liquid carbon-dioxide under liquid-vapor-equilibrium conditions. *Fluid Phase Equilibria* 49, 115–126.

Nalini, N., Sabitha, K., Vishwanathan, P. and Menon, V.P. (1998) Influence of spices on the bacterial (enzyme) activity in experimental colon-cancer. *Journal of Ethnopharmacology* 62, 15–24.

Peter, K.V. (2000) Diversification vital. In: Ram, N. (ed.) *The Survey of Indian Agriculture*. Kasturi and Sons, Madras, India, pp. 83–84.

Potty, S.N. and Krishnakumar, V. (2001) Organic farming in seed spices in relation to quality and export. In: Agrawal, S., Sastry, E.V.D. and Sharma, R.K. (eds) *Seed Spices – Production, Quality, Export*. Pointer Publishers, Jaipur, India, pp. 154–171.

Pruthi, J.S. and Misra, B.D. (1963) Physico-chemical and microbiological changes in curry powders during drying, milling and mixing operations. *Spices Bulletin* 3(3–5), 8.

Singh, G. and Upadhyay, R.K. (1991) Fungitoxic activity of cumaldehyde, main constituent of the *Cuminum cyminum* oil. *Fitoterapia* 62(1), 86.
Singh, G., Kapoor, I.P.S., Pandey, S.K. and Singh, G. (1998) Studies on essential oils – XIII: natural antioxidant for sunflower oil. *Journal of Scientific and Industrial Research* 57(3), 139–142.
Singh, G., Marimuthu, P., Lampasona, M.P. de and Catalan, C.A.N. (2006) *Cuminum cyminum* L. Chemical constituents, antioxidant and antifungal studies on its volatile oil and acetone extract. *Indian Perfumer* 50(3), 31–39.
Srivastava, K.C. (1989) Extracts from two frequently consumed spices – cumin (*Cuminum cyminum*) and turmeric (*Curcuma longa*) – inhibit platelet aggregation and alter eicosanoid biosynthesis in human blood platelets. *Prostaglandins, Leukotrienes and Essential Fatty Acids* 37(1), 57–64.
Takayanagi, T., Ishikawa, T. and Kitajima, J. (2003) Sesquiterpene lactone glucosides and alkyl glycosides from the fruit of cumin. *Phytochemistry* 63(4), 479–484.
Tunç, I., Berger, B.M., Erler, F. and Dagli, F. (2000) Ovicidal activity of essential oils from five plants against two stored product insects. *Journal of Stored Product Research* 36, 161–168.
Vedamuthu, P.G.B., Khader, Md.A. and Rajan, F.S. (1994) Improvement of seed spices. In: Chadha, K.L. and Rethinam, P. (eds) *Advances in Horticulture – Plantation and Spice Crops*, Volume 9, Part 1. Malhotra Publishing House, New Delhi, pp. 345–374.
Wang, Z., Ding, L., Li, T., Zhou, X., Wang, L., Zhang, H., Liu, L., Li, Y., Liu, Z., Wang, H., Zeng, H. and He, H. (2006) Improved solvent-free microwave extraction of essential oil from dried *Cuminum cyminum* L. and *Zanthoxylum bungeanum* Maxim. *Journal of Chromatography* 1102(1–2), 11–17.
Weiss, E.A. (2002) Umbelliferae. In: *Spice Crops*. CAB International, Wallingford, UK, pp. 261–268.

12 Fennel

Shamina Azeez

12.1. Introduction

Fennel (*Foeniculum vulgare* Mill.) belongs to the family Apiaceae (formerly the Umbelliferae). It is native to southern Europe and the Mediterranean region and is cultivated mainly in India, Rumania, Russia, Germany, France, Italy, Japan, Argentina and the USA. India's export of fennel has improved slightly in the years 2001/02, 2002/03 and 2003/04, the value of which is given in Table 12.1.

Etymologically, the word fennel developed from Middle English *fenel*, *feny*; Anglo-Saxon *fenol*, *finol*, from Latin *feniculum*, *fœniculum*, diminutive of *fenum*, *fœnum*, meaning 'hay'. In Ancient Greek, fennel was called *marathon* and is attested in Linear B tablets as *ma-ra-tu-wo*. This is the origin of the place name, Marathon (meaning 'place of fennel'), site of the Battle of Marathon in 490 BC. Greek mythology claims Prometheus used the stalk of a fennel plant to steal fire from the gods. In medieval times, fennel was used in conjunction with St John's wort to keep away witchcraft and other evil things. This might have originated because fennel can be used as an insect repellent. Fennel is thought to be one of the nine herbs held sacred by the Anglo-Saxons (Duke, 2000).

12.2. Botany and Uses

Botany

Weiss (2002) describes the botany of the species in detail, the salient features of which are given here. *Foeniculum* is stated to have three species, *F. vulgare* (fennel), *F. azoricum* Mill. (Florence fennel) and *F. dulce* (sweet fennel). The basic chromosome number of the species is 11, thus fennel is a diploid with 2n = 22. It is a highly aromatic perennial herb, erect, glaucous green and grows to 2 m tall. The leaves grow up to 40 cm long; they are finely dissected, with the ultimate segments filiform, about 0.5 mm wide. The flowers are produced in terminal compound umbels 5–15 cm wide, each umbel section with 20–50 tiny yellow flowers on short pedicels. The fruit is a dry seed from 4–9 mm long, half as wide or less, and grooved.

Uses

Fennel is widely cultivated, both in its native habitat and elsewhere, for its edible, strongly flavoured leaves and seeds. The flavour is similar to that of anise and star anise, though usually not so strong. The taste of fennel

(eds V.A. Parthasarathy, B. Chempakam and T.J. Zachariah)

Table 12.1. Export of fennel from India.

		Value	
Year	Qty (Mt)	(Rs. Lakhs)	(US$ million)
2001/02	4374.41	1695.82	3.56
2002/03	4159.63	1783.75	3.69
2003/04	5200.00	2143.00	4.67

Source: www.indianspices.com.

varies from sweet to slightly bitter, without the anise flavour of wild fennel and closely related local types grown in Central Europe and Russia. The Florence fennel (*F. vulgare* Azoricum Group) is smaller than the wild type and is a selection with inflated leaf bases which form a sort of bulb that is eaten as a vegetable, both raw and cooked. It comes mainly from India and Egypt and has a mild anise-like flavour, but is sweeter and more aromatic. Its flavour comes from anethole, an aromatic compound also found in anise and star anise. There are several cultivars of Florence fennel, which is also known by several other names, notably the Italian name, *finocchio*. In North America, it is often mislabelled as 'anise' (Wetherilt and Pala, 1994).

Fennel has become naturalized along roadsides, in pastures and in other open sites in many regions, including northern Europe, Cyprus, the USA, southern Canada and in much of Asia and Australia. It is propagated by seed and is considered a weed in Australia and the USA (Bown, 2001).

12.3. General Composition

Extraction

In a comparative study on hydrodistillation and supercritical CO_2 (SC-CO_2) extraction of ground fennel seeds, the former possessed a less intense fennel seed aroma than extracts obtained by SC-CO_2 from organoleptic tests (Damjanović *et al.*, 2005). Optimal conditions of SC-CO_2 extraction (high percentage of *trans*-anethole, with significant content of fenchone and reduced content of methylchavicol and co-extracted cuticular waxes), as calculated by these researchers, are: pressure, 100 bar; temperature, 40°C; extraction time, 120 min.

Composition of oils

Bernath *et al.* (1994) analysed the fruit chemical composition and found it contained, on average, per 100 g edible portion: 8.8 g water; 15.8 g protein; 14.9 g fat; 36.6 g carbohydrates; 15.7 g fibre; and 8.2 g ash (containing 1.2 g Ca, 19 mg Fe, 1.7 g K, 385 mg Mg, 88 mg Na, 487 mg P and 28 mg Zn). The contents of vitamin A were: 135 IU; niacin 6 mg; thiamine 0.41 mg; riboflavin 0.35 mg; and energy value about 1440 kJ per 100 g. The fruit contains mucilage, sugars, starch, tannin, fixed oil and essential oil. The main components of the fixed oil are petroselenic, oleic, linoleic and palmitic acids.

The fruit contains a fixed oil from 15 to 30% and a volatile essential oil up to 12%. The fruit also contains flavonoids, iodine, kaempferols, umbelliferone and stigmasterol and ascorbic acid; traces of aluminium, barium, lithium, copper, manganese, silicon and titanium. A non-destructive method of determining oil constituents has been described by Fehrmann *et al.* (1996).

The chemical composition of fennel extracts obtained from supercritical fluid extraction (SFE) of dry-harvested, hydrodistilled and low-pressure solvent-extracted fennel seeds was determined by gas chromatography (Moura *et al.*, 2005). The SFE maximum global yield (12.5%, dry basis) was obtained with dry-harvested fennel seeds. Anethole and fenchone were the major constituents of the extract. The fatty acids, palmitic ($C_{16}H_{32}O_2$), palmitoleic ($C_{16}H_{30}O_2$), stearic ($C_{18}H_{36}O_2$), oleic ($C_{18}H_{34}O_2$), linoleic ($C_{18}H_{32}O_2$) and linolenic ($C_{18}H_{30}O_2$), were also detected.

Parejo *et al.* (2004) identified caffeoylquinic acids, dicaffeoylquinic acids, flavonoids and rosmarinic acid among ten main antioxidant phenolic compounds from bitter fennel, *F. vulgare*, using a simple high-performance liquid chromatography (HPLC). Distilled fennel was found to contain a higher proportion of antioxidant phenolic compounds than non-distilled plant material.

Muckensturm *et al.* (1997) characterized different populations of *F. vulgare* containing 10-nonacosanone as a specific chemical marker. *F. vulgare* subsp. *piperitum* is characterized by the presence of rotundifolone. *p*-Butylanisole is present in traces in fennel which contains a large amount of *trans*-anethole. A chemotaxonomic classification based on the amount of estragole, *trans*-anethole, limonene and fenchone was proposed by the authors for the different varieties and chemotypes of *F. vulgare* subsp. *Vulgare*.

Harborne and Saleh (1971) confirmed the presence of quercetin 3-arabinoside in the leaves of fennel and three other flavonol glycosides, kaempferol 3-arabinoside, kaempferol 3-glucuronide and quercetin 3-glucuronide. A chemotypic characterization of populations of fennel based on the occurrence of glycosides has been attempted. The dried distillation residue of fennel fruits contains 14–22% protein and 12–18% fat and is suitable for stock feed (Weiss, 2002).

12.4. Chemistry

Volatiles

Extraction

The largest quantity of herbal essential oil is obtained by hydrodistilling fresh or slightly wilted foliage just before flowering (Bellomaria *et al.*, 1999). Fruits can be distilled any time after harvest, but they must be milled or crushed and distilled immediately to avoid oil loss by evaporation. The temperature must be high enough to prevent the oil from congealing. Essential oil from different plant parts and between different regional cultivars tends to be very variable (Karaca and Kevseroglu, 1999; Piccaglia and Marotti, 2001). In European and Argentinean types of *F. vulgare*, limonene concentration in the whole plant does not exceed 10%, but α-phellandrene in leaves is between 23 and 25% and in stems between 22 and 28%. In contrast, the limonene content in young leaves and stems of European and Indian types of *F. dulce* is 37–40% and 28 and 34%, respectively, decreasing with age. The α-phellandrene content is low (1–4%) and remains constant with age. The composition of sweet and bitter fennel oil is given in Table 12.2.

Table 12.2. Composition of sweet and bitter fennel oil.

	Fennel oil (%)	
Component	Sweet fennel	Bitter fennel
α-Phellandrene	–	12.98
α-Pinene	4.03	18.10
Anethole	52.03	47.97
Estragole	2.53	8.31
Fenchol	3.18	–
Fenchone	2.67	2.84
Limonene	28.92	–

Source: Karlsen *et al.* (1969).

In the mature fruit, up to 95% of the essential oil is located in the fruit, greater amounts being found in the fully ripe fruit. Hydrodistillation yields 1.5–35.0%. Generally, anethole and fenchone are found more in the waxy and ripe fruits than in the stems and leaves, whereas α-pinene is found more in the latter. A comparison of the composition of fennel oils from flowers and seeds is given in Table 12.3. Wide variations are seen in the content and composition of the oils based on cultivar and geographical origin (Akgül, 1986; Kruger and Hammer, 1999). Miraldi (1999) reported inverse proportions

Table 12.3. Composition of fennel oils from flowers and seeds.

	Fennel oil (%)	
Component	Flowers	Seeds
α-Pinene	5.0	1.4
Anethole	55.5	72.0
Anisaldehyde	1.8	0.5
β-Pinene	1.2	0.3
Estragole	14.6	12.0
Fenchone	5.6	10.5
Limonene	9.0	1.4
Myrcene	3.0	1.3
p-Cymene	4.0	0.6
Unidentified	0.3	–

Source: Retamar (1986).

of *trans*-anethole and estragole, suggesting a common precursor.

Gámiz-Gracia and de Castro (2000) devised a subcritical extractor equipped with a three-way inlet valve and an on/off outlet valve to perform subcritical water extractions in a continuous manner for the isolation of fennel essential oil. The target compounds were removed from the aqueous extract by a single extraction with 5 ml hexane, determined by gas-chromatography-flame ionization and identified by mass spectrometry. This extraction method is superior to both hydrodistillation and dichloromethane manual extraction in terms of rapidity, efficiency, cleanliness and the possibility of manipulating the composition of the extract.

Composition of oil

In India, small seeds generally had higher oil content than larger seeds and the main characteristics were: specific gravity (15°C), 0.9304; refractive index (15°C), 1.4795; optical rotation, +35°; saponification value, 181.2; iodine value (Wijs), 99; unsaponified material, 3.7%. The expressed oil is classified as semi-drying and is a source of lauric and adipic acids (Weiss, 2002). Table 12.4 gives the average physico-chemical properties of fennel volatile oil.

Approximately 45 constituents have been determined from fennel seed oil (Fig. 12.1), the main constituents being *trans*-anethole (60–65%, but up to 90%), fenchone (2–20%), estragol (methyl chavicol), limonene, camphene, α-pinene and other monoterpenes, fenchyl alcohol and anisaldehyde. The major compounds in supercritical

Table 12.4. Physico-chemical properties of fennel volatile oil.

Parameter	Value
Colour of oil	Colourless or pale yellow
Specific gravity	0.889–0.921
Refractive index	1.484–1.568
Optical rotation	+20° to + 58°

Source: Agrawal (2001).

CO_2 and hydrodistilled extracts of ground fennel seeds were *trans*-anethole (68.6–75.0 and 62.0%, respectively), methylchavicol (5.09–9.10 and 4.90%, respectively), fenchone (8.4–14.7 and 20.3%, respectively), respectively (Damjanović *et al.*, 2005).

The yield and composition of the volatile fraction of the pentane extracts of leaves, stems and seeds of *F. vulgare* Mill. have been studied by Guillén and Manzanos (1996). The yield obtained from seeds was much higher than that obtained from leaves and stems. The volatile fraction of the pentane extract of the latter two has a higher concentration of terpene hydrocarbons and a smaller concentration of oxygenated terpene hydrocarbons than that of the seeds. Sesquiterpenes and the antioxidant vitamin E have been detected in the leaves and petroselinic acid in the seeds. Saturated aliphatic hydrocarbons with 25 or more carbon atoms have been found in all the plant parts.

Akgül and Bayrak (1988) reported the volatile oil composition of various parts of bitter fennel (*F. vulgare* var. *vulgare*) growing as wild Turkish plants, investigated by gas-liquid chromatography. The major component of all oil samples was *trans*-anethole (29.70, 37.07, 54.22, 61.08 and 64.71% in leaf, stem, flowering umbel, flower and fruit, respectively). The other main components were α-pinene (in leaf, stem, flowering umbel and flower), α-phellandrene (in leaf, stem and flowering umbel) and fenchone (fruit oil). The volatile oils of flowering umbel, flower and fruit contained high amounts of oxygenated compounds, in gradually increasing percentages. Harborne *et al.* (1969) reported for the first time that the psychotropic aromatic ether myristicin occurred in the seed of cultivated fennel but was absent from wild collections of this species.

The root essential oil contains (on average) α-pinene (1.0%), *p*-cymene (0.3%), β-fenchylacetate (1.0%), *trans*-anethole (1.6%), eugenol (0.2%), myristicin (3%) and dillapiole (87%). On the other hand, the root and bulbous stem base of Florence fennel contains less than 1% of dillapiole but 70% of *trans*-anethole, giving a very different taste. The herbage contains 1.00–2.55% essential oil, up to 75% of which is *trans*-anethole. Anethole and fenchone

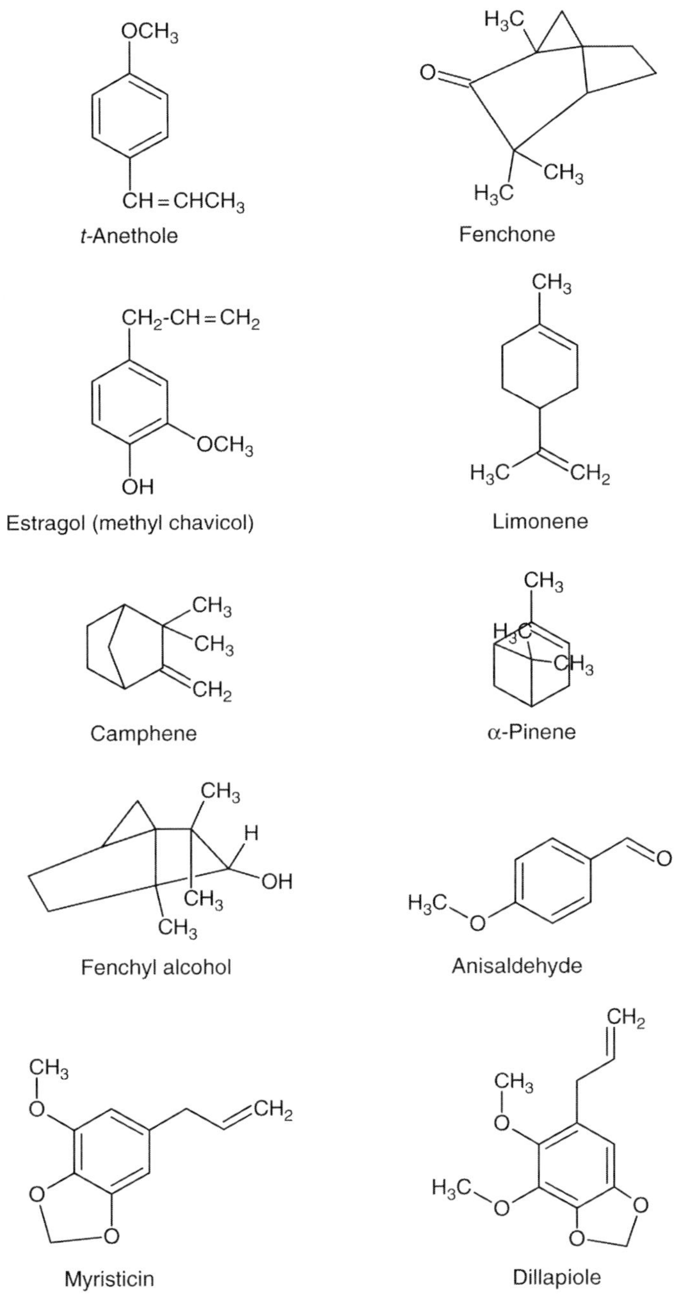

Fig. 12.1. Volatile components in fennel.

concentrations increase from bud stage to fruit ripening, α-pinene and limonene concentrations decrease and estragole concentration remains constant.

Kapoor *et al.* (2004) reported that two arbuscular mycorrhizal (AM) fungi – *Glomus macrocarpum* and *G. fasciculatum* – improved growth and essential oil concentration of fennel significantly (the latter registered a 78% increase in essential oil concentration over non-mycorrhizal control); AM inoculation of plants along with phosphorus fertilization enhanced growth, P-uptake and essential oil content of plants significantly compared with either of the components applied separately. The essential oil characterization by gas-liquid chromatography revealed that the level of anethol was enhanced significantly on mycorrhization.

Biosynthesis

The synthesis of the major essential oil components, estragole and anethole, has been elucidated. Cell-free extracts from bitter fennel tissues display *O*-methyltransferase activities able to methylate chavicol and *t*-anol *in vitro* to produce estragole and *t*-anethole, respectively, using *S*-adenosyl-L-methionine as a methyl group donor (Gross *et al.*, 2002). An association between estragole accumulation and chavicol *O*-methyltransferase activity during the development of different plant parts was found. Young leaves had greater *O*-methyltransferase activity than old leaves. In developing fruits, *O*-methyltransferase activity levels increased until the wasting stage and then decreased drastically.

The metabolism of *l*-*endo*-fenchol to *d*-fenchone in fennel has been studied in quite some detail by Croteau and co-workers (Croteau and Felton, 1980). Croteau *et al.* (1980a) later reported a soluble enzyme preparation from the leaves of fennel which catalysed the cation-dependent cyclization of both geranyl pyrophosphate and neryl pyrophosphate to the bicyclic rearranged monoterpene *l*-*endo*-fenchol. Croteau *et al.* (1980b) found that (+)-(1*S*)-fenchone, an irregular bicyclic monoterpene ketone thought to be derived via rearrangement of a bicyclic precursor, was one of the major terpenoids of the volatile oil of fennel. They could provide strong evidence that fenchone was derived by the cyclization of geranyl pyrophosphate or neryl pyrophosphate to *endo*-fenchol, followed by dehydrogenation of this bicyclic alcohol, and demonstrated the biosynthesis of a rearranged monoterpene in a cell-free system. Croteau *et al.* (1989) elaborated on the biosynthesis of monoterpenes in fennel, geranyl pyrophosphate: (−)-*endo*-fenchol cyclase catalyses the conversion of geranyl pyrophosphate to (−)-*endo*-fenchol by a process thought to involve the initial isomerization of the substrate to the tertiary allylic isomer, linalyl pyrophosphate, and the subsequent cyclization of this bound intermediate.

Quantitative and qualitative assay

Many techniques are followed to identify and quantify the components of fennel essential oil. Križman *et al.* (2006) developed a headspace-gas chromatography method for analysing the major volatile constituents in fennel fruits and leaves – α-pinene, α-phellandrene, limonene, fenchone, estragole and *trans*-anethole.

Betts (1993) reported that 3% bis-methoxybenzilidinebitoluidine $(MBT)_2$ on 'Graphpac' was preferable for assaying sweet fennel oil by providing a more reliable melted liquid crystal stationary phase, with low temperature versatility. Betts (1992) reported earlier that the toroid (or a liquid crystal) phase might be useful for resolving some terpene hydrocarbons in sweet fennel and mace oils and identifying peaks by mass spectra and retention times; and the liquid crystal, the choice for some aromatics, which include minor toxic oil constituents, compared with conventional phases. Betts *et al.* (1991) used the liquid crystal bismethoxybenzilidine-bitoluidine (BMBT) initially as the stationary phase for the gas chromatographic study of some aromatics and a monoterpenoid constituent of fennel volatile oils, which gave best results when used below its melting point of about 180°C. Changes in the sequence of retentions (terpineol-estragole and anetholethymol 'shifts') suggested this liquid crystal might operate by three different mechanisms, dependent on the column treatment.

Pope *et al.* (1991) applied chemical-shift-selective imaging at microscopic resolution of various plant materials, including dried and undried fruits of fennel, to the study of selective imaging of aromatics and carbohydrates, water and oil. The non-invasive nature of the method gives it advantages over established methods of plant histochemistry, which involve sectioning and staining to reveal different chemical constituents.

Chemistry of non-volatiles

Oleoresins

Fennel oleoresin is prepared by solvent extraction of whole seeds and normally contains a volatile oil of 50% or a guaranteed content in the range of 52–58%. Only small quantities are produced for specific uses as it is not a substitute for fennel oil. Chemical analysis by Barazani *et al.* (2002) of the volatile fraction of oleoresins from fruits of seven natural populations of *F. vulgare* var. *vulgare* (bitter fennel) from the wild and after cultivation indicated the presence of two groups of populations. Chemotypic differentiation (relative contents of estragole and *trans*-anethole) or phenotypic plasticity increases within-species chemical variability, but the specific ecological roles of these essential oils remain to be uncovered.

Fixed oils

Of the fatty acid in the fixed oil, most of which is contained in the polygonal cells in the seed endosperm, total monounsaturated acids account for 10% and total polyunsaturated fatty acids 2%. The main components of an expressed oil are petroselinic acid (up to 75%), oleic acid (up to 25%), linoleic acid (up to 15%) and palmitic acid (up to 5%) (Weiss, 2002).

12.5. Culinary, Medicinal and Other Uses

Culinary uses

The bulb, foliage and seeds of the fennel plant all have secure places in the culinary traditions of the world, especially in India and the Middle East. Fennel pollen is the most potent form of fennel, but it is exceedingly expensive. Dried fennel seed is an aromatic, anise-flavoured spice; the seeds are brown or green in colour when fresh and turn slowly to a dull grey as the seed ages. Green seeds are optimal for cooking.

Fennel seeds are sometimes confused with aniseed, which is very similar in taste and appearance, though smaller. Indians often chew fennel seed as a mouth-freshener. Fennel is also used as a flavouring in natural toothpaste. Some people employ it as a diuretic, while others use it to improve the milk supply of breastfeeding mothers.

In India, it is an essential ingredient in the Bengali spice mixture *panch phoron* and in Chinese five-spice powders. In the west, fennel seed is a very common ingredient in Italian sausages and northern European rye breads. Many egg, fish and other dishes employ fresh or dried fennel leaves. Florence fennel is a key ingredient in some Italian and German salads, often tossed with chicory and avocado, or it can be braised and served as a warm side dish. One may also blanch and/or marinate the leaves, or cook them in risotto. In all cases, the leaves lend their characteristically mild, anise-like flavour.

Pharmacological properties

Fennel contains anethole, an antispasmatic, along with other pharmacologically active substances. The various scientifically documented medicinal effects of fennel are listed below.

Antioxidant activity

Water and ethanol extracts of fennel seeds show strong antioxidant activity *in vitro*

(Oktay *et al.*, 2003). One hundred μg of water and ethanol extracts exhibit 99.1% and 77.5% inhibition of peroxidation in the linoleic acid system, respectively, which is greater than the same dose of α-tocopherol (36.1%), a natural antioxidant. Both extracts of fennel have effective reducing power, free radical scavenging, superoxide anion radical scavenging, hydrogen peroxide scavenging and metal-chelating activities, which are directly proportional to the concentration of the sample. Indications are that the fennel seed is a potential source of natural antioxidant.

Anticancer property

Anetholes from fennel, anise and camphor are among the several dietary factors that have the potential to be used to prevent and treat cancer (Aggarwal and Shishodia, 2006). Essential oil of fennel is included in some pharmacopoeias. It is used traditionally in drugs to treat chills and stomach problems.

Antimicrobial property

Croci *et al.* (2002) evaluated the capacity of various fresh vegetables that generally are eaten raw to adsorb hepatitis A virus (HAV) on the surface, and the persistence of the virus. Of the vegetables studied – lettuce, fennel and carrot – lettuce consistently was found to contain the highest quantity of virus; of the other two vegetables, a greater decrease was observed and complete inactivation had occurred at day 4 for carrot and at day 7 for fennel. For all three vegetables, washing did not guarantee a substantial reduction in the viral load.

A combination of oils of fennel, anise or basil with either benzoic acid or methyl-paraben was tested against *Listeria monocytogenes* and *Salmonella enteriditis*. *S. enteriditis* was more sensitive to inhibition by a combination of oil of anise, fennel or basil with methyl-paraben where there was < 10 CFU/ml after 1 h. *L. monocytogenes* was less sensitive to inhibition by each combination; however, there was a significant reduction in growth. Synergistic inhibition by one or more combinations was evident against each microorganism (Fyfe *et al.*, 1998).

Effect on muscles

The effect of commercial essential oils of celery, sage, dill, fennel, frankincense and nutmeg on rat skeletal muscles involved a contracture and inhibition of the twitch response to nerve stimulation, at final bath concentrations of 2×10^{-5} and 2×10^{-4} g/ml (Lis-Balchin and Hart, 1997).

As a relief from nausea

Gilligan (2005) used a variety of aromatherapy treatments on patients suffering from the symptom of nausea in a hospice and palliative care programme, using a synergistic blend of *Pimpinella anisum* (aniseed), *F. vulgare* var. *dulce* (sweet fennel), *Anthemis nobilis* (Roman chamomile) and *Mentha x piperita* (peppermint). The majority of patients who used the aromatherapy treatments reported relief, as per measurements on the Bieri scale, a visual-numeric analogue. Since the patients were also on other treatments for their symptoms, it was impossible to establish a clear scientific link between the aromatherapy treatments and nausea relief, but the study suggested that the oils used in this aromatherapy treatment were successful complements to the relief of this symptom.

Hepatoprotective effect

The hepatotoxicity produced by acute carbon tetrachloride-induced liver injury was found to be inhibited by essential oil from fennel, as evidenced by decreased levels of serum aspartate aminotransferase, alanine aminotransferase, alkaline phosphatase and bilirubin (Özbek *et al.*, 2003).

A greater amount of biliary solids and pronouncedly higher rate of secretion of bile acids were caused by various spices including fennel, probably contributing to the digestive stimulant action of the test spices (Patel and Srinivasan, 2000).

Gershbein (1977) reported increases in the liver increment (the amount of tissue regenerated) in partially hepatectomized rats, by subcutaneous (sc) injection of oils of anise, fennel, tarragon, parsley seed, celery seed and oleoresin, nutmeg, mace, cumin and sassafras and of the aromatic principles, 4-allylanisole, 4-propenylanisole, *p*-isopropylbenzaldehyde, safrole and isosafrole. Many of the agents effective by the sc route were also active when added to the diet.

Reduction in food transit time

Patel and Srinivasan (2001) reported a significant shortening of the food transit time when some prominent dietary spices including fennel were added to the diet.

As a treatment for primary dysmenorrhoea

In a study comparing the efficacy of the drug mefenamic acid against the essence of fennel seeds, Jahromi *et al.* (2003) found that the latter could be used as a safe and effective herbal drug for primary dysmenorrhoea; however, it may have a lower potency than mefenamic acid in the dosages used for this study (2% concentration). Both drugs relieved menstrual pain effectively; the mean duration of initiation of action was 67.5 ± 46.06 min for mefenamic acid and 75 ± 48.9 min for fennel.

Increased ectopic uterine motility is the major reason for primary dysmenorrhoea and its associated symptoms, like pain. Treatments include long-term therapy, where a combination of oestrogens and progestins is used; in short-term therapy, non-steroidal anti-inflammatory drugs (NSAIDs) are sometimes used. Most NSAIDs in long-term therapy show severe adverse effects. Ostad *et al.* (2001) used fennel essential oil (FEO) in an attempt to find agents with less adverse effect. Administration of different doses of FEO reduced the intensity of oxytocin and PGE_2-induced contractions significantly (25 and 50 μg/ml for oxytocin and 10 and 20 μg/ml PGE_2, respectively). FEO also reduced the frequency of contractions induced by PGE_2 but not with oxytocin. The estimated LD_{50} was 1326 mg/kg. No obvious damage was observed in the vital organs of the dead animals.

Antihirsutism activity

Idiopathic hirsutism is the occurrence of excessive male-pattern hair growth in women who have a normal ovulatory menstrual cycle and normal levels of serum androgens. It may be a disorder of peripheral androgen metabolism. Javidnia *et al.* (2003) evaluated the clinical response of idiopathic hirsutism to topical application of creams containing 1 and 2% of fennel extract, which has been used as an oestrogenic agent, by measuring the hair diameter and rate of growth. The efficacy of the cream containing 2% fennel was better than the cream containing 1% fennel and these two were more potent than the placebo. The mean values of hair diameter reduction were 7.8, 18.3 and −0.5% for patients receiving the creams containing 1, 2 and 0% (placebo), respectively.

Acaricidal activity

Lee *et al.* (2006) reported the acaricidal activities of components derived from fennel seed oils against *Tyrophagus putrescentiae* adults using direct contact application and compared with compounds such as benzyl benzoate, dibutyl phthalate and *N,N*-diethyl-*m*-toluamide. The bioactive constituent of the fennel seeds was characterized as (+)-carvone by spectroscopic analyses. The most toxic compound to *T. putrescentiae* was naphthalene, followed by dihydrocarvone, (+)-carvone, (−)-carvone, eugenol, benzyl benzoate, thymol, dibutyl phthalate, *N,N*-diethyl-*m*-toluamide, methyl eugenol, myrcene and acetyleugenol, on the basis of LD_{50} values.

Is fennel teratogenic?

The need to clarify the safety of the use of FEO was addressed by Ostad *et al.* (2004), since its use as a remedy for the control of primary dysmenorrhoea increased concern about its potential teratogenicity due to its oestrogen-like activity. The authors used limb bud mesenchymal cells (which have been used extensively for *in vitro* studies of chondrogenesis since, when grown in high-density cultures, these cells can differentiate into a number of cell types) and the Alcian blue staining method (which is specific for staining cartilage proteoglycan) to determine the teratogenic effect of FEO. Limb bud cells obtained from day 13 rat embryo were cultivated and exposed to various concentrations of FEO for 5 days at 37°C and the number of differentiated foci were counted, against a positive standard control – retinoic acid. The differentiation was also evaluated using limb bud micromass culture using immunocytochemical techniques and BMP-4 antibody. The results showed that FEO at concentrations as low as 0.93 mg/ml produced a significant reduction in the number of stained differentiated foci. However, this reduction was due to cell loss, determined by neutral red cell viability assay, rather than due to decrease in cell differentiation. These findings suggest that the FEO at the studied concentrations may have a toxic effect on fetal cells, but there was no evidence of teratogenicity.

Estragole, a natural constituent of tarragon, sweet basil and sweet fennel, is used widely in foodstuffs as a flavouring agent. Several studies, as detailed in the review by De Vincenzi *et al.* (2000), have shown the carcinogenicity of estragole. The 1-hydroxy metabolites are stronger hepatocarcinogens than the parent compound. Controversial results are reported for the mutagenicity of estragole. However, the formation of hepatic DNA adducts *in vivo* and *in vitro* by metabolites of estragole has been demonstrated.

Sekizawa and Shibamoto (1982) reported the mutagenicity of anethole present in fennel from their studies. Stich *et al.* (1981) examined the clastogenic activities (substances or processes which cause breaks in chromosomes) of quercetin from fennel seeds and the ubiquitous transition metal Mn^{2+} – individually and in various combinations. The clastogenic effects of the simultaneous application of arecoline from betel nut, plus quercetin, were greater than the action of quercetin alone.

Fennel as a food allergen

Changes in dietary habits and the internationalization of foods have led to the increasingly frequent use of spices. Children with allergy symptoms to spices were evaluated, by prick tests using the basic foodstuff, crushed or diluted in saline, for aniseed, cinnamon, coriander, cumin, curry, fennel, nutmeg, paprika, sesame and vanilla; labial and/or challenge tests were performed for certain spices (mustard, fennel) by Rancé *et al.* (1994). The spices responsible for sensitization (found in 46% of cases) were mustard, fennel, coriander, cumin and curry. Fennel was responsible for a case of recurrent angio-oedema (positive labial challenge test). Mustard and fennel are incriminated most frequently and are also responsible for clinical manifestations. Avoidance of these allergens in the diet is made difficult by masking in mixtures of spices or in prepared dishes.

12.6. Quality Aspects

Of the 15 spices marketed in India and screened by Saxena and Mehrotra (1989) for the mycotoxins, aflatoxin, rubratoxin, ochratoxin A, citrinin, zearalenone and sterigmatocystin, samples of coriander and fennel were found to contain the largest number of positive samples and mycotoxins. Other spices like cinnamon, clove, yellow mustard and Indian mustard did not contain detectable amounts of the mycotoxins tested. Aflatoxins are the most common contaminants in the majority of samples, levels being higher than the prescribed limit for human consumption.

The main products from fennel are the green or dried herb, dried fruit or fennel seed, herb and seed oils. The products are elaborated upon below.

Herb

The green herb is used for flavour during cooking or prior to serving. The dried herb is inferior in quality compared with the freeze-dried or frozen ones. The major flavour component is anethole, which gives the herb the odour and flavour of anise.

Herb oil

The use of steam-distilled herb oil from whole plants is declining and few recent reports are available. The oil from fresh or wilted herbage is a nearly colourless to pale yellow mobile liquid, which may darken with time; it lacks the anise odour and the taste is bitter. The main characteristics are: specific gravity (15°C), 0.893–0.925; refractive index (20°C), 1.484–1.508; optical rotation, +40° to +68°; soluble in 0.5–1.0 volumes 90% alcohol (Guenther, 1982).

Seed

Fennel seed is a major culinary and processing spice, used whole or ground, for culinary purposes. The highest average maximum level in the USA is about 0.12% (1190 ppm) in meat and meat products. Quality seeds have a bitter, camphoraceous taste and a pungent odour. It is also used widely in Arab, Chinese and Ayurvedic medicine; its various clinical effects have been detailed in the relevant section above.

Table 12.5. Quality specifications for fennel.

Parameter	Specifications
ASTA Cleanliness Specifications[1]	
Whole insects, dead (by count)	*
Mammalian excreta (mg/lb)	*
Other excreta (mg/lb)	*
Mould (% by weight)	1
Insect-defiled/infested (% by weight)	1
Extraneous foreign matter (% by weight)	0.5
Food and Drug Administration (FDA) Defect Action Levels (DAL)	
Adulteration with mammalian excreta (mg/lb)	3
Volatile oil (% min)	1.5
Moisture[2] (% max)	10.0
Ash (% max)	9.0
Acid-insoluble ash (% max)	1.0
Average bulk index (mg/100 g)	210.0
Defect Action Levels prescribed by USFDA[3]	
Insects (MPM-V32)	20% or more subsamples contain insects
Mammalian excreta	20% or more subsamples or average of more than 3 mg of mammalian excreta per pound

[1]*Source*: Anon. (1991);
[2]ASTA suggested minimum level;
[3]*Source*: Potty and Krishnakumar (2001).
Note: *If more than 20% of the subsamples contain rodent, excreta or whole insects, or an average of 3 mg/lb of mammalian excreta, the lot must be reconditioned.

Seed oil

Fennel seed oil is usually obtained by steam distilling whole or crushed fruit, yielding 1.5–6.5% oil or, more recently, by supercritical carbon dioxide extraction. Generally, there is more oil in European varieties and less in Asian varieties. The oil is almost colourless to pale yellow and crystallizes on standing, so may require warming before use. The congealing temperature should not be below 3°C. The oil has a pleasant, aromatic, anise odour and a characteristic camphor-like taste, spicy and mildly bitter; Arctander (1960) placed the oil in the warm-phenolic, fresh herbaceous group. The oil is used mainly for flavouring food, tobacco and pharma products, in liqueurs, and in industrial perfumery to mask the odour of aerosols, disinfectants, insecticides, etc. The maximum permitted level in food is about 0.3%, but usually less than 0.1%; in perfumery and cosmetics it is 0.4%.

The major characteristics of commercial-grade fennel oil are: specific gravity (25°C), 0.953–0.973; refractive index (20°C),1.528–1.538; optical rotation (23°C), +12° to +24°; slightly soluble in water, soluble in 1.0 volume 90% or 8 volumes 80% alcohol, very soluble in chloroform and ether.

Sweet fennel oil

This is distilled from the fruit of *F. dulce*, its main constituents being limonene (20–25%), fenchone (7–10%) and *trans*-anethole (4–6%). Arctander (1960) placed the oil in the sweet, non-floral, candy-flavoured group. In the USA, the regulatory status generally recognized as safe has been accorded to fennel oil, GRAS 2481, and sweet fennel oil, GRAS 2483.

Anethole

Fennel oil, star anise and anise are natural sources of anethole, although synthetic substitutes are readily available. In many countries, the use of synthetic anethole in food products is illegal. Anethole can also be synthesized from estragole extracted from *Pinus* oil (Weiss, 2002).

The ASTA, FDA and USFDA standards for cleanliness in fennel are given in Table 12.5 and the quality specifications for whole and ground fennel in Table 12.6.

Table 12.6. Quality specifications for whole and ground fennel.

Parameter	Specification
Odour	It should have a warm, agreeable, sweet odour
Volatile oil	A minimum value of 1% in Germany, 3% in the Netherlands, 2% in the UK
Appearance	It should be a free-flowing seed
Colour	In Germany, the colour should be light green and light brownish-green
Aroma	Sweet aroma compared with a herby camphoraceous note
Packing	Whole seed is packed in jute bags; fennel powder is packed either in polywoven or jute bags with inner polylining

Source: Potty and Krishnakumar (2001).

12.7. Conclusion

In summary, *Foeniculum* is stated to have three species, *F. vulgare* (fennel), *F. azoricum* Mill. (Florence fennel) and *F. dulce* (sweet fennel). Fennel is widely cultivated, both in its native habitat and elsewhere, for its edible, strongly flavoured leaves and seeds. The flavour is similar to, but milder than, that of anise and star anise. Anethole and fenchone are the major constituents of the solvent extract of seed; phenols, free fatty acids, carbohydrates, proteins, vitamins and minerals have been reported in varying proportions. In the mature fruit, up to 95% of the essential oil is located in the fruit, greater amounts being found in the fully ripe fruit. Approximately 45 constituents have been determined from fennel seed oil, the main constituents being *trans*-anethole, fenchone, estragol (methyl chavicol), limonene, camphene, α-pinene and other monoterpenes, fenchyl alcohol and anisaldehyde. Fennel is an essential ingredient in the culinary traditions of the world. Many egg, fish and other dishes employ fresh or dried fennel leaves. It is also used in aromatherapy. Of the medicinal properties, it is recognized as antioxidant, hepatoprotective, anticancer, antimicrobial and as a treatment against nausea and primary dysmenorrhoea, among others; but the concern also remains of its teratogenic, mutagenic and food allergen properties. These properties are still to be reconfirmed, but the role of fennel in our culinary tradition is already firmly established. The main products from fennel are the seed, seed oil, herb, herb oil and anethole, for all of which quality specifications exist.

References

Aggarwal, B.B. and Shishodia, S. (2006) Molecular targets of dietary agents for prevention and therapy of cancer. *Biochemical Pharmacology* 71(10), 1397–1421.

Agrawal, S. (2001) Seed spices – an introduction. In: Agrawal, S., Sastry, E.V.D. and Sharma, R.K. (eds) *Seed Spices – Production, Quality, Export*. Pointer Publishers, Jaipur, India, pp. 1–18.

Akgül, A. (1986) Studies on the essential oils from Turkish fennel seeds. In: Brunke, J.E. (ed.) *Progress in Essential Oil Research*. W. de Gruyter and Co., New York, pp. 487–489.

Akgül, A. and Bayrak, A. (1988) Comparative volatile oil composition of various parts from Turkish bitter fennel (*Foeniculum vulgare* var. *vulgare*). *Food Chemistry* 30(4), 319–323.

Anon. (1991) *Clean Spices: A Guide Book for Shippers of Products to the U.S. Spice Trade*. ASTA, Washington, DC, pp. 4.

Arctander, S. (1960) *Perfume and Flavour Materials of Natural Origin*. Mrs. G. Arctander, 6665 Valley View Boulevard, Las Vegas, Nevada.

Barazani, O., Cohen, Y., Fait, A., Diminshtein, S., Dudai, N., Ravid, U., Putievsky, E. and Friedman, J. (2002) Chemotypic differentiation in indigenous populations of *Foeniculum vulgare* var. *vulgare* in Israel. *Biochemical Systematics and Ecology* 30(8), 721–731.

Bellomaria, B., Valentini, G. and Arnold, N. (1999) The essential oil of *F. vulgare*. *Rivista Italiano EPPOS* 27, 43–48.

Bernath, J., Kattaa, A., Nemeth, E. and Frank, R. (1994) Production-biological investigation of fennel (*Foeniculum vulgare*) populations of different genotypes. *Atti del Convegno Internazionale*, 287–292.

Betts, T.J. (1992) Possible value for the gas chromatographic analysis of essential oils of some unusual phase commercial capillaries. *Journal of Chromatography A* 626(2), 294–300.

Betts, T.J. (1993) Capability of a carbon support to improve the gas chromatographic performance of a liquid crystal phase in a packed column for some volatile oil constituents. *Journal of Chromatography A* 641(1,2), 189–193.

Betts, T.J., Charmaine, M.M. and Tassone, A.I. (1991) Use of a liquid crystal stationary phase at temperatures below its melting point for the gas chromatographic study of some volatile oil constituents. *Journal of Chromatography A* 547, 335–344.

Bown, D. (2001) *New Encyclopedia of Herbs and Their Uses*. The Herb Society of America, Dorling Kindersley Publishing, New York.

Croci, L., De Medici, D., Scalfaro, C., Fiore, A. and Toti, L. (2002) The survival of hepatitis A virus in fresh produce. *International Journal of Food Microbiology* 73(1), 29–34.

Croteau, R. and Felton, N.M. (1980) Substrate specificity of monoterpenol dehydrogenases from *Foeniculum vulgare* and *Tanacetum vulgare*. *Phytochemistry* 19(7), 1343–1347.

Croteau, R., Felton, M. and Ronald, R.C. (1980a) Biosynthesis of monoterpenes: conversion of the acyclic precursors geranyl pyrophosphate and neryl pyrophosphate to the rearranged monoterpenes fenchol and fenchone by a soluble enzyme preparation from fennel (*Foeniculum vulgare*). *Archives of Biochemistry and Biophysics* 200(2), 524–533.

Croteau, R., Felton, M. and Ronald, R.C. (1980b) Biosynthesis of monoterpenes: preliminary characterization of *l-endo*-fenchol synthetase from fennel (*Foeniculum vulgare*) and evidence that no free intermediate is involved in the cyclization of geranyl pyrophosphate to the rearranged product. *Archives of Biochemistry and Biophysics* 200(2), 534–546.

Croteau, R., Miyazaki, J.H. and Wheeler, C.J. (1989) Monoterpene biosynthesis: mechanistic evaluation of the geranyl pyrophosphate: (–)-*endo*-fenchol cyclase from fennel (*Foeniculum vulgare*). *Archives of Biochemistry and Biophysics* 269(2), 507–516.

Damjanović, B., Lepojević, Ž., Živković, V. and Tolić A. (2005) Extraction of fennel (*Foeniculum vulgare* Mill.) seeds with supercritical CO_2: comparison with hydrodistillation. *Food Chemistry* 92(1), 143–149.

De Vincenzi, M., Silano, M., Maialetti, F. and Scazzocchio, B. (2000) Constituents of aromatic plants: II. Estragole. *Fitoterapia* 71(6), 725–729.

Duke, J.A. (2000) *Handbook of Medicinal Herbs*, 2nd edition. CRC Press, Boca Raton, Florida.

Fehrmann, A., Schulz, H. and Pank, F. (1996) Non-destructive NIRS-measurement in caraway and fennel fruits. *Bundenanstalt fur Zuchtungforschund und Kulturpflazen* 2(1), 418–421.

Fyfe, L., Armstrong, F. and Stewart, J. (1998) Inhibition of *Listeria monocytogenes* and *Salmonella enteriditis* by combinations of plant oils and derivatives of benzoic acid: the development of synergistic antimicrobial combinations. *International Journal of Antimicrobial Agents* 9(3), 195–199.

Gámiz-Gracia, L. and de Castro, M.D.L. (2000) Continuous subcritical water extraction of medicinal plant essential oil: comparison with conventional techniques. *Talanta* 51(6), 1179–1185.

Gershbein, L.L. (1977) Regeneration of rat liver in the presence of essential oils and their components. *Food and Cosmetics Toxicology* 15(3), 173–181.

Gilligan, N.P. (2005) The palliation of nausea in hospice and palliative care patients with essential oils of *Pimpinella anisum* (aniseed), *Foeniculum vulgare* var. *dulce* (sweet fennel), *Anthemis nobilis* (Roman chamomile) and *Mentha x piperita* (peppermint). *International Journal of Aromatherapy* 15(4), 163–167.

Gross, M., Friedman, J., Dudai, N., Larkov, O., Cohen, Y., Ravid, E.B.U., Putievsky, E. and Lewinsohn, E. (2002) Biosynthesis of estragole and *t*-anethole in bitter fennel (*Foeniculum vulgare* Mill. var. *vulgare*) chemotypes. Changes in SAM:phenylpropene *O*-methyltransferase activities during development. *Plant Science* 163(5), 1047–1053.

Guenther, E. (1982) *The Essential Oils, Individual Essential Oils of the Plant Families Gramineae, Lauraceae, Burseraceae, Myrtaceae, Umbelliferae and Greraniaceae*, Volume 4. Van Nostrand, New York, pp. 634–645.

Guillén, M.D. and Manzanos, M.J. (1996) A study of several parts of the plant *Foeniculum vulgare* as a source of compounds with industrial interest. *Food Research International* 29(1), 85–88.

Harborne, J.B. and Saleh, N.A.M. (1971) Flavonol glycoside variation in fennel, *Foeniculum vulgare*. *Phytochemistry* 10(2), 399–400.

Harborne, J.B., Heywood, V.H. and Williams, C.A. (1969) Distribution of myristicin in seeds of the Umbelliferae. *Phytochemistry* 8(9), 1729–1732.

Jahromi, B.N., Tartifizadeh, A. and Khabnadideh, S. (2003) Comparison of fennel and mefenamic acid for the treatment of primary dysmenorrhea. *International Journal of Gynecology and Obstetrics* 80(2), 153–157.

Javidnia, K., Dastgheib, L., Samani, S.M. and Nasiri, A. (2003) Antihirsutism activity of fennel (fruits of *Foeniculum vulgare*) extract – a double-blind placebo controlled study. *Phytomedicine* 10(6–7), 455–458.

Kapoor, R., Giri, B. and Mukerji, K.G. (2004) Improved growth and essential oil yield and quality in *Foeniculum vulgare* mill on mycorrhizal inoculation supplemented with P-fertilizer. *Bioresource Technology* 93(3), 307–311.

Karaca, A. and Kevseroglu, K. (1999) The research of some important agricultural characteristics of coriander (*C. sativum*) and fennel (*F. vulgare*) varieties of Turkish origin. *Onodokuzmayis Universitesi Fakultesi Dergisi* 14(2), 65–77.

Karlsen, J., Baerheim, S.A., Chingova, B. and Zolatovich, G. (1969) Fruits of Foeniculum species and their essential oils. *Planta Medicine* 1793, 281–293.

Križman, M., Baričevič, D. and Prošek, M. (2006) Fast quantitative determination of volatile constituents in fennel by headspace-gas chromatography. *Analytica Chimica Acta* 557(1–2), 267–271.

Kruger, H. and Hammer, K. (1999) Chemotypes of fennel. *Journal of Essential Oil Research* 11(1), 79–82.

Lee, C.-H., Sung, B.-K. and Lee, H.-S. (2006) Acaricidal activity of fennel seed oils and their main components against *Tyrophagus putrescentiae*, a stored-food mite. *Journal of Stored Products Research* 42(1), 8–14.

Lis-Balchin, M. and Hart, S. (1997) A preliminary study of the effect of essential oils on skeletal and smooth muscle *in vitro*. *Journal of Ethnopharmacology* 58(3), 183–187.

Miraldi, E. (1999) Comparison of the essential oil from ten *F. vulgare* fruit samples of different origins. *Flavour and Fragrance Journal* 14(6), 379–382.

Moura, L.S., Carvalho, R.N. Jr., Stefanini, M.B., Ming, L.C. and Meireles, M.A.A. (2005) Supercritical fluid extraction from fennel (*Foeniculum vulgare*): global yield, composition and kinetic data. *The Journal of Supercritical Fluids* 35(3), 212–219.

Muckensturm, B., Foechterlen, D., Reduron, J.-P., Danton, P. and Hildenbrand, M. (1997) Phytochemical and chemotaxonomic studies of *Foeniculum vulgare*. *Biochemical Systematics and Ecology* 25(4), 353–358.

Oktay, M., Gülin, I. and Küfreviolu, Ö. (2003) Determination of *in vitro* antioxidant activity of fennel (*Foeniculum vulgare*) seed extracts. *Lebensmittel-Wissenschaft und-Technologie* 36(2), 263–271.

Ostad, S.N., Soodi, M., Shariffzadeh, M., Khorshidi, N. and Marzban, H. (2001) The effect of fennel essential oil on uterine contraction as a model for dysmenorrhea, pharmacology and toxicology study. *Journal of Ethnopharmacology* 76(3), 299–304.

Ostad, S.N., Khakinegad, B. and Sabzevari, O. (2004) Evaluation of the teratogenicity of fennel essential oil (FEO) on the rat embryo limb buds culture. *Toxicology in Vitro* 18(5), 623–627.

Özbek, H., Ura, S., Dülger, H., Bayram, İ., Tuncer, İ., Öztürk, G. and Öztürk, A. (2003) Hepatoprotective effect of *Foeniculum vulgare* essential oil. *Fitoterapia* 74(3), 317–319.

Parejo, I., Viladomat, F., Bastida, J. and Codina, C. (2004) Development and validation of a high-performance liquid chromatographic method for the analysis of antioxidative phenolic compounds in fennel using a narrow bore reversed phase C_{18} column. *Analytica Chimica Acta* 512(2), 271–280.

Patel, K. and Srinivasan, K. (2000) Stimulatory influence of select spices on bile secretion in rats. *Nutrition Research* 20(10), 1493–1503.

Patel, K. and Srinivasan, K. (2001) Studies on the influence of dietary spices on food transit time in experimental rats. *Nutrition Research* 21(9), 1309–1314.

Piccaglia, R. and Marotti, M. (2001) Characterization of some Italian types of wild fennel (*Foeniculum vulgare*). *Journal of Agricultural and Food Chemistry* 49(1), 239–244.

Pope, J.M., Rumpel, H., Kuhn, W., Walker, R., Leach, D. and Sarafis, V. (1991) Applications of chemical-shift-selective NMR microscopy to the non-invasive histochemistry of plant materials. *Magnetic Resonance Imaging* 9(3), 357–363.

Potty, S.N. and Krishnakumar, V. (2001) Organic farming in seed spices in relation to quality and export. In: Agrawal, S., Sastry, E.V.D. and Sharma, R.K. (eds) *Seed Spices – Production, Quality, Export*. Pointer Publishers, Jaipur, India, pp. 154–171.

Rancé, F., Juchet-Gibon, A., Fejji, S., Brémont, F. and Dutau, G. (1994) Sensibilisation aux épices chez l'enfant. *Revue Française d'Allergologie et d'Immunologie Clinique* 34(6), 475–479.

Retamar, J.A. (1986) Essential oils from aromatic species. In: Varghese, J. (ed.) *On Essential Oils*. Synthite Industries and Chemicals Private Ltd., Kolenchery, India, pp. 220–221.

Saxena, J. and Mehrotra, B.S. (1989) Screening of spices commonly marketed in India for natural occurrence of mycrotoxins. *Journal of Food Composition and Analysis* 2(3), 286–292.

Sekizawa, J. and Shibamoto, T. (1982) Genotoxicity of safrole-related chemicals in microbial test systems. *Mutation Research/Genetic Toxicology* 101(2), 127–140.

Stich, H.F., Stich, W. and Lam, P.P.S. (1981) Potentiation of genotoxicity by concurrent application of compounds found in betel quid: arecoline, eugenol, quercetin, chlorogenic acid and Mn^{2+}. *Mutation Research/Genetic Toxicology* 90(4), 355–363.

Weiss, E.A. (2002) Umbelliferae. In: Weiss, E.A. (ed.) *Spice Crops*. CAB International, Wallingford, UK, pp. 284–298.

Wetherilt, H. and Pala, M. (1994) Herbs and spices indigenous to Turkey. In: Charalambous, G. (ed.) *Spices, Herbs and Edible Fungi*. Elsevier Science, Amsterdam, pp. 292–294.

13 Fenugreek

N.K. Leela and K.M. Shafeekh

13.1. Introduction

Fenugreek, or methi (*Trigonella foenum-graecum* L.), belongs to the subfamily Papilionacae of the family Leguminosae (bean family, Fabaceae). The plant is an aromatic herbaceous annual, widely cultivated in Mediterranean countries and Asia. It is believed to have originated in south-eastern Europe or south-western Asian countries; an independent centre of origin exists in Ethiopia. In India, its cultivation is concentrated mainly in Rajasthan, which contributes 80% of the total area, as well as production.

Trigonella is a latinized diminutive of Greek *trigonon* (triangle), composed of *treis* (three) and *gony* (knee, angle); it probably refers to the triangular shape of the flowers. The Latin species name *foenum graecum* means 'Greek hay', referring to both the intensive hay fragrance of dried fenugreek herb and its eastern Mediterranean origin (http://www.uni-graz.at/~katzer/engl/Trig_foe.html).

The area and production of fenugreek in India for the period 1994–2004 is shown in Table 13.1. It shows a slight increase in area under cultivation and production of fenugreek during this period, but production had doubled during 2001/02 and has since declined (DASD, 2007).

13.2. Botany and Uses

Fenugreek is a self-pollinated crop. The plants are weak spreading and moderately branched, attaining a height of 30–50 cm. It flowers 30–50 days after sowing and matures in 110–140 days. The leaves are pinnate and trifoliate, with leaflets 2.0–2.5 cm long, oblanceolate-oblong and obscurely dentate. The flowers are white or yellowish white (1 or 2 auxiliary) and the fruit pod is 3–15 cm long with a long persistent beak. Each pod contains 10–20 seeds, which are greenish-brown, along with a deep groove across one corner, giving the seeds a hooded appearance.

Fenugreek requires a moderately cool climate for proper growth and high yield. It can be grown in all types of soils rich in organic matter content and with good drainage. It can also tolerate a salinity condition, as compared with other leguminous crops.

Fenugreek is used both as a herb (the leaves) and a spice (the seed). The seed is used frequently in Indian cuisine in the preparation of pickles, curry powders and pastes. The young leaves and sprouts are eaten as greens and the fresh or dried leaves are used to flavour dishes. In India, fenugreek seeds are mixed with yoghurt and used as a conditioner for hair. It is also one of the ingredients in the making of *khakhra*, a type of bread. Fenugreek is used

Table 13.1. Area and production of fenugreek.

Year	Area (ha)	Production (t)
1994/95	45,733	57,146
1995/96	39,035	47,494
1996/97	33,421	43,741
1997/98	33,590	31,413
1998/99	35,732	35,737
1999/2000	37,250	40,480
2000/01	35,450	52,020
2001/02	115,600	136,640
2002/03	50,600	64,220
2003/04	50,600	64,220

Source: DASD (2007).

also in a type of bread unique to Ethiopian and Eritrean cuisine. It is used as a natural herbal medicine in the treatment of diabetes. Fenugreek also finds use as an ingredient in the production of clarified butter, which is similar to Indian *ghee*. In Yemen, it is the main condiment and an ingredient added to the national dish called *saltah*. It is used widely as a galactogogue, as a digestive aid and also for treating sinus and lung congestion. It reduces inflammation and fights infection. The seeds are used in the preparation of hair tonic and recommended as a cure for baldness in men. Seed powder is used as a yellowish dye in the Far East. The fixed oil in the seed has a celery-like odour, is tenacious and has attracted the interest of the perfume trade (http://en.wikipedia.org/wiki/Fenugreek).

Fenugreek mixed with cottonseed is fed to cows to increase milk flow. Mildewed or sour hay is made palatable to cattle when it is mixed with fenugreek herbage. It is used as a conditioning powder to produce a glossy coat on horses.

13.3. General Composition

Seeds

Fenugreek seed is used as a spice in culinary preparations. In most cases, the whole seeds are used. When separated into testa and albumen, fenugreek has completely different functions. The seeds consist of 75% testa and 25% albumen; the testa contains fragrant essential oil, saponin, protein and it functions as a spice. On the other hand, the albumen consists of 80% water-soluble substance and 20% water-insoluble substance. The water-soluble substance is galactomannan (http://en.wikipedia.org/wiki/Fenugreek). Fenugreek seeds also contain gums (23.06%) and mucilage (28%). The seeds are a rich source of the polysaccharide galactomannan (Pruthi, 1976).

Dried seeds of fenugreek contain moisture (6.3%), protein (9.5%), fat (10%), crude fibre (18.5%), carbohydrates (42.3%), ash (13.4%), calcium (1.3%), phosphorus (0.48%), iron (0.011%), sodium (0.09%), potassium (1.7%) and vitamins – vitamin A (1040 i.u./100 g), vitamin B_1 (0.41 mg/100 g), vitamin B_2 (0.36 mg/100 g), vitamin C (12.0 mg/100 g) and niacin (6.0 mg/100 g).

Another study on the composition of fenugreek indicated the following values: moisture, 7–11% (average 8.7%); crude protein, 27.7–38.6% (average 31.6%); mineral matter (total ash) 3.35–6.80% (average 4.9%); acid insoluble ash, 0.2–2.3% (average 1%); petroleum ether extract, 5.2–8.2% (average 6.3%); alcohol extract, 16.6–24.8% (average 22.4%); and hot water extract, 29.0–39.7% (average 34.0%). The vitamins present in the seeds are: carotene (Vitamin A), 96 μg; thiamine (Vitamin B_1), 0.34 mg; riboflavin (Vitamin B_2), 0.29 mg; and nicotinic acid, 1.1 mg/100 g. The seeds contain folic acid (total 84 μg/100 g; free 14.5 mg/100 g). Germinating seeds contain pyridoxine, cyanocobalamine, calcium pantothenate, biotin and vitamin C. Exposure of the germinating seeds to β- and γ-radiation reduces the vitamin C content.

Young seeds of the plant contain small amounts of sucrose, glucose, fructose, myoinositol, galactinol (1-0-α-D-galactopyranosyl-D-myoinositol), stachyose and traces of galactose and raffinose. Two galactose-containing compounds, verbascose (6^G-*C*6-α-galactosyl)$_3$-sucrose) and digalactosylmyoinositol, have been reported in the seeds. Very little myoinositol is present in mature seeds. The seeds contain small quantities of xylose and arabinose.

The endosperm of the seed contains 14–15% galactomannan. The seeds contain 30% protein. The yield of protein depends on the extractant used. Extraction of the seed with distilled water gave 15% yield, whereas extraction with saline solution and 70% alcohol yielded 25 and 5% of the total protein of the seed, respectively. The content of albumin, globulin and prolamine in these extractants is as follows (% of proteins): lysine, 4.9, 1.7, 0.5; histidine, 2.8, 11.6, 0.4; arginine, 9.3, 11.2, 2.3; cystine, 1.2, 0.6, 3.0; tyrosine, 2.1, 5.7, 4.3 and tryptophan trace, 0.5, 2.4, respectively. Globulin is characterized by high histidine content and the prolamine contains a low percentage of basic nitrogen and a high percentage of cystine and tryptophan.

Seeds extracted by 0.2% NaOH had the following amino acid composition (% of proteins): lysine, 8.0; histidine, 1.1; arginine, 8.0; tyrosine, 3.0; aspartic acid, 9.0; glutamic acid, 9.0; serine, 6.0; glycine, 9.5; threonine, 5.0; alanine, 5.9; phenyalanin, 1.0; leucines, 11.0; proline, –1.0; and valine + methionine, 6.0.

Aqueous extract of the seed contains the amino acids serine, valine + aspartic acid, glutamic acid, threonine, β-alanine, γ-aminobutyric acid and histidine, while extracts of germinating seeds contain methionine 15 μg/g.

Nazar and El Tinay (2007) reported that seeds contained 28.4% protein, 9.3% crude fibre and 7.1% crude fat. Maximum protein solubility was observed at pH 11 (91.3%) and minimum at pH 4.5 (18.5%).

Fenugreek leaves

The proximate composition of fenugreek leaves is as follows (g/100 g of edible matter): moisture, 86.1; protein, 4.4; fat, 0.9; fibre, 1.1; other carbohydrates, 6.0; and ash, 1.5. The mineral components are (mg/100 g edible matter): Ca, 395; Mg, 67; P, 51 (Phytin P, O); Fe, 16.5; ionizable Fe, 2.7; Na, 76.1; K, 31.0; Cu, 0.26; S, 167.0; and Cl, 165.0. Traces of strontium and lead have been reported in some samples (Anon., 1976).

About two-fifths of the total nitrogen of the leaves occurs as non-protein nitrogen. The free amino acids present are: lysine, histidine, arginine, threonine, valine, tryptophan, phenylalanine, isoleucine, leucine, cystine and tyrosine. The non-protein nitrogen fraction is a good source of dietary lysine The analysis of total leaf proteins for essential amino acids gave the following values (g/g N): arginine, 0.35; histidine, 0.11; lysine, 0.3; tryptophan, 0.08; phenyl alanine, 0.30; methionine, 0.09; threonine, 0.20; leucine, 0.39; isoleucine, 0.30; and valine, 0.32 (Anon., 1976). Microwave drying moderately affected the sensory characteristics of fenugreek leaves (Fathima *et al.*, 2001).

13.4. Chemistry

Volatiles

Fenugreek contains 0.02–0.05% volatile oil (Pruthi, 1976; Sankarikutty *et al.*, 1978; Ramachandraiah *et al.*, 1986). It is brown in colour, having a specific gravity of 0.871 at 15.5°C (Pruthi, 1976).

Girardon *et al.* (1985) identified 39 compounds, including *n*-alkanes, sesquiterpenes and some oxygenated compounds, in the volatile oil of fenugreek seeds. The major components are *n*-hexanol, heptanoic acid, dihydroactiniolide, dihydrobenzofuran, tetradecane, α-muurolene, β-elemene and pentadecane (Table 13.2). The dominant aroma component is a hemiterpenoid-γ-lactone, sotolon (3-hydroxy-4,5-dimethyl-2(5*H*)-furanone), which is present in concentrations up to 25 ppm (Girardon *et al.*, 1989). The sensory evaluation, along with aroma quality, is shown in Table 13.3. Blank *et al.* (1997) reported that sotolon (Fig. 13.1) was formed by oxidative deamination of 4-hydroxy-L-isoleucine. There is chemical similarity between sotolon and the phthalides responsible for the quite similar flavour of lovage leaves (http://en.wikipedia.org/wiki/Sotolon). Toasted fenugreek seeds owe their flavour to another type of heterocyclic compound, called pyrazines.

Table 13.2. Volatile components of fenugreek.

Component	Identification A	B	C
n-Hexanol	+	+ +	+ +
2-Heptanone		+	
n-Heptanal		+	+
Aniline		+	
Phenol		+	+
Heptanoic acid		+	+ +
3-Octen-2-one		+	+
1,8-Cineol			+
Undecane	+		
Camphor		+	
5-Methyl-δ-caprolactone	+		
1-Dodecene	+	t	+
Methyl cyclohexyl acetate	+	+	
Dihydrobenzofuran		+ +	+
Dodecane	+		
Decanoic acid		t	+
Thymol		+	
2-Hexylfuran		+	+
Tridecane	+		
γ-Nonalactone	+	+	+
Eugenol	t	+	
δ-Elemene	+	+	
1-Tetradecene	+		+
Tetradecane	+ +	+	
Calarene	+	+	
β-Ionone		t	
α-Muurolene	+		+
Dihydroactinidiolide	+	+ +	
ε-Muurolene	+ +	+	+
β-Elemene	+ +	+	+
β-Selinese	+		
γ-Elemene	+		
γ-Muurolene	+	+	+
Calamenene	+	+	+
Pentadecane	+ +	+	+
Dodecanoic acid		+	+
Diphenyl amine		+	
1-Hexadecene	+	+	
Hexadecane	+	+	+

Note: A: extract from headspace vacuum treatment; B: steam distillation of seeds; C: steam distillation of oleoresin; +, 0.5–5%; ++, 5–10%; t: trace.
Source: Girardon *et al.* (1985).

Lawrence (1987) reviewed the volatile oil composition of fenugreek seeds. Mazza *et al.* (2002) determined the volatile oil composition of Sicilian fenugreek seeds extracted by different methods (Table 13.4). Headspace analysis of solid phase micro extract (SPME) of fenugreek seeds indicated the presence of carbonyl compounds (hexanal, 2-methyl-2-butenal, 3-octen-2-one, *trans-cis-* and *trans-trans-*3,5-octadien-2-one), sesquiterpene hydrocarbons (*δ*-elemene, *γ*-cadinene and *α*-muurolene), alcohols (pentanol, hexanol, 2-methyl-2-buten-1-ol, 1-octen-3-ol), heterocyclic compounds [3-hydroxy-4,5-dimethyl-2(5*H*)-furanone (*γ*-nonalactone), dihydro-5-ethyl-2(3*H*)-furanone (*γ*-caprolactone)] and other furan compounds particularly involved in the aroma. Methanolic extract, as well as aqueous and dichloromethane extracts, contained higher-boiling compounds, such as C_6–C_{18}, saturated acids and long-chain unsaturated acids, such as oleic, linoleic and linolenic. Two isomers of 3-amino-4,5-dimethyl-3,4-dihydro-2(5*H*)-furanone, the precursor of sotolon, were found in all the extracts (Mazza *et al.*, 2002) The aerial parts of fenugreek yield light yellow oil in 0.3% yield. The chief constituents are: *δ*-cadinene (27.6%); *α*-cadinol (12.1%); *γ*-eudesmol (11.2%); *α*-bisabolol (10.5%); *α*-muurolene (3.9%); liguloxide (7.6%); cubenol (5.7%); *α*-muurolol (4.2%); and *epi-α*-globulol (5.7%) (Ahmadiani *et al.*, 2004). Other low-boiling compounds found in fenugreek are indicated in Table 13.5.

Non-volatiles

The non-volatile constituents isolated from fenugreek include steroids, fatty acids and flavonoids. Among these, the furostanol glycosides are probably responsible for the bitter taste. Sterols, diosgenin derivatives and trigonellin (*N*-methyl-pyridinium-3-carboxylate) are the most important among the non-volatiles. Sterols and diosgenin derivatives are of potential interest to the pharmaceutical industry (http://www.uni-graz.at/~katzer/engl/Trig_foe.html).

Lipids

Seeds contain 7.5% total lipids, of which neutral lipids constituted 84.1%, glycolipids 5.4% and phospholipids 10.5%. Neutral lipids consisted mostly of triacylglycerols (86%), diacylglycerols (6.3%) and small

Table 13.3. Odour-active compounds detected in an aroma extract of fenugreek seeds.

Compound	Aroma quality (GC-olfactometry)	Flavour dilution factor (FD factor)
(1) Diacetyl	Buttery	1
(2) 1-Octen-3-one	Mushroom-like	4
(3) (*Z*)-1,5-Octadiene-3-one	Metallic, geranium-like	5
(4) 3-Isopropyl-2-methoxy pyrazine	Roasty, earthy	3
(5) Acetic acid	Acidic, pungent	7
(6) 3-Isobuty-2-methoxy pyrazine	Roasty, paprika-like	3
(7) Linalool	Flowery	4
(8) Butanoic acid	Sweaty, rancid	3
(9) Isovaleric acid	Sweaty, rancid	4
(10) Caproic acid	Musty	3
(11) Eugenol	Spicy	4
(12) 3-Amino-4,5-dimethyl 3, 4-dihydro-2-(5*H*)-furanone	Seasoning-like	5
(13) Sotolon	Seasoning-like	14

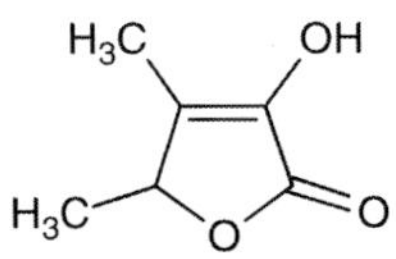

Sotolon

Fig. 13.1. The major aroma component in fenugreek.

Table 13.4. Volatiles from Sicilian fenugreek.

Carbonyl compounds	Hexanal 2-Methyl-2-butenal 3-Octen-2-one *trans-cis*-3, 5-Octadien-2-one *trans-trans*-3, 5-Octadien-2-one
Sesquiterpene hydrocarbons	δ-Elemene γ-Cadinene α-Muurolene
Alcohols	Pentanol Hexanol 2-Methyl-2-buten-1-ol 1-Octen-3-ol
Heterocyclics	Sotolon [3-hydroxy-4, 5-dimethyl-2-(5*H*)-furanone] γ-Nonalactone [dihydro-5-pentyl-2-(5*H*)-furanone] γ-Caprolactone [dihydro-5-ethyl-2-(3*H*)-furanon]

Source: Mazza *et al.* (2002).

quantities of monoacylglycerols, free fatty acids and sterols. Acylmonogalactosyl diacylglycerol and acylated sterylglycoside were the major glycolipids, while sterylglucoside, monogalactosylmonoacylglycerol and digalactosyldiacylglycerol were present in small amounts. The phospholipids consisted of phosphatidylcholine and phosphatidylethanolamine as major components and phosphatidylserine, lysophosphatidylcholine, phosphatidylinositol, phosphatidylglycerol and phosphatidic acid as minor phospholipids (Hemavathy and Prabhakar, 1989).

Fixed oil

The seeds contain about 7% fixed oil consisting mainly of linoleic, oleic and linolenic acids. Fenugreek seeds from Andhra Pradesh contained 5.00–6.45% fatty oil (Ramachandraiah *et al.*, 1986). Hot alcohol was reported as the best solvent for extracting maximum oleoresin (29.02%) from fenugreek (Sankarikutty *et al.*, 1978).

The seeds of fenugreek contain 6–8% of fatty oil with a fetid odour and a bitter taste. Oil samples from Egypt had the following range of characteristics: specific gravity (25°C),

0.9100–0.9142; n_0 (25°C), 1.4741–1.4749; acid value, 1.0–2.0; saponin value, 178.0–183.0; iodine value, 115.0–116.2; thiocyanogen value, 77.2–77.7; RM value, 0.10–0.15; and unsaponifiable matter, 3.9–4.0%. The component fatty acids of the oil are (weight of total acids): palmitic acid, 9.6%; stearic acid, 4.9%; arachidic acid, 2.0; behenic acid, 0.9; oleic acid, 35.1; linoleic acid, 3.7%; and linolenic acid, 13.8%. Lightly toasted fenugreek seeds (150°C) were superior to the medium (175°) and dark-roasted (200°C) seeds with regard to flavour and nutritive value. No appreciable loss in total nitrogen and crude protein was noticed during roasting, but there was a considerable decrease in total and free sugars as the temperature of the roasting increased. Fenugreek leaves contain Vitamin C (~43.10 mg/100 g). By boiling in water, or steaming and frying, the vegetable loses 10.8 and 7.4% of the vitamin, respectively.

Pressure-cooking causes the least loss of ascorbic acid, while stir-frying of vegetable fenugreek causes the greatest loss.

Table 13.5. Volatiles from fenugreek.

α-Pinene	1-Pentanol
β-Pinene	1-Hexanol
Sabinene	2-Methyl-2-butene-1-ol
3-Carene	2-Methyl-2-butenal
Menthol	2-Pentyl furan
β-Terpineol	Formic acid
Cineol	Propanoic acid
Anethol	γ-Butyrolactone
β-Terpinyl acetate	
1-*p*-Menthen-8-yl-acetate	
Carvone	
Linalool	

Steroid glycosides and saponins

The seeds contain mainly two steroidal saponins which, on hydrolysis, give two steroidal sapogenins, diosgenin and gitogenin, in a 9:1 ratio. Tigogenin is reported to be present in traces. Samples of seeds from Algeria, Morocco, Ethiopia and India yielded 0.35, 0.25, 0.20 and 0.10% of diosgenin, respectively. The total saponin content of the seed is reported to be 1% and it can be increased up to 20 times by incubation of seeds with water at 37°C for 1–96 h. The diosgenin levels in fenugreek seeds from Canadian origin ranged from 0.28 to 0.92% (28–92 μg/mg; Taylor *et al.*, 2002).

Several furostanol glycosides have been isolated from fenugreek, which are indicated in Table 13.6. Yoshikawa *et al.* (1997) isolated the furostanol saponins, trigoneosides Ia, Ib, IIa, IIb, IIIa and IIIb from Indian fenugreek. The furostanol glycosides trigofoenosides A and D, F and G have been isolated and reported as their methyl ethers (Gupta *et al.*, 1984; 1985). From the ethanol extract of fenugreek seeds a furostanol saponin, trigoneoside VIII (26-*O*-*β*-D-glucopyranosyl-25 (*R*)-52-furostan-20 (22)-*en*-2 *α*, 3*β*, 26-triol-3-*O*-*β*-D-xylopyranosyl (1→6)-*β*-D-glucopyranoside), has been isolated. The saponins isolated from the leaves include diosgenin, tigogenin and gitogenin, the major one being diosgenin. Saponins isolated from fenugreek are indicated in Table 13.6.

Alkaloid

Seeds contain the alkaloid, trigonelline (0.38%, methyl betaine of nicotinic acid). It yields nicotinic acid on heating with hydrochloric acid at 260–270°C.

Dry fenugreek contains trigonelline during roasting; two-thirds of trigonelline is converted into niacin or nicotinic acid. Nicotinic acid is almost absent in the former. It is reported that fermentation increases both free and total niacin.

Flavonoids

Shang *et al.* (1998) isolated five flavonoids, vitexin, tricin, naringenin, quercetin and tricin-7-*O*-*β*-D-glucopyranoside, from fenugreek seeds. The seeds contain the flavonoid components querticetin, luteolin and their glycosides (Anon., 1976). Han *et al.* (2001) isolated kaempferol glycoside, lilyn (kaempferol-3-*O*-*β*-D-glucosyl-(1→2)-*β*-D-galactoside), kaempferol-3-*O*-*β*-D-glucosyl-(1→2)-*β*-D-galactoside-7-*O*-*β*-D-glucoside, kaempferol-3-*O*-*β*-D-glucosyl-1→(6^{11}-*O*-acetyl)-*β*-D-galactoside-7-*O*-*β*-D-glucoside

Table 13.6. Steroid saponins from fenugreek.

Compound	Reference
Diosgenin	Gupta *et al.*, 1986a
Yamogenin	Taylor *et al.*, 1997
Tigogenin	
Neotigogenin	
Smilagenin	
Sarsa sapogenin	
Yuccagenin	
Gitogenin	
Neogitogenin	Taylor *et al.*, 1997
Protodioscin	Hibasami *et al.*, 2003
Methyl protodioscin	Yang *et al.*, 2005
Methylprotodeltonin	
Trigoneoside Ia [26-*O*-β-D-glucopyranosyl-(25*S*)-5-α-furostan-2-α, 3 β, 22 ζ, 26-tetraol 3-*O*-[β-D-xylopyranosyl (1→6)]-β-D-glucopyranoside]	Yoshikawa *et al.*, 1997
Trigoneoside Ib [26-*o*-β-D-glucopyranosyl-(25*R*)-5-α-furostan-2 α, 3 β, 22 ζ, 26-tetraol 3-*O*-[β-D-xylopyranosyl (1→6)]-β-D-glucopyranoside]	
Trigoneoside IIa [26-*O*-β-D-glucopyranosyl-(25*S*)-5-β-furostan-3 β, 22 ζ, 26-triol 3-*O*-[β-D-xylopyranosyl (1 →6)]-β-D-glucopyranoside]	
Trigoneoside IIb [26-*O*-β-D-glucopyranosyl-(25*R*)-5-β-furostan 3 β, 22 ζ, 26-triol 3-*O*-[β-D-xylopyranosyl (1 →6)]-β-D-glucopyranoside]	
Trigoneoside IIIa [26-*O*-beta-D-glucopyranosyl-(25*S*)-5-α-furostan-3 β, 22 ζ, 26-triol 3-*O*-[α-L-rhamnopyranosyl (1→2)]-β-D-glucopyranoside]	
Trigoneoside IIIb [26-*O*-β-D-glucopyranosyl-(25*R*)-5-α-furostan 3 β, 22 ζ, 26-triol 3-*O*-[α-L-rhamnopyranosyl (1→2)]-β-D-glucopyranoside]	
Trigoneoside IVa	Yoshikawa *et al.*, 1998
Trigoneoside Va	
Trigoneoside Vb	
Trigoneoside VI	
Trigoneoside VIIb	
Trigoneoside VIIIb	
Trigoneoside IX	
Trigoneoside Xa [26-*O*-β-D-glucopyranosyl-(25*S*)-5-α-furostan-2 α, 3 β, 22 ζ, 26 tetraol-3-*O*-α-L rhamnopyranosyl (1→2)-β-D-glucopyranoside]	Murakami *et al.*, 2000
Trigoneoside Xb [26-*O*-β-D-glucopyranosyl-(25*R*)-5-α-furostan-2 α, 3 β, 22 ζ, 26 tetraol-3-*O*-α-L-rhamnopyranosyl (1→2)-β-D-glucopyranoside]	Murakami *et al.*, 2000
Trigoneoside XIb [26-*O*-β-D-glucopyranosyl-(25*R*)-5α-furostan-2 α, 3 β, 22 ζ, 26 tetraol-3-*O*-β-D-xylopyranosyl (1→4)-β-D-glucopyranoside]	Murakami *et al.*, 2000
Trigoneoside XIIa [26-*O*-β-D-glucopyranosyl-(25*S*)-furost-4-en-3 β, 22 ζ, 26 triol-3-*O*-α-L-rhamnopyranosyl (1→2)-β-D-glucopyranoside]	
Trigoneoside XIIb [26-*O*-β-D-glucopyranosyl-(25*R*)-furost-4-en-3 β, 22 ζ, 26 triol-3-*O*-β-L-rhamnopyranosyl (1→2)-β-D-glucopyranoside]	
Trigoneoside XIIIa [26-*O*-β-D-glucopyranosyl-(25*S*)-furost-5-en-3 β, 22 ζ, 26 triol-3-*O*-α-L-rhamnopyranosyl (1→2)-[β-D-glucopyranosyl (1→3)-β-D-glucopyranosyl-(1→4)-[β-D-glucopyranoside]	
Fenugreekine	Ghosal *et al.*, 1974
Trigoneoside A	Gupta *et al.*, 1985
Trigoneoside B	Gupta *et al.*, 1986b
Trigoneoside C	
Trigoneoside D	Gupta *et al.*, 1985
Trigoneoside F	Gupta *et al.*, 1984
Trigoneoside G	

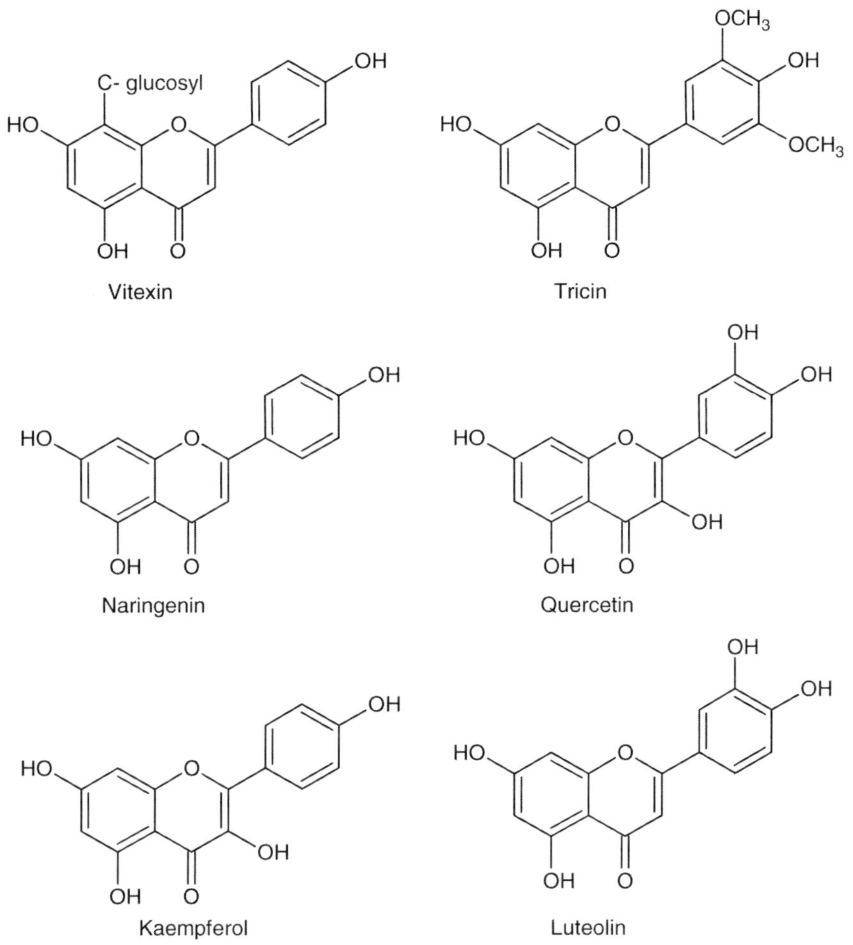

Fig. 13.2. Flavonoids from fenugreek.

and querceticin-3-*O*-*β*-D-glucosyl-(1→2)-*β*-D-galactoside-7-*O*-*β*-D-glucoside from the stem of fenugreek. The seeds of fenugreek contained the flavone glycosides, orientin (0.259%) and vitexin (0.184%) (Huang and Liang 2000). Figure 13.2 shows some of the flavonoids isolated from fenugreek.

Miscellaneous compounds

Fowden *et al.* (1973) isolated 4-hydroxy leucine from the seeds of fenugreek. Later, Alcock *et al.* (1989) determined its absolute configuration as (2*S*, 3*R*, 4*S*). From the leaves and stems, *γ*-schizandrin and scopoletin (7-hydroxy-6-methoxycoumarin) were isolated by Wang *et al.* (1997). Shang *et al.* (2002) isolated *N,N′*-dicarbazyl, glyceryl monopalmitate, stearic acid, β-sitosteryl glucopyranoside, ethyl α-glucopyranoside, D-3-*O*-methyl chiroinsitol and sucrose from seeds. Methylprotodioscin and methylprotodeltonin were isolated from the plant by Yang *et al.* (2005). The non-volatiles from fenugreek are indicated in Fig. 13.3.

13.5. Medicinal and Pharmacological Uses

Fenugreek seeds and leaves have been used extensively in various medicinal preparations. The leaves are refringent and

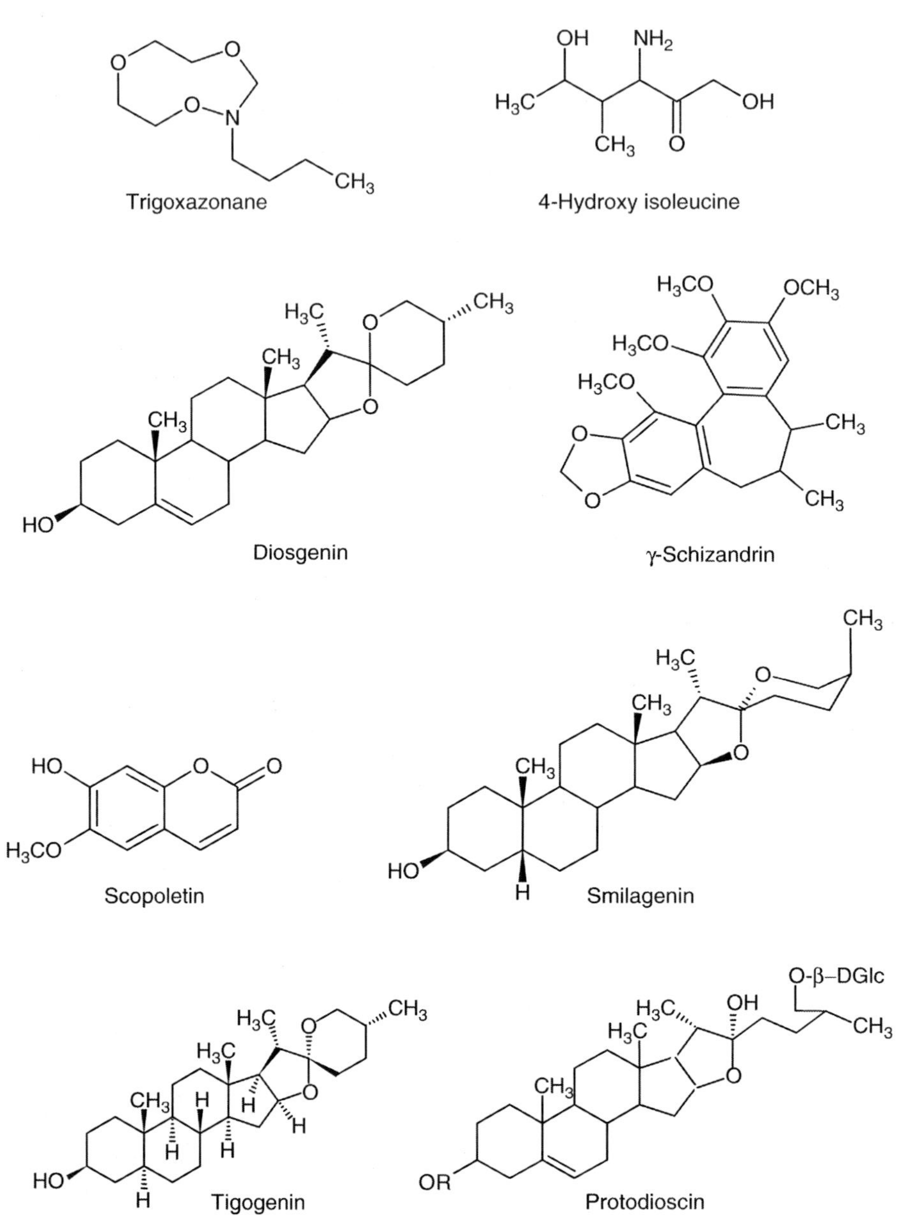

Fig. 13.3. Non-volatiles from fenugreek.

aperients and are given internally for vitiated conditions of *Pitta* in Ayurveda medicine. The seeds are bitter, mucilaginous, aromatic, carminative, tonic, thermogenic, galactogogue, astringent, emollient and an aphrodisiac. They are good for fever, vomiting, anorexia, cough, bronchitis and callosities. Externally, in the form of poultices, they are used for boils, abscesses and ulcers. An infusion of seeds is given to smallpox patients as a cooling drink. Seeds are also used in enlargement of liver and spleen and rickets. Women use the seeds to induce lactation during the post-natal period. Fenugreek seeds contain diosgenin, a steroidal substance, which is used as a starting material in the production of sex hormones and oral contraceptives (Anon., 1976).

The seeds are hot, tonic, antipyretic, anthelmintic, astringent to the bowels, cure leprosy, *vata*, vomiting, bronchitis, piles, remove bad taste from the mouth and are useful in heart disease (Ayurveda). In Unani medicine, the plants and seeds are considered to be suppurative, aperient, diuretic, emmenagogue and useful in dropsy and chronic cough. The leaves are useful in external and internal swelling, burns and prevent hair falling out (Kirtikar and Basu, 1984).

Hypoglycaemic activity

Modern clinical studies have investigated the hypocholesterolaemic and hypoglycaemic actions of fenugreek in normal and diabetic humans. Injection of whole seed extracts for 21 days improved plasma glucose and insulin responses and 24-h urinary concentrations reduced. In diabetic insulin-dependent subjects, daily administration of 25 gm fenugreek seed powder reduced fasting plasma glucose profile, glycosuria and daily insulin requirement (56 to 20 units) after 8 weeks. It also resulted in significant reductions in serum cholesterol concentrations (Sharma, 1986).

Oral administration of methanolic and aqueous extracts of seeds at the dose of 1 g/kg body weight produced a hypoglycaemic effect in mice (Zia *et al.*, 2001a). In non-insulin-dependent diabetic patients, incorporation of 100 g of defatted fenugreek seed powder in the diet for 10 days produced a fall in fasting food-glucose levels and improvement in the glucose tolerance test. Urinary glucose excretion was reduced by 64% in 2 h. Serum total cholesterol, LDL and VLDL cholesterol and triglyceride levels decreased without alteration in the HDL cholesterol fraction (Sharma and Raghuram, 1990).

Furostanol-type steroid saponins in fenugreek increased food intake in normal rats significantly, while modifying the circadian rhythm of feeding behaviour in diabetic rats resulted in a progressive weight gain in contrast to untreated diabetic controls. In normal and diabetic rats, steroid saponins decreased total plasma cholesterol without any change in triglycerides (Petit *et al.*, 1995). Fenugreek improves peripheral glucose utilization, contributing to improvement in glucose tolerance. It exerts its hypoglycaemic effect by acting at the insulin receptor level as well as at the gastrointestinal level. An intravenous glucose tolerance test indicated that fenugreek in the diet reduced the area under the plasma glucose curve significantly and shortened the half-life of plasma glucose, due to increased metabolic clearance. Fenugreek also increased erythrocyte insulin reception (Raghuram *et al.*, 1994). The soluble dietary fibre fraction from fenugreek seeds improves glucose homeostasis in animal models of type I and type II diabetic rats (Hannan *et al.*, 2007).

Fenugreek seeds have hypoglycaemic and hypocholesterolaemic effects on type I and type II diabetes mellitus patients and experimental diabetic animals. Xue *et al.* (2007) reported that rats treated with *T. foenum-graecum* extract had lower blood glucose, glycated haemoglobin, triglycerides, total cholesterol and higher high-density lipoprotein cholesterol compared with diabetic rats.

Narender *et al.* (2006) reported that 4-hydroxyisoleucine, isolated from the seeds, decreased plasma triglyceride levels by 33%, total cholesterol (TC) by 22% and free fatty acids by 14%, accompanied by an increase in the HDL-C/TC ratio by 39% in the dyslipidaemic hamster model.

Broca *et al.* (1999) reported that, in non-insulin-dependent diabetic (NIDD) rats, a single intravenous administration of 4-*OH*-isoleucine (50 mg/kg) partially restored glucose-induced insulin response without affecting glucose tolerance; a 6-day subchronic administration of 4-*OH*-Ile (50 mg/kg, daily) reduced basal hyperglycaemia, decreased basal insulinaemia and improved glucose tolerance. *In vitro*, 4-*OH*-Ile (200 μM) potentiated glucose (16.7 mM)-induced insulin release from NIDD rat-isolated islets.

Feeding the seed mucilage alleviated the reduction in maltase activity during diabetes, but the activities of sucrase and lactase were not changed on feeding. It also showed 30% improvement in urine sugar and urine volume profiles and 26% improvement in fasting blood glucose levels (Kumar *et al.*, 2005a,b). Oral administration of alcoholic extract of fenugreek seeds lowered the blood glucose in alloxan diabetic rats significantly (Vats *et al.*, 2003).

Hypocholesterolaemic activity

Supplements of fenugreek seeds have been shown to lower serum cholesterol, triglyceride and low-density lipoprotein in human patients and experimental models of hypercholesterolaemia and hypertriglyecridaemia (Basch *et al.*, 2003). The hypocholesterolaemic effects of fenugreek seeds were also reported by Singhal *et al.* (1982). The ethanol extract from fenugreek seeds contain hypocholesterolaemic components, saponins which interact with bile salts in the digestive tract (Stark and Madar, 1993).

Ingestion of fenugreek powder reduces total cholesterol and triglyceride levels. Fenugreek is thus considered a dietary supplement for hyperlipidaemia and atherosclerosis in diabetic subjects (Sharma *et al.*, 1996a). The antidiabetic effects of fenugreek seeds in type I and type II diabetes in both human and animal models have been well established (Basch *et al.*, 2003).

Currently, fenugreek is available commercially in encapsulated form and is being prescribed as a dietary supplement for the control of hypercholesterolaemia and diabetes by practitioners of complementary and alternative medicine. It can be found in capsule form in many health food stores. Raju and Bird (2006) reported that supplementation of fenugreek through diet reduced triglyceride accumulation in the liver, without affecting the plasma insulin or glucose levels in obese rats. Administration of sodium orthovanadate and fenugreek seed powder resulted in the normalization of hyperglycaemia, together with glyoxalase I activity, in diabetic rats (Raju *et al.*, 1999).

Anticarcinogenic activity

The anticarcinogenoic activity of fenugreek has been reported by several workers. Devasena and Menon (2003) observed that fenugreek seeds in the diet inhibited colon carcinogensis, by modulating the activities of β-glucoronidase and mucinase. The seed powder in the diet decreased the activity of β-glucoronidase significantly and prevented the free carcinogens from acting on colonocytes. Mucinase helped in hydrolysing the protective mucin. This was attributed to the presence of fibre, flavanoids and saponins (Devasena and Menon, 2003). Sur *et al.* (2001) reported antineoplastic activity of the seed extract. Intraperitoneal administration of the alcohol seed extract before and after inoculation of Ehrlich ascites carcinoma cell in mice inhibited tumour cell growth. Treatment with the extract enhanced both the peritoneal exudates cell and macrophage cell counts (Sur *et al.*, 2001).

Continuous feeding of rats with 1% fenugreek seed powder (FSP) and 0.05% and 0.10% diosgenin suppressed total colonic aberrant crypt foci (ACF) by up to 32, 24 and 42%, respectively, in azoxymethane-induced carcinogenesis in rats. During the promotional stages, FSP inhibited total ACF.

Diosgenin also inhibited the growth of human osteosarcoma 1547 cell line (Moalic *et al.*, 2001; Corbiere *et al.*, 2003). Protodioscin, isolated from fenugreek, displays a strong inhibitory effect against leukaemic cell line HL-60 and a weak growth inhibitory effect on gastric cell line KATO-III (Hibasami *et al.*, 2003).

Devasena and Menon (2007) reported that fenugreek seeds had a modulatory effect on colon tumour incidence, as well as hepatic lipid peroxidation (LPO) during DMH (1,2-dimethylhydrazine) colon carcinogenesis in male Wistar rats. In DMH-treated rats, 100% colon tumour incidence was accompanied by enhanced LPO and a decrease in reduced glutathione (GSH) content, as well as a fall in glutathione peroxidase (GPx), glutathione *S*-transferase (GST), superoxide dismutase (SOD) and catalase (CAT) activities. Inclusion of fenugreek seed powder in the diet of DMH-treated rats reduced the colon tumour incidence to 16.6%, decreased the lipid peroxidation and increased the activities of GPx, GST, SOD and CAT in the liver.

Polyphenolic extract of fenugreek seed acts as a protective agent against ethanol-induced abnormalities in the liver and the effects are comparable with those of a known hepatoprotective agent, silymarin (Kaviarasan and Anuradha, 2007).

Fenugreek seeds showed a protective effect against 7,12-dimethylbenz (alpha) anthracene (DMBA)-induced breast cancer in rats, at 200 mg/kg body weight (Amin *et al.*, 2005). The hepatoprotective properties of fenugreek seeds have also been reported (Thirunavukkarasu *et al.*, 2003; Kaviarasan *et al.*, 2006; Raju and Bird, 2006). Raju *et al.* (2004) reported that diosgenin in fenugreek inhibited cell growth and induced apoptosis in the H-29 human colon cancer cell line. The beneficial effect of fenugreek and diosgenin as a cancer preventive agent has been reported by several workers.

Fenugreek in complementary cancer therapy

Cyclophosphamide (CP) is a commonly used anticancer drug which causes toxicity by its reactive metabolites, such as acrolein and phosphoramide mustard. Fenugreek extract exhibits a protective effect by reversing the cyclophosphamide-induced apoptosis and free radical-mediated lipid peroxidation in the urinary bladder of mice (Bhatia *et al.*, 2006). Hence, fenugreek is suggested as a promising protective medicinal herb for complementary therapy in cancer patients under chemotherapeutic interventions. Diosgenin in fenugreek is one of the molecules identified for the prevention and therapy of cancer due to its ability to interfere with multiple-cell signalling pathways (Aggarwal and Shishodia, 2004).

Diosgenin in fenugreek has been identified as one of the molecular targets that potentially can be used for the prevention and treatment of cancer (Aggarwal and Shishodia, 2006). Diosgenin, a steroidal saponin present in fenugreek, suppresses proliferation, inhibits invasion and suppresses osteoclastogenesis through inhibition of necrosis factor NF-kappaB-regulated gene expression and enhances apoptosis induced by cytokines and chemotherapeutic agents (Shishodia and Aggarwal, 2006).

Anti-inflammatory and antipyretic activity

Fenugreek leaves possess anti-inflammatory and antipyretic effect. The leaf extract reduces formalin-induced oedema in single dose (fenugreek 1000 and 2000 mg/kg, sodium salicylate 300 mg/kg). It also reduces hyperthermia induced by Brewer's yeast 1–2 h after administration (Ahmadiani *et al.*, 2001).

Antioxidant activity

The aerial parts and seeds of fenugreek showed antioxidant and free radical scavenging activities (Bajpai *et al.*, 2005). Supplementation of diet with fenugreek seeds lowered lipid peroxidation (Kaviarasan *et al.*, 2004). A polyphenol-rich extract from the seeds of fenugreek reduced the oxidative haemolysis and lipid peroxidation in normal and diabetic human erythrocytes. Dixit *et al.* (2005) also found that the aqueous fraction of fenugreek exhibited higher antioxidant activity compared with other fractions. Kaviarasan *et al.* (2007) reported the radical scavenging of hydroxyl radicals and inhibition of H_2O_2-induced lipid peroxidation by fenugreek extract in rat liver mitochondria. These studies show the *in vivo* beneficial effect of fenugreek seeds. Supplementation of diet with fenugreek seeds in alloxan-treated diabetic rats resulted in the lowering of lipid peroxidation (Ravikumar and Anuradha, 1999).

Effect on enzyme activities

Fenugreek seed powder reduces the activities of glucose-6-phosphatase and fructose-1, 6-biphosphatase in the liver and kidneys of diabetic rats. The inclusion of fenugreek powder overcomes the toxicity of vanadium encountered when given alone (Gupta *et al.*, 1999). Studies on the *in vivo* effects of insulin, vanadate and fenugreek seed powder on changes in the activity of creatine kinase in the heart, skeletal muscles and liver of rats show that the effects of insulin and vanadate are comparable in restoring normoglycaemia and creatine kinase activities, while fenugreek is slightly less effective (Solomon-Genet *et al.*, 1999).

Effect on ovaries and liver tissues

Fenugreek oil has a stimulating effect on the ovarian activity of mice. Administration of fenugreek oil in mice showed that the total number and quality of cumulus–oocyte complexes increased and the oil stimulated the oocytes to progress in meiosis, but the levels of nucleic acid contents were unaffected (Hassan *et al.*, 2006).

Antifertility effect

The seeds of fenugreek produced an antifertility effect in female rabbits and a toxicity effect in male rabbits (Kassem *et al.*, 2006). Feeding diets containing 30% fenugreek seeds resulted in a reduction of testis weight in males and damage to the seminiferous tubules and interstitial tissues. In addition, the plasma concentration of the antrogen hormone and sperm concentrations was halved in treated animals. In the case of females, development of the fetus was reduced.

Immunomodulatory effect

An aqueous extract of fenugreek at 50–200 mg/kg of body weight showed a stimulatory effect on the immune system of Swiss albino mice (Bin Hafeez *et al.*, 2003).

Nematicidal activity

The aqueous, methanol and chloroform extracts of fenugreek seeds cause significant mortality of *Meloidogyne javanica* larvae (Zia *et al.*, 2001b). Further soil amendments with powered seeds of fenugreek cause soil suppression against *M. javanica*. Decomposed seeds of fenugreek cause marked reduction in nematode population densities and subsequent root-knot development as compared with the aqueous extract of the seeds (Zia *et al.*, 2003). Aqueous leaf extract of fenugreek leaves shows nematicidal activity against J_2 of *M. incognita* (Saxena and Sharma, 2004).

Larvicidal activity

The larvicidal activity of acetone and petroleum ether extracts of *T. foenum-graceum* in combination with *Murraya koenigii*, *Coriandrum sativum* and *Ferula asafetida* produced potential synergistic larvicidal activity against *Aedes aegypti* larvae, although they exhibited comparatively poor larvicidal activity when tested individually (Harve and Kamath, 2004).

Wound healing activity

Aqueous extract of fenugreek seeds promoted significant wound healing activity and the seed suspension was more potent than the aqueous extract (Taranalli and Kuppast, 1996).

Allelopathic effect

Orobanche crenata is a major threat to grain legume production. When intercropped with grain legumes such as pea plants, fenugreek reduced *O. crenata* infection. This is attributed to the allelopathic effect of the fenugreek. The application of root exudates of fenugreek inhibited germination of the seeds of *O. crenata*. The main inhibitory metabolite in this case was characterized as 2-butyl-(1,4,7,2) trioxazonane (trigoxazonane) by Antonio *et al.* (2007).

Toxicity studies of fenugreek

Fenugreek seeds are used to treat dysentery, diarrhoea, dyspepsia, cough, enlargement of liver and spleen, rickets and gout. No renal or hepatic toxicity was observed in patients ingesting an experimental diet containing fenugreek seed powder (25 g/day), even after 24 weeks (Sharma *et al.*, 1996b). An acute intraperitoneal and oral toxicity study of the glycosidic extract of fenugreek leaves concluded that the extract was considered to be safe and have a minimal adverse effect. The intraperitoneal study was aimed at four target organs, e.g. liver, kidney, stomach, small and large intestine, and found that the liver was the only organ affected, where early degeneration with infiltration of mononuclear and mild hepatitis was found in some animals treated with toxic doses of the glycosidic extract (Abdel Barry *et al.*, 2000). Fenugreek seed extract administered twice a week for

4 weeks, at dosages of 1.0, 1.5 and 2.0 g/kg of body weight, exhibited a necrotic effect on the liver and the kidney tissues of male albino Wistar rats. Spermatogenesis was observed in the testes at dosage levels of 1.5 g/kg and 2.0 g/kg of the extract (Effraim *et al.*, 1999).

Debitterized fenugreek powder does not produce any significant acute and cumulative toxicity in mice and rats up to 10% level (Muralidhara *et al.*, 1999).

13.6. Conclusion

Fenugreek is a rich source of various phytochemicals, especially the steroidal saponins. Due to its antioxidant, hypoglycaemic and hypocholesterolaemic activities, fenugreek has great potential for use in the complementary medicines for cancer therapy and diabetes. The saponins present in fenugreek, chiefly diosgenin, are the starting compound for the manufacture of over 60% of the total steroid drugs by the pharmaceutical industry. Currently, diosgenin requirement is met by the Dioscorea species. Fenugreek, being easy to cultivate, might one day replace the present commercial sources. The use of herbs as hypoglycaemic is a major avenue in Indian perspectives, particularly for treating diabetes, and is to be explored more effectively as much information is available on these aspects.

References

Abdel Barry, J.A. and Al Hakiem, M.H.H. (2000) Acute intraperitoneal and oral toxicity of the leaf glycosidic extract of *Trigonella foenum-graecum* in mice. *Journal of Ethnopharmacology* 70(1), 65–68.

Aggarwal, B.B. and Shishodia, S. (2004) Suppression of the nuclear factor kappa B activation pathway by spice derived hytochemicals: reasoning for seasoning. *Annals of New York Academy of Sciences* 1030, 434–441.

Aggarwal, B.B. and Shishodia, S. (2006) Molecular targets of dietary agents for prevention and therapy of cancer. *Biochemical Pharmacology* 71, 1397–1421.

Ahmadiani, A., Javan, M., Semnanian, S., Barat, E. and Kamalineyad, M. (2001) Antiinflammatory and antipyretic effects of *Trigonella foenum graecum* leaf extracts in rats. *Journal of Ethnopharmacology* 75, 283–286.

Ahmadiani, A., Rustaiyan, A., Karimian, M. and Kamalinejad, M. (2004) Volatile constituents from the oil of *Trigonella foenum graecum* L. *Journal of Essential Oil Research* 16(4), 356–357.

Alcock, N.W., Crout, D.H., Gregorio, M.V.M., Lee, E., Pike, G. and Samuel, C.J. (1989) Stereochemistry of 4-hydroxy leucine from *Trigonella foenum graecum*. *Phytochemistry* 28, 1835–1841.

Amin, A., Alkaabi, A., Al-Falaasi, S. and Daoud, S.A. (2005) Chemopreventive activities of *Trigonella foenum-graecum* (Fenugreek) against breast cancer. *Cell Biology International* 29(8), 687–694.

Anon. (1976) *The Wealth of India Raw Materials*, Volume X. CSIR, New Delhi, pp. 298–306.

Antonio, E., Monica, F.A., Anna, A., Diego, R. and Andrea, M. (2007) Trigoxazonane, a mono-substituted trioxazonane from *Trigonella foenum-graecum* root exudates, inhibits *Orobanche crenata* seed germination. *Phytochemistry* 68, 2487–2492.

Bajpai, M., Mishra, A. and Prakash, D. (2005) Antioxidant and free radical scavenging activities of some leafy vegetables. *International Journal of Food Sciences and Nutrition* 56(7), 473–481.

Basch, E., Ulbricht, C., Kuo, G., Szapary, P. and Smith, M. (2003) Therapeutic applications of fenugreek. *Alternative Medicine Review* 8(1), 20–27.

Bhatia, K., Kaur, M., Atif, F., Ali, M., Rehman, H., Rahman, S. and Raisuddin, S. (2006) Aqueous extract of *Trigonella foenum-graecum* L. ameliorates additive urotoxicity of buthionine sulfoximine and cyclophosphamide in mice. *Food and Chemical Toxicology* 44, 1744–1750.

Bin-Hafeez, Haque, R., Parvez, S., Pandey, S., Sayeed, I. and Raisuddin, S. (2003) Immunomodulatory effect of fenugreek (*Trigonella foenum-graecum* L.) extract in mice. *International Immunopharmacology* 3, 257–265.

Blank, I., Jianming Lin, Devaud, S., Fumeaux, R. and Fay, L.B. (1997) The principal flavour components of fenugreek (*Trigonella foenum graecum*). In: Risch, S.J. and Chi, T.H. (eds) *Spices: Flavour Chemistry and Antioxidant Properties*. ACS, Washington, DC, pp. 12–28.

Broca, C., Gross, R., Petit, P., Sanvaire, Y., Manteghetti, M., Tournies, M., Masiello, P., Gonnis, R. and Riber, G. (1999) 4-Hydroxy isoleucine – experimental evidence of its insulinotropic and antidiabetic properties. *American Journal of Physiology* 277(4), E617–E623.

Corbiere, C., Liagre, B., Bianchi, A., Bordji, K., Dauca, M., Netter, P. and Beneytout, J.L. (2003) Different contribution of apoptosis to the antiproliferative effects of diosgenin and other plant steroids, hecogenin and tigogenin, on human 1547 osteosarcoma cells. *International Journal of Oncology* 22(4), 899–905.

DASD Directorate of Arecanut and Spices Development (2007) Naseema Printers and Publishers, Kochi, India, pp. 110.

Devasena, T. and Menon, V.P. (2003) Fenugreeek affects the activity of β-glucoronidase and mucinase in the colon. *Phytotherapy Research* 17(9), 1088–1091.

Devasena, T. and Menon, V.P. (2007) Fenugreek seeds modulate 1, 2–dimethyl hyrdrazine-induced hepatic oxidative stress during colon carcinogenesis. *Italian Journal of Biochemistry* 56, 28–34.

Dixit, P., Ghaskadbi, S., Mohan, H. and Devasagayam, T.P. (2005) Antioxidant properties of germinated fenugreek seeds. *Phytotherapy Research* 19(11), 977–983.

Effraim, K.D., Jaxks, T.W. and Nwafor, P.A. (1999) Investigation of the toxicity potential of *Trigonella foenum graecum. Pakistan Veterinary Journal* 19(1), 13–16.

Fathima, N., Begum, K. and Rajalakshmi, D. (2001) Microwave drying of selected greens and their sensory characteristics. *Plant Foods Human Nutrition* 56, 303–311.

Fowden, L., Pratt, H.M. and Smith, A. (1973) 4-Hydroxy leucine from seed of *Trigonella foenum graecum. Phytochemistry* 12, 1707–1711.

Ghosal, S., Srivastava, R.S., Chatterjee, C.D. and Dutta, K.S. (1974) Fenugreek, a new steroidal sapogenin-peptide ester of *Trigonella foenum. Phytochemistry* 13(10), 2247–2251.

Girardon, P., Bessiere, J.M., Baccou, J.C. and Sauvaire, Y. (1985) Volatile constituents of fenugreek seeds. *Planta Medica* 6, 533–534.

Girardon, P., Sauvaire, Y, Baccou, J.C. and Bessiere, J.M. (1989) Identification de la 3-hydroxy-4,5-dimethyl-2(5H)-furanone dans l'arome des graines Fenugrec (*Trigonella foenum graecum* L.). *Lebensmittel-Wissenschaft Und-technologie* 19, 44–46.

Gupta, D., Raju, J. and Baquer, N.Z. (1999) Modulation of some gluconeogenic enzyme activities in diabetic rat liver and kidney: effect of anti-diabetic compounds. *Indian Journal of Experimental Biology* 37(2), 196–199.

Gupta, R.K., Jain, D.C. and Thakur, R.S. (1984) Furostanol glycosides from *Trigonella foenum-graecum* seeds. *Phytochemistry* 23, 2605–2607.

Gupta, R.K., Jain, D.C. and Thakur, R.S. (1985) Furostanol glycosides from *Trigonella foenum-graecum* seeds. *Phytochemistry* 24(10), 2399–2401.

Gupta, R.K., Jain, D.C. and Thakur, R.S. (1986a) Minor steroidal sapogenins from fenugreek seeds, *Trigonella foenum graecum. Journal of Natural Products* 49(6), 1153.

Gupta, R.K., Jain, D.C. and Thakur, R.S. (1986b) Two furostanol saponins from *Trigonella foenum graecum. Phytochemistry* 25(9), 2205–2207.

Han, Y., Nishibe, S., Noguchi, Y. and Zhexiong, J. (2001) Flavonol glycosides from glycosides from the stems of *Trigonella foenum graecum. Phytochemistry* 49, p.1153.

Hannan, J.M., Ali, L., Rokeya, B., Khaleque, J., Akhter, M., Flatt, P.R. and Abdel-Wahab, Y.H. (2007) Soluble dietary fibre fraction of *Trigonella foenum-graecum* (fenugreek) seed improves glucose homeostasis in animal models of type 1 and type 2 diabetes by delaying carbohydrate digestion and absorption, and enhancing insulin action. *British Journal of Nutrition* 97(3), 514–521.

Harborne, J.B. and Marbry, T.J. (1975) *The Flavonoids*, Vols I and II. Academic Press, New York.

Harve, G. and Kamath, V. (2004) Larvicidal activity of plant extracts used alone and in combination with known synthetic larvicidal agents against *Aedes aegypti. Indian Journal of Experimental Biology* 42(12), 1216–1219.

Hassan, A.M., Wagdy, K.B.K. and Kawkab, A.A. (2006) Genetic and histopathology studies on mice: effect of fenugreek oil on the efficiency of ovarian and liver tissues. *African Journal of Biotechnology* 5(5), 477–483.

Hemavathy, J. and Prabhakar, J.V. (1989) Lipid composition of fenugreek (*Trigonella foenumgraecum* L.) seeds. *Food Chemistry* 31(1), 1–7.

Hibasami, H., Moteki, H., Ishikawa, H., Imai, K., Yoshioka, K., Ishii, M.Y. and Komiya, T. (2003) Protodioscin isolated from fenugreek (*Trigonella foenum graecum*) induces cell death and morphological change indicative of apoptosin in leukemic cell line H-60, but not in gastric cancer cell line KATO III. *International Journal of Molecular Medicine* 11(1), 23–26.

http://www.uni-graz.at/~katzer/engl/Trig_foe.html.

http://en.wikipedia.org/wiki/Fenugreek.

http://en.wikipedia.org/wiki/Sotolon.

Huang, W. and Liang, X. (2000) Determination of 2 flavone glycosides in the seeds of *Trigonella foenum graecum* L. from various production localities. *Journal of Plant Resources and Environment* 9(4), 53–54.

Kassem, A., Al-Aghbari, A., Al-Habori, M. and Al-Mamary, M. (2006) Evaluation of the potential antifertility effect of fenugreek seeds in male and female rabbits. *Contraception* 73, 301–306.

Kaviarasan, S. and Anuradha, C.V. (2007) Fenugreek seed polyphenols protect liver from alcohol toxicity: a role on hepatic detoxification system and apoptosis. *Pharmazie* 62, 299–304.

Kaviarasan, S., Vilayalakshmi, K. and Anuradha, C.V. (2004) Polyphenol-rich extract of fenugreek seeds protects erythrocytes from oxidative damage. *Plant Foods for Human Nutrition* 59(4), 143–147.

Kaviarasan, S., Ramamurty, N., Gunasekaran, P., Varalakshmi, E. and Anuradha, C.V. (2006) Fenugreek (*Trigonella foenum graecum*) seed extract prevents ethanol-induced toxicity and apoptosis in Chang liver cells. *Alcohol and Alcoholism* 41(3), 267–273.

Kaviarasan, S., Naik, G.H.,Gangabhagirathi, R., Anuradha, C.V. and Priyadarsini, K.I. (2007) *In vitro* studies on antiradical and antioxidant activities of fenugreek (*Trigonella foenum-graecum*) seeds. *Food Chemistry* 103(1), 31–37.

Kirtikar, and Basu (1984) *Indian Medicinal Plants*, Text 1. Lalit mohan basu, Allahabad, India, pp. 699–702.

Kumar, G.S., Shetty, A.K. and Salimath, P.V. (2005a) Modulatory effect of fenugreek seed mucilage and spent turmeric on intestinal and renal disaccharidases in streptozotocin induced diabetic rats. *Journal of Medicinal Food* 8(3), 382–385.

Kumar, G.S., Shetty, A.K., Sambaiah, K. and Salimath, P.V. (2005b) Antidiabetic property of fenugreek seed mucilage and spent turmeric in streptozotocin-induced diabetic rats. *Nutrition Research* 25, 1021–1028.

Lawrence, B.M. (1987) Progress in essential oils. *Perfumer and Flavourist* 12(5), 54–62.

Mazza, G., Tommaso, D. di and Foti, S. (2002) Volatile constituents of Sicilian fenugreek (*Trigonella foenum-graecum*) seeds. *Sciences des Aliments* 22(3), 249–264.

Moalic, S., Liagre, B., Corbiere, C., Bianchi, A., Dauca, M., Bordji, K. and Beneytout, J.L. (2001) A plant steroid diosyesis induces apoptasis, cell cycle arrest and cox activity in osterosarcoma cells. *FEBS Letters* 506(3), 225–230.

Murakami, T., Kishi, A., Matsuda, H. and Yoshikawa, M. (2000) Medicinal foodstuffs. XVII. Fenugreek seed (3). Structures of new furostanol-type steroid saponins, trigoneosides Xa, Xb, XIb, XIIa, XIIb and XIIa, from the seeds of Egyptian *Trigonella foenum-graecum*. *Chemical and Pharmaceutical Bulletin* 48(7), 994–1000.

Muralidhara, K., Narasimha murthy, K., Viswanathan, S. and Ramesh, B.S. (1999) Acute and subchronic toxicity assessment of debitterised fenugreek powder in the mouse and rat. *Food and Chemical Toxicology* 37, 831–838.

Narender, T., Puri, A., Shweta Khaliq, T., Saxena, R., Bhatia, G. and Chandra, R. (2006) 4-Hydroxyisoleucine, an unusual amino acid as antidyslipidemic and antihyperglycemic agent. *Bioorganic and Medicinal Chemistry* 16(2), 293–296.

Nazar, A.E. and El Tinay, A.H. (2007) Functional properties of fenugreek (*Trigonella foenum graecum*) protein concentrate. *Food Chemistry* 103, 582–589.

Petit, P.R., Sauvaire, Y.D., Hillaire-Buys, D.M., Leconte, O.M., Baissac, Y.G., Ponsin, G.R. and Ribes, G.R. (1995) Steroid saponins from fenugreek seeds: extraction, purification, and pharmacological investigation on feeding behavior and plasma cholesterol. *Steroids* 60(10), 674–680.

Pruthi, J.S. (1976) *Spices and Condiments*. National Book Trust, New Delhi, India, pp. 117–121.

Raghuram, T.C., Sharma, R.D., Sivakumar, B. and Sahay, B.K. (1994) Effect of fenugreek seeds on intravenous glucose disposition in non-insulin dependent diabetic patients. *Phytotherapy Research* 8(2), 83–86.

Raju, J. and Bird, R.P. (2006) Alleviation of hepatic steatosis accompanied by modulation of plasma and liver TNF-alpha levels by *Trigonella foenum-graecum* (fenugreek) seeds in Zucker obese (fa/fa) rats. *International Journal of Obesity* 30(8), 1298–1307.

Raju, J., Gupta, D., Rao, A.R. and Baquer, N.Z. (1999) Effect of antidiabetic compounds on glyoxalase I activity in experimental diabetic rat liver. *Indian Journal of Experimental Biology* 37(2), 193–195.

Raju, J., Patlolla, J.M., Swamy, M.V. and Rao, C.V. (2004) Diosgenin, a steroid saponin of *Trigonella foenum graecum* (Fenugreek), inhibits azoxymethane-induced aberrant crypt foci formation in F344 rats and induces apoptosis in HT-29 human colon cancer cells. *Annals of New York Academy of Sciences* 13(8), 1392–1398.

Ramachandraiah, O.S., Reddy, P.N., Azeemoddin, G., Ramayya, D.A. and Rao, S.D.T. (1986) Essential and fatty oil content in umbelliferous and fenugreek seeds of Andhra Pradesh habitat. *Indian Cocoa, Arecanut and Spices Journal* 10(1), 12.

Ravikumar, P. and Anuradha, C.V. (1999) Effect of fenugreek seeds on blood lipid peroxidation and antioxidants in diabetic rats. *Phytotherapy Research* 13(3), 197–201.

Sankarikutty, B., Sumathykutty, M.A., Bhat, A.V. and Mathew, A.G. (1978) Studies on extractions of oils and oleoresins from cumin, fennel and fenugreek. *Indian Arecanut, Spices and Cocoa Journal* 2(2), 25–30.

Saxena, R. and Sharma, R. (2004) Effect of aq. extracts of leaves and roots of some edible plants on J_2 mortality of *M. incognita*. *Nematology* 15, 91–96.

Shang, M., Cai, S., Han, J., Li, J., Zhao, Y., Zheng, J., Namba, T., Kadota, S., Tezuka, Y. and Fan, W. (1998) Studies on flavonoids from fenugreek (*Trigonella foenum-graecum*). *Zhongguo Zhong Yao Za Zhi [China Journal of Chinese Materia Medica]* 23(10), 614–616.

Shang, M., Cai, S.O., Lin, W.H., Wang, M.C. and Park, J.H. (2002) Studies on the chemical constituents from the seed of *Trigonella foenum graecum*. *Zhaong guo Zhong Yao Za Zhi* 27, 277–279.

Sharma, R.D. (1986) Effect of fenugreek leaves and seeds on blood glucose and serum insulin response in human subjects. *Nutrition Research* 6, 1353–1364.

Sharma, R.D. and Raghuram, T.C. (1990) Hypoglycemic effect of fenugreek seeds on non-insulin dependent subjects. *Nutrition Research* 10, 731–739.

Sharma, R.D., Sarkar, A., Hazra, D.K., Misra, B., Singh, J.B., Maheswaryi, B.B. and Sharma, S.B. (1996a) Hypolipidaemic effect of fenugreek seeds. A chronic study in non-insulin dependent diabetic patients. *Phytotherapy Research* 10(4), 332–334.

Sharma, R.D., Sarkar, A., Hazra, D.K., Misra, B., Singh, J.B. and Maheswaryi, B.B. (1996b) Toxicological evaluation of fenugreek seeds. A long-term feeding experiment in diabetic patients. *Phytotherapy Research* 10(6), 519–520.

Shishodia, S. and Aggarwal, B.B. (2006) Diosgenin inhibits osteoclastogenesis, invasion, and proliferation through the downregulation of Akt, I kappa B kinase activation and NF-kappa B-regulated gene expression. *Oncogene* 25(10), 1463–1473.

Singhal, P.C., Gupta, R.K. and Joshi, L.D. (1982) Hypocholesterolmic effect of seeds. *Current Science* 51, 136–137.

Solomon Genet, Kale, R.K. and Baquer, N.Z. (1999) Effect of vanadate, insulin and fenugreek (*Trigonella foenum-graecum*) on creative kinase level in tissues of diabetic rat. *Indian Journal of Experimental Biology* 37(2), 200–202.

Stark, A. and Madar, Z. (1993) The effect of an ethanol extract derived from fenugreeek (*Trigonella foenum-graecum*) on bile acid absorption and cholesterol levels in rats. *British Journal of Nutrition* 69, 277–287.

Sur, P., Das, M., Gomes, A., Vedasiromony, J.R., Sahu, N.P., Banerjee, S., Sharma, R.M. and Ganguly, D.K. (2001) *Trigonella foenum-graecum* (fenugreek) seed extract as an antineoplastic agent. *Phytotherapy Research* 15(3), 257–259.

Taranalli, A.D. and Kuppast, I.J. (1996) Study of wound healing activity of seeds of *Trigonella foenum-graecum* in rats. *Indian Journal of Pharmaceutical Sciences* 58(3), 117–119.

Taylor, W.G., Zaman, M.S., Mir, Z., Mir, P.S., Acharya, S.N., Mears, G.J. and Elder, J.L. (1997) Analysis of steroidal sapogenins from amber fenugreek (*Trigonella foenum-graecum*) by capillary gas chromatography and combined gas chromatography/mass spectrometry. *Journal of Agricultural and Food Chemistry* 45(3), 753–759.

Taylor, W.G., Zulywak, H.J., Richards, K.W., Acharya, S.N., Bittman, S. and Elder, J.L. (2002) Variation in diosgenin levels among 10 accessions of fenugreek seeds produced in Western Canada. *Journal of Agricultural and Food Chemistry* 50, 5994–5997.

Thirunavukkarasu, V., Anuradha, C.V. and Viswanathan, P. (2003) Protective effect of fenugreek (*Trigonella foenum-graecum*) seeds in experimental ethanol toxicity. *Phytotherapy Research* 17(7), 737–743.

Vats, V., Yadav, S.P. and Grover, J.K. (2003) Effect of *Trigonella foenum-graecum* on glycogen content of tissues and the key enzymes of carbohydrate metabolism. *Journal of Ethnopharmacology* 85(2/3), 237–242.

Wang, D., Sun, H., Han, Y., Wang, X. and Yuan, C. (1997) Chemical constituents of the stems and leaves of *Trigonella foenum-graecum* L. *Zhongguo Zhong Yao Za Zhi* 22(8), 486–487.

Xue, W.L., Li, X.S., Zhang, J., Liu, Y.H., Wang, Z.L. and Zhang, R.J. (2007) Effect of *Trigonella foenum-graecum* (fenugreek) extract on blood glucose, blood lipid and hemorheological properties in streptozotocin-induced diabetic rats. *Asia Pacific Journal of Clinical Nutrition* 16 (Suppl. 1), 422–426.

Yang, W.X., Huang, H.Y., Wang, Y.J., Jia, Z.Y. and Li, L.L. (2005) Study on chemical constituents in total saponin from *Trigonella foenum-graecum*. *Zhongguo Zhong Yao Za Zhi* 30(18), 1428–1430.

Yoshikawa, M., Murakami, T., Komatsu, H., Murakami, N., Yamahara, J. and Matsuda, H. (1997) Medicinal foodstuffs. IV. Fenugreek seed. (1) Structures of trigoneosides Ia, Ib, IIa, IIIa, IIIb, new furostanol saponins from the seeds of Indian *Trigonella foenum-graecum*. *Chemical and Pharmaceutical Bulletin* 45(1), 81–87.

Yoshikawa, M., Murakami, T., Komatsu, H., Murakami, N., Yamahara, J. and Matsuda, H. (1998) Medicinal foodstuffs. VIII. Fenugreek seed. (2) Structures of six new furostanol saponins, trigoneosides IVa, Va, Vb, VI, VIIb, and VIIIb, from the seeds of Indian *Trigonella foenum-graecum* l. *Heterocycles* 47, 397–405.

Zia, T., Nazrul Hasnain, S. and Hassan, S.K. (2001a) Evaluation of the oral hypoglycemic effect of *Trigonella foenum-graecum* (methi) in normal mice. *Journal of Ethnopharmacology* 75, 191–195.

Zia, T., Siddiqui, I.A. and Nazrul Hasnain, S. (2001b) Nematicidal activity of *Trigonella foenum-graecum* L. *Phytotherapy Research* 15(6), 538–540.

Zia, T., Siddiqui, I.A., Shaukat, S.S. and Nazrul Hasnain, S. (2003) *Trigonella foenum-graecum* (fenugreek) mediated suppression of *Meloidogyne javanica* in mungbean. *Archives of Phytopathology and Plant Protection* 36(1), 23–31.

14 Paprika and Chilli

T. John Zachariah and P. Gobinath

14.1. Introduction

Paprika is defined in the USA as a sweet, dried, red powder. This mild powder can be made from any type of *Capsicum annuum* that is non-pungent and has a brilliant red colour. Paprika may be pungent in Hungary, but paprika is always non-pungent in the international trade. Paprika comes from milling dry fruits of different varieties of *C. annum* L. In Europe, it is produced principally in Hungary, Turkey and Spain. Spanish powdered paprika again is classified, based on its pungency, into three categories, such as hot (*picante*), sweet (*dulce*) and an intermediate called *ocal* or *agridulce* (Mateo *et al.*, 1997).

It is estimated that world production of chillies is about 2.5 million t and paprika accounts for one-third of the total world consumption of chilli (red pepper). India tops the list, with about one million t from 8.28 million ha. India has emerged as the major producer and supplier of chillies in the international market (Thampi, 2003). Production details of chilli from different countries are listed in Table 14.1.

14.2. Botany and Uses

Capsicum fruits in different forms are popular food additives in most parts of the world. The genus *Capsicum* is a member of the Solanaceae family that includes tomato, potato, tobacco and petunia. The genus *Capsicum* consists of approximately 22 wild species and five domesticated species, *C. annuum*, *C. baccatum*, *C. chinense*, *C. frutescens* and *C. pubescens*. *Capsicum* is endemic to the Western hemisphere and the pre-Columbian distribution extended from the southernmost border of the USA to the temperate zone of southern South America. Despite their vast trait differences, most chilli cultivars cultivated commercially in the world belong to the species *C. annuum*. The tabasco (*C. frutescens*) and habanero (*C. chinense*) are the best-known exceptions (Bosland, 1996).

The terminology *Capsicum* is confusing. Pepper, chili, chile, chilli, aji, paprika and *Capsicum* are used interchangeably for plants in the genus *Capsicum*. *Capsicum* is reserved for taxonomic discussion. The word *chile* is a variation of *chil* derived from the Nahuatl (Aztec) dialect, which referred to plants now known as *Capsicum*, whereas *aji* is a variation of *axi* from the extinct Arawak dialect of the Caribbean (Bosland, 1996). Chile pepper has come to mean pungent chilli cultivars. However, chile means pepper (*Capsicum*), whether the fruits are pungent or not. Generally, chili is used to identify the state dish of Texas, which is a combination of pungent chile cultivars and

Table 14.1. World chilli production.

Country	Production (t)
India	850,000
China	400,000
Pakistan	300,000
South Korea	150,000
Mexico	300,000
Bangladesh	100,000
Other countries	400,000

Source: Thampi (2003).

meat. Bell pepper generally refers to non-pungent blocky chile types. Additional confusion is present within species designation because *C. annuum* was sometimes called *C. frutescens* in the scientific literature. The five domesticated species are *C. annuum* L., *C. baccatum* L., *C. chinense* Jacq., *C. frutescens* L. and *C. pubescens* R. & P. (IBPGR, 1983). A peculiar chilli category is paprika. It is not a pod-type in the USA, but it is a product. In Europe, there are chilli pod-types that are paprika. This is because in the Hungarian language *paprika* means *Capsicum* (Bosland, 1996).

The paprika-type chilli, which was evolved in the temperate regions around the Mediterranean and some parts of the USA, is used widely as a table spice and also in the meat processing industry as a natural colourant. This is valued principally for the brilliant red colour it gives to pale foods and also for its delicate aroma. Colour is very important in paprika and chilli powder. Paprika and paprika oleoresin are used currently in a wide assortment of foods, drugs and cosmetics, as well as for improving the feather colour of flamingoes in zoos (Bunnell and Bauernfeind, 1962). Technological innovations have already made great strides in separating colour components and pungent constituents.

Paprika is, in some cases, classified as non-pungent (Bosland, 1996), even though it may contain low or high levels of pungent compounds. There are many hot, pungent types that vary in pungency; for example, New Mexico, Jalapeno, Cayenne, etc. Thai and Habanero types increase in pungency in this order. Chillies generally are smaller and lower in red colour than non-pungent fruit. Paprika, like chillies and capsicums, is always a ground product. Oleoresin paprika is prepared from varieties of *C. annuum* L., from which paprika is produced (Purseglove *et al.*, 1981).

14.3. General Composition

Chilli contains many chemicals, including water, fixed (fatty) oils, steam-volatile oil, carotenoids, capsaicinoids, resin, protein, fibre and mineral elements. Many of these chemicals have importance for nutritional value, taste, colour and aroma. The two most important groups of chemicals found in chilli are the carotenoids and capsaicinoids (Bosland and Votava, 2000).

Water is the main constituent in peppers. In chilli, the amount of water is dependent on the age and type of pod harvested. Spice varieties allowed to dry on the plant may contain 70% water. Chilli fruits contain sugar, pentosans and raw fibre. Glucose accounts for 90–98% of the sugar content of red mature paprika pod. The amount of sugar in a pod varies by cultivar, agroclimatic conditions and type. Total and reducing sugars are at maximum levels in red succulent fruits.

Cellulose and other fibrous material may account for up to 20% of the dry weight of pericarp tissue. The skins contain 77% soluble fibre and 80% total dietary fibre. This amount of fibre is greater than in either rice or oats (Adeyeyei and Otokiti, 1999). Investigation of polar extracts from ripe fruits of *C. annuum* L. var. *acuminatum* yielded three new glycosides, capsosides A (1) and B (2) and capsianoside VII (3), along with seven known compounds (Iorizzi *et al.*, 2001).

Lysine, arginine, proline, tyrosine, tryptophan, methionine, valine, phenylalanine, leucine, glutamic acid, glycine, asparagines, threonine and alanine are found in chilli. Asparagine, glutamine, glutamic acid and tryptophan account for 95% of the free amino acids. A small amount of aspartic acid was detected. The total amount of ascorbic acid in fruits was 121 mg/100 g fresh weight (FW) (Kim *et al.*, 1997). *C. annuum* is a rich source of vitamins (Anu and Peter, 2000).

Fatty acid carotenoid esters and unesterified hypophasic and epiphasic carotenoids were extracted from paprika fruit at different stages of ripening and processing. Monoesters of capsanthin contained mostly unsaturated fatty acids ($C_{18:2}$), while diesters of capsanthin and capsorubin contained saturated fatty acids such as C_{12}, C_{14} and C_{16}. The carotenoid esters were more stable, toward lipoxygenase-catalysed linoleic acid oxidation, than free pigments. Capsanthin esters containing saturated fatty acids resisted the enzymic oxidation better than the others (Biacs *et al.*, 1989). Studies by Bekker *et al.* (2002) on lipids of *C. annuum* fruit pulp identified the presence of isoprenes (19% of unsaponified mass), triterpenes (30%) and sterols (38%). The fatty acid content of the saponified part of the extract ranged from 0.6 to 45.0% of the saponified mass. Linoleic acid is the principal component in seeds (54%) and pulp (45%) and linolenic acid is about 10% in the pulp.

Hungarian studies have shown that the pericarp has 16–17% protein and the seeds contain 18% protein. When the microelements were investigated it was found that iron was present in the largest concentration, followed by bromide and manganese. Other microelements found were cadmium, calcium, cobalt, copper, magnesium, phosphorus, potassium, sodium and zinc. Fruits of the *Capsicum* species have a relatively low volatile oil ranging from about 0.1 to 2.6% in paprika. The characteristic aroma and flavour of fresh fruit is imparted by the volatile oil (Pruthi, 2003). The comparative chemical composition of chilli and paprika is given in Table 14.2.

Analysis of chemical constituents in fruits of red pepper (cv. Bugang) revealed five natural capsaicinoids. They were capsaicin, nordihydrocapsaicin, dihydrocapsaicin, vanillyl decanamide and homodihydrocapsaicin. The concentration of total capsaicinoids in fruits was 5.4 mg/100 g FW. Eleven carotenoids were identified, with a total concentration of 65 mg/100 g FW. The concentration of free amino acids in fruits was 0.9 g/100 g FW (Kim *et al.*, 1997).

Table 14.2. Composition of paprika and chilli.

Constituents	Paprika (Indian)	Pepper (Chilli)
Moisture (g)	7.90	6.50
Protein (g)	13.80	14.0
Fat (g)	10.40	14.10
Total carbohydrate (g)	60.30	58.20
Fibre carbohydrate (g)	19.00	15.60
Total ash (g)	7.60	7.20
Minerals		
Calcium (g)	0.20	0.10
Phosphorus (g)	0.30	0.32
Sodium (g)	0.02	0.01
Potassium (g)	2.40	2.10
Iron (mg)	23.10	9.90
Vitamins		
Thiamine (mg)	0.60	0.59
Riboflavin (mg)	1.36	1.66
Niacin (mg)	15.30	14.20
Ascorbic acid (mg)	58.80	63.70
Vitamin A (IU)	4915	6165

Maturity-related changes

Changes in the weight and composition of pepper (sweet) and (hot) fruits during their maturation were monitored in a greenhouse trial conducted for two successive seasons. In sweet peppers, FW and DW (dry weight) increased with increasing maturity, peaking 30–40 days after fruit set and then declining (due to senescence and water loss), whereas in hot peppers FW and DW continued to increase to the final sampling date (48 days after fruit set). Total carbohydrate concentrations generally decreased with increasing maturity in both cultivars. Protein concentration fluctuated, but tended to decrease with increasing maturity. Capsaicin concentration and yield (hot cultivar only) increased with increasing maturity, reaching 230 mg/100 g and 4.738 mg/fruit, respectively, 48 days after fruit set. Harvesting dates of 30 and 36 days after fruit set are recommended for the sweet and hot cultivars, respectively, when grown in plastic greenhouses (El Saeid, 1995).

Genotype-related changes

Physico-chemical characteristics of 12 cultivars (Konkan Kirti, Pusa Jwala, Jayanti, Phule Sai, Surkta, RHRC-P, RHRC-E, RHRC-16–5, BC-30, KDCS-810, LCA-334 and PMR-57) grown in Maharashtra (India) revealed large varietal differences for the content of moisture (80.19–87.45%), protein (1.44–2.16%), ash (0.66–1.20%), fat (0.92–1.56%), fibre (2.54–4.02%) and carbohydrate (6.05–11.11%). Among the cultivars, BC-30 contained the highest amounts of protein, ash, fat and carbohydrate. PMR-57 and Konkan Kirti showed the highest crude fibre content (4.02 and 3.99%, respectively). Pusa Jwala recorded the highest ascorbic acid content in fruits (162 mg/100 g). KDCS-810 contained the highest capsaicin (180 mg/100 g). Konkan Kirti, Pusa Jwala, Jayanti, Phule Sai, LCA-334, RHRC-16–5 and PMR-57 were superior to other cultivars as flavouring in food products. LCA-334 recorded the highest total chlorophyll content in its fruits (11.9 mg/100 g). Phosphorus content ranged from 38.6 to 68.8 mg/100 g. Potassium content varied from 0.30 to 0.53 g/100 g, while calcium and iron contents ranged from 7.5 to 19.2 and 0.8 to 2.42 mg/100 g, respectively. BC-30 had the highest contents of phosphorus, potassium, calcium and iron. Fruit weight varied from 1.02 to 3.06 g and pericarp weight ranged from 0.71 to 2.44 g. Phule Sai showed higher fruit weight and pericarp weight than all other cultivars, except BC-30, which was on par with it. Seed number in the chilli cultivars ranged from 26 to 76 (Gupta and Tambe, 2003).

14.4. Chemistry

Colour and pigments

The colour of chilli spice powder is due to the presence of red-pigmented carotenoids. The main pigments are capsanthin, capsorubin, zeaxanthin and cryptoxanthin. Carotenoids are very stable in intact plant tissue. However, when chillies are processed by drying and then grinding into spice powder, the carotenoids auto-oxidize easily, due to the effects of heat, light and oxygen. This leads to a more orange and less intense coloration that devalues the spice powder. In addition, carotenoids have provitamin A activity (Mosquera and Mendez, 1993; Wall and Bosland, 1993; Daood *et al.*, 2006).

Carotenoid compounds

Carotenoids control pod colour, with approximately 20 carotenoids contributing to the colour of the powder. Carotenoid compounds are yellow-to-red pigments of aliphatic or alicyclic structures composed of isoprene units, which are normally fat-soluble colours. The keto-carotenoids, capsanthin, capsorubin and cryptocapsin are unique *Capsicum* carotenoids. The major red colour in chilli comes from the carotenoids capsanthin and capsorubin, while the yellow-orange colour is from β-carotene and violaxanthin. Capsanthin, the major carotenoid in ripe fruits, contributes up to 60% of the total carotenoids. Capsanthin and capsorubin increase proportionally with advanced stages of ripeness, with capsanthin being the more stable of the two. The amount of carotenoids in fruit tissue depends on factors such as cultivar, maturity stage and growing conditions (Reeves, 1987). Deli and Toth (1997) observed the changes in carotenoid pigment composition of *Capsicum* cv. Bovet 4 fruits (grown in Hungary) during ripening. In the chromatograms, 56 peaks were detected and 34 carotenoids were identified. In ripe fruits, capsanthin, capsorubin, zeaxanthin, cucurbitaxanthin A and β-carotene were the main carotenoids, the remainder being capsanthin 5,6-epoxide, capsanthin 3,6-epoxide, karpoxanthin, cucurbitaxanthin B, violaxanthin, cycloviolaxanthin, antheraxanthin, capsanthone, nigroxanthin, β-cryptoxanthin and several *cis* isomers and furanoid oxides.

The carotenoid composition of traditional sweet cultivars of paprika (*C. annuum*) from Szeged (Sz-20) and Mihalyteleki (MT) was compared with that of cultivars produced by cross-breeding MT and the Spanish cultivar Negral, which has an intense brownish-red colour. Cultivars Sz-20 and MT were characterized by their high red xanthophyll content, as well as by their high capsanthin/capsorubin ratios. Negral and its F1 and F5 generation hybrids had a lower red pigment content but a higher capsorubin level. The principal difference between the cultivars appeared to be the fatty acid esters of the major red xanthophylls. Crossing Negral with the Hungarian cultivar Red Longum produced a hybrid of relatively higher red pigment content than that of the Spanish origin. F1 hybrids had a carotenoid composition similar to that of the Spanish parents, but the F5 generation showed improved characteristics, such as high colour intensity and high capsanthin/capsorubin ratio (Biacs *et al.*, 1993).

Deli *et al.* (1992) studied the carotenoid composition in the fruits of black paprika (*C. annuum* variety *longum nigrum*) during ripening. In the chromatograms, 58 peaks were detected and 34 carotenoids (92–95% of the total carotenoid content) were identified completely or tentatively. The total carotenoid content of the ripe fruits was about 3.2 g/100 g DW, of which capsanthin constituted 42%, zeaxanthin 8%, cucurbitaxanthin A (3,6-epoxy-5,6-dihydro-β, β-carotene-5,3′-diol) 6.6%, capsorubin 3.2% and β-carotene 7%. The remainder was composed of capsanthin 5,6-epoxide, capsanthin 3,6-epoxide (3,6-epoxy-5,3′-dihydroxy-5,6-dihydro-β,κ-caroten-6′-one), karpoxanthin, violaxanthin, antheraxanthin, zeaxanthin, β-cryptoxanthin, lutein and several *cis* isomers and furanoid oxides. Molnar *et al.* (2005) studied carotenoids in the hypophasic and epiphasic fractions from red paprika. The hypophasic and epiphasic carotenoids of paprika (PM1) and (PM2) were obtained by extraction, saponification and partition between MeOH-H_2O (9:1) (hypophasic) and hexane (epiphasic). A high content of capsanthin was quantified in hypophasic carotenoids (PM1) from red paprika. On the other hand, a high content of β, β-carotene and β-cryptoxanthin was found in epiphasic fractions. Structures of major carotenoids are illustrated in Fig. 14.1.

Quantification of colour value

The colour of chilli powder can be measured either as extractable red colour or surface colour. Extractable colour is the official method used by the American Spice Trade Association (ASTA, 1985) and in international trade. Generally, in trade, the lower limit allowable for chilli powder is 120 ASTA units and for non-pungent paprika, 160–180 ASTA units. The higher the colour level, the better the quality of the spice. The loss of red coloration during storage needs to be considered to allow the spice to be of acceptable colour when it reaches the consumer (Govindarajan and Sathyanarayana, 1986; Hari *et al.*, 2005).

Surface colour measurements will give some indication as to how the chilli powder will look to the eye. The lightness (L) value can give some indication of colour differences, as powder of higher colour intensity will have a lower 'L' value. However, high-temperature drying has other quality defects, such as darkening of powder, and the 'L' value may therefore be low. For chilli powder, a hue angle (h°) of 0° is red and 90° is yellow; therefore, the closer the value to 90°, the more orange a powder will appear (Jorge *et al.*, 1997). As it is difficult to interpret complex 'L' and 'h°' data, the standard technique used by the spice industry is to measure extractable colour and to observe the powder visually for defects.

Quantitative estimation of the red components of Hungarian paprika indicated total pigment ranging from 4.07 to 5.49 g and capsanthin between 2.19 and 3.49 g, capsorubin 0.42 to 0.98 g per kg of pericarp. Thus, this major pigment accounts for 65–80% of the total colour (Govindarajan and Sathayanarayana, 1986). Composition of these pigments varies with maturity stage and is also related to the cultivar.

Red bell peppers contain 280 ug/gm total carotenoids. Capsanthin accounts for 60% of the total carotenoids. They also contain 11% β-carotene and 20% capsorubin. Capsanthin is acylated with C_{12} to C_{18} saturated fatty acids (Schweiggert *et al.*, 2005).

Processing of paprika and carotenoids

Mosquera *et al.* (1993) studied the effect of the processing of paprika on the main carotenes and esterified xanthophylls present in the fresh fruit. Over-ripe fruits of pepper

Capsanthin: (3*R*,3′*S*,5′*R*)-3,3′-dihydroxy-ß,-caroten-6′-one(orange-red)$C_{40}H_{56}O_3$

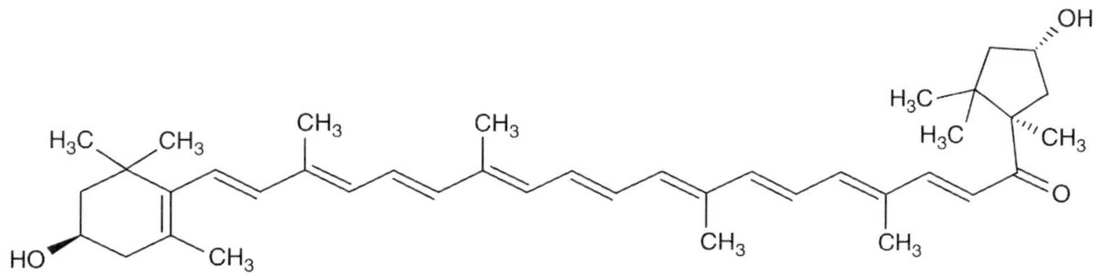

Capsorubin: 3,3′-dihydroxy-carotene-66′dione(orange-red)$C_{40}H_{56}O$

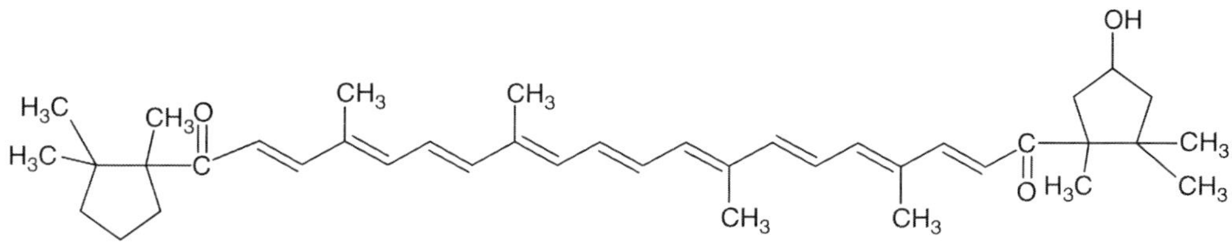

Cryptoxanthin: (*R*)-3,5,5-trimethyl-4-[3,7,12,16-tetramethyl-18-(2,6,6-trimethylcyclohex-1-enyl)-octadeca-1,3,5,7,9,11,13,15,17-nonaenyl]-cyclohex-3-enol

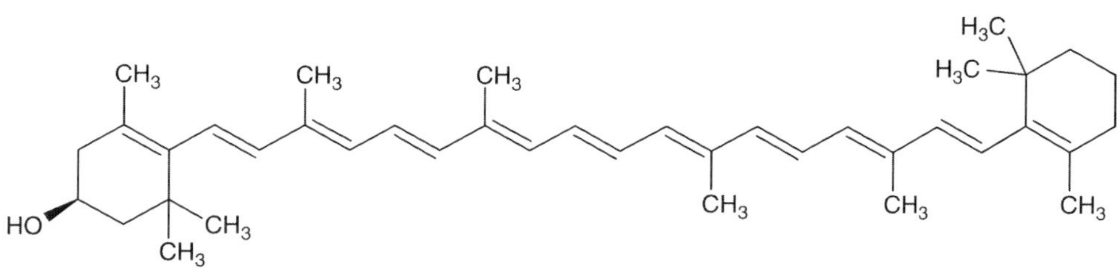

Zeaxanthin: 4-[18-(4-hydroxy-2,6,6-trimethyl-1-cyclohexenyl)-3,7,12,16-tetramethyl-octadeca-1,3,5,7,9,11,13,15,17-nonaenyl]-3,5,5-trimethyl-cyclohex-3-en-1-ol.

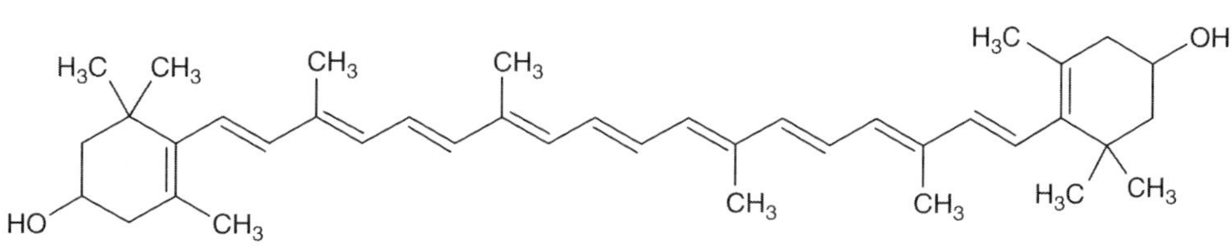

Fig. 14.1. Structures of major carotenoids from paprika.

cultivars Bola and Agridulce were dried for 8 h at 60°C in industrial drying tunnels. The paprika obtained was analysed at the end of the process and after 9 months of storage in darkness at 10°C. Capsanthin was the major pigment at all stages, followed by β-carotene in fresh fruits and capsorubin. Once fruit processing had begun, Agridulce always showed a higher pigment content than Bola. During milling, there was a marked decrease in all pigments. The loss in concentration of pigments caused by the addition of seed during milling was greater than that caused by degradation. During drying and milling, the yellow pigments, particularly β-carotene, were the most unstable. The red pigments were highly stable, with minimal degradation due to processing (Rodrigues *et al.*, 1998).

Localization of colouring matter

The colouring matter of paprika is present in nature as fatty acid esters. These carotenoid pigments are present in the thylakoid membranes of the chromoplasts. Using thin layer chromatography and column chromatography, the total extracts of pericarp after hydrolysis showed capsanthin as the major pigment. In plants, carotenoids are synthesized in both the chloroplasts of photosynthetic tissues and the chromoplast of flowers, fruits and roots. Chemically, carotenoids are lipid-soluble, symmetrical hydrocarbons with a series of conjugated double bonds. The double bond structure is responsible for the absorption of visible light. Carotenoids function as accessory pigments for photosynthesis but, more importantly, as photoprotectants in the plant. The primary function of β-carotene and other carotenoids is to protect the chloroplasts from photo-oxidative damage. However, carotenoids are unstable when exposed to light, oxygen or high temperatures. The carotenoids in the fruits are important for attracting seed dispersers (birds).

The green, yellow, orange and red colours originate from the carotenoid pigments of fruits during ripening. More than 30 different pigments have been identified in chilli fruits. These pigments include the green chlorophylls a and b, the yellow-orange lutein, zeaxanthin, violaxanthin, antheraxanthin, β-cryptoxanthin and β-carotene. The red pigments, capsanthin, capsorubin and cryptocapsin, are found only in chilli fruits (Deli and Molnar, 2002).

In chilli, 95% of the total provitamin A is in green pod and β-carotene accounts for 93% in mature red pods. When mature red pods were measured, the cultivars with the highest and the lowest provitamin A activity were both yellow max pod types. In the matured red pods, the α-, β-carotene and provitamin A activity increased by 344, 255 and 229%, respectively, as the pods matured.

Free and bound carotenoids

A protocol for extraction and chromatographic separation with a C30-reverse-phase column for analysis of non-saponified lipid extracts of paprika fruits by liquid chromatography-mass spectrometry (LC-MS) was developed by Breithaupt and Schwack (2000). Using this procedure, it was possible to identify the main mono- and diesterified derivatives of capsanthin, capsorubin, β-cryptoxanthin and zeaxanthin occurring naturally in red peppers. LC-MS analyses proved that xanthophylls of red peppers were acylated exclusively with saturated C_{12}, C_{14} and C_{16} fatty acids, whereas unsaturated, as well as C_{18} fatty acids, generally were absent. However, saponification experiments on paprika lipid extracts showed that approximately 75% of the total fatty acids of red peppers were C_{18} fatty acids. In contrast, direct extracts of green peppers comprised only free carotenoids, while the fatty acid distributions of green and red peppers did not differ significantly.

Accumulation of carotenoids due to ethephon

Perucka (1996) studied ethephon-induced changes in the accumulation of carotenoids in the red pepper fruit. Ethephon application stimulated fruit maturation. The mass of ripe fruits from the second and third harvests was increased by an average of > 44% with ethephon application. Treated fruits from the third harvest had higher concentrations of capsanthin (11% on average), β-carotene (14%) and β-cryptoxanthin-provitamin A (18%) than control fruits. The increase in these pigments was accompanied by a decrease in the amount of zeaxanthin and the disappearance of neoxanthin.

Carotenoids and maturity

Deli *et al.* (1992) studied the carotenoid composition in the fruits of black paprika (*C. annuum* variety *longum nigrum*) during ripening. During ripening, an increase in capsanthin and, to a lesser extent, an increase in carotenoids with kappa and oxabicyclo [2.2.1] end groups, was observed.

A detailed analysis of the carotenogenesis was performed throughout the ripening process, with special emphasis on the ripe stage, for selecting the best cultivar for paprika production (Mendez *et al.*, 2002). Most of the cultivars showed a total carotenoid content in the range of 7000–9700 mg/kg DW. In general, the chlorophyll-retaining character was related to high carotenoid content. The general trend of the cultivation cycle was that the shorter the cycle, the higher the total carotenoid content. The lowest total carotenoid content was 4856.77 mg/kg DW, which showed the longest cultivation cycle. The carotenogenic capacity of the cultivars has been discussed relative to total carotenoid content and the R/Y and Caps/Zeax ratios, the main quality traits for breeding cultivars for the production of high quality paprika. The cultivar with the highest total carotenoid content, high R/Y (2.11) ratio, and the highest Caps/Zeax (9.85) ratio was found to be the most suitable cultivar for paprika production in terms of carotenoid pigment biosynthesis capacity (Mendez *et al.*, 2002).

Ripening index and pigments

Mendez and Mosquera (2000) indicated the usefulness of xanthophyll esterification accompanying carotenoid overaccumulation in chromoplasts of *C. annuum* in ripening fruits as a ripeness index. Changes in the degree of xanthophyll esterification during capsicum fruit ripening were studied in five cultivars. Esterification of xanthophylls with fatty acids was linked to the transformation of chloroplasts (present in green fruits) into chromoplasts (present in red fruits). Changes in the fractions of free, and partially and totally esterified, carotenoids were similar between cultivars, reflecting the constitutive nature of esterification as part of the ripening process and being controlled by it. From the first stages of ripening, the fraction of totally esterified pigments (zeaxanthin diester, β-cryptoxanthin diester, capsanthin diester and capsorubin diester) made up almost 50% of the total carotenoid content. The proportion of the partially esterified pigment fraction (zeaxanthin monoester, capsanthin monoester and capsorubin monoester) in the total carotenoid content increased, with a gradual decrease in the fraction of free pigments (β-cryptoxanthin, β-carotene, zeaxanthin, capsanthin and capsorubin). In fully ripe fruits, a balance was reached between the three esterification fractions (free, partially esterified and totally esterified), with mean values of 24.17 ± 4.06, 31.48 ± 4.61 and 44.36 ± 5.05, respectively, which is almost independent of cultivar. This suggested a marked control of the carotenoid composition of the totally developed chromoplast, indicating its use as an index of ripeness (Krajayklang *et al.*, 2000).

New carotenoids

The structure of a new carotenoid, isolated from fruits of the red tomato-shaped paprika, was elucidated to be (3*S*,5*R*,6*S*,5϶P)-3,6-epoxy-5,6-dihydro-5-hydroxy-β,κ-carotene-3′,6′-dione by spectroscopic analyses and mass spectrometry and was designated as capsanthone 3,6-epoxide. Capsanthone 3,6-epoxide is assumed to be an oxidative metabolite of capsanthin 3,6-epoxide in paprika (Maoka *et al.*, 2001a).

Eleven apo-carotenoids (1–11), including five new compounds, 4, 6, 9, 10 and 11, were isolated from the fruits of the red paprika collected from Japan by Maoka *et al.* (2001b). The structures of new apocarotenoids were determined to be apo-14′-zeaxanthinal (4), apo-13-zeaxanthinone (6), apo-12′-capsorubinal (9), apo-8′-capsorubinal (10) and 9,9′-diapo-10,9′-retro-carotene-9,9′-dione (11) by spectroscopic analysis. The other six known apocarotenoids were identified to be apo-8′-zeaxanthinal (1), apo-10′-zeaxanthinal (2), apo-12′-zeaxanthinal (3), apo-15-zeaxanthinal (5), apo-11-zeaxanthinal (7) and apo-9-zeaxanthinone (8), which had not been found previously in paprika. These apocarotenoids were assumed to be oxidative cleavage products of C40 carotenoid, such as capsanthin in paprika.

Biosynthesis of carotenoids

The amount of carotenoids formed by biosynthesis was quantified by the accumulation of the colourless carotenoid phytoene in the presence of the inhibitor, norflurazon. When applied, substantial amounts of this rather photostable intermediate were formed in the light. However, carotenoid biosynthesis was completely stalled in darkness. This switch-off in the absence of light is related to the presence of very low messenger levels of the phytoene synthase gene, *psy*, and the phytoene desaturase gene, *pds*. It could be demonstrated that light-independent degradation or conversion of carotenoids to abscisic acid was a minor process (Simkin *et al.*, 2003).

Changes in the biosynthesis of individual carotenoid pigments were investigated by Mendez *et al.* (2000) during fruit ripening of five cultivars of red pepper. Ripening fruit of the five cultivars showed the typical and characteristic pattern of carotenoid biosynthesis for the *Capsicum* genus. In the five cultivars, lutein and neoxanthin, both characteristic chloroplast pigments, decreased in concentration with ripening, and eventually disappeared. β-Carotene, antheraxanthin and violaxanthin increased in concentration, and other pigments such as zeaxanthin, β-cryptoxanthin, capsanthin, capsorubin, capsanthin-5,6-epoxide and cucurbitaxanthin A were biosynthesized *de novo.* A pool of zeaxanthin stands out from the rest of pigments during ripening, which reveals the importance of this pigment as a branching point in the carotenoid biosynthesis in *Capsicum.* In all the red varieties, there was an inverse relationship between total carotenoid content and the red-to-yellow isochromic pigment fraction ratio (R/Y) and the capsanthin-to-zeaxanthin ratio (Caps/Zeax) (Mendez *et al.*, 2000). Deli *et al.* (2001) suggested possible biosynthetic routes for the formation of minor carotenoids containing 3,5,6-trihydroxy-β, 3,6-epoxy-β and 6-hydroxy-γ end groups.

Studies on the temperature profile simulating the traditional slow-drying process of red pepper fruits for paprika production indicate that the evolution of carotenoid concentration, the main quality trait for paprika, depends directly on the physical conditions imposed (Galvez *et al.*, 2004). During the drying process, three different stages could be observed in relation to the carotenoids. The first stage corresponded to a physiological adaptation to the new imposed conditions with a slight reduction in colour value, followed by a sharp increase in the carotenoid content in the final stage.

The study demonstrated that fine control of the temperature and moisture content would help to modulate carotenogenesis positively and minimize catabolism, making it possible to adjust the drying process to the ripeness stage of fruits to improve carotenoid retention and the quality of the resulting product (Galvez *et al.*, 2004).

Molecular genetics of biosynthesis

Classical genetic studies have determined that the yellow fruit colour in pepper (*Capsicum*) is recessive to red in the locus *y*. The relation of the *y* locus with the gene coding for capsanthin–capsorubin synthase (CCS), which synthesizes the red carotenoid pigments in the mature fruit, was studied by Popovsky and Paran (2000). Co-segregation of *y* and CCS in populations derived from crosses between plants bearing red × white and red × yellow fruits indicated the correspondence of the two genes. They obtained indications for the occurrence of a deletion in the CCS gene in plants containing the recessive *y* allele. This deletion did not contain the distal 220 bp of the 3′ end of the gene. They used the CCS gene to determine the genotype of peppers with different fruit colours at the *y* locus. In BC1 segregants from a red × white cross, the red- and peach-fruited progenies had the wild-type allele at the CCS locus, while the orange-, yellow- and white-fruited progenies had the mutant allele. Screening orange-fruited cultivars with CCS, as well as segregation analysis of CCS in an additional red × white cross, indicated two possible genotypes of the orange fruit colour in this locus.

Neoxanthin, a precursor of the plant hormone abscisic acid, is an allenic xanthophyll recognized as the last product of carotenoid synthesis in green plants. A cDNA for neoxanthin synthase (NSY) was isolated from tomato cv. Philippino using a molecular approach based on the mechanistic and structural similarities of NSY to two other closely related carotenogenic enzymes, lycopene cyclase (LCY) and capsanthin–capsorubin synthase (CCS) (Bouvier *et al.*, 2000).

The differentiation of chloroplasts into chromoplasts involves a series of biochemical changes that culminate in the intense accumulation of long-chain chromophore carotenoids, such as lycopene, rhodoxanthin, astaxanthin, anhydroeschsoltzxanthin, capsanthin and capsorubin. The signal pathways mediating these transformations are unknown. Chromoplast carotenoids are known to accumulate in green tissues experiencing stress conditions and studies indicate that they provide efficient protection against oxidative stress. The addition of reactive oxygen species (ROS) progenitors, such as menadione, *tert*-butylhydroperoxide or paraquat and pro-oxidants such as diamide or buthionine sulphoximine, to green pericarp discs of pepper fruits rapidly and dramatically induces the simultaneous expression of multiple carotenogenic gene mRNAs that give rise to capsanthin. Similarly, down-regulation of catalase by amitrole induces expression of carotenogenic gene mRNAs, leading to the synthesis of capsanthin in excised green pericarp discs. ROS signals from plastids and mitochondria also contribute significantly to this process. Analysis of the capsanthin–capsorubin synthase promoter (database accession number Y14165), in combination with a β-glucuronidase reporter gene, reveals strong activation in transformed pepper protoplasts challenged with the above ROS. Collectively, these data demonstrate that ROS act as a novel class of second messengers that mediate intense carotenoid synthesis during chromoplast differentiation (Bouvier *et al.*, 1998).

Chen *et al.* (1998) conducted a study to investigate the regulation of the fibrillin (*fib*) gene, along with two carotenoid biosynthesis genes, namely those encoding geranylgeranyl pyrophosphate synthase (*ggpps*) and capsanthin–capsorubin synthase (*ccs*) from bell pepper, whose expression is greatly induced during fruit ripening. A homologous transient expression assay has shown that high expression of these genes in pepper fruit is regulated essentially at the transcriptional level. Transcription of *ccs* is mainly fruit-specific and that of *ggpps* is highly induced in fruits. Expression of *fib* is more complex: it is induced not only by a developmental process in fruits but also, in pepper and tobacco leaves, by diverse environmental factors such as drought and mechanical wounding. The wound-induced transcriptional activation of *fib* is light- and oxygen-dependent.

A geranylgeranyl pyrophosphate synthase (*ggpps*) [farnesyl transtransferase] gene from *C. annuum* (bell pepper) was cloned by screening with an internal fragment of the *ggpps* cDNA, which was labelled by random priming (Badillo *et al.*, 1995). The nucleotide sequence (EMBL/GenBank/DDBJ accession number X80267) showed that this gene, like the capsanthin–capsorubin gene but unlike the phytoene synthase gene from *C. annuum*, was not interrupted by an intron. Southern blot analysis of *C. annuum* genomic DNA suggested the presence of a single gene highly similar to the cDNA and also of additional related sequences.

The fruit colour of *C. annuum* is a trait of great economic importance in breeding. Lang *et al.* (2004) evaluated molecular genetic analysis of six genes involved in the carotenoid biosynthetic pathway in order to develop a system of molecular marker-assisted selection (MAS) for breeding pepper cultivars with various fruit colours. The capsanthin–capsorubin synthase (*ccs*) gene showed a polymorphism in the PCR pattern in the segregation population derived from a cross between a pepper accession (cv. msGTY-1) with orange fruits and a pepper accession (cv. 277long) with red fruits. There was a deletion in the upstream region of the *ccs* gene in the plants with orange fruits. Southern hybridization analysis and sequencing analysis indicated that 211-bp of the downstream region of the gene was conserved in the plants with orange fruits, while no transcript of the *ccs* gene was detected by RT-PCR in the mature orange fruits. Carotenoid composition analysis using the TLC method showed that one of the major pepper carotenoids, capsanthin, was present in the red fruits but not in the orange fruits. The PCR polymorphism of the *ccs* gene and TLC pattern of carotenoid composition were completely co-segregated with the fruit colour in the F2 population, suggesting that the *ccs* gene determines the fruit colour by changing the carotenoid composition.

Paprika oleoresin

Paprika oleoresin is prepared industrially by solvent extraction of the dried fruit and the subsequent removal of the solvent. It is a dark red liquid of oily appearance with the characteristic odour and flavour of paprika. It has a density of 0.935–0.945 g/ml with allowed residual solvent of < 25 ppm. The solvent generally used for extracting oleoresin is hexane, ethyl acetate or ethylene dichloride. One kg of paprika oleoresin of 100,000 c.u. is of equal value to approximately 30 kg of a good grade of Spanish paprika. Generally, it is transported in metallic drums with an epoxy-phenolic inside inlay and it is stored invariably in airtight containers in a cool, dark and dry place. The shelf life with consistent quality is about 1 year. It is not recommended to freeze paprika oleoresin. Pigment composition was determined by HPLC. The system hexane/acetone/isopropyl alcohol (3:2:1) gave the best yield and colour value and a chromatographic pattern identical to that of the commercial oleoresin (Lemos *et al.*, 1993).

Paprika oleoresin is fractioned by extraction with supercritical carbon dioxide (SCF-CO_2). Higher extraction volumes, increasing extraction pressures and the use of solvents such as 1% ethanol or acetone resulted in greater pigment yields. Within the 2000–7000 psi range, total oleoresin yield always approached 100%. Pigments isolated at lower pressure consisted almost exclusively of β-carotene, while pigments obtained at higher pressures contained a greater proportion of red carotenoids (capsanthin, capsorubin, zeanthaxin and β-cryptoxanthin) and small amounts of β-carotene (Reilly *et al.*, 2001).

This technique (extraction using super critical carbondioxide) removes the paprika oil and β-carotene during the first extraction step, allowing for second-stage oleoresin extracts with a high pigment concentration (200% relative to the reference) and a red/yellow pigment ratio of 1.8 (Jaren Galan *et al.*, 1999).

Standardization of solvent

Mini *et al.* (1998) conducted a study to standardize the solvent for chilli oleoresin extraction. Extraction of oleoresin from *Capsicum* using six different solvents was carried out in order to find the most suitable one for standardizing the procedure of extraction. The solvents used were acetone, ethyl alcohol, dichloroethane, hexane, benzene and ethyl acetate. Extraction was fastest using ethyl acetate and took the minimum number of siphonings (9.2), which was similar to that for acetone (11.2) and benzene (11.0). Ethyl alcohol had the highest siphoning number (22.2) and time (40.2 min) for extraction. The yield of oleoresin ranged from 10.6% for dichloroethane to 37.2% for ethyl alcohol. Acetone produced maximum efficiency for colour value, with ethyl acetate having a similar value. The lowest colour value was produced by ethyl alcohol. Ethyl acetate produced maximum efficiency for capsaicin extraction (75.79 mg/g). Ethyl acetate was standardized as the best solvent for oleoresin extraction.

Genotype variation

Sankari *et al.* (2000) conducted a study on the per se performance of individual progenies of F3 generations of the cross Ramanathapuram Local X Jalapeno hot pepper with respect to oleoresin and related constituents. Oleoresin content ranged from 9.1 to 11.5%, with a mean value of 10.6%, along with variability in colour value.

Capsanthin esters

Goda *et al.* (1995) separated capsanthin esters from oleoresin of paprika fruits from Spain and determined their chemical structures without saponification. The major monoesterified capsanthin was identified as 3'-*O*-myristoylcapsanthin. It is suggested that the rate of esterification of fatty acid to the hydroxyl group on the cyclopentane ring of capsanthin is different to that on the cyclohexene ring.

Priya *et al.* (2002) noted the effect of seasons and growth environments on the growth and yield of paprika. Shading, especially during winter, increased the contents of capsanthin, oleoresin and ascorbic acid of paprika.

Pungency and capsaicinoids

Borges (2001) describes in detail why chilli is pungent. Pungency is produced by the capsaicinoids, a group of alkaloid compounds that are found only in the plant genus, *Capsicum*. The nature of the pungency has been established as a mixture of seven homologous branched-chain alkyl vanillylamides. They are often called capsaicin after the most prevalent compound. Dihydrocapsaicin is usually the second most prevalent capsaicinoid, while the other five compounds, norcapsaicin, nordihydrocapsaicin, nornordihydrocapsaicin, homocapsaicin and homodihydrocapsaicin, are considered minor capsaicinoids because of their relatively low abundance in most natural products. Capsaicin is a powerful and stable alkaloid that can be detected by human taste buds in solutions of ten parts per million. Capsaicin's composition ($C_{18}H_{27}NO_3$) is similar to piperine ($C_{17}H_{19}NO_3$), which gives black pepper its bite.

The capsaicinoids present in the capsicum fruit are predominantly capsaicin and dihydrocapsaicin, making up 80–90% (Govindarajan and Sathyanarayana, 1991). Dihydrocapsaicin accounts for about 22% of the total capsaicinoids mixture and has about the same pungency as capsaicin. Capsaicinoids are ingested mainly as naturally occurring, pungency-producing components of the *Capsicum* species (chilli, cayenne pepper, red pepper). They range typically from 0.1 mg/g in chilli pepper to 2.5 mg/g in red pepper and 60 mg/g in oleoresin red pepper. Pepper varieties from *C. frutescens*, *C. annuum* and *C. chinense* were found to contain 0.22–20.00 mg total capsaicinoids/g DW. Cayenne pepper samples had mean capsaicin and dihydrocapsaicin contents of 1.32 and 0.83 mg/g DW, respectively.

Capsaicinoids are controlled by a range of factors – genetics, position of fruit and interaction of the plant and environment. For example, cool growing conditions reduce heat levels dramatically. Capsaicinoids are synthesized mainly in the placenta and are also found in the seed (Purseglove *et al.*, 1981). Heat is perceived to be due to the interaction of the capsaicinoids with the nerve cells of the skin and mucous membranes, which triggers responses similar to the ones caused by thermal heat. For trade, it is important to measure heat levels adequately. This allows standardization of products. However, responses are very subjective, due to tasters getting used to capsaicinoids over time, showing reduced responses to similar heat stimuli. The commercial method of determining heat levels in chillies is to measure pungency as Scoville heat units (SHU) (discussed later in this chapter). In organoleptic terms, the main discriminative factor between the dried products is the degree of pungency exhibited. This is highest in chillies, moderate to mild in capsicums and absent in most forms of paprika (Purseglove *et al.*, 1981). Chilli pungency is expressed in SHU (Scoville, 1912). The Scoville organoleptic test was the first reliable measurement of the pungency of chilli. This test has now been replaced by instrumental methods. The most common instrumental method is high-performance liquid chromatography (HPLC). It provides accurate and efficient analysis of the content and type of capsaicinoids present in a chilli sample. HPLC analysis has become the standard method for routine analysis by the processing industry (ISO7543.2:1993). The method is rapid and can handle a larger number of samples. A common practice today is to multiply capsaicinoid ppm by 15 to convert to SHU. The capsaicinoids are produced in glands on the placenta of the fruit. While seeds are not the source of pungency, occasionally they absorb capsaicin because of their proximity to the placenta. No other plant part produces capsaicinoids. Ishikawa *et al.* (1998) evaluated the contents of capsaicinoids and their phenolic intermediates in the various tissues. Capsaicinoids accumulate primarily in the placenta of fruits (average of 33 and 38 μmol/g of capsaicin and dihydrocapsaicin, respectively). Approximately 58 and 49% of vanillylamine and phenolic intermediates, respectively, accumulate in the placenta. Their distribution is not correlated with the production of capsaicinoids (Reilly *et al.*, 2001).

Yao *et al.* (1994) extracted and quantified capsaicin and dihydrocapsaicin using SCF-CO_2 and organic solvent extraction from Scotch Bonnet (*C. annuum* L.). The super critical CO_2 extract afforded 3.2 and 0.58% capsaicin and dihydrocapsaicin, while the combined organic extract yielded 0.5 and 0.09%, respectively, per g (DW).

Sato *et al.* (1999) developed a rapid quantitative method for estimating capsaicinoids in placentas of *Capsicum*. This was achieved by employing a direct connection of supercritical fluid extraction and supercritical fluid chromatography. They suggested this method was useful as a rapid (20 min) and safe screening test for the pungency of *C. annuum* fruits.

Effect of seasonal changes and other parameters on capsaicin

Yun *et al.* (2002) observed the changes in fruit component by temperature treatment after harvest of unripened fruit in hot pepper. The capsaicinoid contents of 100% coloured red fruits were highest at 30°C and lowest at 25°C. The capsaicin and β-carotene contents of 100% coloured red fruits of hot pepper were highest when stored at 15°C, while there were no significant differences in those contents among the other temperature treatments. In addition, the contents of cryptoflavin and cryptocapsin were highest at 15 and 25°C, respectively.

Lindsay and Bosland (1995) reported that pungency increased with increased environmental stress. More specifically, any stress to the chilli plant increased the amount of capsaicinoid level in the pods. A few hot days could increase the capsaicinoid content significantly. In New Mexico, it was observed that even after furrow irrigation, the heat level increased in the pods. Anthropopathically, the plant has sensed the flooding of its root zone as a stress and has increased the capsaicinoid level in its pods. If the same cultivar was grown in both a hot semi-arid region and a cool coastal region, the fruits harvested from the hot semi-arid region would be higher in capsaicinoids than the fruits harvested from the cool coastal climate.

The effect of water stress on phenyl propanoid metabolism was studied in Padron pepper plants growing in a greenhouse. The soluble phenolic and lignin contents of fruits from control plants were higher than those of water-stressed plants. The amount of capsaicinoids (capsaicin and dihydrocapsaicin) in the fruits of water-stressed plants was higher than that in control plants (Estrada *et al.*, 1999a).

The effect of seasonal changes on the pungency level of Padron pepper fruits was studied. The pattern of capsaicinoid accumulation was the same during the different months, but there was a considerable increase in capsaicinoid levels in August and September in all the growth stages studied (Estrada *et al.*, 1999b).

The kinetics of degradation of both the green- and total-colour of green chilli purée was studied at selected temperatures (50–90°C) by Ahmed *et al.* (2002). They reported that pungency of green chilli purée decreased during thermal processing as the capsaicin content was reduced from 559 to 441 μg/g, while the SHU decreased from 8500 to 7480.

Robi and Sreelathakumary (2004) noted the influence of maturity at harvest on capsaicin and ascorbic acid content in hot chilli (*C. chinense* Jacq.). Analysis of variance for capsaicin and ascorbic acid content over different maturity stages showed that maturity x genotype interaction was significant. Significant variation was observed among genotypes for capsaicin content at the colour changing (1.26–3.02%), red ripe (1.32–3.18%) and withering stages (1.48–3.36%), with certain genotypes giving the highest contents at all stages. Similarly, significant variation was observed among genotypes for ascorbic acid content at the colour changing (89.40–130.12 mg/100 g fresh fruit), red ripe (95.23–136.45 mg/100 g fresh fruit) and withering stages (89.26–136.59 mg/100 g fresh fruit). In general, genotypes had the highest capsaicin content at the withering stage and the highest ascorbic acid content at the red ripe stage.

Iwai *et al.* (1977) observed the formation of the pungent principles in the fruits of sweet pepper, *C. annuum* L. var. *grossum*, during postharvest ripening under continuous light. Marked accumulation of carotenoid was also observed during postharvest ripening under continuous light.

Gnayfeed *et al.* (2001) examined the level of bioactive compounds in red pepper (paprika) as affected by ripening and genotype. Results indicated that all the examined compounds were at a low level in mature green fruits and the onset of climacteric ripening caused their content to grow. With the advance in ripening, carotenoids were being formed even at the overripening stage, while tocopherols, capsaicinoids and ascorbic acid reached their maximum level at the colour break or red stage and then declined.

Cultivar- and location-specific variability

The capsaicin content of red and green chilli varieties from different parts of India, namely, Tamilnadu, Kashmir, Simla and Sankeshwari, was determined using HPLC. The percentage of capsaicin in red chillies from Madras, Kashmir, Sankeshwari and Kolhapur (Lavangi) was 0.210, 0.310, 0.127 and 0.220, respectively, and green chillies did not interfere with the analysis of capsaicin by HPLC (Krishnamurthy *et al.*, 1999).

The capsaicinoid concentrations of the developing (20–100 days after flowering (DAF)) fruits of *C. annuum* and *C. frutescens–chinense* complex cultivars were determined by HPLC. Three capsaicinoids were detected (capsaicin, dihydrocapsaicin and nordihydrocapsaicin), and intercultivar and interspecific differences in capsaicinoid concentrations were observed as early as 20 DAF. Capsaicinoid concentrations were highest between 20 and 40 DAF (Rani, 1996; Minami *et al.*, 1998).

Zewdie and Bosland (2001) found that capsaicinoid profiles were not good chemotaxonomic indicators for *Capsicum* species. They also found that it was not always true to state that capsaicin and dihydrocapsaicin were the major capsaicinoids. Changes in the mineral elements and capsaicin content of chilli (*C. annuum* L. and *C. frutescence* L.) fruits during development were studied in chilli cultivars, Krishna, Pusa Jawala and Pusa Sadabahar. Of the two edible stages (20 and 30 DAF), the mineral and capsaicin contents of fruits were highest at 20 DAF. Fruit capsaicin content of cv. Krishna was similar to that of Pusa Sadabahar and significantly higher than that of Pusa Jawala (Das *et al.*, 1996).

Immature and ripe fruits of Japanese hot and sweet peppers were analysed for total phenolic, flavonoid and capsaicinoid contents. The amount of phenolic compounds was higher in hot and ripe pepper fruits than in sweet and immature pepper fruits. Capsaicinoids, which were detected in hot peppers only, were highest in immature 'and ripe' fruits. The data reveal that the total phenolics content is a good indicator of the antioxidant activity of pepper fruits and that the antioxidant levels in sweet peppers, especially in the ripe fruits, are higher than they are in hot peppers (Saga and Sato, 2003).

High-capsaicin chilli

Attempts have been made to develop high-capsaicin chillies and eventually to develop hybrids which combine the advantage of high capsaicin content. Capsaicin content is a Mendelian character; pungency is heritable and is simple monogenic dominant to non-pungency. Selection based on parental performance per se has been found useful, with high × high crosses on average giving superior performance in comparison with high × low or low × low crosses. One of the first results was the cv. Pusa Jwala (*C. annuum*). It is virus resistant, has a high potential yield, wide adaptability, high pungency, good colour and 6–8% capsaicin in the oleoresin. Pusa Sadabahar, a superior line selected in a population of bird chillies, has 12% capsaicin content in the oleoresin. The cross Pusa Jwala × IC31339 (*C. frutescens*) has given exceptionally good transgressant lines for oleoresin quality over both parental varieties, having over 15% capsaicin. The transgressant lines PS31–3 and LG1 possessed 20 and 27.5% capsaicin in the oleoresin, respectively (Tewari, 1990).

Mathur *et al.* (2000), in their search for the hottest chilli variety in India, identified the Tezpur cultivar (*C. frutescens* var. nagahari) as having the highest amounts of capsaicin and dihydrocapsaicin (4.28 and 1.42% w/w, respectively). As per the authors, it contributes a pungency rating of 855,000 SHU (Scoville heat units), which is the hottest chilli known to date (see below).

Scoville heat units

In 1912, Wilbur Scoville developed a method to measure the heat level of a chilli pepper. This test is named after him, is called the Scoville organoleptic test and is a dilution-taste procedure (Scoville, 1912). The test used a panel of five human representatives, who tasted a chilli sample and then recorded the heat level. A sample was diluted, until the pungency no longer could be detected. Scoville blended pure ground chillies with a sugar–water solution and a panel of testers then sipped the concoctions, in increasingly diluted concentrations, until they reached the point at which the liquid no longer burned the mouth. A number was then assigned to each chilli based on how much it needed to be diluted before heat could no longer be tasted. The organoleptic method or taste test has been the standard method for pungency analysis. Although this method is used widely, it has limitations. Tasters must be trained and their ability to test many samples is restricted by the heat of the test solution. Taster fatigue is a real phenomenon and tasters are also unable to distinguish between the different capsaicinoids.

The pungency of chilli is measured in multiples of 100 units, from the bell pepper at zero Scoville units to the incendiary Habanero at 300,000 Scoville units. One part of chilli 'heat' per 1,000,000 drops of water rates as only 1.5 Scoville units. The substance that makes a chilli so hot is capsaicin. Pure capsaicin rates over 15,000,000 Scoville units! The 'Red Savina' Habanero has been tested at over 577,000 Scoville units (Padilla and Yahia, 1998). Tiwari *et al.* (2005) identified the hottest chilli variety in the world. A special variety of chilli, Nagarhari, grown in Tezpur (Assam) has been found to possess a pungency of 855,000 SHU with 5% capsaicinoids. Table 14.3 give the SHU of major capsaicinoids and Table 14.4 illustrates heat units of some of the international chilli varieties.

Capsaicin, also known as *N*-vanillyl-8-methyl-6-(*E*)-nonenamide, is the most pungent among capsaicinoids that can be isolated. It is sparingly soluble in water but very soluble in fats, oils and alcohol.

It has been shown organoleptically that humans not only note intensity of pungency but also perceive each capsaicinoid differently. The investigations of Krajewska and Powers (1988) revealed that nordihydrocapsaicin (NDC) was the least irritating and burning was located in the front of the mouth and palate. It causes a 'mellow warming effect'. The heat sensation developed immediately after swallowing and receded rapidly. In comparison, capsaicin and dihydrocapsaicin were more irritating and were described as having a 'typical' heat sensation. Both compounds produced heat in the mid-mouth and mid-palate, as well as the throat and the back of the tongue. In contrast, homodihydrocapsaicin was very irritating, harsh and very sharp. Heat did not develop immediately and it affected the throat, back of the tongue and the palate for a prolonged period. The heat sensation can last for up to 6 h after ingestion. Different combinations of these capsaicinoids produce the different pungency characteristics of individual chilli varieties (Cooper *et al.*, 1991).

The analytical method for determining chilli colour and pungency can be categorized into five basic groups: (i) organoleptic; (ii) colorimetric methods: chromogenic reagents react directly with the phenolic hydroxyl of the vanillyl moiety in the extracts of fruits; (iii) thin-layer chromatography (TLC) and paper chromatography; (iv) gas chromatography; and (v) high-performance liquid chromatography (HPLC) (Kocsis *et al.*, 2002).

Biosynthesis of capsaicin

Capsaicinoids, responsible for hot flavour, are synthesized through the cinnamic acid pathway and their degradation is aided by the action of peroxidases. The evaluation of capsaicinoids during the development, maturation and senescence of fruits of Habanero (*C. chinense* Jacq.), De Arbol (*C. annuum* var.

Table 14.3. Major capsaicinoids and their SHU (Dray, 1992).

Capsaicinoid	Abbreviation	Typical relative amount (%)	Scoville heat units (SHU)	Chemical structure
Capsaicin	C	69	16,000,000	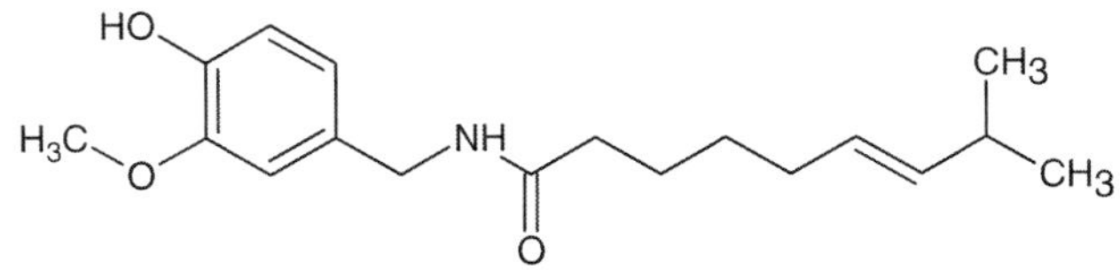
Dihydrocapsaicin	DHC	22	16,000,000	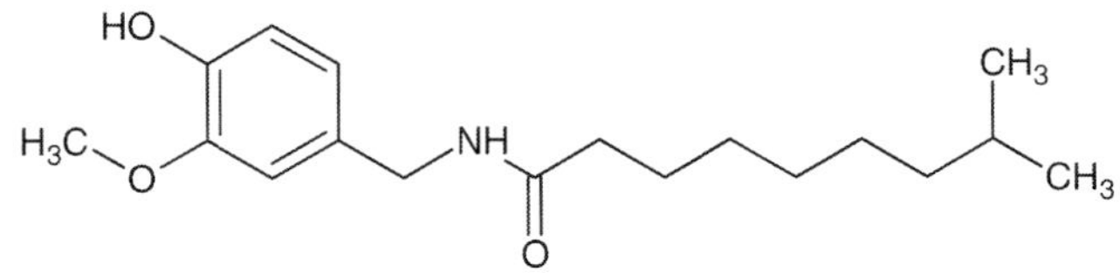
Nordihydrocapsaicin	NDHC	7	9,100,000	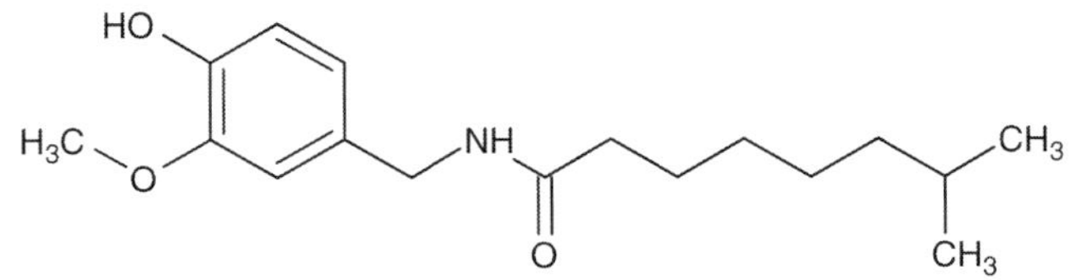
Homodihydrocapsaicin	HDHC	1	8,600,000	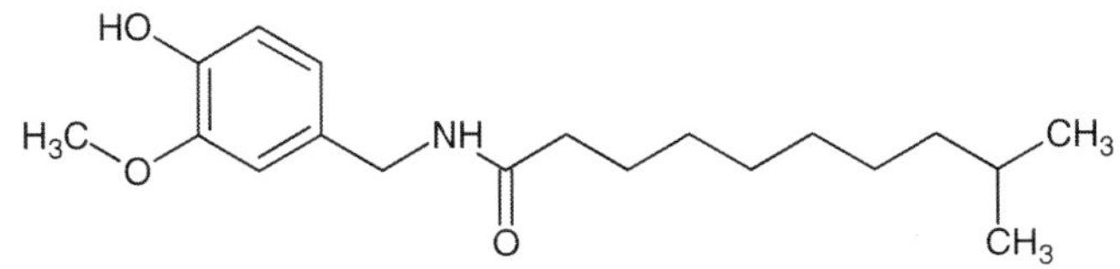
Homocapsaicin	HC	1	8,600,000	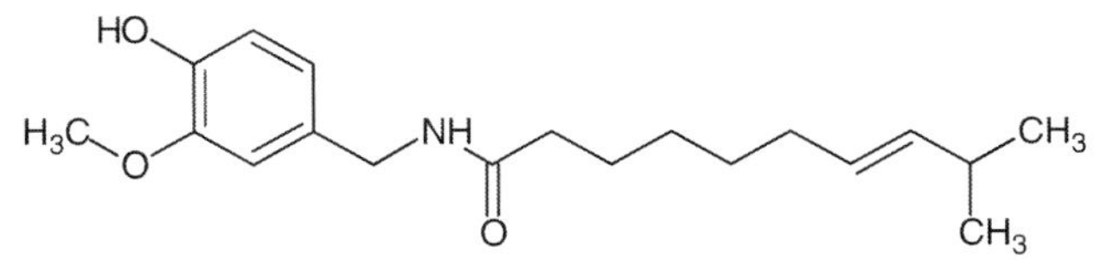

Table 14.4. Heat units of major international chilli varieties.

Chilli type	Scoville heat units (SHU)
Bell/sweet	0–100
New Mexican	500–1,000
Espanola	1,000–1,500
Ancho & Pasilla	1,000–2,000
Cascabel & Cherry	1,000–2,500
Jalapeno & Mirasol	2,500–5,000
Serrano	5,000–15,000
De Arbol	15,000–30,000
Caynne & Tabasco	30,000–50,000
Chiltepin	50,000–100,000
Scotch Bonnet & Thai	100,000–350,000
Habanero	200,000–350,000
Nagarhari (Assam)	855,000

annuum) and Piquin (*C. annuum* var. *aviculare)* was investigated. The relationship between capsaicinoid content and peroxidase activity was determined. Capsaicinoids were most abundant in Habanero, followed by De Arbol and Piquin. There was a higher content of capsaicin than dihydrocapsaicin in all three fruits. The contents of capsaicinoids (particularly capsaicin and dihydrocapsaicin) increased continuously and reached a peak 45–50 days after fruit set in Habanero and De Arbol and 40 DAF in Piquin, and then declined. Peroxidase activity increased at the time when the concentrations of capsaicinoids started to decrease. There was an inverse relationship between the evolution of capsaicinoids and peroxidase activity that might indicate that this enzyme was involved in capsaicin degradation (Padilla and Yahia, 1998).

Capsaicin is a bioactive molecule synthesized by enzymatic (putative capsaicin synthase) condensation of vanillylamine, a phenyl propanoid intermediate with 8-methyl-nonenoic acid, a fatty acid derivative from the leucine/valine pathway. Analysis of levels of 8-methyl-nonenoic acid and phenyl propanoid intermediates in high-, medium- and low-pungent *Capsicum* genotypes revealed that the 8-methyl-nonenoic acid pool plays a crucial role in determining the efficacy of capsaicin levels. Cerulenin-mediated inhibition of 8-methyl-nonenoic acid synthesis decreased the capsaicin biosynthesis in *Capsicum* cell suspension cultures. Similarly, amino oxy acetate inhibited vanillylamine synthesis but failed to reduce capsaicin production. The mRNA transcript analysis of keto acyl synthase (KAS), a crucial enzyme involved in 8-methyl-nonenoic acid, and an amino transferase (AMT) involved in vanillylamine biosynthesis, was studied. The mRNA transcript analysis revealed the progressive developmental expression of the *kas* gene in the placenta during the ontogeny of the fruit, whereas AMT transcript levels did not show significant differences. Hence, the study demonstrates the influence of 8-methyl-nonenoic acid and its possible regulatory role in capsaicin biosynthesis in *Capsicum* spp. (Prasad *et al.*, 2006). Alejo (2006) described various techniques for studying the biosynthetic mechanism of capsaicinoids. Several efforts have been focused on investigating the conditions.

Quantitative variation in the accumulation of two major capsaicinoids responsible for pungency in the fruit of chilli peppers, capsaicin and dihydrocapsaicin, was analysed in a cross between the non-pungent *C. annuum* parent cv. Maor and a pungent *C. frutescens* parent, accession BG 2816. QTL (quantitative fruit linkage) interval analysis for individual and total capsaicinoid content identified a major QTL, termed cap, which explained 34–38% of the phenotypic variation for this trait in two growing environments. For all measurements, the allele of the pungent parent BG 2816 contributed to the increased level of pungency. To determine whether known structural genes in the pathway could define a candidate for this QTL, 12 clones obtained from differentially expressed transcripts from placental tissue in pungent peppers were also mapped. None of them had a significant effect on this trait, nor did the allelic state at the locus C, the on/off switch for pungency in pepper, located on chromosome-2 (Blum *et al.*, 2004).

Studies carried out by Bernal *et al.* (1995) on the ability of hot pepper (*C. annuum*) peroxidase and the phenolic precursors of capsaicin biosynthesis showed that hot pepper peroxidase, and especially hot pepper peroxidase iso enzyme B6 (Prx B6), was capable of oxidizing the phenolic precursors of capsaicin, with caffeic acid and ferulic acid being the best substrates. Vanillylamine was the only precursor that did not act as a substrate for peroxidase catalysed oxidations. Since the basic peroxidase isoenzyme B6 is located in cell walls, this isoenzyme may be involved in the insolubilization of cell wall-bound phenyl propanoid precursors. These results lend weight to the biochemical evidence for supporting the existence of an oxidative competitive sink for phenyl propanoid intermediates of capsaicin biosynthesis, which probably competes with capsaicin itself to yield lignin-like substances in the cell wall of *C. annuum* (Bernal *et al.*, 1995).

Phenylalanine ammonia lyase (PAL), cinnamic acid 4-hydroxylase (CA4H), *p*-coumaric acid 3-hydroxylase (CA3H), caffeic acid *O*-methyl transferase (CAOMT) and capsaicinoid synthase (CS) activities involved in the capsaicin biosynthetic pathway were investigated in developing or mature fruits or callus cultures of *Capsicum* cv. Tampiqueno 74. The maximum activities for PAL, CA4H and CA3H in callus cultures were similar to those in fruits, but values for CAOMT and CS activities in callus tissues were six fold lower than those in fruits. In general, fruit tissue showed low PAL, CA4H and CS activities during the period from 0 to 22 DAF and high enzyme activities at the time of maximum fruit growth (day 30) in calluses, peak activities for all enzymes except CA4H observed at day 7 of culture, which corresponded to the period of initial growth (Alejo and Peralta, 1993).

Alejo and Garciglia (1992) conducted studies relating PAL activity and capsaicin-precursor compounds in *p*-fluorophenylalanine-resistant and -sensitive variant cells of chilli pepper. PFP (*p*-fluorophenylalanine)-resistant cells exhibited the highest PAL activity and free phenylalanine and phenolic compound concentrations, even after a minimum of nine subcultures (15 days each), in the absence of PFP, indicating that the selected trait was stable. PFP-resistant cells accumulated greater amounts of capsaicin precursors (cinnamic, caffeic and ferulic acids) than either the non-selected cells or the sensitive cell line. *p*-Coumaric acid was not detected at significant levels in any of the cell cultures. Overall, accumulation of free phenylalanine correlated well with PAL activity and the concentrations of phenolic compounds, phenylpropanoids and capsaicin, suggesting an active flow through the phenylpropanoid pathway in PFP-resistant cells.

Studies carried out by Kopp and Jurenitsch (1981) on the biosynthesis of capsaicinoids in *C. annuum* var. *annuum* established that valine, isoleucine and leucine were direct precursors of the respective even- and odd-numbered branched fatty acid moieties of the capsaicinoids.

Microscopic investigation of the structure of the placenta of *C. annuum* var. *annuum* cv. Karayastubusa was undertaken for the intracellular localization of capsaicin and its analogues in *Capsicum* fruit. The epidermal tissue of the placenta appeared to be the site of capsaicinoid accumulation.

Yu *et al.* (2005) evaluated capsaicin biosynthesis in water-stressed hot pepper fruits. The concentration of capsaicin in the placenta of fruits in the water deficit treatment began to increase rapidly 10 DAF. It reached maximum at 30 DAF and was 3.84-fold higher than in the placenta of control treatment plants. In the pericarp, the concentration of capsaicin reached maximum at 50 DAF and was 4.52-fold higher than in the control treatment. PAL activity was higher in the placenta of fruits in the water deficit treatment than in the fruits of control plants at 50 DAF. At 40 or 50 DAF, cinnamic acid 4-hydroxylase (*trans*-cinnamate-4-monooxygenase) (C4H) activity was higher in plants subjected to the water deficit treatment than in control plants. Capsaicinoid synthase (CS) activity 40 DAF was 1.45- to 1.58-fold higher in fruits in the water deficit treatment than in fruits in the control treatment.

Diaz *et al.* (1998) and Estrada *et al.* (2000) conducted detailed studies on the biosynthesis of phenolics in capsicum and phenolic metabolism with reference to capsaicin biosynthesis.

Yu *et al.* (2005) found increase in capsaicin biosynthesis in water-stressed hot pepper fruits. 'Hungariana', 'Beauty Zest' and 'Home Flavor' hot pepper plants (*C. annuum* var. *annuum*) were grown with ample or limited water supply. The fruits of plants in the water deficit treatment were small, had a proportionally heavier placenta and had a higher concentration of capsaicin. The concentration of capsaicin in the placenta of 'Beauty Zest' fruits in the water deficit treatment began to increase rapidly 10 DAF. It reached maximum at 30 DAF and was 3.84-fold higher than in the placenta of the control treatment plants. In the pericarp, the concentration of capsaicin reached maximum at 50 DAF and was 4.52-fold higher than in the control treatment. In 'Hungariana' fruits, the concentration of capsaicin in the placenta was not significantly different among treatments. PAL activity was higher in the placenta of 'Beauty Zest' fruits in the water deficit treatment than in the fruits of control plants at 50 DAF. At 40 or 50 DAF, cinnamic acid-4-hydroxylase [*trans*-cinnamate 4-monooxygenase] (C_4H) activity was higher in plants subjected to the water deficit treatment than in the control plants. In both treatments, C_4H activity in the placenta was 1.4- to 1.5-fold greater than in the pericarp 40 DAF. CS activity 40 DAF was 1.45- to 1.58-fold higher in fruits in the water deficit treatment than in fruits in the control treatment. Although peroxidase activity was lower in plants in the water deficit treatment than in the control treatment, the difference was not significant.

Flavour and aroma

In paprika, the volatile oil in the fruits ranges from 0.1 to 2.6%. The characteristic aroma and flavour of the fresh fruit is imparted by volatile oil (Pruthi, 2003).

Chilli or paprika is used for flavour, not heat, in some cuisines. Flavour is a complex sensation determined in the mouth. One of the most potent volatiles known to humans is found in chilli, the pyrazine 2-methoxy-3-isobutyl-pyrazine, the 'green bell pepper' smell. Reports indicate that humans can detect this odour at two parts per trillion. Luning *et al.* (1994) reported the presence of more than 80 odour compounds in *C. annuum*. *C. frutescens* cv. 'Tabasco' contained 125 compounds whose relative abundance changed with the season of harvest. The composition of aroma compounds of Tabasco differed significantly from that of the green bell pepper. The Tabasco sample contained no pyrazine compounds. To reconstitute the Tabasco aroma, it took three main chemicals: 4-methyl-1-pentyl-2-methylbutyrate, 3-methyl-1-pentyl-3-methylbutyrate and iso-hexyl-isocaproate (Kocsis *et al.*, 2002).

Van Ruth *et al.* (1995) described the flavour components in bell peppers. Out of 47 compounds identified, 12 could be detected by assessors at a sniffing port on a gas chromatograph. 3(2220) 2-Methylpropanal,9 2-methylbutanal, 10 3-methylbutanal, 13(2370) 2,3-butanedione, 15(3382) 1-penten-3-one, 19(2557) hexanal, 25(2540) heptanal, 29 β-ocimene, 35 *trans*-3-hepten-2-one, 39 dimethyltrisulphide, 45(3132) 2-isobutyl-3-methoxypyrazine and 47(3639) β-cyclocitral were the major compounds.

Other flavouring compounds found in reasonable quantities are 14(3098) pentanal, 21 1-methyl-1*H*-pyrrole, 23(3584) 1-penten-3-ol, 26(2633) (*R*)-(+)-limonene, 36 *cis*-2-heptenal, 40(2782) nonanal, 42(2805) 1-octen-3-ol and 44(2362) decanal. The compound described as the character-imparting compound in bell peppers is 2-isobutyl-3-methoxypyrazine. This very powerful odourant has a threshold in water of 1 part in 10^12 (Govindarajan and Sathyanarayana, 1986).

14.5. Medicinal and Pharmacological Properties

Medicinal use of *Capsicum* has a long history, dating back to the Mayan civilization, for treating asthma, coughs and sore throats. The Aztecs used chilli pungency to relieve toothache. The pharmaceutical industry uses capsaicin as a counter-irritant balm for external application. It is the active ingredient in Heet and Sloan's Liniment, two rubdown liniments used for sore muscles. The capsaicin is used to alleviate pain (Rashid *et al.*, 2003). Capsaicin-stimulated release of substance P was studied in depth by Purkiss *et al.* (2000). It involves two distinct mechanisms. Its mode of action is thought to be from nerve endings releasing a neurotransmitter called substance 'P'. Substance P informs the brain that something painful is occurring. Capsaicin causes an increase in the amount of substance P released. Eventually, substance P is depleted and further releases from the nerve ending are reduced. Purkiss *et al.* (2000) found that the capsaicin (10 μM) stimulated a twofold increase in the release of substance P in the absence of extracellular Ca^{2+}. It was concluded that capsaicin induced the release of substance P from dorsal root ganglion neurons via two mechanisms, one requiring extracellular Ca^{2+} and the intact synaptosomal-associated protein 25 kDa (SNAP-25) and the other independent of extracellular Ca^{2+} and not involving SNAP-25 (Dray, 1992).

The physiological, pharmacological and toxicological interests of chilli and of their active principle – capsaicin – have been studied extensively. Because of its specific excitatory and neurotoxic properties on C-fibres, capsaicin has been used extensively as the main tool for neuropharmacological studies concerning pain and thermoregulation. Hence, this molecule is emerging to target disease associated with deep pain. The molecule action of capsaicin is mediated by neuropeptides such as calcitonin-related peptide, nitric oxide, neurokinins and glutamate. Use of capsazepine, a standard capsaicin antagonist and resiniferatoxin, an extremely potent capsaicin analogue, has led to the discovery of VR1, a specific receptor mediating capsaicin action (Coulibaly *et al.*, 1998).

Creams containing capsaicin are used to reduce post-operative pain for mastectomy patients and for amputees suffering from phantom limb pain. Prolonged use of the cream has also been found to help reduce the itching of dialysis patients, the pain from shingles and cluster headaches. Further research has indicated that capsaicin cream reduces pain associated with arthritis. The repeated use of the cream apparently counters the production of substance P in the joint, hence less pain. Reducing substance P also helps reduce long-term inflammation, which can cause cartilage breakdown. Carotenoids also have antioxidative properties that may prevent cancer (Yoshimura *et al.*, 2000).

Antioxidant activity

Perucka *et al.* (2001) studied the antioxidant activity of capsaicin, dihydrocapsaicin content, fractions containing flavonoids, as well as L-phenylalanine ammonia-lyase (PAL) in the fruit of the hot pepper, *C. annuum* cv. Cyklon and Bronowicka Ostra. The fruit of the cv. Bronowicka Ostra was characterized by a higher PAL activity, capsaicinoid level and higher antioxidant activity than that found in Cyklon. It was found that the red fruit of both cultivars had higher antioxidant activity in the capsaicinoid and flavonoid fractions. The PAL activity was much higher in the red than in the green (unripe) fruits. The antioxidant activity under the same experimental conditions was similar in both capsaicinoid and flavonoid fractions of cv. Bronowicka Ostra.

Rizvi and Srivastava (1999) studied the protective effect of capsaicin on the osmotic fragility of human erythrocytes. Capsaicin (from capsicum) exerted a stabilizing effect on human erythrocytes, making them more resistant to lysis under hypotonic stress. The protective effect of capsaicin was due probably to a direct interaction of capsaicin with the erythrocyte membrane rather than due to any alteration in the intracellular metabolism of erythrocytes.

Chilli and its pungent ingredient, capsaicin, has been shown to protect against experimental gastric mucosal injury induced by various necrotizing agents such as ethanol and aspirin and stress. The protective action of capsaicin and chilli on haemorrhagic shock-induced gastric mucosal injury in the rat and the gastrointestinal injury and immunity of capsaicin was reported by Choo *et al.* (1998).

Yeoh *et al.* (1995) demonstrated that chilli could protect humans against aspirin-induced gastroduodenal mucosal injury. A survey conducted by Kang *et al.* (1995) established the protective effect of chilli against peptic ulcer. Red pepper and natural and synthetic capsaicin in the diet significantly decreased cholesterol in the liver (Sambaiah and Satyanarayana, 1980).

Anticancerous activity

Zhang *et al.* (2003) found the potential of capsaicin in inhibiting the growth of adult T-cell leukaemia cells. Human T-cell leukaemia virus type 1 (HTLV-1)-associated adult T-cell leukemia (ATL) has been found to be resistant to conventional chemotherapy. Capsaicin treatment inhibited the growth of ATL cells in both a dose- and time-dependent manner. The inhibitory effect was due mainly to the induction of cell cycle arrest and apoptosis.

Aflatoxigenic activity

Masood *et al.* (1994) tested capsanthin and capsaicin, the colouring and pungent principles of *C. annuum*, respectively, against growth and aflatoxin production of *Aspergillus flavus* in SMKY liquid medium. Capsanthin inhibited both the growth and toxin production of *A. flavus* completely at all concentrations tested (0.2, 0.6 and 1.0 mg/ml) until the 4th day of incubation.

Non-lethal use of capsaicin

Capsaicin is also the active ingredient in the chemical riot control agent, pepper spray. When the spray comes in contact with skin, especially eyes or mucous membranes, it is very painful. In large quantities, capsaicin can be a lethal poison. Symptoms of overdose include difficulty in breathing, blue skin and convulsions and uncontrollable, painful nipple erections. Even though the concentration required to kill a human is quite large, and the low concentration of capsaicin in chilli makes accidental poisoning by chilli consumption exceedingly unlikely, capsaicin has been implicated in some cases of infanticide in India.

14.6. International Specifications, Desirable Limits

Defining quality

Yadav *et al.* (2000) established the association between morphological and quality traits of chilli fruit germplasm. Good pungency, bright red colour, high oleoresin concentration and few seeds in the fruit are the main characters on which quality is based and priced.

The final quality of chilli spice powder is assessed by a number of different parameters. Colour and pungency levels are the most obvious parameters assessed, but sweetness and flavour of non-pungent paprika powders are also important. In addition, the spice trade may specify limits of impurity, levels of microbial counts of, for example, fungi, yeasts, *Salmonella* and coli forms, particle size and moisture content, among others. The main desired characters are colour, pungency and less extraneous contamination in relation to their biochemistry, assessment and desired levels. Table 14.5

Table 14.5. Commercial specification for chilli (as sold in whole, cracked and powdered forms).

Trait	Specification
Colour	It shall possess a characteristic yellowish-red to red colour, with yellow seeds for whole chillies
Taste	It shall have a mild odour and a perceptible sense of pungency
Pungency	20,000–30,000 SHU (crushed chillies) (HPLC method) 30,000–45,000 SHU (chilli powder); In the UK and the Netherlands, ground chillies of pungency less than 20,000 SHU are preferred. In Germany, whole chillies of pungency 20,000–40,000 SHU are accepted
Texture	Any granulation ranging between 300 and 500 μ is accepted (for ground chillies)
Iron filings	In the UK, consignments are required to be put through magnets. Detection of iron filings is considered to be a hazard. A maximum limit of 10 ppm is allowed
Packing	Net wt 25 kg in multilayer paper bag Some prefer cloth bags with inner polylinings for crushed chillies For chilli powder, packing in consumer packs is preferred Packing in polywoven bags is being encouraged

Table 14.6. International (ISO) standards for *Capsicum*, chillies, paprika products and methods of testing.

ISO Standard No.	Title of ISO specification
ISO standards for whole and ground capsicums, etc.	
ISO 972: 1997	Chillies and capsicums, whole or ground (powdered) – specification
ISO 7540: 1984	Ground (powdered) paprika (*Capsicum annuum* L.) – specification
ISO 3588: 1977	Determination of degree of fineness of grinding – hand-sieving method (reference method)
ISO 7542: 1984	Ground (powdered) paprika (*Capsicum annuum* L.) – microscopical examination
ISO 3513: 1995	Chillies – determination of Scoville index
ISO 7541: 1989	Ground (powdered) paprika – determination of total natural colouring matter content
ISO 7543–1: 1994	Chillies and chilli oleoresins – determination of total capsaicinoid content; Part 1: Spectrometric method
ISO 7543: 1993	Chillies and chilli oleoresins – determination of total capsaicinoid content; Part 2: Method using high-performance liquid chromatography (HPLC method)
International (ISO) methods of testing spices and spice products	
ISO 2825: 1981	Preparation of a ground sample for analysis
ISO 948: 1980	Method of sampling
ISO 927: 1982	Determination of extraneous matter
ISO 1298: 1982	Determination of moisture content – entrainment method
ISO 941: 1980	Determination of cold water-soluble extract
ISO 6571: 1984	Determination of volatile oil content
ISO 928: 1997	Determination of total ash
ISO 1108: 1992	Determination of non-volatile ether extract
ISO 930: 1997	Determination of acid-insoluble ash

Source: compiled by Pruthi (2003).

Capsanthin: (3*R*, 3′*S*, 5′*R*)-3, 3′-Dihydroxy-ß-caroten-6′-one (orange-red) $C_{40}H_{56}O_3$

Capsorubin: 3,3′-dihydroxy-carotene-6 6' dione (orange-red) $C_{40}H_{56}O$

Cryptoxanthin: (*R*)-3,5,5-Trimethyl-4-[3,7,12,16-tetramethyl-18-(2,6,6-trimethylcyclohex-1-enyl)-octadeca-1,3,5,7,9,11, 13,15,17-nonaenyl]-cyclohex-3-enol

Zeaxanthin: 4-[18-(4-hydroxy-2,6,6-trimethyl-1-cyclohexenyl)-3,7,12,16-tetramethyl-octadeca-1,3,5,7,9,11,13,15, 17-nonaenyl]-3,5,5-trimethyl-cyclohex-3-en-1-ol.

gives the commercial specification for chilli. International (ISO) standards for *Capsicum*, chillies and paprika products and methods of testing are given in Table 14.6.

14.7. Conclusion

It is estimated that the world production of chillies is about 2.5 million t and paprika accounts for about one-third of the total world consumption of chilli. Capsicum fruits in different forms are popular food additives in most part of the world. Paprika and chilli are consumed worldwide, either as a spice or a natural colourant. It is valued principally for the brilliant red colour it gives to pale foods and also for its delicate aroma. Paprika and paprika oleoresin are used currently in a wide assortment of foods, drugs and cosmetics, as well as for improving the feather colour of flamingoes in zoos. The colour of chilli powder is due to the presence of red-pigmented carotenoids. The main pigments are capsanthin, capsorubin, zeaxanthin and cryptoxanthin. The pungency of chilli is attributed to the five natural capsaicinoids, which are capsaicin, dihydrocapsaicin, nor-dihydrocapsaicin, vanillyl decanamide and homodihydrocapsaicin. Tremendous developments are taking place in the utilization of these compounds, paving the way for future exploitation.

References

Adeyeyei, E.I. and Otokiti, M.K.O. (1999) Proximate composition and some nutritionally valuable minerals of two varieties of *Capsicum annum* L. (bell and cherry peppers). *Discovery and Innovation* 11(1/2), 75–81.

Ahmed, J., Shivhare, U.S. and Debnath, S. (2002) Colour degradation and rheology of green chilli puree during thermal processing. *International Journal of Food Science and Technology* 37(1), 57–63.

Alejo, N.O. (2006) Capsaicin accumulation in *Capsicum* spp. suspension cultures. *Methods in Molecular Biology* 318, 327–334.

Alejo, N.O. and Garciglia, R.S. (1992) Phenylalanine ammonia-lyase activity and capsaicin-precursor compounds in *p*-fluorophenylalanine-resistant and -sensitive variant cells of chili pepper (*Capsicum annuum*). *Physiologia Plantarum* 85(2), 173–179.

Alejo, N. and Peralta, J.E.G. (1993) Activity of enzymes involved in capsaicin biosynthesis in callus tissue and fruits of chilli pepper (*Capsicum annuum* L.). *Journal of Plant Physiology* 141(2), 147–152.

Anu, A. and Peter, K.V. (2000) The chemistry of paprika. *Capsicum and Eggplant Newsletter* 19, 19–22.

ASTA (1985) *Official Analytical Methods for Spices*, 3rd edn. American Spices Trade Association, New York.

Badillo, A., Steppuhn, J., Deruere, J., Camara, B. and Kuntz, M. (1995) Structure of a functional geranylgeranyl pyrophosphate synthase gene from *Capsicum annuum*. *Plant Molecular Biology* 27(2), 425–428.

Bekker, N.P., Ul'chenko, N.T. and Glushenkova, A.I. (2002) Lipids of *Capsicum annuum* fruit pulp. *Chemistry of Natural Compounds* 38(5), 466.

Bernal, M.A., Calderon, A.A., Ferrer, M.A., Merino de Caceres, F. and Barcelo, R.A. (1995) Oxidation of capsaicin and capsaicin phenolic precursors by the basic peroxidase isoenzyme B6 from hot pepper. *Journal of Agricultural and Food Chemistry* 43(2), 352–355.

Biacs, P.A., Daood, H.G., Pavisa, A. and Hajdu, F. (1989) Studies on the carotenoid pigments of paprika (*Capsicum annuum* L. var Sz-20). *Journal of Agricultural and Food Chemistry* 37(2), 350–353.

Biacs, P.A., Daood, H.G., Huszka, T.T. and Biacs, P.K. (1993) Carotenoids and carotenoid esters from new cross-cultivars of paprika. *Journal of Agricultural and Food Chemistry* 41(11), 1864–1867.

Blum, E., Mazourek, M., O'Connell, M., Curry, J., Thorup, T., Liu, K., Jahn, M. and Paran, I. (2004) Molecular mapping of capsaicinoid biosynthesis genes and quantitative trait loci analysis for capsaicinoid content in Capsicum. *Theoritical and Applied Genetics* 108(1), 79–86.

Borges, R.M. (2001) Why are chillies pungent? *Journal of Biosciences* 26(3), 289–291.

Bosland, P.W. (1996) Capsicums: innovative uses of an ancient crop. In: Janick, J. (ed.) *Progress in New Crops*. ASHS Press, Arlington, Virginia, pp. 479–487.

Bosland, P.W. and Votava, E.J. (2000) *Peppers: Vegetable and Spice Capsicums*. CAB International, Wallingford, UK, pp. 204.

Bouvier, F., Backhaus, R.A. and Camara, B. (1998) Induction and control of chromoplast specific carotenoid genes by oxidative stress. *Journal of Biological Chemistry* 273(46), 30651–30659.

Bouvier, F., D'Harlingue, A., Ackhaus, R.A., Kumagai, M.H. and Camara, B. (2000) Identification of neoxanthin synthase as a carotenoid cyclase paralog. *European Journal of Biochemistry* 267(21), 6346–6352.

Breithaupt, D.E. and Schwack, W. (2000) Determination of free and bound carotenoids in paprika (*Capsicum annuum* L.) by LC/MS. *European Food Research and Technology* 211(1), 52–55.

Bunnell, R.H. and Bauernfeind, J.C. (1962) Chemistry, uses, and properties of carotenoids in foods. *Food Technology* 16, 36–43.

Chen, H.C., Klein, A., Xiang, M.H., Backhaus, R.A. and Kuntz, M. (1998) Drought and wound induced expression in leaves of a gene encoding a chromoplast carotenoid associated protein. *Plant Journal* 14(3), 317–326.

Choo, H.T., Jin, Y.K., Wee, A. and Kok, O.L. (1998) Gastrointestinal injury and immunity. Protective action of capsaicin and chilli on haemorrhagic shock induced gastric mucosal injury in the rat. *Journal of Gastroenterology and Hepatology* 13(10), 1007–1014.

Cooper, T.H., Guzinski, J.A and Fisher, C. (1991) Improved high performance liquid chromatography method for the determination of major capsaicinoids in capsicum oleoresin. *Journal of Agricultural and Food Chemistry* 39, 2253–2256.

Coulibaly, S.K., Coxam, V. and Barlet, J.P. (1998) Physiological, pharmacological and toxicological interests of chillies and of their active principle: Capsaïcin. *Médecine et Nutrition* 34, 236–245.

Daood, H.G, Tömösközi-Farkas, R. and Kapitány, J. (2006) Antioxidant content of bio and conventional spice red pepper (*Capsicum annuum* L) as determined by HPLC. *Acta Agronomica Hungarica* 54(2), 133–140.

Das, P., Borua, I., Sarkar, C.R. and Phookan, A.K. (1996) Changes in mineral elements and capsaicin content of chilli (*Capsicum annum* L. and *Capsicum frutescens* L.) fruits during development. *Journal of the Agricultural Science Society of North East India* 9(1), 104–106.

Deli, J. and Molnar, P. (2002) Paprika carotenoids: analysis, isolation, structure elucidation. In: *Current Organic Chemistry*, Volume 6(13). Bentham Science Publishers, Oak Park, Illinois, pp. 1197–1219.

Deli, J. and Toth, G. (1997) Carotenoid composition of the fruits of *Capsicum annuum* cv. Bovet 4 during ripening. *Zeitschrift fur Lebensmittel Untersuchung und Forschung* 205(5), 388–391.

Deli, J., Matus, Z. and Szabolcs, J. (1992) Carotenoid composition in the fruits of black paprika (*Capsicum annuum* variety longum nigrum) during ripening. *Journal of Agricultural and Food Chemistry* 40(11), 2072–2076.

Deli, J., Molnar, P., Matus, Z. and Toth, G. (2001) Carotenoid composition in the fruits of red paprika (*Capsicum annuum* var. lycopersiciforme rubrum) during ripening biosynthesis of carotenoids in red paprika. *Journal of Agricultural and Food Chemistry* 49(3), 1517–1523.

Diaz, J., Bernal, A., Merino, F. and Ros Barcelo, A. (1998) Phenolic metabolism in *Capsicum annuum* L. *Recent Research Developments in Phytochemistry* 2(2), 155–169.

Dray, A. (1992) Mechanism of action of capsaicin like molecules on sensory neurons. *Life Science* 51(23), 1759–1765.

El Saeid, H.M. (1995) Chemical composition of sweet and hot pepper fruits grown under plastic house conditions. *Egyptian Journal of Horticulture* 22(1), 11–18.

Estrada, B., Pomar, F., Diaz, J., Merino, F. and Bernal, M.A. (1999a) Pungency level in fruits of the Padron pepper with different water supply. *Scientia Horticulturae* 81(4), 385–396.

Estrada, B., Diaz, J., Merino, F. and Bernal, M.A. (1999b) The effect of seasonal changes on the pungency level of Padron pepper fruits. *Capsicum and Eggplant Newsletter* 18, 28–31.

Estrada, B., Bernal, M.A., Díaz, J., Pomar, F. and Merino, F. (2000) Fruit development in *Capsicum annuum*: changes in capsaicin, lignin, free phenolics, and peroxidase patterns. *Journal of Agricultural and Food Chemistry* 48(12), 6234–6239.

Galvez, P.A., Mendez, H.D. and Mosquera, M.M.I. (2004) Changes in the carotenoid metabolism of capsicum fruits during application of modelized slow drying process for paprika production. *Journal of Agricultural and Food Chemistry* 52(3), 518–522.

Gnayfeed, M.H., Daood, H.G., Biacs, P.A. and Alcaraz, C.F. (2001) Content of bioactive compounds in pungent spice red pepper (paprika) as affected by ripening and genotype. *Journal of the Science of Food and Agriculture* 81(15), 1580–1585.

Goda, Y., Sakamoto, S., Nakanishi, T., Maitani, T. and Yamada, T. (1995) Identification of monoesterified capsanthin in paprika (*Capsicum annuum*): the nature of esterification of capsanthin. *Chemical and Pharmaceutical Bulletin* 43(7), 1248–1250.

Govindarajan, V.S. and Sathyanarayana, M.N. (1986) *Capsicum* production, technology, chemistry, and quality. Part III. Chemistry of the color, aroma, and pungency stimuli. *Critical Reviews in Food Science and Nutrition* 24(3), 245–355.

Govindarajan, V.S. and Sathyanarayana, M.N. (1991) *Capsicum* production, technology, chemistry, and quality. Part V. Impact on physiology, pharmacology, nutrition, and metabolism structure, pungency, pain, and desensitization sequences. *Critical Reviews in Food Science and Nutrition* 29(6), 435–473.

Gupta, D.N. and Tambe, N.G. (2003) Physicochemical characteristics of some promising varieties of chilli grown in Konkan Region. *Journal of Maharashtra Agricultural Universities* 28(3), 327–328.

Hari, G.S., Rao, P.V. and Reddy, Y.N. (2005) Evaluation of paprika (*Capsicum annuum* L.) germplasm and its utility in breeding for higher yield and better quality grown under irrigated conditions. *Research on Crops* 6(2), 266–269.

IBPGR (1983) *Genetic resources of* Capsicum. International Board for Plant Genetic Resources, Rome (now International Plant Genetic Resources Institute (IPGRI), Italy.

Iorizzi, M., Lanzotti, V., Marino, S. de, Zollo, F., BlancoMolina, M., Macho, A. and Munoz, E. (2001) New glycosides from *Capsicum annuum* L. var. acuminatum. Isolation, structure determination, and biological activity. *Journal of Agricultural and Food Chemistry* 49(4), 2022–2029.

Ishikawa, K., Janos, T., Sakamoto, S. and Nunomura, O. (1998) The contents of capsaicinoids and their phenolic intermediates in the various tissues of the plants of *Capsicum annuum* L. *Capsicum and Eggplant Newsletter* 17, 22–25.

Iwai, K., Lee, K.R., Kobashi, M. and Suzuki, T. (1977) Formation of pungent principles in fruits of sweet pepper, *Capsicum annuum* L. var. grossum during post-harvest ripening under continuous light. *Agricultural and Biological Chemistry* 41(10), 1873–1876.

Jaren Galan, M., Nienaber, U. and Schwartz, S.J. (1999) Paprika (*Capsicum annuum*) oleoresin extraction with supercritical carbon dioxide. *Journal of Agricultural and Food Chemistry* 47(9), 3558–3564.

Jorge, A., Osuna, G., Marisa, M.W. and Cynthia, A.W. (1997) Natural antioxidants for preventing color loss in stored paprika. *Journal of Food Science* 62(5), 1017–1021.

Kang, J.Y., Yeoh, K.G., Chia, H.P., Lee, H.P., Chia, Y.W., Guan, R. and Yap, I. (1995) Chili protective factor against peptic ulcer? *Digestive Diseases and Sciences* 40(3), 576–579.

Kim, S.H., Kim, Y.H., Lee, Z.W., Kim, B.D. and Ha, K.S. (1997) Analysis of chemical constituents in fruits of red pepper (*Capsicum annuum* L. cv. Bugang). *Journal of the Korean Society for Horticultural Science* 38(4), 384–390.

Kocsis, N., Amtmann, M., Mednyanszky, Z. and Korany, K. (2002) GC-MS investigation of the aroma compounds of Hungarian red paprika (*Capsicum annuum*) cultivars. *Journal of Food Composition and Analysis* 15, 195–203.

Kopp, B. and Jurenitsch, J. (1981) Biosynthesis of capsaicinoids in *Capsicum annuum* var. *annuum*. II. Formation of the fatty acid moiety of the capsaicinoids from L-valine, L-leucine and L-isoleucine. *Planta-Medica* 43(3), 272–279.

Krajayklang, M., Klieber, A. and Peter, R.D. (2000) Colour at harvest and post-harvest behaviour influence paprika and chilli spice quality. *Postharvest Biology and Technology* 20, 269–278.

Krajewska, A.M. and. Powers, J.J. (1988) Sensory properties of naturally occurring capsaicinoids. *Journal of Food Science* 53(3), 902–905.

Krishnamurthy, R., Malve, M.K. and Shinde, B.M. (1999) Evaluation of capsaicin content in red and green chillies. *Journal of Scientific and Industrial Research* 58(8), 629–630.

Lang, Y.Q., Yanagawa, S., Sasanuma, T. and Sasakuma, T. (2004) Orange fruit color in capsicum due to deletion of capsanthin capsorubin synthesis gene. *Breeding Science* 54(1), 33–39.

Lemos, A.L.D.C., Rusig, O. and Carvalho, P.R.N. (1993) Evaluation of different solvents for paprika oleoresin extraction. *Arquivos de Biologia e Tecnologia* 36(4), 659–667.

Lindsay, K. and Bosland P.W. (1995) A field study of environmental interaction on pungency. *Capsicum Eggplant Newsletter* 14, 36–38.

Luning, P.A., Rijk, T., Withers, H.J. and Roozent, J.P. (1994) Gas chromatography, mass spectrometry, and sniffing port analyses of volatile compounds of fresh bell peppers *(Capsicum annuum)* at different ripening stages. *Journal of Agricultural and Food Chemistry* 42, 977–983.

Maoka, T., Fujiwara, Y. and Hashimoto, A.N. (2001a) Capsanthone 3,6-epoxide, a new carotenoid from the fruits of the red paprika *Capsicum annuum* L. *Journal of Agricultural and Food Chemistry* 49(8), 3965.

Maoka, T., Fujiwara, Y., Hashimoto, K. and Akimoto, N. (2001b) Isolation of a series of apocarotenoids from the fruits of the red paprika *Capsicum annuum* L. *Journal of Agricultural and Food Chemistry* 49(3), 1601–1606.

Masood, A., Dogra, J.V.V. and Jha, A.K. (1994) The influence of colouring and pungent agents of red chilli (*Capsicum annuum*) on growth and aflatoxin production by *Aspergillus flavus*. *Letters in Applied Microbiology* 18(4), 184–186.

Mateo, J., Aguirrezabal, M., Dominguez, C. and Zumalacarregui, J.M. (1997) Volatile compounds in Spanish paprika. *Journal of Food Composition and Analysis* 10, 225–232.

Mathur, R., Dangi, R.S., Dass, S.C. and Malhotra, R.C. (2000) The hottest chilli variety in India. *Current Science* 79(3), 287–288.

Mendez, H.D. and Mosquera, M.M.I. (2000) Xanthophyll esterification accompanying carotenoid overaccumulation in chromoplast of *Capsicum annuum* ripening fruits is a constitutive process and useful for ripeness index. *Journal of Agricultural and Food Chemistry* 48(5), 1617–1622.

Mendez, H.D., Ladronde, R.G. and Mosquera, M.I. (2000) Carotenoid biosynthesis changes in five red pepper (*Capsicum annuum* L.) cultivars during ripening. Cultivar selection for breeding. *Journal of Agricultural and Food Chemistry* 48, 3857–3864.

Mendez, H.D., Garcia, C.J. and Mosquera, M.M.I. (2002) Characterization of carotenoid high-producing *Capsicum annuum* cultivars selected for paprika production. *Journal of Agricultural and Food Chemistry* 50(20), 5711–5716.

Minami, M., Toyota, M., Inoue, T., Nemoto, K. and Ujihara, A. (1998) Changes of capsaicinoid contents during maturing stage in chilli pepper (*Capsicum* spp.). *Journal of the Faculty of Agriculture, Shinshu University* 35(1), 45–49.

Mini, C., Vahab, M.A., Augustine, A. and Rajan, S. (1998) Standardisation of solvent for chilli oleoresin extraction. *South Indian Horticulture* 46(3–6), 309–311.

Molnar, P., Kawase, M., Satoh, K., Sohara, Y., Tanaka, T., Tani, S., Sakagami, H., Nakashima, H., Motohashi, N., Gyemant, N. and Molnar, J. (2005) Biological activity of carotenoids in red paprika, Valencia orange and golden delicious apple. *Phytotherapy Research* 19(8), 700–707.

Mosquera, M.I. and Mendez, H.D. (1993) Separation and quantification of the carotenoid pigments in red peppers (*Capsicum annuum* L.) paprika and oleoresin by reversed phase HPLC. *Journal of Agricultural and Food Chemistry* 41, 1616–1620.

Mosquera, M.I., Galan, M.J. and Fernandez, G.J. (1993) Effect of processing of paprika on the main carotenes and esterified xanthophylls present in the fresh fruit. *Journal of Agricultural and Food Chemistry* 41(11), 2120–2124.

Padilla, C.M. and Yahia, E.M. (1998) Changes in capsaicinoids during development, maturation, and senescence of chile peppers and relation with peroxidase activity. *Journal of Agricultural and Food Chemistry* 46(6), 2075–2079.

Perucka, I. (1996) Ethephon induced changes in accumulation of carotenoids in red pepper fruit (*Capsicum annuum* L.). *Polish Journal of Food and Nutrition Sciences* 5(4), 61–68.

Perucka, I., Materska, M. and Jongen, W.M.F. (2001) Phenylalanine ammonia-lyase and antioxidant activities of lipophilic fraction of fresh pepper fruits *Capsicum annum* L. In: Knorr, D. (ed.) *Nutritionists Meet Food Scientists and Technologists*. Proceedings of a Conference, Porto, Portugal, 12–14 April 2000, Part 2. *Innovative Food Science and Emerging Technologies* 2(3), 189–192.

Popovsky, S. and Paran, I. (2000) Molecular genetics of the *y* locus in pepper: its relation to capsanthin-capsorubin synthase and to fruit color. *Theoretical and Applied Genetics* 101(1–2), 86–89.

Prasad, B.C.N., Gururaj, H.B., Kumar, V., Giridhar, P., Parimalan, R., Sharma, A. and Ravishankar, G.A. (2006) Influence of 8-methyl-nonenoic acid on capsaicin biosynthesis in *in vivo* and *in vitro* cell cultures of *Capsicum* spp. *Journal of Agricultural and Food Chemistry* 54(5), 1854–1859.

Priya, N., Vijayakumar, M., Veeraragavathatham, D., Jeyakumar, P. and Chezhian, N. (2002) Effect of seasons and growth environments on paprika (*Capsicum annuum* var. longum) growth and yield. *South Indian Horticulture* 50(4/6), 463–471.

Pruthi, J.S. (2003) Chemistry and quality control of capsicums and capsicum products. In: De, A.K. (ed.) *Capsicum. The genus Capsicum*. Taylor and Francis, London and New York, pp. 25–70.

Purkiss, J., Welch, M., Doward, S. and Foster, K. (2000) Capsaicin stimulated release of substance P from cultured dorsal root ganglion neurons: involvement of two distinct mechanisms. *Biochemical Pharmacology* 59(11), 1403–1406.

Purseglove, J.W., Brown, E.G., Green, C.L. and Robbins, S.R.J. (1981) *Spices*, Volume I. Longman, New York, pp. 331–439.

Rani, P.U. (1996) Evaluation of chilli germplasm for capsanthin and capsaicin contents and effect of storage on ground chilli. *Madras Agricultural Journal* 83(5), 288–291.

Rashid, M., Inoue, M., Bakoshi, S. and Ueda, H. (2003) Increased expression of vanilloid receptor 1 on myelinated primary afferent neurons contributes to the antihyperalgesic effect of capsaicin cream in diabetic neuropathic pain in mice. *Journal of Pharmacology And Experimental Therapeutics* DOI: 10.1124/jpet.103.050948.

Reeves, M.J. (1987) Re-evaluation of *Capsicum* color data. *Journal of Food Science* 52, 1047–1049.

Reilly, C.A., Crouch, D.J. and Yost, G.S. (2001) Quantitative analysis of capsaicinoids in fresh peppers, oleoresin capsicum and pepper spray products. *Journal of Forensic Sciences* 46(3), 502–509.

Rizvi, S.I. and Srivastava, P.P. (1999) Protective effect of capsaicin on the osmotic fragility of human erythrocytes. *Indian Journal of Experimental Biology* 37(8), 835–836.

Robi, R. and Sreelathakumary, I. (2004) Influence of maturity at harvest on capsaicin and ascorbic acid content in hot chilli (*Capsicum chinense* Jacq.). *Capsicum and Eggplant Newsletter* 23, 13–16.

Rodrigues, P., Morais, H., Mota, T., Olivera, S., Forgaes, E. and Cserhati, T. (1998) Use of HPLC and multivariate methods for the evaluation of the stability of colour pigments of paprika (*Capsicum annuum*) powder. *Analytica Chimica Acta* 372, 411–416.

Saga, K. and Sato, G. (2003) Varietal differences in phenolic, flavonoid and capsaicinoid contents in pepper fruits (*Capsicum annuum* L.). *Journal of the Japanese Society for Horticultural Science* 72(4), 335–341.

Sambaiah, K. and Satyanarayana, M.N. (1980) Hypocholesterolemic effect of red pepper and capsaicin. *Indian Journal of Experimental Biology* 8(8), 898–899.

Sankari, A., Veeraragavathatham, D. and Sathiyamurthy, V.A. (2000) Per se performance of individual progenies of F3 generations of the cross Ramanathapuram Local × Jalapeno hot pepper (*Capsicum annuum* L.).

In: *Spices and Aromatic Plants: Challenges and Opportunities in the New Century*. Contributory papers. Centennial Conference on Spices and Aromatic Plants, Calicut, Kerala, India, 20–23 September 2000. Indian Society for Spices, Calicut, India, pp. 64–67.

Sato, K., Sasaki, S.S., Goda, Y., Yamada, T., Nunomura, O., Ishikawa, K. and Maitani, T. (1999) Direct connection of supercritical fluid extraction and supercritical fluid chromatography as a rapid quantitative method for capsaicinoids in placentas of Capsicum. *Journal Agricultural and Food Chemistry* 47(11), 4665–4668.

Schweiggert, U., Kammerer, R.D., Carle, R. and Schieber, A. (2005) Characterization of carotenoids and carotenoid esters in red pepper pods (*Capsicum annuum* L.) by high performance liquid chromatography/atmospheric pressure chemical ionization mass spectrometry. *Rapid Communications in Mass Spectrometry* 19, 2617–2628.

Scoville, W.L. (1912) Note on *Capsicum*. *Journal of the American Pharmaceutical Association* 1, 453 (cited by Bosland, P.W. (1996)).

Simkin, A.J., Zhu, C., Kuntz, M. and Sandmann, G. (2003) Light dark regulation of carotenoid biosynthesis in pepper (*Capsicum annuum*) leaves. *Journal of Plant Physiology* 160(5), 439–443.

Tewari, V.P. (1990) Development of high capsaicin chillies (*Capsicum annuum* L.) and their implications for the manufacture of export products. *Journal of Plantation Crops* 18(1), 1–13.

Thampi, P.S.S. (2003) A glimpse of the world trade in capsicum. In: De, A.K. (ed.) *Capsicum. The Genus Capsicum*. Taylor and Francis, London and New York, pp. 16–24.

Tiwari, A., Kaushik, M.P., Pandey, K.S. and Dangi, R.S. (2005) Adaptability and production of hottest chilli variety under Gwalior agroclimatic conditions. *Current Science* 88(10), 1545–1546.

Van Ruth, S.M., Roozen, J.P. and Cozijnsen, J.L. (1995) Changes in flavour release from rehydrated diced bell peppers (*Capsicum annuum* L.) by artificial saliva components in three mouth model systems. *Journal of the Science of Food and Agriculture* 67, 189–196.

Wall, M.M. and Bosland, P.W. (1993) The shelf-life of chiles and chile containing products. In: Charalambous, G. (ed.) *Shelf Life Studies of Foods and Beverages, Chemical, Biological, and Nutritional Aspects*. Elsevier Publishing, New York, pp. 487–500.

Yadav, S., Mandal, S., Verma, S.K., Muneem, K.C. and Singh, M. (2000) Association between morphological and quality traits of chilli fruit germplasm. In: Ramana, K.V., Eapen, S.J., Babu, K.N., Krishnamurthy, K.S. and Kumar, A. (eds) *Spices and Aromatic Plants: Challenges and Opportunities in the New Century*. Contributory papers. Centennial Conference on Spices and Aromatic Plants, Calicut, Kerala, India. Indian Society for Spices, Calicut, India, pp. 23–25.

Yao, J., Nair, M.G. and Chandra, A. (1994) Supercritical carbon dioxide extraction of Scotch Bonnet (*Capsicum annuum*) and quantification of capsaicin and dihydrocapsaicin. *Journal of Agricultural and Food Chemistry* 42(6), 1303–1305.

Yeoh, K.G., Kang, J.Y., Yap, I., Guan, R., Tan, C.C., Wee, A. and. Teng, C.H. (1995) Chili protects against aspirin-induced gastroduodenal mucosal injury in humans. *Digestive Diseases and Sciences* 40(3), 580–583.

Yoshimura, M., Yonehara, N., Ito, T., Kawai, Y. and Tamura, T. (2000) Effects of topically applied capsaicin cream on neurogenic inflammation and thermal sensitivity in rats. *Japanese Journal of Pharmacology* 82(2), 116–121.

Yu, S., Chang, Y.Y. and Ting, N.L. (2005) Capsaicin biosynthesis in water stressed hot pepper fruits. *Botanical Bulletin of Academia Sinica* 46(1), 35–42.

Yun, H.K., Kim, K.Y., Kim, Y.C., Seo, T.C., Kim, I.S., Yoo, K.C. and Higashio, H. (2002) Changes of color and fruit component by temperature treatment after harvest of unripened fruit in hot pepper. *Journal of the Korean Society for Horticultural Science* 43(3), 289–292.

Zewdie, Y. and Bosland, P.W. (2001) Capsaicinoid profiles are not good chemotaxonomic indicators for *Capsicum* species. *Biochemical Systematics and Ecology* 29(2), 161–169.

Zhang, J., Nagasaki, M., Tanaka, Y. and Morikawa, S. (2003) Capsaicin inhibits growth of adult T-cell leukemia cells. *Leukemia Research* 27(3), 275–283.

15 Vanilla

Shamina Azeez

15.1. Introduction

Vanilla, a vine known since the time of the Aztecs, is a genus of about 110 species in the orchid family. *Vanilla planifolia* yields the popular, commercial flavouring agent vanillin. Vanilla is the world's third most expensive spice, next to saffron and cardamom. *V.* Mill. is the only genus among orchids whose members produce a spice. The name originates from the Spanish word 'vainilla', the diminutive form of 'vaina' (meaning 'sheath'), which is in turn derived from Latin 'vagina'. It belongs to the family Orchidaceae, subfamily Epidendroideae, tribe Vanilleae and subtribe Vanillinae (Weiss, 2002.)

This evergreen genus occurs in tropical and subtropical regions, from tropical America to tropical Asia, New Guinea and West Africa. The main species harvested for vanillin is *V. planifolia*, a native of Mexico, though now grown widely throughout the tropics. Additional sources include *V. pompona* and *V. tahitiensis*, grown in Tahiti.

The total area of vanilla cultivation in the world during the year 2001 was 40,846 ha and production was 5583 t. There has been no appreciable increase in area under vanilla cultivation in the traditional vanilla-growing countries, according to the UN Food and Agriculture Organization (2005). The major vanilla-producing countries are Madagascar, The Comoro Islands, Indonesia, Mexico and Réunion. Among these countries Madagascar holds the prominent position, having a cultivated area of 25,550 ha under vanilla. Of late, Indonesia has started to produce more, with a production of 2102 t from 9700 ha. Others are Mexico, China, The Comoro Islands, Réunion, Tonga, French Guiana, Malawi, Uganda, Zimbabwe, Guadeloupe, Kenya, Fiji Islands, Cook Islands and Turkey. Madagascar stands as the world's largest producer of vanilla, e.g. 3.0 t, of a total world production of 7.3 t (Table 15.1).

The import of vanillin and ethyl vanillin together to India during 2000/01 was 404 t. Even if only 10% of import of these synthetic substitutes was replaced by natural product, the requirement of vanilla beans would be 2020 t at the rate of 2% vanillin content. This is almost one half of the entire global production of vanilla beans, indicating the great potential for vanilla development in India (http://www.hinduonnet.com/businessline/2003/03/03/).

Vanilla was highly regarded in pre-Columbian Mesoamerica and was brought back to Europe, and from there to the rest of the world, by the Spanish Conquistadors. In ancient Mexico, the Totonac people (present state of Veracruz (Papantla), Mexico) were regarded as the producers of the best vanilla.

(eds V.A. Parthasarathy, B. Chempakam and T.J. Zachariah)

Table 15.1. Main vanilla-producing countries.

Country	Production, 2005 (t)
Madagascar	3.0
India	2.4
China	1.0
Mexico	0.2
Turkey	0.2
The Comoro Islands	0.1
World total	7.3

Source: UN Food and Agriculture Organization (2005).

They continued to be the world's chief producers through the mid-19th century. At that time, French vanilla growers in Mexico traded their knowledge of artificial pollination of flowers for the Totonac knowledge of preparing the beans (Correll, 1953; Purseglove *et al.*, 1981).

Vanillin, the crystalline component, was first isolated from vanilla pods by Gobley in 1858. By 1874, vanillin had been obtained from glycosides of pine tree sap, temporarily causing an economic depression in the natural vanilla industry. When vanillin, a cheaper synthetic substitute of vanilla, was used for New Coke in 1985 by the Coca-Cola Corporation, the world's largest customer of natural vanilla extract, the Madagascan economy crashed. It recovered only when New Coke flopped. The world market price of vanilla has been at the mercy of cartels, typhoons and political instability, apart from the ever-fluctuating demand and supply pressures. Production of secondary vanilla metabolites, particularly vanillin from *V. planifolia* cell suspension cultures from various plant parts, remains experimental (Havkin-Frenkel *et al.*, 1997.) This technology could reduce the cost of growing vanilla beans greatly, but could seriously affect the economy of vanilla-producing countries such as Madagascar, Java and Tahiti (Simpson and Ogarzaly, 1986).

15.2. Botany and Uses

The basic chromosome number of the genus is x = 16; *V. planifolia*, *V. fragrans*, *V. pompona*, *V. tahitensis* and other species are diploid, with 2n = 32. Vanilla is a tropical climbing orchid, with a long, green, fleshy stem that sprouts roots and clings to trees parasitically. Vanilla climbs over supports and can grow as high as possible. When cultivated, the vines are trained on to posts and support trees up to a height of about 130–135 cm, to facilitate trailing of the vines and artificial hand pollination. Suitable live supports are *Plumaria alba*, *Erythrina lithosperma*, *Jatropha carcas* and *Glyricidia maculata* (http://www.kissankerala.net), but *E. lithosperma* is highly susceptible to wasp attack and hence is not an ideal live support. Just before the plant flowers, the grower usually prunes 10–15 cm from the vine tip; this stops linear growth and seems to benefit flowering (Childers *et al.*, 1959). Its yellow or orange orchidaceous flowers grow in bunches, which bloom one flower each day, opening one by one, during the 2-month season.

Vanilla is a tropical crop and requires a warm climate with frequent rains (annual rainfall of 150–300 cm). Uncleared jungle areas with natural shade or filtered sunlight and soil or rich humus layer undisturbed on the top are ideal for vanilla plantations. The spice is cultivated on various types of soils from sandy loam to laterites and is propagated by planting shoot cuttings *in situ*. Rooted cuttings of 60 cm length (or even tissue culture-derived plants) are planted; longer cuttings bear earlier. The cutting is planted with the onset of the monsoon rains.

V. planifolia is the only orchid used for industrial purposes (in the food and cosmetic industries). *Vanilla* species are used as food plants by the larvae of some Lepidopteran species, including *Hypercompe eridanus* and *H. icasia*. The seeds will not germinate in normal soil; they need a certain symbiotic fungus.

The racemose inflorescences, short-lived flowers, arise successively on short peduncles from the leaf axils or scales (Weiss, 2002). The small lily-like, greenish-yellow vanilla flowers, 3.6 × 5.2 cm long, develop in axillary racemes (Woebse, 1963). There may be 20–100 flowers on a single raceme. Each flower opens up in the morning and closes late in the afternoon, never to re-open

(Childers *et al.*, 1959). If pollination has not occurred meanwhile, the flower will be shed. The flowers are self-fertile but need pollinators to perform this task. They are pollinated by stingless bees and certain hummingbirds, which visit the flowers primarily for its nectar (DeVarigny, 1894; Correll, 1953), but hand pollination is the best method in commercially grown vanilla as flowering is not synchronous. Practically all vanilla is produced now by hand pollination, a labour-intensive task, which accounts for 40% of the total labour cost in vanilla production (Gregory *et al.*, 1967). Approximately 6–8 months after pollination, the green vanilla beans are harvested.

The fruit ('vanilla bean') is an elongated, fleshy seed pod, 10–20 cm long. A single fruit is formed from the pollination of one flower and from this is derived the characteristic flavour compound. It ripens gradually (8–9 months after flowering), eventually turning black and giving off a strong aroma. Each pod contains thousands of minute seeds, but it is the pod that is used for vanilla flavouring. The vanillin in the green beans is present exclusively in conjugated form, principally as the *β*-D-glucoside and, at this stage, the beans display no trace of the characteristic vanilla flavour. This only develops during the fermentation or 'curing' process, which can take more than 6 months to occur. During curing, vanillin *β*-D-glucoside and related *β*-D-glucosides come into contact with *β*-D-glucosidases, resulting in the release of free vanillin and related substances (notably 4-hydroxybenzaldehyde) (Kanisawa *et al.*, 1994; Rao and Ravishankar, 2000b; Dignum *et al.*, 2001). The vanillin content of cured pods is usually *c.* 2–2.5% and, in addition, the number of minor components present is around 200.

Species with common names include:

- *V. aphylla*: leafless vanilla
- *V. barbellata*: small bearded vanilla, wormvine orchid, leafless vanilla, snake orchid
- *V. chamissonis*: Chamisso's vanilla
- *V. claviculata*: green withe
- *V. dilloniana*: leafless vanilla
- *V. edwallii*: Edwall's vanilla
- *V. mexicana*: Mexican vanilla
- *V. odorata*: inflated vanilla
- *V. phaeantha*: leafy vanilla
- *V. planifolia*: vanilla, flat plane-leaved vanilla, West Indian vanilla
- *V. poitaei*: Poit's vanilla
- *V. siamensis*: Thai vanilla.

Vanilla fruits are harvested when fully mature, but before they are too ripe. If harvested immature, the full-bodied aroma and requisite colour do not develop and the beans are more prone to fungal infection. The pods take about 9 months to mature. They are harvested when the tips (the thickest portion of the fruit, the blossom end) begin to turn pale yellow, overall pod colour changes from dark green to light green, the fruits lose their shine and two distinct lines appear from end to end (David, 1950; Purseglove, 1985). The flavouring comes from the vanilla bean. The prepared beans are very dark brown, slender, pleated and about 20 cm long, but tough and pliable. Quality vanilla has a frosting of crystal called *givre* (French for 'frost'), which contains the active ingredient 'vanillin' that produces the characteristic fragrance and is produced during the process of induced fermentation. These pods are called 'fine vanilla'. 'Woody vanilla' is shorter, lighter coloured, uncrystallized, stronger and slightly bitter.

Processing for vanillin

Unlike most spices, the processing or curing of vanilla is quite complicated, since fresh vanilla pods do not have any taste; this, and the need for manual pollination (outside Mexico), makes vanilla one of the most expensive spices. Vanillin is bound as a glycoside and must be set free by enzymatic reaction, normally induced by a sequence of blanching (Bourbon) or steaming (Mexico) operations. During the curing process, the flavour precursors, which are glucosides, are broken down by glucosidase into vanillin and glucose and some other minor aromatic substances. Odoux *et al.* (2003) have purified beta-D-glucosidase from beans of

V. planifolia and have found the enzyme to be a tetramer (201 kDa) made up of four identical subunits (50 kDa).

Weiss (2002) described in detail the stepwise procedure for curing vanilla by the traditional method. This includes killing the vegetative tissue of the vanilla pod to prevent it from growing further after harvest; sweating to allow enzymes to process the compounds in the beans into vanillin and other compounds important to the final vanilla flavour; and drying to prevent rotting and to lock in the aroma in the pods. After the final step of conditioning, the beans are graded for quality. At this point, the beans are dark, oily and pliable. One kg of cured beans is derived from about 6 kg of green pods. Theodose (1973) gives details of the two main traditional forms of curing employed in Mexico. The Bourbon curing technique developed on the island of Réunion is slightly different from the above and is described by Correll (1953). Steps for vanilla curing under laboratory conditions have been described by Dignum *et al.* (2002).

Uses

The predominant commercial use of vanilla is for its flavour compound, vanillin. It finds use not only as a flavouring agent in ice creams, bakery products and puddings, etc., but is also important in the perfumery and cosmetic industry. A few studies on its medicinal properties have also been reported, which are detailed in a subsequent section on culinary and medicinal uses.

15.3. General Composition

Cured vanilla beans contain vanillin, organic acids, fixed fatty oil, wax, gum, resins, tannins, pigments, sugars, cellulose and minerals. The relative amounts of these depend on the species, the environmental factors during growth, harvesting, processing and grading procedures (Purseglove *et al.*, 1981). Mature, fresh, green pods contain about 20% water and each 100 g of dried pod contains, on average, 3–5 g protein, 11 g fat, 7–9 g sugar, 15–20 g fibre, 5–10 g ash, 1.5–3.0 g vanillin, 2 g of a soft resin and an odourless vanillic acid (Weiss, 2002).

Vanilla contains 25% of sugars, 15% fat, 15–30% cellulose and 6% minerals (Uhl, 2000). Water content is unusually high (35%). The nutritional content of vanilla extract in 34.4% ethanol is given in Table 15.2 (USDA National Nutrient Database for Standard Reference, 2002).

Sagrero-Nieves and Schwartz (1988) studied the phenolic content of vanilla, vanillic acid and 4-hydroxybenzaldehyde (HBA) in *V. planifolia* with respect to the harvest period of beans (August–December). The moisture content of the beans decreased from 87.6 to 81.4%. Vanillic acid remained constant and both vanillin and HBA

Table 15.2. Nutrient content of vanilla extract in 34.4% ethanol.

Nutrient	Value per 100 g edible portion
Proximates	
Water (g)	52.58
Energy (kcal)	288
Protein (g)	0.06
Total lipid (g)	0.06
Ash (g)	0.26
Carbohydrate by difference (g)	12.65
Minerals	
Ca (mg)	11
Fe (mg)	0.12
Mg (mg)	12
P (mg)	6
K (mg)	148
Na (mg)	9
Zn (mg)	0.11
Cu (mg)	0.072
Mn (mg)	0.23
Vitamins	
Thiamin (mg)	0.011
Riboflavin (mg)	0.095
Niacin (mg)	0.425
Pantothenic acid (mg)	0.035
Vitamin B_6 (mg)	0.026
Lipids	
Total saturated fatty acids (g)	0.010
Monounsaturated fatty acids (g)	0.010
Polyunsaturated fatty acids (g)	0.004

Source: USDA National Nutrient Database for Standard Reference (2002), www.nal.usda.gov.

increased from 0.20 and 0.05 to 11.30 and 1.03 mg/g dry weight, respectively. The higher phenolic content could be attributed to fermentation on the vine. No correlation was observed between the extract colour (i.e. green versus brown) and the vanillin content. Funk and Brodelius (1990a) established a cell suspension culture of *V. planifolia* in MS-medium. 2,4-D suppressed, while NAA enhanced the formation of extractable phenolics and cytokinins appeared to favour lignin biosynthesis. Treatment of the culture with chitosan resulted in the induction of various enzymes of phenylpropanoid metabolism, while the amount of extractable phenolics decreased due to their rapid incorporation into polymeric ligneous material.

Bouquet: highly fragrant and aromatic.

Flavour: rich, full, aromatic and powerful. Madagascar and Mexico make the best quality. Indonesian and Tahitian vanilla is weaker and considered inferior.

Hotness scale: in the scale devised to relate the 'hotness' of spices, a value on the arbitrary scale ranging from 0 to 10, vanilla scores 1. Generally, most spices fall in the middle, with only the hottest of Mexican chillies scoring 9 or 10.

15.4. Chemistry

Volatiles

The main aroma compound in vanilla is vanillin. Several other volatile constituents are responsible for its characteristic aroma with sweet, balsamic, creamy, woody, spicy, fruity, herbaceous, phenolic and cinnamon-like notes. A compilation of volatile substances identified has been reported by Maarse *et al.* (1994) and Ranadive (1994).

Vanillin

The characteristic aroma of vanilla is obtained after a curing process of green fruits, which contain many different glucosidic compounds. The curing process is required to hydrolyse the glucosides and to release the aroma compounds. *β*-Glucosidases are believed to play an important role in this process (Arana, 1943). The *β*-glucosidase activity in vanilla beans is highest at 6–7 months after pollination (Arana, 1943; Wild-Altamirano, 1969) and the amount of glucosides is also at its highest level then (Kanisawa, 1993). The curing process kills the *β*-glucosidase activity (Ranadive *et al.*, 1983; Dignum *et al.*, 2002), indicating that the aroma formation might not be a completely enzymatic process.

Extraction

Aroma compounds from vanilla beans have been extracted using several extraction procedures, using alcohols and organic solvents (Galletto and Hoffman, 1978; Dignum *et al.*, 2002), direct thermal desorption (Hartman *et al.*, 1992; Adedeji *et al.*, 1993) and solid-phase microextraction (SPME) (Sostaric *et al.*, 2000), followed by identification of the compounds by gas chromatography-mass spectrometry (GC-MS).

Waliszewski *et al.* (2007a) found that a combination of hydration process in 5% ethanol for 48 h and enzymatic pretreatment with stable cellulolytic preparations up to 12 h could double vanillin content in the ethanolic extract and yield a product of excellent sensory properties. Extraction of glucovanillin from green pods of *V. fragans* (*V. planifolia*) and simultaneous conversion to vanillin by a combination of pectinase (polygalacturonase) and cellulase enzyme activities involving cell wall degradation and glucovanillin hydrolysis, in the presence of 47.5% ethanol for 8 h at 70°C, was found to be highly efficient (Ruiz *et al.*, 2001). Extracted vanillin was 3.13 times higher than the one obtained with the Soxhlet method. The classical curing/extraction process results in 1.1–1.8 g of vanillin/100 g of dry pods. Thus, it was concluded that the enzymatic reaction might substitute the microbial process involved in tissue fermentation previous to vanillin extraction with the simultaneous hydrolysis of glucovanillin.

Longares and Canizares (2006) devised a new method for the quick extraction of vanillin and *p*-hydroxybenzaldheyde (PHB) of vanilla beans from *V. fragans* by irradiating with microwave energy to accelerate the extraction process. Combined with simultaneous determination (using the Vierordt's method) and photometric monitoring (at 348 and 329 nm), this resulted in a 62-fold decrease in the extraction time and a 40–50% increase in the vanillin and PHB concentrations compared with the official Mexican extraction method.

Nguyen *et al.* (1991) extracted oleoresin from vanilla beans with 2–62 g CO_2/g dried bean at 306–309 K and 10–13 MPa. Vanillin yields of up to 95% were attained. Vanillin purity was higher with supercritical CO_2 extraction than with conventional aqueous ethanol extraction, with vanillin representing 74–97% of the flavour and fragrance compounds, compared with 61% using alcohol extraction.

Other aroma compounds

Aroma compounds in cured vanilla beans from different countries, e.g. Madagascar, Tonga, Costa Rica, Java, Indonesia and Mexico, have been documented. Over 100 volatile compounds have been detected, including aromatic carbonyls, aromatic alcohols, aromatic acids, aromatic esters, phenols and phenol ethers, aliphatic alcohols, carbonyls, acids, esters and lactones, of which the aldehyde vanillin is the most abundant. The level of the aldehydes, e.g. vanillin and *p*-hydroxybenzaldehyde and their respective acids (vanillic acid and *p*-hydroxybenzoic acid), in cured vanilla beans is used as an indicator of cured vanilla bean quality for commercial purposes (Klimes and Lamparsky, 1976; Adedeji *et al.*, 1993; Ranadive, 1994).

Silva *et al.* (2006) showed from sensory analysis that aromatic extracts obtained with a pentane/ether (1/1 v/v) solvent mixture, from cured vanilla beans, provided the flavour most representative of vanilla bean. They found clear differences between the numbers of aroma compounds identified in different organic aroma extracts: 65 volatiles were identified in a pentane/diethyl ether extract by GC-MS analysis; ether extraction gave 54 volatiles; the pentane/dichloromethane solvent yielded only 41 volatiles. The volatile compounds identified included 25 acids, 15 phenolic compounds, ten alcohols, four aldehydes, four heterocyclic compounds, four esters, two hydrocarbons and one ketone (Table 15.3). The tentatively identified compounds 2-heptenal, (*E*)-2-decenal and 2-heptenoic acid, were reported for the first time. Aromatic acids, aliphatic acids and phenolic compounds were the major volatiles. Quantification of the aroma compounds revealed that vanillin, vanillic acid, *p*-hydroxybenzaldehyde and *p*-hydroxybenzoic acid were the major compounds (Table 15.4). Vanillin is reported to represent 50% of the total quantified volatiles in Bourbon vanilla and 30% in Mexican vanilla whereas, in the study by Silva *et al.* (2006), the vanillin concentration (19,118 ppm) represented 85% of the volatile compounds.

When the pentane/diethyl ether extract was subjected to GC-O analysis, two of the 26 compounds detected were found at concentrations of less than 1 ppm, 13 at < 4 ppm, six at < 20 ppm, three at < 150 ppm and two at > 150 ppm. The compounds guaiacol, 4-methylguaiacol, acetovanillone and vanillyl alcohol, found at much lower concentrations in vanilla beans than vanillin, proved to be as intense as vanillin (Table 15.5). Ten phenolic compounds were detected in vanilla extracts as being aroma-active. Guaiacol, 4-methylguaiacol and acetovanillin, occurring at concentrations of 3.8–13.7 ppm, were similar in intensity to vanillin, which was detected at a concentration of more than 1000 times that of these compounds. Methyl salicylate, detected at a level of less than 1 ppm, was perceived as being as intense as vanillin. *p*-Cresol, methyl cinnamate and anisyl alcohol, occurring at concentrations of 1.1–2.6 ppm, were of medium intensity. Sweet, woody, balsamic, spicy, vanilla-like and toasted notes were attributed to phenolic compounds. Vanillic acid was not perceived by panellists because its elution required a high temperature, which caused a burnt odour in the sniffing port. The aldehydes 2-heptenal and (*E*)-2-decenal, identified here for the first time in vanilla beans, were perceived as being of medium intensity, with green, oily and herb-like floral notes. Aliphatic, acetic, isobutyric, isovaleric and valeric acids were

Table 15.3. Volatile compounds detected in aroma extracts from cured vanilla beans obtained using various organic solvents.

Compound	Ether	Pentane/ether	Pentane/dichloromethane
Phenols			
Guaiacol	*	*	*
4-Methylguaiacol		*	*
Phenol	*	*	
p-Cresol	*	*	*
4-Vinylguaiacol		*	
Vanillyl methyl ether		*	*
4-Vinylphenol	*	*	
Vanillin	*	*	*
Acetovanillone	*	*	*
Vanillyl alcohol	*	*	*
Vanilloylmethyl cetone	*	*	
p-Hydroxybenzaldehyde	*	*	*
p-Hydroxybenzyl alcohol	*	*	
Vanillic acid	*	*	*
p-Hydroxybenzoic acid	*	*	*
Aliphatic acids			
Acetic acid	*	*	*
Propanoic acid	*	*	*
Isobutyric acid	*	*	
Butyric acid	*	*	
Isovaleric acid	*	*	*
Valeric acid	*	*	*
Hexanoic acid	*	*	*
Heptanoic acid	*	*	*
Octanoic acid	*	*	*
2-Heptenoic acid	*	*	*
Nonanoic acid	*	*	*
Dodecanoic acid	*	*	
Myristic acid	*	*	
Pentadecanoic acid	*	*	
Hexadecanoic acid	*	*	*
9-Hexadecanoic acid		*	
Heptadecanoic acid	*	*	
Stearic acid	*	*	
Oleic acid	*	*	
Linoleic acid	*	*	
Aromatic acids			
Benzoic acid	*	*	
Benzene propanoic acid	*	*	*
Cinnamic acid (isomer 1)	*	*	*
Cinnamic acid (isomer 2)	*	*	*
Anisic acid	*	*	
Alcohols			
1-Octen-3ol	*	*	
2,3-Butanediol (isomer 1)	*	*	*
1-Octanol	*	*	
2,3-Butanediol (isomer 2)	*	*	*
1,2-Propanediol	*	*	
Benzyl alcohol	*	*	*
2-Phenylethanol	*	*	*
Benzene propanol	*	*	

Continued

Table 15.3. *Continued*

Compound	Ether	Pentane/ether	Pentane/dichloromethane
Anisyl alcohol	*	*	*
Cinnamyl alcohol	*	*	*
Aldehydes			
2-Heptenal	*	*	*
(*E*)-2-Decenal		*	*
(*E*,*Z*)-2,4-Decadienal	*	*	*
(*E*,*E*)-2,4-Decadienal	*	*	*
Esters			
Methyl salycilate		*	
Methyl cinnamate		*	
Anisyl formate		*	
Ethyl linolenate	*	*	*
Hydrocarbons			
Tricosane		*	*
Pentacosane		*	*
Heterocyclics			
Furfural	*	*	*
γ-Butyrolactone	*	*	*
Pantolactone		*	*
1H-pyrrole-2,5-dione, ethyl-4-methyl	*	*	*
Ketone			
3-Hydroxy-2-butanone	*	*	*

Souce: Silva *et al.* (2006).

perceived by the panellists as having sour, buttery and oily notes.

In the overall vanilla aroma, minor compounds like *p*-cresol, creosol, guaiacol and 2-phenylethanol have a high impact. This is shown by GC-olfactometry analysis of cured vanilla beans. Dignum *et al.* (2004) investigated the presence of *β*-D-glucosides of these compounds in order to determine whether these compounds were derived from glucosides or if they were formed during the curing process via different pathways. Glucosides of vanillin, vanillic acid, *p*-hydroxybenzaldehyde, vanillyl alcohol, *p*-cresol, creosol and bis[4-(*β*-D-glucopyranosyloxy)-benzyl]-2-isopropyltartrate and bis[4-(*β*-D-glucopyranosyloxy)-benzyl]-2-(2-butyl)tartrate have been identified in a green bean extract. Glucosides of 2-phenylethanol and *p*-cresol were not hydrolysed. *β*-Glucosidase does not have a high substrate specificity for the naturally occurring glucosides compared with the synthetic *p*-nitrophenol glucoside.

Werkhoff and Güntert (1997) reported for the first time the identification of some ester components in Bourbon vanilla beans – pentyl salicylate and citronellyl isobutyrate – as constituents of food aroma or essential oils (Table 15.6).

Adedeji *et al.* (1993) used a direct thermal desorption technique (220°C) to analyse the volatiles from beans that might cause the thermal degradation and transformation of sugar into common volatile compounds such as 3,5-dimethyl-2,4(3H,5H)-furandione and 3,5-dihydroxy-6-methyl-2,3-dihydro-4H-pyran-4-one. This last compound was detected at a high concentration (3880 ppm) in Mexican vanilla, being the third most abundant compound after vanillin and 2-furfural, and far more abundant than vanillic acid, *p*-hydroxybenzaldehyde or *p*-hydroxybenzoic acid.

Vanillin structure and properties

The compound predominantly responsible for the characteristic flavour and smell of vanilla is vanillin (4-hydroxy-3-methoxybenzaldehyde, MW 152.14), which is an aromatic aldehyde (Fig. 15.1). The fermented fruit

Table 15.4. Concentrations of volatile compounds in pentane/ether extract from cured vanilla beans.

Compound (quantified HPLC)	Concentration, ppm (mg/kg of cured vanilla)	Compound (quantified by HPLC)	Concentration, ppm (mg/kg of cured vanilla)
Guaiacol	9.3	Benzene propanoic acid	3.9
4-Methylguaiacol	3.8	Cinnamic acid (isomer 1)	3.4
Phenol	1.8	Cinnamic acid (isomer 2)	9.5
p-Cresol	2.6	Anisic acid	
4-Vinylguaiacol	1.2	*Alcohols*	
Vanillyl methyl ether	< 1	1-Octen-3ol	< 1
4-Vinyl phenol	1.8	2,3-Butanediol (isomer 1)	16.5
Vanillin	19,118	1-Octanol	1.1
Acetovanillone	13.7	2,3-Butanediol (isomer 2)	8.0
Vanillyl alcohol	83.8	1,2-Propanediol	< 1
Vanilloyl-methyl cetone	2.2	Benzyl alcohol	2.7
p-Hydroxybenzaldehyde	873.3	2-Phenylethanol	1.0
p-Hydroxybenzyl alcohol	65.1	Benzene propanol	< 1
Vanillic acid	1315	Anisyl alcohol	2.4
p-Hydroxybenzoic acid	255	Cinnamyl alcohol	< 1
Aliphatic acids		*Aldehydes*	
Acetic acid	124.3	2-Heptenal	2.1
Propanoic acid	1.7	(*E*)-2-Decenal	1.8
Isobutyric acid	1.7	(*E*,*Z*)-2,4-Decadienal	1.4
Butyric acid	< 1	(*E*,*E*)-2,4-Decadienal	1.2
Isovaleric acid	3.8	*Esters*	
Valeric acid	1.5	Methyl salycilate	< 1
Hexanoic acid	< 1	Methyl cinnamate	1.1
Heptanoic acid	1.9	Anisyl formate	2.3
Octanoic acid	5.5	Ethyl linolenate	13.5
2-Heptenoic acid	1.7	*Hydrocarbons*	
Nonanoic acid	15.7	Tricosane	15.9
Dodecanoic acid	2.2	Pentacosane	19.9
Myristic acid	12.4	*Heterocyclics*	
Pentadecanoic acid	13.4	Furfural	< 1
Hexadecanoic acid	126.6	γ-Butyrolactone	< 1
9-Hexadecanoic acid	5.7	Pantolactone	1.4
Heptadecanoic acid	5.7	1H-pyrrole-2,5-dione, ethyl-4-methyl	1.8
Stearic acid	13.9		
Oleic acid	16.3	*Ketone*	
Linoleic acid	225.6	3-Hydroxy-2-butanone	14.6
Aromatic acids			
Benzoic acid	2.6		

Source: Silva *et al.* (2006).

contains about 2% vanillin, depending on provenance (Mexico 1.75%, Sri Lanka 1.50% and Indonesia 2.75%). Vanillin is also found in gum benzoin, Peru balsam and clove oil. In clove bud oil, vanillin probably originates by air-oxidation of eugenol (Guenther, 1982).

According to Gildemeister and Hoffmann (1899), vanillin crystallizes from hot water in the form of colourless needles at 81–82°C. It possesses the strong and intensely sweet odour characteristic of vanilla. On careful heating, vanillin can be sublimated without decomposition; by prolonged heating at 105°C, vanillin decomposes with the formation of non-volatile products.

Vanillin is readily soluble in alcohol, ether, chloroform and hot water; relatively insoluble in cold water, for which reason vanillin can be recrystallized from water. At 75–80°C, one part of vanillin dissolves

Table 15.5. Aroma-active compounds detected by GC-O analysis of an aroma extract from cured vanilla beans.

Compound	Concentration, ppm	Odour quality	Intensity[a]
Phenols			
Guaiacol	9.3	Chemical, sweet spicy	+ + +
4-Methylguaiacol	3.8	Sweet, woody	+ + +
p-Cresol	2.6	Balsamic, woody, spicy	++
4-Vinylguaiacol	1.2	Chemical, phenolic	+
4-Vinylphenol	1.8	Sweet, woody	++
Vanillin	19,118	Vanilla, sweet	+ + +
Acetovanillone	13.7	Vanilla, sweet, honey	+ + +
Vanillyl alcohol	83.8	Vanilla-like	+ + +
p-Hydroxybenzaldehyde	873	Vanilla-like, biscuit	++
p-Hydroxybenzyl alcohol	65.1	Vanilla-like, sweet	++
Aliphatic acids			
Acetic acid	124	Sour, vinegar	++
Isobutyric acid	1.7	Buttery	++
Butyric acid	< 1	Buttery, oily	+
Isovaleric acid	3.8	Buttery, oily	++
Valeric acid	1.5	Cheese	+ + +
Alcohols			
2,3-Butanediol (isomer 2)	8.0	Floral, oily	+
Anisyl alcohol	2.4	Herbal	++
Aldehydes			
2-Heptenal	2.1	Green, oily	+
(*E*)-2-Decenal	1.8	Herb-like, floral	++
(*E*,*Z*)-2,4-Decadienal	1.4	Herb-like, fresh	++
(*E*,*E*)-2,4-Decadienal	1.2	Fatty, wood	++
Esters			
Methyl salicylate	< 1	Chalk	+ + +
Methyl cinnamate	1.1	Sweet	++
Ethyl linolenate	13.5	Sweet	++
Ketones			
3-Hydroxy-2-butanone	14.6	Buttery	+
Unknown[b]	6.2	Vanilla-like, chemical	+ + +

Note: [a](+) Weak, (++) medium, (+ + +) strong.
Source: Silva *et al.* (2006).

in 20 parts of water, at 14°C in 90–100 parts of water. At 7–8°C, most of the vanillin will crystallize from the solution gradually. Vanillin is soluble in sodium carbonate solution, but not in sodium bicarbonate solution (Guenther and Althausen, 1978).

About 170 other volatile constituents, mostly present at below 1 ppm levels, have been reported in vanilla by Klimes and Lamparsky (1976). Besides vanillin (85% of total volatiles), other important aroma components are glucovanillin, anisic acid, aldehyde, *p*-hydroxybenzoic acid, *p*-hydroxybenzaldehyde (up to 9%), vanillic acid, *p*-hydroxybenzyl alcohol, vanillyl alcohol and *p*-hydroxybenzyl methyl ether (1%), phenols, lactones, furans and esters (Uhl, 2000.)

The characteristic fragrance of Tahiti vanilla is due to the presence of piperonal (heliotropin, 3,4-dioxymethylenbenzaldehyde) and diacetyl (butandione). Piperonal is an aromatic aldehyde with a floral odour and is used in flavouring and perfumes. Separation of a fragrant 5-piperidone compound, containing three methyl groups, and also of methylbenzoate from Bourbon vanilla bean extract was achieved by liquid chromatography and gas-liquid partition

Table 15.6. Newly identified volatile ester components in Bourbon vanilla beans and their sensory impressions.

Component	Sensory impression
Hexyl butanoate	Fruity, green, sweet
Butyl hexanoate	Heavy-fruity, sweet
Fenchyl acetate	Fir needle oil, sweet, conifer-like
Menthyl acetate	Minty, woody-herbaceous, fruity, floral
α-Terpinyl acetate	Herbaceous, sweet, spicy, bergamot, lavender
Bornyl acetate	Herbaceous-piney, sweet, balsamic, camphory
Isobornyl acetate	Balsamic-camphoraceous, pine needle-like
Linalyl acetate	Sweet, floral-fruity, bergamot-like, lavender-like
Citronellyl isobutyrate	Fruity-rosy, sweet, bergamot-like
Phenethyl formate	Green-herbaceous, floral
Anisyl formate	Herbaceous-green, sweet, spicy, vanilla-like
Ethyl salicylate	Sweet, floral-fruity, heavy-fruity
Pentyl salicylate isoamyl	Floral, green, oil of winter-green, rose oxide-like
Salicylate hexyl	Sweet, clover-like
Salicylate	Sweet-herbaceous, floral, green, spicy

Source: Werkhoff and Güntert (1997).

chromatography, respectively (Feyertag and Hutchins, 1981).

The molecular structures of vanillin (4-hydroxy-3-methoxybenzaldehyde), isovanillin (3-hydroxy-4-methoxybenzaldehyde) and ethylvanillin(3-ethoxy-4-hydroxybenzaldehyde) were determined by Egawa *et al.* (2006) by means of gas electron diffraction. Among them, vanillin and ethylvanillin have a vanilla odour but isovanillin smells different. Vanillin and isovanillin have two stable conformers and ethylvanillin has four.

Non-volatiles

Oleoresin

Vanilla is available in three physical forms, whole beans, splits and cuts. Vanilla powder is a mixture of ground vanilla in a carrier such as 30% sugar (Purseglove *et al.*, 1981). Vanilla extract is made by cutting the cured beans into small pieces and percolating in successive quantities of hot 65–70% alcohol. The extract is very concentrated, a few drops sufficing for most uses. Vanilla oleoresin involves solvent extraction of chopped beans and later evaporation of the vanilla extract under vacuum, leaving a dark, viscous mass (Cowley, 1973). The oleoresin is diluted with solvents to give one- two- or 10-fold strengths (Uhl, 2000). Vanilla absolute and tincture are very concentrated ethanol or benzene extracts of vanilla aroma, used for perfumery purposes.

Vanilla essence comes in two forms: the actual extract of the seedpods and the far cheaper synthetic essence, basically consisting of a solution of synthetic vanillin in ethanol. Natural vanilla is an extremely complicated mixture of several hundred different compounds, versus synthetic vanillin which is derived from phenol and is of high purity. Many commercial vanilla extracts are now actually blends of natural and synthetic vanillin. The occurrence of several non-vanillin aroma and flavour components in minor or trace amounts in beans is the reason for their organoleptic superiority over synthetic vanilla and blends. Natural vanilla has a delicate, rich and mellow aroma and aftertaste, while the synthetic material is quite heavy, grassy and less pleasant.

Vanillin synthesis

Although more than 12,000 t of vanillin is produced each year, less than 1% of this is natural vanillin from *Vanilla*; the remainder is synthesized much more cheaply via chemical processes.

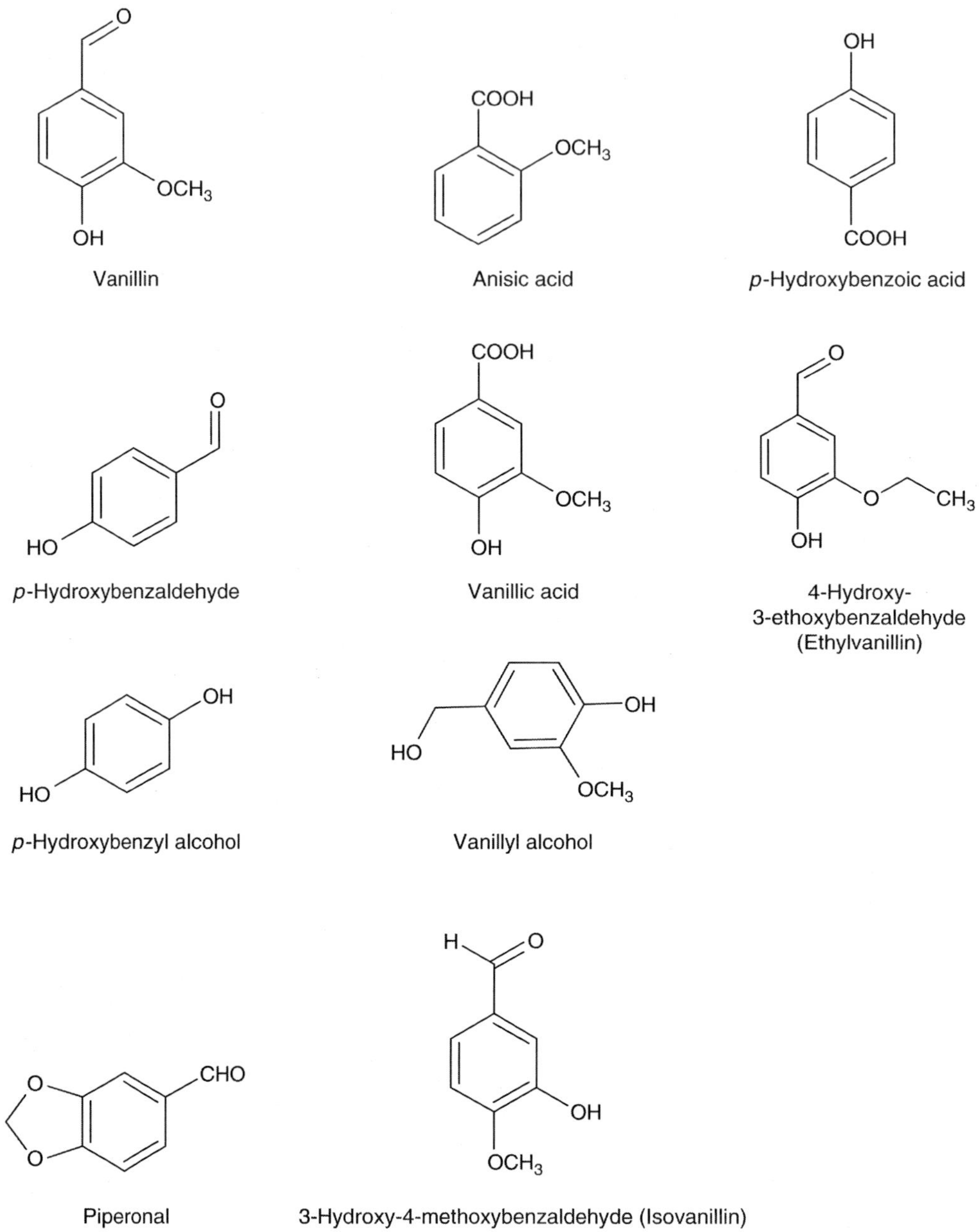

Fig. 15.1. Volatile compounds of vanilla.

The 'classical' synthesis of vanillin from eugenol or isoeugenol was developed in 1896 and it remained the preferred method for about 50 years. Vanillin is now prepared industrially in large amounts by the Reimer–Tiemann reaction, starting with guaiacol (catechol monomethylether), from which it is formed along with *o*-vanillin. Another source of vanillin is lignin, a by-product of paper pulp manufacture. Bioconversion of ferulic acid to vanillin is also possible by liquid cultures of various fungi, which may be an economical route as well.

The value of vanillin extracted from *Vanilla* pods is calculated variously as being between US$1200/kg and US$4000/kg, in contrast to the price of synthetic vanillin, <US$15/kg (Muheim and Lerch, 1999). Synthetic vanillin is used in both food and non-food applications, in fragrances and as a flavouring in pharmaceutical preparations. Currently, approximately 50% of the worldwide production of synthetic vanillin is used as an intermediate in the chemical and pharmaceutical industries for the production of herbicides, antifoaming agents or drugs such as papaverine, L-dopa, L-methyldopa and the antimicrobial agent, trimethoprim (Hocking, 1997). Synthetic vanillin is also used in household products, such as air fresheners and floor polishes.

Biosynthesis

Biosynthesis of vanillin has been attempted by various workers, as reviewed by Walton *et al.* (2003).

In 1965, Zenk, using radioactively labelled ferulic and vanillic acids, proposed a route by which both vanillin and vanillic acid were derived from ferulic acid. A CoA-dependent *β*-oxidative cleavage of feruloyl-CoA led to the formation of vanilloyl-CoA, which was further reduced to vanillin, or alternatively deacylated to vanillic acid (Fig. 15.2). The addition of ferulic acid to callus cultures or tissue cultures resulted in increased levels of vanillin production, suggesting that ferulic acid might indeed be a precursor of vanillin (Romagnoli and Knorr, 1988; Labuda *et al.*, 1993).

A more complex pathway was proposed by Funk and Brodelius (1990a,b, 1992), where caffeic acid was first methylated at the 4-position, followed by the 3-position to produce 3,4-dimethoxycinnamic acid, which was then demethylated at the 4-position, prior to a glucosylation step. Side-chain cleavage was proposed to occur at a late stage to produce vanillic acid (or its *β*-D-glucoside), which was then reduced to vanillin (Fig. 15.3).

Enzyme preparations of *β*-glucosidase have been used to release vanillin from vanilla pods as an alternative to conventional curing (Dignum *et al.*, 2001; Ruiz *et al.*, 2001).

A new model (Fig. 15.4) for hydroxycinnamate chain-shortening and vanillin formation in plants was revealed with the isolation of 4-hydroxycinnamoyl-CoA hydratase/lyase (HCHL) and its gene from a soil bacterium, *Pseudomonas fluorescens* strain AN103, which had been isolated by growth on ferulic acid as a sole carbon source (Gasson *et al.*, 1998; Narbad and Gasson, 1998; Mitra *et al.*, 1999).

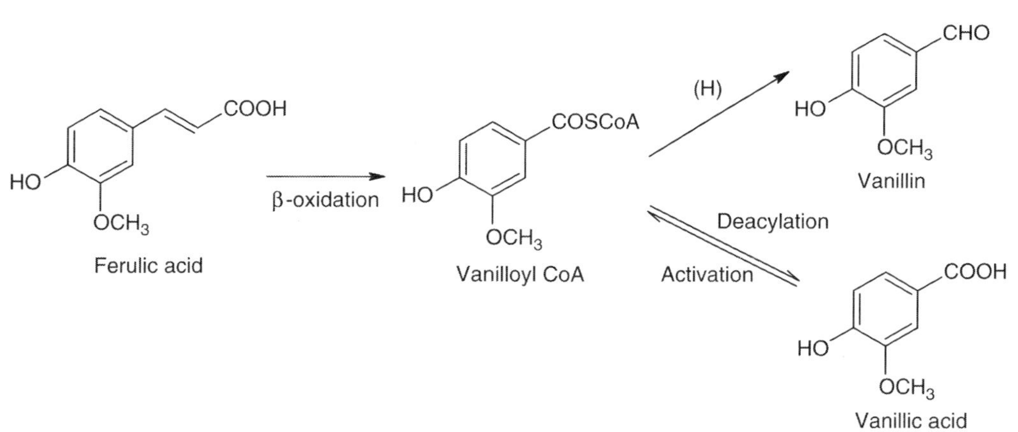

Fig. 15.2. Vanillin biosynthesis (Zenk, 1965).

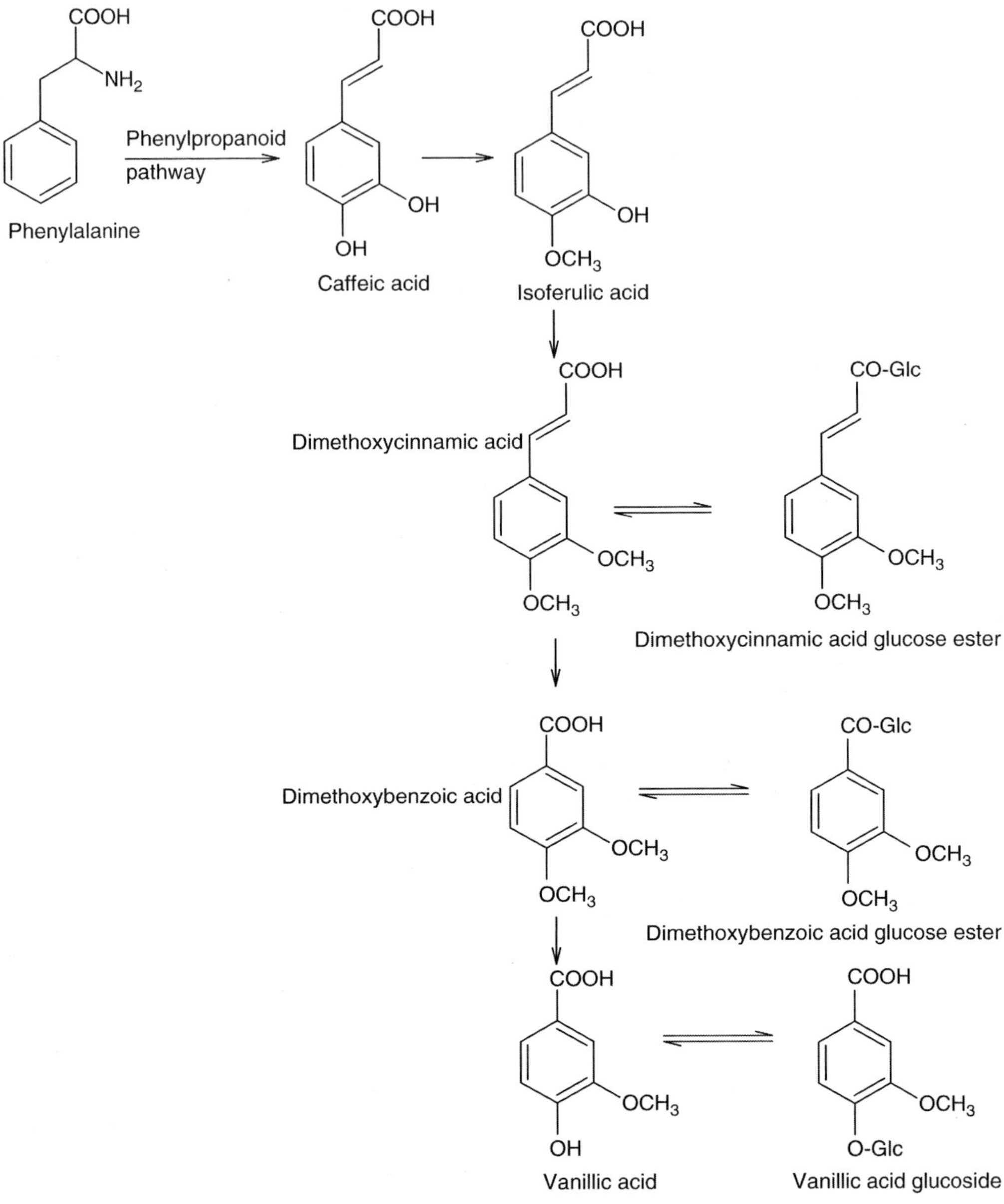

Fig. 15.3. A proposed route to vanillic acid via isoferulic acid (Funk and Brodelius, 1990a,b, 1992).

The biosynthetic origins of vanillin can be determined by the analysis of naturally occurring isotope ratios (in practice, chiefly $^{2}H/^{1}H$ and $^{13}C/^{12}C$), using isotope ratio-mass spectrometry (IR-MS) and nuclear magnetic resonance (site-specific natural isotope fractionation: SNIF-NMR®). Isotopic ratio characteristics are influenced by the biosynthetic route, as well as by the environmental (including climatic) conditions under which biosynthesis occurs (Martin *et al.*, 1992; Jamin *et al.*, 1997; Martin, 1998). The extent of the overall incorporation of naturally occurring ^{13}C, denoted by the $\delta_{PDB}{}^{13}C$

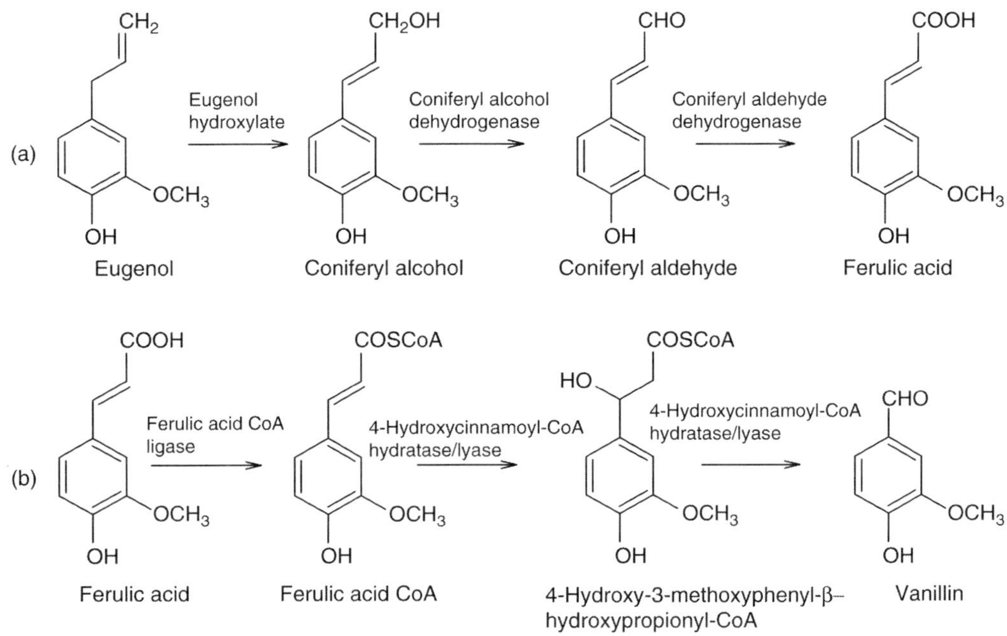

Fig. 15.4. The routes (a) from eugenol to ferulic acid and (b) from ferulic acid to vanillin in *Pseudomonas* strains (Gasson *et al.*, 1998; Mitra *et al.*, 1999).

(delta Pee Dee Belemnite) value, for vanillin and 4-hydroxybenzaldehyde samples isolated from *Vanilla* spp. falls within a characteristic range (around −21.0 for vanillin) that reflects the crassulacean acid metabolism (CAM) pathway of photosynthesis by which *Vanilla* fixes CO_2 (Lamprecht *et al.*, 1994; Remaud *et al.*, 1997). This is quite different from values determined for samples of vanillin that have been produced from the degradation of lignin, or chemically from fossil fuel sources, where CO_2 is not fixed originally by CAM metabolism (between *c.* −25 and −37). Vanillin samples produced by other means, for example by microbial fermentation or by metabolic engineering in plants (or microbes), or enzymatically, will display $\delta_{PDB}{}^{13}C$ values that will reflect the mechanism of the pathway involved.

Biotransformation

It is possible to generate vanillin from other plant-derived materials by biotransformation.

- Isorhapontin, a monoglucosylated stilbene constituent of spruce bark, can be cleaved by a dioxygenase isolated from *Pseudomonas* strain TMY1009 (Kamoda *et al.*, 1989).
- Soybean lipoxygenase can produce vanillin from esters of coniferyl alcohol (Markus *et al.*, 1992).
- van den Heuvel *et al.* (2001) used a broad-specificity *Penicillium* flavoenzyme, vanillyl alcohol oxidase, to produce vanillin by the biotransformation of vanillylamine (obtainable by the hydrolysis of capsaicin) and of creosol (a major component of creosote obtained from heating wood or coal tar).

Yoshida *et al.* (1997) achieved the production of vanillin by oxidation of vanillylamine using amine oxidase (AO) from *Aspergillus niger* and monoamine oxidase (MAO) from *Escherichia coli*. Enzyme kinetic studies have revealed that AO is a more efficient producer of vanillin than MAO. Continuous production of vanillin with immobilized AO was also investigated, and industrial synthesis of vanillin through AO from *A. niger* was suggested as a possibility by the authors. Such approaches are, in principle, attractive since the technologies are reproducible, predictable and acceptable and, given adequate demand, scale-up and stability would also be cost effective.

Tripathi *et al.* (2002) studied the biotransformation of phenylpropanoid intermediates – ferulic acid, coniferyl aldehyde and *p*-coumaric acid in free and immobilized cell cultures of *Haematococcus pluvialis*, which accumulated vanilla flavour metabolites – vanillin, vanillic acid, vanillyl alcohol, protocatechuic acid, *p*-hydroxybenzoic acid, *p*-hydroxybenzaldehyde and *p*-coumaric acid when treated with these precursors, to a range corresponding to vanilla flavour metabolites.

Tissue cultures

Tissue or organ cultures of *Vanilla* to produce vanillin and related flavour compounds have been explored by Knorr *et al.* (1993), Rao and Ravishankar (2000b), Dignum *et al.* (2001) and Priefert *et al.* (2001). Such cultures have the potential to produce *c.* 200 compounds that reportedly are present in (cured) vanilla pods; *Vanilla* cells and organs, and cells of *Capsicum frutescens* (Rao and Ravishankar, 2000a), have been cultured and successfully demonstrated to accumulate vanillin and associated metabolites (vanillic acid and ferulic acid), but production is low. Rao and Ravishankar (1999) reported that suspended and immobilized cell cultures of *C. frutescens* accumulated vanilla (and capsaicin) flavour metabolites when fed with isoeugenol. The addition of β-cyclodextrin and isoeugenol increased the accumulation of vanillin. Isoeugenol-treated immobilized cells, when challenged with aqueous mycelial extract of *A. niger*, yielded maximum vanillin concentrations, whereas the addition of a medium filtrate of *A. niger* led to a marginal increase in the vanillin.

A novel process for producing natural vanillin flavour from ferulic acid precursor has been developed by Westcott *et al.* (1993) using vanilla plant aerial roots as the biocatalyst. The charcoal used in the process acts as a product reservoir for the vanillin produced, thus relieving possible product inhibition and/or further metabolism. The vanillin is then removed from the charcoal by selective solvent extraction. The remaining unreacted ferulic acid remains adsorbed to the charcoal and can be recycled for further reaction. The aerial root tissue can be reused several times, but its activity gradually declines with reuse. Vanillin productivities of 400 mg/kg dry weight tissue/day and concentrations of 7 g/kg of root tissue can be obtained regularly. This concentration is *c.* 35-fold greater than the concentrations of vanillin originally present in the aerial root tissue and is about 40% of that present in matured vanilla beans. Using aerial roots supplied with ferulic acid, vanillin is produced five to ten times faster than its normal synthesis in vanilla beans, or in aerial roots not supplied with precursor. The composition of the vanilla flavour produced using the aerial root method is comparatively close to that of vanilla beans; in particular, it contains *p*-hydroxybenzaldehyde, at a vanillin:pHB ratio of 7.8:1, as compared with a ratio of 12.8:1 for bean-derived vanilla. This may impart a superior organoleptic value and make the product of this aerial root process more valuable.

Metabolic engineering

An innovative approach towards an enhanced capacity for vanillin formation would be to introduce an enzyme or pathway to generate vanillin from a mainstream intermediate of the plant phenylpropanoid pathway. The isolation of the gene encoding the bacterial vanillin-forming enzyme HCHL (detailed earlier) raised this possibility (Gasson *et al.*, 1998). In principle, feruloyl-CoA, an intermediate of the plant monolignol pathway (Whetten and Sederoff, 1995), could be converted directly to vanillin and acetyl-CoA in a single step.

In summary, of the alternatives available for introducing a pathway of vanillin production *de novo,* or for enhancing vanillin production in *Vanilla*, HCHL presents the most attractive option of generating vanillin from a phenylpropanoid precursor (feruloyl-CoA) naturally present in plants (Whetten and Sederoff, 1995).

Qualitative and quantitative analysis of vanilla

Waliszewski *et al.* (2007b) described a simple and rapid HPLC technique for vanillin determination in alcohol vanilla extract, and the method has been applied successfully for the determination of vanillin in some commercial extracts for routine analysis. de Jager *et al.* (2007) developed a LC-MS method for the determination of vanillin, coumarin and ethyl vanillin in vanilla products using LC-electrospray ionization in the positive ionization mode. The limits of detection for the method ranged from 0.051 to 0.073 μg/ml.

Bettazzi *et al.* (2006) developed a disposable electrochemical sensor for the detection of vanillin in vanilla extracts and in commercial products. An analytical procedure based on square-wave voltammetry (SWV) was optimized and a detection limit of 0.4 μM for vanillin was found. The method was applied to the determination of vanillin in natural concentrated vanilla extracts and in final products such as yoghurt and compote. The results obtained with electrochemical quantification of vanillin in the extract samples correlated well with the HPLC results.

It has been shown by Tenailleau *et al.* (2004) that the $^{13}C/^{12}C$ ratio can be used to determine the origin of vanillin by quantitative measurements of the ^{13}C NMR signals for each of the eight C atoms. Thus, attempts to substitute cheaper synthetic vanillin fraudulently can be detected even when ^{13}C-substituted materials are used.

Boyce *et al.* (2003) reported the use of the mixed micellar electrokinetic capillary chromatography (MECC) method for the qualitative and quantitative determination of key components, including vanillin, 4-hydroxybenzaldehyde, 4-hydroxybenzoic acid, vanillic acid and 3-methoxybenzaldehyde in natural vanilla extracts, nature identical extracts and synthetic flavourings. The limits of detection (LOD) ranged between 5–10 μg/ml. Bütehorn and Pyell (1996) had earlier demonstrated the potential use of micellar elektrokinetic chromatography (MEKC) in food analysis and a rapid method for the determination of vanillin and related compounds and possible synthetic additives to vanilla flavourings by MEKC as a screening method for quality control.

15.5. Culinary and Medicinal Uses

There are three main commercial forms of natural vanilla:

- Whole bean
- Powder
- Extract (alcoholic solution; as per Food and Drug Administration requirements, at least 35% vol. of alcohol).

Culinary uses

Vanilla has been coveted over the ages for culinary and medicinal reasons alike. While traditional medical uses of vanilla have faded away, its culinary traditions have changed little (Bythrow, 2005). Vanilla's high status in the culinary world comes from a long history of flavouring and its mellow fragrance enhances a variety of sweets and desserts, such as chocolates, custards, creams, soufflés, liqueurs, ice cream, sugar cookies, puff pastries and butter creams; its usage in salty foods is very uncommon.

Vanilla, being native to Central America and having a long record of pre-Columbian usage, was used by both the Mayas and, later, the Aztecs to flavour a special drink prepared from water, cocoa beans and spices: *chacau haa* (or *chocol haa*) in the Mayan and *cacahuatl* in the Aztec tongue (Náhuatl). Mayan chocolate, as still drunk in southern Mexico (Yucatán), Guatemala and Belize, is often spicy, containing chillies and other spices, native (allspice, annatto) or imported (black pepper, cinnamon).

Vanilla was first used in Europe, mainly for the same purpose as earlier in America, to flavour drinking chocolate, a very popular drink among the 17th century European nobility. European drinking chocolate was almost exclusively sweet and might have used a lot of additional flavourings, e.g. anise, cinnamon, but also exotic animal products like musk and ambergris.

Vanilla flavour in foodstuffs may be achieved by adding vanilla essence or by cooking vanilla beans in the liquid preparation. A stronger aroma may be attained if the beans are split in two; in this case, the innards of the beans (the aroma-filled seeds – the tiny black grains) are mixed into the preparation. Natural vanilla gives a brownish to yellowish colour to preparations, depending on concentration. Good-quality vanilla has a strong aromatic flavour, but foodstuffs with small amounts of low-quality vanilla or artificial vanilla-like flavourings are far more common, since true vanilla is much more expensive. Methyl, as well as ethyl, vanillin is used by the food industry; the latter is more expensive and has a stronger note.

Medicinal properties

From the time of the Aztecs, vanilla was considered an aphrodisiac. This reputation was much enhanced in 1762 when a German study found that a medication based on vanilla extract cured impotence. It was also once believed that vanilla was a febrifuge, i.e. used to reduce fevers, though it is used rarely for any medicinal purposes other than as a pharmaceutical flavouring. Essential oil of vanilla and vanillin were and are sometimes used in aromatherapy.

Antimicrobial property

In common with many other low-molecular weight phenolic compounds, vanillin displays antioxidant and antimicrobial properties and hence has the potential for use as a food preservative (Burri *et al.*, 1989; Davidson and Naidu, 2000). It is active against both Gram-positive and Gram-negative food-spoilage bacteria and has been shown to be effective against both yeasts and moulds in fruit purées and laboratory growth media (Cerrutti *et al.*, 1997; López-Malo *et al.*, 1998; Fitzgerald *et al.*, 2003). When used at a concentration of ~ 13 mM, vanillin inhibited the growth of *Saccharomyces cerevisiae*, *Zygosaccharomyces bailii*, *Debaryomyces hansenii* and *Z. rouxii* in culture medium and apple purée for 40 days. However, vanillin was less effective in banana purée, where ~ 20 mM was insufficient to inhibit the growth of *Z. bailii*; the authors concluded that the higher lipid/protein levels in bananas interfered with vanillin's antimicrobial activity (Cerrutti and Alzamora, 1996). The inhibition was biostatic in nature. During fermentation, the bioconversion of sub-MIC levels of vanillin to vanillyl alcohol and low levels of vanillic acid were demonstrated in the culture medium, presumably catalysed by constitutively expressed, non-specific dehydrogenases, neither of which was antagonistic to yeast cell growth (Fig. 15.5).

The results indicate the importance of the aldehyde moiety in the vanillin structure for its antimicrobial activity and that the bioconversion of vanillin could be advantageous for the yeasts, but only at levels below MIC. It was observed that increased vanillin concentrations inhibited its own bioconversion, suggesting that the activity required intact cells with metabolic capacity.

One limitation is the strong flavour of vanillin at the minimal inhibitory concentrations required, but this may be partially overcome by using it in combination with other, synergistic, antimicrobials, thus lowering the effective concentrations that are necessary (Gould, 1996).

Studies by López-Malo *et al.* (1995, 1997) showed the incorporation of vanillin (~ 3–7 mM) into fruit-based agars (apple, banana, mango, papaya and pineapple) inhibited the growth of *Aspergillus flavus*, *A. niger*, *A. ochraceus* and *A. parasiticus* for 2 months. Furthermore, synergistic effects

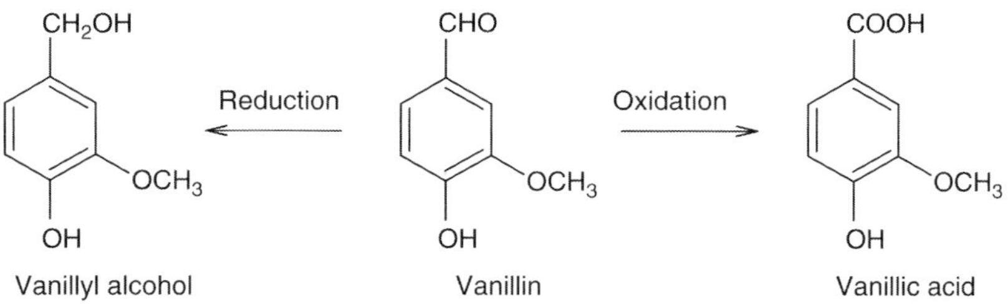

Fig. 15.5. The bioconversion pathway of vanillin in *Saccharomyces cerevisiae*.

were observed when vanillin and potassium sorbate were used in combination. Matamoros-León *et al.* (1999) established that, with a slight reduction in pH and water activity (a_w), ~ 3 mM vanillin in combination with ~ 2 mM potassium sorbate could inhibit the growth of *Penicillium digitatum*, *P. glabrum* and *P. italicum* for 1 month.

Antioxidant property

Vanillin has been reported to act as an antioxidant in complex foods containing polyunsaturated fatty acids (Burri *et al.*, 1989).

Antigenotoxic effect

There is some evidence for the antimutagenic effects of vanillin; for example, in suppressing chromosomal damage induced by methotrexate in the Chinese hamster V79 cell line (Keshava *et al.*, 1998). Inouye *et al.* (1988) reported the suppression of the induction of micronuclei by mitomycin C (MMC) in mouse bone marrow cells by post-treatment with vanillin. Post-treatment with vanillin at 500 mg/kg caused about 50% decrease in the frequency of micronucleated polychromatic erythrocytes (MN-PCEs). The suppressant effect was not due to a delay in the formation of MN-PCEs but to the cytotoxic action of vanillin. Vanillin acts as an anticlastogenic factor *in vivo*.

However, a study by Salih (2006) demonstrated that when a food additive was present in *E. coli* cell suspension during sunlight exposure, the number of induced mutations increased to varying extents over that seen with sunlight alone. Vanilla produced mutations in an additive fashion, while flavoured colourants like raspberry and peach increased the number of mutations in a dose-dependent manner. The impact of this investigation reflects the significance of the combination of sunlight and chemical food additives as a potential risk, which requires special attention and necessitates further investigations to evaluate this risk.

15.6. International Specifications for Quality and Desirable Limits

Everything expensive gets adulterated and faked – vanilla is no exception. Synthetic vanillin is an obvious choice to 'spice up' beans of low quality, or beans that have been extracted to yield the expensive vanilla extract (obtained by macerating vanilla pods in a mixture of water and alcohol). Synthetic vanillin could appear in the extract itself. Tonka bean extract features regularly in vanilla extract, especially in Mexico.

A high-performance liquid chromatographic procedure was developed for the isolation and quantitation of coumarin from vanilla-based liquid flavourings of Mexican origin by Thompson and Hoffmann (1988) in 40 products representing 14 different Mexican brands, which were assayed for coumarin, vanillin and ethyl vanillin. The procedure has been adapted to the analysis of other products including domestic vanilla extracts and imitation vanilla flavourings for 37 compounds, including vanillin, ethyl vanillin, 4-hydroxybenzaldehyde and piperonal.

The quality of the cured vanilla beans depends on a whole range of factors, starting from the agroclimatic condition during cultivation, through raw material production to curing. According to the ISO 3493 International Standards for vanilla, four commercial forms have been established:

1. Vanilla pods, consisting of whole pods which may be split.
2. Cut vanilla, consisting of parts of pods, split or not, and deliberately cut or broken.
3. Vanilla in bulk, consisting of vanilla in pods and cut vanilla.
4. Vanilla powder; obtained by grinding vanilla pods with permitted additives after drying.

Whole vanilla

Vanilla pods

The general characteristics desired for vanilla pods are that they are whole, sound, supple and full, of typical flavour, of uniform chocolate brown to dark brown colour and without any other stain for the non-split and split pods. The pods must have been cured suitably to develop their flavour and contain optimal moisture content. The pods may be rimy, with a mark at the bottom one-third of their length.

Cut vanilla

Cut vanilla must be prepared from vanilla pods as specified above, be sound and of good flavour and be chocolate brown to dark brown in colour.

Vanilla in bulk

Vanilla in bulk is obtained from either the pods or cut vanilla; it must be sound and of good specific flavour, chocolate brown to dark brown in colour, generally wooded and have several large stains.

Vanilla powder

Vanilla powder is obtained from either of the above three forms and must be able to pass through a mesh of 1.25 mm, be brown to dark brown in colour and have the characteristic flavour of vanilla.

Non-split and split vanilla

Non-split and split vanilla may fall into any of the four categories, depending on their quality. Their moisture contents may be 38% for categories 1 and 2, 30% for category 3 and cut and bulk vanilla pods, 25% for category 4 and 20% for vanilla powder (ISO 5565–2).

Vanilla extract

The quality of vanilla extract is defined by Winton's analytical values (Merory, 1960) (Table 15.7). The concentration of vanillin is a major criterion, although organoleptic quality does not depend on it entirely. The various characteristic flavour notes that define vanilla are woody, pruney, resinous, leathery, floral and fruity aromatics (Gillette and Hoffman, 1992). Bourbon vanilla serves as the standard by which the chemical and sensory qualities

Table 15.7. Quality parameters for vanilla extract.

Quality factor	Minimum	Maximum	Average
Vanillin (g/100 ml extract)	0.11	0.35	0.19
Ash (g/100 ml extract)	0.220	0.432	0.319
Soluble ash (g/100 ml extract)	0.179	0.357	0.265
Lead number (Winton)	0.40	0.74	0.54
Alkalinity of total ash (N/10 acid/100 ml extract)	30.00	54.00	30.00
Alkalinity of soluble ash (N/10 acid/100 ml extract)	22.00	40.00	42.00
Total acidity (N/10 alkali/100 ml extract)	30.00	52.00	30.00
Acidity other than vanillin (N/10 alkali/100 ml extract)	14.00	42.00	

Source: Merory (1960).

of other types of vanilla can be assessed. Imitation vanilla, when spiked with vanillin, is inferior in quality compared with the natural extract, as the characteristic flavour components are missing. No modern processing technology can improve the quality of a poor bean; deterioration in quality can also result from improper curing and handling.

Specifications

The standards defined by the Food and Drug Administration of the USA for vanilla products are given below.

21CFR 169.3
Sec. 169.3 Definitions

a) The term vanilla beans mean the properly cured and dried pods of *Vanilla planifolia* Andrews and of *Vanilla tahitensis* Moore.
b) The term unit weight of beans means, in the case of vanilla beans, containing not more than 25% moisture; it means the weight of such beans equivalent in content of moisture-free vanilla-bean solids to 13.35 ounces of vanilla beans containing 25% moisture.
c) The term unit of vanilla constituent means the total sapid and odorous principles extractable from one unit weight of vanilla beans, as defined in paragraph (b) of this section, by an aqueous alcohol solution in which the content of ethyl alcohol by volume amounts to not less than 35%.

21CFR 169.175
Sec. 169.175 Vanilla extract

a) Vanilla extract is the solution in aqueous ethyl alcohol of the sapid and odorous principles extractable from vanilla beans. In vanilla extract, the content of ethyl alcohol is not less than 35% by volume and the content of vanilla constituent, as defined in Sec. 169.3 (c), is not less than one unit per gallon. The vanilla constituent may be extracted directly from vanilla beans or it may be added in the form of concentrated extract or concentrated vanilla flavouring or vanilla flavouring concentrated to the semi-solid form called vanilla oleoresin. Vanilla extract may contain one or more of the following optional ingredients:
(1) Glycerin
(2) Propylene glycol
(3) Sugar (including invert sugar)
(4) Dextrose
(5) Maize syrup (including dried maize syrup).

Source: US Food and Drug Administration (2002), Code of Federal Regulations, 21CFR Part 169.

Packing and marking

Vanilla pods are to be put in packets of pods of the same length and then packaged in clean, sound, watertight containers of a material that will not affect the product in any way (e.g. tin-plate boxes). The containers must have vanilla of the same category and homogeneous. The same specifications hold good for different forms of vanilla. The packets of the above products of vanilla must contain the following information: name of the product (botanical species), commercial form, producing country, code, batch or test certificate number, or similar means of identification, any other information required by the customer, and reference to the international standard (ISO, 1997.)

15.7. Conclusion

To conclude, it is worth reiterating that the only orchid used as a spice, the vanilla pod, has been recognized for its culinary and medicinal uses since the time of the Aztecs. Vanilla is the world's third most expensive spice, from which is obtained the popular commercial flavouring agent, vanillin. The characteristic aroma of vanilla is obtained only after a time-consuming and labour-intensive curing process. The main aroma compound in vanilla is vanillin; over 100 volatile compounds have been detected, including aromatic carbonyls, aromatic alcohols, aromatic acids, aromatic esters, phenols and phenol ethers, aliphatic alcohols, carbonyls, acids, esters and lactones. The level of the aldehydes, vanillin and *p*-hydroxybenzaldehyde and their respective acids (vanillic acid and *p*-hydroxybenzoic acid) in cured vanilla beans is used as an indicator of bean quality for commercial purposes. Vanilla's high status in the culinary world comes from a long history of flavouring chocolates, sweets and desserts. Its medicinal uses are demonstrated in its antimicrobial, antioxidant and antigenotoxic effects. The ISO 3493 International Standards for vanilla dictates the quality standards for the different forms of vanilla: vanilla pods, cut vanilla, vanilla in bulk and vanilla powder.

References

Adedeji, J., Hartman, T.G. and Ho, C. (1993) Flavor characterization of different varieties of vanilla beans. *Perfumer and Flavors* 18, 115–133.

Arana, F.E. (1943) Action of β-glucosidase in the curing of vanilla. *Food Research* 8, 343–351.

Bettazzi, F., Palchetti, I., Sisalli, S. and Mascini, M. (2006) A disposable electrochemical sensor for vanillin detection. *Analytica Chimica Acta* 555(1), 134–138.

Boyce, M.C., Haddad, P.R. and Sostaric, T. (2003) Determination of flavour components in natural vanilla extracts and synthetic flavourings by mixed micellar electrokinetic capillary chromatography. *Analytica Chimica Acta* 485(2), 179–186.

Burri, J., Graf, M. Lambelet, P. and Löliger, J. (1989) Vanillin: more than a flavouring agent – a potent antioxidant. *Journal of the Science of Food and Agriculture* 48, 49–56.

Bütehorn, U. and Pyell, U. (1996) Micellar electrokinetic chromatography as a screening method for the analysis of vanilla flavourings and vanilla extracts. *Journal of Chromatography A* 736(1–2), 321–332.

Bythrow, J.D. (2005) Historical perspective: vanilla as a medicinal plant. *Seminars in Integrative Medicine* 3(4), 129–131.

Cerrutti, P. and Alzamora, S.M. (1996) Inhibitory effects of vanillin on some food spoilage yeasts in laboratory media and fruit purées. *International Journal of Food Microbiology* 29, 379–386.

Cerrutti, P., Alzamora, S.M. and Vidales, S.L. (1997) Vanillin as an antimicrobial for producing shelf-stable strawberry purée. *Journal of Food Science* 62, 608–610.

Childers, N.E., Cibes, H.R. and Hernandez-Medina, E. (1959) Vanilla – the orchid of commerce. In: Whithner, C.L. (ed.) *The Orchids – A Scientific Survey*. Ronald Press Co., New York, pp. 477–508.

Correll, D.S. (1953) Vanilla. Its botany, history, cultivation and economic importance. *Economic Botany* 7(4), 291–358.

Cowley, E. (1973) Vanilla and its uses. In: *Proceedings of the Conference on Spices*, Tropical Products Institute, London, p. 79.

David, P.A. (1950) Vanilla culture in the College of Agriculture at Los Banos. *Phillippines Agriculture* 33, 239–349.

Davidson, P.M. and Naidu, A.S. (2000) Phyto-phenols. In: Naidu, A.S. (ed.) *Natural Food Antimicrobial Systems*. CRC Press LLC, Boca Raton, Florida, pp. 265–294.

de Jager, L.S., Perfetti, G.A. and Diachenko, G.W. (2007) Determination of coumarin, vanillin, and ethyl vanillin in vanilla extract products: liquid chromatography mass spectrometry method development and validation studies. *Journal of Chromatography A* 1145(1–2), 83–88.

DeVarigny. C. (1894) Fertilization of the vanilla flower by bees. *Bombay Natural History Society* 4, 555–556.

Dignum, M.J.W., Kerler, J. and Verpoorte, R. (2001) Vanilla production: technological, chemical, and biosynthetic aspects. *Food Review International* 17, 199–219.

Dignum, M.J.W., Kerler, J. and Verpoorte, R. (2002) Vanilla curing under laboratory conditions. *Food Chemistry* 79(2), 165–171.

Dignum, M.J.W., van der Heijden, R., Kerler, J., Winkel, C. and Verpoorte, R. (2004) Identification of glucosides in green beans of *Vanilla planifolia* Andrews and kinetics of vanilla β-glucosidase. *Food Chemistry* 85(2), 199–205.

Egawa, T., Kameyama, A. and Takeuchi, H. (2006) Structural determination of vanillin, isovanillin and ethylvanillin by means of gas electron diffraction and theoretical calculations. *Journal of Molecular Structure* 794(1–3), 92–102.

Feyertag, E. and Hutchins, R. (1981) Isolation and tentative identification of a new constituent of bourbon vanilla bean extract. *Food Chemistry* 7(4), 311–315.

Fitzgerald, D.J., Stratford, M. and Narbad, A. (2003) Analysis of the inhibition of food spoilage yeasts by vanillin. *International Journal of Food Microbiology* 23rd International Specialized Symposium on Yeasts (ISSY-23) 86(1–2), 113–122.

Funk, C. and Brodelius, P.E. (1990a) Influence of growth regulators and an elicitor on phenylpropanoid metabolism in suspension cultures of *Vanilla planifolia*. *Phytochemistry* 29(3), 845–848.

Funk, C. and Brodelius, P.E. (1990b) Phenylpropanoid metabolism in suspension cultures of *Vanilla planifolia* Andr. II. Effects of precursor feeding and metabolic inhibitors. *Plant Physiology* 94, 95–101.

Funk, C. and Brodelius, P.E. (1992) Phenylpropanoid metabolism in suspension cultures of *Vanilla planifolia* Andr. IV. Induction of vanillic acid formation. *Plant Physiology* 99, 256–262.

Galletto, W.G. and Hoffman, P.G. (1978) Some benzyl ethers present in the extract of vanilla (*Vanilla planifolia*). *Journal of Agricultural and Food Chemistry* 26, 195–197.

Gasson, M.J., Kitamura, Y., McLauchlan, W.R., Narbad, A., Parr, A.J., Parsons, E.L.H., Payne, J., Rhodes, M.J.C. and Walton, N.J. (1998) Metabolism of ferulic acid to vanillin. *Journal of Biological Chemistry* 237, 4163–4170.

Gildemeister, E. Von and Hoffmann, Fr. (1899) *Die Ätherischen Öle*, 3rd edn., Volume III. Julius Springer, Schimmel & Co., Leipzeig, Berlin, pp. 288.

Gillette, M.H. and Hoffman, P.G. (1992) Vanilla extract. In: Hui, Y.H. (ed.) *Encyclopedia of Food Science and Technology*, Volume 4. Wiley, New York, pp. 2641–2657.

Gould, G.W. (1996) Industry perspectives on the use of natural antimicrobials and inhibitors for food applications. *Journal of Food Protection* (suppl.), 82–86.

Gregory, L.E., Gaskins, M.H. and Colberg, C. (1967) Parthenocarpic pod development by *Vanilla planifolia* Andrews induced with growth-regulating chemicals. *Economic Botany* 21, 351–357.

Guenther, E. (1982) Oil of clove. In: *The Essential Oils*, Volume 4. Robert E. Krieger Publishing Co., Florida, pp. 435.

Guenther, E. and Althausen, D. (1978) Aromatic aldehydes. In: *The Essential Oils*, Volume 2. Robert E. Krieger Publishing Co., Huntington, New York, pp. 361–362.

Hartman, T.G., Karmas, K., Chen, J., Shevade, A., Deagro, M. and Hwang, H. (1992) Determination of vanillin, other phenolic compounds, and flavours in vanilla beans. In: Ho, C.-T., Lee, C.Y. and Huang, M.T. (eds) *Phenolic Compounds in Food and Their Effects on Health I, ACS Symposium Series No. 506*. American Chemical Society, Washington, DC, pp. 60–76.

Havkin-Frenkel, D., Dorn, R. and Leuster, T. (1997) Plant tissue culture for production of secondary metabolites. *Food Technology (Chicago)* 51(11), 56–58.

Hocking, M.B. (1997) Vanillin: synthetic flavoring from spent sulfite liquor. *Journal of Chemical Education* 74, 1055–1059.

http://www.hinduonnet.com/businessline/2003/03/03/.

http://www.kissankerala.net.

Inouye, T., Sasaki, Y.F., Imanishi, H., Watanabe, M., Ohta, T. and Shirasu, Y. (1988) Suppression of mitomycin C-induced micronuclei in mouse bone marrow cells by post-treatment with vanillin. *Mutation Research/Fundamental and Molecular Mechanisms of Mutagenesis* 202(1), 93–95.

Jamin, E., Naulet, N. and Martin, G.J. (1997) Multi-element and multi-site isotopic analysis of nicotine from tobacco leaves. *Plant, Cell and Environment* 20, 589–599.

Kamoda, S., Habu, N., Samejima, M. and Yoshimoto, T. (1989) Purification and some properties of lignostilbene-α,β-dioxygenase responsible for the C_α–C_β cleavage of a diarylpropane type lignin model compound from *Pseudomonas* sp. TMY1009. *Agricultural and Biological Chemistry* 53, 2757–2761.

Kanisawa, T. (1993) Flavor development in vanilla beans. *Kouryou* 180, 113–123.

Kanisawa, T., Tokoro, K. and Kawahara, S. (1994) Flavor development in the beans of *Vanilla planifolia*. In: Kurihara, K., Suzuki, N. and Ogawa, H. (eds) *Olfaction Taste XI, Proceedings of the International Symposium*. Springer, Tokyo, pp. 268–270.

Keshava, C., Keshava, N., Whong, W.-Z., Nath, J. and Ong, T.-M. (1998) Inhibition of methotrexate-induced chromosomal damage by vanillin and chlorophyllin in V79 cells. *Teratogenesis, Carcinogenesis, and Mutagenesis* 17, 313–326.

Klimes, I. and Lamparsky, D. (1976) Vanilla volatiles – a comprehensive analysis. *International Flavours and Food Additives* 7, 272–291.

Knorr, D., Caster, C., Dörneburg, H., Dorn, R., Gräf, S., Havin-Frenkel, D., Podstolski, A. and Werrman, U. (1993) Biosynthesis and yield improvement of food ingredients from plant cell and tissue cultures. *Food Technology* 47, 57–63.

Labuda, I.M., Keon, K.A. and Goers, S.K. (1993) Progress in flavour precursor studies: analysis, generation, biotechnology. In: Schreier, P. and Winterhalter, P. (eds) *Proceedings of the International Conference, Würzburg, Germany*. Allured Publishing Corp., Carol Stream, Illinois, pp. 477.

Lamprecht, G., Pichlmayer, F. and Schmid, E.R. (1994) Determination of the authenticity of vanilla extracts by stable isotope ratio analysis and component analysis by HPLC. *Journal of Agricultural and Food Chemistry* 42, 1722–1727.

Longares, P.A. and Canizares, M.M.P. (2006) Focused microwave-assisted extraction and simultaneous spectrophotometric determination of vanillin and *p*-hydroxybenzaldehyde from vanilla fragans. *Talanta* 69(4), 882–887.

López-Malo, A., Alzamora, S.M. and Argaiz, A. (1995) Effect of natural vanillin on germination time and radial growth of moulds in fruit-based agar systems. *Food Microbiology* 12, 213–219.

López-Malo, A., Alzamora, S.M. and Argaiz, A. (1997) Effect of vanillin concentration, pH and incubation temperature on *Aspergillus flavus, Aspergillus niger, Aspergillus ochraceus* and *Aspergillus parasiticus* growth. *Food Microbiology* 14, 117–124.

López-Malo, A., Alzamora, S.M. and Argaiz, A. (1998) Vanillin and pH synergistic effects on mould growth. *Journal of Food Science* 63, 143–146.

Maarse, H., Visscher, C.A., Willlemsen, L.C. and Boelens, M.H. (1994) *Volatile Compounds in Food. Qualitative and Quantitative Data*, Supplement 4. TNO, Zeist, Netherlands, pp. 219–227.

Markus, P.H, Peters, A.L.J. and Roos, R. (1992) Process for the preparation of phenylaldehydes. European Patent Application, EP 0 542 348 A2.

Martin, G.J. (1998) Recent advances in site-specific natural isotope fractionation studied by nuclear magnetic resonance. *Isotopes in Environmental and Health Studies* 34, 233–243.

Martin, G.J., Martin, M.L. and Zhang, B.-L. (1992) Site-specific natural isotope fractionation of hydrogen in plant products studied by nuclear magnetic resonance. *Plant, Cell and Environment* 15, 1037–1050.

Matamoros-León, B., Argaiz, A. and López-Malo, A. (1999) Individual and combined effects of vanillin and potassium sorbate on *Penicillium digitatum, Penicillium glabrum*, and *Penicillium italicum* growth. *Journal of Food Protection* 62, 540–542.

Merory, J. (1960) *Food Flavorings: Composition, Manufacture and Use*. AVI Publ. Co. Inc., Westport, Connecticut.

Mitra, A., Kitamura, Y., Gasson, M.J., Narbad, A., Parr, A.J., Payne, J., Rhodes, M.J.C., Sewter, C. and Walton, N.J. (1999) 4-hydroxycinnamoyl-CoA hydratase/lyase (HCHL) – an enzyme of phenylpropanoid chain cleavage from *Pseudomonas*. *Archives of Biochemistry and Biophysics* 365, 10–16.

Muheim, A. and Lerch, K. (1999) Towards a high-yield conversion of ferulic acid to vanillin. *Applied Microbiology and Biotechnology* 51, 456–461.

Narbad, A. and Gasson, M.J. (1998) Metabolism of ferulic acid via vanillin using a novel CoA-dependent pathway in a newly-isolated strain of *Pseudomonas fluorescens*. *Microbiology* 144, 1397–1404.

Nguyen, K., Barton, P. and Spencer, J.S. (1991) Supercritical carbon dioxide extraction of vanilla. *The Journal of Supercritical Fluids* (Proceedings of the Symposium on Supercritical Fluids) 4(1), 40–46.

Odoux, E., Chauwin, A. and Brillouet, J.M. (2003) Purification and characterization of vanilla bean (*Vanilla planifolia* Andrews) beta-D-glucosidase. *Journal of Agricultural and Food Chemistry* 51(10), 3168–3173.

Priefert, H., Rabenhorst, J. and Steinbüchel, A. (2001) Biotechnological production of vanillin. *Applied Microbiology and Biotechnology* 56, 296–314.

Purseglove, J.W. (1985) *Tropical Crops: Monocotyledons*. Longman Group Ltd., Harlow, UK.

Purseglove, J.W., Brown, E.G., Green, C.L. and Robbins, S.R.J. (1981) *Spices*, Volume 2. Longman, Inc., New York, pp. 736–813.

Ranadive, A.S. (1994) Vanilla cultivation, curing, chemistry, technology and commercial products. In: Charalambous, G. (ed.) *Spices, Herbs and Edible Fungi*. Elsevier, Amsterdam, pp. 517–577.

Ranadive, A.S., Szkutnica, K., Guerrera, J.G. and Frenkel, C. (1983) Vanillin biosynthesis in vanilla beans. In: *Proceedings of the IX International Congress of Essential Oils*, Singapore, pp. 147–154.

Rao, S.R. and Ravishankar, G.A. (1999) Biotransformation of isoeugenol to vanilla flavour metabolites and capsaicin in suspended and immobilized cell cultures of *Capsicum frutescens*: study of the influence of β-cyclodextrin and fungal elicitor. *Process Biochemistry* 35(3–4), 341–348.

Rao, S.R. and Ravishankar, G.A. (2000a) (Review) Vanilla flavour: production by conventional and biotechnological routes. *Journal of the Science of Food and Agriculture* 80, 289–304.

Rao, S.R. and Ravishankar, G.A. (2000b) Biotransformation of protocatechuic aldehyde and caffeic acid to vanillin and capsaicin in freely suspended and immobilized cell cultures of *Capsicum frutescens*. *Journal of Biotechnology* 76, 137–146.

Remaud, G.S., Martin, Y.-L., Martin, G.G. and Martin, G.J. (1997) Detection of sophisticated adulterations of natural vanilla flavors and extracts: application of the SNIF-NMR method to vanillin and *p*-hydroxybenzaldehyde. *Journal of Agricultural and Food Chemistry* 45, 859–866.

Romagnoli, L.G. and Knorr, D. (1988) Effects of ferulic acid treatment on growth and flavour development of cultured *Vanilla planifola* cells. *Food Biotechnology* 2, 83–104.

Ruiz, T.F., Perez, A.I. and Lopez, M. (2001) An enzymatic extraction and transformation of glucovanillin to vanillin from vanilla green pods. *Journal of Agricultural and Food Chemistry* 49(11), 5207–5209.

Sagrero-Nieves, L. and Schwartz, S.J. (1988) Phenolic content of *Vanilla planifolia* as affected by harvest period. *Journal of Food Composition and Analysis* 1(4), 362–365.

Salih, F.M. (2006) Risk assessment of combined photogenotoxic effects of sunlight and food additives. *Science of the Total Environment* 362(1–3), 68–73.

Silva, A.P., Odoux, E., Brat, P., Ribeyre, F., Rodriguez-Jimenes, G., Robles-Olvera, V., García-Alvarado, M.A. and Günata, Z. (2006) GC-MS and GC-olfactometry analysis of aroma compounds in a representative organic aroma extract from cured vanilla (*Vanilla planifolia* G. Jackson) beans. *Food Chemistry* 99(4), 728–735.

Simpson, B.B. and Ogarzaly, M.C. (1986) *Economic Botany: Plants in Our World*. McGraw-Hill, New York.

Sostaric, T., Boyce, M.C. and Spickett, E.E. (2000) Analysis of the volatile compounds in vanilla extracts and flavorings by solid-phase microextraction and gas chromatography. *Journal of Agricultural and Food Chemistry* 48, 5802–5807.

Tenailleau, E.J., Lancelin, P., Robins, R.J. and Akoka, S. (2004) Authentication of the origin of vanillin using quantitative natural abundance ^{13}C NMR. *Journal of Agricultural and Food Chemistry* 52, 7782–7787.

Theodose, R. (1973) Traditional methods of vanilla preparation and their improvement. *Tropical Science* 15, 47–57.

Thompson, R.D. and Hoffmann, T.J. (1988) Determination of coumarin as an adulterant in vanilla flavoring products by high-performance liquid chromatography. *Journal of Chromatography A* 438, 369–382.

Tripathi, U., Rao, S.R. and Ravishankar, G.A. (2002) Biotransformation of phenylpropanoid compounds to vanilla flavor metabolites in cultures of *Haematococcus pluvialis*. *Process Biochemistry* 38(3), 419–426.

Uhl, S.R. (2000) Emerging flavor contributors. In: *Handbook of Spices, Seasonings and Flavorings*. Technomic Publishing Co., Inc., Pennsylvania, pp. 217–219.

UN Food and Agriculture Organization (2005) http://www.hinduonnet.com/businessline/2003/03/03/.

US Food and Drug Administration (2002) Code of Federal Regulations, 21CFR Part 169.

USDA National Nutrient Database for Standard Reference (2002) www.nal.usda.gov.

van den Heuvel, R.R.H., Fraaije, M.W., Laane, C. and van Berkel, W.J.H. (2001) Enzymatic synthesis of vanillin. *Journal of Agriculture and Food Chemistry* 49, 2954–2958.

Waliszewski, K.N., Ovando, S.L. and Pardio, V.T. (2007a) Effect of hydration and enzymatic pretreatment of vanilla beans on the kinetics of vanillin extraction. *Journal of Food Engineering* 78(4), 1267–1273.

Waliszewski, K.N., Pardio, V.T. and Ovando, S.L. (2007b) A simple and rapid HPLC technique for vanillin determination in alcohol extract. *Food Chemistry* 101(3), 1059–1062.

Walton, N.J., Mayer, M.J. and Narbad, A. (2003) Molecules of interest – vanillin. *Phytochemistry* 63(5), 505–515.

Weiss, E.A. (2002) Orchidaceae. In: *Spice Crops*. CAB International, Wallingford, UK, pp. 136–154.

Werkhoff, P. and Güntert, M. (1997) Identification of some ester compounds in Bourbon vanilla beans. *Lebenomittel-Wissenschaft und Technology* 30, 429–431.

Westcott, R.J., Cheetham, P.S.J. and Barraclough, A.J. (1993) Use of organized viable vanilla plant aerial roots for the production of natural vanillin. *Phytochemistry* 35(1), 135–138.

Whetten, R. and Sederoff, R. (1995) Lignin biosynthesis. *Plant Cell* 7, 1001–1013.

Wild-Altamirano, C. (1969) Enzymic activity during growth of vanilla fruit. 1. Proteinase, glucosidase, peroxidase and polyphenoloxidase. *Journal of Food Science* 34, 235–238.

Woebse, A. (1963) Pollinating the vanilla flower. *American Orchid Society Bulletin* 32(12), 1009–1010.

Yoshida, A., Takenaka, Y., Tamaki, H., Frebort, I., Adachi, O. and Kumagai, H. (1997) Vanillin formation by microbial amine oxidases from vanillylamine. *Journal of Fermentation and Bioengineering* 84(6), 603–605.

Zenk, M.H. (1965) Biosynthese von vanillin in *Vanilla planifolia* Andr. *Z. Pflanzenphysiology* 53, 404.

16 Ajowan

T. John Zachariah

16.1. Introduction

Ajowan, or bishop's weed (*Trachyspermum ammi* L. Sprague ex. Turill, syn. *T. copticum* Linn and *Carum copticum* Hiern), belonging to the family Apiaceae, is an important seed spice. It is used as a spice in certain areas of Asia only. It is a small seed-like fruit, egg-shaped and greyish in colour. The plant has a similarity to parsley. Because of their seed-like appearance, the fruit pods are sometimes called ajwain seeds, or bishop's weed seeds. It is found mostly in Indian cooking, where it is also known as bishop's weed or carom. It is particularly suited to the delicate vegetarian cuisine from the state of Gujarat. Table 16.1 illustrates the export of ajowan to various countries from India. Pakistan and Saudi Arabia are the leading importers (Malhotra and Vijay, 2004).

The greyish-green ajowan seeds are striped and curved (similar to cumin or caraway seeds in appearance), often with a fine silk stalk attached. They are usually sold whole. The seeds are often chewed on their own for medicinal value, tasting bitingly hot and bitter, leaving the tongue numb for a while. Cooking is found to mellow ajowan; when crushed, the seeds have a strong and distinctive thyme-like fragrance.

Ajowan possesses a harsh thyme-like flavour with a bit of a kick, leaving a milder, pleasant aftertaste (http://www.theepicentre.com/Spices/ajowan.html; Malhotra and Vijay, 2004).

16.2. Botany and Uses

Botany

Ajowan is an annual herbaceous, 30–70 cm (1–2 ft) in height, bearing feathery leaves and red flowers. When the seeds are ripe, they are dried and threshed. Ajowan is native to India, but is also cultivated in Iran, Egypt, Pakistan and Afghanistan.

The plant is profusely branched, having a height of 60–90 cm, erect, with soft fine hair. It has feather-like leaves, pinnately divided into two or three with linear segments. The flowers appear as a terminal compound umbel. The minute greyish-white fruits are ovoid in nature. The fruits are similar to parsley in size and shape and measure 1.7–3.0 mm long, 1.5–2.4 mm broad and 0.5–1.4 mm thick. Each mericarp has five ridges and the odour is similar to thyme. The diploid chromosome number of ajowan is 2n = 18. The flowers are protandrous and cross-pollination occurs through insects.

Table 16.1. Export of ajowan seed from India from 1996/97 to 2000/01 (quantity in tonnes and value in US$ million).

	1996/97		1997/98		1998/99		1999/2000		2000/01	
	Quantity	Value	Quantity	Value	Quantity	Value	Quantity	Value	Quantity	Value
Pakistan	–	–	–	–	–	–	–	–	335	1.68
Saudi Arabia	401	3.5	207	1.9	283	3.06	236	3.31	159	1.48
USA	41	0.4	21	0.24	33	0.55	39	3.56	55	0.66
UAE	46	0.37	15	0.11	28	0.24	5	0.04	79	0.64
Malaysia	35	0.2	29	0.28	20	0.11	–	–	62	0.55
Indonesia	40	0.28	35	0.22	–	–	–	–	45	0.35
Nepal	2	0.006	35	0.13	4	0.02	1	0.15	31	0.28
South Africa	29	0.31	11	0.13	6	0.11	13	0.22	14	0.22
Kenya	22	0.17	0.7	0.006	44	0.37	9	0.08	25	0.22
Bangladesh	–	–	–	–	–	–	–	–	44	0.2
Canada	16	0.17	3	0.04	29	0.2	9	0.15	22	0.2
UK	60	0.53	42	0.4	63	0.71	50	0.068	20	0.17
Other countries	212	1.37	99	0.71	150	1.48	53	0.75	71	0.52
Total	904	7.4	498	4.22	660	6.88	465	8.95	962	7.17

Source: Malhotra and Vijay (2004).

Uses

Raw ajowan smells almost exactly like thyme because it also contains thymol, but is more aromatic and less subtle in taste, as well as slightly bitter and pungent. Even a small amount of raw ajowan will dominate the flavour of a dish completely.

Ajowan has a particular affinity to starchy foods like savoury pastries and breads, especially parathas. Snacks like Bombay mix and potato balls get an extra kick from ajowan. It is also good with green beans and root vegetables, lentil dishes and recipes using chickpea flour ('Besan' or Bengal gram flour). It is occasionally an ingredient of curry powder (http://www.theepicentre.com/Spices/ajowan.html).

In Indian cuisine, ajwain is almost never used raw, but either dry-roasted or fried in ghee or oil. This develops a much more subtle and complex aroma, somewhat similar to caraway but 'brighter'.

Flatulence caused by beans is reduced when ajowan is cooked with the beans. It may be used as a substitute for cumin as well. It is also known traditionally as a digestive aid and an anti-emetic (http://en.wikipedia.org/wiki/Ajwain). The whole ajowan seed, powder and oil are used as adjuncts for flavouring foods, as antioxidants and as a preservative in confectionary, beverages and pan mixtures. Ajowan oil is also used in the preparation of lotions and ointments in the cosmetic industries (Malhotra and Vijay, 2004; Malhotra, 2006).

Products

The characteristic odour of ajowan oil is due to the high content of thymol (61%). Thymol easily crystallizes out from the oil and is sold in Indian markets as ajowan *kaphool*, or *sat-ajowan*, and is much valued in medicines. Thymol is used as an ingredient of deodorants, mouthwashes, toothpastes and many pharmaceutical preparations. The leftover residue after distillation contains 15–17% protein and 20–25% fat and is valued as cattle feed. The major processed products are ajowan oil, oleoresin, thymol, thymol crystals, dethymolized oil (thymene) and fatty oils (Malhotra and Vijay, 2004).

Ajowan oleoresin prepared from seeds gives a warm, aromatic and pleasing flavour to food products and is used in processed foods, snacks, sauces and various vegetable preparations. Ajowan oil can be treated with aqueous alkaline solution to extract thymol (Pruthi, 2001). Fatty oils produced from ajowan seed have their use in various pharmaceutical and cosmetic industries, are used in the soap industry for flavouring and as a deodorant. They are also used for perfuming disinfectant soaps and as an insecticide. A thymol-free fraction of the oil, known as 'thymene', finds application in soap perfumes (Malhotra and Vijay, 2004).

Table 16.2. Chemical composition of ajowan (ground spice per 100 g).

Composition	Content
Carbohydrate (g)	24.6
Protein (g)	17.1
Fibre (g)	21.2
Water (g)	7.4
Food energy (calorie)	363
Minerals (g)	7.9
Ca (g)	1.525
P (g)	0.443
Na (mg)	56
K (mg)	1.38
Fe (mg)	27.7
Thiamine (mg)	0.21
Riboflavin (mg)	0.28
Niacin (mg)	2.1

Source: Malhotra and Vijay (2004).

16.3. General Composition

Ajowan seed contains generally 8.9% moisture, 15.4% protein, 18.1% fat (ether extract), 11.9% crude fibre, 38.6% carbohydrates, 7.1% mineral matter, 1.42% calcium, 0.30% phosphorus and 14.6 mg/100 g iron, with a calorific value of 379.4 per 100 g. The percentage of seed oil extracted with *n*-hexane is 31.80%, while that with ethanol is 28%. The neutral lipid component of the oil includes hydrocarbons, esters, sterol esters, triglycerides, free fatty acids, diglycerides, sterols and monoglycerides, whereas the polar lipid components are phosphatidyl ethanolamines and phosphatidyl cholines (Qasim and Khan, 2001).

The oleoresin yield of ajowan is 24.66%, containing 12.15% volatile oil and 87.85% non-volatile material. The oleoresin samples can be kept cold (8–10°C), as well as at ambient temperature (25–30°C), for 60 days without any significant changes in their quality (Nagalakshmi *et al.*, 2000).

The chemical composition of ajowan (ground spice) is given in Table 16.2.

16.4. Chemistry

Ajowan oil is extracted from the seed by the steam distillation method. The two kinds of oils, i.e. essential oil (volatile oil) and non-volatile fatty oils, are extracted. Two integrated methods are available to recover both these oils from the crushed seeds: the sequential method and the combined extraction method. Ajowan seeds contain 3–4% essential oil and 26% fatty oils (Malhotra and Vijay, 2004).

The composition of the volatile oil, which determines the odour and flavour characters, has been of particular interest to chemists. Ajowan oil is composed of phenols, terpenes and *p*-cymene. The essential oil contains more than 27 compounds, of which thymol (61%) is the major one, the others being paracymene (15.6%), γ-terpinene (11.9%), β-pinene (4–5%), dipentene (4–6%), camphene and myrcene. The essential oil composition of ajowan seed is given in Table 16.3.

The oil also contains, in addition to thymol, a liquid hydrocarbon, which is called cymol or crymene. It is also probable that the oil contains another hydrocarbon, which is isomeric with oil of turpentine (Malhotra and Vijay, 2004).

Saharkhiz *et al.* (2005) studied the effects of different harvesting stages on the essential oil content and composition of ajowan cultivated in Iran. The essential oils of the fruits were extracted by hydrodistillation and analysed by capillary gas chromatography (GC) and GC-MS. The essential oil content of fruits harvested at pasty and ripe

Table 16.3. Essential oil composition of ajowan seed.

Component	Essential oil (%)
Phenolic part	
Safrole	0.10
Thymol	87.75
Carvacrol	11.17
Non-phenolic part	
α-Thujene	0.27
α-Pinene	0.28
β-Pinene	2.38
Myrcene	0.81
p-Cymene	60.78
Limonene	8.36
γ-Terpinene	22.26
Terpinolene	0.13
Linalool	0.27
Camphor	0.28
(*Z*)-β-Terpineol	0.19
(*E*)-β-Terpineol	1.35
Borneol	0.49
Terpinen-4-ol	0.12
α-Terpineol	0.22
Carvone	0.15
Safrole	0.16

Source: Malhotra and Vijay (2004).

stages was 7.1 (w/w) and 3.2% (w/w) based on dry weights, respectively. The major components of the oil of fruits harvested at the pasty stage were γ-terpinene (43.2%), thymol (32.4%) and *p*-cymene (20.7%), while the main components of the oil of ripe fruits were thymol (36.7%), γ-terpinene (36.5%) and *p*-cymene (21.1%).

Nagalakshmi *et al.* (2000) indicated the proximate composition of ajowan seeds and the physico-chemical characteristics of the volatile oil. The GC-MS analysis of the volatile oil shows the presence of 17 constituents, of which thymol (39.36%), γ-terpinene (30.97%), *p*-cymene (19.47%), β-pinene (5.45%) and α-pinene (1.48%) were the major constituents. Malhotra and Vashishtha (2005) reported the effect of seed rate and row spacing on growth, yield and essential oil of ajowan genotypes. Figure 16.1 illustrates the structures of the major volatiles of ajowan.

Mineral nutrition and oil composition

Studies on the effect of the interaction of nitrogen and phosphorus on the seed and essential oil yield of ajowan show an increase in the essential oil yield, with increasing levels of P and N interactions. (Krishnamoorthy and Madalageri, 2000).

Akbarinia *et al.* (2004) reported the effect of different rates of chemical fertilizer, manure and a mixture of both on seed yield and composition of the essential oil of ajowan. An increase in N and P fertilizer rate to 90 and 60 kg/ha, respectively, increased seed yield but had no effect on the essential oil content. An increase in the application rate of green manure also increased seed yield and oil content. The integrated treatments give the highest seed and essential oil yields. Nitrogen at 60 kg/ha, 40 kg P/ha and 20 t manure/ha increased thymol content, while it decreased *p*-cymene content. The integrated treatments also gave the highest essential oil and thymol contents (Akbarinia

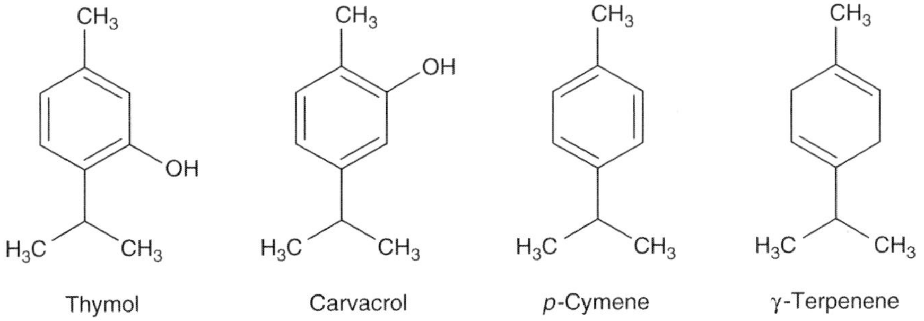

Fig. 16.1. Major volatiles of Ajowan.

et al., 2005). Nitrogen and P application had no effect on the essential oil content in comparison with the control, whereas the manure applied, separately or with NP, increased the essential oil content. The highest oil content was recorded in the N:P (30:20) kg/ha + 35 t manure/ha treatment, followed by the 30 t manure/ha and N:P (60:40) kg/ha + 25 t/ha treatments. Nine components were identified in the oil of ajowan seed, with thymol, δ-terpinene and *p*-cymene as the major constituents. Application of N:P (90:60) kg/ha or 20 t manure/ha increased thymol but decreased *p*-cymene. Application of N and P or manure had no significant effect on δ-terpinene. Compared with separate application of N and P or manure, the NP + manure treatment recorded the highest thymol. There were no significant differences between the NP + manure treatments regarding the major constituents (Akbarinia, *et al.*, 2005).

16.5. Medicinal and Pharmacological Properties

Ajowan seed has been popular from ancient times for its use in folk medicines. The seeds contain an essential oil with 50% thymol, which is a strong germicide, antispasmodic and fungicide. Thymol is also used in toothpaste and perfumery. It is used in a steeped liquid form against diarrhoea and flatulence. In India, the seeds are used as a household remedy for indigestion and colic and are used in poultices to relieve asthma and arthritis. It is also reported to have aphrodisiac properties (http://en.wikipedia.org/wiki/Ajwain; http://www.theepicentre.com/Spices/ajowan.html). Thymol isolated from the oil is a powerful antiseptic and an ingredient in a number of skin ointments/powders, deodorants, mouthwashes, toothpastes and gargles.

Ajowan has long been used in indigenous medicines for the treatment of diarrhoea, dysentery, atonic dyspepsia, cholera, flatulence and indigestion. The oil has properties for using as an expectorant in emphysema, bronchial pneumonia and respiration ailments, and also possesses an antidiuretic effect. Alcoholic extract of ajowan exhibited potent antimicrobial effects, inhibiting the growth of *Bacillus subtilis*, *Escherichia coli* and *Saccharomyces cerevisiae*. The methanolic extracts of ajowan seeds possess natural antioxidant properties. The aqueous portion left after the separation of essential oil from ajowan is known as *omum*-water (ajowan water) in India, which is used against flatulence and in gripe water preparations for children. Hajare *et al.* (2005) described the aflatoxin inactivation potential of aqueous extract of ajowan. Sharangi and Datta (2005) also described the antihelmintic, carminative, laxative, diuretic, flatulence preventive and dyspepsia preventive properties of ajowan.

Antibacterial activity

The antibacterial activity of homogenized seed oil and residues indicates that the powdered seeds of *C. copticum* exhibit antibacterial activity against *Staphylococcus aureus* only and not *E. coli*. The oil extracted with *n*-hexane exhibited antibacterial activity against both organisms, while the oil extracted with ethanol and *n*-hexane did not. The residue left after ethanol extraction exhibits antibacterial activity against *E. coli* only (Qasim and Khan, 2001).

The antimicrobial activity of ajowan oils against *Aspergillus niger*, *S. cerevisiae*, *Mycoderma* sp., *Lactobacillus acidophilus* and *B. cereus*, as estimated by the paper disc agar diffusion method, has been reported by Meena and Sethi (1994).

Antioxidant activity

The oils of ajowan show excellent antioxidant effects (better than those of the synthetic antioxidant and butylated hydroxytoluene; Gurdip *et al.*, 1998). Mehta *et al.* (1994) demonstrated ajowan as a source of natural lipid antioxidant. Soybean oil treated with methanolic extracts has been subjected to storage and heating tests, which showed a marked decrease in oxidation of the oil as measured using peroxide values, conjugated diene values and GC analysis of oxidized fatty acid methyl esters. The formation of primary and secondary oxidation products of oxidized soybean oil was significantly lower for oil treated with ajowan extracts than control.

Mehta and Zayas (1995) showed antioxidant properties in a methanolic extract of ajowan using linoleic acid.

Phytomedicine

Ajowan is known to traditional healers to have hypotensive properties. Bioassay-directed fractionation of seeds results in the isolation of thymol. In anaesthetized rats, thymol (1–10 mg/kg, i.v.) produces dose-dependent reductions in blood pressure and heart rate (Aftab *et al.*, 1995).

Among other products, ajowan salt is prepared commercially by mixing finely ground rock salt and is used mostly for folk remedies of digestive problems (Malhotra and Vijay, 2004). Ajowan seeds are reported to be useful in flatulence, colic, atonic dyspepsia, diarrhoea, cholera, hysteria and spasmodic affections of the bowel. The seed produces a feeling of warmth and relieves the sinking and fainting feelings which accompany bowel disorders. Ajowan seed in conjunction with asafoetida, myrobalan and rock salt proved beneficial in stomach ache problems. A hot poultice of seed is used as a dry fomentation to the chest in asthma and expectoration from bronchitis (Malhotra and Vijay, 2004).

Indian folk remedies suggest that ajowan seed with a little rock salt mixture daily after meals improves indigestion and irregular diet. Buch *et al.* (1988) report the effect of ajowan volatile oils on ejaculated human spermatozoa. The volatile oils are instantaneously spermicidal in varying dilutions and their action is dependent on concentration and time.

Ajowan seed and its extract do not appear to have any significant toxicity. Normally, the concentrations of compounds in ajowan do not pose a health threat for consumption or to fieldworkers handling the plants (Malhotra and Vijay, 2004).

16.6. Specifications

Saxena *et al.* (2004) established the quality standards of whole ajowan and ajowan powder. The whole and powdered forms of ajowan (dried fruits) being sold in Indian markets have been evaluated. The moisture content, organic extraneous matter, inorganic extraneous matter, damaged/shrivelled/immature seeds and volatile oil content of whole ajowan and the moisture, total ash, acid-insoluble ash, non-volatile ether extract, crude fibre and volatile oil content of powdered ajowan are used as quality parameters. Agmark grade specifications of ajowan seed are given in Table 16.4.

16.7. Conclusion

Ajowan, or bishop's weed (*Trachyspermum ammi*), is an important seed spice particularly suitable to delicate vegetarian cuisine. It has a strong and distinctive thyme-like

Table 16.4. Agmark grade specifications of ajowan seed.

	Special characteristics			
Grade designation	Inorganic matter (% by foreign weight maximum)	Organic damaged, discoloured matter (% by weight weight maximum)	Shrivelled, immature, and weevilled (% by maximum)	Moisture
Special	0.24	0.50	1.0	11
Good	0.50	0.75	2.0	11
Fair	1.00	1.00	3.0	11

Source: Malhotra and Vijay (2004).

fragrance, leaving a milder, pleasant after-taste to the food. The characteristic odour of ajowan oil is due to the high content of thymol in the oil. It is used as an ingredient in deodorants, mouthwashes, toothpastes and many other pharmaceutical preparations. The major processed products are ajowan oil, oleoresin, thymol crystals, dethymolized oil and fatty oils. In the new era of ethnic foods, ajowan possesses a great future.

References

Aftab, K., Atta ur-Rahman and Usmanghani, K. (1995) Blood pressure lowering action of active principle from *Trachyspermum ammi* (L.) Sprague. *Phytomedicine* 2(1), 35–40.

Akbarinia, A., Ghalavand, A., Sefidcon, F., Rezaee, M.B. and Sharifi, A. (2004) Study on the effect of different rates of chemical fertilizer, manure and mixture of them on seed yield and composition of essential oil of Ajowan (*Trachyspermum copticum*). *Pajouhesh va Sazandegi in Agronomy and Horticulture* (61), 32–41.

Akbarinia, A., Sefidkon, F., Ghalavand, A., Sarvestani, Z.T. and Ashorabadi, E.S. (2005) Effect of nitrogen phosphorus fertilizer and manure on content and chemical composition of *Trachyspermum ammi* L. *Journal of Essential Oil Bearing Plants* 8(1), 82–86.

Buch, J.G., Dikshit, R.K. and Mansuri, S.M. (1988) Effect of certain volatile oils on ejaculated human spermatozoa. *Indian Journal of Medical Research* 87, 361–363.

Gurdip, S., Kapoor, I.P.S., Pandey, S.K. and Singh, G. (1998) Studies on essential oils – part thirteen: natural antioxidant for sunflower oil. *Journal of Scientific and Industrial Research* 57(3), 139–142.

Hajare, S.S., Hajare, S.N. and Sharma, A. (2005) Aflatoxin inactivation using aqueous extract of ajowan (*Trachyspermum ammi*) seeds. *Journal of Food Science* 70(1), C29–C34.

http://en.wikipedia.org/wiki/Ajwain.

http://www.theepicentre.com/Spices/ajowan.html.

Krishnamoorthy, V. and Madalageri, M.B. (2000) Effect of interaction of nitrogen and phosphorus on seed and essential oil yield of ajowan (*Trachyspermum ammi* L.) genotypes. *Journal of Spices and Aromatic Crops* 9(2), 137–139.

Malhotra, S.K. (2006) Minor seed spices: ajowan, dill, celery, aniseed. In: Ravindran, P.N., Nirmal Babu, K., Shiva, K.N. and Kallupurackal, J.A. (eds) *Advances in Spices Research*. Agrobios, Jodhpur, India, pp. 785–801.

Malhotra, S.K. and Vashishtha, B.B. (2005) Effect of seed rate and row spacing on growth, yield and essential oil of ajowan (*Trachyspermum ammi* L.) genotypes. *Indian Journal of Arecanut, Spices and Medicinal Plants* 7(1), 18–20.

Malhotra, S.K. and Vijay, O.P. (2004) Ajowan. In: Peter, K.V. (ed.) *Handbook of Herbs and Spices*, Volume 2. Woodhead Publishing Limited, Cambridge, UK, pp. 107–116.

Meena, M.R. and Sethi, V. (1994) Antimicrobial activity of essential oils from spices. *Journal of Food Science and Technology* Mysore 31(1), 68–70.

Mehta, R.L. and Zayas, J.F. (1995) Antioxidative effect of ajowan in a model system. *Journal of the American Oil Chemists' Society* 72(10), 1215–1218.

Mehta, R.L., Zayas, J.F. and Yang, S.S. (1994) Ajowan as a source of natural lipid antioxidant. *Journal of Agricultural and Food Chemistry* 42(7), 1420–1422.

Nagalakshmi, S., Shankaracharya, N.B., Naik, J.P. and Rao, L.J.M. (2000) Studies on chemical and technological aspects of ajowan (*Trachyspermum ammi* (L.) Syn. *Carum copticum* Hiern) seeds. *Journal of Food Science and Technology* 37(3), 277–281

Pruthi, J.S. (2001) *Minor Spices and Condiments*. ICAR, New Delhi, pp. 659–660.

Qasim, M. and Khan, M.R. (2001) Biochemical and antibacterial studies of Ajowan (*Carum copticum*) oil. *Pakistan Journal of Scientific and Industrial Research* 44(3), 184–185.

Saharkhiz, M.J., Omidbaigi, R. and Sefidkon, F. (2005) The effects of different harvest stages on the essential oil content and composition of ajowan (*Trachyspermum ammi* Sprague) cultivated in *Iran. Journal of Essential Oil Bearing Plants* 8(3), 300–303.

Saxena, V.B., Paul, T.M., Vaidya, P.S. and Jaiswal, P.K. (2004) Quality standards of ajowan whole and ajowan powder. *Indian Journal of Arecanut, Spices and Medicinal Plants* 6(3), 95–97.

Sharangi, A.B and Datta, S. (2005) Medicinal properties of spices. *Indian Journal of Arecanut, Spices and Medicinal Plants* 7(2), 42–49.

17 Star Anise

B. Chempakam and S. Balaji

17.1. Introduction

Star anise (*Illicium verum* Hook) is a spice that closely resembles anise in flavour, obtained from the star-shaped pericarp. It is native to southern China and northern Vietnam and is grown almost exclusively in southern China, Indochina and Japan. The spice was first introduced into Europe in the 17th century. The oil, produced by a process of steam distillation, is substituted for European aniseed (*Pimpinella anisum* L.) in commercial drinks (Morton, 2004). The fruit is star-shaped and consists of eight to 13 carpels joined centrally and is a well-known spice used in Vietnamese cuisine (Loi and Thu, 1970). It is so named from the stellate form of its fruit. The essential oil of star anise fruits is used in the confectionary trade to flavour liquorice and other sweets and in the baking trade to flavour cakes, cookies and biscuits. It has a volatile oil content of 2.5–3.5% in the fresh fruit and 8–9% in the dried material. The fixed oil content is about 20% (Heath, 1981). This small tree, belonging to the family Iliciaceae, and which grows in the evergreen forests of southern China and the mountainous regions of Indochina, is cultivated in the Vietnamese province of Lang Son and in the mountainous regions of Eastern Laos.

Japanese star anise (*Illicium anisatum*), a similar tree, is not edible because it is highly toxic. Cases of illness, including serious neurological effects such as seizures, which are reported after using star anise tea, may be the result of using this species (Biessels *et al.*, 2002). It is similar to *I. verum*, but its fruit is smaller and with a weaker odour, which is said to be more similar to cardamom than to anise. While it is poisonous, and therefore unsuitable for internal use, the Chinese use the fruits to treat some skin problems (Lai *et al.*, 1997).

Vietnam produces more than 5000 t of star anise seeds per annum. It is estimated that the combined production of China and Vietnam is more than 25,000 t per annum. In addition, 200–250 t of essential oil are shipped to France and the Czech Republic. In China, which is the largest supplier of star anise to the world market, Vietnamese star anise is blended and then exported to France. In France, it is used as a raw material in the production of alcoholic beverages (FAO, 2003).

17.2. Botany and Uses

Botany

The plant belongs to the genus of the family Illiciaceae, order Illiciales, subclass Magnoliidae, class Magnoliopsida. It is a

 Chemistry of Spices (eds V.A Parthasarathy, B. Chempakam and T.J. Zachariah)

small to medium-sized evergreen tree reaching up to 8m (26ft) in height. The trees have evergreen, aromatic leaves and bisexual flowers. The leaves are lanceolate and the axillary flowers are yellow, the female portion of the flower consisting of seven to 15 carpels (Rosengarten, 1969). The fruits are star-shaped, reddish-brown, consisting of 6–8 carpels arranged in a whorl. Each carpel is 10mm long, boat-shaped, hard and wrinkled, containing a seed. The seeds are brown, compressed, ovoid, smooth, shiny and brittle.

They are harvested before they ripen, then sun dried. Star anise, as the name suggests, is star-shaped, radiating between five and ten pointed boat-shaped sections, about eight on average. These hard sections are seedpods. Tough-skinned and rust-coloured, they measure up to 3cm (1.25in) long. The fruit is picked before it ripens and is then dried (Lust, 1974). The tree is propagated by seed and cultivated mainly in China and Japan for export and home markets.

Uses

The stars are available whole, or ground to a red-brown powder. The bulk of the oil in commerce is obtained from the star anise fruit in China. Apart from its use in sweetmeats and confectionery, it contributes to meat and poultry dishes, combining especially well with pork and duck. It is also one of the ingredients used to make the broth for the Vietnamese noodle soup called *pho'*.

Star anise is an ingredient of the traditional five-spice powder of Chinese cooking. Chinese stocks and soups very often contain the spice. In the West, star anise is added in fruit compotes and jams and in the manufacture of anise-flavoured liqueurs, the best known being anisette (Morton, 2004). The water-soluble extract of *I. anisatum* promotes hair growth and may be a useful additive in hair growth products (Sakaguchi *et al.*, 2004). Star anise is also used in different Indian curry powders for making meat preparations.

17.3. General Composition

The seeds contain some volatile oil, resin and a large amount of fixed oil (Meisner, 1818). The fruit (without the seeds) contains volatile oil, resin, fat, tannin, pectin and mucilage. The volatile oil (*oil of star-anise*) amounts to about 4–5% and is almost identical with *oil of anise* (from *P. anisum*, Linné). Star-anise oil from Chinese fruit has a specific gravity at 15°C (59°F) of 0.980–0.990 and its known constituents are anethol, phellandrene, safrol and hydro-quinone-ethyl-ether (Flückiger, 1879). Poisonous *sikimin* has been detected in the fruit (Eykmann, 1881), while Schlegel (1885) found a crystalline principle of a pronounced odour of musk. He also found *saponin* in the watery extract.

The closely related Japanese star anise, *I. anisatum*, is highly toxic. It contains a poisonous sesquiterpene lactone, called anisatin, and also shikimin and sikimitoxin, which causes severe inflammation of the kidneys, urinary tract and digestive organs, as well as affecting the nervous system (Lederer *et al.*, 2006). Other compounds present in this toxic species of *Illicium* are safrole and eugenol, which are not present in *I. verum* and are used to identify its adulteration.

Anisatin and its derivates are suspected of acting as strong GABA antagonists. It is impossible to recognize Chinese and Japanese star anise in its dried or processed form by its appearance only, due to morphological similarities between the species. There are cases of product recalls when products containing star anise were found to be contaminated by Japanese anise (Biessels *et al.*, 2002; Johanns *et al.*, 2002; Lee *et al.*, 2003b; Vandenberghe *et al.*, 2003; Ize-Ludlow *et al.*, 2004).

17.4. Chemistry

Chinese star anise is an evergreen bush of the magnolia order grown in Vietnam and southern China. The ripe, strongly anise-smelling fruits open up in a star. They are used as a spice and for the production of star anise oil by steam distillation. Star anise oil is a colourless to pale yellow liquid which solidifies on cooling.

Volatiles

Physical properties

Table 17.1 presents the physical properties of the star anise essential oil. The colour of the steam-distilled oil samples was greenish-yellow and that of oils from liquid CO_2 extraction was yellow. This can also be seen by the difference between the average dominant wavelength for liquid CO_2 extraction and that for steam distillation.

The specific gravity of steam-distilled oil was a little higher than that of the oil from liquid CO_2 extraction. The specific gravities of the samples from the two methods were similar to the standard values of anise essential oil reported by the FCC (Food Chemicals Codex), as cited by Heath (1981). The refractive indices, as well as the optical rotation of the oils from steam distillation and those from liquid CO_2 were not significantly different. All values of optical rotation were of *levo* orientation.

Chemical composition

The essential oil, which ranges from 2.5 to 5.0%, includes the following chemical compounds: α-pinene, camphene, β-pinene, linalool, *cis*-anethole, *trans*-anethole, safrole, anisaldehyde and acetoanisole. The chemical structures of these compounds are given in Fig. 17.1. The oil is of medium viscosity and will solidify at low temperatures and needs to be hand-warmed before use. The oil is extracted by steam distillation from the dried ripe fruit and seeds (WHO Institute of Materia Medica, 1990).

The main component (80–90%) is (*E*)-anethole. Star anise oil and (*E*)-anethole isolated from it are used in anise liqueur (Anisette, Sambuca) and anise brandy (Pernod, Ouzo, Raki, Arak), liquorice sweets, toothpaste, etc. It has almost completely replaced the original anise seed oil, obtained from the umbellifer *P. anisum*. Shikimic acid (Wang *et al.*, 2001), used in the production of the antiviral drug Tamiflu® (Roche), is extracted from the fruits of Chinese star anise and related species (Rahway, 1989).

Star anise contains primarily anethole and fatty oil. Essential oil of star anise has a sweetish, burning flavour and a highly aromatic odour. It is located primarily in the woody shell and, to a lesser extent, in the seed. An elevated moisture content and excessively high temperature create a risk of self-heating.

Zhou *et al.* (2005) determined anethole composition quantitatively by GC in the fruit of *I. verum* from various places in the Guangxi province. The average recovery rate and the RSD were 102.31 and 1.78%, respectively. The content of anethole in the fruit of *I. verum* from various places was more than 4.5%.

Table 17.1. Physical properties, extraction yield and anethole content of star anise essential oil.

Character	Steam distillation	Liquid CO_2 extraction
Extraction yield (% r.m.)	10.2	11.2
Anethole content in essential oil (%)	92.2	89
Total anethole content (% r.m.)	9.4	10.0
Physical properties		
Colour	Greenish-yellow	Yellow
Dominant wavelength (nm)	571.8	575.2
Specific gravity	0.9873 (25,125)	0.9859 (25/25)
Refractive index	1.5553 (25°C)	1.5517 (25°C)
Optical rotation	0.3167 (20°C)	0.3333 (20°C)

Source: Tuan and Hangantileke (1997).

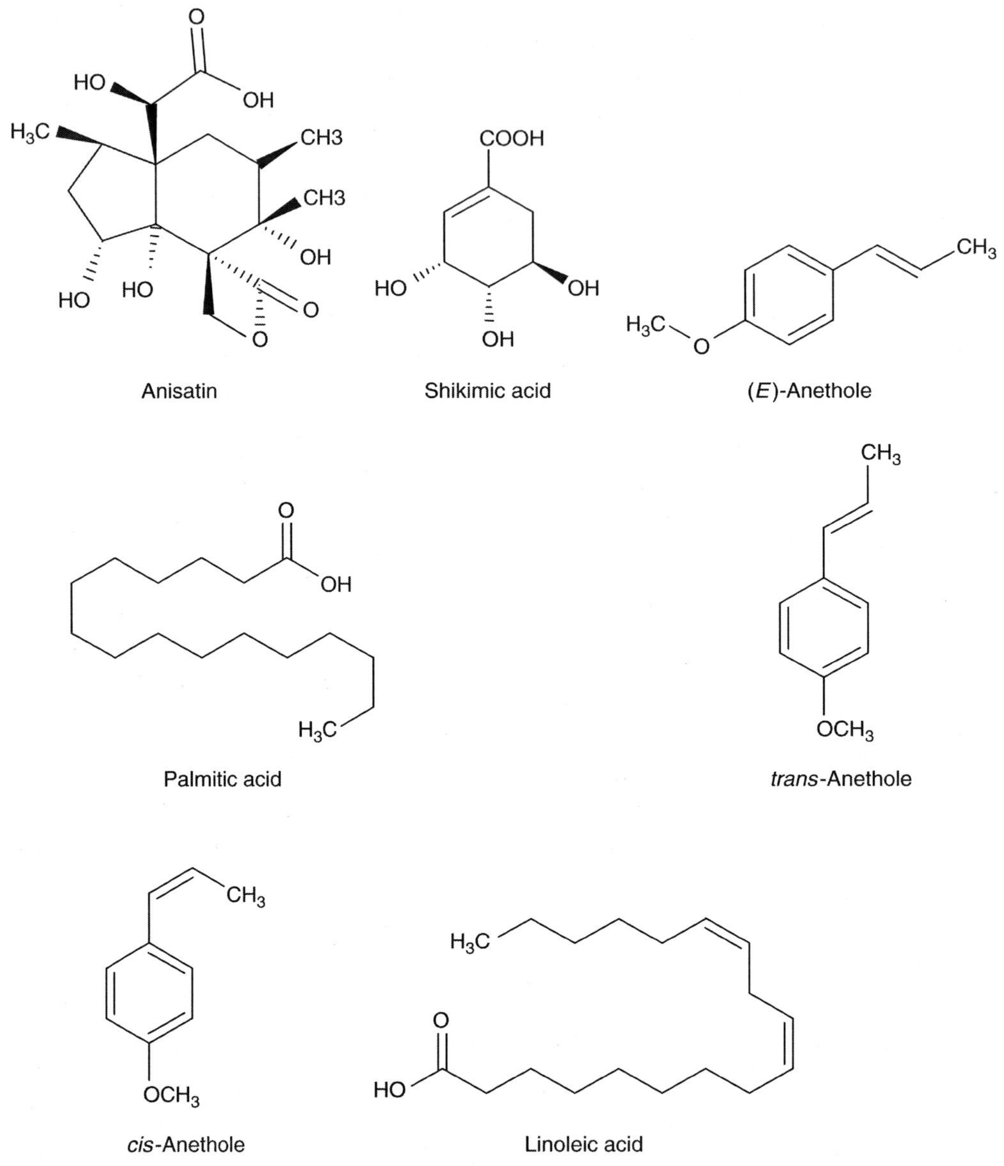

Fig. 17.1. Components of essential oil from star anise (*I. verum*).

Extraction techniques

The essential oil from star anise fruits traditionally is extracted by steam distillation. This process is not expensive but can induce thermal degradation, hydrolysis and water solubilization of some fragrance constituents (Reverchon, 1997).

In recent years, extraction of oils from plant materials with liquid and supercritical CO_2 has drawn increasing attention from researchers in the food and other industries (Stahl and Gerard, 1985; Rizvi *et al.*, 1986). The composition of star anise essential oil isolated by supercritical extraction is qualitatively similar to the composition of hydrodistilled anise oil reported in the literature (Cu, 1986; Cu *et al.*, 1990). From a quantitative point of view, essential oil obtained by supercritical extraction contains a higher percentage of anethole with respect to hydrodistilled oil (Table 17.1). Indeed, the percentage of this antioxidant compound ranged from 71.6 to 85.9% in the hydrodistilled oil against 94.2% found in the supercritical extract (Tuan and Hangantileke, 1997).

Star anise volatile oils can also be isolated by supercritical CO_2 extraction coupled to a fractional separation technique. Gas chromatography-mass spectrometry analysis of the various fractions obtained in different extraction and fractionation conditions allowed the identification of the best operating conditions for the isolation of essential oil. A good extraction performance was obtained operating at 90 bar and 50°C (for 630 min) for both treated materials. Optimum fractionation was achieved in both cases by operating at 90 bar and −10°C in the first separator and at 15 bar and 10°C in the second (Della Porta *et al.*, 1998).

Results of detailed GC-MS analysis of these two fractions are reported in Table 17.2 (volatile oil column and waxes column). Anise waxes were formed mainly by *n*-pentacosane (35.7%), *n*-heneicosane (25.8%), *n*-tricosane (10.3%), *n*-docosane (9.0%) and *n*-tetracosane (6.2%). Volatile oil contained 94.2% of anethole (*cis* and *trans*). In the oil, estragole (1.4%), limonene (1.7%), linalool (0.3%), two terpineol isomers (0.3%) and linalyl acetate (0.3%) were also present. Caryophyllene (0.5%) and *trans*-bergamotene (0.7%) were the main compounds among sesquiterpenes.

A relatively simple apparatus was described by Lucchesi *et al.* (2004) for extracting essential oils from aromatic plant material by atmospheric solvent-free microwave extraction (SFME) without the addition of any solvent or water. The essential oils from spices like star anise extracted by SFME for 30 min and 1 h were similar to those obtained by conventional hydrodistillation (HD) for (respectively) 4 and 8 h.

Another extraction technique for volatile oil by online coupled packed capillary high-performance liquid chromatography-capillary gas chromatography (micro-HPLC-CGC) yielded new compounds that had neither been found nor separated before by conventional capillary gas chromatography-mass spectrometry method (Wang *et al.*, 2004).

Composition

The GC-MS pattern of fruit volatile oil of star anise (*I. verum* Hook) shows the presence of 25 components, which account for 99.9% of the total amount (Padmashree *et al.*, 2007). The major components are *trans*-anethole (93.9%), estragole (1.05%) and limonene (1.05%) (Table 17.3). Fifteen components are identified from its acetone extract, accounting for 80.27% of the total amount. *trans*-Anethole (51.81%) is found as a major component, along with linoleic acid (11.6%), 1-(4-methoxyphenyl)-prop-2-one (6.71%), foeniculin (5.29%) and palmitic acid (1.47%).

Phenylpropanoids

Two new phenylpropanoid glucosides and an alkyl glucoside from the fruits of *I. verum* were isolated and their chemical structures were elucidated on the basis of spectroscopic studies (Lee *et al.*, (2003a,b). The two racemic mixtures of phenylpropanoids [1-(4′-methoxyphenyl)-(1*R*, 2*S* and 1*S*, 2*R*)-propanediol and 1-(4′-methoxyphenyl)-(1*R*, 2*R* and 1*S*, 2*S*)-propanediol] were isolated, along with two known phenylpropanoids.

Sesquiterpenes

A summary of the different sesquiterpenes isolated from the extracts of various plant parts of star anise is shown in Table 17.4.

Three neurotropic sesquiterpenoids, veranisatins A, B and C, were isolated from star anise (*I. verum* Hook. fil., Illiciaceae) (Nakamura *et al.*, 1996). A dichloromethane extract of *I. tsangii* yielded murolane sesquiterpenes and menthane monoterpenes (Ngo *et al.*, 1999).

Sy and Brown (1998) isolated prezizaane sesquiterpene angustisepalin from the aerial parts of *I. angustisepalum*. Angustisepalin is formally the 10-benzoyl ester of neomajucin, previously reported from *I. majus*.

Table 17.2. Area percentages of the compounds found in star anise extracts.

Compounds	SFE oil (%)[a]	Waxes (%)[a]	HP (%)[a]
α-Thujene	t	–	–
α-Pinene	0.09	–	–
β-Pinene	0.03	–	–
Myrcene	0.04	–	–
α-Phellandrene	0.02	–	–
3-Carene	0.09	–	–
α-Terpinene	0.01	–	–
para-Cymene	0.04	–	–
Limonene	1.74	–	–
cis-Ocimene	t	–	–
γ-Terpinene	0.06	–	–
Terpinolene	0.02	–	–
Linalool	0.31	–	–
4-Terpineol	0.20	–	–
α-Terpineol	0.09	–	–
Estragole	1.45	–	–
cis-Anethole	0.15	–	–
trans-Anethole	94.05	–	60.53
Lynalyl acetate	0.11	–	–
α-Cubebene	0.21	–	–
β-Elemene	0.01	–	–
Caryophyllene	0.53	–	0.36
α-*trans*-Bergamotene	0.72	–	0.53
α-Humulene	0.02	–	–
β-*cis*-Farnesene	0.01	–	–
δ-Cadinene	–	–	0.11
Spathulenol	–	–	0.15
$C_{12}H_{15}N_3O_2$	–	–	0.87
Torreyol	–	–	0.07
α-Cadinol	–	–	0.46
$C_{16}H_{14}O$	–	–	2.56
trans-cis-Farnesol	–	–	2.11
$C_{15}H_{26}O$	–	–	1.97
$C_{15}H_{20}O_3$	–	–	2.23
Palmitic acid	–	0.72	0.33
Methyl palmitate	–	1.37	0.80
$C_{20}H_{28}O$	–	–	0.53
Methyl linoleate	–	–	0.54
n-Heneicosane	–	25.85	3.15
Methyl heneicosane	–	3.44	–
n-Docosane	–	9.03	7.10
n-Tricosane	–	10.35	7.37
n-Tetracosane	–	6.22	4.27
n-Pentacosane	–	35.71	1.70
n-Hexacosane	–	2.18	1.92
n-Heptacosane	–	1.73	0.34
n-Octacosane	–	0.64	t
n-Nonacosane	–	2.77	t

Note: [a]Percentages are expressed as gas chromatograph areas without any correction factor.
t = Percentages lower than 0.01; – = not detectable.
SFE oil column: compounds recovered in the second separator.
Waxes column: compounds recovered in the first separator (90 bar and 50°C for 510 min).
HP column: composition of the extract (300 bar and 50°C for 180 min) recovered in the second separator on a previously treated matter.
Source: Della Porta *et al.* (1998).

Table 17.3. Flavour profile of star anise volatile oil.

Compound	Identity	Peak (%)
α-Pinene	KI,MS	0.12 ± 0.020
β-Pinene	KI,MS	0.03 ± 0.020
Myrcene	KI,MS	0.02 ± 0.003
α-Phellandrene	KI,MS	0.04 ± 0.001
3-Carene	KI,MS	0.15 ± 0.020
α-Terpinene	KI,MS	0.02 ± 0.001
p-Cymene	KI,MS	0.05 ± 0.003
Limonene	KI,MS	1.05 ± 0.040
trans-Ocimene	KI,MS	0.09 ± 0.010
cis-β-Ocimene	KI,MS	0.01 ± 0.001
γ-Terpinene	KI,MS	0.04 ± 0.001
Terpinolene	KI,MS	0.03 ± 0.003
Linalool	KI,MS	0.29 ± 0.020
γ-Terpineol	KI,MS	0.12 ± 0.030
4-Terpineol	KI,MS	0.09 ± 0.020
α-Terpineol	KI,MS	0.08 ± 0.010
Estragole	KI,MS	1.05 ± 0.120
cis-Anethole	KI,MS	0.14 ± 0.020
trans-Anethole	KI,MS	93.9 ± 1.560
α-Cubebene	KI,MS	0.10 ± 0.010
β-Clemene	KI,MS	0.01 ± 0.001
Caryophyllene	KI,MS	0.10 ± 0.010
Bergamotene	KI,MS	0.01 ± 0.002
Δ-Cardinene	KI,MS	0.04 ± 0.002
α-Cadinol	KI,MS	0.02 ± 0.001

Source: Padmashree *et al.* (2007).

Three novel seco-prezizaane sesquiterpenes were isolated from leaves of *I. parviflorum* (swamp star anise, yellow star anise), a species occurring endemically in central Florida. The compound cycloparvifloralone possesses a hitherto unknown ring system with a cage-like acetal/hemiketal structure. Lactones (cycloparviflorolide) and parviflorolide, which were obtained as an inseparable mixture, coexist in hemiketal/keto equilibrium. It could be shown that a 4,7-hemiketal occurs in an analogous fashion to pseudoanisatin, a known constituent of other *Illicium* species. From the fruits of *I. floridanum* the novel ortholactone was also isolated (Schmidt, 1999).

17.5. Medicinal and Pharmacological Properties

Star anise essential oil, *I. verum*, is often used as a substitute for anise seed oil in perfumery because it shares similar chemistry. It is used in aromatherapy to help relieve coughing, colic, cramping, hiccups and indigestion. It should be used in moderation to avoid skin irritation (Rosengarten, 1969; Lust, 1974; Malcolm, 1987).

Like anise, star anise has been assigned the following pharmacological properties:

- Carminative
- Stomachic
- Stimulant and diuretic
- Antirheumatic
- Antimicrobial

Table 17.4. Sesquiterpenes isolated from the extracts of various plant parts of star anise.

Pericarp of *I. merrillianum* (Huang *et al.*, 2002, 2004)	Bark of *I. difengi* (Huang *et al.*, 1997)	Pericarp of *I. minwanense* (Yokoyama *et al.*, 2002, 2003)	Fruit of *I. floridanum* Ellis (Schmidt *et al.*, 2001)
3-Deoxypseudoanisatin 2β-Hydroxy-3,6-dedioxypseudoanisatin 8-α-Hydroxy-10-deoxycyclomerrillianolide 10-β-Hydroxypseudoanisatin 10-β-Hydroxy cyclopseudoanisatin 1,6-Dihydroxy-3-deoxyminwanensin 8-Deoxy merrilliortho-lactone	3 β-*O*-acetyl-mangiferolic acid Mangiferonic acid Mangiferolic acid Butulinic acid	(1*S*)- and (1*R*)-minwanenone, 1-α-Hydroxy-6-deoxypseudoanisatin (2*S*)-Hydroxy-6-deoxypseudoanisatin 3-Oxopseudoanisatin (3*S*, 6*R*)-4, 7-epoxy-6-deoxypseudoanisatin 7-*O*-methylpseudomajucin (+)-8,11,13,15-abietatetraene)	Debenzoyl-7-deoxo-1-α 7-α-Dihydroxytashironin Debenzoyl-7-deoxo-7 α-hydroxytashironin Debenzoyl-7-deoxo-7 α-hydroxy-3-oxotashironin (1–3)

- Chemopreventive
- Insecticidal
- Anti-flu.

Antimicrobial property

This spice has got potent antimicrobial properties. Chemical studies indicate that a major portion of this antimicrobial property is due to anethole, present in the dried fruit. Studies with isolated anethole (compared with standard anethole) indicated that it was effective against bacterial, yeast and fungal strains (De *et al.*, 2002). The recent findings of Singh *et al.* (2006) showed that the volatile oil inhibited the growth of *Fusarium moniliforme* completely at a dose of 6 µl. In the case of extract, 50% mycelial zone inhibition was obtained for *Penicillium citrinum* and *P. viridicatum*. Moreover, the volatile oil was found to be effective for controlling the growth of *F. moniliforme* and *Aspergillus niger*, whereas the extract has been found to be highly effective for *A. flavus*. The extract has shown better activity for *Staphylococcus aureus* and *Bacillus cereus* in comparison with volatile oil and commercial bactericide, i.e. Ampicillin. However, volatile oil has shown better activity for *Salmonella aeruginosa* and *B. subtilis*.

Phenylpropanoid glucosides from the fruits are preventive agents against sepsis (Lee *et al.*, 2003b). Anethol extract from star anise seeds inhibits fungal growth (Hitokoto *et al.*, 1980). It is also effective against dermatitis and the oil does not give cross-reactions and pseudo-cross-sensitivity (Rudzki and Grzywa, 1976).

Antioxidant activity

The extract has shown excellent activity for the inhibition of primary and secondary oxidation products in rapeseed oil and could be considered as a natural antioxidant, which may be used for the chemoprevention of diseases occurring due to oxidative deterioration (Anon., 1992). The antioxidant activity is due to the high percentage of anethole, which is more than 80% (Padmashree *et al.*, 2007).

Chemopreventive property

Phenylpropanoids and phytoquinoids isolated from *Illicium* plants showed inhibitory activities against Epstein-Barr virus early antigen (EBV-EA), even at 1 × 10 mol ratio, and the inhibitory activity of their compounds was found to be more than that of β-carotene (Itoigawa *et al.*, 2004). Two phenylpropanoids having a prenyl group, 4-allyl-2-methoxy-6-(3-methyl-2-butenyl) phenol and 4-allyl-2, 6-dimethoxy-3-(3-methyl-2-butenyl) phenol, showed more potent activities as antitumour promoters. The presence of a prenyl moiety in the phenylpropanoids plays an important role in antitumour-promoting activity. Hence, the prenylated phenylpropanoids might be valuable as potential cancer chemopreventive agents.

Insecticidal property

Thirteen seco-prezizaane terpenoids isolated from star anise species (*I. floridanum*, *I. parviflorum* and *I. verum*) were found to possess insecticidal activity (Kuriyama *et al.*, 2002). Anisatin and pseudoanisatin exhibited moderate insecticidal activity against German cockroaches (*Blattella germanica* L.). The insecticidal activities of phenylpropene and (*E*)-anethole, derived from the fruit of star anise, *I. verum*, were examined by Chang and Ahn (2002) against adults of *B. germanica*. As naturally occurring insect-control agents, the *I. verum* fruit-derived materials could be useful for managing populations of *B. germanica*. Insecticidal properties were also observed in non-polar crude extracts of star anise against eggs, larvae and adults of *Tribolium castaneum* (Herbst) and *Sitophilus zeamais* Motsch (Ho *et al.*, 1995).

Anti-flu activity

Star anise is the industrial source of shikimic acid, a primary ingredient used to create the anti-flu drug, Tamiflu (Goodman, 2005). Tamiflu is regarded as the most promising drug to mitigate the severity of the bird flu H5N1 strain of virus. Currently, Tamiflu is the only drug available which may reduce the severity of bird flu (also known as avian flu).

17.6. International Specifications, Desirable Limits

Packaging and storage

Star anise is harvested and shipped all year round. It is packaged in, among other things, bast bales (50 kg) and bags (40 kg).

Conditions for storage

TEMPERATURE Favourable travel temperature range: 5–25°C. Star anise should be transported in areas which exhibit the lowest temperatures during the voyage and are dry. In any event, storage beneath the weather deck or, in the case of shipping, in containers in the uppermost layer on deck, must be avoided as the deck or container is strongly heated by the intense tropical sun and, at temperatures of > 25°C, essential oils may be lost.

HUMIDITY/MOISTURE Star anise should be stored away from goods which are sensitive to moisture/humidity or release moisture (e.g. copra).

Designation	Humidity/water content (%)
Relative humidity	60–70
Water content	8–12
Maximum equilibrium moisture content	65

VENTILATION Star anise requires particular temperature, humidity/moisture and possibly ventilation conditions. In order to avoid formation of mould, the storage space should be cool, dry and, most particularly, easy to ventilate.

Recommended ventilation conditions: air exchange rate: six changes/h (airing).

MECHANICAL INFLUENCES

1. Star anise easily becomes fragile and must therefore be handled with appropriate care.
2. Breakage may amount to as much as 25%.
3. With bagged cargo, point loads applied, for example, by hooks may result in damage (tears) to the bags, and thus in loss of volume. Plate or bag hooks, which, due to their shape, distribute the load and reduce the risk of damage, should thus be used.

Adulteration and its identification

Chinese star anise (*I. verum* Hook, F.) is a well-known spice used in the treatment of infant colic. Japanese star anise (*I. anisatum* L.), however, has been documented to have both neurologic and gastrointestinal toxicities.

A methodological approach for an effective and reliable quality control of Chinese star anise (*I. verum* Hook. F.) was developed and validated by Lederer *et al.* (2006). A combined method of TLC and HPLC-MS/MS was used for differentiation of various *Illicium* species, especially Chinese and Japanese star anise. Species can be distinguished by their TLC flavonoid pattern. A sensitive and selective HPLC/ESI-MS/MS method was developed for the detection and quantification of lower admixtures of *I. anisatum* and of further toxic *Illicium* species at a low concentration range using the sesquiterpene, lactone anisatin, as a marker. This assay includes a solid-phase extraction clean-up procedure with a high recovery (> 90%).

Star anise herbal tea may be adulterated with *I. anisatum* Linn. A short and rapid method using microscopy and gas chromatography (GC) was developed to detect *I. anisatum* Linn., an adulterant in the powdered mixture of *I. verum*. Anatomical differences in the epicarp cells of *I. verum* and *I. anisatum* fruits can be defined clearly under fluorescent microscopy and scanning electron microscopy. A GC method can also be used for quick identification of possible *I. anisatum* adulteration with *I. verum* (Joshi *et al.*, 2005).

17.7. Conclusion

Star anise belongs to a family of spices with a rich history. In addition to its traditional uses, it has multiple applications in botany, chemistry, pharmacology and therapy. The spice is back in the region as an ingredient of the drug to fight bird flu. It is also the primary source of shikimic acid used to produce oseltamivir phosphate, sold under the brand name, Tamiflu. Thus, it has become a major weapon against global influenza. Although only limited human-to-human transmission has been confirmed, scientists fear a worldwide pandemic could erupt if the virus mutates to a highly pathogenic form that humans can pass efficiently among themselves. Now scientists are finding faster, cheaper ways to produce more of the only drug proven capable of combating avian flu. Tamiflu reduces flu mortality by inhibiting the virus from spreading.

Moreover, star anise contains bioactive compounds possessing insecticidal properties, which can be exploited for evolving natural grain protectants. More attention also has to be focused on the antimicrobial, antioxidant and chemopreventive properties of the spice.

References

Anon. (1992) Les agents antioxygenes. *Pour la Science* 179, 10–12.

Biessels, G.J., Vermeij, F.H. and Leijten, F.S. (2002) Epileptic seizure after a cup of tea: intoxication with Japanese star anise. *Ned Tijdschr Geneeskd* 146(17), 813–816.

Chang, K.S. and Ahn, Y.J. (2002) Fumigant activity of (*E*)-anethole identified in *Illicium verum* fruit against *Blattella germanica*. *Pest Management Science* 58(2), 161–166.

Cu, J.Q. (1986) The kingdom of essential oil. In: Lawrence, B.M., Mookherjee, B.D. and Willis, B.J. (eds) *Flavour and Fragrances: A World Perspective*, Proceedings of the 10th International Congress of Essential Oils, Washington, DC, 16–20 November. Elsevier, Amsterdam.

Cu, J.Q., Perineau, F. and Goepfert, G. (1990) GC/MS analysis of star anise oil. *Journal of Essential Oil Research* 2, 91–92.

De, M., De, A.K., Sen, P. and Banerjee, A.B. (2002) Antimicrobial properties of star anise (*Illicium verum* Hook f). *Phototherapeutic Research* 16(1), 94–95.

Della Porta, G., Taddeo, R., D'Urso, E. and Reverchon, E. (1998) Isolation of clove bud and star anise essential oil by supercritical CO_2 extraction. *Lebensmittel–Wissenschaft und Technologie* 31, 454–460.

Eykmann, J.F. (1881) Isolation of Shikimin from star anise. *Pharmacology Journal and Transactions* 11, 1046.

FAO (1993) Beer, J.H. de (ed.) *Non-wood Forest Products in Indochina – Focus: Vietnam*. FAO Regional office for Asia and Pacific, Bangkok.

Flückiger, F.A. (1879) *Pharmacographia: A History of the Principal Drugs of Vegetable Origin, Met with in Great Britain and British India*. Macmillan, London.

Goodman, P.S. (2005) Star rises in fight against bird flu. Demand for a Chinese fruit skyrockets. *Washington Post Foreign Service* November 18, p. D01.

Heath, H.B. (1981) *Source Book of Flavours*. The AVI Publishing Company, Westport, Connecticut, pp. 221–222.

Hitokoto, H., Morozumi, S., Wauke, T., Sakai, S. and Kurata, H. (1980) Inhibitory effects of spices on growth and toxin production of toxigenic fungi. *Applied and Environmental Microbiology* 39(4), 818–822.

Ho, S.H., Ma, Y., Goh, P.M. and Sim, K.Y. (1995) Star anise, *Illicium verum* Hook f. as a potential grain protectant against *Ttibolium castaneum* (Herbst) and *Sitophilus zeamais* Motsch. *Postharvest Biology and Technology* 6, 341–347.

Huang, J.M., Nakade, K., Kondo, M., Yang, C.S. and Fukuyama, Y. (2002) Brine shrimp lethality test active constituents and new highly oxygenated seco-prezizaane-type sesquiterpenes from *Illicium merrillianum*. *Chemical and Pharmaceutical Bulletin* (Tokyo) 50(1), 133–136.

Huang, J.M., Yang, C.S., Zhao, R., Takahashi, H. and Fukuyama, Y. (2004) Seven novel seco-prezizaane-type sesquiterpenes from the pericarps of *Illicium merrillianum*. *Chemical and Pharmaceutical Bulletin* (Tokyo) 52(1), 104–107.

Huang, P., Nishi, M., Zheng, X.Z., Lai, M.X. and Naknishi, T. (1997) Triterpene acids from the barks of *Illicium difengpi*. *Yao Xue Xue Bao* 32(9), 704–707.

Itoigawa, M., Ito, C., Tokuda, H., Enjo, F., Nishino, H. and Furukawa, H. (2004) Cancer chemopreventive activity of phenylpropanoids and phytoquinoids from *Illicium* plants. *Cancer Letters* 214(2), 165–169.

Ize-Ludlow, D., Ragone, S., Bruck, S.I., Jeffrey, N., Bernstein, J.N., Duchowny, M. and Garcia Peña, B.M. (2004) Neurotoxicities in infants seen with the consumption of star anise tea. *Pediatrics* 114(5), e653–e656.

Johanns, E.S., van der Kolk, L.E., van Gemert, H.M., Sijben, A.E., Peters, P.W. and de Vries, I. (2002) An epidemic of epileptic seizures after consumption of herbal tea. *Ned Tijdschr Geneeskd* 146, 813 –816.

Joshi, V.C., Srinivas, P.V. and Khan, I.A. (2005) Rapid and easy identification of *Illicium verum* Hook. f. and its adulterant *Illicium anisatum* Linn. by fluorescent microscopy and gas chromatography. *Journal of the Association of Official Analytical Chemists International* 88(3), 703–706.

Kuriyama, T., Schmidt, T.J., Okuyama, E. and Ozoe, Y. (2002) Structure–activity relationships of seco-prezizaane terpenoids in gamma-aminobutyric acid receptors of houseflies and rats. *Bioorganic and Medicinal Chemistry* 10(6), 1873–1881.

Lai, M., Yao, W., Yang, M., Huang, P., Liu, B. and Zheng, X. (1997) Studies on the pharmacognosy of *Cortex Illicii* and its adulterants. *Zhong Yao Cai* 20(12), 601–604.

Lederer, I., Schulzki, G., Gross, J. and Steffen, J.P. (2006) Combination of TLC and HPLC-MS/MS methods. Approach to a rational quality control of Chinese star anise. *Journal of Agricultural and Food Chemistry* 54, 1970–1974.

Lee, S.W., Li, G., Lee, K.S., Song, D.K. and Son, J.K. (2003a) A new phenylpropanoid glucoside from the fruits of *Illicium verum*. *Archives of Pharmacological Research* 26(8), 591–593.

Lee, S.W., Li, G., Lee, K.S., Jung, J.S., Xu, M.L., Seo, C.S., Chang, H.W., Kim, S.K., Song, D.K. and Son, J.K. (2003b) Preventive agents against sepsis and new phenylpropanoid glucosides from the fruits of *Illicium verum*. *Planta Medica* 69(9), 861–864.

Loi, D.T. and Thu, N.V. (1970) *Medicinal Trees and Pharmaceutical Material and Herbs in Vietnam*. Pharmacy College Publisher, Hanoi, pp. 209–211.

Lucchesi, M.E., Chemat, F. and Smadja, J. (2004) Solvent-free microwave extraction: an innovative tool for rapid extraction of essential oil from aromatic herbs and spices. *The Journal of Microwave Power and Electromagnetic Energy* 39(3–4), 135–139.

Lust, J. (1974) *The Herb Book*. Bantam Books, New York, pp. 344.

Malcolm, S. (1987) *The Encyclopedia of Herbs and Herbalism*. Macdonald and Company, Turin, Italy.

Meisner, W. (1818) Beitrag zur Geschichte des Chrom. *Annalen der Physik* 60(12), 366–394.

Morton, M. (2004) *Cupboard Love, a Dictionary of Culinary Curiosities*. Insomniac Press, Toronto, Canada.

Nakamura, T., Okuyama, E., and Yamazaki, M. (1996) Neurotropic components from star anise (*Illicium verum* Hook fil.). *Chemical and Pharmaceutical Bulletin* (Tokyo) 44(10), 1908–1914.

Ngo, K.S., Wong, W.T. and Brown, G.D. (1999) Murolane sesquiterpenes from *Illicium tsangii*. *Journal of Natural Products* 62(4), 549–553.

Padmashree, A., Roopa, N., Semwal, A.D., Sharma, G.K., Agathian, G. and Bawa, A.S. (2007) Star-anise (*Illicium verum*) and black caraway (*Carum nigrum*) as natural antioxidants. *Food Chemistry* 104(1), 59–66.

Rahway, N.J. (1989) *The Merck Index*. Eleventh edition. Merck & Co., Inc., Whitehouse Station, New Jersey.

Reverchon, E. (1997) Supercritical fluid extraction and fractionation of essential oils and related product. *Journal of Supercritical Fluids* 10, 1–37.

Rizvi, S.S.H., Benado, A.L., Zollweg, J.A. and Daniels, J.A. (1986) Supercritical fluid extraction operating principles and food applications. *Food Technology* (Chicago) 40(7), 57–64.

Rosengarten, F., Jr. (1969) *The Book of Spices*. Livingston Publishing Co., Wynnewood, Pennsylvania.

Rudzki, E. and Grzywa, Z. (1976) Sensitizing and irritating properties of star anise oil. *Contact Dermatitis* 2(6), 305–308.

Sakaguchi, I., Ishimoto, H., Matsuo, M., Ikeda, N., Minamino, M. and Kato, Y. (2004) The water-soluble extract of *Illicium anisatum* stimulates mouse vibrissae follicles in organ culture. *Experimental Dermatology* 13(8), 499–504.

Schlegel, C.E. (1885) The fruit of *Illicium anisatum*, Loureiro. *The American Journal of Pharmacy* 57, 426–427.

Schmidt, T.J. (1999) Novel seco-prezizaane sesquiterpenes from North American *Illicium* species. *Journal of Natural Products* 62(5), 684–687.

Schmidt, T.J., Muller, E. and Fronczek, F.R. (2001) New allo-cedrane type sesquiterpene hemiketals and further sesquiterpene lactones from fruits of *Illicium floridanum*. *Journal of Natural Products* 64(4), 411–414.

Singh, G., Maurya, S., deLampasona, M.P. and Catalan, C. (2006) Chemical constituents, antimicrobial investigations and antioxidative potential of volatile oil and acetone extract of star anise fruits. *Journal of the Science of Food and Agriculture* 86(1), 111–121.

Stahl, E. and Gerard, D. (1985) Solubility behaviour and fractionation of essential oils in dense carbon dioxide. *Perfume Flavor* 10, 29–37.

Sy, L.K. and Brown, G.D. (1998) A prezizaane sesquiterpene from *Illicium angustisepalum*. *Phytochemistry* 49(6), 1715–1717.

Tuan, D.Q. and Ilangantileke, S.G. (1997) Liquid CO_2 extraction of essential oil from star anise fruits (*Illicium verum* H.). *Journal of Food Engineering* 311, 47–57.

Vandenberghe, N., Pittion-Vouyouvitch, S., Flesch, F., Wagner, M. and Godet, E. (2003) An inaugural generalized tonic-clonic convulsive crisis following ingestion of Japanese star anise. *Presse Medicale* 11 32(1), 27–28.

Wang, X.Q., Guo, Y.J. and Yang, C.S. (2001) Determination of shikimic acid in fruit of *Illiciaceae* plants by HPLC with diode-array detection. *Zhongguo Zhong Yao Za Zhi* 26(7), 447–449.

Wang, X., Wang, H., Ma, J., Xu, Y. and Guan, Y. (2004) Analysis of volatile oil in *Illicium verum* fruit by on-line coupled packed capillary liquid chromatography/capillary gas chromatography. *Se Pu* 22(2), 101–105.

WHO (Manila), Institute of Materia Medica (1990) *Medicinal Plants in Vietnam*. WHO Regional Publications, Western Pacific Series No. 3, Hanoi and Manila, pp. 211–212.

Yokoyama, R., Huang, J.M., Yang, C.S. and Fukuyama, Y. (2002) New seco-prezizaane-type sesquiterpenes, jiadifenin with neurotrophic activity and 1,2-dehydroneomajucin from *Illicium jiadifengpi*. *Journal of Natural Products* 65(4), 527–531.

Yokoyama, R., Huang, J.M., Hosoda, A., Kino, K., Yang, C.S. and Fukuyama, Y. (2003) Seco-prezizaane-type sesquiterpenes and an abietane-type diterpene from *Illicium minwanense*. *Journal of Natural Products* 66(6), 799–803.

Zhou, J., Lu, G., Zhong, X. and Wen, H. (2005) Quantitative determination of anethole in the fruit of *Illicium verum* from various places of Guangxi province. *Zhong Yao Cai* 28(2), 106–107.

18 Aniseed

N.K. Leela and T.M. Vipin

18.1. Introduction

Anise or aniseed (*Pimpinella anisum* L.) is a flowering plant in the family *Apiaceae*, native to the eastern Mediterranean region and South-west Asia. It is widely cultivated in southern and central Europe, the former USSR, North Africa and, to a lesser extent, Mexico and South America (Ross, 2001). In India, it is grown to a small extent as a culinary herb or as a garden plant. The spice aniseed is the fruit of *P. anisum*.

18.2. Botany and Uses

The aniseed plant grows to 30–60 cm high with ternately pinnate leaves. The flowers are small, white and produced in compound umbels. The fruit is ovoid or pyriform, laterally compressed, 3–5 mm in length and 2–3 mm wide, greyish-green to greyish-brown, with a peculiar sweet smell. Each fruit contains two carpels, both containing an anise seed. The seed is small and curved, about 0.5 cm long and greyish-brown. It usually contains hair-like protrusions from each end of the seed. The pericarp is broadly ovoid, five-ridged with short hairs and numerous vittae (Ross, 2001). The ripe dry fruits are harvested between July and September.

Anise leaves are used to treat digestive problems and to relieve toothache, and its essential oil is used to treat lice and scabies. In aromatherapy, aniseed essential oil is used to treat colds and flu. It is also being researched for the treatment of bird flu (Waumans *et al.*, 2006).

In India, aniseed is also used as mouth freshener, for flavouring some foods and in confectionaries. Anise was used as a cure for sleeplessness, chewed with alexanders (*Smyrnium olusatrum*) and a little honey in the morning to freshen the breath and, when mixed with wine, as a remedy for scorpion stings. In the Middle East, aniseed is used in producing alcoholic beverages. In Thailand, it is used to flavour tea.

Anise oil is used for both hunting and fishing. It smells similar to liquorice and is put on fishing lures to attract fish (http://en.wikipedia.org/wiki/Anise). Anethole (4-methoxyphenyl-1-propene), the principal component of anise oil, is a precursor that can produce 2,5-dimethoxybenzaldehyde, which is used in the synthesis of psychedelic drugs such as DOB (2,5-dimethoxy-4-bromoamphetamine) (Waumans *et al.*, 2006). Anise is also the main flavour in several types of liquors. It has a particular effect on some dogs that parallels the effect of catnip on house cats. The residue left after extraction of oil is used as cattle feed

(Nath *et al.*, 1966). It contains 17–19% protein and 16–12% fat (Anon., 1969).

18.3. General Composition

Aniseed contains moisture (9–13%), protein (18%), fatty oil (8–23%), essential oil (2–7%), sugars (35%), starch (5%), N-free extract (22–28%) and crude fibre (12–25%) (Pruthi, 1976). The nutritional, fatty acid and elemental composition of aniseed is given in Table 18.1.

Table 18.1. Nutritional, fatty acid and elemental composition of aniseed.

Composition	USDA Handbook 8–21
[1]*Nutrients* (%)	
Water (g)	9.54
Food energy (Kcal)	337
Protein (g)	17.6
Fat (g)	15.90
Carbohydrate (g)	50.02
Ash (g)	6.95
Calcium (g)	0.646
Phosphorus (mg)	440
Sodium (mg)	16
Potassium (mg)	1441
Iron (mg)	36.96
[2]*Fatty acids* (%)	
Petroselinic acid	23.5
Oleic acid	56.0
Linoleic acid	17.1
Palmitic acid	3.2
[3]*Elements*	
Mg	–
Al (mg/kg)	147
Si (mg/kg)	269
P (mg/kg)	570
S (mg/kg)	517
Cl (%)	0.11
K (%)	0.54
Ca (%)	0.25
Ti	–
Mn (mg/kg)	19
Fe (mg/kg)	156
Cu (mg/kg)	53
Zn (mg/kg)	35
Br	–
Rb	–
Sr (mg/kg)	6

Source: [1]Tainter and Grenis (1993); [2]Anon. (1969); [3]Al-Bataina *et al.* (2003).

18.4. Chemistry

Aniseed contains volatile oil, furanocoumarins, flavonoids, fatty acids, phenylpropanoids, sterols and proteins. Anethole has an observed oestrogenic effect, and the seeds as a whole are mildly oestrogenic. This effect may substantiate the herb's use as a stimulant of breast-milk production (http://www.herbs2000.com/herbs/herbs_anise.htm).

Volatiles

Aniseed contains 1.5–4.0% essential oil. Up to 6% oil has been reported from Syrian aniseed. The physico-chemical properties of the oil are given in Table 18.2 (Anon., 1969). The primary constituent of the oil is anethole (80%) (http://www.purplesage.org.uk/profiles/aniseed.htm). Other components in the volatile oil are anisaldehyde, anisketone and methyl chavicol (estragole). El-Wakeil *et al.* (1986) found an increase in *t*-anethole and a decrease in the other oil components of anise oil during extended storage. The flavour of anise is similar to liquorice and fennel, the latter also containing a high level of anethole. The essential oil of anise is commonly used, although the oleoresin is not (Tainter and Grenis, 1993). The volatile oil composition as analysed by various workers is given in Table 18.3 and Fig. 18.1.

The extraction of seed with ether yields a dark green, fatty oil. The major fatty acid is oleic acid, which is followed by petroselinic acid. The fatty acid composition is indicated in Table 18.1.

Rodrigues *et al.* (2003) extracted 3.13–10.67% essential oil by the supercritical fluid extraction (SFE) method. The oil contained anethole (~90%), γ-himachalene (2–4%), *p*-anisaldehyde (< 1%), methylchavicol (0.9–1.5%), *cis*-pseudoisoeugenyl-2-methylbutyrate (~3%) and *t*-pseudoisoeugenyl-2-methylbutyrate (~ 1.3%) as the major constituents (Rodrigues *et al.*, 2003).

Tabanca *et al.* (2006) analysed essential oils from 15 *Pimpinella* species by gas chromatography (GC) and gas chromatography-

Table 18.2. Physico-chemical properties and fatty acid composition of aniseed oil.

Physico-chemical properties	Aniseed volatile oil	Ether extract
Specific gravity (at 20°C)	0.978–0.992	0.922
Refractive index (at 20°C)	1.553–1.560	1.472
Optical rotation (at 20°C)	–2°C to +1°C	–

Source: Anon. (1969).

Table 18.3. Volatiles from aniseed.

Compound	Leaf	Fruit	Reference(s)
cis-Anethole	+	–	Embong *et al.*, 1977
t-Anethole	+	+	Tam *et al.*, 1979 Embong *et al.*, 1977
Anisic acid	–	+	Embong *et al.*, 1977
Anisyl alcohol	–	+	Embong *et al.*, 1977
Anisyl ketone	–	+	Embong *et al.*, 1977
Benzene, 2-hydroxy-5-methoxy-*trans*-propenyl 2-methyl-butyrate	–	+	Kubeczka *et al.*, 1976
Bergamotene	–	+	De Maack *et al.*, 1982
Bisabolene, β: callus tiss	–	+	Reichling *et al.*, 1985
Caffeic acid	–	+	Schultz and Herrmann 1980
Camphene			Mekhtieva, 1991
Camphor	–	+	Embong *et al.*, 1977
Carvone	–	+	Embong *et al.*, 1977; De Maack *et al.*, 1982
Carvone dihydro acetate	–	+	Embong *et al.*, 1977
β-Caryophyllene	–	+	Embong *et al.*, 1977
Cinnamaldehyde			De Maack *et al.*, 1982
Cinnamyl alcohol			De Maack *et al.*, 1982
β-Elemene			De Maack *et al.*, 1982
Estragole	–	+	Embong *et al.*, 1977
Eugenol (2-methyl-butyrate), pseudo-iso			Reichling *et al.*, 1985
Eugenol, (2-methyl-butyryl ester), iso pseudo epoxy	–	+	Reichling *et al.*, 1985
Eugenol, (2-methyl-butyryl ester), iso pseudo			Reichling *et al.*, 1985
Eugenol, iso pseudo 2-methyl-butyrate	–	+	Schultz *et al.*, 1986
Eugenol iso pseudo epoxy 2-methyl-butyrate	–	+	Kleiman *et al.*, 1988
Eugenol	–	+	Embong *et al.*, 1977
Fenchone	–	+	Embong *et al.*, 1977
Limonene	–	+	Embong *et al.*, 1977
Linalool	–	+	Embong *et al.*, 1977
Phellandrene			Rutovskii and Leonov, 1924
4,4-Dimethoxy stilbene	–	+	Miething *et al.*, 1990

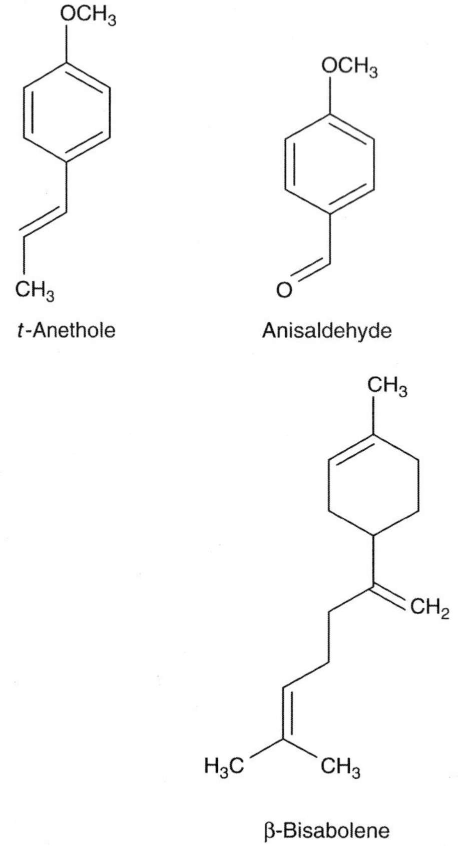

Fig. 18.1. Volatiles from aniseed.

mass spectrometry (GC-MS) techniques. A total of 140 compounds were identified, which included mono-, sesqui- and trinorsesquiterpenoids, propenylphenols and pseudoisoeugenols. Trinorsesquiterpenoids and phenylpropanoids are the chemical markers of the *Pimpinella* species. The essential oils obtained from *Pimpinella* roots share the same principal compound, epoxypseudoisoeugenyl-2-methybutyrate, at concentrations from 20 to 82.6% (Tabanca *et al.*, 2006).

Headspace solvent microextract (HSME) of aniseed contained 90% *t*-anethole (Abbas *et al.*, 2005).

The major components of the essential oils from the hairy root cultures were *trans*-epoxypseudoisoeugenyl 2-methylbutyrate, geijerene, pregeijerene, zingiberene and *β*-bisabolene (Santos *et al.*, 1998). *t*-Epoxypseudoisoeugenyl 2-methylbutyrate, *β*-bisabolene and geijerene were the major components of the essential oil from the roots of the plant, whereas the main component of the fruit oil was *t*-anethole. Geijerenes were not detected in the fruit oil (Santos *et al.*, 1998; Table 18.4).

Non-volatiles

The non-volatile constituents isolated from aniseed can be divided into phenolic acids, flavonoids, furanocoumarins, sterols and glucosides (Table 18.5).

Phenolic acids

The phenolic acids isolated from aniseed include caffeic acid, hydroxybenzoic acid, hydroxycinnamic acid and *p*-coumaric

Table 18.4. Percentage composition of essential oils of *P. anisum* isolated from fruits, roots and hairy root cultures grown in different media.

			B5/2		MS/2		SH		B5	
Component	Fruits	Roots	Light	Dark	Light	Dark	Light	Dark	Light	Dark
α-Heptanal	–	0.2	t	t	t	t	t	t	t	t
Benzaldehyde	–	0.7	–	–	–	–	–	–	–	–
Sabinene	–	0.2	–	–	–	–	–	–	–	–
n-Octanal	–	0.1	0.4	1.3	3.0	2.3	1.7	1.4	3.3	0.2
Myrcene	–	0.1	–	–	–	–	–	–	–	–
Decane	–	–	t	t	t	t	t	t	t	t
α-Terpinene	–	t	–	–	–	–	–	–	–	–
Phenylacetaldehyde	–	–	t	t	t	t	t	t	t	t
p-Cymene	–	t	t	t	t	t	t	t	t	t
Limonene	–	t	t	t	t	t	t	t	t	t
γ-Terpinene	–	0.1	–	–	–	–	–	–	–	–
m-Cresol	–	–	0.2	t	1.5	t	0.4	0.2	0.5	t
n-Nonanal	–	–	t	t	t	t	t	t	t	t
Undecane	–	t	0.3	0.2	2.8	1.9	2.2	1.0	2.5	0.2
Geijerene isomer	–	0.2	0.2	0.2	t	t	0.4	0.3	t	t
Geijerene	–	1.8	3.0	2.6	4.7	3.9	9.3	5.3	8.9	1.1
Borneol	–	t	–	–	–	–	–	–	–	–
Terpinen-4-ol	–	t	–	–	–	–	–	–	–	–
α-Terpineol	–	0.5	–	–	–	–	–	–	–	–
Estragole	2.2	–	–	–	–	–	–	–	–	–
Anisaldehyde	1.9	–	–	–	–	–	–	–	–	–
cis-Anethole	t	–	–	–	–	–	–	–	–	–
Isogeijerene C	–	–	t	t	0.4	t	t	t	t	t
trans-Anethole	92.5	2.0	t	t	t	0.9	0.8	1.1	0.6	t
Pregeijerene	–	6.0	3.9	5.0	15.1	12.1	24.3	12.7	15.6	2.8
Carvacrol	–	–	t	t	1.7	1.4	t	0.5	1.2	t
Dodecanol	–	0.3	1.7	1.3	9.9	7.0	10.7	6.9	7.1	0.2
β-Caryophyllene	–	0.2	t	t	0.3	1.2	0.3	t	t	t
t-α-Bergamotene	–	–	t	t	0.8	0.5	0.4	0.5	t	t
t-β-Farnesene	–	t	0.3	0.2	t	0.8	0.4	0.6	1.2	0.2
Dodecanol	–	–	0.6	0.3	2.9	0.9	9.0	2.5	2.6	0.6
ar-Curcumene	–	–	t	t	3.0	1.5	2.9	0.7	2.4	t
Zingiberene	–	0.2	3.4	3.5	12.0	9.6	10.1	7.5	8.8	2.3
β-Bisabolene	–	6.4	4.6	8.2	14.4	14.6	9.2	7.8	7.7	3.1
Sesquiphellandrene	–	–	0.3	0.3	1.7	3.0	0.3	0.3	0.5	t
Elemol	–	–	t	t	0.8	0.4	t	t	0.6	t
t-Pseudoisoeugenyl 2-methylbutyrate	0.1	4.7	3.6	2.8	t	1.9	4.0	9.0	t	4.0
cis-Epoxypseudoisoeugenyl 2-methylbutyrate	–	t	t	0.1	t	t	0.4	0.2	0.7	0.2
t-Epoxypseudoisoeugenyl 2-methylbutyrate	t	70.2	52.8	56.4	2.4	16.7	5.7	21.3	0.8	78.1

t = trace.
Source: Santos *et al.* (1998).

Table 18.5. Non-volatiles from aniseed.

Compound	Leaf	Fruit	Reference
Abscisic acid	–	+	
Benzoic acid, 4-β-D-glucopyranosyl-oxy	–	+	Dirks and Herrmann, 1984
Benzoquinone, 1,4-callus tissue	–	+	Lichtenthaler and Straub, 1975
Caffeic acid	–	+	Schultz and Herrmann, 1980
Bergapten:callus tissue	–	+	De Maack *et al.*, 1982
p-Coumaric acid	–	+	Schultz and Herrmann, 1980

acid (Schultz and Herrmann, 1980). 4-Methoxy-2-(*trans*-1-propenyl) phenyl (±)-2-methylbutanoate was isolated from anise plants (Carter *et al.*, 1977).

Flavonoids

The flavonoids, luteolin, luteolin 7-*O*-β-D-xyloside and cynaroside, were isolated by El-Moghazi *et al.* (1979) and orientin and vitexin by Akunzemann and Herrmann (1977). Quercetin-3-glucuronide, rutin, isoorientin, isovitexin and apigenin-7-glucoside have been reported from aniseed (Ozguven, 2000). Figure 18.2 depicts the structure of flavonoids in aniseed.

Furanocoumarins

5-Methoxy psoralen and xanthotoxin were reported by Ceska *et al.* (1987). Zobel *et al.* (1991) isolated psoralen, bergapten, scoparon, scopoletin and seslin from the leaves of aniseed and Zobel and Brown (1991) isolated psoralen and bergapten from the fruit.

Sterols

Aniseed oil contains 0.7% sterols. Sitosterol and stigmasterol were identified as the major components (Zlatanov and Ivanov, 1995). Aniseed also contains coumarins and glycosides.

Glucosides

From the polar portion of the methanolic extract of the fruit of anise (*P. anisum* L.), Fujimatu *et al.* (2003) isolated aromatic compound glucosides and an alkyl glucoside. The new compounds identified were (*E*)-3-hydroxyanethole β-D-glucopyranoside, (*E*)-1′-(2-hydroxy-5-methoxyphenyl)propane-β-D-glucopyranoside, 3-hydroxyestragole-β-D-glucopyranoside, methyl syringate 4-*O*-β-D-glucopyranoside, hexane-1,5-diol-1-*O*-β-D-glucopyranoside and 1-deoxy-D-erythritol-3-*O*-β-D-glucopyranoside (Fujimatu *et al.*, 2003).

From anise seeds, a phenolic glucoside, namely 4-(β-D-glucopyranosyloxy) benzoic acid, was isolated by Dirks and Herrmann (1984). This compound is widely distributed among the Apiaceae and is also present in star anise (Illiciaceae).

Ishikawa *et al.* (2002) isolated 17 glucosides of phenylpropanoids, including four stereoisomers of anethole glycol 2′-*O*-β-D-glucopyranoside and four stereoisomers of 1′-(4-hydroxyphenyl) propane-1′, 2′-diol-2′-*O*-β-D-glucopyranoside, together with anethole glycols and guaiacyl glycerol, the water-soluble portion of the methanolic extract of the fruit of anise.

18.5. Medicinal and Pharmacological Properties

The volatile oil in aniseed provides the basis for its internal use to ease griping, intestinal colic and flatulence. It also has a marked expectorant and antispasmodic action and may be used in bronchitis and in tracheitis, where there is persistent irritable coughing, and to reduce the symptoms of whooping cough. Externally, the oil may be used in an ointment base for the treatment of scabies and lice infestations. Aniseed's mild oestrogenic effects, thought

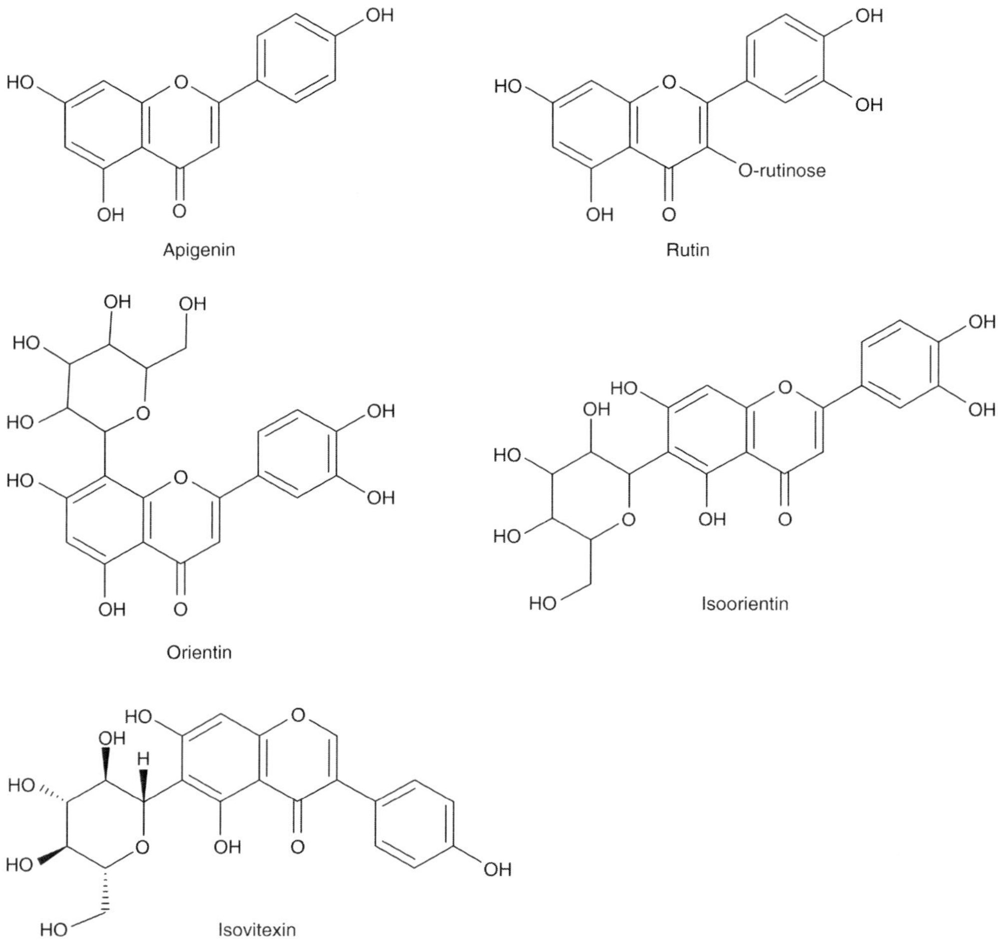

Fig. 18.2. Flavonoids from aniseed.

to be due to the presence of diantheole and photoantheole, explain the use of this plant in folk medicine to increase milk secretion, facilitate birth and increase libido (http://www.purplesage.org.uk/profiles/aniseed.htm).

The essential oil of aniseed is reported to be antiseptic, antispasmodic, carminative, diuretic, expectorant and stimulant. It is good for bronchitis, colds, cramps, emotional balancing, headlice, muscular aches and pains, muscular spasm, rheumatism and stress (http://www.tigerlillys.co.nz/Properties.htm).

Gilligan (2005) suggested that aniseed oil in combination with *Foeniculum vulgare* var. *dulce* (sweet fennel), *Anthemis nobilis* (Roman chamomile) and *Mentha x piperita* (peppermint) oils used in aromatherapy treatment were successful complements to the relief of nausea in a hospice and palliative care programme.

Antispasmodic

Aniseed is used in folk medicine as an antispasmodic agent. Tirapelli *et al.* (2007) report that ethanol:water (40:60) extract of aerial parts of anise (50μg/ml) inhibited acetylcholine-induced contraction in rat smooth muscle.

The *Pimpinella* species are used as food plants by the larvae of some *Lepidoptera* species, including the lime-speck pug and wormwood pug.

Anti-inflammatory

Aniseed possesses anti-inflammatory properties. Topical application of ethyl acetate and hexane extracts of aniseed, at a dose of 20 μl/animal, produced an anti-inflammatory effect in mouse treated with 12-*O*-tetradecanoyl phorbol-13-acetate (Okuyama *et al.*, 1995).

Anti-ulcer effect

Aqueous suspension of anise possesses significant cytoprotective and anti-ulcer activities against experimentally induced gastric lesions. Al-Mofleh *et al.* (2007) reported that in pylorus-ligated Shay rats anise suspension reduced basal gastric acid secretion and acidity significantly and completely inhibited ruminal ulceration. The suspension replenished ethanol-induced depleted levels of gastric mucosal NP-SH and gastric wall mucus concentration significantly. The anti-ulcer effect of anise is possibly prostaglandin-mediated and/or through its antisecretory and antioxidative properties.

Antifungal activity

Aniseed possesses varying levels of fungitoxicity. Aniseed fluid extract shows antimycotic activity against *Candida albicans, C. parapsilosis, C. tropicalis, C. pseudotropicalis* and *C. krusei* with MIC values between 17 and 20% (v/v). Extract of fruits of anise inhibits the growth of dermatophyte species (*Trichophyton rubrum, T. mentagrophytes, Microsporum canis* and *M. gypseum*) with MIC values between 1.5 and 9% (v/v). The essential oil of anise shows strong antifungal activity against yeasts with MIC lower than 1.56% (v/v) and dermatophytes with MIC lower than 0.78% (v/v) (Kosalec *et al.*, 2005).

Insecticidal activity

The essential oil of *P. anisum* is highly effective as both larvicidal and ovicidal against three mosquito species, *Anopheles stephensi, Aedes aegypti* and *Culex quinquefasciatus*. The oil showed toxicity against 4th instar larvae of *A. stephensi* and *A. aegypti* with LD_{95} values of 115.7 μg/ml, whereas it was 149.7 μg/ml against *C. quinquefasciatus* larvae (Prajapati *et al.*, 2005). Recently, larvicidal activity of the essential oil against the seaside mosquito, *Ochlerotatus caspius*, has been reported by Knio *et al.* (2007).

Essential oil from anise shows potent fumigant activity against the larvae of *Lycoriella ingénue* (Dufour). *trans*-Anethole, the chief constituent of the anise oil, was toxic with an LC_{50} value of 0.20 μl/l (Park *et al.*, 2006). Essential oils extracted from the seeds of anise exhibit significant repellency against the adult females of the mosquito, *C. pipiens* (Erler *et al.*, 2006).

Antibacterial activity

Methanol extract of *P. anisum* seeds is effective against the Gram-negative bacterium *Helicobacter pylori* at MIC of 100 μg/ml (Mahady *et al.*, 2005). This bacterium is recognized as the primary etiological factor associated with the development of gastritis and peptic ulcer disease. HP infections are also associated with chronic gastritis, gastric carcinoma and primary gastric B-cell lymphoma.

18.6. ISO Specifications

ASTA recommends a moisture limit of 10% in the whole spice. Ash and acid-insoluble ash should be no greater than 6 and 1%, respectively. Tables 18.6 and 18.7 show the chemical and physical specifications for the whole and ground spice, including the FDA DALs for the whole spice.

Table 18.6. Whole anise: chemical and physical specifications.

Specification	Suggested limits
ASTA cleaniless specifications	
Whole dead insects, by count	4
Mammalian excreta (mg/lb)	3
Other excreta (mg/lb)	5
Mould, % by weight	1
Insect defiled/infested, % by weight	1
Extraneous, % by weight	1
FDA DALs (condimental seed)	
Adulteration with mammalian excreta (mg/lb)	Average of 3
Volatile oil (% min.)	2.5
Moisture (% max.)	10
Ash (% max.)	6
Acid-insoluble ash (% max.)	1
Average bulk index (mg/100 g)	230

Source: Tainter and Grenis (1993).

Table 18.7. Ground anise: chemical and physical specifications.

Specification	Suggested limits
FDA DALs	None
Volatile oil (% min.)	2
Moisture (% max.)	10
Total ash (% max.)	6
Acid-insoluble ash (% max.)	1
Military specifications (EE-S-63lJ, 1981)	None
Bulk index 2 (ml/100 g)	175

Source: Tainter and Grenis (1993).

18.7. Conclusion

Aniseed contains a volatile oil, furanocoumarins, flavonoids, fatty acids, phenylpropanoids, sterols and proteins. The chief component of aniseed oil is *t*-anethole. Aniseed is reported to have anti-ulcer, anti-inflammatory, antimicrobial and insecticidal properties.

References

Abbas, B.S., Jabbari, A. and Yamini, Y. (2005) Headspace solvent microextraction: a very rapid method for identification of volatile components of Iranian *Pimpinella anisum* seed. *Analytica Chimica Acta* 530(1), 155–161.

Akunzemann, J. and Herrmann, K. (1977) Isolation and identification of flavon (ol)-*O*-glycosides in caraway (*Carum carvi*), fennel (*Foeniculum vulgare* Mill.), anise (*Pimpinella anisum* L.) and coriander (*Coriandrum sativum* L.) and of flavone-*C*-glycosides in anise. I. Phenolics of spices. *Z Lebensm Unters Forsch* 164, 194–200.

Al-Bataina, B.A., Maslat, A.O. and Al-Kofahi, M.M. (2003) Element analysis and biological studies on ten oriental spices using XRF and Ames test. *Journal of Trace Elements in Medicine and Biology* 17(2), 85–90.

Al-Mofleh, I.A., Alhaider, A.A., Mossa, J.S., Al-Soohai bani, M.O. and Rafatullah, S. (2007) Aqueous suspension of anise '*Pimpinella anisum*' protects rats against chemically induced gastric ulcers. *World Journal of Gastroenterology* 13(7), 1112–1118.

Anon. (1969) *The Wealth of India Raw Materials*, Volume VIII. CSIR, New Delhi.

Carter, G.T., Schinoes, H.K. and Lichtenstein, E.P. (1977) 4-Methoxy-2-(*trans*-1-propenyl)phenyl (±)-2-methylbutanoate was isolated from anise plants. *Phytochemistry* 16, 615–616.

Ceska, O., Chaudhary, S.K., Warrington, P.J. and Ashwood-Smith, M.J. (1987) Photoactive furocoumarins in fruits of some umbellifers. *Phytochemistry* 26, 165–169.

De Maack, F., Frunet, D., Malnati, J.C. and Estienne, J. (1982) Study of minor constituents in anethole samples obtained from anise oil. I. Study of the origin of an anethole by the identification of sesquiterpenes. *Ann Falsif Expert Chim Toxicology* 75, 357–367 (quoted from Ross, I.A., 2001).

Dirks, U. and Herrmann, K. (1984) 4-(β-D-glucopyranosyloxy) benzoic acid, a characteristic phenolic constituent of the Apiaceae. *Phytochemistry* 23, 1811–1812.

El-Moghazi, A.M., Ali, A.A., Ross, K. and Mottaleb, B.U.T. (1979) Flavonoids of *Pimpinella anisum* L. growing in Egypt. *Fitoterapia* 50, 267–268.

El-Wakeil, F., Khairaj, M., Mosri, S., Farag, R.S., Shihata, A.A. and Badel, A. (1986) Biochemical studies on the essential oil of some fruits of Umbelliferae family. *Vereinigung der Seifen-, Parfüm- und Waschmittelfachleute*. 112, 77–80.

Embong, M.B., Hadziyev, D. and Molnar, S. (1977) Essential oils from spices grown in Alberta – anise oil (*Pimpinella anisum*). *Canadian Journal of Plant Science* 57, 829–837.

Erler, F., Ulug, I. and Yalcinkaya, B. (2006) Repellent activity of five essential oils against *Culex pipiens. Fitoterapia* 77(7–8), 491–494.

Fujimatu, E., Ishikawa, T. and Kitajima, J. (2003) Aromatic compound glucosides, alkyl glucoside and glucide from the fruit of anise. *Phytochemistry* 63(5), 609–616.

Gilligan, N.P. (2005) The palliation of nausea in hospice and palliative care patients with essential oils of *Pimpinella anisum* (aniseed), *Foeniculum vulgare* var. *dulce* (sweet fennel), *Anthemis nobilis* (Roman chamomile) and *Mentha x piperita* (peppermint). *International Journal of Aromatherapy* 15(4), 163–167.

http://en.wikipedia.org/wiki/Anise.

http://www.herbs2000.com/herbs/herbs_anise.htm.

http://www.purplesage.org.uk/profiles/aniseed.htm.

http://www.tigerlillys.co.nz/Properties.htm.

Ishikawa, T., Fujimatu, E. and Kitajima, J. (2002) Water-soluble constituents of anise: new glucosides of anethole glycol and its related compounds. *Chemical and Pharmacological Bulletin* 50(11), 1460–1466.

Kleiman, R., Plattner, R.D. and Weisleder, D. (1988) Antigermination activity of phenylpropenoids from the genus *Pimpinella. Journal of Natural Products* 51(2), 249–256.

Knio, K.M., Usta, J., Dagher, S., Zournajian, H. and Kreydiyyeh, S. (2007) Larvicidal activity of essential oils extracted from commonly used herbs in Lebanon against the seaside mosquito, *Ochlerotatus caspius. Bioresource Technology* 99 (4): 763–768.

Kosalec, I., Pepelinjak, S. and Kustrak, D. (2005) Antifungal activity of fluid extract and essential oil from anise fruits (*Pimpinella anisum* L., Apiaceae). *Acta Pharmaceutica* 55(4), 377–385.

Kubeczka, K.H., Massow, F.V., Formacek, V. and Smith, M.A.R. (1976) A new type of phenylpropane from the essential fruit oil of *Pimpinella anisum. Zeitschrift für Natur forsch Ser B* 31, 283.

Lichtenthaler, H.K. and Straub, V. (1975) The formation of lipoquinones in tissue cultures. *Planta Medica Supplement* 1975, 198.

Mahady, G.B., Pendland, S.L., Stoia, A., Hamill, F.A., Fabricant, D., Dietz, B.M. and Chadwick, L.R. (2005) *In vitro* susceptibility of *Helicobacter pylori* to botanical extracts used traditionally for the treatment of gastrointestinal disorders. *Phototherapy Research* 19(11), 988–991.

Mekhtieva, N.P. (1991) Essential oils of *Pimpinella aromatica. Chemistry of Natural Compounds* 27(2), 249–250.

Miething, H., Seger, V. and Hansel, R. (1990) Determination of photo anethole from a stored essential oil of anise fruits as 4,4´-di-methoxystilbene by high performance liquid chromatography ultraviolet coupling. *Phytotherapy Research* 4(3), 121–123.

Nath, K., Agarwala, O.N. and Mahadevan, V. (1966) 'Spent aniseed' (*Pimpenella anisum*) as a cattle feed. *Indian Veterinary Journal* 43(10), 908–912.

Okuyama, T., Matsuda, M., Matsuda, Y., Baba, M., Masubuchi, H., Adachi, M., Okada, Y., Hashimoto, T., Zou, L.B. and Nishino, H. (1995) Studies on cancer biochemoprevention of natural resources. X. Inhibitory effect of spices on TPA- enhanced 3H-choline incorporation in phospholipid of C3 H 10 T1/2 cells and on-TPA induced ear edema. *Zhonghua Yaoxue Zashi* 47, 421–430.

Ozguven, M. (2000) Aniseed. In: Peter, K.V. (ed.) *Handbook of Herbs and Spices*. Woodhead Publishing Ltd, Cambridge, UK, pp. 39–51.

Park, I.K., Choi, K.S., Kim, D.H., Choi, I.H., Kim, L.S., Bak, W.C., Choi, J.W. and Shin, S.C. (2006) Fumigant activity of plant essential oils and components from horseradish (*Armoracia rusticana*), anise (*Pimpinella anisum*) and garlic (*Allium sativum*) oils against *Lycoriella ingenua* (Diptera: Sciaridae). *Pest Management Science* 62(8), 723–728.

Prajapati, V., Tripathi, A.K., Aggarwal, K.K. and Khanuja, S.P. (2005) Insecticidal, repellent and oviposition-deterrent activity of selected essential oils against *Anopheles stephensi, Aedes aegypti* and *Culex quinquefasciatus. Bioresource. Technology* 96(16), 1749–1757.

Pruthi, J.S. (1976) *Spices and Condiments*. National Book Trust, India.

Reichling, J., Becker, H., Martin, R. and Burkhardt, G. (1985) Comparative studies on the production and accumulation of essential oil in the whole plant and in the cell culture of *Pimpinella anisum* L. *Zeitschrift für Naturforschung B* 40(7/8), 465–468.

Rodrigues, V.M., Rosa, P.T.V., Marques, M.O.M., Petenate, A.J. and Meireles, M.A.A. (2003) Supercritical extraction of essential oil from aniseed (*Pimpinella anisum* L) using CO_2: solubility, kinetics, and composition data. *Journal of Agricultural and Food Chemistry* 51(6), 1518–1523.

Ross, I.A. (2001) *Medicinal Plants of the World: Chemical Constituents, Traditional and Modern Medicinal Uses*, Volume 2. Humana Press, Totowa, New Jersey, pp. 363–374.
Rutovskii, B.N. and Leonov, P. (1924) Crimean oil of anise. *Trav Sci Inst Chim Pharm Moscon* 10, 64–68.
Santos, P.M., Figueiredo A.C., Oliveira, M.M., Barroso, J.G., Pedro, L.G., Deans, S.G., Younus, A.K.M. and Scheffer, J.J.C. (1998) Essential oils from hairy root cultures and from fruits and roots of *Pimpinella anisum*. *Phytochemistry* 48, 455–460.
Schultz, J.M. and Herrmann, K. (1980) Occurrence of hydroxybenzoic acid and hydroxycinnamic acid in spices. IV. Phenolics of spices. *Zeitschrift für Lebensmittel-Untersuchung und -Forschung* 171, 193–199.
Schultz, W., Lange, G. and Heinrich, G. (1986) Mass spectrometric study of medicinal plants. II. Direct mass spectrometric analysis of *Anisi fructus*. *Deutsche Apotheker-Zeitung* 126 (51/52), 2787–2793.
Tabanca, N., Demirci, B., Ozek, T., Kirimer, N., Baser, K.H.C., Bedir, E., Khan, I.A. and Wedge, D.E. (2006) Gas chromatographic-mass spectrometric analysis of essential oils from *Pimpinella* species gathered from Central and Northern Turkey. *Journal of Chromatography A* 1117(2), 194–205.
Tainter, R.D. and Grenis, T.A. (1993) *Spices and Seasonings. Food Science and Technology*. VCH Publishers, New York, p. 226.
Tam, N.T., Thanh, H.T.K. and Hoa, L.C. (1979) Contributions to the study of essence of the anise leaf grown in Lang Son (Vietnam). *Duoc Hoc* 1979(2), 18–20.
Tirapelli, C.R., Andrade, C.R. de, Cassano, A.O., De Souza, F.A., Ambrosio, S.R., Costa, F.B. da and Oliveira, A.M. de (2007) Antispasmodic and relaxant effects of the hydroalcoholic extract of *Pimpinella anisum* (Apiaceae) on rat anococcygeus smooth muscle. *Journal of Ethnopharmacology* 110(1), 23–29.
Waumans, D., Bruneel, N. and Tytgat, J. (2006) *Anise Oil as a Precursor for 2-Alkoxy-5-methoxybenzaldehydes. DEA Microgram Journal* 2 (1). Retrieved on 9 December 2006.
Zlatanov, M. and Ivanov, S.A. (1995) Studies on sterol composition of the seed oil of some representatives of the family Apiaceae. *Fett issenschaft Technologie* 97(10), 381–383.
Zobel, A.M. and Brown, S.A. (1991) Psoralens on the surface of seeds of Rutaceae and fruits of Umbelliferae and Leguminosae. *Canadian Journal of Botany* 169, 485–488.
Zobel, A.M., Wang, J.Y., March, R.E. and Brown, S.A. (1991) Identification of eight coumarins occurring with psoralen, xanthotoxin, and bergapten on leaf surfaces. *Journal of Chemical Ecology* 17, 1859–1870.

19 Garcinia

K.S. Krishnamurthy and V.P. Sapna

19.1. Introduction

The genus *Garcinia* belongs to the family Clusiaceae (Guttiferae). It consists of about 180 species, of which ~ 30 species are found in India. The genus is also found in Africa. The well-known and widespread species in Asia are *G. mangostana*, *G. cambogia*, *G. dulcis* and *G. tinctoria*. About 40 species of *Garcinia* produce edible fruits (Yapwattanaphum *et al.*, 2002). *G. dulcis* has a wider potential as a home garden fruit in the tropics, along with *G. mangostana*, but other species are not suitable (Martin *et al.*, 1987).

Maximum density of the *Garcinia* species is seen in the north of the Malaysian Archipelago, with approximately 28 species in Malaysia, about 23 in Thailand, about 20 in Indonesia and 19 in the Philippines. Myanmar and Indochina have fewer species. The Malaysian distribution is linked to the 18 tropical species found in the Andaman and Nicobar Islands. In other parts of India also, e.g. the north-east, Assam, Tamil Nadu and Western Ghats, the species are linked to the Malaysian distribution (Bin Osman and Rahman Milan, 2006). Table 19.1 shows the area and production of *G. mangostana* in major producing countries.

Of the *Garcinia* species, only three appear to have been moved to another region or continent. *G. cochinensis* Choisy, native to Indochina, moved to Brazil and Florida and is often cultivated in Cambodia; *G. livingstonei* T. Anders moved from Africa to Florida and *G. tinctoria* moved to Australia and Madagascar. There is evidence of incipient domestication and primitive cultivation for the fruits of *G. atroviridis* and *G. hombroniana* Pierre in Malaysia, *G. indica* Choisy in north-east India, *G. multiflora* Champ. in Vietnam and Laos and *G. pedunculata* Roxb. in Assam (India) and Bangladesh. Domestication to some extent is also noted for *G. dulcis* in Indonesia, Malaysia and the Phillippines and *G. tinctoria* in India and Malaysia (Bin Osman and Rahman Milan, 2006).

19.2. Botany and Uses

Botany

Garcinia species are evergreen trees or shrubs with a straight trunk tapering to a conical canopy and branches in alternating pairs at an acute angle on the trunk. These species are mostly understorey trees of lowland evergreen forests. The trees resemble *Eugenia* in shape but the branching habit

Table 19.1. Area and production of *Garcinia mangostana*.

Country	Year	Area (ha)	Production (t)
Indonesia	Undated[1]	10,750	NA
Malaysia	1998[2]	7,632	NA
Philippines	2000[3]	1,354	4,692
Thailand	2000[4]	11,000	46,000

Source: [1]BAPPEDA (2001); [2]MOA (2001); [3]DA-AMAS (2004); [4]Maneesin (2002).

and presence of latex distinguish it from *Eugenia*. The leaves are simple and entire, opposite or in whorls of three, coriaceous, often with glandular and resinous cells. The flowers are unisexual on separate trees but occasionally bisexual, borne in tufts or singly in the axils of leaves, regular with four persistent sepals and four petals in red, pink, yellow or white. The male flowers have seven or more stamens inserted on a receptacle and the female flowers have a large hypogynous ovary mounted on a receptacle. The ovaries are many chambered. The fruits are fleshy berries and contain one to four flattened seeds in a pulpy mass. Among the species, *G. cambogia*, *G. indica* and *G. atroviridis* have attracted wide attention all over the world due to the presence of hydroxy citric acid, which is generally known as an antiobesity factor.

G. cambogia is a small to medium-sized tree with a rounded crown. It has horizontal or drooping branches; its leaves are dark green and shiny, elliptic obviate, fruits are ovoid, yellow or red when ripe, with six to eight grooves and the fruits have six to eight seeds surrounded by a succulent aril. The tree is commonly found in the evergreen forests of Western Ghats, from Konkan southward to Travancore, and in the Shola forests of Nilgiris up to an altitude of about 1800 m (6000 ft). It flowers during the hot season and the fruits ripen during the rainy season. The seeds of *G. cambogia* contain 31% edible fat.

G. indica is a slender evergreen tree with drooping branches. It has ovate or oblong lanceolate leaves; the fruits are globose or spherical, dark purple when ripe and enclosing five to eight large seeds. The tree is found in the tropical rainforests of Western Ghats, from Konkan southward to Mysore, Coorg and Wynad. It flowers in November–February and the fruits ripen in April–May. The root is astringent. The seeds of the fruit have edible fat, commercially known as Kokam butter.

G. atroviridis is a moderate-sized graceful tree, 9–15 m (30–50 ft) high, found in the north-eastern districts of upper Assam. The leaves are glabrous, large, glossy green and the base is contracted. It has terminal flowers; the female flower is solitary and large. The fruits are orange-yellow, subglobose, fluted, with a firmly textured outer rind and a rather thin and translucent pulp surrounding the seeds.

Uses

The fruit of *G. indica* has an agreeable flavour and a sweetish acid taste. It is used as a garnish to give an acid flavour to curries and also for preparing syrups during hot months. In Ceylon, the dried fruit rinds of *G. cambogia* are used along with salt in the curing of fish. The fruit rinds of *G. atroviridis* are also too acidic to be eaten raw, but the taste is excellent when stewed with sugar. In Malay, the sun-dried rinds of underripe fruits of *G. atroviridis* are sold for the dressing of fish and as a sour relish for use in curries in place of tamarind. The acidic pulps of *G. mangostana*, *G. atroviridis*, *G. parviflora*, *G. cambogia* and *G. indica* have been reported as a substitute for tamarind to impart flavour. The dried rinds of *G. cambogia* and *G. indica* are used as a condiment for flavouring curries in place of tamarind or lemon. Kokam butter extracted from the seeds of *G. indica* is used in confectionery and cooking as a substitute for ghee. Oil extracted from *G. mangostana* is used as a substitute for kokam butter. *G. indica* seed contains 23–26% oil, which is used in the preparation of chocolates, cosmetics and medicines. The rind of mangosteen is reported to contain tannins and is used to tan leather and to dye fabric black. The fruit of *G. atroviridis* is used as a fixative with alum in the dyeing of silk. It is also used as an ingredient in soap and shampoo preparations. Gamboge, a gum resin collected from *Garcinia* after making incisions in the bark, is used as a pigment and traded in the world market. *Garcinia* species are also used in the paint and lacquer industries.

19.3. General Composition

The nutritional composition of ripe edible aril indicates that mangosteen contains a high percentage of carbohydrates, mostly in sugar form. It is relatively low in minerals and vitamins. The calcium and phosphorus content is high. The percentage of total soluble solids ranges from 13 to 15.2% in immature and 18.3 to 19% in mature fruits (Table 19.2).

A gum resin obtained from *G. hanburyi*, gamboges, consists mainly of resin (71.6–74.2%), gum (21.8–24%), moisture (4.8%), traces of starch and woody fibre. The resin has the nature of an acid (gambogic acid) and is the active principle of the gum. Its specific gravity is 1.221. It forms soluble salts with alkalis and insoluble precipitates with salts of heavy metals; this class of compound has been called gambogiates.

Table 19.2. Composition of 100 g ripe mangosteen aril.

Parameter	Range
Moisture (%)	79.2–84.9
Calories	60–81
Protein (%)	0.5–0.7
Fat (%)	0.1–0.8
Carbohydrates (%)	14.3–19.8
Total sugars (%)	17.5
Reducing sugars (%)	4.3
Fibre (%)	0.3–5.1
Ash (%)	0.20–0.23
Calcium (%)	0.01–18.00
Phosphorus (%)	0.02–17.00
Iron (%)	0.2–0.9
Vitamin A (IU)	0–14
Thiamine (%)	0.03–0.09
Riboflavin (%)	0–0.06
Niacin (%)	0–0.1
Ascorbic acid (%)	1–66
Acid	0.49
Citric acid (g/100 g)	0.63

Source: Kanchanapom and Kanchanapom (1998).

19.4. Chemistry

Volatiles

The aroma of mangosteen is contributed by 52 volatile compounds, 28 of which have been identified. In terms of quantity, the major compounds are (*Z*)-hex-3-en-1-ol (27%), octane (15%), hexyl acetate (8%) and α-copaene (7%). The main contributors to the mangosteen flavour are hexyl acetate, (*Z*)-hex-3-enyl-acetate (*cis*-hex-3-enyl-acetate) and (*Z*)-hex-3-en-1-ol (MacLeod and Pieris, 1982). The major groups of compounds found in mangosteen aril are alcohols, aldehydes and ketones, esters, hydrocarbons, terpenes, etc. The compounds present are given in Table 19.3.

The main essential oil components of the edible fruits of *G. huillensis* Welw. ex. Oliv. growing wild in the Gutu and Rusape areas of Zimbabwe are α-humulene (23.0%), valencene (18.2%), caryophyllene (12.6%), caryophyllene oxide (6.3%) and δ-selinene (5.0%) (Chagonda Lameck and Chalchat, 2005).

Table 19.3. Volatile flavour components of mangosteen aril.

Group	Compounds
Alcohols	Hexan-1-ol, (*Z*)-hex-3-en-1-ol
Aldehydes and ketones	Acetone, benzaldehyde, hexanal, (*E*)-hex-2-enal, 2-furaldehyde, furfuryl methyl ketone, 5-methyl-2-furaldehyde, nonanal, phenylacetaldehyde
Esters	Hexyl acetate, (*Z*)-hex-3-enyl-acetate
Hydrocarbons	Ethyl cyclohexane, heptane, octane, toluene, *o*-, *m*-, *p*-xylene
Terpenes	α-Copane, α-terpineol, guaiene, valencene, δ-cadinene, γ-cadinene
Miscellaneous	Dichloromethane, pyridine

Source: MacLeod and Pieris (1982).

Non-volatiles

In general, the *Garcinia* species contains xanthones and phenolic compounds. Xanthones and xanthone derivatives have been isolated from the various species of *Garcinia* (Rama Rao *et al.*, 1980; Minami *et al.*, 1994). However, the isolation of (–)-hydroxycitric acid [(–)-HCA] from a few species of *Garcinia* and its biological properties has attracted the attention of biochemists and health practitioners. (–)-HCA is found in the fruit rinds of *G. cambogia*, *G. indica* and *G. atroviridis* (Lewis and Neelakantan, 1965; Lewis, 1969), which are abundant in the Indian subcontinent and western Sri Lanka (CSIR, 1956; Watt, 1972).

Hydroxycitric acid ((–)-HCA)

The dried rind of the fruit of *G. cambogia*, popularly known as 'Malabar tamarind', is used extensively all over the west coast of South India for culinary purposes and in Colombo for the curing of fish. The organic acids responsible for the bacteriostatic effect of the pickling medium in the 'Colombo curing' of fish (Lewis *et al.*, 1964; Lewis and Neelakantan, 1965) were identified mistakenly as tartaric and citric acids (Sreenivasan and Venkataraman, 1959). Lewis and Neelakantan (1965) isolated the principal acid in the fruit rinds of *G. cambogia* and identified it as (–)-HCA on the basis of chemical and spectroscopic studies. This is known to have a body-trimming effect and hence has become a very valuable commodity for health practitioners. The fruit rinds of *G. cambogia* and *G. indica* contain 20–30% (–)-HCA. The structure of hydroxycitric acid and its derivatives is given in Fig. 19.1.

Regulation of fatty acid synthesis through (–)-HCA

The biological effect of (–)-HCA stems from the inhibition of extramitochondrial cleavage of citrate to oxaloacetate and acetyl-CoA catalysed by ATP:citrate lyase. This limits the availability of acetyl-CoA units required for fatty acid synthesis and lipogenesis (Sullivan, 1984). The inhibition of ATP:citrate lyase by (–)-HCA leads to less dietary carbohydrate utilization for the synthesis of fatty acids, resulting in more glycogen storage in the liver and muscles. Many *in vitro*

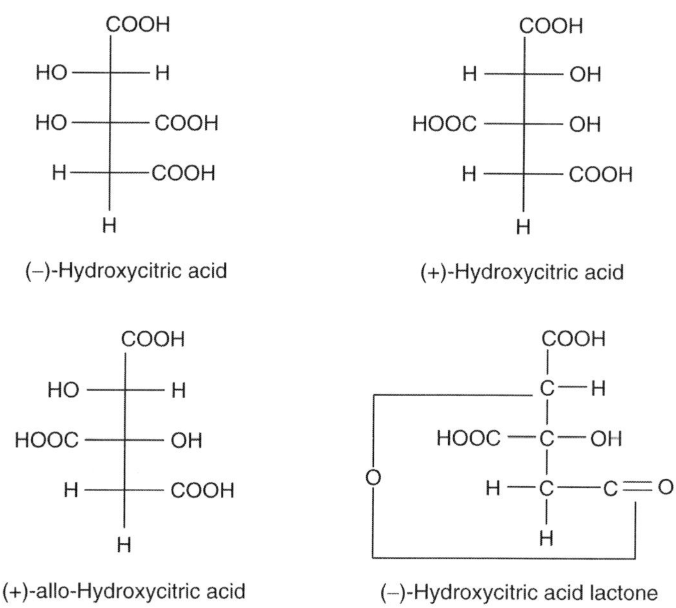

Fig. 19.1. Structure of hydroxy citric acid and its derivatives.

and *in vivo* studies demonstrated that (–)-HCA suppresses *de novo* fatty acid synthesis and lipogenesis. Increase in the rate of *in vivo* hepatic glycogen synthesis with the administration of (–)-HCA has been reported by Sullivan *et al.* (1974). Increased hepatic fatty acid oxidation leading to increased levels of acetyl-CoA and ATP is responsible for the food intake-suppressive effects of (–)-HCA.

Though (–)-HCA is safe to consume, it has impacts on the production of fatty acids and cholesterol, which may influence the production of sterols directly, thus restricting the production of steroid hormones. Hence, (–)-HCA should not be recommended during pregnancy. Excessive exposure of tissues to fatty acids is likely to be the chief cause of the various anomalies that lead to sustained hyperglycaemia in type-2 diabetes. Disinhibition of hepatic fatty acid oxidation and inhibition of fatty acid synthesis with (–)-HCA and camitine (McCarty, 1995) has wide scope as a new weight-loss strategy, but diabetic patients may suffer from enhanced excessive hepatic gluconeogenesis.

Xanthone

The *Garcinia* species produce a range of xanthones, biflavonoids and lactones which have been isolated from the fruit rind, bark and roots. Most of these compounds are xanthones and xanthone derivatives (Rama Rao *et al.*, 1980; Bennet and Lee, 1989; Minami *et al.*, 1994). Xanthones are a unique class of biologically active compounds possessing numerous bioactive capabilities, such as antioxidant properties, maintenance of intestinal health, strengthening the immune system, neutralizing free radicals, supporting cartilage and joint function and promoting a healthy seasonal respiratory system (http://www.xango.com/learn/xanthones.html). Various xanthones, xanthone derivatives, biflavonoids and lactones have been isolated from different vegetative parts of the *Garcinia* species, the details of which are summarized below. Structures of some of the compounds are given in Fig. 19.2.

G. BRACTEATA Leaves and bark: 5-*O*-methylxanthone VI, bracteaxan-thones I and II, nemorosonol, simple-xanthones, garcibracteatone, neoiso-bractanins A and B, xerophenone C (Thoison *et al.*, 2005).

G. MANGOSTANA Heartwood: mangoxanthone and a new benzophenone(3′,6-dihydroxy-2,4,4′ trimethoxybenzophenone) (Nilar *et al.*, 2005).

Root bark, stem bark and latex: α-mangostin, β-mangostin, γ-mangostin, garcinone E, methoxy-β-mangostin and a new geranylated biphenyl derivative 3-hydroxy-4geranyl-5 methoxybiphenyl (Dharmaratne *et al.*, 2005). Pericarp: mangostinone, α-mangostin, β-mangostin, γ-mangostin, garcinone E, 1,5-dihydroxy-2-(3methylbut-2-enyl)-3-methoxy-xanthone and 1,7-dihydroxy-2-(3-methylbut-2-enyl) 3-methoxyxanthone (Asai *et al.*, 1995).

Green fruit hulls: mangostenol, mangostenone A and B, trapezifolixanthone, tovophyllin B, α- and β-mangostins, garcinone B, mangostinone, mangostanol and the flavonoid epicatechin (Suksamrarn *et al.*, 2002). Fruit hull: α- and γ-mangostins, procyanidins A-2 and B-2, (–)-epicatechin (Yoshikawa *et al.*, 1994). Three new tetraoxygenated xanthones (garcinones A, B and C) (Sen *et al.*, 1982). Phenolics, P_1 (1,3,6,7-tetrahydroxy-2,8-(3-methyl-2-butenyl)), P_2 [1,3,6-trihydroxy-7-methoxy-2,8-(3-methyl-2-butenyl) xanthone] and P_3 (epicatechin) (Yu *et al.*, 2006).

Dry fruit hull: two new xanthones, a *bis*-pyrano xanthone, BR-xanthone-A and 1-methoxy-2,4,5-trihydroxyxanthone, BR-xanthone-B (Balasubramanian and Rajagopalan, 1988).

Leaves: 2-ethyl-3-methylmaleimide *N*-β-D-glucopyranoside (Krajewski *et al.*, 1996). A new triterpene, 3*β*-hydroxy-26-*nor*-9,19-cyclolanost-23-en-25-one (Parveen *et al.*, 1991).

G. GRIFFITHII Stem bark: guttiferone I, cambogin, 1,7-dihydroxy-xanthone, 1,3,6,7-tetrahydroxyxanthone, 1,3,5,6-tetrahydroxyxanthone (Nilar *et al.*, 2005). Griffipavixanthone (Xu *et al.*, 1998).

G. KOLA Root: 3″,4′,4′,5,5″,7,7″-heptahydroxy-3,8″ biflavanone (Han *et al.*, 2005).

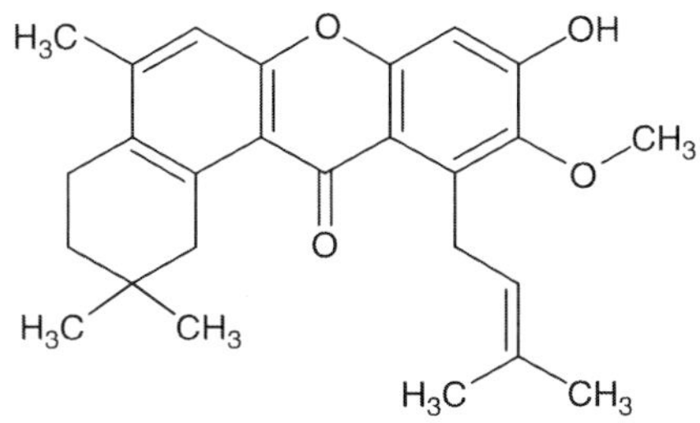

α-Mangostin

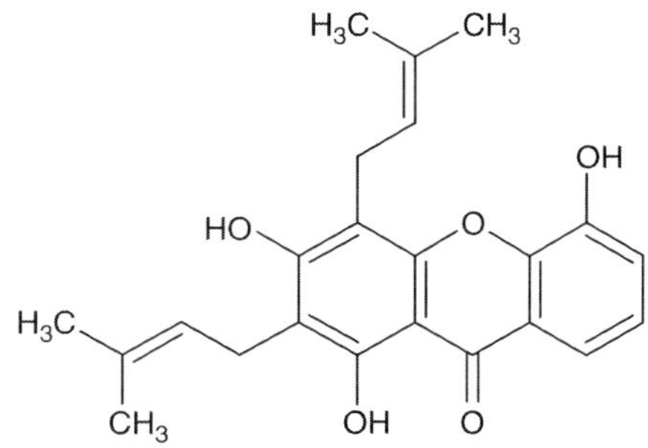

8-Desoxygaranin

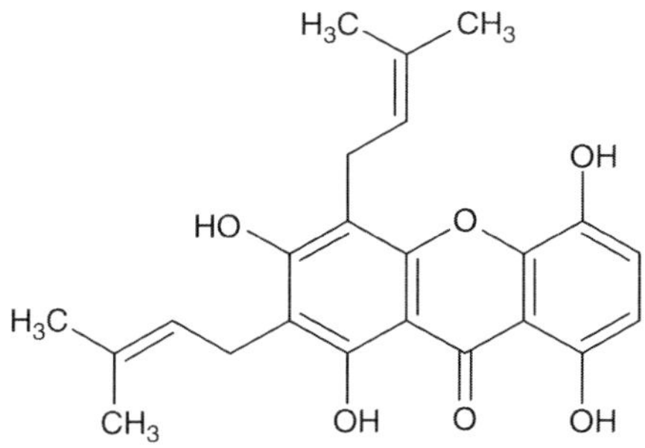

Gartanin

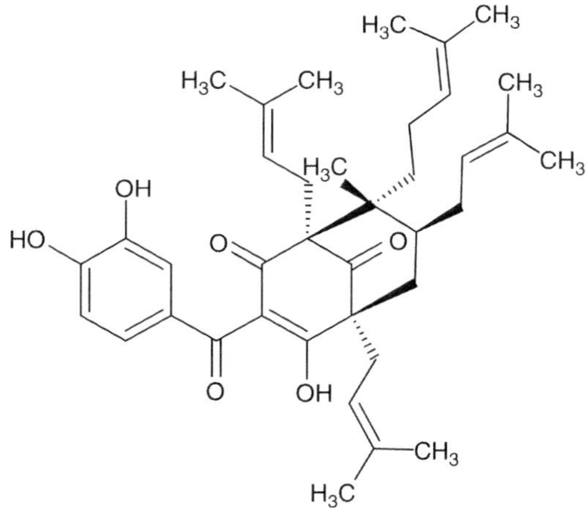

Guttiferone A

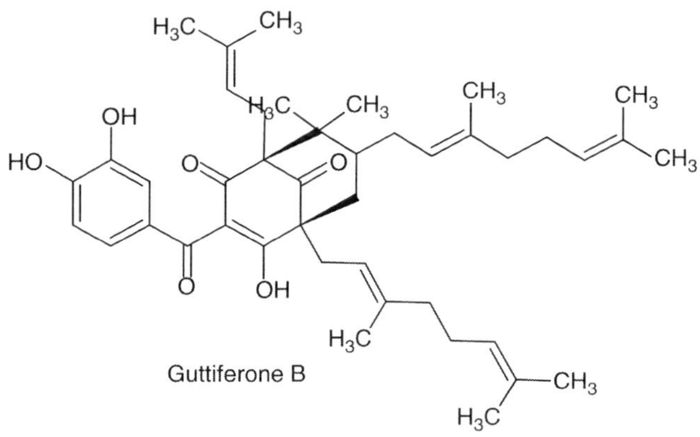

Guttiferone B

Fig. 19.2. Structures of some of the compounds isolated from *Garcinia* sp. *Continued*

Biflavnonoids (Iwu *et al.*,1990). 6-Asyl-1,2 benzopyran derivative, garcipyran (Niwa *et al.*, 1994).

Leaf: glycosides, flavonoids and tannins (Obuekwe and Onwukaeme, 2004).

G. COWA Latex: cowa xanthone A–E and six previously reported xanthones (Mahabusarakam *et al.*, 2005). Cowanin, cowanol, cowa-xanthone, 1,3,6-trihydroxy-7-methoxy-2,5-*bis*(3-methyl-2-butenyl)

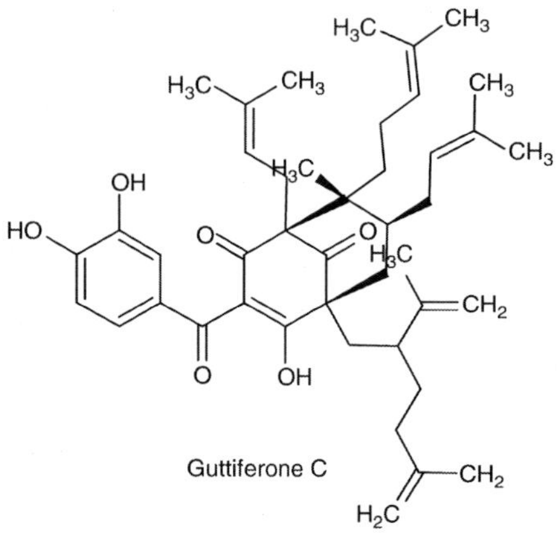

Guttiferone C

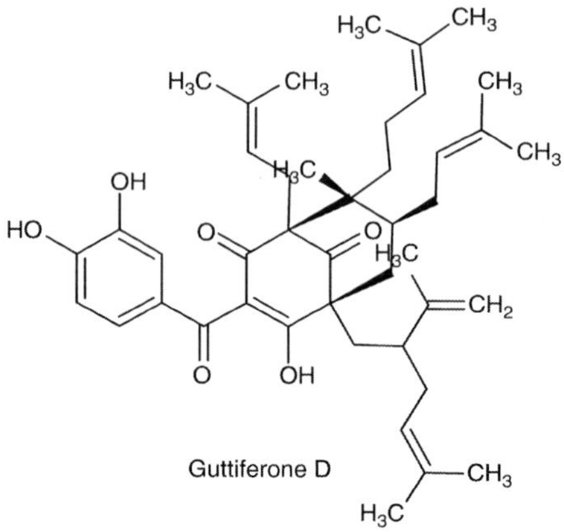

Guttiferone D

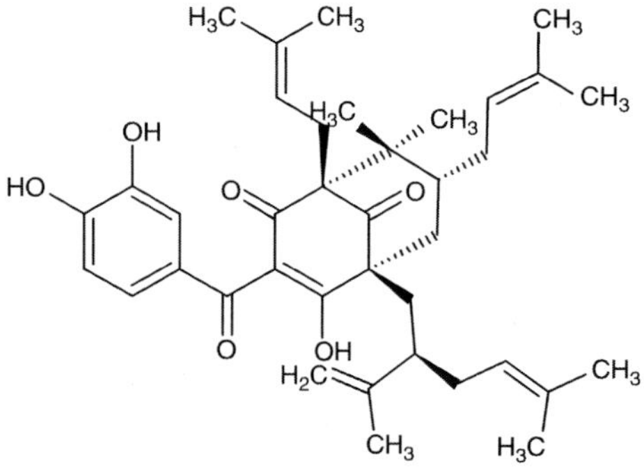

Guttiferone F

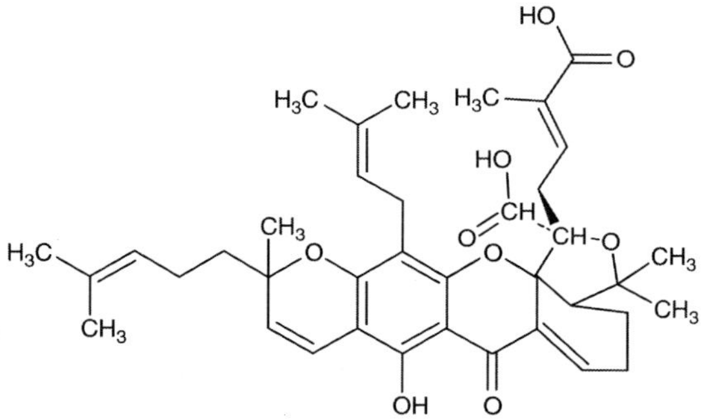

trans-Garcinolic acid

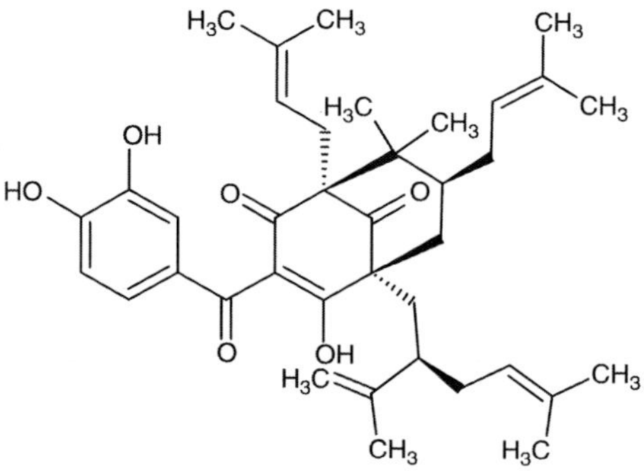

Xanthochymol

Fig. 19.2. *Continued*

xanthone and norcowanin (Pattalung *et al.*, 1994).

Stem: 1,3,6-trihydroxy-7-methoxy-8-(3,7-dimethyl-2,6-octadienylxanthone (Lee and Chan, 1977).

Stem bark: new xanthone, 7-*O*-methylgarcinone E (Likhitwitayawuid *et al.*, 1997).

Fruit: tetraoxygenated xanthones, cowaxanthones A–E, together with ten previ-

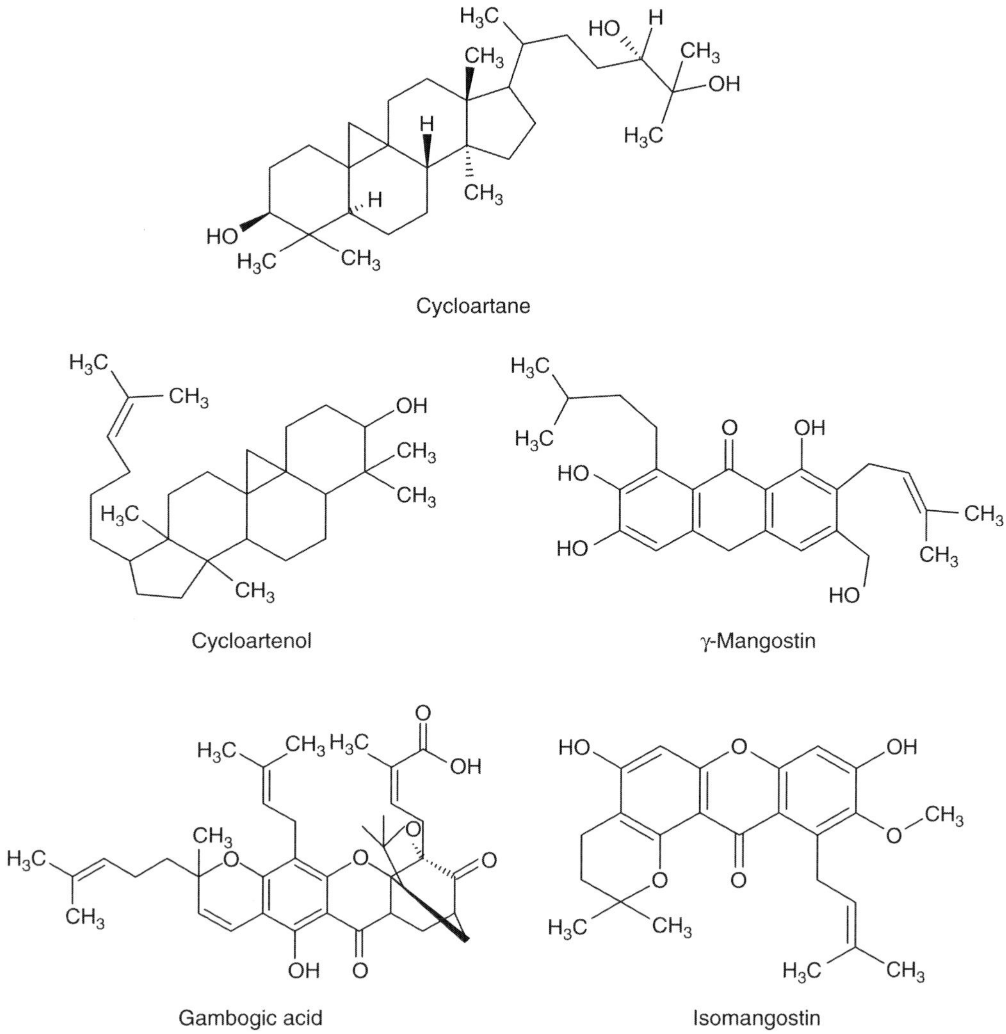

Fig. 19.2. *Continued*

ously reported tetraoxygenated xanthones (Panthong *et al.*, 2006).

G. ATROVIRIDIS Fruit: atrovirisidone, naringenin and 3,8 binaringenin (Permana et al., 2005), 2-(butoxy carbonyl methyl)-3-butoxy carbonyl-2-hydroxy-3-propanolide and 1′,1″-dibutyl methyl hydroxycitrate (Mackeen et al., 2002).

Roots: benzoquinone atrovirinone and the depsidone atrovirisidone (Permana *et al.*, 2001).

Stem bark: a new xanthone, atroviridin (Kosin *et al.*, 1998).

G. SUBELLIPTICA Root bark: subelliptinones E and F, 1,5-dihydroxy-3-methoxy xanthone (Iinuma *et al.*, 1995a). Subelliptenone H and I (with 1,1-dimethylallyl group) (Iinuma *et al.*, 1995b).

Wood: *Garcinia*xanthone D and 1,4,5-trihydroxyxanthone (Minami *et al.*, 1995). 2,5-Dihydroxy-1-methoxyl xanthone, 1-*o*-methylsympho-xanthone, *Garcinia*xanthone E, symphoxanthone and subelliptenone A (Minami *et al.*, 1996). A new benzophenone derivative, 2′,3′,6-trihydroxy-2,4-dimethoxy-benzo-phenone and a new xanthone, 1,6-*O*-

dimethylsymphoxanthone (Minami *et al.*, 1998). Four new phloroglucinol derivatives, named garsubellins B–E (Fukuyama *et al.*, 1998). Subellinone, a novel polyisoprenylated phloroglucinol (Fukuyama *et al.*, 1993).

Pericarp: garcinielliptone FA, benzoylphloroglucinol, garcinielliptone FB (Wu *et al.*, 2005).

Heartwood: two new prenylated xanthones, *Garcinia*xanthones A and B (Fukuyama *et al.*, 1991).

G. HANBURYI Latex: 11 cytotoxic xanthones, e.g. gambogin, morellin dimethyl acetal, isomoreollin B, moreollic acid, gambogenic acid, gambogenin, isogambogenin, desoxygambogenia, gambogenin dimethyl acetal, gambogellic acid, hanburin and gambogic acid, isomorellin, morellic acid, desoxymorellin (Asano *et al.*, 1996).

G. DULCIS Bark: a new xanthone dulciol A, 12b-hydroxydes-D-garcigerin and toxyloxanthone B (Iinuma *et al.*, 1996).

Roots: four novel xanthones with a 1,1-dimethyl allyl group (dulciols B–E) (Iinuma *et al.*, 1996).

Leaves: pyranoxanthone, dulxanthone E (5,9,10,12-tetramethoxy-2,2-dimethyl-2H-pyrano[5,6-b]xanthen-6-one) (Kosela *et al.*, 1999a).

Green fruits: dulcinoside, dulcisisoflavone, dulcisxanthone A and sphaerobioside acetate (Deachathai *et al.*, 2005).

Ripe fruits: dulcisflavan, dulcisxanthone B and isonormangostin (Deachathai *et al.*, 2005).

G. FUSCA Stem bark: fusca xanthones A–H (Ito et al., 2003b).

G. XANTHOCHYMUS Wood: two prenylated-xanthones 1,4,5,6-tetrahydroxy-7,8-di(3-methylbut-2-enyl) xanthone and 1,2,6-trihydroxy-5-methoxy-7-(3-methylbut-2-enyl) xanthone (Chanmahasathien et al., 2003).

G. CUNEIFOLIA Stem bark: a new xanthone cuneifolin (Ee *et al.*, 2003).

G. ASSUGU Two new benzo phenones corresponding to the 13-*O*-methyl ethers of isogarcinol and garcinol (Ito *et al.*, 2003a).

G. SPECIOSA Bark and stems: protostane triterpenes (Rukachaisirikul *et al.*, 2003c).

Bark: four 17,14-friedolanostanes and five lanostanes, as well as friedelin (Vieira *et al.*, 2004).

G. NIGROLINEATA Leaves: ten new 1,3,5-trioxygenated xanthones and one new quinone derivative, nigrolineaquinone A (Rukachaisirikul *et al.*, 2003a).

Bark: nine xanthones, nigrolineaxanthones A–I and nine known xanthones (Rukachaisirikul *et al.*, 2003b).

G. SCORTECHINII Fruits: four caged tetraprenylated xanthones (scortechinones Q-T, 1–4), four rearranged xanthones (scortechinones U-X, 5–8), two sesquiterpene derivatives, two triflavanoids and 11 caged polyprenylated compounds (Sukpondma *et al.*, 2005).

G. SMEATHMANNII Bark: smeathxanthone A and B (Komguem *et al.*, 2005).

G. HUMILIS Bark: guttiferone I (Herath *et al.*, 2005).

G. CAMBOGIA Fruit rind: hydroxy citric acid (Jayaprakasha and Sakariah, 1998).

Root: a new xanthone, garbogiol (Iinuma *et al.*, 1998).

Bark: known xanthone (rheediaxanthone A) and two known benzophenones (garcinol and isogarcinol) (Iinuma *et al.*, 1998).

G. VIEILLARDII Stem bark: 6-*O*-methyl-2-deprenylrheediaxanthoneBandvieillardixanthone (Hay *et al.*, 2004a). 1,6-Dihydroxyxanthone pancixanthone A, isocudraniaxanthone B, isocudraniaxanthone A, 2-deprenylrheediaxanthone B and 1,4,5-trihydroxyxanthone (Hay *et al.*, 2004b).

G. GAUDICHAUDII Leaf: cytotoxic gaudichaudiones A–D (penta-cyclic tetra-isoprenylated xanthonoids) (Cao *et al.*, 1998a). Fifteen novel cytotoxic compounds, gaudichaudiones A–H and gaudichaudiic acids A–E, including the known morellic acid and forbesione (Cao *et al.*, 1998b).

Bark: gaudispirolactone and 7-isoprenylmorellic acid (Wu *et al.*, 2001).

G. FORBESII Branch: a new chromenoxanthone, forbexanthone, as well as the known compounds pyranojacareubin and 1,3,7-trihydroxy-2,3-methylbut-2-enyl-xanthone (Harrison *et al.*, 1993).

G. VOLKENSII Heartwood: known biflavonoids GB-1a, GB-2a and morelloflavone and a new flavanone, volkensiflavone, whose constituent units are naringenin and apigenin (Herbin *et al.*, 1970).

G. ANDAMANICA Leaves: sorbifolin 6-galactoside and scutellarein 7-diglucoside (Alam *et al.*, 1986). A new flavone glycoside 4′-hydroxywogonin 7-neohesperidoside (Alam *et al.*, 1987).

G. NERVOSA; G. POLYANTHA; G. PYRIFERA Stem bark: isocowanin(8-geranyl-4-(3,3-dimethylallyl)-7-methoxy-1,3,6-trihydroxyxanthone), isocowanol (8-geranyl-4-(3-hydroxymethyl-3-methyl-allyl)-7-methoxy-1,3,6-trihydroxy-xanthone) and nervosaxanthone (4,8-di(3,3-dimethylallyl)-2-(1,1-dimethyl-allyl)-1,3,5,6-tetrahydroxyxanthone) (Ampofo and Waterman, 1986).

G. THWAITESII *β*-Amyrin and tirucallol, four biflavonoids and a new xanthone, 2,5-dihydroxy-1,6-dimethoxyxanthone (Gunatilaka *et al.*, 1983).

G.DENSIVENIA Stembark:1,3,5,6-tetraoxygenated xanthone pyranojacareubin (1,5-dihydroxy-6′,6′-dimethylpyrano (2′,3′:3,2)-6″,6″-dimethylpyrano (2″,3″:6,7)-xanthone) and the biflavonoids morelloflavone and *O*-methyl fukugetin (Waterman and Crichton, 1980b).

G. PEDUNCULATA Heartwood: 2,4,6,3′,5′-pentahydroxybenzophenone and 1,3,5,7-tetrahydroxyxanthone (Rama Rao *et al.*, 1974).

G. LIVINGSTONII Heartwood, bark and leaves: moarelloflavone (BGH-II) and a new biflavonyl, BGH-111, along with optically active amentoflavone and podocarpusflavone A (Pelter *et al.*, 1971).

Root bark: five prenylated xanthones (Diserens *et al.*, 1992a). Three xanthone dimers garcilivin A–C that are structurally related to 1,4,5-trihydroxy-3-(3-methylbut-2-enyl)-9*H*-xanthen-9-one (Diserens *et al.*, 1992b).

G. NERVOSA Leaves: I-5,II-5,I-7,II-7,I-3′,I-4′,II-4′-hepta-hydroxy-[I-3, II-8]-flavanonylflavone (Babu *et al.*, 1988). A new isoflavone, 5,7,4′-trihydroxy-2′,3′,6′-trimethoxyisoflavone, nervosin, along with two known isoflavones, irigenin (5,7,3′-trihydroxy-6,4′,5′-trimethoxyisoflavone) and 7-methyltectorigenin(5,4′-dihydroxy-6,7-di-methoxyisoflavone) (Ilyas *et al.*, 1994).

G. PSEUDOGUTTIFERA Heartwood: benzophenones, e.g. 6-hydroxy-2,4-dimethoxy-3,5-*bis*(3-methyl-2-butenyl)benzophenone (myrtiaphenone-A); 2,2-dimethyl-8-benzoyl-7-hydroxy-5-methoxy-6-(3-methyl-2-butenyl)benzopyran (myrtiaphenone-B); 2,6-dihydroxy-4-methoxy-3,5-*bis*(3-methyl-2-butenyl)benzophenone (vismiaphenone-C) and a new benzophenone, 2,2-dimethyl-8-benzoyl-3,7-dihydroxy-5-methoxy-6-(3-methyl-2-butenyl)-3,4-dihydrobenzopyran (pseudoguttiaphenone-A) and a triterpene, eupha-8,24-dien-3*β*-ol (Ali *et al.*, 2000).

G. POLYANTHA Stem bark: bangangxanthone A [1,5,8-trihydroxy-6′-methyl-6′-(4-methylpent-3-enyl)-pyrano[2′,3′:3,4]xanthone] and B [1,4,8-trihydroxy-2-prenylxanthone], along with two known xanthones, 1,5-dihydroxyxanthone, 2-hydroxy-1,7-dimethoxyxanthone and the pentacyclic triterpenoids, friedelin, oleanolic acid and lupeol (Lannang *et al.*, 2005).

G. VIRGATA Stem bark: two prenylated xanthones and two formyl-*δ*-tocotrienol derivatives (Merza *et al.*, 2004).

G. XANTHOCHYMUS Wood: two prenylated xanthones, 1,4,5,6-tetrahydroxy-7,8-di(3-methylbut-2-enyl)xanthone and 1,2,6-trihydroxy-5-methoxy-7-(3-methylbut-2-enyl)xanthone and a known xanthone,

12b-hydroxy-des-D-garcigerrin A (Chanmahasathien *et al.*, 2003).

G. MERGUENSIS Bark: xanthones, e.g. merguenone, 1,5-dihydroxy-6′-methyl-6′-(4-methyl-3-pentenyl)-pyrano(2′,3′:3,2)-xanthone, subelliptenone H, 8-deoxygartanin, rheediaxanthone A, morusignin G, 6-deoxyjacareubin, 1,3,5-trihydroxy-4,8-di(3-methylbut-2-enyl)-xanthone, rheediachromenoxanthone and 6-deoxyisojacareubin (Nguyen *et al.*, 2003).

G. VILERSIANA Bark: four triterpenoids (olean-12-ene-3β,11α-diol, lupeol, β-amyrin and oleanolic acid) and six xanthones (globuxanthone, subelliptenone H, subelliptenone B, 12b-hydroxy-des-D-garcigerrin A, 1-*O*-methylglobu-xanthone and symphoxanthone) (Nguyen and Harrison, 2000).

G. CONRAUANA Stem bark: 3-(3,3″-dimethylallyl)-conrauanalactone [4-hydroxy-3-(3″,3″-dimethylallyl)-6-pentadecylpyran-2-one] (Hussain and Waterman, 1982). Bark: conrauanalactone (4-hydroxy-6-pentadecyl-2-pyrone),5,7-dihydroxychromone and eriodictyol (Waterman and Crichton, 1980a).

G. LATERIFLORA Lateriflorone, a cytotoxic natural product with spiroxalactone skeleton (Kosela *et al.*, 1999b).

G. QUADRIFARIA Stem bark: xanthone 1,3,5-trihydroxy-4,8-di(3,3-dimethylallyl)xanthone and the biflavonoids, *O*-methylfukugetin and morelloflavone (Waterman and Hussain, 1982).

G. STAUDTII Stem bark: rheedia xanthone-A and xanthochymol (Waterman and Hussain, 1982).

G. MANNII Stem bark: a new biflavanone, I-3′-II-3,3′-I-4′-II-4′-I-5-II-5-I-7-II-7-nonahydroxy-I-3-II-8-biflavanone (Crichton and Waterman, 1979).

G. OPACA Leaf: macluraxanthone, 1,3,5-trihydroxy-6′,6′-dimethylpyrano-(2′,3′:6,7)-4-(1,1-dimethylprop-2-enyl)xanthone and two new prenylated xanthones, 1,3,5-trihydroxy-6′,6′-dimethylpyrano-(2′,3′:6,7)-2-(3-methylbut-2-enyl)-4-(1,1-dimethylprop-2-enyl)xanthone and 4″,5″-dihydro-1,5-dihydro-1,5-dihy-droxy-6′,6′-dimethylpyrano(2′,3′:6,7)-2-(3-methylbut-2-enyl)-4″,4″,5″-trimethylfurano(2″,3″:3,4) xanthone (Goh *et al.*, 1992).

G. QUAESITA Bark: hermonionic acid and a new phenol, quaesitol (Gunatilaka *et al.*, 1984).

A new method based on reversed-phase high-performance liquid chromatography with photodiode array detector (LC–PDA) enabled simultaneous analysis of six naturally occurring xanthones (3-isomangostin, 8-desoxygartanin, gartanin, α-mangostin, 9-hydroxycalabaxanthone and β-mangostin). Separation was performed on a Phenomenex Luna C18 (2) (150 mm × 3 mm, 5 μm) column. The xanthones were identified by retention time, ultraviolet (UV) spectra and quantified by LC–PDA at 320 nm. The precision of the method was confirmed by the relative standard deviation (RSD), which was ≤ 4.6%. The recovery was in the range of 96.58–113.45%. A good linear relationship was established in over two orders of magnitude range. The limits of detection (LOD) for six xanthone compounds were ≤ 0.248 μg/ml. The identity of the peaks was further confirmed by high-performance liquid chromatography with time-of-flight mass spectrometry (LC–TOF MS) system coupled with electrospray ionization (ESI) interface. The developed methods were applied to the determination of six xanthones in *G. mangostana* products. The methods also are effective for the analysis of real samples (Ji *et al.*, 2006).

A sensitive liquid chromatography/electrospray ionization tandem mass spectrometrical (LC/ESI–MS/MS) method was developed for the identification and quantification of two polyisoprenylated benzophenones, xanthochymol and isoxanthochymol, in the extracts of the fruit rinds, stem bark, seed pericarps and leaves of *G. indica* and in the fruit rinds of *G. cambogia* (Chattopadhyay and Kumar, 2006).

A recycling counter-current chromatographic system was first set up with a high-speed counter-current chromatography instrument coupled with a column-switching valve. This system was first applied successfully to the preparative separation of epimers, gambogic acid and epigambogic acid from *G. hanburyi* using *n*-hexane:methanol: water (5:4:1, v/v/v) as the two-phase solvent system (Han *et al.*, 2006).

19.5. Medicinal and Pharmacological Uses

Traditionally, *Garcinia* species have been used as anti-inflammatory, anti-immunosuppressive, antiprotozoal and antimicrobial agents. The various medicinal properties attributed to *Garcinia* are:

- Antioxidant
- Anti-inflammatory agent
- Analgesic
- Astringent
- Hepatotic tonic
- Cancer suppressant
- Anti-HIV agent
- Antibacterial agent
- Antiobesity factor
- Veterinary medicine
- Other medicinal uses.

Antioxidant

Kolaviron isolated from *G. kola* seed extract appears to act as an *in vivo* natural antioxidant and an effective hepatoprotective agent and is as effective as butylated hydroxyanisole (Farombi *et al.*, 2000). The *G. indica* extract which contains garcinol has both antifungal and antioxidant properties and has potential for use as a biopreservative in food applications and nutraceuticals (Selvi *et al.*, 2003). The growth of *Aspergillus flavus* was inhibited completely by the hexane and chloroform extracts from *G. cowa* and the chloroform extract from *G. pedunculata*. The antiaflatoxigenic activities of the extracts from *G. cowa* and *G. pedunculata* may be due to their effective antioxidative properties, which could suppress the biosynthesis of aflatoxin (Joseph *et al.*, 2005). Superoxide anions in xanthine and xanthine oxidase systems were scavenged by the xanthones isolated from *G. subelliptica* (Minami *et al.*, 1995). Xanthone present in the hulls of *G. mangostana* exhibits a potent radical scavenging activity (Yoshikawa *et al.*, 1994).

Anti-inflammatory agent

G. mangostana fruit hulls are used as an anti-inflammatory agent (Chairungsrilerd *et al.*, 1996). In Thai medicine, the fruit hull is used to heal skin infections and wounds (Parveen *et al.*, 1991; Yaacob and Tindall, 1995; Ohizumi, 1999). The rind is also used to treat respiratory disorders (Wahyuono *et al.*, 1999).

Analgesic

A decoction from the leaves and roots of *G. atroviridis* is used in the treatment of earache (CSIR, 1956). A decoction of the fruit rind of *G. cambogia* is given for rheumatism and bowel complaints (Jena *et al.*, 2002).

Astringent

G. mangostana fruit hulls are used as an astringent and to treat diarrhoea (Chairungsrilerd *et al.*, 1996). The bark and leaves are also used as an astringent for the cure of aphtha or thrush (Coronel, 1983).

Hepatotic tonic

Syrup from the juice of *G. indica* fruit is given in bilious infections (Jena *et al.*, 2002). *G. kola* extracts are used extensively in traditional

African medicine for the treatment of coughs, inflammation of the respiratory tract and liver cirrhosis (Iwu *et al.*, 1990).

Cancer suppressant

Among the nine Thai medicinal plants tested for antiproliferative activity against SKBR3 human breast adenocarcinoma cell line using MTT assay, ethanolic extracts of *G. mangostana* had strong antiproliferation, potent antioxidation and induction of apoptosis (Moongkarndi *et al.*, 2004b). Thus, this indicates that this substance can show different activities and has potential for cancer chemoprevention which is dose-dependent as well as exposure time-dependent (Moongkarndi *et al.*, 2004a). Investigations on the induction of apoptosis in human leukaemia HL-60 cells, the inhibition of NO generation and the inhibition of LPS-induced iNOS gene expression by Western blot analysis suggest the possible chemopreventive ability of garcinol (purified from *G. indica* fruit rind) and its oxidation products (Sang *et al.*, 2002). Dietary administration of garcinol inhibited 4-NQO-induced tongue carcinogenesis through suppression of increased cell proliferation activity in the target tissues and/or COX-2 expression in tongue lesions (Yoshida *et al.*, 2005).

Human telomerase reverse transcriptase gene expression was inhibited by gambogic acid in human hepatoma SMMC-7721 cells, indicating the gambogic acid's potent anticancer activity (Guo *et al.*, 2006). The high potency of gambogic acid, a natural product isolated from the resin of the *G. hurburyi* tree as an inducer of apoptosis, its novel mechanism of action, easy isolation and abundant supply, as well as the fact that it is amenable to chemical modification, makes gambogic acid an attractive molecule for the development of anticancer agents (Zhang *et al.*, 2004). A new benzoylphosphoglucinol, garcinielliptone FB, isolated from the pericarp of *G. subelliptica,* exhibited cytotoxic activity against several human cancer cells (Wu *et al.*, 2005). Atrovirisidone B isolated from *G. atroviridis* showed cytotoxic activity against human breast, prostate and large cells. Cancer chemopreventive activity was exhibited by *G. assugu* plants (Ito *et al.*, 2003a) and also by *G. fusca* plants (Ito *et al.*, 2003b).

Anti-HIV agent

Ethanol extract of the fruit peel of *G. mangostana* showed potent inhibiting activity against HIV-1 protease; the compound responsible was isolated and established as mangostin (Chen *et al.*, 1996). Protostane triterpenes, e.g. garciosaterpenes A, B and C, obtained from methanol extracts of bark and stems of *G. speciosa*, showed anti-HIV-1 activity (Rukachaisirikul *et al.*, 2003c).

Antibacterial agent

The polyoxygenated xanthones present in the rind of *G. mangostana* act as antibacterial agents. Antibacterial biphenyl derivatives have been isolated from *G. bancana* (Rukachaisirikul *et al.*, 2005). Nigrolineaxanthone N isolated from the leaves of *G. nigrolineata* showed significant antibacterial activity against methicillin-resistant *Staphylococcus aureus* (Rukachaisirikul *et al.*, 2003a). α-Mangostin alone or in combination with gentamicin against vancomycin-resistant enterococci (VRE) and in combination with VCM (vancomycin hydrochloride) against methicillin-resistant *S. aureus* (MRSA) might be useful in controlling VRE and MRSA infections (Sakagami *et al.*, 2005). GBI, a hydroxybiflavanol present in the seed extract of *G. kola*, exhibited activity against Gram-positive and Gram-negative bacteria, *Candida albicans* and *A. flavus* (Madubunyi, 1995). Cowanol and cowaxanthone from *G. cowa* also have moderate antibacterial activity against *S. aureus* (Pattalung *et al.*, 1994). Compounds extracted from the roots of *G. atroviridis* showed mild inhibitory activity towards *Bacillus cereus* and *S. aureus* (Permana *et al.*, 2001).

Antiobesity factor

(–)-Hydroxycitric acid (HCA) is found in the fruit rinds of certain species of *Garcinia*, including *G. cambogia*, *G. indica* and *G. atroviridis*. These are in great demand by health practitioners, as HCA is known to induce weight loss (Jena *et al.*, 2002). Preliminary research based on laboratory and animal experiments suggests that (–)-HCA may be a useful weight-loss aid (Lowenstein, 1971; Triscari and Sullivan, 1977). Animal research also indicated that (–)-HCA suppressed appetite and food intake to induce weight loss (Greenwood *et al.*, 1981; Rao and Sakariah, 1988). Heymsfield *et al.* (1998) noted that, although (–)-HCA appeared to be a promising experimental weight control agent, studies in humans were limited (Thom, 1996; Rothacker and Waitman, 1997) and results have been contradictory. Supporting evidence of human (–)-HCA efficacy for weight control is based largely on studies with small sample sizes. Their study failed to detect either the weight-loss or fat-mobilizing effects of (–)-HCA. (–)-Hydroxycitric acid (HCA-SX) and, to a greater degree, the combination of HCA-SX, niacin-bound chromium and *Gymnema sylvestre*, can serve as safe weight management supplements (Preuss *et al.*, 2004). HCA from *G. cambogia* resulted in a reduction in body weight but did not cause any changes in major organs or in the haematology, clinical chemistry and histopathology in rats (Shara *et al.*, 2004). *G. cambogia* extracts containing a high concentration of hydroxycitric acid were effective in suppressing fat accumulation in developing male Zucker obese rats, but were highly toxic to the testis (Saito *et al.*, 2005). Burdock *et al.* (2005) expressed that the toxicity, as reported by Saito *et al.* (2005), was misleading and was dependent on other factors such as dose, frequency of administration, etc.

Veterinary medicine

A decoction of the fruit rind of *G. cambogia* is employed in veterinary medicine as a rinse for diseases of the mouth in cattle (Jena *et al.*, 2002). Gamboge resin extracted from *G. Morella* is used as a strong purgative in veterinary medicine and gamboges from *G. hanburyi* is also used similarly in Indochina (Howes, 1949; Dastur, 1964).

Other medicinal uses

The fruit of *G. indica* is anthelmintic and useful in piles, dysentery, tumour, pain and heart complaints (Jena *et al.*, 2002). *G. dulcis* is used in traditional medicine to treat lymphatitis, parotitis and struma (Iinuma *et al.*, 1996). *G. kola* also is used in traditional medicine to treat a variety of ailments (Madubunyi, 1995). In Nigerian medicine, *G. kola* is used to treat hepatitis, laryngitis and gastroenteritis (Braide, 1993). The latex of *G. cowa* is used in Thai folk medicine as an antifever agent (Pattalung *et al.*, 1994). The bark of *G. lucida* is used by traditional healers in Cameroon to treat gastric infections and as an antidote against poison (Nyemba *et al.*, 1990). The seeds of *G. kola* enjoy a folk reputation in Africa as a poison antidote (Iwu *et al.*, 1987). *G. mangostana* rind is used as a cure for dysentery and chronic intestinal catarrh (Coronel, 1983) and is used as a lotion and medicine for menstruation (Burkill, 1966). The stem bark of *G. epunctata* has medicinal value in Cameroon (Mbafor *et al.*, 1989).

19.6. Conclusion

The major flavouring compound in *G. cambogia*, *G. indica* and *G. atroviridis* is (–)-hydroxycitric acid. Though this is emerging as a handy tool to treat obesity, more evidence needs to be gathered to prove its potential as an antiobesity factor satisfactorily. The various naturally occurring xanthones in different species of *Garcinia* also have medicinal use as radical scavenging agents and also are employed to treat infections and some respiratory disorders. Anti-HIV, anticancer and antibacterial activities have been reported from some species of *Garcinia*. Strong systematic research including clinical trials is needed to prove the reported claims, though, traditionally, *Garcinia* species have been used to treat all kinds of ailments.

References

Alam, M.S., Qasim, M.A., Kamil, M. and Ilyas, M. (1986) Sorbifolin 6-galactoside from *Garcinia andamanica*. *Phytochemistry* 25(12), 2900–2901.

Alam, M.S., Kamil, M. and Ilyas, M. (1987) 4′-Hydroxywogonin 7-neohesperidoside from *Garcinia andamanica*. *Phytochemistry* 26(6), 1843–1844.

Ali, S., Goundar, R., Sotheeswaran, S., Beaulieu, C. and Spino, C. (2000) Benzophenones of *Garcinia pseudoguttifera* (Clusiaceae). *Phytochemistry* 53(2), 281–284.

Ampofo, S.A. and Waterman, P.G. (1986) Xanthones from three *Garcinia* species. *Phytochemistry* 25(10), 2351–2355.

Asai, F, Tosa, H., Tanaka, T. and Iinuma, M. (1995) A xanthone from pericarps of *Garcinia mangostana*. *Phytochemistry* 39(4), 943–944.

Asano, J., Chiba, K., Tada, M. and Yoshii, T. (1996) Cytotoxic xanthones from *Garcinia hanburyi*. *Phytochemistry* 41(3), 815–820.

Babu, V., Ali, S.M., Sultana, S. and Ilyas, M. (1988) A biflavonoid from *Garcinia nervosa*. *Phytochemistry* 27(10), 3332–3335.

Balasubramanian, K. and Rajagopalan, K. (1988) Novel xanthones from *Garcinia mangostana*, structures of BR-xanthone-A and BR-xanthone-B. *Phytochemistry* 27(5), 1552–1554.

BAPPEDA (2001) Proyek Perkebunan Manggis, Kabupaten Belitung.

Bennet, G. and Lee, H.H. (1989) Xanthones from Guttiferae. *Phytochemistry* 28, 967–998.

Bin Osman, M. and Rahman Milan, A. (2006) *Mangosteen – Garcinia mangostana*. Southampton Centre for Underutilised Crops, University of Southampton, Southampton, UK, p. 170.

Braide, V.B. (1993) Anti-inflamatory effect of kola iron, a biflavonoid extract of *Garcinia kola*. *Fitoterapia* 64(5), 433–436.

Burdock, G., Bagchi, M. and Bagchi, D. (2005) *Garcinia cambogia* toxicity is misleading. *Food and Chemical Toxicology* 43(11), 1683–1684.

Burkill, I.H. (1966) *A Dictionary of the Economic Products of the Malay Peninsula*. Ministry of Agriculture and Cooperative, Kuala Lumpur.

Cao, S.G., Sng, V.H.L., Wu, X.H., Sim, K.Y., Tan, B.H.K., Pereira, J.T., Wong, W.H., Hew, N.F. and Goh, S.H. (1998a) Cytotoxic caged tetraprenylated xanthonoids from *Garcinia gaudichaudii* (Guttiferae). *Tetrahedron Letters* 39(20), 3353–3356.

Cao, S.G., Sng, V.H.L., Wu, X.H., Sim, K.Y., Tan, B.H.K., Pereira, J.T. and Goh, S.H. (1998b) Novel cytotoxic polyprenylated xanthonoids from *Garcinia gaudichaudii* (Guttiferae). *Tetrahedron* 54(36), 10915–10924.

Chagonda Lameck, S. and Chalchat, J.C. (2005) The essential oil of the fruit of *Garcinia huillensis* Welw. ex. Oliv. from Zimbabwe. *Flavour and Fragrance Journal* 20(3), 313–315.

Chairungsrilerd, N., Takeuchi, K., Ohizumi, Y., Nozoe, S. and Ohta, T. (1996) Mangostanol, a prenyl xanthone from *Garcinia mangostana*. *Phytochemistry* 43(5), 1099–1102.

Chanmahasathien, W., Li, Y., Satake, M., Oshima, Y., Ruangrungsi, N. and Ohizumi, Y. (2003) Prenylated xanthones with NGF-potentiating activity from *Garcinia xanthochymus*. *Phytochemistry* 64(5), 981–986.

Chattopadhyay, S.K. and Kumar, S. (2006) Identification and quantification of two biologically active polyisoprenylated benzophenones, xanthochymol and isoxanthochymol, in *Garcinia* species using liquid chromatography–tandem mass spectrometry. *Journal of Chromatography B* 844(1), 67–83.

Chen, S.X., Wan, M. and Loh, B.N. (1996) Active constituents against HIV 1 protease from *Garcinia mangostana*. *Planta Medica* 62(4), 381–382.

Coronel, R.E. (1983) Mangosteen. In: *Promising Fruits of the Philippines*. College of Agriculture, UPLB, Los Banos, Philippines, pp. 307–322.

Crichton, E.G. and Waterman, P.G. (1979) Manniflavanone, a new 3,8-linked flavanone dimer from the stem bark of *Garcinia mannii*. *Phytochemistry* 18(9), 1553–1557.

CSIR (1956) *The Wealth of India (Raw Materials)*, Volume IV. CSIR, New Delhi, pp. 99–108.

DA-AMAS (2004) *Mangosteen Industry Situation Report*. Agribusiness and Marketing Assistance Service, Diliman, Quezon City, Philippines.

Dastur, J.F. (1964) *Useful Plants of India and Pakistan*. Reprint 1985, Taraporevala, Bombay, India.

Deachathai, S., Mahabusarakam, W., Phongpaichit, S. and Taylor, W.C. (2005) Phenolic compounds from the fruit of *Garcinia dulcis*. *Phytochemistry* 66(19), 2368–2375.

Dharmaratne, H.R.W., Piyasena, K.G.N.P. and Tennakoon, S.B. (2005) A geranylated biphenyl derivative from *Garcinia mangostana*. *Natural Product Research* 19(3), 239–243.

Diserens, I.S., Rogers, C., Sordat, B. and Hostettmannt, K. (1992a) Prenylated xanthones from *Garcinia livingstonii*. *Phytochemistry* 31(1), 313–316.

Diserens, I.S., Hamburger, M., Rogers, C. and Hostettmannt, K. (1992b) Dimeric xanthones from *Garcinia livingstonii*. *Phytochemistry* 31(10), 3589–3593.

Ee, G.C.L., Phong, K.H., Mong, X.H., Shaari, K. and Sukari, M.A. (2003) Cuneifolin, a new xanthone from *Garicinia cuneifolia*. *Natural Product Sciences* 9(3), 174–176.

Farombi, E.O., Tahnteng, J.G., Agboola, A.O., Nwankwo, J.O. and Emerole, G.O. (2000) Chemoprevention of 2-acetylaminofluorene-induced hepatotoxicity and lipid peroxidation in rats by kolaviron-A *Garcinia kola* seed extract. *Food and Chemical Toxicology* 38(6), 535–541.

Fukuyama, Y., Kamiyama, A., Mima, Y. and Kodama, M. (1991) Prenylated xanthones from *Garcinia subelliptica*. *Phytochemistry* 30(10), 3433–3436.

Fukuyama, Y., Kaneshi, A., Tani, N. and Kodama, M. (1993) Subellinone, apolyisoprenylated phloroglucinol derivative from *Garcinia subelliptica*. *Phytochemistry* 33(2), 483–485.

Fukuyama, Y., Minami, H. and Kuwayama, A. (1998) Garsubellins, polyisoprenylated phloroglucinol derivatives from *Garcinia subelliptica*. *Phytochemistry* 49(3), 853–857.

Goh, S.H., Jantan, I., Gray, A.I. and Waterman, P.G. (1992) Prenylated xanthones from *Garcinia opaca*. *Phytochemistry* 31(4), 1383–1386.

Greenwood, M.R., Cleary, M.P., Gruen, R., Blasé, D., Stern, J.S., Triscari, J. and Sullivan, A.C. (1981) Effect of (–)-hyroxycitrate on development of obesity in the Zucker obese rat. *American Journal of Physiology* 240, E72–78.

Gunatilaka, A.A.L., Sriyani, H.T.B., Sotheeswaran, S. and Waight, E.S. (1983) 2,5-Dihydroxy-1,6-dimethoxyxanthone and biflavonoids of *Garcinia thwaitesii*. *Phytochemistry* 22(1), 233–235.

Gunatilaka, A.A.L., Sriyani, H.T.B. and Sotheeswaran, S. (1984) Quaesitol, a phenol from *Garcinia quaesita*. *Phytochemistry* 23(11), 2679–2681.

Guo, Q.L., Lin, S.S., You, Q.D., Gu, H.Y., Yu, J., Zhao, L., Qi, Q., Liang, F., Tan, Z. and Wang, X. (2006) Inhibition of human telomerase reverse transcriptase gene expression by gambogic acid in human hepatoma SMMC-7721 cells. *Life Sciences* 78(11), 1238–1245.

Han, Q.B., Lee, S.F., Qiao, C.F., He, Z.D., Song, J.Z., Sun, H.D. and Xu, H.X. (2005) Complete NMR assaignments of the antibacterial biflavonoid GB1 from *Garicinia kola*. *Chemical and Pharmaceutical Bulletin* 53(8), 1034–1036.

Han, Q.B., Song, J.Z., Qiao, C.F., Wong, L. and Xu, H.X. (2006) Preparative separation of gambogic acid and its C-2 epimer using recycling high-speed counter-current chromatography. *Journal of Chromatography A* 1127(1–2), 298–301.

Harrison, L.J., Leong, L.S., Sia, G.L., Sim, K.Y. and Tan, H.T.W. (1993) Xanthones from *Garcinia forbesii*. *Phytochemistry* 33(3), 727–728.

Hay, A.E., Aumond, M.C., Mallet, S., Dumontet, V., Litaudon, M., Rondeau, D. and Richomme, P. (2004a) Antioxidant xanthones from *Garcinia viellardii*. *Journal of Natural Products* 67(4), 707–709.

Hay, A.E., Hélesbeux, J.J., Duval, O., Labaïed, M., Grellier, P. and Richomme, P. (2004b) Antimalarial xanthones from *Calophyllum caledonicum* and *Garcinia viellardii*. *Life Sciences* 75(25), 3077–3085.

Herath, K., Jayasuriya, H., Ondeyka, J.G., Guan, Z.Q., Borris, E.P., Stijfhoorn, E., Sevenson, D., Wang, J.H., Sharma, N., MacNaul, K., Menke, J.G., Ali, A., Schulman, M.J. and Singh, S.B. (2005) Guttiferone I, a new prenylated benzophenone from *Garcinia humilis* as a liver X receptor ligand. *Journal of Natural Products* 68(40), 617–619.

Herbin, G.A., Jackson, B., Locksley, H.D., Scheinmann, F. and Wolstenholme, W.A. (1970) The biflavonoids of *Garcinia volkensii* (Guttiferae). *Phytochemistry* 9(1), 221–226.

Heymsfield, S.B., Allison, D.B., Vasselli, J.R., Pietobelli, D.G. and Nunthe, C. (1998) *Garcinia cambogia* (hydroxycitric acid) as a potential antiobesity agent. A randomized control trial. *Journal of the American Medical Association* 280, 1596–1600.

Howes, F.N. (1949) Vegetable gums and resins. *Chronica Botanica*. Waltham, Massachusetts.

http://www.xango.com/learn/xanthones.html.

Hussain, R.A. and Waterman, P.G. (1982) Lactones, flavonoids and benzophenones from *Garcinia conrauana* and *Garcinia mannii*. *Phytochemistry* 21(6), 1393–1396.

Iinuma, M., Tosa, H., Tanaka, T., Asai, F. and Shimano, R. (1995a) Three xanthones from root bark of *Garcinia subelliptica*. *Phytochemistry* 38(1), 247–249.

Iinuma, M., Tosa, H., Tanaka, T., Asai, F. and Shimano, R. (1995b) Two xanthones with a 1,1-dimethylallyl group in root bark of *Garcinia subelliptica*. *Phytochemistry* 39(4), 945–947.

Iinuma, M., Ito, T., Tosa, H., Tanaka, T. and Riswan, S. (1996) Five new xanthones from *Garcinia dulcis*. *Journal of Natural Products* 59(5), 472–475.

Iinuma, M., Ito, T., Miyake, R., Tosa, H., Tanaka, T. and Chelladurai, V. (1998) A xanthone from *Garcinia cambogia*. *Phytochemistry* 47(6), 1169–1170.

Ilyas, M., Kamil, M., Parveen, M. and Khan, M.S. (1994) Isoflavones from *Garcinia nervosa*. *Phytochemistry* 36(3), 807–809.

Ito, C., Itoigawa, M., Miyamoto, Y., Onoda, S., Rao, K.S., Mukainaka, T., Tokuda, H., Nishino, H. and Furukawa, H. (2003a) Polyprenylated benzophenones from *Garcinia assugu* and their potential cancer chemopreventive activities. *Journal of Natural Products* 66(2), 206–209.

Ito, C., Itoigawa, M., Takakura, T., Ruangrunsi, N., Enjo, F., Tokuda, H., Nishino, H. and Furukawa, H. (2003b) Chemical constituents of *Garcinia fusca*: structure elucidation of eight new xanthones and their cancer chemopreventive activity. *Journal of Natural Products* 66(2), 200–205.

Iwu, M., Igboko, O.A., Onwuchekwa, U.A. and Okunji, C.O. (1987) Evaluation of the antihepatotoxic activity of the biflavonoids of *Garcinia kola* seeds. *Journal of Ethnopharmacology* 21(2), 127–138.

Iwu, M.M., Igboko, O.A. and Tempesta, M.S. (1990) Biflavonoid constituents of *Garcinia kola* roots. *Fitoterapia* 61(2), 178–181.

Jayaprakasha, G.K. and Sakariah, K.K. (1998) Determination of organic acids in *Garcinia cambogia* (Desr.) by high performance liquid chromatography. *Journal of Chromatography A* 806(2), 337–339.

Jena, B.S., Jayaprakasha, G.K., Singh, R.P. and Sakariah, K.K. (2002) Chemistry and biochemistry of (–)-hydroxycitric acid from *Garcinia*. *Journal of Agricultural and Food Chemistry* 50, 10–22.

Ji, X., Avula, B. and Khan, I.A. (2006) Quantitative and qualitative determination of six xanthones in *Garcinia mangostana* L. by LC–PDA and LC–ESI-MS. *Journal of Pharmaceutical and Biomedical Analysis* 43(4), 1270–1276.

Joseph, G.S., Jayaprakasha, G.K., Selvi, A.T., Jena, B.S. and Sakariah, K.K. (2005) Antiaflatoxigenic and antioxidant activities of *Garcinia* extracts. *International Journal of Food Microbiology* 101(2), 153–160.

Kanchanapom, K. and Kanchanapom, M. (1998). Mangosteen. In: Shaw, P.E., Jr., Chan, H.T. and Nagi, S. (eds) *Tropical and Subtropical Fruits*. AgScience Inc., Auburndale, Florida, pp. 191–216.

Komguem, J., Meli, A.L., Manfouo, R.N., Lontsi, D., Ngounou, F.N., Kuete, V., Kamdem, H.W., Tane, P., Ngadjui, B.T. and Sondengam, B.L. (2005) Xanthones from *Garcinia smeathmannii* (Oliver) and their antimicrobial activity. *Phytochemistry* 66(14), 1713–1717.

Kosela, S., Hu, L.H., Yip, S.C., Rachmatia, T., Sukri, T., Daulay, T.S., Tan, G.K., Vittal, J.J. and Sim, K.Y. (1999a) Dulxanthone E: a pyranoxanthone from the leaves of *Garcinia dulcis*. *Phytochemistry* 52(7), 1375–1377.

Kosela, S., Cao, S.G., Wu, X.H., Vittal, J.J., Sukri, T., Masdianto, Goh, S.H. and Sim, K.Y. (1999b) Lateriflorone, a cytotoxic spiroxalactone with a novel skeleton, from *Garcinia lateriflora BL. Tetrahedron Letters* 40(1), 157–160.

Kosin, J., Ruangrungsi, N., Ito, C. and Furukawa, H. (1998) A xanthone from *Garcinia atroviridis*. *Phytochemistry* 47(6), 1167–1168.

Krajewski, D., Tóth, G. and Schreier, P. (1996) 2-Ethyl-3-methylmaleimide from the leaves of mangosteen (*Garcinia mangostana*). *Phytochemistry* 43(1), 141–143.

Lannang, A.M., Komguem, J., Ngninzeko, F.N., Tangmouo, J.G., Lontsi, D., Ajaz, A., Choudhary, M.I., Ranjit, R., Devkota, K.P. and Sondengam, B.L. (2005) Bangangxanthone A and B, two xanthones from the stem bark of *Garcinia polyantha* Oliv. *Phytochemistry* 66(19), 2351–2355.

Lee, H.H. and Chan, H.K. (1977) 1,3,6-Trihydroxy-7-methoxy-8-(3,7-dimethyl-2,6-octadienyl) xanthone from *Garcinia cowa*. *Phytochemistry* 16(12), 2038–2040.

Lewis, Y.S. (1969) Isolation and properties of hydroxycitric acid. In: Colowick, S.P. and Kaplan, N.O. (eds) *Methods in Enzymology*, Volume 13. Academic Press, New York, pp. 613–619.

Lewis, Y.S. and Neelakantan, S. (1965) (–)-Hydroxycitric acid – the principal acid in the fruits of *Garcinia cambogia*. *Phytochemistry* 4, 619–625.

Lewis, Y.S., Neelakantan, S. and Anjana Murthy, C. (1964) Acids in *Garcinia cambogia*. *Current Science* 33, 82–83.

Likhitwitayawuid, K., Phadungcharoen, T., Mahidol, C. and Ruchirawat, S. (1997) 7-*O*-Methylgarcinone e from *Garcinia cowa*. *Phytochemistry* 45(6), 1299–1301.

Lowenstein, J.M. (1971) Effect of (–)-hydroxycitrate on fatty acid synthesis by rat liver *in vivo*. *Journal of Biological Chemistry* 246, 629–632.

Mackeen, M.M., Ali, A.M., Lajis, N.H., Kawazu, K., Kikuzaki, H. and Nakatani, N. (2002) Antifungal *Garcinia* acid esters from the fruits of *Garcinia atroviridis*. *Zeitschrift fur Naturforschung Section C, Biosciences* 57(3/4), 291–295.

MacLeod, A.J. and Pieris, N.M. (1982) Volatile flavour components of mangosteen, *Garcinia mangostana*. *Phytochemistry* 21, 117–119.

Madubunyi, I.I. (1995) Antimicrobial activities of the constituents of *Garcinia kola* seeds. *International Journal of Pharmacognosy* 33(3), 232–237.

Mahabusarakam, W., Chairerk, P. and Taylor, W.C. (2005) Xanthones from *Garcinia cowa* Roxb. Latex. *Phytochemistry* 66(10), 1148–1153.

Maneesin, P. (2002) Roles of packaging and post harvest technology on export of fresh fruits. *Thai Packaging Newsletter* Jan–Mar, 47–58 (in Thai).

Martin, F.W., Campbell, C.W. and Ruberte, R.M. (1987) Perennial edible fruit of the tropics – an inventory. *Agriculture Handbook No. 642*. USDA, ARS, Washington, DC.

Mbafor, J.J., Fomum, Z.T., Promsattha, R., Sanson, D.R. and Tempesta, M.S. (1989) Isolation and characterization of taxifolin 6-C-glucoside from *Garcinia epunctata*. *Journal of Natural Products* 52(2), 417–419.

McCarty, M.F. (1995) Promotion of hepatic lipid oxidation and gluconeogenesis – a strategy for appetite control. *Medical Hypotheses* 44, 278–286.

Merza, J., Aumond, M.C., Rondeau, D., Dumontet, V., Ray, A.M.L., Séraphin, D. and Richomme, P. (2004) Prenylated xanthones and tocotrienols from *Garcinia virgata*. *Phytochemistry* 65(21), 2915–2920.

Minami, H., Kinoshita, M., Fukuyama, Y., Kodama, M., Yoshizawa, T., Sugiura, M., Nakagawa, K. and Tago, H. (1994) Antioxidant xanthones from *Garcinia subelliptica*. *Phytochemistry* 36, 501–506.

Minami, H., Takahashi, E., Fukuyama, Y., Kodama, M., Yoshizawa, T. and Nakagawa, K. (1995) Novel xanthones with superoxide scavenging activity from *Garcinia subelliptica*. *Chemical and Pharmaceutical Bulletin* 43(2), 347–349.

Minami, H., Takahashi, E., Kodama, M. and Fukuyama, Y. (1996) Three xanthones from *Garcinia subelliptica*. *Phytochemistry* 41(2), 629–633.

Minami, H., Hamaguchi, K., Kubo, M. and Fukuyama, Y. (1998) A benzophenone and a xanthone from *Garcinia subelliptica*. *Phytochemistry* 49(6), 1783–1785.

MOA (2001) Pakej Teknologi Manggis. Jabatan Pertanian, Malaysia (in Malay).

Moongkarndi, P., Kosem, N., Kaslungka, S., Luanratana, O., Pongpan, N. and Neungton, N. (2004a) Antiproliferation, antioxidation and induction of apoptosis by *Garcinia mangostana* (mangosteen) on SKBR3 human breast cancer cell line. *Journal of Ethnopharmacology* 90(1), 161–166.

Moongkarndi, P., Kosem, N., Luanratana, O., Jongsomboonkusol, S. and Pongpan, N. (2004b) Antiproliferative activity of Thai medicinal plant extracts on human breast adenocarcinoma cell line. *Fitoterapia* 75(3–4), 375–377.

Nguyen, L.H.D. and Harrison, L.J. (2000) Xanthones and triterpenoids from the bark of *Garcinia vilersiana*. *Phytochemistry* 53(1), 111–114.

Nguyen, L.H.D., Vo, H.T., Pham, H.D., Connolly, J.D. and Harrison, L.J. (2003) Xanthones from the bark of *Garcinia merguensis*. *Phytochemistry* 63(4), 467–470.

Nilar, Nguyen, L.H.D., Venkatraman, G., Yeow, S.K. and Harrison, L.J. (2005) Xanthones and benzophenones from *Garcinia griffithii* and *Garcinia mangostana*. *Phytochemistry* 66(14), 1718–1723.

Niwa, M., Ito, J., Terashima, K. and Aqil, M. (1994), Garcipyran, a novel 6-aryl-1,2-benzopyran derivative from *Garcinia kola*. *Heterocycles* 38(8), 1927–1932.

Nyemba, A.M., Mpondo, T.N., Connolly, J.D. and Rycroft, D.S. (1990) Cycloartane derivatives from *Garcinia lucida*. *Phytochemistry* 29(3), 994–997.

Obuekwe, I.F. and Onwukaeme, N.D. (2004) Phytochemical analyses and antimicrobial activities of the leaf and stem bark extracts of *Garcinia kola* Herkel (family Guttiferae). *Pakistan Journal of Scientific and Industrial Research* 47(2), 160–162.

Ohizumi, Y. (1999) Search for antogonists of luztanin and serotonin from the Thai medicinal plant *Garcinia mangostana* and their pharmacological studies. *Bioenvironment* 2, 215.

Panthong, K., Pongcharoen, W., Phongpaichit, S. and Taylor, W.C. (2006) Tetraoxygenated xanthones from the fruits of *Garcinia cowa*. *Phytochemistry* 67(10), 999–1004.

Parveen, M., Khan, N.U.D., Achari, B. and Dutta, P.K. (1991) A triterpene from *Garcinia mangostana*. *Phytochemistry* 30(1), 361–362.

Pattalung, P.N., Thongtheeraparp, W., Wiriyachitra, P. and Taylor, W.C. (1994) Xanthones of *Garcinia cowa*. *Planta Medica* 60(4), 365–368.

Pelter, A., Warren, R., Chexal, K.K., Handa, B.K. and Rahman, W. (1971) Biflavonyls from guttiferae – *Garcinia livingstonii*. *Tetrahedron* 27(8), 1625–1634.

Permana, D., Lajis, N.K., Mackeen, M.M., Ali, A.M., Aimi, N., Kitajima, M. and Takayama, H. (2001) Isolation and bioactivities of constituents of the roots of *Garcinia atroviridis*. *Journal of Natural Products* 64(7), 976–979.

Permana, D., Abas, F., Maulidiani, Shaari, K., Stanslas, J., Ali, A.M. and Lajis, N.H. (2005) Atrovirisidone B, a new prenylated depsidone with cytotoxic property from the roots of *Garcinia atroviridis*. *Zeitschrift-fur-Naturforschung Section C, Biosciences* 60(7/8), 523–526.

Preuss, H.G., Bagchi, D., Bagchi, M., Rao, C.V.S., Satyanarayana, S. and Dey, D.K. (2004) Efficacy of a novel, natural extract of (–)-hydroxycitric acid (HCA-SX) and a combination of HCA-SX, niacin-bound chromium and *Gymnema sylvestre* extract in weight management in human volunteers: a pilot study. *Nutrition Research* 24(1), 45–58.

Rama Rao, A.V., Sarma, M.R., Venkataraman, K. and Yemul, S.S. (1974) A benzophenone and xanthone with unusual hydroxylation patterns from the heartwood of *Garcinia pedunculata*. *Phytochemistry* 13(7), 1241–1244.

Rama Rao, A.V., Venkataswamy, G. and Yemul, S.S. (1980) Xanthochymol and isoxanthochymol; two polyisoprenylated benzophenones from *Garcinia xanthoxhymus*. *Indian Journal of Chemistry* 19, 627–633.

Rao, R.N. and Sakariah, K.K. (1988) Lipid-lowering and antiobesity effect of (–)-hydroxycitric acid. *Nutrition Research* 8, 209–212.

Rothacker, D.Q. and Waitman, B.E. (1997) Effectiveness of a *Garcinia cambogia* and natural caffeine combination in weight loss: a double-blind placebo-controlled pilot study. *International Journal of Obesity* 21(Suppl. 2), 53.

Rukachaisirikul, V., Kamkaew, M., Sukavisit, D., Phongpaichit, S., Sawangchote, P. and Taylor, W.C. (2003a) Antibacterial xanthones from the leaves of *Garcinia nigrolineata*. *Journal of Natural Products* 66(12), 1531–1535.

Rukachaisirikul, V., Ritthiwigrom, T., Pinsa, A., Sawangchote, P. and Taylor, W.C. (2003b) Xanthones from the stem bark of *Garcinia nigrolineata*. *Phytochemistry* 64(6), 1149–1156.

Rukachaisirikul, V., Pailee, P., Hiranrat, A., Tuchinda, P., Yoosook, C., Kasisit, J., Taylor, W.C. and Reutrakul, V. (2003c) Anti-HIV 1 protostane triterpenes and digeranylbenzophenone from trunk bark and stems of *Garcinia speciosa*. *Planta Medica* 69(12), 1141–1146.

Rukachaisirikul, V., Naklue, W., Sukpondma, Y. and Phongpaichit, S. (2005) An antibacterial biphenyl derivative from *Garcinia bancana* Miq. *Chemical and Pharmaceutical Bulletin* 53(3), 342–343.

Saito, M., Ueno, M., Ogino, S., Kubo, K., Nagata, J. and. Takeuchi, M. (2005) High dose of *Garcinia cambogia* is effective in suppressing fat accumulation in developing male Zucker obese rats, but highly toxic to the testis. *Food and Chemical Toxicology* 43(3), 411–419.

Sakagami, Y., Iinuma, M., Piyasena, K.G.N.P. and Dharmaratne, H.R.W. (2005) Antibacterial activity of α-mangostin against vancomycin resistant *Enterococci* (VRE) and synergism with antibiotics. *Phytomedicine* 12(3), 203–208.

Sang, S., Liao, C.H., Pan, M.H., Rosen, R.T., Shiau, S.Y.L., Lin, J.K. and Ho, C.T. (2002) Chemical studies on antioxidant mechanism of garcinol: analysis of radical reaction products of garcinol with peroxyl radicals and their antitumor activities. *Tetrahedron* 58(51), 10095–10102.

Selvi, A.T., Joseph, G.S. and Jayaprakasha, G.K. (2003) Inhibition of growth and aflatoxin production in *Aspergillus flavus* by *Garcinia indica* extract and its antioxidant activity. *Food Microbiology* 20(4), 455–460.

Sen, A.K., Sarkar, K.K., Mazumder, P.C., Banerji, N., Uusvuori, R. and Hase, T.A. (1982) The structures of garcinones a, b and c: three new xanthones from *Garcinia mangostana*. *Phytochemistry* 21(7), 1747–1750.

Shara, M., Ohia, S.E., Schmidt, R.E., Yasmin, T., Zardetto Smith, A., Kincaid, A., Bagchi, M., Chatterjee, A., Bagchi, D. and Stohs, S.J. (2004) Physico-chemical properties of a novel (–)-hydroxycitric acid extract and its effect on body weight, selected organ weights, hepatic lipid peroxidation and DNA fragmentation, hematology and clinical chemistry, and histopathological changes over a period of 90 days. *Molecular and Cellular Biochemistry* 260(1/2), 171–186.

Sreenivasan, A. and Venkataraman, R. (1959) Chromatographic detection of the organic constituents of Gorikapuli (*Garcinia cambogia* Desr.) used in pickling fish. *Current Science* 28, 151–152.

Sukpondma, Y., Rukachaisirikul, V. and Phongpaichit, S. (2005) Xanthone and sesquiterpene derivatives from the fruits of *Garcinia scortechinii*. *Journal of Natural Products* 68(7), 1010–1017.

Suksamrarn, S., Suwannapoch, N., Ratananukul, P., Aroonlerk, N. and Suksanrarn, A. (2002) Xanthones from the green fruit hulls of *Garcinia mangostana*. *Journal of Natural Products* 65(5), 761–763.

Sullivan, A.C. (1984) Effect of (–)-hydroxycitrate on lipid metabolism. In: Perks, E.G. and Witting, L. (eds) *Modification of Lipid Metabolism*. Academic Press, New York, pp. 143–74.

Sullivan, A.C., Triscari, J. and Miller, O.N. (1974) The influence of (–)-hydroxycitrate on *in vivo* rates of hepatic glycogenesis and cholesterogenesis. *Federal Proceedings* 33, 656.

Thoison, O., Cuong, D.D., Gramain, A., Chiaroni, A., Hung, N.V. and Sevenet, T. (2005) Further rearranged prenylxanthones and benzophenones from *Garcinia bracteata*. *Tetrahedron* 61(35), 8529–8535.

Thom, E. (1996) Hydroxycitrate (HCA) in the treatment of obesity. *International Journal of Obesity* 20 (Suppl. 2), 53.

Triscari, J. and Sullivan, A.C. (1977) Comparative effects of (–)-hydroxycitrate and (+)allo-hydroxycitrate on acetyl CoA carboxylase and fatty acid and cholesterol synthesis *in vivo*. *Lipids* 12, 357–363.

Vieira, L.M.M., Kijjoa, A., Silva, A.M.S., Mondranondra, I.O., Kengthong, S., Gales, L., Damas, A.M. and Herz, W. (2004) Lanostanes and friedolanostanes from the bark of *Garcinia speciosa*. *Phytochemistry* 65(4), 393–398.

Wahyuono, S., Astuti, P. and Artama, W.T. (1999) Characterization of a bioactive substance Alpha mangostin isolated from the hull of *G. mangostana*. *Indonesian Journal of Pharmacology* 10(3), 8.

Waterman, P.G. and Crichton, E.G. (1980a) Pyrones from the bark of *Garcinia conrauana*: conrauanalactone, a novel C_{20} derived α-pyrone. *Phytochemistry* 19(6), 1187–1189.

Waterman, P.G. and Crichton, E.G. (1980b) Xanthones and biflavonoids from *Garcinia densivenia* stem bark. *Phytochemistry* 19(12), 2723–2726.

Waterman, P.G. and Hussain, R.A. (1982) Major xanthones from *Garcinia quadrifaria* and *Garcinia staudtii* stem barks. *Phytochemistry* 21(8), 2099–2101.

Watt, G. (1972) *Dictionary of the Economic Products of India: Periodical Export.* Volume III. Cosmo publications, Delhi, India, p. 464.

Wu, C.C., Weng, J.R., Won, S.J. and Lin, C.N. (2005) Constituents of the pericarp of *Garicinia subelliptica*. *Journal of Natural Products* 68(7), 1125–1127.

Wu, J., Xu, Y.J., Cheng, X.F., Harrison, L.J., Sim, K.Y. and Goh, S.H. (2001) A highly rearranged tetraprenyl xanthonoid from *Garcinia gaudichaudii* (Guttiferae). *Tetrahedron Letters* 42(4), 727–729.

Xu, Y.J., Cao, S.G., Wu, X.H., Lai, Y.H., Tan, B.H.K., Pereira, J.T., Goh, S.H., Venkatraman, G., Harrison, L.J. and Sim, K.Y. (1998) Griffipavixanthone, a novel cytotoxic bixanthone from *Garcinia griffithii* and *G. pavifolia*. *Tetrahedron Letters* 39(49), 9103–9106.

Yaacob, O. and Tindall, H.D. (1995) *Mangosteen Cultivation*. FAO Plant Production and Protection Paper No. 129, FAO, Rome, p. 100.

Yapwattanaphum, C., Subhadrabandhu, S., Sugiura, A., Yonemori, K. and Utsunomiya, N. (2002) Utilization of some *Garcinia* species in Thailand. *Acta Horticulturae* 575(2), 563–570.

Yoshida, K., Tanaka, T., Hirose, Y., Yamaguchi, F., Kohno, H., Toida, M., Hara, A., Sugie, S., Shibata, T. and Mori, H. (2005) Dietary garcinol inhibits 4-nitroquinoline 1-oxide-induced tongue carcinogenesis in rats. *Cancer Letters* 221(1), 29–39.

Yoshikawa, M., Harada, E., Miki, A., Tsukamoto, K., Liang, S.Q., Yamahara, J. and Murakami, N. (1994) Antioxidant constituents from the fruit hulls of mangosteen (*Garcinia mangostana* L.) originating in Vietnam. *Yakugaku Zasshi Journal of the Pharmaceutical Society of Japan* 114(2), 129–133.

Yu, L., Zhao, M., Yang, B., Zhao, Q. and Jiang, Y. (2006) Phenolics from hull of *Garcinia mangostana* fruit and their antioxidant activities. *Food Chemistry* 104(1), 176–181.

Zhang, H.Z., Kasibhatla, S., Wang, Y., Herich, J., Guastella, J., Tseng, B., Drewe, J. and Cai, S.X. (2004) Discovery, characterization and SAR of gambogic acid as a potent apoptosis inducer by a HTS assay. *Bioorganic and Medicinal Chemistry* 12(2), 309–317.

20 Tamarind

K.S. Krishnamurthy, V.P. Sapna and V.A. Parthasarathy

20.1. Introduction

The tamarind, *Tamarindus indica* L. (family Leguminosae), is a widely distributed tree spice which is also grown as a shade tree on highways. It is one of the most important multi-purpose tree species in the Indian subcontinent. It is a large evergreen tree with an exceptionally beautiful spreading crown and is cultivated throughout almost the whole country, except in the Himalayas. It is cultivated in more than 53 countries in the world.

Tamarind is thought to have originated in Madagascar (Von Maydell, 1986; Hockin, 1993). The tree grows wild throughout the Sudan and was introduced into and adopted in India so long ago that it has often been reported as also being indigenous there. The tamarind fruit was at first thought to be produced by an Indian palm, as the name tamarind comes from a Persian word *Tamar-I-hind*, meaning date of India. Its name *amlika* in Sanskrit indicates its ancient presence in the country (Mishra, 1997). In Myanmar, it is reported as one of the commonest village trees in the dry zone (Troup, 1921). The fruit was well known to the ancient Egyptians and to the Greeks in the 4th century BC. It is now cultivated throughout semi-arid Africa and South Asia, where it has become naturalized in several regions. The tree has long been naturalized in the East Indies and the islands of the Pacific. The tamarind was introduced into tropical America, Bermuda, the Bahamas and the West Indies certainly much earlier. It has been planted extensively in Bangladesh, India, Myanmar, Malaysia, Sri Lanka, Thailand and several African, Australian, Central American and South American countries. The fruit became known in Europe during the Middle Ages (http://www.hort.purdue.edu/newcrop/morton/tamarind.html). Commercial plantations are reported in Belize, Central American countries and in north Brazil (Sharma and Bhardwaj, 1997). In all tropical and near-tropical areas, including South Florida, it is grown as a shade and fruit tree, along roadsides and in backyards and parks. India has 61,700 ha of cultivated tamarind and produces 179,310 t (Premaja and Manojkumar, 2007).

20.2. Botany and Uses

Botany

Tamarind is an evergreen, moderate to large tree, growing up to 24 m in height and 7 m in girth. The most useful part is the pod, but all the tree parts are used in one way or another. The tree is highly wind-resistant, with strong, supple branches, drooping at the ends, and has a dark grey, rough, fissured bark. The flowers are borne in small racemes, very small, five-petalled, yellow coloured with orange or red streaks. The flower buds are distinctly pink. The sepals, four in number, shed when the flower opens. The pods are 7.5–20 cm long, 2.5 cm broad and 1 cm thick, more or less constricted between the seeds, slightly curved, brownish-ash coloured and scurfy. There are three to 12 seeds in each pod contained in loculi, enveloped by a tough, leathery parchment-like membrane, called the endocarp. Outside the endocarp is the light-brownish, red, sweetish acidic, edible pulp, traversed by a number of branched, ligneous strands. The outermost covering of the pod is fragile and easily separable. The pods begin to ripen from February to April. At first, the pods are tender-skinned with green, highly acid flesh and soft, whitish, underdeveloped seeds. As they mature, the pods fill out somewhat and the juicy, acidulous pulp turns brown or reddish-brown. Thereafter, the skin becomes a brittle, easily cracked shell and the pulp dehydrates naturally to a sticky paste enclosed by a few coarse strands of fibre extending lengthwise from the stalk (http://www.hort.purdue.edu/newcrop/morton/tamarind.html; http://www.uni-graz.at/~katzer/engl/Tama_ind.html).

Uses

All parts of the tamarind tree are useful in one way or another. Tamarind is valued mostly for its fruit and pulp, which is used for a wide variety of domestic and industrial purposes (Kulkarni *et al.*, 1993), in particular to prepare juice, jam, syrup and sweets. Tamarind juice concentrate (TJC) is a convenient product due to the ease with which it can be dissolved and reconstituted in warm water. The specific heat of TJC increases with temperature and the glass transition temperature of the product is −70.74°C (Ahmed *et al.*, 2007). Tamarind intake appears to have an additional beneficial effect on the mobilization of deposited fluoride from bone, by enhancing urinary excretion of fluoride (Khandare *et al.*, 2004). It can also be eaten raw and is used in southern India to induce an acidic flavour to curries. The combinations of acidulants such as tamarind and antioxidant spices improves the retention of β-carotene during cooking (Gayathri *et al.*, 2004). Tamarind extract can be used as a replacement for citric acid, phosphoric acid and other acids that are added to soft drinks. Beverages containing tamarind extract have an improved shelf life due to higher pH and also yield a flavour profile equivalent to or better than beverages sweetened with aspartame (Zablocki and Perore, 1996).

The pulp, when mixed with seawater, cleans silver, copper and brass. The leaves are eaten by cattle and goats and furnish fodder for silkworms, *Anaphe* sp. in India and *Hypsoides vuilletii* in West Africa. Tamarind leaves and flowers are useful as mordants in dyeing. Tamarind seeds yield amber oil, useful as an illuminant and as a varnish, especially preferred for painting dolls and idols. The powder made from tamarind kernels has been adopted by the Indian textile industry as 300% more efficient and more economical than cornstarch for sizing and finishing cotton, jute and spun viscose, as well as having other technical advantages. It is commonly used for dressing homemade blankets. Other industrial uses include employment in the colour printing of textiles, paper sizing, leather treating, etc. Tamarind seed husk is a natural source of tannin that can be used beneficially to manipulate rumen fermentation (Bhatta *et al.*, 2001). At low concentration, this tannin has a beneficial effect on the performance of crossbred lactating cows (Bhatta *et al.*, 2000).

The heartwood is highly prized for furniture, panelling, wheels, axles, gears for mills, ploughs, planking for the sides of boats, wells, mallets, knife and tool handles, rice pounders, mortars and pestles. Tamarind twigs are sometimes used as 'chewsticks' and the bark of the tree as a masticatory, alone or in place of lime with betelnut. The bark is often employed in tanning hides and in dyeing, and is burnt to make ink. The lac produced by the lac insect parasitizing on a tamarind tree may be harvested and sold as stick-lac for the production of lacquers and varnish.

20.3. General Composition

The tamarind fruit (pod) has mainly pulp and seeds. The seeds are covered by a thin parchment, membrane-like structure. The pulp constitutes 30–50% of ripe fruit (Purseglove, 1987; Shankaracharya, 1998). The shell and fibre account for 11–30% and the seed constitutes around 25–40% (Chapman, 1984). The fruit pulp (both ripe and dried) contains mainly tartaric acid, reducing sugars, pectin, tannin, fibre and cellulose. The general composition of tamarind fruits is given in Table 20.1.

Both pulp and seed are a good source of potassium, calcium and phosphorus. The kernel and testa are very rich in potassium, sodium, zinc and iron. The whole seed has 13–27% protein, 4.5–16% fat, 11–25% total sugars and 50–57% carbohydrates. The pulp is considered a promising source of tartaric acid, alcohol (12%) and pectin (21–22%). The red pulp of some types contains the pigment, chrysanthemin. The seeds contain approximately 63% starch, 14–18% albuminoids and 4.5–6.5% of a semi-drying oil. The seed kernel also has similar fat and protein contents, while the carbohydrate content is slightly higher (65–72%). The testa is rich in fibre, e.g. ~ 21% (Ishola *et al*., 1990; Bhattacharya *et al*., 1993). The mineral composition of tamarind is given in Table 20.2.

The tender leaves and flowers also show high calcium and phosphorus content. The leaves have 16–18% carbohydrates. Thiamine, riboflavin and niacin were also detected in both leaves and flowers. Table

Table 20.1. Composition of tamarind fruits.

Constituent	Content
Moisture (%)	15–30 (62.5–69.2)
Proteins (%)	2.00–8.79 (1.4–3.3)
Fat/oil/crude lipid (%)	0.50–2.53 (0.27–0.81)
Carbohydrates, total (%)	56.70–70.70
Fibre, crude (%)	2.20–18.30
Tartaric acid, total (%)	8.00–18.00 (8.40–12.40)
Reducing sugars (%)	25.00–45.00
Total ash (%)	2.10–2.90 (1.20–1.72)
Pectin (%)	2.00–4.00
Cellulose (%)	19.40 (1.80–3.20)
Albuminoids (%)	3.00–4.00
Total available carbohydrates (%)	41.77
Alcohol-insoluble solids (%)	22.70
Water-insoluble solids (%)	22.70
Non-reducing sugars (%)	16.52
Total sugars (%)	41.20 (21.40–30.85)
Starch (%)	5.70
Tannin (mg)	600.00
Ascorbic acid a (mg)	3.00–9.00
β-carotene equivalent (μg)	10.00–60.00
Thiamine (mg)	0.18–0.22
Riboflavin (mg)	0.07–0.09
Niacin (mg)	0.60
pH	(3.15)
Pentoses (%)	(4.20–4.80)
Sucrose (%)	(0.10–0.80)

Note: The values given in the parantheses are for ripe fruit.
Source: Meillon (1974); Duke (1981); Ishola *et al.* (1990).

Table 20.2. Mineral composition of tamarind (mg/kg).

Mineral	Pulp	Seed	Kernel	Testa
Calcium	81–466	9.3–786.0	120	100
Phosphorus	86–190	68.4–165.0	–	–
Magnesium	72.03	17.5–118.3	180	120
Potassium	62–570	272.8–610.0	1020	240
Sodium	3.0–76.7	19.2–28.8	210	240
Copper	21.83	1.6–19.0	–	–
Iron	1.3–10.9	6.5	80	80
Zinc	1.06	2.8	100	120
Nickel	0.52	–	–	–

Source: Marangoni *et al.* (1988); Ishola *et al.* (1990); Bhattacharya *et al.* (1993).

Table 20.3. Composition of tender leaves and flowers of tamarind.

Constituent	Tender leaves	Flowers
Moisture (%)	70.5–78.0	80.0
Protein (%)	4.0–5.8	2.8
Fat/oil (%)	1.2–2.1	1.5
Fibre (%)	1.9–3.0	1.5
Carbohydrates (total) (%)	16.0–18.0	
Ash/minerals (%)	1.0–1.5	0.7
Calcium (mg)	101–250	35.5
Magnesium (mg)	71.0	
Phosphorus (mg)	140.0	45.6
Iron (mg)	2.0–5.2	1.5
Thiamine (mg)	0.1–0.2	0.07
Riboflavin (mg)	0.1–0.2	0.14
Niacin (mg)	1.5–4.1	1.14
Vitamin C (mg)	3.0–6.0	13.80
Carotenes (mg)		0.31

Source: Lewis and Neelakantan (1964); Duke (1981).

20.3 gives the composition of the tender leaves and flowers.

Among fatty acids, linoleic acid was the major constituent (nearly 50%), followed by oleic (~ 24%) and palmitic acids (~ 15%). Table 20.4 gives the fatty acid composition of the seed oil of tamarind. Among sterols, betasitosterol constituted 66–72%, followed by campesterol (16–19%) and stigmasterol (Andriamanantena *et al.*, 1983).

Tamarind kernel powder has around 15% dietary fibre and 14% crude protein. Crude lipid is around 8%, 4.5% ash and has a calorific value of 1511kj/100g dry matter. Total protein fractionation revealed that 100g of seed flour yields around 7g protein, of which albumins and globulins each constituted ~ 2.5g, prolamines 0.7g and glutelins 1.3g.

Table 20.4. Fatty acid composition of seed oil (%).

Fatty acid	Content
Lauric acid (C12:0)	NP
Myristic acid (C14:0)	NP
Palmitic acid (C16:0)	14.67
Stearic acid (C18:0)	5.27
Oleic acid (C18:1)	23.67
Linoleic acid (C18:2)	49.13
Linolenic acid (C18:3)	2.23
Behenic acid (C22:0)	5.03

NP = not present.
Source: Pugalenthi *et al.* (2004).

20.4. Chemistry

Volatiles

Analysis of the volatile compounds of tamarind revealed the presence of more than 80 compounds. Aromatic and furan derivatives were dominant. The major constituents were 2-phenyl acetaldehyde (25.4% of total volatiles), which has a fruity and honey-like odour, 2-furfural (20.7%), having a caramel-like flavour, followed by hexadecanoic acid (18.1%) and limonene, which has a citrus flavour. A list of the volatile compounds detected in tamarind is given in Table 20.5.

The composition of volatile constituents in tamarind varies with climate and also between varieties. Grollier *et al.* (1998) reported that the main flavour compound of the tamarind pulp was 2-acetylfuran. An analysis of the volatile constituents of the fruit pulp of tamarinds grown in Malaysia indicated the presence of 66 compounds; furan derivatives and carboxylic acids were dominant, accounting for 44.4 and 33.3% of the total volatiles, respectively. The major components were furfural (38.2%), palmitic acid (14.8%), oleic acid (8.1%) and phenylacetaldehyde (7.5%) (Wong *et al.*, 1998). Sagrero *et al.* (1994) identified 16 volatile compounds from the fruit pulp of tamarind collected from Australia through GC and GC-MS and reported that aromadendrene was the major constituent in the oil, corresponding to 90% of the flavour constituents. Structures of some of the volatile components are given in Fig. 20.1.

Composition of the leaf oil of Tamarindus indica *L.*

The major constituents of the leaf oil of tamarind are linalool anthranilate, benzyl benzoate and limonene. α-Pinene, β-pinene, nerol, etc., were noticed in minor quantity. The components are listed in Table 20.6.

Table 20.5. Volatile compounds of tamarind (mg/kg).

Compound	Content	Compound	Content
Acetaldehyde	< 0.01	γ-Terpinene	0.01
Ethanol	< 0.01	Acetophenone	< 0.01
Diacetyl	< 0.01	Methylbenzoate	0.02
Ethyl acetate	0.08	*cis*-Linalool oxide	0.01
Isovaleraldehyde	0.09	4-Methylbenaldehyde	0.02
2-Methylbutanal	0.03	Terpinolene	< 0.01
1-Penten-3-ol	< 0.01	*trans*-Linalool oxide	< 0.01
2-Ethylfuran	0.01	α-Dimethylstyrene	< 0.01
Isoamyl alcohol	0.01	Nonanal	0.02
2-Methylbutanol	< 0.01	2-Phenylethanol	0.02
1-Methyl-1H-pyrrole	0.03	4-Methylacetophenone	< 0.01
1*H*-Pyrrole	0.02	(*E*,*Z*)-2,6-Nonadienal	< 0.01
Pyrrolidine	< 0.01	2-Methylacetophenone	< 0.01
Toluene	0.01	Terpinen-4-ol	< 0.01
3-Methyl-2-butenol	< 0.01	Octanoic acid	< 0.01
Hexanal	0.02	Al-terpineol	0.02
1-Ethyl-1h-pyrrole	< 0.01	Methyl salicylate	0.03
2-Furfural	0.62	Safranal	< 0.01
(*E*)-2-Hexenal	0.01	Al-ionene	0.02
Ethylbenzene	0.02	2,3-Dihydrobenzofuran	< 0.01
p-Xylene	0.13	2-Phenylethyl butyrate	< 0.01
o-Xylene	0.04	(*E*)-2-Decenal	0.02
Heptanal	< 0.01	Vitispirane	0.02
Methional	0.01	(*E*)-Anethole	< 0.01
2-Acetylfuran	< 0.01	1*H*-Indole	0.04
α-Pinene	< 0.01	Methyl decanoate	< 0.01
Propylbenzene	< 0.01	Eugenol	< 0.01
Benzaldehyde	0.03	(*E*)-2-Undecenal	< 0.01
5-Methylfurfural	0.01	Decanoic acid	< 0.01
β-Pinene	< 0.01	Al-cedrene	0.01
6-Methyl-5-hepten-2-one	< 0.01	Geranyl acetone	< 0.01
Sabinene	< 0.01	Dodecanoic acid	0.02
Octanal	< 0.01	Tetradecanoic acid	0.06
α-Phellandrene	< 0.01	Methyl hexadecanoate	< 0.01
(*E*,*E*)-2,4-Heptadienal	0.01	Hexadecanoic acid	0.54
α-Terpinene	< 0.01	Ethyl linoleate	0.02
1,2,4-Trimethylbenzene	< 0.01	Octadecanoic acid	0.02
p-Cymene	0.01	2-Butoxyethanol acetate	0.02
1,8-Cineole	< 0.01	Cyclohexyl acetate	< 0.01
Benzyl alcohol	< 0.01	Limonene	0.15
2-Phenylacetaldehyde	0.76		

Source: Pino *et al.* (2004).

Non-volatiles

Tamarind has tartaric acid as its major organic acid component. Many polyphenols are found in the coat of the tamarind fruit. Tamarind kernel powder (TKP) has a xyloglucan, which has a variety of uses. TKP, a crude extract of tamarind seeds, has been used as a replacement for starch in cotton sizing and as a wet-end additive in the paper industry, where it replaces starch and galactomannans (Glicksman, 1986). Refined tamarind seed polysaccharide is used as a thickening, stabilizing and gelling agent in the food industry, particularly in Japan where it is a permitted food additive (Glicksman, 1986; Gidley *et al.*, 1991).

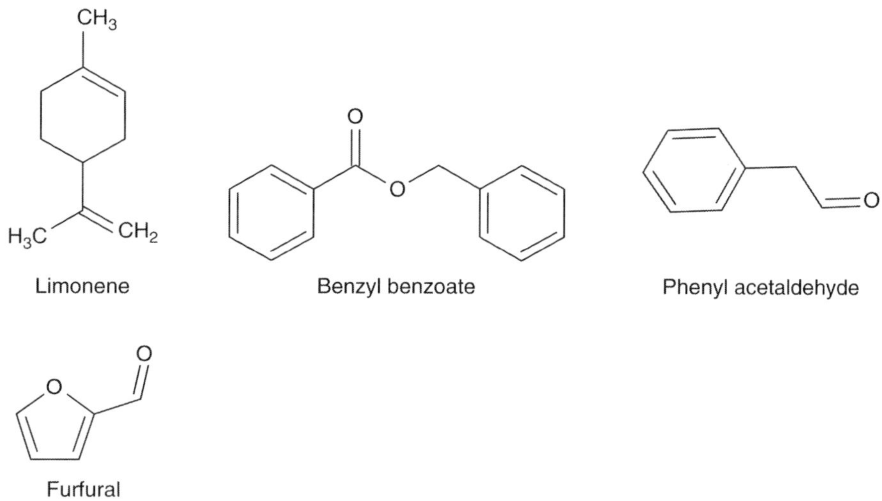

Fig. 20.1. Structures of major volatile compounds of tamarind.

Composition of tamarind kernel powder

Tamarind kernel powder, on fractionation, affords a homogeneous polysaccharide composed of D-glucose, D-xylose, D-galactose and L-arabinose in the molar ratios of 8:4:2:1 (Srivastava and Singh, 1967). The polysaccharide is composed of (1 $\rightarrow$ 4)-β-D-glucan backbone substituted with side-chains of α-D-xylopyranose and β-D-galactopyranosyl (1 $\rightarrow$ 2)-α-D-xylopyranose linked (1 $\rightarrow$ 6) to glucose residues. The glucose, xylose and galactose units are present in the ratio of 2.8:2.25:1.0 (Glicksman, 1986; Gidley *et al.*, 1991).

Table 20.6. Leaf oil components of tamarind (%).

Compound	Content
(*E*)-2-Hexanal	1.7
α-pinene	1.0
β-pinene	1.4
p-Cymene	0.6
Limonene	24.4
(*E*)-*b*-Ocimene	t
Linalool	1.0
Linalool anthranilate	4.7
Al-terpineol	0.7
Nerol	1.0
Benzyl benzoate	40.6
Pentadecanol	8.2
Hexadecanol	12.4

t = trace.
Source: Pino *et al.* (2002).

Extraction and isolation techniques of tamarind xyloglucans and polyphenols

The major polyphenolic compounds of tamarind pericarp were extracted using organic solvents and the metabolites were isolated by semi-preparative high-performance liquid chromatography. Their structures were elucidated by liquid chromatography–electrospray-ionization-mass spectrometry (LC–ESI-MS), nano-electrospray-ionization mass spectrometry (ESI-MS) and, where possible, by gas chromatography-mass spectrometry (GC-MS) and ^{1}H and ^{13}C NMR (Sudjaroen *et al.*, 2005). Solvent extraction experiments showed that ethanol had a higher selectivity than ethyl acetate for extraction of (–)-epicatechin; yields of (–)-epicatechin using ethanol were about 150mg/100g. The antioxidant mixture extracted from sweet Thai tamarind seedcoat using solvent extraction with ethanol was found to be the most active in terms of peroxide value (Luengthanaphol *et al.*, 2004).

Several oligosaccharide fragments, ranging from two to nine contiguous residues, have been isolated from purified tamarind xyloglucan using enzymatic digestion and partial acid hydrolysis. Structures were determined using matrix-assisted laser adsorption ionization-time of flight (MALDI-TOF) mass spectrometry, gas chromatography (GC), gas chromatography-mass spectrometry (GC-MS) and Dionex high-pH anion exchange–high performance liquid chromatography (HPAE–HPLC) (Marry *et al.*, 2003).

The profile (%) of polyphenolics in tamarind pericarp was dominated by proanthcyanidins in various forms, such as (+)-catechin, procyanidin B_2, (−)-epicatechin (9.4), procyanidin trimer (11.3), procyanidin tetramer (22.2), procyanidin pentamer (11.6) and procyanidin hexamer (12.8), along with taxifolin (7.4), apigenin (2.0), eriodictyol (6.9), luteolin (5.0) and naringenin (1.4) of total phenols, respectively. Tamarind seeds comprised procyanidins only, represented (%) mainly by oligomeric procyanidin tetramer (30.2), procyanidin hexamer (23.8), procyanidin trimer (18.1), procyanidin pentamer (17.6), with lower amounts of procyanidin B_2 (5.5) and (−)-epicatechin (Sudjaroen *et al.*, 2005).

Properties of TKP

TKP is one of the cheapest gums available, but it has an unpleasant odour, dull colour and a presence of water insolubles. Low solubility in cold water and fast biodegradability are some other characteristics. Humidification and compression of the composite material prepared from the tamarind seed gum and the cellulosic-rich sisal plant fibre increased its adhesive strength. This material has potential industrial applications, such as in false roofing and room partitioning (Veluraja *et al.*, 1997). A number of chemical modifications of tamarind seed polysaccharide have been described, including acetyl (Rao and Beri, 1955), hydroxyalkyl (Schiavio and Maderno, 1958; Shimohiro *et al.*, 1983) and carboxymethyl (Shimohiro *et al.*, 1983; Omya and Tabuchi, 1985) derivatives. Carboxymethylation of tamarind kernel powder increased its solubility in cold water and the stability of its paste to microorganisms (Goyal *et al.*, 2006). The functional, as well as the nutritional, properties of the meal and concentrate from tamarind kernels (raw and roasted) revealed that the *in vitro* digestibility was 71.3; the kernel protein was rich in lysine, glutamic acid, aspartic acid, glycine and leucine, but deficient in sulphur-containing amino acids (Bhattacharya *et al.*, 1994).

Tamarind xyloglucan imparted more viscous, liquid-like rheological properties and heat stability to the gelatinized tapioca starch/xyloglucan mixtures (Pongsawatmanit *et al.*, 2006). Tamarind seed powder can also be used for the production of tannase under solid-state fermentation (SSF) using *Aspergillus niger* ATCC 16620 (Sabu *et al.*, 2005). The tamarind foam mats prepared from the binary combinations of the foaming agents ovalbumin, mesquite gum and a low-molecular weight, surface-active blend yielded foams that exhibited longer drainage mean times, higher yield stress, apparent plastic viscosity, critical drying time, instantaneous elastic modulus and a shorter onset of the critical drying time to yield tamarind powders with better sensory flavour perception (Carter *et al.*, 2001).

20.5. Medicinal and Pharmacological Uses

Tamarind has many medicinal uses. It is reported to have the following medicinal properties:

- Digestive
- Hepatic tonic
- Anti-inflammatory
- Corneal wound healing
- Antioxidant
- Other medicinal uses.

Digestive

Due to its medicinal value, tamarind is used as an Ayurvedic medicine for gastric and/or digestion problems. Tamarind pulp alone, or in combination with limejuice, honey, milk, dates, spices or camphor, is used as a digestive, even for elephants (Morton, 1987).

Hepatic tonic

The flowers are used to cure jaundice and bleeding piles (Brown, 1954; de Padua *et al.*, 1978). Tamarind pulp alone, or in combination with limejuice, honey, milk, dates, spices or camphor, is used as a remedy for biliousness and bile disorders and as an antiscorbutic (Morton, 1987).

Anti-inflammatory

Tamarind leaves and flowers, dried or boiled, are used as poultices for swollen joints, sprains and boils (http://www.kingtutshop.com/Egyptian-Herb/Tamarind.htm). The pulp is used as an anti-inflammatory agent and as a gargle for sore throats. Lotions and extracts made from the leaves are used in treating conjunctivitis and are also used as antiseptics (Morton, 1987). Lotions and poultices from the bark are applied on open sores and caterpillar rashes. The powdered seed paste/seedcoat is also used to cure boils (http://www.haryana-online.com/Flora/imli.htm). The bark is used as a tonic or in poultices to treat ulcers, wounds, boils, sores and rashes in eastern Sudan and also in the Philippines (Dalziel, 1937). The bark of the tree is regarded as an effective astringent, tonic and febrifuge.

Corneal wound healing

Bacterial keratitis is a serious infectious ocular disease requiring prompt treatment to prevent frequent and severe visual disabilities. The tamarind seed polysaccharide (TSP) appears to be a promising candidate as a vehicle for the topical treatment of bacterial keratitis. TSP prolongs the precorneal residence times of antibiotics and enhances drug accumulation in the cornea, probably by reducing the washout of topically administered drugs (Ghelardi *et al.*, 2004). The ability of the TSP to promote corneal wound healing may depend on its influence on the integrin recognition system (Burgalassi *et al.*, 2000a). It can be a potentially useful adjuvant for ophthalmic delivery systems (Burgalassi *et al.*, 2000b). Timolol in association with TSP has a prolonged duration of action and is suitable for ocular administration in cases of elevated intraocular pressure (D'Amico *et al.*, 1999).

Antioxidant

Methanol and aqueous acetone extracts of dry-heated tamarind seedcoat sample showed hydroxyl radical scavenging activity, as well as exhibiting good antioxidant activity against the linoleic acid emulsion system, and the values were lower and higher than the synthetic antioxdiant, BHA, and ascorbic acid, respectively (Siddhuraju, 2006). The seeds showed a much higher antioxidant activity and phenolic content than the edible portions (Soong and Barlow, 2004). Treatment of hypercholesterolaemic hamsters with the *T. indica* fruit pulp extract led to a decrease in the levels of serum total cholesterol, non-HDL cholesterol and triglyceride and to an increase of high-density lipoprotein (HDL) cholesterol levels. It also led to decreased lipid peroxidation in serum and improved the efficiency of the antioxidant defence system, indicating the potential of tamarind extracts in diminishing the risk of atherosclerosis development in humans (Martinello *et al.*, 2006). Tamarind may be an important source of cancer chemopreventive natural products in tropical regions (Sudjaroen *et al.*, 2005). Tamarind xyloglucans were immunoprotective at low picogram doses. The tamarind xyloglucans also blocked UV-activated phosphorylation of SAPK/JNK protein (Strickland *et al.*, 1999).

Other medicinal uses

Tamarind preparations are recognized universally as refrigerants in fevers and as laxatives and carminatives (Morton, 1987). The laxative properties of the pulp and the diuretic properties of the leaf sap have been confirmed by modern medicinal science (Bueso, 1980). In South-east Asia, the fruit is prescribed to counteract the ill effects of overdoses of false chaulmoogra, *Hydnocarpus anthelmintica* Pierre, given in leprosy. The pulp is said to aid the restoration of sensation in cases of paralysis (Morton, 1987). In Colombia, an ointment made of tamarind pulp, butter and other ingredients is used to cure domestic animals of vermin. The pulp is mixed with salt, as a liniment for rheumatism (Morton, 1987). It is also administered to alleviate sunstroke, *Datura* poisoning and alcoholic intoxication (Benthal, 1933; Dalziel, 1937; Eggeling and Dale, 1951; Chaturvedi, 1985), etc. It is reported to aid in the care of malarial fever also (Timyan and Bwa, 1996). In Mauritius, the Creoles mix salt with the pulp and use it as a liniment for rheumatism and make a decoction of the bark for asthma (http://www.kingtutshop.com/Egyptian-Herb/Tamarind.htm). Lotions and extracts made from tamarind preparations are used in treating dysentry, jaundice, erysipelas, haemorrhoids and various other ailments (Morton, 1987).

Tamarind leaves, seeds and fruits are also used in traditional Indian medicine (Jayaweera, 1981). The fruit shells are burnt and reduced to an alkaline ash, which enters into medicinal formulae. Powdered seeds/seedcoat, with or without cumin seeds and palm sugar, are prescribed for chronic diarrhoea and dysentery. Root infusion is used to cure chest complaints and is an ingredient in prescriptions for leprosy (http://www.haryana-online.com/Flora/imli.htm). The leaves and roots contain the glycosides, vitexin, isovitexin, orientin and isoorientin. Tamarind root bark is used for abortion. The root bark is ground into a powder, mixed with hot water and administered 3 days prior to an abortion and for the prevention of pregnancies (Lakshmanan and Narayanan, 1994). A decoction of the bark is used in cases of gingivitis, asthma and eye inflammation. The bark yields the alkaloid, hordenine. The bark has also been used to recover loss of sensation due to paralysis and to heal urinary discharges and gonorrhoea. Bark is also one of the ingredients of *Abayalavana*, used to cure enlarged spleen in India. In South India, tamarind-pepper rasam is also considered an effective home remedy for a cold.

20.6. Quality Specifications

There are quality specifications to be followed for seedless tamarind and dry tamarind, as well as tamarind concentrate, a product resulting from the concentration of hot water extract of soluble solids of tamarind pulp under vacuum, which is marketed by private agencies. Seeds are removed from the matured fruit, pressed and sold as tamarind pulp. This pulp generally is devoid of rind, seed and fibre. Seed content should not exceed 10% for average, 7% for fair, 5% for good or 3% for special-quality tamarind pulps. As per Agmark grades, a higher moisture content of up to 20% is allowed for pulp. Agmark specifications for seedless tamarind vary from Special to Grade C.

Special, Grade A and Grade B are awarded to dry tamarind under Agmark rules based on the percentage of rind, fibre, moisture and insect damage. For tamarind seed, there are only two grades, e.g. Special and Grade A (Table 20.7). Quality specifications are listed for undecorticated and decorticated tamarind seeds, as well as tamarind powder.

Prescription for requirement, methods of sampling and test for tamarind concentrate

Tamarind concentrate shall be obtained by hot water extraction of clean tamarind pulp, with subsequent concentration under vacuum. The tamarind fruits shall be mature, sound, fresh and shall be free from insect and fungal attack or any other blemish which affects quality. There should not be any added colouring or flavouring agents.

It shall be manufactured under hygienic conditions, light to dark brown in colour, and the flavour should be characteristic of tamarind fruit. No burnt flavour should be present. It shall be free from harmless extraneous vegetable materials (fibre and rind common to tamarind, and stems up to

Table 20.7. Agmark specifications for tamarind seedless, dry tamarind and tamarind seed.

Character/grade	Special	A	B	C
Tamarind seedless (%/weight max.)				
Moisture	15	17	20	20
Seed content	5	10	15	20
Foreign matter (organic)	4	6	8	10
Foreign matter (inorganic)	1	1.5	2	2
Dry tamarind (%/weight max.)				
Seed content	35	40	45	–
Fibre	6	8	10	–
Rind	3	4	6	–
Insect damage	2	3	5	–
Moisture	15	20	25	–
Tamarind seed (%/weight max.)				
Extraneous matter	1	2	–	–
Damaged and discoloured	2	5	–	–
Weight/l	900	800	–	–
Moisture	9	10	–	–

Source: Saideswara Rao and Mary Mathew (2001).

10 mm in length and sepal bracts aggregating an area of 5 cm^2, pits (a whole pit (stone) or one half seed of the tamarind fruit) and pit fragments (piece of tamarind seed less than the equivalent of one half pit and which weighs at least 5 mg). It should be free from living insects, moulds, insect fragments and rodent contamination visible to the naked eye.

Packing

Tamarind preparations shall be packed in tin plate or glass containers which should be sealed appropriately. The tin plate container shall be lacquered with acid-resistant lacquer. Each container should be labelled with the name of the material, name and address of the manufacturer, net mass of the contents of the container in grams, date of manufacture, list of additives and manufacturer's licence number. Requirements for tamarind concentrate and limits of heavy metals are given in Table 20.8.

Patents

A general patents search revealed as many as 161 web pages for green tamarind, 77 for tamarind powder, 53 for tamarind pulp, 51 for tamarind paste, 35 for tamarind juice

Table 20.8. Requirements for tamarind concentrate and limits for heavy metals (as per IS 5955:1993).

Tamarind concentrate	Requirement
Characteristics	
Headspace of the can in mm, max.	6
Microbiological requirements	To satisfy the test
Moisture,% by mass, max.	15
Total soluble solids, % by mass, min.	65
Total insoluble pulp, % by mass, max.	2
Total tartaric acid, %, min.	9
Acid insoluble ash (on dry basis), % by mass, max.	0.5
Total reducing sugar, % by mass, max.	35
Limits for heavy metals	
Arsenic, ppm, max.	1.0
Lead, ppm, max.	2.5
Copper, ppm, max.	30
Zinc, ppm, max.	50
Tin, ppm, max.	250

concentrate, 26 for tamarind jam, 21 for tamarind kernel powder, 15 for tamarind pickles and three for tamarind chutney (http://www.scopus.com/scopus/home.url).

As many as 15 Indian patents for tamarind were filed from 1995 to 2004. Patent areas included process of tamarind paste and concentrate; tamarind pickles and tamarind garlic chutney preparations; preparation of Indian traditional tokku-like product from green tamarind; preparation of tamarind extract in the form of paste/jam; improved process of the production of tamarind powder; preparation of carboxy methyl tamarind kernel powder for printing polyester fabrics; recovery of potassium bitartarate; pectin sugar fruit acids by by-products from tamarind pulp; process of extraction of polysaccharides from tamarind seed kernel; recovery of tartaric acid from tamarind pulp; and a method of improving jute yarn sizing by the application of modified tamarind kernel powder (http://www.indianpatents.org.in/db/test).

20.7. Conclusion

Tamarind is rich in calcium, potassium, zinc and iron. The major flavour compounds in tamarind are furfural and phenyl acetaldehyde. Benzyl benzoate, linalool anthranilate and limonene are the major leaf oil components. Quality specifications are available for a few tamarind products, but a large number of tamarind products are emerging and there is scope for developing quality specifications for other products as well. Tamarind is used traditionally as an astringent, an anti-inflammatory and anti-diuretic agent, a liniment for rheumatism, a laxative, a carminative and a digestive agent. Scientific basis for such claims is lacking and hence warrants detailed investigation.

References

Ahmed, J., Ramaswamy, H.S. and Sashidhar, K.C. (2007) Rheological characteristics of tamarind (*Tamarindus indica* L.) juice concentrates. *LWT* 40, 225–231.

Andriamanantena, R.W., Artuad, J., Gaydofn, E.M., Iatrides, M.C. and Chavalier, J.L. (1983) Fatty acid and sterol composition of Malagasy tamarind kernels. *Journal of the American Oil Chemists Society* 60(7), 1318–1321.

Benthal, A.P. (1933) *The Trees of Calcutta and its Neighbourhood.* Thacker Spink and Company, London, p. 513.

Bhatta, R., Krishnamoorthy, U. and Mohammed, F. (2000) Effect of feeding tamarind (*Tamarindus indica*) seed husk as a source of tannin on dry matter intake, digestibility of nutrients and production performance of crossbred dairy cows in mid-lactation. *Animal Feed Science and Technology* 83(1), 67–74.

Bhatta, R., Krishnamoorthy, U. and Mohammed, F. (2001) Effect of tamarind (*Tamarindus indica*) seed husk tannins on *in vitro* rumen fermentation. *Animal Feed Science and Technology* 90(3–4), 143–152.

Bhattacharya, S., Bal, S., Mukherjee, R.K. and Bhattacharya, S. (1993) Some physical and engineering properties of tamarind (*Tamarindus indica*) seed. *Journal of Food Engineering* 18, 77–79.

Bhattacharya, S., Bal, S., Mukherjee, R.K. and Suvendu Bhattacharya (1994) Functional and nutritional properties of tamarind (*Tamarindus indica*) kernel protein. *Food Chemistry* 49(1), 1–9.

Brown, W.H. (1954) *Useful Plants of the Philippines.* Republic of the Philippines. Department of Agriculture and Natural Resource, Technical Bulletin. Bur Printer, Manila.

Bueso, C.E. (1980) Sour sop, tamarind and chironka. In: Nagy, S. and Shaw, P.E. (eds) *Tropical and Subtropical Fruits.* AVI Publishing Westport, Connecticut, pp. 375.

Burgalassi, S., Raimondi, L., Pirisino, R., Banchelli, G., Boldrini, E. and Saettone, M.F. (2000a) Effect of xyloglucan (tamarind seed polysaccharide) on conjunctival cell adhesion to laminin and on corneal epithelium wound healing. *European Journal of Ophthalmology* 10(1), 71–76.

Burgalassi, S., Chetoni, P., Panichi, L., Boldrini, E. and Saettone, M.F. (2000b) Xyloglucan as a novel vehicle for timolol: pharmacokinetics and pressure lowering activity in rabbits. *Journal of Ocular Pharmacology and Therapeutics* 16(6), 497–509.

Carter, E.J.V., Paredes, G.E., Beristain, C.I. and Tehuitzil, H.R. (2001) Effect of foaming agents on the stability, rheological properties, drying kinetics and flavour retention of tamarind foam mats. *Food Research International* 34, 587–598.

Chapman, K.R. (1984) Tamarind. In: Page, P.E. (ed.) *Tropical Tree Fruits for Australia*. Queensland Department of Primary Industries, Information Series Q 183018, Brisbane, Australia, pp. 83–86.

Chaturvedi, A.N. (1985) Firewood farming on the degraded lands of the Gangetic plain. *Uttar Pradesh Forest Bulletin*15(1). Government of India Press, Lucknow, India, p. 286.

Dalziel, J.M. (1937) *The Useful Plants of West Tropical Africa*. Crown Agents for Overseas Governments and Administrations, London, p. 612.

D'Amico M., Di Filippo C., Lampa, E., Boldrini, E., Rossi, F., Ruggiero, A. and Filippelli, A. (1999) Effects of timolol and of timolol with tamarind seed polysaccharide on intraocular pressure in rabbits. *Pharmacy and Pharmacology Communications* 5(5), 361–364.

De Padua, L.S., Lugod, G.C. and Pancho, J.V. (1978) *Handbook of Philippines Medicinal Plants*. Los Banos Technical Bulletin, College, Laguna, Philippines, 3.

Duke, J.A. (1981) *Handbook of Legumes of World Economic Importance*. Plenum Press, New York, pp. 228–230.

Eggeling, W.J. and Dale, I.R. (1951) *The Indigenous Tree of the Uganda Protectorate*. The Government Printer, Entebbe, Uganda, p. 491.

Gayathri, G.N., Platel, K., Prakash, J. and Srinivasan, K. (2004) Influence of antioxidant spices on the retention of β-carotene in vegetables during domestic cooking processes. *Food Chemistry* 84(1), 35–43.

Ghelardi, E., Tavanti, A., Davini, P., Celandroni, F., Salvetti, S., Parisio, E., Boldrini, E., Senesi, S. and Campa, M. (2004) A mucoadhesive polymer extracted from tamarind seed improves the intraocular penetration and efficacy of rufloxacin in topical treatment of experimental bacterial keratitis. *Antimicrobial Agents and Chemotherapy* 48(9), 3396–3401.

Gidley, M.J., Lillford, P.J., Rowlands, D.W., Lang, P., Dentini, M. and Crescenzi, V. (1991) Structure and solution properties of tamarind-seed polysaccharide. *Carbohydrate Research* 214, 299–314.

Glicksman, M. (1986) *Food Hydrocolloids*. CRC Press, Boca Raton, Florida, pp. 191–202.

Goyal, P., Kumar, V. and Sharma, P. (2006) Carboxymethylation of tamarind kernel powder. *Carbohydrate Polymers* 69(2), 251–255.

Grollier, C., Debien, C., Dornier, M. and Reynes, M. (1998), Prominent characteristics and possible uses of the tamarind. *Fruits-Paris* 53(4), 271–278.

Hockin, G.D. (1993) *Trees for Drylands*. Tamarindus indica *L., Family Leguminosae: Caesalpinoideae*. Oxford and IBH Publishing Co., New Delhi, pp. 305–307.

http://www.haryana-online.com/Flora/imli.htm.

http://www.hort.purdue.edu/newcrop/morton/tamarind.html.

http://www.indianpatents.org.in/db/test.

http://www.kingtutshop.com/Egyptian-Herb/Tamarind.htm.

http://www.scopus.com/scopus/home.url.

http://www.uni-graz.at/~katzer/engl/Tama_ind.html.

Ishola, M.M., Agbaji, E.B. and Agbaji, A.S. (1990) A chemical study of *Tamarindus indica* (Tsamiya) fruits grown in Nigeria. *Journal of the Science of Food and Agriculture* 51, 141–143.

Jayaweera, D.M.A. (1981) *Medicinal Plants (Indigenous and Exotic) Used in Ceylon. Part II. Flacourtiaceae-Lytharaceae*. National Science Council of Sri Lanka, Colombo, pp. 244–246.

Khandare, A.L., Uday Kumar, P., Shanker, R.G., Venkaiah, K. and Lakshmaiah, N. (2004) Additional beneficial effect of tamarind ingestion over defluoridated water supply to adolescent boys in a fluorotic area. *Nutrition* 20(5), 433–436.

Kulkarni, R.S., Gangaprasad, S. and Swamy, G.S.K. (1993) Tamarind: economically an important minor forest produce. *Minor Forest Products News* 3(3), 6.

Lakshmanan, K.K. and Narayanan, A.S.S. (1994) Antifertility herbals used by the tribals in Anaikkatty hills. *Journal of Economic and Taxonomic Botany* 14(1), 171–173.

Lewis, Y.S. and Neelakantan, S. (1964) The chemistry, biochemistry and technology of tamarind. *Journal of Scientific and Industrial Research* 23, 204–206.

Luengthanaphol, S., Mongkholkhajornsilp, D., Douglas, S., Douglas, P.L., Pengsopa, L.I. and Pongamphai, S. (2004) Extraction of antioxidants from sweet Thai tamarind seed coat—preliminary experiments. *Journal of Food Engineering* 63(3), 247–252.

Marangoni, A., Alli, I. and Kermasha, S. (1988) Composition and properties of seeds of the true legume *Tamarindus indica* L. *Journal of Food Science* 53, 1452–1455.

Marry, M., Cavalier, D.M., Schnurr, J.K., Netland, J., Yang, Z., Pezeshk, V., York, W.S., Pauly, M. and White, A.R. (2003) Structural characterization of chemically and enzymatically derived standard oligosaccharides isolated from partially purified tamarind xyloglucan. *Carbohydrate Polymers* 51(3), 347–356.

Martinello, F., Soares, S.M., Franco, J.J., Santos, A.C., Sugohara, A., Garcia, S.B., Curti, C. and Uyemura, S.A. (2006) Hypolipemic and antioxidant activities from *Tamarindus indica* L. pulp fruit extract in hypercholesterolemic hamsters. *Food and Chemical Toxicology* 44(6), 810–818.

Meillon, S. (1974) Process for making drinks, syrups, juice and liquor and solid extracts based on tamarind products thus obtained. French Patent No. 2231322 (English summary).

Mishra, R.N. (1997) Tamarindus indica *L.: An Overview of Tree Improvement.* Proceedings of National Symposium on *Tamarindus indica* L., 27–28 June, Tirupati, Andhra Pradesh, Government of Andhra Pradesh, India, pp. 51–58.

Morton, J. (1987). Tamarind. In: Morton, J.F. (ed.) *Fruits of Warm Climates*. Florida Flair Books, Miami, Florida, pp. 115–121.

Omya, T. and Tabuchi, K. (1985) Tamarind gum carboxymethyl ether alkali metal salts. Japanese Patent 60, 166, 302.

Pino, J.A., Escalona, J.C., Licea, I., Perez, R. and Aguero, J. (2002) Leaf oil of *Tamarindus indica* L. *Journal of Essential Oil Research* 14, 187–188.

Pino, J.A., Marbot, R. and Vazquez, C. (2004) Volatile components of tamarind (*Tamarindus indica* L.) grown in Cuba. *Journal of Essential Oil Research* 16, 318–320.

Pongsawatmanit, R., Temsiripong, T., Ikeda, S. and Nishinari, K. (2006) Influence of tamarind seed xyloglucan on rheological properties and thermal stability of tapioca starch. *Journal of Food Engineering* 77(11), 41–50.

Premaja, P. and Manojkumar, K. (2007) *Arecanut and Spices Database*. Directorate of Arecanut and Spices Development, Government of India, Calicut, Kerala, pp. 102.

Pugalenthi, M., Vadivel, V., Gurumoorthi, P. and Janardhanan, K. (2004) Composative nutritional evaluation of little known legumes, *Tamarindus indica, Erythirna indica* and *Sesbania bispinosa. Tropical and Subtropical Agroecosystems* 4, 107–123.

Purseglove, J.W. (1987) *Tropical Crops. Dicotyledons*. Longman Science and Technology, London, pp. 204–206.

Rao, P.S. and Beri, R.M. (1955) Acetylation of tamarind seed jellose. *Proceedings of the Indian Academy of Sciences A* 42, 199–203.

Sabu, A., Pandey, A., Jaafar Daud, M. and Szakacs, G. (2005) Tamarind seed powder and palm kernel cake: two novel agro residues for the production of tannase under solid state fermentation by *Aspergillus niger* ATCC 16620. *Bioresource Technology* 96(11), 1223–1228.

Sagrero, L., Bartley, J.P. and Provis Schwede, A. (1994) Supercritical fluid extraction of the volatile constituents from tamarind (*Tamarindus indica* L.). *Journal of Essential Oil Research* 6, 547–548.

Saideswara Rao, Y. and Mary Mathew, K. (2001). Tamarind. In: Peter, K.V. (ed.) *Handbook of Herbs and Spices*. Woodhead Publishing Limited, Cambridge, UK, pp. 287–296.

Schiavio, A. and Maderno, C. (1958) Soluble derivatives of the glucogalactoxylans of tamarind flour. Italian Patent 574, 928.

Shankaracharya, N.B. (1998) Tamarind – chemistry, technology and uses: a critical appraisal. *Journal of Food Science and Technology* 35(3), 193–208.

Sharma, S. and Bhardwaj, R. (1997) *Tamarind – A Suitable Fruit Crop for Dry Arid Regions*. Proceedings of National Symposium on *Tamarindus indica* L., 27–28 June, Tirupati, Andhra Pradesh, Government of Andhra Pradesh, India, pp. 4–6.

Shimohiro, Y., Ogawa, M. and Ohnishi, T. (1983) Thickening agent for paste. European Patent Application 130, 385.

Siddhuraju, P. (2006) Antioxidant activity of polyphenolic compounds extracted from defatted raw and dry heated *Tamarindus indica* seed coat. *Food Science and Technology* 40(6), 982–990.

Soong, Y.Y. and Barlow, P.J. (2004) Antioxidant activity and phenolic content of selected fruit seeds. *Food Chemistry* 88(3), 411–417.

Srivastava, H.C. and Singh, P.P. (1967) Structure of the polysaccharide from tamarind kernel. *Carbohydrate Research* 4(4), 326–342.

Strickland, F.M., Darvill, A., Albersheim, P., Eberhard, S., Pauly, M. and Pelley, R.P. (1999) Inhibition of UV-induced immune suppression and interleukin-10 production by plant oligosaccharides and polysaccharides. *Photochemistry and Photobiology* 69(2), 141–147.

Sudjaroen, Y., Haubner, R., Würtele, G., Hull, W.E., Erben, G., Spiegelhalder, B., Changbumrung, S., Bartsch, H. and Owen, R.W. (2005) Isolation and structure elucidation of phenolic antioxidants from Tamarind (*Tamarindus indica* L.) seeds and pericarp. *Food and Chemical Toxicology* 43(11), 1673–1682.

Timyan, J. and Bwa, Y.O. (1996) *Important Trees in Haiti*. Southeast Consortium for International Development, Washington, DC.

Troup, R.S. (1921) *The Silviculture of Indian Trees. Leguminosae (Carsalpinieae) to Vorbenaceae. 7.* Tamarindus indica *L.* Clarendon Press, Oxford, UK, 263–363.

Veluraja, K., Ayyalnarayanasubburaj, S. and Paulraj, A.J. (1997) Preparation of gum from tamarind seed and its application in the preparation of composite material with sisal fibre. *Carbohydrate Polymers* 34, 377–379.

Von Maydell, M.J. (1986) *Trees and Shrubs of Sahel; Their Characteristics and Uses.* Deutsche. Geseuschft Fuer Techhische Zusammenarboit, Eschborn, Germany, pp. 111.

Wong, K.C., Tan, C.P., Chow, C.H. and Chee, S.G. (1998) Volatile constituents of the fruit of *Tamarindus indica* L. *Journal of Essential Oil Research* 10(2), 219–221.

Zablocki, L. and Perore, S. (1996) Tamarind extract (NutraSweet Co., Deerfield, Illinois) United States Patents US 5 474791. In: *Trends in Food Science and Technology* 7, 307.

21 Parsley

Shamina Azeez and V.A. Parthasarathy

21.1. Introduction

Parsley, *Petroselinum crispum* (Syn. *Apium petroselinum* Linn.; *P. lativum* Hoffm.; *Carum petroselinum* Benth), is a biennial herb belonging to the family Apiaceae. It is native to southern Europe and western Asia and in many parts of the world is cultivated commercially as an annual for its attractive and aromatic leaves. In America, parsley is used mostly as a garnish, while in Europe and the Middle East it is used almost as often as salt (http://www.chilipaper.com/). Chopped parsley leaves are a popular decoration in Central Europe (similar to the use of coriander leaves in China, South-east Asia and parts of India), mostly for soups and vegetables. The Latin name, *Petroselinum*, was derived from Greek *pétros*, rock, stone. *Selinum* was the Latin name of celery. The species name was given because of the crispate leaf shape. Parsley has been known for over 1000 years in the Mediterranean (http://www.uni-graz.at/).

Parsley is of European (probably Western Mediterranean) origin. The plant was introduced into England from Sardinia in 1548. According to Linnaeus, the wild habitat of parsley is Sardinia, from where it was brought to England and apparently first cultivated there in 1548, while Bentham considered it a native of the Eastern Mediterranean regions and De Candolle reported Turkey, Algeria and the Lebanon to be its home (Charles, 2000). Since its introduction into the British Isles in the 16th century it has been completely naturalized in various parts of England and Scotland. The Greeks held parsley in high esteem, crowning the victors with chaplets of parsley at the Isthmian games and making wreaths with it for adorning the tombs of their dead. Homer relates that warriors fed the leaves to chariot horses (http://www.botanical.com/). European colonists brought parsley to the USA in the 17th century. It is grown throughout Florida as a commercial crop of minor importance in the vegetable-producing areas of central and southern Florida (http://edis.ifas.ufl.edu/).

Petroselinum, the specific name of parsley, from which the English name is derived, is of classic origin. This last name in the Middle Ages was corrupted into *Petrocilium* – this was anglicized into petersylinge, persele, persely and finally parsley. Linnaeus in 1764 named it *A. petroselinum*, and later to the genus *Carum* (http://www.botanical.com/botanical/mgmh/p/parsle09. html).

21.2. Botany and Uses

Botany

The erect-growing parsley reaches a height of 0.30–0.46 m (1 to 1.5 ft) and has green leaves and greenish-yellow flowers in compound umbels. The seeds are smooth, ribbed and ovate. Two different varieties are commonly grown, e.g. root parsley (var. *tuberosum*), which has a tender, edible root (used as an aromatic vegetable) and leaf parsley, cultivated solely for its leaves (var. *latifolium* – broad-leaved; var. *crispum* – curly-leaved) for use as a garnish (http://www.uni-graz.at/).

It is propagated by planting seeds, which are sown about 6 mm deep and covered with a thin mulch layer until germination, which occurs in 7–12 days. The seedlings may be transplanted later. The plants are spaced 5–8 cm apart in rows, 0.30 m (1 ft) apart. Parsley requires a very moist soil and careful weeding is necessary.

Cultivars

There are no fewer than 37 varieties reported and the most valuable is 'Curled Leaf,' a compact type with close, perfectly curled leaves and very finely divided leaf type. 'Italian' (or plain-leaf) is a less decorative but flavourful parsley that most closely resembles the original non-curly plants of Europe. It is not cultivated much now, the leaves being less attractive than those of the curled, is of a less brilliant green and coarser in flavour. The 'Hamburg', or turnip-rooted parsley, is grown only for the sake of its enlarged fleshy parsnip-like and turnip-shaped taproot. 'Neapolitan' (or celery leaf) is grown for its leaf stalks, which are blanched and eaten like celery; and 'Dwarf' is suitable both for ornamental and culinary purposes. Both the crowded, dense-leaved type and the broad, open-growing type were described as early as the 4th century BC (http://www.botanical.com/).

Production

The estimated number of hectares of parsley cultivated in North America is 25,091, while worldwide it is 250,905. The yield per acre is reported to be 4238 kg dry herb and 32 kg oil, with oil on a fresh-weight basis being 0.26% (http://www.ag.montana.edu).

Parsley is a biennial plant but is usually produced as an annual crop. It can be grown from seeds or divisions in fertile soils in full or partial sunlight. Parsley matures in 70–90 days; the harvest begins in October and continues through March, depending on weather and location. Parsley plants form a healthy rosette in the first year, winter mortality being low. Plant growth and seed production are excellent in the second year. Parsley leaves can be hand-harvested three to four times in a season and the plants yield approximately 2.24–6.72 t/ha. Fresh parsley can be stored for up to 2.5 months at 0°C (http://www.ams.usda.gov/). The roots, leaves and seeds of parsley are used either fresh or as dried oil. The components derived are starch, mucilage, sugar, volatile oil, terpenes, apiin and apiole.

Uses

Parsley leaves are ready for use about 3 months after seeding. A few leaves at a time may be removed from each plant, or the entire bunch of leaves may be removed for use. Although parsley leaves are used most commonly in the fresh green condition as a garnish, their characteristic flavour and green colour can be retained if the leaves are dried rapidly. Dehydrated parsley flakes are produced from parsley grown in commercial fields. Green parsley leaves have a mild, agreeable flavour and are an excellent source of vitamin C, iodine, iron and other minerals. Quite often, parsley is left on the plate to become the last bite, as it tends to sweeten the breath (http://edis.ifas.ufl.edu/).

The finely chopped leaves are used as flavouring in sauces, soups, stuffing, rissoles, minces, etc., and are also sprinkled over vegetables or salads. The leaves are also dried and powdered as a culinary flavouring when fresh leaves are not available. In addition to the leaves, the stems are also dried and powdered, both as a culinary colouring and as a dye. The roots of the turnip-rooted variety are used as a vegetable and flavouring. The 2-year-old roots are used for medicinal purposes, the leaves are dried, for making parsley tea, and the seeds are used for the extraction of an oil called apiole, which is of considerable curative value. The best seed for medicinal purposes is that obtained from the Triple Moss curled variety, which is grown for producing apiole (http://www.botanical.com/).

Parsley leaves, which are strongly diuretic, can jump-start weight loss, and their high vitamin C content makes them useful against colds and flu. Their invigorating, mild flavour is a key ingredient in tabbouleh, a Middle Eastern salad (http://findarticles.com/). The powdered seeds of parsley are a folk remedy for hair growth and scalp stimulation, when massaged into the scalp. It also has strong antioxidant properties (Pizzorno and Murray, 1985).

21.3. General Composition

Extraction

Soysal (2004) determined the effects of microwave output power on drying time, drying rate and the dried product quality in terms of the colour of the parsley leaves when dried in a domestic microwave oven. The value of the drying constant increases with increased microwave output power. Microwave drying does not affect the colour parameters of the leaves, except for some decrease in whiteness. Although some darkening may occur, microwave drying maintains a good green colour close to that of the original fresh parsley leaves.

Composition

A rich source of iron and vitamins C and A, parsley also yields fatty acids and an essential or volatile oil. The essential oil of the leaves is considered superior to that from the seeds and is used in condiments and seasonings. Parsley seed oil is used in fragrances for perfumes, soaps and creams. Parsley has a very high content of vitamins (β-carotene, thiamin, riboflavin and vitamins C and E) and is a rich source of calcium, iron and folate (Athar *et al.*, 1999). A high proportion of the carotene is 9-*cis*-β-carotene, which is considered effective against cancer and cardiovascular disease (Ben-Amotz and Fishier, 1998).

Factors affecting composition

CULTIVAR The turnip-rooted variety, Halblange [Half-long], had the lowest ratios of myristicin to apiole (Franz and Glasl, 1976). Parsley cultivars belonging to the *vulgare* group had the highest content of sugar, crude protein and carotene, and those of radicosum the highest content of ascorbic acid. Madzharova *et al.* (1973) crossed the celery cvs Pioneer and Prolet with the parsley cvs Listen and Berlinski and with Festival 68 (parsley × celery). New leaf forms were obtained which had tender leaves, were rich in vitamin C, minerals, protein and sugars, had a celery aroma and could be used like parsley. Certain lines relatively resistant to *Septoria apiicola* were selected. The essential oil of the celery × parsley hybrid, named Festival 68, was similar to that of parsley. The principal constituents in the hybrid essential oil were γ-terpene and heptanol, in parsley myrcene and in celery myrcene and limonene. There were variations in the essential oil content of different varieties. Franz and Glasl (1976) found that Hamburger Schmitt [Hamburg Cutting] and Enface Schmitt [Plain Cutting] had a relatively high percentage of oil in the fruits, and the fruit oil in the former contained 22% 2,3,4,5-tetramethoxyallylbenzene. The turnip-rooted variety Challenge [Half-long] had the lowest ratios of myristicin to apiole. A correlation was noted between the aromatic properties and root shape, cvs with long, thin and evenly tapering roots having the highest aromatic rating. Simon and Quinn (1988) detected thymol in seven accessions, at 2% or less, and this was claimed to be the first report of this compound in parsley leaf oil.

CLIMATE The quantity of vitamin C was increased by large temperature changes, especially by low night temperatures; thus, the contents were most frequently highest in the north (Hardh, 1975). However, Moore *et al.* (1997) reported that, when grown at high CO_2, leaf ribulose-1,5-bisphosphate carboxylase/oxygenase content was not affected in parsley that produces mannitol.

FERTILIZER Fertilizer application has been reported to influence the quality of parsley. Studies on the application of NPK at 45 g N + 100 g P_2O_5 + 55 kg K_2O/m^3, or two or three times this rate of NPK, indicated that N and P rates, but not the K rate, had a significant effect on the total chlorophyll content, which increased as the rate increased in both cases (Gurgul *et al.*, 1996). There was no change in the total sugar and ascorbic acid contents in response to N, P or K application, but in the case of P only there was a clear increasing trend in both as the application rate increased. Ascorbic acid content increased in response to increasing the K rate. Applying N, P or K increased the activity of peroxidase and catalase, particularly during the early phases of growth (Gurgul *et al.*, 1996).

ORGANIC FERTILIZERS Franken and Gnadinger (1994) studied the molecular aspects of the symbiosis between plants and arbuscular endomycorrhizal fungi in parsley cv. Hamburger Schnitt. Phosphate nutrition and low light conditions influenced plant–fungal interactions negatively in different ways. Without chemical fertilizers, legume green manure crops, particularly sunn hemp and hyacinth bean, can increase the yield of culinary herbs like parsley in a crop rotation system (Palada *et al.*, 2004).

NUMBER OF CUTS Essential oil yield and other chemical parameters are also influenced by the number of cuts. Essential oil was greatest at 0.02% chlorophyll contents, which increased gradually from the first to the third cut (El Sherbeny and Hussein, 1993).

STORAGE CONDITIONS The least decay in storage at room temperature occurs when parsley is harvested at 70 days. Storage at 0°C and 84% RH doubled the shelf life compared with storage at room temperature. For root parsley, irrespective of cultivar, the vitamin C content was highest in roots from the May sowings. The best storage can be obtained with roots of plants sown in April and root contents of vitamin C, dry matter, sugars and nitrates decreased during storage (Bakowski *et al.*, 1994).

DISTANT HYBRIDIZATION A hybrid parsley, cultivars Festival 68, was derived from a parsley × celery cross, producing 50–80% more foliage than standard parsley cultivars (Madjarova and Bubarova, 1978). Festival 68 has the morphological characteristics resembling parsley. The leaves have a high content of ascorbic acid, sugars and essential oils. The leaf yields are 50–80% higher than those of ordinary commercial varieties. New root forms from the same interspecific combination, having intermediate characters in leaf rosette and higher contents of ascorbic acid, carotene, chlorophyll, essential oils and amino acids than either parent, have been obtained, together with others which have larger roots than those of their parents and a long storage period, similar to that of parsley. No differences were observed in total β-carotene levels or in its *cis*-isomer fractions at the doses of ionizing radiation required for the preservation of foods, nor did it contribute to a decrease of vitamin A (Sebastião *et al.*, 2002).

Table 21.1. The composition of parsley leaves (per 100 g edible material).

Parameter	Content
Water (g)	79–89
Fibre (g)	0.9–9.1
Starch (g)	0
Sugar (g)	t
Total acidity (meq)	–
Ash (g)	1.4–2.4
Fat (g)	t–1.0
Protein (g) (N × 6.25)	3.7–5.2
Calories (Kcal)	21–60
Ascorbic acid (mg)	110–200
Carotene (mg)	4.4–8.8
Thiamine (mg)	0.09–0.2
Riboflavin (mg)	0.18–0.6
Niacin (mg)	0.53–1.8
Folic acid (pg)	40
Calcium (mg)	139–325
Iron (mg)	2.3–19.0

t = trace.
Source: http://aggie-horticulture.tamu.edu/.

Toxic compounds, such as the photosensitizing furocoumarines including psoralen, bergaptene and isoimperatorin (Manderfeld *et al.*, 1997), which can induce dermatitis, have been found in parsley roots, though in very low concentrations (Lagey *et al.*, 1995). The composition of parsley leaves is given in Table 21.1.

21.4. Chemistry of Volatiles

Extraction

Fischer *et al.* (1991) have applied a high-speed counter-current chromatography with an Ito multi-layer coil separator-extractor to perform efficient separations of aroma-relevant constituents, such as phthalides, from celery and parsley roots.

Different plant materials including parsley were extracted with liquid carbon dioxide under liquid-vapour conditions by Naik and Lentz (1989). The yields of the CO_2 extractions were 10–360% larger than the yields of the steam distillations, while the extraction time was only 1/2 to 1/10 of the time needed for distillation. The energy consumption of the extraction process was approximately a factor of three lower than the energy required for steam distillation.

Composition

The major constituents of parsley leaves are 1,3,8-*p*-menthatriene, followed by β-phellandrene, myristicin and myrcene. Parsley accessions high in the specific constituents (percentage of essential oil) 1,3,8-*p*-menthatriene (68%), myristicin (60%), β-phellandrene (33%), apiole (22%), myrcene (16%), terpinolene and 1-methyl-4-isopropenylbenzene (13%), and a compound of molecular weight 268 (dimer) (10%), were identified by Simon and Quinn (1988). Lamarti *et al.* (1991) reported that the curly-leaved parsley cultivars could be distinguished by their light, dark-brown mericarps, the essential oil of which was rich in monoterpenes, particularly α-pinene (15.7–24.1%) and β-pinene (9.6–15.1%). The structure of the volatile components of parsley is given in Fig. 21.1.

In root parsley, only α-pinene (10.6%), β-pinene (7.1%), myristicin (2.5%) and apiole (79.8%) have been found, while apiole (30.4–67.5%) is the principal constituent of Giant Italian parsley (Table 21.2). Myristicin (0.7–62.3%) was present in all parsley specimens analysed. Volatile chemicals obtained from the leaves of parsley, *P. sativum*, by steam distillation, isopentane extraction and headspace analysis were identified by Kasting *et al.* (1972) using GLC-MS. The presence in leaf oil of α-pinene, β-pinene, myrcene, β-phellandrene, *trans*-β-ocimene, γ-terpinene, 1-methyl-4-isopropenyl benzene and 1,3,8-*p*-menthatriene, as shown by earlier investigators, has been confirmed and the number of volatile chemicals detected in the leaves increased by an additional 42. Sniffing tests of effluent from a gas chromatograph of a concentrate from parsley leaves has shown that 1,3,8-*p*-menthatriene is only one of several compounds that give a parsley-like aroma.

The 45 aroma volatiles of desert parsley were identified by MacLeod *et al.* (1985), including 11 previously not reported as parsley leaf volatiles. The major constituents of the

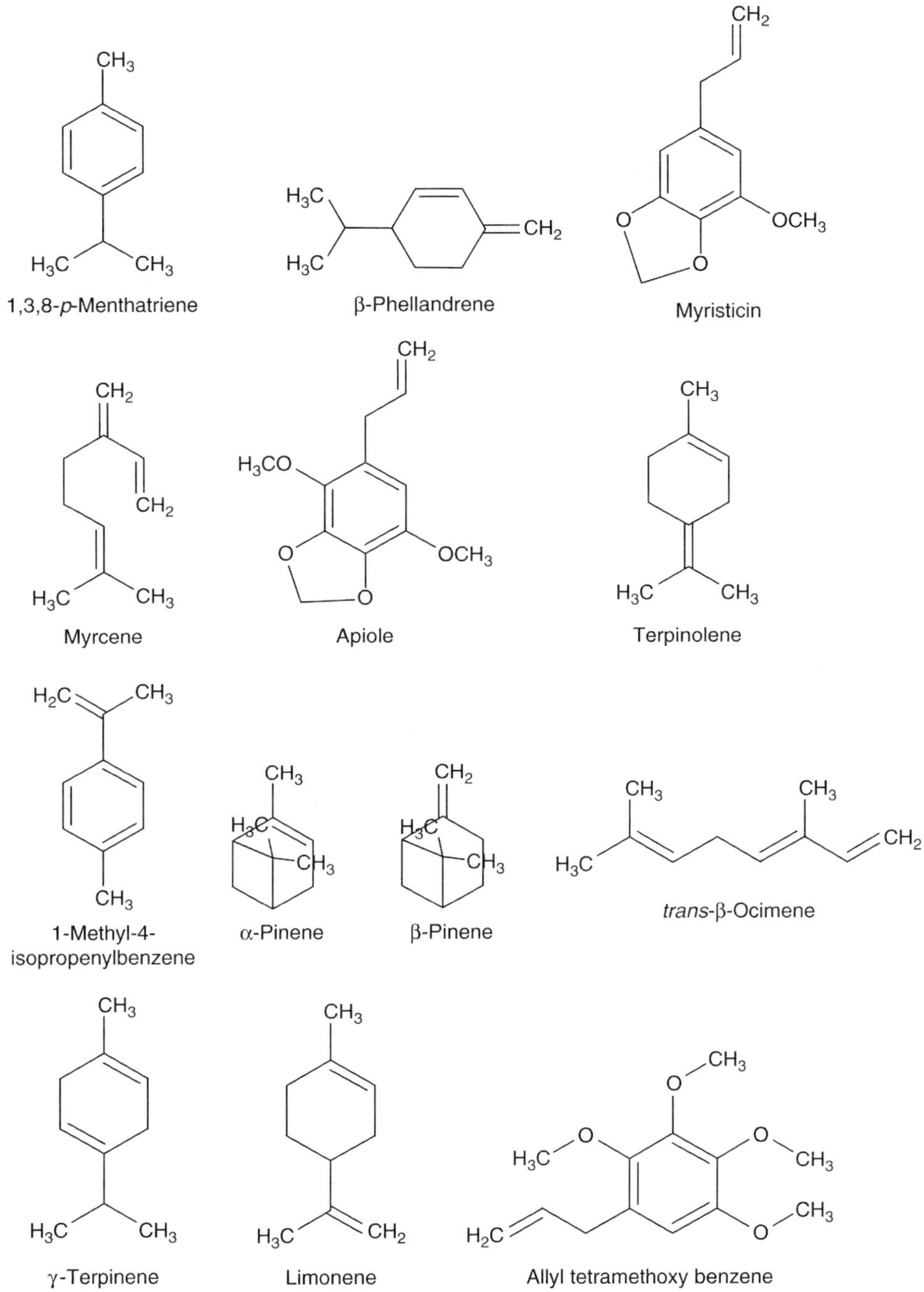

Continued

Fig. 21.1. Volatile constituents of parsley.

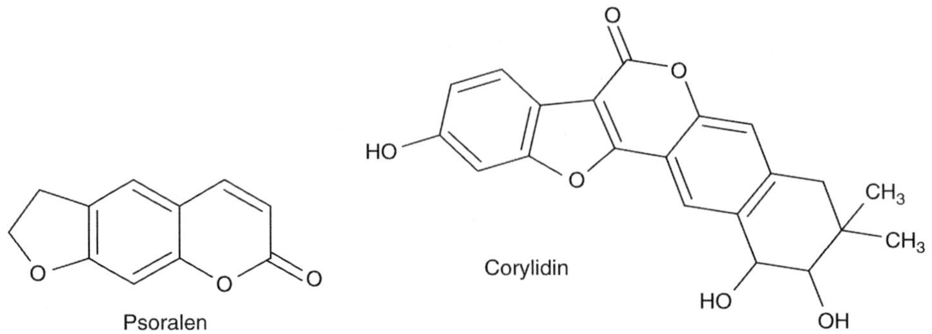

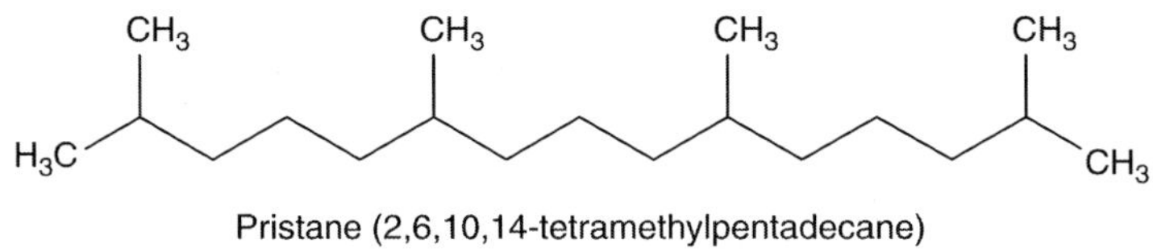

Fig. 21.1. *Continued*

Table 21.2. Major volatile constituents of parsley (% of essential oil).

Parsley leaves	Curly-leaved parsley	Root parsley	Giant Italian parsley
1,3,8-*p*-Menthatriene (68)	α-Pinene (15.7–24.1)	α-Pinene (10.6)	Apiole (30.4–67.5)
β-Phellandrene (33)	β-Pinene (9.6–15.1)	β-Pinene (7.1)	
Myristicin (60)		Myristicin (2.5)	
Myrcene (16)		Apiole (79.8)	
Apiole (22)			
1-Methyl-4-isopropenylbenzene (13)			

sample are 4-methoxy-6-(prop-2-enyl)benzo-1,3-dioxolan (myristicin) 4,7-dimethoxy-5-(prop-2-enyl)benzo-1,3 dioxolan (apiole), β-phellandrene, *p*-mentha-1,3,8-triene and 4-isopropenyl-1-methylbenzene. Aroma assessments with GC show that apiole, in particular, has a desirable parsley odour character. One component, 2-(*p*-tolyl)propan-2-ol, is a new aroma volatile and, together with *p*-mentha-1,3,8-triene, may be unique to parsley.

The essential oils of leaves and roots have approximately the same composition. The main components (10–30%) are myristicin, limonene and 1,3,8-*p*-menthatriene; minor components are mono- and sesquiterpenes. The curly varieties (var. *crispum*) tend to be richer in myristicin, but contain much less essential oil than var. *latifolium* (0.01 and 0.04%, respectively) (http://www.botanical.com/botanical/). In contrast, the essential oil from the fruits (3–6%) is dominated either by myristicin (60–80%; mostly var. *tuberosum* and var. *crispum*) or by apiole (70%, mostly var. *latifolium*). A third chemical race shows allyl tetramethoxy benzene (55–75%), which can also appear in apiole-dominated oils (up to 20%).

Spraul *et al.* (1992) revealed the presence of two novel compounds in parsley, $C_{20}H_{30}O_4$ and $C_{20}H_{32}O_3$, isolated and purified by means of high-speed counter-current chromatography and their structures elucidated by spectroscopic methods and some chemical transformations. Their systematic names according to the chemical abstract nomenclature are:

[1*S*-[1α,2β(*Z*),4α,8aβ]]-[1,2,4a,5,6,7,8,8a-octahydro-l-hydroxy-4,4a-dimethyl-l-(l-methylethyl)-7-oxo-2-naphthalenyl]-2-methyl-2-butenoate for compound 1 and [1*S*-[1α,2β(*Z*),4aα,8aβ]]-[1,2,4a,5,6,7,8,8a-octahydro-l-hydroxy-4,4a-dimethyl-1-(l-methylethyl)-2-naphthalenyl]-2-methyl-2-butenoate for compound 2. The trivial names crispanone and crispane, respectively, were proposed. Later, the structure of the sesquiterpenes, crispanone and crispane, was revised to that of siol angelate and lasidiol angelate, respectively, and a novel phenylpropanoid (apional) was also isolated by Appendino *et al.* (1998).

Factors affecting the quality of parsley

FERTILIZERS Growing parsley in nickel (Ni)-supplemented clay soils, at low levels (50 mg/kg soil), increases leaf essential oil content and quality without affecting leaf chlorophyll and iron contents, but reduces total soluble solids, L-ascorbic acid, nitrate and ammonium levels. Increasing Ni levels up to 100 mg/kg soil results in visible symptoms of leaf chlorosis, which coincides with a sudden drop in leaf chlorophyll content and reduced N and Mg levels relative to that of the control. The main aroma constituent of parsley leaves, 1,3,8-*p*-menthatriene, which forms about 62% of the essential oil, showed a 10–25% increase over that of the control with 25 mg or higher levels of Ni fertilization per kg soil. It is suggested that low levels of Ni fertilization, particularly 50 mg/kg clay soil, strongly improve not only parsley leaf yield and quality (i.e. leaf area, mineral content, oil yield and flavour) but also the leaves are safer for human consumption since their nitrate and ammonium contents are reduced significantly (Atta-Aly, 1999).

STRESS Changes in the quantity and quality of the volatile oils from parsley in response to certain stress agents, infection with *Cercospora petroselini* and treatment of this infection with Cuprosan, have been reported (Hashem and Sahab, 1999), e.g. increase in the concentration of Cu ions in the leaves; isolation of psoralen in samples treated with Cu salts; isolation of corylidin, angladin and pereflorin B from parsley infected with *C. petroselini*. Of these compounds, corylidin, angladin and psoralen inhibited growth of *Pseudomonas putida*, *Escherichia coli* and *Rhizobium meloloti* (Gram-negative bacteria), while pereflorin B inhibited *Streptococcus lactis* and *Bacillus subtilis* (Gram-positive bacteria).

Of the 55 common fruits and vegetables assessed for their concentration of pristane, a natural saturated terpenoid alkane (2,6,10,14-tetramethylpentadecane), by quantitative gas-liquid chromatography, the highest content was observed in parsley, which contained 124 μg/g of fresh sample; pristane levels in the remaining 54 foodstuffs analysed ranged from 0.02 to 1.70 μg of pristine/g of fresh sample. The amounts of pristane in average serving sizes of representative samples ranged from 1.5 to 107 μg. On the basis of the data obtained from this study by Chung *et al.* (1989), it appears that we are exposed to appreciable amounts of pristane in diets that include parsley.

The irradiation of parsley with doses as high as 5 M rad does not bring about any distinct qualitative and quantitative changes, as per the study of Josimovic (1983).

21.5. Chemistry of Non-volatiles

Composition

The nutrient content in 100 g fresh leaves of Hamburg and leafy-type parsley is summarized in Table 21.3. Parsley blanched before freezing showed significant losses in the contents of vitamin C (47–51%), nitrates (22–33%) and nitrites (43–55%), and lesser but significant loss of dry matter. During freezing and storage of frozen products, there were losses in vitamin C, β-carotene and chlorophyll, while the levels of nitrates and nitrites were variable. Particularly great losses of vitamin C and β-carotene were observed in non-blanched frozen leaves stored at –20°C. After 9 months' storage, frozen products preserved 10–44% of vitamin C, 37–91% of β-carotene, 78–95% of

Table 21.3. Proximate composition of Hamburg and leafy-type parsley.

Parameter	Hamburg	Leafy type
Dry matter (g)	20.0	17.3
Vitamin C (mg)	310	257
β-Carotene (mg)	7.5	9.4
Chlorophyll (mg)	203	–
N-NO_3 (mg)	30.8	68.5
N-NO_2 (mg)	0.078	0.077

Source: Lisiewska and Kmiecik (1997).

chlorophyll and 78–153% of nitrates. Of the types of parsley analysed, the Hamburg type was a better raw material for freezing because of a significantly higher content of vitamin C and chlorophyll and significantly less nitrates in frozen products. When the storage temperature was −30°C, the blanching of leaves was not necessary, although it helped their pressing into cubes (Lisiewska and Kmiecik, 1997).

Natural antioxidants

With the increasing interest in the food industry for natural sources of antioxidants for their beneficial effects on health, new potential sources have been screened among edible aromatic plants and microalgae. The α-tocopherol content (a potent antioxidant) in parsley was reported to be 3.45 mg/100 g of fresh leaves obtained through supercritical fluid extraction (Diego *et al*., 2004).

Recent epidemiological studies have directed the attention from the synthetic all-*trans* β-carotene to natural carotenoids predominant in fruits and vegetables as possible active ingredients for the prevention of cancer and cardiovascular diseases. Fruits and vegetables commonly consumed in Israel were analysed by Ben-Amotz and Fishier (1998) for their carotenoid content, with emphasis on 9-*cis* β-carotene using reversed-phase, 3D photodiode array HPLC. Fourteen carotenoids were eluted in order of decreasing polarity, from polar oxycarotenoids to lipophilic hydrocarbons. The richest sources of total carotenoids (> 100 µg/g dry weight) in Israeli vegetables were carrot, dill, parsley, tomato, lettuce, sweet potato and red pepper. The green vegetables had high contents of xanthophylls and hydrocarbon carotenes. Relatively high ratios (9-*cis* to all-*trans* β-carotene) of above 0.2 g/g were noted in sweet potato, papaya, parsley, lettuce, dill, apricot, pepper, prune and pumpkin. The authors are of the opinion that the high content of 9-*cis* β-carotene in certain fruits and vegetables and the wide variety of carotenoids and stereoisomers of carotenoids in all plants should shift nutritional and medical attention from the synthetic all-*trans* β-carotene towards natural carotenoids as potential candidates for chemoprevention.

Earlier reports by Hart and Scott (1995) showed that parsley is a good source (> 1000 µg/100 g) of lutein and lycopene. There was little or no loss of carotenoids on cooking. Green vegetables showed an average increase in lutein levels of 24% and of 38% in β-carotene levels.

Rontani *et al*. (2005) reported the results of experiments that supported the significance of the photo-oxidation of the unsaturated components of higher plant cutins in the natural environment. Visible light-induced senescence experiments carried out with parsley resulted in the formation of 9-hydroperoxy-18-hydroxyoctadec-10(*trans*)-enoic and 10-hydroperoxy-18-hydroxyoctadec-8(*trans*)-enoic acids derived from type II (i.e. involving $^1O_2$) photo-oxidation of 18-hydroxyoleic acid and subsequent cutin depolymerization. These results showed that, in senescent plants, where the $^1O_2$ formation rate exceeds the quenching capacity of the photoprotective system, $^1O_2$ can migrate outside the chloroplasts and affect the unsaturated components of cutins.

Flavonoids

Plant species of the family Apiaceae are known to accumulate flavonoids, mainly in the form of flavones and flavonols (Fig. 21.2). Kreuzaler and Hahlbrock (1973) isolated 24 different flavonoid glycosides from illuminated cell suspension cultures of parsley (*P. hortense*). The chemical structures of 14 of these compounds were further

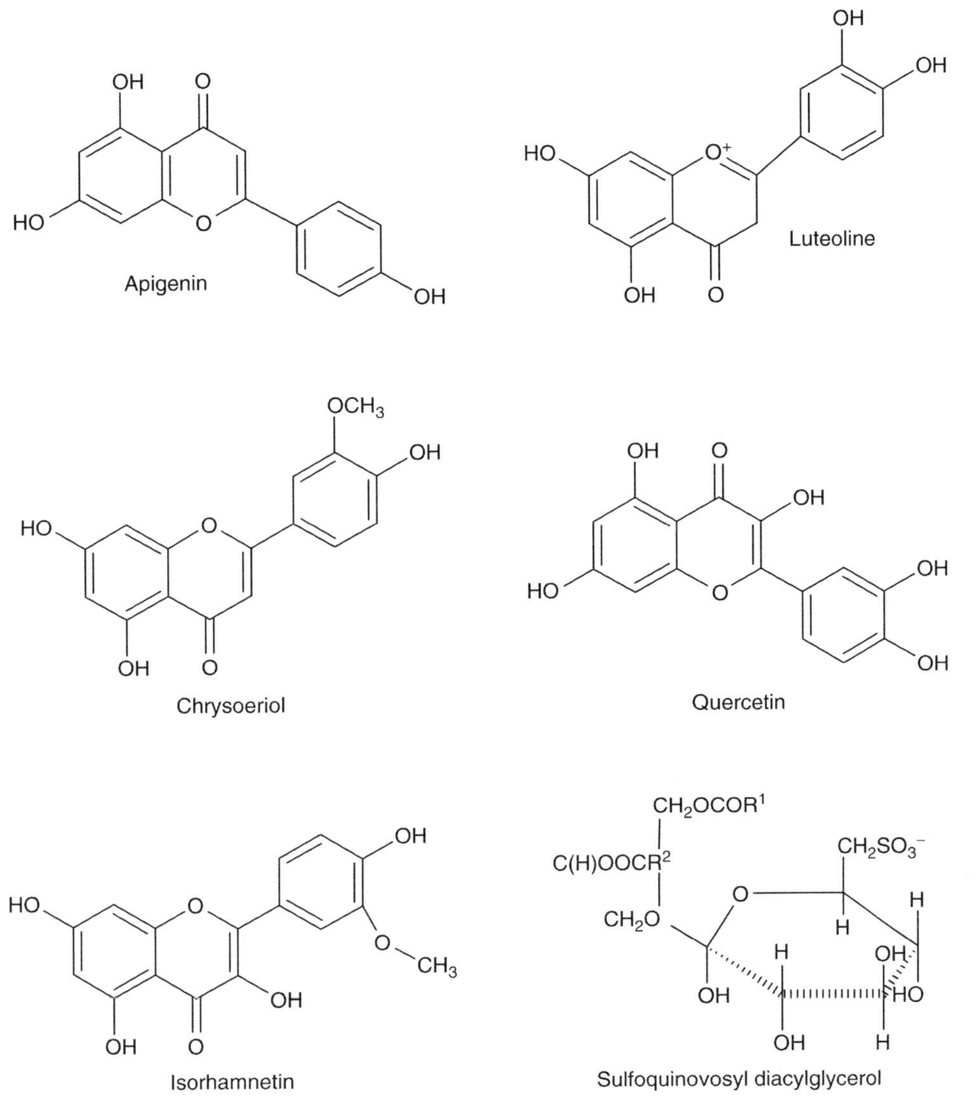

Fig. 21.2. Major flavonoids in parsley.

characterized. The aglycones identified were the flavones apigenin, luteolin and chrysoeriol, and the flavonols quercetin and isorhamnetin. The flavones occurred either as 7-*O*-glucosides or as 7-*O*-apioglucosides, while the flavonols were 3-*O*-monoglucosides or 3,7-*O*-diglucosides. One-half of these glycosides were electrophoretically mobile and substituted with malonate residues. Kuriyama *et al.* (2005) purified the major glycolipids in monogalactosyl diacylglycerol, digalactosyl diacylglycerol and sulphoquinovosyl diacylglycerol (SQDG) from dried vegetables and examined their anticancer property (elaborated in the subsequent section on medical uses).

Justesen and Knuthsen (2001) quantified, by HPLC and mass spectrometry, flavonoids in commonly eaten fresh herbs, including parsley. Five major flavonoid aglycones were detected and quantified by HPLC after acid hydrolysis: apigenin, isorhamnetin, kaempferol, luteolin and quercetin. The highest levels of flavonoids were found in parsley (510–630mg apigenin/100g).

The key reaction of flavonoid biosynthesis, the condensation of the acyl residues from one molecule of 4-coumaroyl-CoA and three molecules of malonyl-CoA, previously had been assumed to be catalysed by a 'flavonone synthase'. Studies by Heller and Hahlbrock (1980) indicated that the immediate product of the synthase reaction was not the flavonone but the isomeric chalcone. The term 'chalcone synthase' was therefore suggested for the enzyme.

The genes of 2-oxoglutarate-dependent dioxygenases (2-*ODD*), flavone synthase (*FNS*) or flavonone 3 β-hydroxylase (*FHT*) and flavonol synthase (*FLS*), which are involved in the biosynthesis of these secondary metabolites, were cloned from parsley leaves by Gebhardt *et al.* (2005). A cDNA encoding flavone synthase I (*FNS I*) was amplified by RT-PCR from leaflets of *P. crispum* cv. Italian Giant seedlings and functionally expressed in yeast cells. The identity of the recombinant, 2-oxoglutarate-dependent enzyme was verified in assays converting (2*S*)-naringenin to apigenin (Martens *et al.*, 2001).

Furanocoumarins

Beier *et al.* (1994) first reported the isolation of the linear furanocoumarin, saxalin, from fresh parsley leaves and dried parsley flakes. Psoralen, graveolone, bergapten, xanthotoxin, isoimperatorin, isopimpinellin, oxypeucedanin and oxypeucedanin hydrate were found and the levels of psoralen, bergapten, xanthotoxin and isopimpinellin were quantified by HPLC – fresh parsley leaves had 112μg/g fresh weight. One brand of parsley flakes had a total of 304μg/g dry weight of the three major photosensitizing linear furanocoumarins (psoralen, bergapten and xanthotoxin). The major flavone glycoside in parsley leaves was identified by Eckey-Kaltenbach *et al.* (1993) as 6′-*O*-malonylapiin.

Parsley's defence mechanism to fungal attack was studied by Kauss *et al.* (1992), who found that pre-incubation of suspension-cultured parsley cells with methyl jasmonate greatly enhanced their ability to respond to fungal elicitors by secretion of coumarin derivatives, especially at relatively low elicitor concentration, and they also observed the incorporation of esterified hydroxycinnamic acids and 'lignin-like' polymers into the cell wall. These three responses correspond to defence reactions induced locally when a fungal pathogen attacks plant cells.

Hagemeier *et al.* (1999) reported the accumulation of furanocoumarins (marmesin and bergapten) and various non-coumarin compounds in parsley cell cultures, as a result of a 25-amino acid oligopeptide (Pep25) elicitor of *Phytophthora sojae*. These compounds were isolated by preparative HPLC and identified by spectroscopic methods (MS, NMR) as 5-hydroxy- and 7-hydroxy-3-butylidenephthalides, including two novel conjugates of the 7-hydroxy derivative, i.e. 7-*O*-glucoside and 7-*O*-(6′-malonylglucoside). With the aid of germination assay with lettuce seeds, Kato *et al.* (1978) identified one of the germination inhibitors in parsley seeds to be heraclenol. Figure 21.3 lists the structures of the major furanocoumarins in parsley.

Guiet *et al.* (2003) demonstrated that deuterium (^2H) distribution in fatty acids was non-statistical and could be related to isotopic discrimination during chain extension and desaturation. Petroselinic acid (C18:1Δ^6) (Fig. 21.4), a fatty acid characteristic of the seeds of the Apiaceae, has been shown to be biosynthesized from palmitoyl-ACP (C16:0) by two steps, catalysed by a dedicated Δ^4-desaturase and an elongase. The isotopic profile resulting from this pathway is similar to the classical plant fatty acid pathway, but the isotopic fingerprint from both the desaturase and elongase steps shows important differences relative to oleic and linoleic acid biosynthesis.

21.6. Uses

Culinary uses

The inclusion of parsley in Portuguese gastronomy is an established tradition. It is often

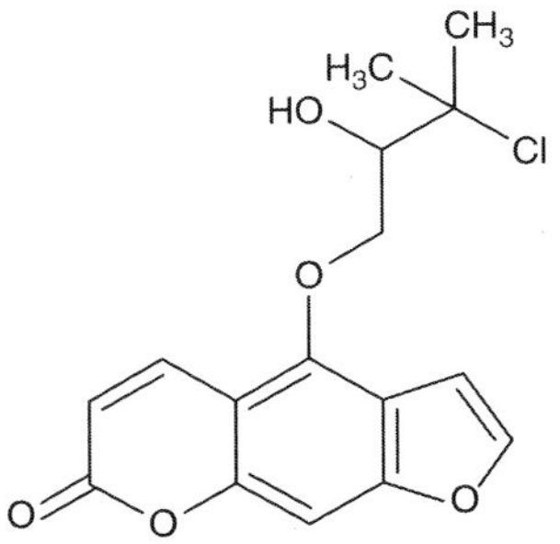

Saxalin

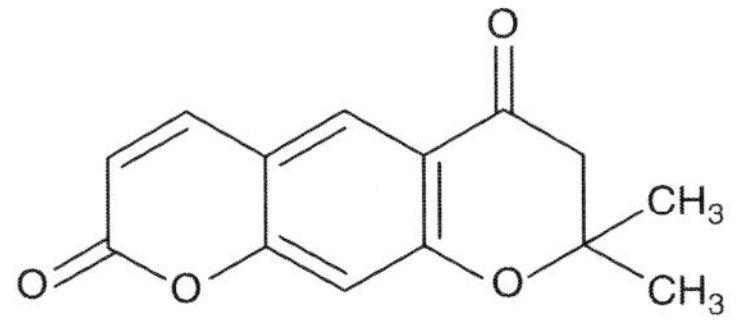

Graveolone

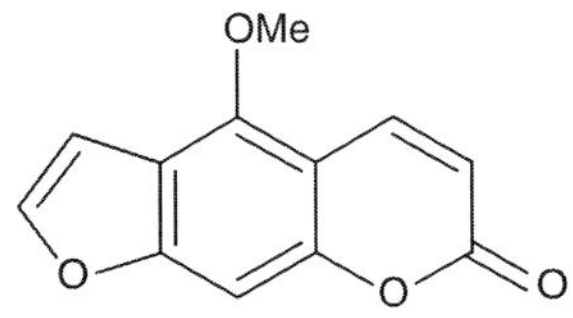

Bergapten

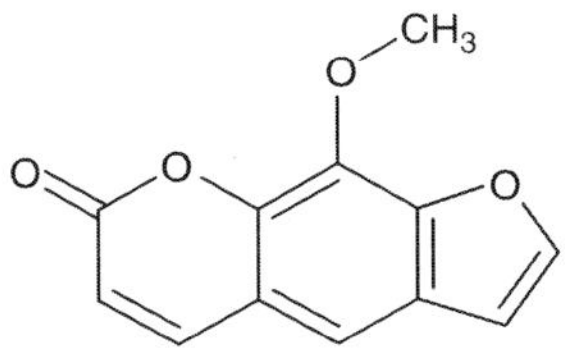

Xanthotoxin

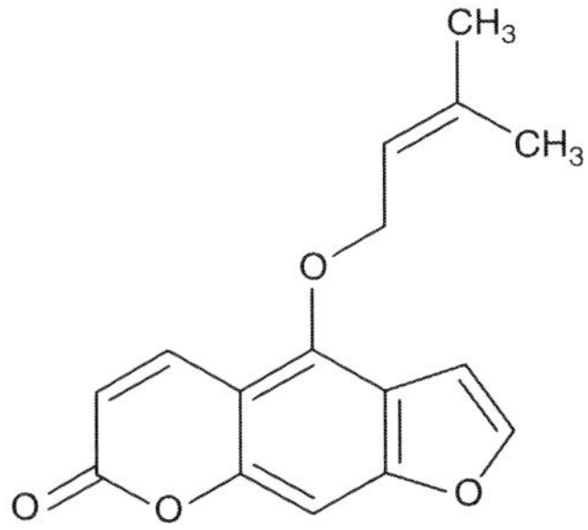

Isoimperatorin

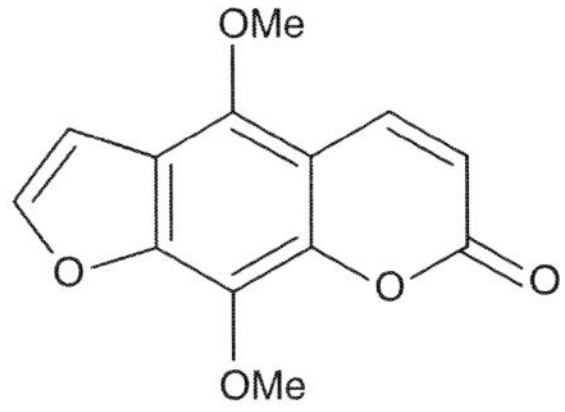

Isopimpinellin

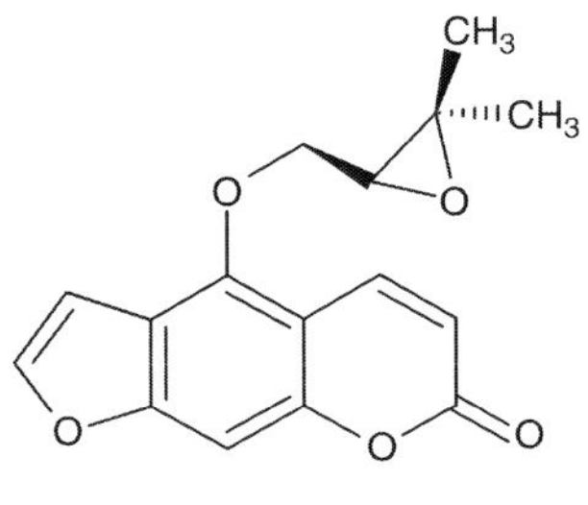

Oxypeucedanin

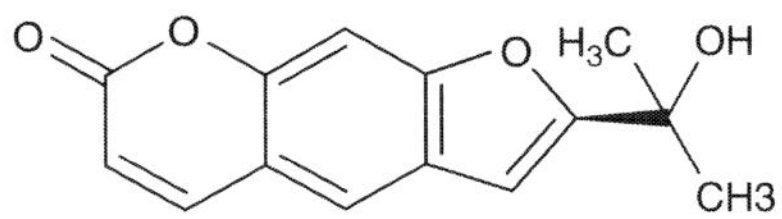

Marmesin

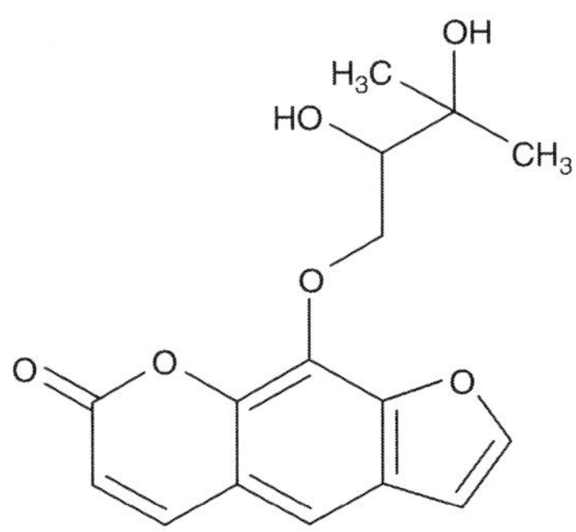

Heraclenol

Fig. 21.3. Major furanocoumarins detected in parsley.

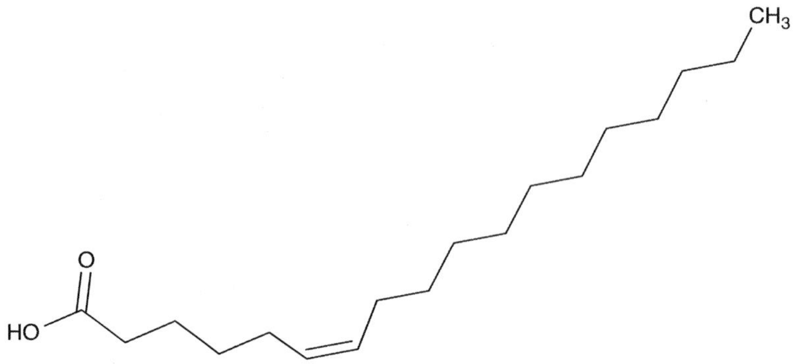

Fig. 21.4. Petroselinic acid.

used finely chopped, either incorporated with all other ingredients or added raw to the final recipe for decoration and its unique flavour (green salads, cold and hot dishes). The use of parsley in this form requires a prior preparation of the herb (selection, washing and disinfecting and fine chopping), which is a time-consuming task, producing a large amount of waste and making the prediction of product quantification difficult, especially in catering. The ideal situation both for final consumers and caterers would be the supply of ready-to-use finely chopped parsley (minimally processed). Such a product requires adequate processing conditions for the maintenance of the sensorial characteristics of parsley and its preservation. Minimally processed products have been available for many years, but the types and quantity have expanded tremendously in past decades. Initially, the food service industry was the main user of fresh-cut products, but use has expanded to restaurants, supermarkets and warehouse stores. The food service industry and restaurants favour fresh-cuts because manpower for preparation and special systems to handle waste are not required and specific forms of fresh-cuts can be delivered at short notice. Fresh-cut products are thus convenience foods with the additional benefit of reduced wastage for retail consumers (Watada *et al.*, 1996).

Fresh-cut products differ from intact fruits and vegetables in terms of their physiology, handling and storage requirements. The fresh-cut process results in tissue and cell integrity disruption, with a concomitant increase in enzymatic, respiratory and microbiological activity, and therefore reduced shelf life (Watada *et al.*, 1996; O'Beirn *et al.*, 1999). This effect might be minimized by the use of adequate temperature management and modified atmosphere packaging (MAP), a technological process which involves either actively or passively controlling or modifying the atmosphere surrounding the product within a package made of various types and/or combinations of films (Farber *et al.*, 2003). Yamauchi and Watada (1993) showed that the decrease in pigments was less in leaves held in a controlled atmosphere with 10% O_2 and 10% CO_2 than when held in fresh air, and that parsley flavour and aroma were retained better in perforated film packages than in sealed film packs (Manzano *et al.*, 1995).

The most important quality parameters in fresh-chopped parsley shelf life determination were the weight loss, exudate and sensory and microbiological criteria. More recent studies by Rosa *et al.* (2007) indicated that fresh-cut parsley packed in a passive atmosphere seemed to result in a better product when compared with that packed in an active modified atmosphere, showing quality and stability for 6 days, suggesting that the application of MAP technology, especially a passive atmosphere, might be able to retard the deterioration of fresh-chopped parsley and have greater potential in food catering units.

For medicinal purposes, the roots are collected in the second year, in autumn or late summer, when the plant has flowered. Parsley leaves can be dried in the oven on muslin trays till thoroughly dry and crisp, after which the leaves are rubbed by hand or passed through a coarse wire sieve and the powder stored in air- and light-tight tins to preserve the good colour. The oil is extracted from the 'seeds', or rather fruits, when fresh (http://www.botanical.com/). Prior to processing, parsley leaves may be kept for > 3 days in a non-cooled store and for up to 15 days in a cold store, assuming > 50% of the material should maintain its quality (Lisiewska *et al.*, 1997).

Medicinal uses and side effects

Several scientific studies provide evidence of the traditional use of parsley in medicine. Food plants of the Apiaceae plant family such as parsley, carrots and celery contain a group of bioactive aliphatic C_{17}-polyacetylenes, which were shown to be highly toxic towards fungi, bacteria and mammalian cells and to display neurotoxic, anti-inflammatory and antiplatelet aggregatory effects and to be responsible for allergic skin reactions in a study by Christensen and Brandt (2006). The effect of these polyacetylenes towards human cancer cells, their human bioavailability and their ability to reduce tumour formation in a mammalian *in vivo* model indicate that they may be beneficial for health.

Anticancer property

In humans, apiaceous vegetables (parsley, carrots, parsnips, celery, etc.) inhibit human cytochrome P-450 1A2 (hCYP1A2), a biotransformation enzyme known to activate several procarcinogens, including aflatoxin B1 (AFB). Peterson *et al.* (2006) reported that the apiaceous constituents psoralen, 5-methoxypsoralen (5-MOP), 8-methoxypsoralen (8-MOP), and apigenin were potent inhibitors of hCYP1A2, whereas quercetin was a modest hCYP1A2 inhibitor. The 2h pretreatment of intact yeast cells with psoralen, 5-MOP and 8-MOP significantly improved cell survival after subsequent 4h AFB treatment and reduced hCYP1A2-mediated mutagenicity of AFB. Apigenin also decreased mutagenicity significantly. These results suggest that *in vivo* CYP1A2 inhibition by apiaceous vegetables may be due to the phytochemicals present and imply that apiaceous vegetable intake may be chemopreventive by inhibiting CYP1A2-mediated carcinogen activation.

Kuriyama *et al.* (2005) found that the glycolipids in monogalactosyl diacylglycerol, digalactosyl diacylglycerol and sulphoquinovosyl diacylglycerol (SQDG) in common dried vegetables, also reported in parsley, were inhibitors of both DNA polymerase α (pol α) *in vitro* and the proliferation of human cancer cells. A significant correlation was found between SQDG content and inhibition of DNA polymerase. Therefore, the inhibition of pol α activity by SQDG may lead to cell growth suppression. Based on these results, Kuriyama *et al.* (2005) concluded that the glycolipid fraction from common vegetables is a potentially novel source of food material for anticancer activity.

Ohyama *et al.* (1987) investigated the urinary mutagenicity of healthy men after strictly defined meals by means of the Ames Salmonella/microsome test. When the subjects ate 150g of fried salmon at one meal, a potent mutagenicity of almost 5000 revertants of TA98 strain was present in all 6h urine samples. On the other hand, fewer than 2500 revertants were present in the urine when the subjects consumed 70g of parsley and 150g of fried salmon simultaneously, being sufficient evidence of parsley's protective effect.

Edenharder *et al.* (2002) reported that the genotoxic activity of 2-amino-1-methyl-6-phenylimidazo[4,5-*b*]pyridine (PhIP) was reduced strongly by green parsley in a dose-dependent manner. It is suggested that the possible mode of mechanism of protection against genotoxicity could be through enzyme inhibition (cytochrome P450 dependent monooxygenase 1A2 and sulphotransferase) by complex mixtures of plant origin. Edenharder *et al.* (1994) had reported previously that parsley was inactive when investigated for antimutagenic potencies with respect to the mutagenic activities induced by 2-amino-3-methyl[4,5-f]quinoline (IQ) and, in part, by 2-amino-3,4-dimethylimidazo[4,5-f]quinoline (MeIQ) or 2-amino-3,8-dimethylimidazo[4,5-f]quinoxaline (MeIQx) in *Salmonella typhimurium* TA98 and TA100.

Antioxidant property

Reactive oxygen species (ROS) are produced in the course of normal metabolism and, because of their high reactivity, accumulation of ROS beyond the immediate needs of the cell may affect cellular structure and functional integrity by bringing about oxidative degradation of biomolecules, such as DNA, proteins and lipids. Although cells possess an intricate network of defence mechanisms to neutralize excess ROS and reduce oxidative stress, some tissues, especially the brain, are more vulnerable to oxidative stress because of their elevated consumption of oxygen and the consequent generation of large amounts of ROS. For the same reason, the mitochondrial DNA (mtDNA) of brain cells is highly susceptible to structural alterations, resulting in mitochondrial dysfunction. Several lines of evidence strongly suggest that these effects of ROS may be related etiologically to a number of neuro-degenerative disorders. Many herbs are known as excellent sources of natural antioxidants, and consumption of fresh herbs in the diet contributes to daily antioxidant intake.

Popović *et al.* (2007) studied the *in vitro* and *in vivo* antioxidant activity of the different extracts of the leaves and root of parsley. All extracts were good scavengers of DPPH and OH^- radicals and reduced the intensity of lipid peroxidation *in vitro*. The *in vivo* effects were evaluated on some antioxidant systems (activities of lipid peroxidase, GSH-peroxidase, peroxidase, catalase and xanthine oxidase and GSH content) in mice liver and blood after treatment with the examined parsley extracts, or in combination with carbon tetrachloride (CCl_4). On the basis of the results obtained, it can be concluded that the examined extracts exhibited a certain protective effect. However, combined treatments with CCl_4 and the examined extracts showed both positive and negative synergism, inducing or suppressing the influence of CCl_4.

Zhang *et al.* (2006) evaluated the antioxidant capacities of the essential oil of parsley using different *in vitro* assays: β-carotene-bleaching assay, DPPH free radical-scavenging assay and Fe^{2+}-metal-chelating assay. Results showed that parsley oil (PO) possessed a certain degree of antioxidant activity in terms of β-carotene-bleaching capacity and free radical-scavenging activity, though much weaker than those of BHT and of α-tocopherol; but its metal-chelating capacity was negligible. Myristicin in PO was found as a dominant compound (32.75%) that exhibited a moderate antioxidant activity, followed by apiole (17.54%), but it might be the major contributor to the antioxidant activity of PO. These results suggest that the PO and its two major components can be potential alternative natural antioxidants.

A comparison of the antioxidant and antibacterial effects of extracts of parsley and cilantro (*Coriandrum sativum*) by Wong and Kitts (2006) revealed that parsley leaves had a higher concentration of phenolic compounds than cilantro. This finding corresponded to a difference in the reducing and scavenging activities of lipid- and water-soluble radicals. The greater antioxidant activity observed in the iron-induced linoleic acid model system occurred with the methanol stem extract from both herbs and was attributed to a greater iron-chelating activity, more so than to reducing or radical scavenging activities. On the contrary, a pro-oxidant activity of the aqueous extracts from both herbs acted to maintain the iron of the iron–ligand complex in an active ferrous state. The greater bacterial cell damage caused by the methanol stem extracts resulted in a greater growth inhibition towards *B. subtilis* and *E. coli*. This study shows that the phenolic compounds extracted from both parsley and cilantro are responsible, in part, for both antioxidant and antibacterial activities. In a study by Hinneburg *et al.* (2006), hydrodistilled extracts from basil, laurel, parsley, juniper, aniseed, fennel, cumin, cardamom and ginger were assessed for their total antioxidant activities by several *in vitro* methods. Although parsley showed the best performance in the iron chelation assay, it was less effective at retarding the oxidation of linoleic acid in the linoleic acid peroxidation assay.

Hepatoprotective effect

In Turkey, parsley is one of the medicinal herbs used by diabetics. Ozsoy-Sacan *et al.* (2006) investigated the effects of parsley and glibornuride on the liver tissue of streptozotocin-induced diabetic rats. In the STZ-diabetic group, blood glucose levels, serum alkaline phosphatase activity, uric acid, sialic acid, sodium and potassium levels, liver lipid peroxidation (LPO) and non-enzymatic glycosylation (NEG) levels increased, while liver glutathione (GSH) levels and body weight decreased. In the diabetic group given parsley, blood glucose, serum alkaline phosphatase activity, sialic acid, uric acid, potassium and sodium levels and liver LPO and NEG levels decreased, but liver GSH levels increased. In the diabetic group given glibornuride, blood glucose, serum alkaline phosphatase activity, serum sialic acid, uric acid, potassium, and liver NEG levels decreased, but liver LPO, GSH, serum sodium levels and body weight increased. Parsley extract has a protective effect against hepatotoxicity caused by diabetes comparable to glibornuride, probably due to its antioxidant property.

The liver increment (the amount of tissue regenerated) in partially hepatectomized rats was increased significantly by sc injection of oils of anise, fennel, tarragon, parsley seed, celery seed and oleoresin, nutmeg, mace, cumin and sassafras and of the aromatic principles, 4-allylanisole, 4-propenylanisole, *p*-isopropylbenzaldehyde, safrole and isosafrole. Most of the essential oils were ineffective in total doses of up to 3000 mg/kg because they contained a high percentage of terpenes, which proved inert. Many of the agents were also effective when added to the diet (Gershbein, 1977).

Diuretic effect of parsley seeds

Kreydiyyeh and Usta (2002) provided substantial evidence of the advocated diuretic effect of parsley in folk medicine and determined the mechanism of action of the herb from animal studies. Rats, when offered an aqueous parsley seed extract to drink, eliminated a significantly larger volume of urine; these findings were supported by the results of other experiments using an *in situ* kidney perfusion technique, which also demonstrated a significant increase in urine flow rate with parsley seed extract. This effect was still apparent in the presence of amiloride and furosemide and in the absence of sodium, but not in the absence of potassium, suggesting that the diuretic effect of the herb is mediated through an increase in K^+ retention in the lumen. Parsley extract was shown, on the other hand, to reduce the activity of the Na^+–K^+ ATPase in both cortex and medulla homogenates. Such an inhibition would decrease apical cellular Na^+ reabsorption, lower K^+ secretion, increase K^+ concentration in the intercellular space and consequently inhibit passive K^+ influx across the tight junctions. The mechanism of action of parsley seems to be mediated through an inhibition of the Na^+–K^+ pump that would lead to a reduction in Na^+ and K^+ reabsorption, leading thus to an osmotic water flow into the lumen, and diuresis.

Laxative property

Kreydiyyeh *et al.* (2001) provided scientific evidence to confirm the laxative property of parsley, as claimed in folk medicine, and explained its mechanism of action. A perfusion technique was used to measure net fluid absorption from rat colon. The addition of an aqueous extract of parsley seeds to the perfusion buffer, and the omission of sodium, both significantly reduced net water absorption from the colon, as compared with the control. Parsley, added to a sodium-free buffer, did not lead to any further significant change in water absorption as compared with parsley alone; suggesting that with parsley, sodium absorption was already inhibited. Since K^+ and Cl^- secretion depends on the activity of the $NaKCl_2$ transporter, the latter was inhibited with furosemide, which increased net water absorption significantly. When parsley and furosemide were added together, net water absorption was significantly higher than with parsley alone and significantly lower than with furosemide alone. In addition, parsley extract was shown to inhibit the *in vitro* activity of the Na^+–K^+ ATPase in a colon homogenate and the activity of a partially purified dog kidney ATPase. The results suggest that parsley acts by inhibiting sodium, and consequently water absorption, through an inhibition of the Na^+–K^+ pump and by stimulating the $NaKCl_2$ transporter and increasing electrolyte and water secretion.

Allergic reactions

Zuskin *et al.* (1988) studied immunological and respiratory findings in spice-factory workers. Intradermal skin testing with mixed spice dust allergen demonstrated positive skin reactions in 73.3% of exposed and in 33.3% of control workers. Increased IgE serum levels were found in 36.8% of exposed and in 9.7% of the control workers. The prevalence of chronic respiratory symptoms was significantly higher in the exposed workers than in the control workers. There was, however, no consistent correlation between skin reactivity and chronic respiratory symptoms. There was a high prevalence of acute symptoms during the work shift. These complaints were more frequent in workers with positive skin tests for the symptoms of cough, chest tightness and irritated and dry throat. Ventilatory capacity was measured by recording maximum expiratory flow-volume curves. There were statistically significant mean reductions during the work shift for all measured lung function parameters in workers with positive skin reactions. Aqueous extracts of different spices, including parsley, caused a dose-related contractile response of isolated guinea pig tracheal smooth muscle. These data suggest that immunologic reactions to spices are frequent in spice workers and may be related to acute symptoms and lung function changes, not to chronic changes; also, in addition to any immunologic response these spices may produce *in vivo*, they probably provoke direct irritant reactions in the airways, as suggested by *in vitro* data.

21.7. Quality Aspects

The quality standards for parsley refer to its freshness, green colour and freedom from defects or seed stems and decays (USDA, 2002). To avoid contamination, it is recommended that parsley be harvested using gloves, though in practice most of the crop is harvested by hand. It is also recommended that parsley be removed from field heat quickly, without excessive drying, to retain maximum green colour and freshness. Parsley can be pre-cooled with ice (Cantwell and Reid, 1992) or by vacuum-cooling (Aharoni *et al.*, 1989); forced-air or hydro-cooling are also used (Joyce *et al.*, 1986).

Optimum storage conditions

The recommended conditions for commercial storage of parsley leaves are 0°C, 95–100% RH (UC-Davis, 2002), under which conditions parsley can be stored for 1–2 months, compared with only 3 days at 18–20°C and 85–90% RH (Lisiewska *et al.*, 1997). The end point of storage at 0°C is the wilting of parsley, at around 20% weight loss (Hruschka and Wang, 1979). MAP is effective in extending storage life, but temperature changes and condensation must be avoided. Aharoni *et al.* (1989) found that non-perforated polyethylene liners delayed yellowing and decay at low temperature. Park *et al.* (1999) achieved 77 days of storage at 0°C or 35 days at 5°C, with good retention of firmness and vitamin C content, using a 40 μm-thick ceramic film. A preharvest spray with gibberellic acid may extend storage life (Lers *et al.*, 1998). Hamburg parsley roots (without leaves) can be stored at 0°C for several months (Bakowski *et al.*, 1994; Elkner *et al.*, 1998). Parsley flavour and aroma were better retained in perforated film packages than in sealed film packs (Manzano *et al.*, 1995).

Controlled atmosphere

Parsley can tolerate 8–10% O_2 + 8–10% CO_2 (Saltveit, 1997), but this may be of little benefit at 0°C. Ten percent O_2 + 11% CO_2 was found optimal for delaying yellowing in parsley stored at 5°C (Apeland, 1971). Storage in 10% O_2 + 10% CO_2 (Yamauchi and Watada, 1993) or 10% CO_2 (Lers *et al.*, 1998) delayed yellowing at room temperature. Parsley is not chilling-sensitive and should be stored as cold as possible without freezing, which occurs at −1.1°C. Parsley produces very little ethylene but is very sensitive to it (Joyce *et al.*, 1986; Tsumura *et al.*, 1993). It has an extremely high respiration rate; young leaves respire at a higher rate than old leaves at harvest but the respiration rate does not decrease as much in older leaves as in younger leaves after harvest, so younger leaves store better (Apeland, 1971). There are no quarantine issues involved. In the USA, only one grade of parsley is available (US No. 1) which is of similar varietal characteristics, i.e. not mixing curly- and flat-leaved varieties that meet quality criteria.

Contamination

The incidence of human pathogens on fresh produce is a serious concern in industrialized countries, Salmonella being one of the most commonly isolated pathogens associated with fresh fruits and vegetables. Outbreaks of salmonellosis have been linked to a wide variety of fresh produce, including parsley (CDC, 2000). Parsley is often eaten raw or with rich dressings, which may result in the regrowth of some pathogenic bacteria, threatening public health, without clear outbreaks. Food safety is a major concern in parsley; the personal hygiene of the staff handling the material is therefore paramount. Contaminated fresh parsley has also been linked to outbreaks of *Shigella sonnei* (Crowe *et al.*, 1999), enterotoxigenic *E. coli* (Naimi *et al.*, 2003), thermotolerant campylobacters (Park and Sanders, 1992) and verotoxinogenic *Citrobacter freundii* (causing gastroenteritis and haemolytic uraemic syndrome; Tschape *et al.*, 1995).

Contamination of fruits and vegetables may occur at various stages during production, harvest, processing and transport. Aycicek *et al.* (2006) found that *E. coli* was significantly more often detected on parsley (21/30) and dill samples (12/30) and, consequently, it was suggested that fresh salad vegetables (especially parsley, dill and cos lettuce) might contain pathogenic microorganisms and represent a risk for consumers regarding foodborne disease. The importance of adequate measures throughout the farm-to-table food chain cannot be over-emphasized.

Erdoğrul and Şener (2005) reported parsley as harbouring *Enterobius vermicularis, Ascaris* eggs and *Entamoeba histolyca* cysts. Kozan *et al.* (2005) detected by light microscopy helminth eggs in 5.9% of unwashed samples but not in any washed samples of raw parsley in Ankara, Turkey. Helminth eggs included *Taenia* spp. (3.5%), *Toxocara* spp. (1.5%) and *Ascaris lumbricoides* (1.0%). Approximately 11% of unwashed lettuce and parsley was contaminated, compared with only 2.5% of carrot samples and none in red cabbage, rocket, tomatoes or green peppers. These results highlight the importance of properly washing/disinfecting raw vegetables before consumption. Ruiz *et al.* (1987) detected a high degree of faecal contamination in vegetables, including parsley, from farms, wholesale markets, supermarkets and small shops in Granada, Spain. *E. coli* were detected in 86.1% of the samples and salmonellae were isolated from 7.5% of samples. The serotype most frequently isolated was *S. typhimurium*.

The irrigation of vegetables with raw wastewater has been practised in many countries. This water is contaminated with different serogroups of *Salmonella* B and C. These same serogroups were detected on vegetables irrigated with wastewater effluents (Melloul *et al.*, 2001). It is necessary to treat wastewater effluents to a level where no residual contaminants can be detected on irrigated crops (Armon *et al.*, 1994). Hence, quality of irrigation is very important to prevent contamination and epidemics, even if *Salmonella* does not persist beyond 3 days after irrigation in parsley, as reported by Melloul *et al.* (2001). In addition, such wastewater contains high levels of trace elements and heavy metals like lead, cadmium, nickel, mercury, uranium, copper, zinc, boron, cobalt, chromium, arsenic, molybdenum, manganese, etc. Many of these are non-essential and are toxic to plants, animals and humans (Kanwar and Sandha, 2000). Where sewage water contaminated by heavy metals has been used for irrigation, lead and cadmium inhibited growth, with plants treated with cadmium having more symptoms of toxicity than those treated with lead (Salim *et al.*, 1995). In areas where natural water used for drinking and irrigation contained high concentrations of fluoride, F-concentrations in plant tissues also increased (Kabasakalis and Tsolaki, 1994).

Attached microorganisms (pathogens and spoilage bacteria) are not removed easily by washing with water or antibacterial agents. Chlorinated water is somewhat beneficial in reducing contamination (Park and Sanders, 1992). Karapinar and Gönül (1992) reported that dipping fresh parsley containing 10^7 *Yersinia enterocolitica* per gram into 2% (v/v) acetic acid or 40% (v/v) vinegar solutions for 15 min exerted pronounced bactericidal effect against this organism. No viable aerobic bacteria were recovered after 30 min dip in 5% (v/v) acetic acid, whereas vinegar led to 3–6 $\log_{10}$ cycles decrease in the number of aerobic bacteria depending on the vinegar concentration and holding time.

Chemical treatments, such as calcium or sodium hypochlorite, hydrogen peroxide, ethanol and a variety of detergents, partially reduced (if any) the populations of the pathogens (Beuchat, 1997; Gandhi *et al.*, 2001). At present, chlorine at a concentration of 50–200 mg/l is the primary post-harvest sanitizing agent in routine use in the fresh produce industry (Beuchat *et al.*, 1998); this is usually ineffective in eliminating pathogens. It is hypothesized that the reduction of the oxidizing power of the chlorine could be due to the high organic load of the plants (Burnett and Beuchat, 2000) or lower accessibility of the target pathogen, achieved by either internalization of the organisms into the plant tissue or aggregation and biofilm production on the plants. Biofilms are assemblages of microorganisms adherent to each other and/or to a surface and embedded in a matrix of exopolymers (Costerton *et al.*, 1999); chemical sanitizers generally are unable to eliminate most biofilm-associated bacteria. Scher *et al.* (2005) revealed that *S. typhimurium* embedded in the biofilm matrix resisted sodium hypochlorite at concentrations above 500 mg/l, while planktonic cells were sensitive to less than 50 mg/l; also, most isolates of *Salmonella* spp. originating from produce were able to synthesize the main components of the biofilm matrix – curli and cellulose (Zogaj *et al.*, 2003; Solomon *et al.*, 2005). Lapidot *et al.* (2006) compared the adhesion and persistence of the wild and its biofilm-deficient isogenic mutant of *S. typhimurium* cells and found that biofilms were likely to influence the effectiveness of strategies to control foodborne pathogens on parsley and that biofilm formation strengthened the adhesion and provided protection against disinfection after storage of the contaminated produce, and not immediately after contamination.

The effect of growing plants in polluted environs has also been reported. Monocyclic aromatic hydrocarbons (MAHs: benzene, toluene, ethylbenzene and xylenes) were extracted from fruit and vegetables and determined by GC-MS (with selected-ion monitoring mode) by Górna-Binkul *et al.* (1996). It was observed that uptake of MAHs depended on the species and took place in different parts of the plant. The highest concentrations of MAHs were found in parsley leaves (*m*- and *p*-xylene).

21.8. Conclusion

Parsley is a biennial herb, native to Southern Europe and Western Asia. The finely chopped leaves are used as flavouring in Central Europe, similar to the use of coriander leaves, in sauces, soups, stuffing, rissoles, minces, etc., and also sprinkled over vegetables or salads. In addition to the leaves, the stems are also dried and powdered, both as a culinary colouring and as a dye. The roots of the turnip-rooted variety are used as a vegetable and flavouring and for medicinal purposes. Parsley is a rich source of Vitamins C, A and E, thiamin, riboflavin and folate; it is high in calcium and iron, and also fatty acids and an essential or volatile oil. The essential oil of the leaves is superior to that of the seeds and is used in condiments and seasonings; seed oil is used in perfumes, soaps and creams. The major constituents of parsley are 1,3,8-*p*-menthatriene, β-phellandrene, myristicin, myrcene, apiole, terpinolene, 1-methyl-4-isopropenylbenzene, α-pinene and β-pinene, trans-β-ocimene and γ-terpinene. Apart from its culinary uses, parsley is known for its anticancer, antioxidant, diuretic and laxative properties; a matter of concern is reports of the allergic reactions induced in workers handling the herb. Photosensitizing toxic furocoumarines, including psoralen, bergaptene and isoimperatorin, which can induce dermatitis, have been found in parsley roots. The quality standards for parsley refer to its freshness, green colour and freedom from defects or seed stems and decays. Optimum storage conditions, including modified atmosphere packaging and controlled atmosphere packaging, have been reported. Since parsley is often eaten raw or with rich dressings, food safety is a major concern. Parsley has been linked to outbreaks of *Shigella sonnei*, *Escherichia coli*, thermotolerant campylobacters and verotoxinogenic *Citrobacter freundii*. Irrigation with raw wastewater has been linked to *Salmonella* B and C and heavy metal contamination. Chemical treatments, such as calcium or sodium hypochlorite, hydrogen peroxide, ethanol and detergents, partially reduced the populations of the pathogens; the reason for the lack of sanitizer effectiveness is thought to be due to reduction of the oxidizing power of chlorine by the high organic load of the plants or lower accessibility of the target pathogen, due either to internalization of the organisms into the plant tissue or aggregation and biofilm production on the plants.

References

Aharoni, N., Reuveni, A. and Dvir, O. (1989) Modified atmospheres in film packages delay senescence and decay of fresh herbs. *Acta Horticulturae* 258, 255–260.

Apeland, J. (1971) Factors affecting respiration and color during storage of parsley. *Acta Horticulturae* 20, 43–52.

Appendino, G., Jakupovic, J. and Bossio, E. (1998) Structural revision of the parsley sesquiterpenes crispanone and crispane. *Phytochemistry* 49(6), 1719–1722.

Armon, R., Dosoretz, C.G., Azov, Y. and Shelef, G. (1994) Residual contamination of crops irrigated with effluent of different qualities: a field study. *Water Science and Technology* 30, 239–248.

Athar, N., Spriggs, T.W. and Liu, P. (1999) *The Concise New Zealand Food Composition Tables*, 4th edn. NZ Institute of Crop and Food Research, Palmerston North, New Zealand.

Atta-Aly, M.A. (1999) Effect of nickel addition on the yield and quality of parsley leaves. *Scientia Horticulturae* 82(1–2), 9–24.

Aycicek, H., Oguz, U. and Karci, K. (2006) Determination of total aerobic and indicator bacteria on some raw eaten vegetables from wholesalers in Ankara, Turkey. *International Journal of Hygiene and Environmental Health* 209(2), 197–201.

Bakowski, J., Michalik, H. and Umiecka, L. (1994) Effect of sowing and harvest date on the quality of root parsley and its storability. *Biuletyn Warzywniczy* 41, 157–167.

Beier, R.C., Ivie, G.W. and Oertli, E.H. (1994) Linear furanocoumarins and graveolone from the common herb parsley. *Phytochemistry* 36(4), 869–872.

Ben-Amotz, A. and Fishier, R. (1998) Analysis of carotenoids with emphasis on 9-*cis* β-carotene in vegetables and fruits commonly consumed in Israel. *Food Chemistry* 62(4), 515–520.

Beuchat, L.R. (1997) Comparison of chemical treatments to kill Salmonella on alfalfa seeds destined for sprout production. *International Journal of Food Microbiology* 34, 329–333.

Beuchat, L.R., Nail, B.V., Adler, B.B. and Clavero, M.R. (1998) Efficacy of spray application of chlorinated water in killing pathogenic bacteria on raw apples, tomatoes, and lettuce. *Journal of Food Protection* 61, 1305–1311.

Burnett, S.L. and Beuchat, L.R. (2000) Human pathogens associated with raw produce and unpasteurized juices, and difficulties in decontamination. *Journal of Industrial Microbiology and Biotechnology* 25, 281–287.

Cantwell, M.I. and Reid, M.S. (1992) Postharvest handling systems: fresh herbs. In: Kader, A.A. (ed.) *Postharvest Technology of Horticultural Crops*, 2nd edn. University of California, Oakland, California, pp. 211–213.

CDC (2000) Report. Food borne outbreak response and surveillance unit, http://www.cdc.gov/foodborneoutbreaks/us_outb/fbo2000/bacterial00.htm). In: Cook, K.A. *et al.* (1998) Outbreak of Salmonella serotype Hartford infections associated with unpasteurized orange juice. *Journal of the American Medical Association* 280, 1504–1509.

Charles, D.J. (2000). Parsley. In: Peter, K.V. (ed.) *Handbook of Herbs and Spices*. CRC Press, Woodhead Publishing Limited, Cambridge, UK, pp. 230–242.

Christensen, L.P. and Brandt, K. (2006) Bioactive polyacetylenes in food plants of the Apiaceae family: occurrence, bioactivity and analysis. *Journal of Pharmaceutical and Biomedical Analysis* 41(3), 683–693.

Chung, J.G., Garrett, L.R., Byers, P.E. and Cuchens, M.A. (1989) A survey of the amount of pristane in common fruits and vegetables. *Journal of Food Composition and Analysis* 2(1), 22–27.

Costerton, J.W., Stewart, P.S. and Greenberg, E.P. (1999) Bacterial biofilms: a common cause of persistent infections. *Science* 284, 1318–1322.

Crowe, L., Lau, W. and McLeod, L. (1999) Outbreaks of *Shigella sonnei* infection associated with eating fresh parsley – United States and Canada. *Journal of the American Medical Association* 281, 1785–1787.

Diego, J.M., Gómez-Coronadoa, Ibañez, E., Rupérez, F.J. and Barbas, C. (2004) Tocopherol measurement in edible products of vegetable origin. *Journal of Chromatography A* (Food Science) 1054(1–2), 227–233.

Eckey-Kaltenbach, H., Heller, W., Sonnenbichler, J., Zetl, I., Schäfer, W., Ernst, D. and Sandermann, H. Jr. (1993) Oxidative stress and plant secondary metabolism: 6'-*O*-malonylapiin in parsley. *Phytochemistry* 34(3), 687–691.

Edenharder, R., Kurz, P., John, K., Burgard, S. and Seeger, K. (1994) *In vitro* effect of vegetable and fruit juices on the mutagenicity of 2-amino-3-methylimidazo[4,5-f]quinoline, 2-amino-3,4-dimethylimidazo[4,5-f]quinoline and 2-amino-3,8-dimethylimidazo[4,5-f]quinoxaline. *Food and Chemical Toxicology* 32(5), 443–459.

Edenharder, R.J., Sager, W.H., Muckel, G.E. and Platt, K.L. (2002) Protection by beverages, fruits, vegetables, herbs, and flavonoids against genotoxicity of 2-acetylaminofluorene and 2-amino-1-methyl-6-phenylimidazo[4,5-*b*]pyridine (PhIP) in metabolically competent V79 cells. *Mutation Research/Genetic Toxicology and Environmental Mutagenesis* 521(1–2), 57–72.

Elkner, K., Horbowicz, M. and Kosson, R. (1998) Effect of storage on contents of dietary fibre and its composition in some cultivars of red beet, root parsley and carrot. *Vegetable Crops Research Bulletin* 49, 107–120.

El Sherbeny, S.E. and Hussein, M.S. (1993) Comparative study on three aromatic plants as affected by cutting frequency. *Egyptian Journal of Physiological Sciences* 17, 95–102.

Erdoğrul, Ö. and Şener, H. (2005) The contamination of various fruit and vegetables with *Enterobius vermicularis, Ascaris* eggs, *Entamoeba histolytica* cysts and *Giardia* cysts. *Food Control* (5th International Meeting of the Noordwijk Food Safety and HACCP Forum) 16(6), 557–560.

Farber, J.N., Harris, L.J., Parish, M.E., Beuchat, L.R., Suslow, T.V. and Gorny, J.R. (2003) Microbiological safety of controlled and modified atmosphere packaging of fresh and fresh-cut produce. *Comprehensive Reviews in Food Science and Food Safety* 2 (Supplement), 142–160.

Fischer, N., Weinreich, B., Nitz, S. and Drawert, F. (1991) Applications of high-speed counter-current chromatography for the separation and isolation of natural products. *Journal of Chromatography A* 538(1), 193–202.

Franken, P. and Gnadinger, F. (1994) Analysis of parsley arbuscular endomycorrhiza: infection development and mRNA levels of defense-related genes. *Molecular Plant Microbe Interactions* 7, 612–620.

Franz, C. and Glasl, H. (1976) Comparative investigations of fruit-, leaf- and root-oils of some parsley varieties. *Qualitas Plantarum* 25, 253–262.

Gandhi, M., Golding, S., Yaron, S. and Matthews, K.R. (2001) Use of green fluorescent protein expressing Salmonella Stanley to investigate survival, spatial location, and control on alfalfa sprouts. *Journal of Food Protection* 64, 1891–1898.

Gebhardt, Y., Witte, S., Forkmann, G., Lukačin, R., Matern, U. and Martens, S. (2005) Molecular evolution of flavonoid dioxygenases in the family Apiaceae. *Phytochemistry* (Evolution of Metabolic Diversity) 66(11), 1273–1284.

Gershbein, L.L. (1977) Regeneration of rat liver in the presence of essential oils and their components. *Food and Cosmetics Toxicology* 15(3), 173–181.

Górna-Binkul, A., Keymeulen, R., Van Langenhove, H. and Buszewski, B. (1996) Determination of monocyclic aromatic hydrocarbons in fruit and vegetables by gas chromatography-mass spectrometry. *Journal of Chromatography A* 734(2), 297–302.

Guiet, S., Robins, R.J., Lees, M. and Billault, I. (2003) Quantitative ^{2}H NMR analysis of deuterium distribution in petroselinic acid isolated from parsley seed. *Phytochemistry* 64(1), 227–233.

Gurgul, E., Herman, B. and Biczak, R. (1996) The effect of NPK fertilization on chlorophyll, sugars, ascorbic acid content and activity of catalase and peroxidase in leaf parsley. *Folia Horticulturae* 8(1), 39–51.

Hagemeier, J., Batz, O., Schmidt, J., Wray, V., Hahlbrock, K. and Strack, D. (1999) Accumulation of phthalides in elicitor-treated cell suspension cultures of *Petroselinum crispum*. *Phytochemistry* 51(5), 629–635.

Hardh, J.E. (1975) The influence of the environment of the nordic latitudes on the quality of vegetables. *Plant Foods for Human Nutrition (formerly) Qualitas Plantarum* 25(1), 43–56.

Hart, D.J. and Scott, K.J. (1995) Development and evaluation of an HPLC method for the analysis of carotenoids in foods, and the measurement of the carotenoid content of vegetables and fruits commonly consumed in the UK. *Food Chemistry* 54(1), 101–111.

Hashem, F.A. El-M. and Sahab, A.F. (1999) Chemical response of parsley and mentha herbs to certain stress agents. *Food Chemistry* 65(1), 29–33.

Heller, W. and Hahlbrock, K. (1980) Highly purified 'flavanone synthase' from parsley catalyzes the formation of naringenin chalcone. *Archives of Biochemistry and Biophysics* 200(2), 617–619.

Hinneburg, I., Dorman, H.J.D. and Hiltunen, R. (2006) Antioxidant activities of extracts from selected culinary herbs and spices. *Food Chemistry* 97, 122–129.

Hruschka, H.W. and Wang, C.Y. (1979) Storage and shelf-life of packaged watercress, parsley and mint. *USDA Market Research Report* No. 1101.

http://aggie-horticulture.tamu.edu/.

http://edis.ifas.ufl.edu/.

http://findarticles.com/.

http://www.ag.montana.edu/.

http://www.ams.usda.gov/.

http://www.botanical.com/.

http://www.botanical.com/ botanical/ mgmh/p/parsle09. html.

http://www.chilipaper.com/.

http://www.uni-graz.at/.

Josimovic, L. (1983) Study on some chemical changes in irradiated pepper and parsley. *The International Journal of Applied Radiation and Isotopes* 34(5), 787–791.

Joyce, D., Reid, M.S. and Katz, P. (1986) Postharvest handling of fresh culinary herbs. *Perishables Handling*, 58, 1–4.

Justesen, U. and Knuthsen, P. (2001) Composition of flavonoids in fresh herbs and calculation of flavonoid intake by use of herbs in traditional Danish dishes. *Food Chemistry* 73(2), 245–250.

Kabasakalis, V. and Tsolaki, A. (1994) Fluoride content of vegetables irrigated with waters of high fluoride levels. *Fresenius Environment Bulletin* 3(6), 377–380.

Kanwar, J.S. and Sandha, M.S. (2000) Wastewater pollution injury to vegetable crops – a review. *Agricultural Reviews* 21, 133–136.

Karapinar, M. and Gönül, S.A. (1992) Removal of *Yersinia enterocolitica* from fresh parsley by washing with acetic acid or vinegar. *International Journal of Food Microbiology* 16(3), 261–264.

Kasting, R., Andersson, J. and von Sydow, E. (1972) Volatile constituents in leaves of parsley. *Phytochemistry* 11(7), 2277–2282.

Kato, T., Kobayashi, M., Sasaki, N., Kitahara, Y. and Takahashi, N. (1978) The coumarin heraclenol as a growth inhibitor in parsley seeds. *Phytochemistry* 17(1), 158–159.

Kauss, H., Krause, K. and Jeblick, W. (1992) Methyl jasmonate conditions parsley suspension cells for increased elicitation of phenylpropanoid defense responses. *Biochemical and Biophysical Research Communications* 189(1), 304–308.

Kozan, E., Gonenc, B., Sarimehmetoglu, O. and Aycicek, H. (2005) Prevalence of helminth eggs on raw vegetables used for salads. *Food Control* 16(3), 239–242.

Kreuzaler, F. and Hahlbrock, K. (1973) Flavonoid glycosides from illuminated cell suspension cultures of *Petroselinum hortense*. *Phytochemistry* 12(5), 1149–1152.

Kreydiyyeh, S.I. and Usta, J. (2002) Diuretic effect and mechanism of action of parsley. *Journal of Ethnopharmacology* 79(3), 353–357.

Kreydiyyeh, S.I., Usta, J., Kaouk, I. and Al-Sadi, R. (2001) The mechanism underlying the laxative properties of parsley extract. *Phytomedicine* 8(5), 382–388.

Kuriyama, I., Musumi, K., Yonezawa, Y., Takemura, M., Maeda, N., Iijima, H., Hada, T., Yoshida, H. and Mizushina, Y. (2005) Inhibitory effects of glycolipids fraction from spinach on mammalian DNA polymerase activity and human cancer cell proliferation. *The Journal of Nutritional Biochemistry* 16(10), 594–601.

Lagey, K., Duinslaeger, L. and Vanderkelen, A. (1995) Burns induced by plants. *Burns* 21, 542–543.

Lamarti, A., Badoc, A. and Bouriquet, R. (1991) A chemotaxonomic evaluation of *Petroselinum crispum* (Mill. Nym. ex A.W. Hill) (parsley) marketed in France. *Journal of Essential Oil Research* 3, 425–433.

Lapidot, A., Romling, U. and Yaron, S. (2006) Biofilm formation and the survival of *Salmonella typhimurium* on parsley. *International Journal of Food Microbiology* 109, 229–233.

Lers, A., Jiang, W.B., Lomanies, E. and Aharoni, N. (1998) Gibberellic acid and CO_2 additive effect in retarding postharvest senescence of parsley. *Journal of Food Science* 63, 66–68.

Lisiewska, Z. and Kmiecik, W. (1997) Effect of freezing and storage on quality factors in Hamburg and leafy parsley. *Food Chemistry* 60(4), 633–637.

Lisiewska, Z., Kmiecik, W. and Budnik, A. (1997) Effect of conditions and time of storage on technological quality changes of parsley leaves. *Folia Horticulturae* 9(2), 21–29.

MacLeod, A.J., Snyder, C.H. and Subramanian, G. (1985) Volatile aroma constituents of parsley leaves. *Phytochemistry* 24(11), 2623–2627.

Madjarova, D.J. and Bubarova, M.G. (1978) New forms obtained by hybridization of *Apium graveolens* and *Petroselinum hortense*. *Acta Horticulturae* 73, 65–72.

Madzharova, D., Benbasat, E., Bubarova, M. and Chavdarov, I. (1973) New forms of celery and celeria. *Gradinarska i Lozarska Nauka* 10(3), 85–93.

Manderfeld, M.M., Schaer, H.W., Davidson, P.M. and Zottola, E.A. (1997) Isolation and identification of antimicrobial furocoumarins from parsley. *Journal of Food Protection* 60, 72–77.

Manzano, M., Citterio, B., Maifreni, M., Paganessi, M. and Comi, G. (1995) Microbial and sensory quality of vegetables for soup packaged in different atmospheres. *Journal of Science and Food Agriculture* 67, 521–529.

Martens, S., Forkmann, G., Matern, U. and Lukacin, R. (2001) Cloning of parsley flavone synthase I. *Phytochemistry* 58(1), 43–46.

Melloul, A.A., Hassani, L. and Rafouk, L. (2001) *Salmonella* contamination of vegetables irrigated with untreated wastewater. *World Journal of Microbiology and Biotechnology* 17, 207–209.

Moore, Bd., Palmquist, D.E. Seemann, J R. (1997) Influence of plant growth at high CO_2 concentrations on leaf content of ribulose-1,5-bisphosphate carboxylase/oxygenase and intracellular distribution of soluble carbohydrates in tobacco, snapdragon, and parsley. *Plant Physiology* 115, 241–248.

Naik, S.N. and Lentz, H. (1989) Extraction of perfumes and flavours from plant materials with liquid carbon dioxide under liquid-vapor equilibrium conditions. *Fluid Phase Equilibria* 49, 115–126.

Naimi, T.S., Wicklund, J.H., Olsen, S.J., Krause, G., Wells, J.G., Bartkus, J.M., Boxrud, D.J., Sullivan, M., Kassenborg, H., Besser, J.M., Mintz, E.D., Osterholm, M.T. and Hedberg, C.W. (2003) Concurrent outbreaks of *Shigella sonnei* and enterotoxigenic *Escherichia coli* infections associated with parsley: implications for surveillance and control of food borne illness. *Journal of Food Protection* 66, 535–541.

O'Beirn, D., Francis, G.A. and Thomas, C. (1999) The microbiological safety of minimally processed vegetables. *International Journal of Food Science and Technology* 34, 1–22.

Ohyama, S., Kitamori, S., Kawano, H., Yamada, T., Inamasu, T., Ishizawa, M. and Ishinishi, N. (1987) Ingestion of parsley inhibits the mutagenicity of male human urine following consumption of fried salmon. *Mutation Research Letters* 192(1), 7–10.

Ozsoy-Sacan, O., Yanardag, R., Orak, H., Ozgey, Y., Yarat, A. and Tunali, T. (2006) Effects of parsley (*Petroselinum crispum*) extract versus glibornuride on the liver of streptozotocin-induced diabetic rats. *Journal of Ethnopharmacology* 104(1–2), 175–181.

Palada, M.C., Davis, A.M., Crossman, S.M.A., Robles, C. and Chichester, E.A. (2004) Sustainable crop management practices for improving production of culinary herbs in the Virgin Islands. *Acta Horticulturae* (ISHS) 629, 289–298.

Park, C.E. and Sanders, G.W. (1992) Occurrence of thermotolerant campylobacters in fresh vegetables sold at farmers' outdoor markets and supermarkets. *Canadian Journal of Microbiology* 38, 313–316.

Park, K.W., Kang, H.M., Yang, E.M. and Jung, J.C. (1999) Effects of film package and storage temperature on quality of parsley in modified atmosphere storage. *Acta Horticulturae* 483, 291–298.

Peterson, S., Lampe, J.W., Bammler, T.K., Gross-Steinmeyer, K. and Eaton, D.L. (2006) Apiaceous vegetable constituents inhibit human cytochrome P-450 1A2 (hCYP1A2) activity and hCYP1A2-mediated mutagenicity of a flatoxin B1. *Food and Chemical Toxicology* 44, 1474–1484.

Pizzorno, J.E. and Murray, M.T. (1985) *A Textbook of Natural Medicine*. John Bastyr College Publications, Seattle, Washington, DC.

Popović, M., Kaurinović, B., Jakovljević, V., Mimica-Dukic, N. and Bursać, M. (2007) Effect of parsley (*Petroselinum crispum* (Mill.) Nym. ex A.W. Hill, Apiaceae) extracts on some biochemical parameters of oxidative stress in mice treated with CCl_4. *Phytotherapy Research* 21(8), 717–723.

Rontani, J.F., de Rabourdin, A., Pinot, F., Kandel, S. and Aubert, C. (2005) Visible light-induced oxidation of unsaturated components of cutins: a significant process during the senescence of higher plants. *Phytochemistry* 66(3), 313–321.

Rosa, C., Sapata, M. and Guerra, M.M. (2007) Chemical and sensory characteristics and microbiological safety of fresh finely chopped parsley packed in modified atmosphere. *Food Control* 18, 1008–1012.

Ruiz, G.V.B., Galves-Vargas, R. and Garcia-Villanova, R. (1987) Contamination of fresh vegetables during cultivation and marketing. *International Journal of Food Microbiology* 4, 285–291.

Salim, R., Isa, M., Al Subu, M.M., Sayrafi, S.A. and Sayrafi, O. (1995) Effect of irrigation with lead and cadmium on the growth and on the metal uptake of cauliflower, spinach and parsley. *Journal of Environmental Science and Health, Part A, Environmental Science and Engineering and Toxic and Hazardous Substance Control* 30(4), 831–849.

Saltveit, M.E. (1997) A summary of CA and MA requirements and recommendations for harvested vegetables. In: Saltveit, M.E. (ed.) *Vegetables and Ornamentals*. CA'97 Proceedings, Volume 4, University of California, Davis, California, pp. 98–117.

Scher, K., Romling, U. and Yaron, S. (2005) Effect of heat, acidification, and chlorination on *Salmonella enterica* serovar *Typhimurium* cells in a biofilm formed at the air–liquid interface. *Applied Environmental Microbiology* 71, 1163–1168.

Sebastião, K.I., Almeida-Muradian, L.B., Romanelli, M.F., Koseki, P.M. and Villavicencio, A.L.C.H. (2002) Effect of gamma-irradiation on the levels of total and *cis/trans* isomers of beta-carotene in dehydrated parsley. *Radiation Physics and Chemistry* 63(3–6), 333–335.

Simon, J.E. and Quinn, J. (1988) Characterization of essential oil of parsley. *Journal of Agricultural and Food Chemistry* 36, 467–472.

Solomon, E.B., Niemira, B.A., Sapers, G.M. and Annous, B.A. (2005) Biofilm formation, cellulose production, and curli biosynthesis by Salmonella originating from produce, animal, and clinical sources. *Journal of Food Protection* 68, 906–912.

Soysal, Y. (2004) Microwave drying characteristics of parsley. *Biosystems Engineering* 89(2), 167–173.

Spraul, M.H., Nitz, S., Drawert, F., Duddeck, H. and Hiegemannt, M. (1992) Crispane and crispanone, two compounds from *Petroselinum crispum* with a new carbon skeleton. *Phytochemistry* 31(9), 3109–3111.

Tschape, H., Prager, R., Streckel, W., Fruth, A., Tietze, E. and Bohme, G. (1995) Verotoxinogenic *Citrobacter freundii* associated with severe gastroenteritis and cases of haemolytic uraemic syndrome in a nursery school: green butter as the infection source. *Epidemiology and Infection* 114, 441–450.

Tsumura, F., Ohsako, Y. and Haraguchi, Y. (1993) Rapid changes in parsley leaves during storage in controlled or ethylene-containing atmosphere. *Journal of Food Science* 58, 619–625.

UC-Davis (2002) http://postharvest.ucdavis.edu/produce/storage/index.html.

USDA (2002) http://www.ams.usda.gov/standards/vegfm.htm.

Ward, J. (200) Richters Catalogue. In: *The Market for Herbs and Essential Oils. Food and Rural Development, Alberta Agriculture, Alberta, Canada* (www.richters.com).

Watada, A.E., Ko, N.P. and Minott, D.A. (1996) Factors affecting quality of fresh-cut horticultural products. *Postharvest Biology and Technology* 9, 115–125.

Wong, Y.Y.P. and Kitts, D.D. (2006) Studies on the dual antioxidant and antibacterial properties of parsley (*Petroselinum crispum*) and cilantro (*Coriandrum sativum*) extracts. *Food Chemistry* 97, 505–515.

Yamauchi, N. and Watada, A.E. (1993) Pigment changes in parsley leaves during storage in controlled or ethylene containing atmosphere. *Journal of Food Science* 56, 616–618.

Zhang, H., Chen, F., Wang, X. and Yao, H.-Y. (2006) Evaluation of antioxidant activity of parsley (*Petroselinum crispum*) essential oil and identification of its antioxidant constituents. *Food Research International* 39, 833–839.

Zogaj, X., Bokranz, W., Nimtz, M. and Romling, U. (2003) Production of cellulose and curli fimbriae by members of the family Enterobacteriaceae isolated from the human gastrointestinal tract. *Infection Immunology* 71, 4151–4158.

Zuskin, E., Kanceljak, B., Skuric, Z., Pokrajac, D., Schachter, E.N., Witek, T.J. and Maayani, S. (1988) Immunological and respiratory findings in spice-factory workers. *Environmental Research* 47(1), 95–108.

22 Celery

K.S. Krishnamurthy

22.1. Introduction

Celery (*Apium graveolens* L.) is a salad crop grown for its long fleshy leaf stalk. It resembles leafy onion. The seed is both a spice and a condiment. It ranks second in importance among salad crops. In the USA, France and in other European countries, it is grown commercially and is available in the market throughout the year, but it is not a commercially popular vegetable in India. There are three commonly used types of celery. The leafy types are referred to as var. *seculinum*, blanched celery as var. *dulce* and celeriac (with swollen edible roots) as var. *rapaceum* (Smith, 1976). The celeriac is also known as the turnip-rooted celery (Pruthi, 1976). Intergeneric hybridization of celery and parsley (*Petroselinum crispum*) has been reported (Smith, 1976).

22.2. Botany and Uses

Botany

Celery belongs to the family Umbelliferae. It is a biennial plant, although grown as an annual crop. The leafstalks are 15–38 cm long; bear three pairs and a terminal leaflet, coarsely serrated and alternately lobed or divided. The flower stalks are 60–100 cm in length, branched and leafy (Bailey, 1950). The flowers are small, white and borne in compound umbels. The fruit is rigid, 1.0–1.5 mm long, 1.0 mm in diameter and contains a small brown-coloured seed (Knott, 1960). The seeds are somewhat bitter in taste and 10 g contains about 26,450 seeds (Choudhury, 1970).

Celery grows wild in Europe, the Mediterranean region and in Asia, west of the Himalayas. The ancient Greeks and Egyptians cultivated celery. It was probably first grown as a medicinal plant, later for the leaves as flavouring. Celery has a long history in China, dating back to at least the 6th century AD. It is reported as being cultivated in several African countries, in Eritrea, Ethiopia, Mozambique and Réunion, and more commonly in South Africa (http://en.wikipedia.org/wiki/Celery). According to Thompson and Kelly (1957), the site of origin of celery extends from Sweden to Algeria, Egypt, Abyssinia, Asia and even to Caucacus, Baluchistan and the mountain regions of India. It has been found growing wild in California and in New Zealand. The first mention of its cultivation as a food plant was in 1623 in France. Janes (1954) reported that it was a native British plant, cultivated there for centuries.

Uses

The succulent leafstalk, often with a part of leafblade, is used for the preparation of sauces, vegetable juices, stews, soups, salads, etc. The type known as Chinese celery has thinner stalks and a stronger flavour and is rarely consumed raw, but is often added to soups and stir-fries. Celeriac, or turnip-rooted celery, is used mainly as a cooked vegetable in stews and soups, but is becoming increasingly popular as a grated raw salad. Leaf celery, also called smallage, is chopped and used as garnish and flavouring, either fresh or in dried powdered form (http://en.wikipedia.org). Sometimes, celery is also used in fried form.

Celery seed is used as a condiment in European countries and a spice in India. The seeds can be used as a flavouring or spice, either as whole seeds or ground and mixed with salt, as celery salt. Celery salt can also be made from an extract of the roots. Volatile oil obtained from the seeds is used in the perfume and pharmaceutical industries (http://en.wikipedia.org). The growing plant is an insect repellent; it repels the cabbage white butterfly so is a good companion for brassicas (Riotte, 1978). Essential oil of celery is most active against *Campylobacter jejuni*, with BA50 values ranging from 0.003 to 0.009 (Friedman *et al.*, 2002). The oil is toxic to cercariae of *Schistosoma mansoni* (96% killed at 40 ppm) and also exerts a chemotactic effect (Saleh *et al.*, 1985). All parts of the plant have medicinal uses.

22.3. General Composition

The leaves of celery are more nutritive than stalks, especially in vitamin A, protein and calcium. Traces of copper and arsenic have been reported in the tuberous root. The herb contains the glucoside apiin (Anon., 1952). A nutritive analysis of celery leaves is given in Table 22.1. Celery seeds are also nutritive. The major composition of the seeds is carbohydrate, followed by fat, protein and ash. They also contain micronutrients and vitamin A. Table 22.2 gives the composition of celery seeds as per ASTA and the *USDA Agricultural Handbook*.

Table 22.1. Nutritive composition of celery leaves (per 100 g of edible portion).

Moisture (g)	88.0	Thiamine (mg)	0.00
Fat (g)	0.6	Riboflavin (mg)	0.11
Protein (g)	6.3	Vitamin C (mg)	62
Carbohydrate (g)	1.6	Calcium (mg)	230
β-carotene (mg)	3990	Iron (mg)	6.3

Table 22.2. Nutritional composition of celery seed (per 100 g).

Composition	Content[1]	Content[2]
Water (g)	6.04	5.00
Food energy (Kcal)	392	450
Protein (g)	18.07	18.00
Fat (g)	25.27	22.80
Carbohydrates (g)	41.35	43.80
Ash (g)	9.27	10.20
Calcium (g)	1.767	1.800
Phosphorus (mg)	547	550
Sodium (mg)	160	170
Potassium (mg)	1400	1400
Iron (mg)	44.9	44.9
Thiamine (mg)	–	0.41
Riboflavin (mg)	–	0.49
Niacin (mg)	–	4.4
Ascorbic acid (mg)	17.14	17.00
Vitamin A activity (RE)	5	5

[1]*USDA Handbook*; [2]ASTA.
Source: Anon. (1977); ASTA (1977).

22.4. Chemistry

Volatiles

The volatiles showed variation in different environments. The composition of the oil from the fresh aerial parts of *A. graveolens* var. *secalinum* (at the flowering stage) obtained from three locations in Egypt revealed that the main components were α- and *β*-pinene, myrcene, limonene, *cis*-*β*-ocimene, γ-terpinene, *cis*-allo-ocimene, *trans*-farnesene, humulene, apiol, *β*-selinene, senkyunolide and neocnidilide (Saleh *et al.*, 1985). The chief component in the essential oils from fruits, and to a lesser extent in the leaves and stems, was limonene, whereas the roots and tubers had more *trans*-ocimene, 3-methyl-

4-ethylhexane and *β*-pinene (Fehr, 1981). Sipailiene *et al.* (2005) reported that the main constituents in the oil from roots were limonene, carvone and 3-*n*-butylphthalide. The structures of some of the volatiles are given in Fig. 22.1.

The volatiles also showed variation between genotypes. The concentration of terpenes and phthalides, the key volatile components, found in various cultivars of both celery and celeriac varied over a wide range (Van *et al.*, 1990). Eleven flavour compounds were identified in cv. Black Celery of Trevi, 17 compounds in cv. D'Elne and 21 compounds in cv. Verde Pascal. The main constituent of these cultivars was limonene. Twenty-one compounds were identified in cv. Dorato d'Asti. The main constituent was *γ*-terpinene. The cv. Black Celery of Trevi had a very low amount of *γ*-terpinene compared with other varieties (Tirillini *et al.*, 2004).

Studies on the qualitative and quantitative changes in the flavour of juices from Monarch and Bergers weiBe Kugel varieties showed the presence of 3-butylphthalide enantiomers. The enantiomeric distribution of 3-butylphthalide lowers the flavour quality of the Bergers weiBe Kugel celery juice (Greule *et al.*, 2005). Among the 12 compounds identified as potent odorants, 3-*n*-butylphthalide 1, sedanenolide 2, and *trans*- and *cis*-sedanolides 3, four were assessed to be most contributive to the overall odour of celery. These three phthalides, (3*E*,5*Z*)-1,3,5-undecatriene, myrcene and (*E*)-2 nonenal, were common to both raw and boiled materials. Two compounds ((*Z*)-3-hexenal and (*Z*)-3-hexenol) were dominant in raw celery and four compounds (2-methylbutanoic acid, sotolon, *β*-damascenone and *β*-ionone) were dominant in boiled celery (Kurobayashi *et al.*, 2006).

GC-MS analysis to study the content and composition of extracts of celery revealed the presence of terpenoids, sesquiterpenoids and phthalides in the essential oils and extracts obtained with organic solvents from two celery cultivars (Wolski *et al.*, 2004). The composition of the essential oil obtained from the fruits of three *A. graveolens* var. *dulce* cultivars, i.e. Helios, Orient and Zefir, showed that the main components of the essential oil were isoprenoids, including monoterpenes and sesquiterpenes. Essential oil content ranged from 2.5 to 3.0%. The percentage

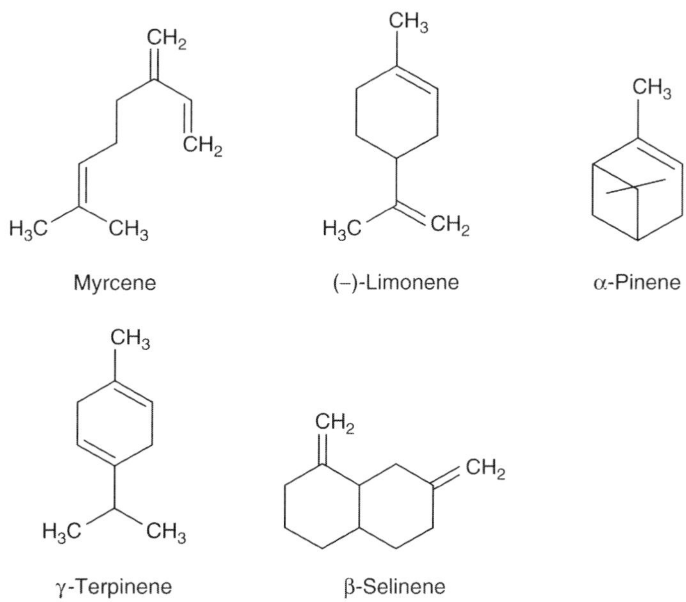

Fig. 22.1. Structures of some of the volatiles of celery.

of monoterpenes in the essential oil ranged from 79.1 to 82.8, while that of sesquiterpenes ranged from 15.7 to 18.8 (Wolski *et al.*, 2001).

Volatile constituents of celery × parsley hybrid

The celery × parsley hybrid inherited all the terpenoids from celery and heptanol from parsley, while synthesizing new compounds of its own. The content of these new compounds was higher than that of the main celery components, limonene and myrcene (Madjarova *et al.*, 1979). The volatile components of celery consisted of monoterpene hydrocarbons (46.0%) and phthalides (42.3%). The major components were limonene and 3-butyl-4,5-dihydrophthalide (or sedanenolide). Celery volatiles contained higher concentrations of γ-terpinene and α-pinene. A celery-like odour was associated during GC elution with each of the 16 phthalides reported (MacLeod and Ames, 1989).

Influence of nitrogen on volatile constituents

Increasing the rate of nitrogen fertilizer application affected the relative proportions of some of the essential oil components (Martin *et al.*, 1985). At higher N rates, limonene and other monoterpene concentrations decreased to very low values, while those of sedanolide and phthalides increased to above 75% (D'Antuono *et al.*, 2002).

Leaf oil

The chief components of the celery leaf oil were limonene, myrcene and *cis*-ocimene (Bubarova, 1973; Fehr, 1974). The sesquiterpene content of the leaf oil (< 5%) was relatively low compared with that of the fruit oil. The essential oil of leaves contained a higher amount of limonene compared with the roots and a very small amount of carvone (Sipailiene *et al.*, 2005). Studies conducted on the leaf essential oil showed that the matured dry leaf contained essential oil having a composition similar to that of seed oil (Thappa *et al.*, 2003). High limonene content in celery oil renders it unfit for direct consumption. The increase in phthalide content is linked with the corresponding decrease in limonene content. The reduced limonene content offers significant advancement towards the development of celery oil with improved quality (Thappa *et al.*, 2003).

Leaf oil in relation to location

The three main constituents of volatiles from leaf stems of a local cv. from Libya were apiole (about 23%), 3-butylphthalide (about 22%) and sedanolide (about 24%). The last two possess a strong characteristic celery aroma (MacLeod *et al.*, 1988). Limonene (40.5%), β-selinene (16.3%), *cis*-ocimene (12.5%) and β-caryophyllene (10.5%) were the major volatile oil constituents of celery leaves collected from Nigeria (Ehiabhi *et al.*, 2003). Analysis of celery leaf oil from Cuba revealed 28 compounds, representing about 94% of the oil. Limonene (18.3%), β-caryophyllene (13.5%) and 3-butyl-4,5-dihydrophthalide (32.1%) were the major constituents (Pino *et al.*, 1997).

Seed oil

Limonene (50.1–65.5%) and β-selinene (11.2–22.2%) were the major components in the seed oil of celery. A sesquiterpene ether, kessane (2.2–7.6%), was also detected (Philippe *et al.*, 2002).

Sterols in seed oil

Sitosterol and stigmasterol were the major components of celery seed oil. The other components were cholesterol, brassicasterol, campesterol, ∂7-campesterol, ∂5-avenasterol, ∂7-stigmasterol and ∂7-avenasterol (Zlatanov and Ivanov, 1995). Oil bodies isolated from celery cell suspension cultures contained at least 60% of the total steryl ester present in the cells. Free sterols comprised < 0.5% of the total lipid in the oil body. Sterylesters constituted 4.5% of the total lipid of celery oil bodies. The proportion of precursor 4-methylsterols in the free sterol fraction of celery was greater in the oil body (Dyas and Goad, 1994). Stigmasterol was either undetected or contributed very little to the steryl ester fraction. The precursor sterols (cycloartenol, obtusifoliol) were esterified in greater proportion to palmitic acid than to the C_{18} fatty acids in suspension cultures (Dyas *et al.*, 1994).

Phospholipids in seed oil

The phospholipid composition of glyceride oils from celery seeds indicated that the phospholipid content was 1.7–3.7% in glyceride oils and 0.2–0.5% in seeds. Phosphatidylcholine (38.5–51.1% in the phospholipids fraction), phosphatidylinositol (18.6–32.0%) and phosphatidylethanolamine (9.3–18.6%) were identified as major components in all the glyceride oils (Zlatanov, 1994).

Volatile production in cell suspension cultures

Two phthalides, 3-isobutylidene-3a,4,5,6-tetrahydrophthalide and 3-isobutylidene-3a,4-dihydrophthalide, were identified in the intact plant and the differentiated callus, but were absent in the undifferentiated callus. The phthalide composition of this culture was comparable to that of the intact plant (Abta *et al.*, 1979). In the green slow growth cultures, flavour compounds, the phthalides (3-butylphthalide and sedanenolide) and other terpenoid compounds were present, both in the media and the cells. Maximum secondary product synthesis and greening occurred in media containing 3,5-dichlorophenoxyacetic acid [3,5-D]. There was no stimulation of phthalide and terpenoid synthesis in cultures grown in a medium containing 2,4-D in darkness and at high (30°C) and intermediate (20°C) temperatures. However, when the cultures were maintained at 4°C for the first 5 days of the growth cycle and then transferred to 25°C, phthalides and other terpenoids, particularly limonene, were released into the medium (Watts *et al.*, 1984).

In young plants, the concentrations of limonene and other terpenoids increased in the petiole and leaf extract as the level of chlorophyll increased. In the cell cultures, greening was induced by transferring to a medium where 2,4-D was replaced by 3,5-dichlorophenoxyacetic acid (3,5-D). During the first subculture, phthalides (the major flavour compounds) were produced in the growth phase, but were present in trace amounts only in the second and third subcultures. Limonene levels, however, increased after the first subculture. Phthalide production was stimulated by the transfer from a 2,4-D- to a 3,5-D-containing medium but, as the level of aggregation and greening increased in the culture, phthalide production was reduced and limonene production increased (Watts *et al.*, 1985).

Non-volatiles

The quantification of the potentially health-defensive and disease-preventive flavonoids, quercetin, kaempferol, myricetin, apigenin and luteolin, in celery revealed that the quercetin level was generally below 10 mg/kg and kaempferol was below 30 mg/kg. Detectable amounts of apigenin, luteolin and myricetin were also present (Lugasi and Hovari, 2000). The leaf stalks of three celery cultivars (Helios, Orient and Tango F1), analysed for the composition of tannins and free phenolic acids, indicated that tannin levels ranged from 5.78 to 7.68%. Phenolic acids, protocatechuic, *p*-hydroxybenzoic, caffeic, *p*-coumaric and ferulic acids were also detected (Najda and Dyduch, 2004). Four linear furocoumarins (psoralen, bergapten, xanthotoxin and isopimpinellin) were isolated from three varieties of healthy, commercially grown celery. Previously, psoralen had not been reported to occur in celery (Beier *et al.*, 1983). The phenolic compounds, particularly the flavonoids, may be responsible, in part, for the antioxidant activity of traditional plant extracts including celery (Pendry *et al.*, 2005).

Five sesquiterpenoid glucosides (celerioside A–E) and three phthalide glycosides (celephtalide A–C), together with six aromatic compound glucosides, two norcarotenoid glucosides and a lignan glucoside, were isolated from the water-soluble portion of the methanol extract of the seed (Kitajima *et al.*, 2003). Cyclooxygenase inhibitory and antioxidant bioassay-directed extraction and purification of celery seeds yielded sedanolide, senkyunolide-N, senkyunolide-J, 3-hydroxymethyl-6-methoxy-2,3-dihydro-1*H*-indol-2-ol, L-tryptophan and 7-[3-(3,4-dihydroxy-4-hydroxymethyl-tetrahydro-furan-2-yloxy)-4,5-dihydroxy-6hydroxymethyl-tetrahydro-pyran-2-yloxy]-5-hydroxy-2-(4-hydroxy-3-methoxy-phenyl)-chromen-4-one (Momin and Nair, 2002).

Extraction of flavour compounds

The best results were obtained with C_{18} Nova-Pak, C_{18} Symmetry and C_{18} Genesis columns for the analysis of free and conjugated flavonoids and the same were used for the quantitative analysis of endogenous flavones and flavonols in acid-hydrolysed extracts from celery (Crozier *et al.*, 1997). A newly developed simultaneous purging and solvent-extraction apparatus was used to isolate and identify 14 terpenes and two aromatic compounds in the headspace sample from celery (Macku and Shibamoto, 1991). The supercritical fluid extraction of oil from milled celery seeds, using CO_2 as a solvent, indicated a significant increase in extraction rate with increase of pressure or decrease of the particle size of the celery seed. A similar effect was observed with the increase of the solvent flow rate and decrease of temperature (Papamichail *et al.*, 2000). High-speed counter-current chromatography (HSCCC) with an Ito multilayer coil separator–extractor was applied to perform efficient separations of natural products such as phthalides (aroma compound) from celery and parsley roots (Fischer *et al.*, 1991).

22.5. Medicinal and Pharmacological Uses

Celery has been used traditionally to treat many disorders. All parts of the plant are known to be a remedy for one or more maladies. Wild celery also is no exception to this. The medicinal value of celery in treating various ailments and its various other medicinal and pharmacological properties is outlined below:

- Treatment for rheumatism
- Treatment for urinary disorders
- Digestive remedy
- Cure for nervous disorders
- Antimicrobial activity
- Other medicinal uses.

Treatment for rheumatism

The herb is used in treating rheumatism (Launert, 1981). Infusions from the seeds are used for rheumatoid arthritis and gout. The essential oil is used in warm water to soak painful, gouty areas of the feet. Root tinctures and also fresh juice from the whole plant are also used in arthritic remedies (http://www.innvista.com/health/herbs/celery.htm).

Treatment for urinary disorders

The diuretic property of celery has been used to prepare herbal medicine (Houghton, 1995). The ripe seeds, herb and root are diuretic (Lust, 1983; Chiej, 1984; Grieve, 1984). The seeds are used mainly as a diuretic and can help clear toxins from the system, especially in cases of gout where uric acid crystals collect in the joints. Root tinctures have been used to cure urinary disorders, such as urinary stones, and used as a kidney stimulant and cleanser. Fresh juice from the whole plant is also used as a cure for urinary tract inflammations and urethritis (http://www.innvista.com/health/herbs/celery.htm). The herb is used against kidney complaints (Launert, 1981).

Digestive remedy

Wild celery relieves indigestion (Bown, 1995). The celery roots act as a bitter digestive remedy and liver stimulant. The seeds also act as a digestive stimulant (http://www.innvista.com/health/herbs/celery.htm).

Cure for nervous disorders

Celery is also used to cure a few nervous disorders (http://www.innvista.com/health/herbs/celery.htm). An essential oil obtained from the plant has a calming effect on the central nervous system. Some of its constituents have antispasmodic, sedative and anticonvulsant actions. It has been shown to be of value in treating high blood pressure (Chavallier, 1996). Wild celery promotes restfulness and sleep in hysteria patients (Grieve, 1984) and is also used to lower blood pressure (Bown, 1995).

Antimicrobial activity

Sesquiterpene lactones from *A. graveolens* showed activity against *Bacillus subtilis* and *Proteus vulgaris* and also tested fungi (Jawad *et al.*, 1985). Essential oil of celery inhibited the growth of *Listeria monocytogenes*, *Staphylococcus aureus*, *Escherichia coli* O:157:H7, *Yersinia enterocolitica*, *Pseudomonas aeruginosa*, *Lactobacillus plantarum*, *Aspergillus niger*, *Geotrichum* and *Rhodotorula* (Elgayyar *et al.*, 2001). The root extracts possessed high activity against *B. cereus* and *Enterococcus faecalis* only (Sipailiene *et al.*, 2005).

Other medicinal uses

The ripe seeds, herb and root are aperient, carminative, emmenagogue, galactogogue, nervine, stimulant and tonic (Lust, 1983; Chiej, 1984; Grieve, 1984). Eating fresh stalks is good for lactating mothers, although wild celery is more effective (http://www.innvista.com/health/herbs/celery.htm). Wild celery also stimulates the uterus and is anti-inflammatory (Bown, 1995). A few compounds, such as L-tryptophan and others isolated from celery seeds, exhibited good antioxidant activity at concentrations of 125 and 250 μg/ml (Momin and Nair, 2002).

The polyacetylenes present in celery have shown to be highly toxic towards fungi, bacteria and mammalian cells, to display neurotoxic, anti-inflammatory and antiplatelet aggregatory effects and to be responsible for allergic skin reactions. The effect of these polyacetylenes towards human cancer cells, their human bioavailability and their ability to reduce tumour formation in a mammalian *in vivo* model indicates that they may also provide benefits for health (Christensen and Brandt, 2006). Celery oil induced significant anti-inflammatory activities 2, 3 and 5 h after administration to rats (Afifi *et al.*, 1994). *A. graveolens* is a potent plant against experimentally induced hepatocarcinogenesis in Wistar rats (Sultana *et al.*, 2005).

22.6. Quality Specifications

Chemical and physical specifications

Celery contains 1.5–3.0% volatile oil, primarily containing about 60–70% *d*-limonene and 10–20% *β*-selinene. The characteristic celery odour is thought to be due to oxygenated compounds present in the oil (sedanolide and sedanonic acid anhydride). Essential oil of celery seed is available; however, the most common extractive form is the oleoresin, due to its fuller flavour. This product contains 12–16% volatile oil.

ASTA suggest maximum moisture levels of 10%. Ash and acid-insoluble ash should be less than 10 and 2%, respectively. Tables 22.3 and 22.4 summarize the typical chemical and physical specifications, including the FDA's DAL, for whole celery and ground celery, respectively.

ISO specifications

The following are the ISO specifications available for celery:

ISO 927, Spices and condiments – determination of extraneous matter content.
ISO 928, Spices and condiments – determination of total ash.
ISO 930, Spices and condiments – determination of acid-insoluble ash.
ISO 939, Spices and condiments – determination of moisture content – entertainment method.
ISO 948, Spices and condiments – sampling.
ISO 2825, Spices and condiments – preparation of a ground sample for analysis.
ISO 6571, Spices, condiments and herbs – determination of volatile oil content.

Table 22.4. Ground celery: chemical and physical specifications.

Specification	Suggested limits
FDA DAL	None
Volatile oil[1] (% min.)	1.0
Moisture[1] (% max.)	10.0
Total ash[1] (% max.)	10.0
Acid-insoluble ash[1]	2.0
Military specifications (EE-S-631J, 1981)	
Volatile oil (ml/100 g, min.)	2.0
Moisture (% max.)	10.0
Total ash (% max.)	14.0
Acid-insoluble ash (% max.)	2.0
Granulation (% min. through a USS No. 55)	95
Non-volatile ether extract (% min.)	12.0
Bulk index (ml/100 g)	190

Note: [1]These are suggested limits that the authors put together from the data collected over the previous 5 years. These numbers are equivalent to the level in to which most quality spices fall.
Source: Tainter and Grenis (2001).

Celery seeds specification (IS 3797:1993)

The following Indian Standards are necessary adjuncts to this standard:
IS No. Title
1070; 1992 Reagent grade water (third revision).
1797; 1985 Methods of test for spices and condiments (second revision).
13145; 1993 Spices and condiments – methods for sampling (first revision).
For the purpose of this standard, the following definitions shall apply.
Extraneous matter: includes dust, dirt, stones, clay particles, chaff and stem or straw.
Damaged seeds: discoloured, shrivelled and immature seeds.
Foreign seeds: seeds other than those of *A. graveolens* L.
Celery seeds shall be of three grades, namely, Special, Good or Fair.

Table 22.3. Whole celery: chemical and physical specifications.

Specification	Suggested limits
ASTA cleanliness specifications	
Whole dead insects, by count	4
Mammalian excreta (mg/lb)	3
Other excreta (mg/lb)	3.0
Mould, % by weight	1.00
Insect-defiled/infested, % by weight	1.00
Extraneous, % by weight	0.50
FDA DAL (condimental seed)	
Adulteration with mammalian excreta (mg/lb)	Average of 3
Volatile oil[1] (% min.)	1.5
Moisture[2] (% max.)	10.0
Ash[1] (% max.)	10.0
Acid-insoluble ash[1] (% max.)	2.0
Average bulk index (mg/100 g)	195

Note: [1]These are suggested limits that the authors put together from the data collected over the previous 5 years. These numbers are equivalent to the level in to which most quality spices fall.
[2]ASTA suggested maximum moisture level.
Source: Tainter and Grenis (2001).

Requirements

DESCRIPTION The celery seeds shall be the dried, ripe fruits of *A. graveolens* L. The seeds shall be light brown to greyish-brown in colour. The shape of the seeds shall be ovoid to hemispherical, their length about 1.0–1.5 mm and width 0.5–1.0 mm. The seeds shall have several raised streaks running along their longitudinal axis, these streaks being lighter than the rest of the seeds.

TASTE AND AROMA OR FLAVOUR The taste and aroma or flavour of celery seeds shall be fresh and characteristic of the type and variety. The material shall be free from foreign taste and aroma or flavour, as well as from any musty odour.

FREEDOM FROM MOULDS AND INSECT ATTACK The celery seeds shall be free from visible insects and moulds and shall be practically free from dead insects and contamination by rodents, visible to the naked eye, corrected, if necessary, in any particular case. In case the magnification exceeds × 10, this fact should be stated in the test report. Table 22.5 lists the grade designations of celery seeds and their requirements as per IS 3797:1993.

Packing

Celery seeds shall be packed in sound, clean and dry containers made of a material which shall protect celery seeds from insect infestation, as well as from any offensive odour.

Marking

The following particulars shall be marked or labelled on each container:

1. Name of the material and grade designation.
2. Variety or trade name.
3. Batch or code number.
4. Net mass.
5. Date of packing.
6. Country of origin.

Sampling

Representative samples of celery seeds shall be drawn according to IS 13145:1991.

Tests

Tests shall be carried out in accordance with Table 22.5.

Quality of reagents

Unless specified otherwise, pure chemicals (chemicals that do not contain impurities which affect the results of analysis) and distilled water (see IS 1070:1992) shall be employed in tests.

22.7. Conclusion

Celery is used both as a vegetable and a spice. It has several medicinal uses. It is believed to cure most urinary disorders. It has diuretic properties. It is also commonly used to relieve pain. Limonene is the major volatile compound in both leaf and seed oil. Phthalides, myrcene, pinene, etc., are also present. Higher nitrogen inhibited limonene

Table 22.5. Grade designations of celery seeds and their requirements (as per IS 3797:1993).

Sl. No.	Characteristic	Requirement			Method of test, ref. to Cl No. IS 1797:1985
		Special	Good	Fair	
i	Moisture (% by mass, max.)	10.0	11.0	11.0	9
ii	Total ash (on dry basis) (% by mass, max.)	10.0	11.0	12.0	6
iii	Acid-insoluble ash (on dry basis) (% by mass, max.)	2.0	2.5	3.0	8
iv	Volatile oil content (on dry basis) (% by mass, min.)	2.0	1.5	1.5	15
v	Extraneous matter (% by mass, max.)	0.5	1.5	3.0	4
vi	Damaged and foreign seeds (% by mass, max.)	1.5	2.5	3.0	4

production but favoured phthalide production. The volatile oil is reported to be inhibitory to some pests. In cell suspension cultures, 3,5-dichlorophenoxyacetic acid stimulated flavour production. There is great scope for *in vitro* production of celery flavour. Also, there is scope for directed studies on some of the medicinal properties, such as its diuretic properties and its use in controlling pain, as well as the treatment of urinary disorders.

References

Abta, A.S., Galpin, I.J. and Collin, H.A. (1979) Flavour compounds in tissue cultures of celery. *Plant Science Letters* 16(2–3), 129–134.

Afifi, N.A., Ramadan, A., El Kashoury, E.A. and El Banna, H.A. (1994) Some pharmacological activities of essential oils of certain umbelliferous fruits. *Veterinary Medical Journal Giza* 42(3), 85–92.

Anon. (1952) *Wealth of India*. CSIR, New Delhi.

Anon. (1977) *Composition of Foods: Spices and Herbs*. USDA Agricultural Handbook 8–2, Washington, DC.

ASTA (1977) *The Nutritional Composition of Spices*. ASTA Research Committee, Washington, DC.

Bailey, L.H. (1950) *The Standard Cyclopedia of Horticulture*. The Macmillan Co., New York.

Beier, R.C., Ivie, G.W., Oertli, E.H. and Holt, D.L. (1983) HPLC analysis of linear furocoumarins (psoralens) in healthy celery (*Apium graveolens*). *Food and Chemical Toxicology* 21, 163–165.

Bown, D. (1995) *Encyclopaedia of Herbs and their Uses*. Dorling Kindersley, London.

Bubarova, M. (1973) Studies on the leaf essential oils of celery, parsley and their hybrids from the taxonomic aspect. *Gradinarska i Lozarska Nauka* 10(1), 23–32.

Chavallier, A. (1996) *The Encyclopedia of Medicinal Plants*. Dorling Kindersley. London.

Chiej, R. (1984) *Encyclopaedia of Medicinal Plants*. MacDonald, London.

Choudhury, B. (1970) *Vegetables* (2nd edn.). National Book Trust, New Delhi.

Christensen, L.P. and Brandt, K. (2006) Bioactive polyacetylenes in food plants of the Apiaceae family: occurrence, bioactivity and analysis. *Journal of Pharmaceutical and Biomedical Analysis* 41(3), 683–693.

Crozier, A., Jensen, E., Lean, M.E.J. and McDonald, M.S. (1997) Quantitative analysis of flavonoids by reversed-phase high-performance liquid chromatography. *Journal of Chromatography A* 761(1–2), 315–321.

D'Antuono, L.F., Neri, R. and Moretti, A. (2002) By-products of vegetable celery (*Apium graveolens* L. var. dulce) as potential source of flavours. *Acta Horticulturae* (576), 327–331.

Dyas, L. and Goad, L.J. (1994) The occurrence of free and esterified sterols in the oil bodies isolated from maize seed scutella and a celery cell suspension culture. *Plant Physiology and Biochemistry Paris* 32(6), 799–805.

Dyas, L., Threlfall, D.R. and Goad, L.J. (1994) The sterol composition of five plant species grown as cell suspension cultures. *Phytochemistry* 35(3), 655–660.

Ehiabhi, O.S., Edet, U.U., Walker, T.M., Schmidt, J.M., Setzer, W.N., Ogunwande, I.A., Essien, E. and Ekundayo, O. (2003) Constituents of essential oils of *Apium graveolens* L., *Allium cepa* L., and *Voacanga africana* Staph. from Nigeria. *Journal of Essential Oil Bearing Plants* 9(2), 126–132.

Elgayyar, M., Draughon, F.A., Golden, D.A. and Mount, J.R. (2001) Antimicrobial activity of essential oils from plants against selected pathogenic and saprophytic microorganisms. *Journal of Food Protection* 64(7), 1019–1024.

Fehr, D. (1974) The essential oil of celery leaves. *Pharmazie* 29(5), 349.

Fehr, D. (1981) Investigations into the aromatic compounds of celery. Part 2. *Pharmazie* 36(5), 374–376.

Fischer, N., Weinreich, B., Nitz, S. and Drawert, F. (1991) Applications of high-speed counter-current chromatography for the separation and isolation of natural products. *Journal of Chromatography A* 538(1), 193–202.

Friedman, M., Henika, P.R. and Mandrell, R.E. (2002) Bactericidal activities of plant essential oils and some of their isolated constituents against *Campylobacter jejuni, Escherichia coli, Listeria monocytogenes* and *Salmonella enterica*. *Journal of Food Protection* 65(10), 1545–1560.

Greule, M., Mosandl, A., Kreck, M. and Dietrich, H. (2005) Quality assessment of celeriac juices using enantioselective flavour analysis. *Deutsche Lebensmittel Rundschau* 101(5), 199–204.

Grieve, M. (1984) *A Modern Herbal*. Penguin, London, 919 p.

Houghton, P. (1995) Herbal products. 9. Bearberry, dandelion and celery. *Pharmaceutical Journal* 255(6855), 272–273.

http://en.wikipedia.org/wiki/Celery.

http://www.innvista.com/health/herbs/celery.htm

Janes, E.R. (1954) *The Vegetable Garden*. Penguin Handbooks, London, pp. 88–100.

Jawad, A L.M., Dhahir, A.B.J., Hussain, A.M., Ali, K.F. and Saleh, H.M. (1985) Antimicrobial activity of sesquiterpene lactones extracted from Iraqi plants. Part II. *Journal of Biological Sciences Research* 16(2), 17–22.

Kitajima, J., Ishikawa, T. and Satoh, M. (2003) Polar constituents of celery seed. *Phytochemistry* 64(5), 1003–1011.

Knott, J.E. (1960) *Hand Book for Vegetable Growers*. Wiley, New York.

Kurobayashi, Y., Kouno, E., Fujita, A., Morimitsu, Y. and Kubota, K. (2006) Potent odorants characterize the aroma quality of leaves and stalks in raw and boiled celery. *Bioscience, Biotechnology and Biochemistry* 70(4), 958–965.

Launert, E. (1981) *Edible and Medicinal Plants*. Hamlyn, London.

Lugasi, A. and Hovari, J. (2000) Flavonoid aglycons in foods of plant origin I. Vegetables. *Acta Alimentaria Budapest* 29(4), 345–352.

Lust, J. (1983) *The Herb Book*. Bantam, New York.

Macku, C. and Shibamoto, T. (1991) Application of simultaneous purging and solvent extraction technique for flavour monitoring of natural products. *Food Chemistry* 42(2), 121–127.

MacLeod, A.J., MacLeod, G. and Subramanian, G. (1988) Volatile aroma constituents of celery. *Phytochemistry* 27(2) 373–375.

MacLeod, G. and Ames, J.M. (1989) Volatile components of celery and celeriac. *Phytochemistry* 28(7), 1817–1824.

Madjarova, D., Bubarova, M., Hristova, I., Madzharova, D. and Khristova, I. (1979) The biochemical nature of forms obtained by the remote hybridization between genera *Apium, Petroselinum*. II. Essential oils. *Herba Hungarica* 18(3), 185–195.

Martin, S., Deiser, E. and Buchloh, G. (1985) Does fertilizer application affect flavour? *Gemuse* 21(9), 356–357.

Momin, R.A. and Nair, M.G. (2002) Antioxidant, cyclooxygenase and topoisomerase inhibitory compounds from *Apium graveolens* Linn. Seeds. *Phytomedicine* 9(4), 312–318.

Najda, A. and Dyduch, J. (2004) Determination of quantitative composition of polyphenolic compounds occurring in leaf stalks of three celery cultivars (*Apium graveolens* L. var. dulce Mill./Pers.). *Folia Universitatis Agriculturae Stetinensis, Agricultura* 95, 259–264.

Papamichail, I., Louli, V. and Magoulas, K. (2000) Supercritical fluid extraction of celery seed oil. *The Journal of Supercritical Fluids* 18(3), 213–226.

Pendry, B., Busia, K. and Bell, C.M. (2005) Phytochemical evaluation of selected antioxidant-containing medicinal plants for use in the preparation of a herbal formula – a preliminary study. *Chemistry and Biodiversity* 2(7), 917–922.

Philippe, J., Suvarnalatha, G., Sankar, R. and Suresh, S. (2002) Kessane in the Indian celery seed oils. *Journal of Essential Oil Research* 14(4), 276–277.

Pino, J.A., Rosado, A. and Fuentes, V. (1997) Leaf oil of celery (*Apium graveolens* L.) from Cuba. *Journal of Essential Oil Research* 9(6), 719–720.

Pruthi, J.S. (1976) *Spices and Condiments*, National Book Trust, New Delhi.

Riotte, L. (1978) *Companion Planting for Successful Gardening*. Garden Way, Vermont, Inc., New York.

Saleh, M.M., Zwaving, J.H., Malingre, T.M. and Bos, R. (1985) The essential oil of *Apium graveolens* var. secalinum and its cercaricidal activity. *Pharmaceutisch Weekblad* 7(6), 277–279.

Sipailiene, A., Venskutonis, P.R., Sarkinas, A. and Cypiene, V. (2005) Composition and antimicrobial activity of celery (*Apium graveolens*) leaf and root extracts obtained with liquid carbon dioxide. *Acta Horticulturae* 677, 71–77.

Smith, P.M. (1976) In: Simmonds, N.W. (ed.) *Evolution of Crop Plants*. Longmans, London.

Sultana, S., Ahmed, S., Jahangir, T. and Sharma, S. (2005) Inhibitory effect of celery seeds extract on chemically induced hepatocarcinogenesis: modulation of cell proliferation, metabolism and altered hepatic foci development. *Cancer Letters* 221(1), 11–20.

Tainter, D.R. and Grenis, A.T. (2001) Spices and Seasonings–a food technology handbook, 2nd edn. Wiley-VCH, New York, 256 p.

Thappa, R.K., Dhar, A.K., Balyan, S.S., Khan, S., Raina, P., Dhar, P.L. and Choudhary, D.K. (2003) Variation in the major components of seed and leaf oil of genetically improved celery (*Apium graveolens* L.). *Journal of Food Science and Technology* 40(4), 426–428.

Thompson, H.C. and Kelly, W.C. (1957) *Vegetable Crops*. McGraw-Hill Book Co. Inc., New York and London, 611 pp.

Tirillini, B., Pellegrino, R., Pagiotti, R., Pocceschi, N. and Menghini, L. (2004) Volatile compounds in different cultivars of *Apium graveolens* L. *Italian Journal of Food Science* 16(4), 477–482.

Van, W.F., Dirinck, P., Vulsteke, G., Schamp, N. and Van Wassenhove, F. (1990) Aromatic volatile composition of celery and celeriac cultivars. *HortScience* 25(5), 556–559.

Watts, M.J., Galpin, I.J. and Collin, H.A. (1984) The effect of growth regulators, light and temperature on flavour production in celery tissue cultures. *New Phytologist* 98(4), 583–591.

Watts, M.J., Galpin, I.J. and Collin, H.A. (1985) The effect of greening on flavour production in celery tissue cultures. *New Phytologist* 100(1), 45–56.

Wolski, T., Najda, A. and Mardarowicz, M. (2001) GC/MS analysis of essential oil of the fruits of three varieties of *Apium graveolens* L. var. dulce Mill. Pers. Annales Universitatis Mariae Curie Skodowska Sectio EEE, *Horticultura* 9 (Supplement), 203–211.

Wolski, T., Najda, A. and Hoderna Kedzia, E. (2004) The content and composition of essential oils and extracts obtained from fruits of some plants of the Umbelliferae (Apiaceae) family and preliminary antibacterial assessment of extracts. *Postepy Fitoterapii* 5(13), 119–125.

Zlatanov, M. (1994) Studies on phospholipid composition of glyceride oils of some representatives of the family Apiaceae. *Fett Wissenschaft Technologie* 96(12), 456–457.

Zlatanov, M. and Ivanov, S.A. (1995) Studies on sterol composition of the seed oil of some representatives of the family Apiaceae. *Fett Wissenschaft Technologie* 97(10), 381–383.

23 Curry Leaf

V.A. Parthasarathy, T. John Zachariah and B. Chempakam

23.1. Introduction

Murraya koenigii, commonly known as the curry leaf tree, is a native of India, Sri Lanka and other South Asian countries. Curry leaves are grown throughout India and they adorn every backyard, especially in the southern states, where most cuisines are prepared with the subtle flavouring of this highly aromatic leafy spice. The leaves are used to flavour a range of dishes and typically these are fried in oil until crisp to impart flavour to all types of curry preparations (Choudhury and Garg, 2007). Fresh leaves release a strong aroma while cooling. The plant has also been used in traditional Indian medicine systems for a variety of ailments. The oil derived from the leaves is also used in the perfume and soap industries (Shanthala and Prakash, 2005).

23.2. Botany

Curry leaf (*M. koenigii* L. Spreng) belongs to the citrus family, Rutaceae. It is a small tree maintained, under cultivation, as a small shrub. Some trees have been observed at a height of more than 5 m. Under cultivation, they are maintained below 2.5 m high. The leaves are exstipulate, bipinnately compound, 30 cm long, each bearing 24 leaflets, having reticulate venation; the leaflets are lanceolate, 4.9 cm long, 1.8 cm broad, having 0.5 cm-long petioles (Parmar and Kaushal, 1982). Lalitha *et al.* (1997) and Lal *et al.* (2000) described the diversity in the genetic resources. Reisch *et al.* (1994c) studied the chemotypes available in Sri Lanka.

23.3. Composition

The leaf extract of curry leaf has been reported to contain moisture (66.3%), protein (1%), fat (1%), carbohydrate (16%), fibre (6.4%) and mineral matter (4.2%). The main minerals per 100 g of leaves are calcium (810 mg), phosphorus (600 mg) and iron (2.1 mg). The vitamins in the leaves are carotene (12,600 i.u.), nicotinic acid (2.3 mg) and vitamin C (4 mg) (Anon., 1962; Kumar *et al.*, 1999). The extract also contains oxalic acid, which reduces the availability of calcium. The contents are total oxalate (1.352%) and soluble oxalate (1.155%) (Ananthasamy *et al.*, 1960; Walde *et al.*, 2005). The effects of storage temperature on the nutritive value were studied by Palaniswamy *et al.* (2002). Reisch *et al.* (1994a) found the furocoumarins in the seeds.

(eds V.A. Parthasarathy, B. Chempakam and T.J. Zachariah)

23.4. Chemistry

Volatile oils

Philip (1981) reported that young leaves contained more volatile oil and oleoresin than mature leaves. Table 23.1 describes the variability in the percentage composition of constituents in young and old leaves (Hiremath *et al.*, 1997; Mallavarapu *et al.*, 1999). The variation in oil recovery and composition during dehydration was described by Madalageri *et al.* (1996).

Terpenes are the main constituents of the volatile essential oil of *M. koenigii* leaves, which are used for curry flavouring (MacLeod and Pieris, 1982). The oil of *M. koenigii* produces less than 4% of other components, with eight monoterpene hydrocarbons (about 16%) and 17 sesquiterpene hydrocarbons (about 80%). The major constituents responsible for aroma are β-caryophyllene, β-gurjunene, β-elemene, β-phellandrene and β-thujene (Kumar *et al.*, 1999). The volatile oils from the leaves of six species of the genus *Murraya* have been studied by GC-MS and about 60 monoterpene and sesquiterpenes components were identified. From these results, and published data on other species, it appears that the oils are either predominantly sesquiterpenoid or monoterpenoid in nature. The distinction between the two oil types coincides with other chemical data that support the division of the genus into two sections, *Murraya* and *Bergera* (Lal *et al.*, 2000; Ramalakshmi *et al.*, 2000).

Table 23.1. Volatile constituents in young and old leaves of *Murraya koenigii*.

Constituent	Content	
	Young leaves	Old leaves
Caryophyllene (%)	26.3	–
Cadinene (%)	18.2	–
Cadinol (%)	12.8	–
D-Sabinene (%)	9.2	31.8–44.8
Dipentene (%)	6.8	–
D-α-Pinene (%)	5.5	19.0–19.7
β-Pinene (%)	–	4.2–4.7
D-1-α-Phellandrene (%)	4.6	–
β-Phellandrene (%)	–	6.5–7.9
D-α-Terpinene (%)	3.2	1.3–4.3
Δ-Terpinene (%)	–	3.9–7.1
Terpinene-4-ol (%)	–	5.2–9.9
Lauric acid (%)	2.7	–
Palmitic acid (%)	3.4	–

Source: Philip (1981); Mallavarapu *et al.* (1999).

The flavour volatile constituents of the seed cotyledons, fruits and leaves of *M. koenigii* L. were analysed by GC-MS and compared with curry leaf flavour constituents (Walde *et al.*, 2005). These consist of monoterpene hydrocarbons, oxygenated monoterpenes and sesquiterpenes. The major constituents of curry leaf are monoterpenes (70%), seed cotyledons (86%) constituting α-pinene (52%) and *cis*-β-ocimene (34%); raw fruit oil containing monoterpenes (80%) and oxygenated monoterpenes (4.8%); and fruit pulp oil containing monoterpenes (61%).

The curry leaf plant is highly valued for its characteristic aroma and medicinal value. A number of leaf essential oil constituents and carbazole alkaloids have been extracted from the plant (Hiremath and Madalageri, 1997). A large number of studies have been carried out on the chemical composition of various parts of the curry leaf plant (Raina *et al.*, 2002). It has been reported that the leaves contain 34 compounds, which constitute about 97.4% of the oil. The major constituents identified were α-pinene (51.7%), sabinene (10.5%), β-pinene (9.8%), β-caryophyllene (5.5%), limonene (5.4%), bornyl acetate (1.8%), terpinen-4-ol (1.3%), γ-terpinene (1.2%) and α-humulene (1.2%) (Rana *et al.*, 2004), while an earlier study by Wong and Tie (1993) identified 62 components, the main constituents being β-phellandrene (24.4%), α-pinene (17.5%), β-caryophyllene (7.3%) and terpinen-4-ol (6.1%). Mallavarapu *et al.* (2000) identified 48 constituents of the leaf essential oil, representing 95% of the essential oil, and 42 constituents of the fruit essential oil, accounting for 98.5%. The major constituents of the leaf essential oil were α-pinene (9%), α-phellandrene (6.1%), β-phellandrene (50.1%), (*E*)-β-ocimene (7.1%) and β-caryophyllene (4.9%). The main constituents of the fruit essential oil were α-pinene (48.1%), β-pinene (7.1%), myrcene (3.1%), β-phellandrene (26%), γ-terpinene (3%) and β-caryophyllene (3%).

Variation in chemical composition has been observed among different agroclimatic locations. Analyses of the essential oil from the leaves of *M. koenigii* growing in southern Nigeria revealed an oil composition of predominantly sesquiterpenes (89.1%). The main components of the oil were β-caryophyllene (20.5%), bicyclogermacrene (9.9%), α-cadinol (7.3%), caryophyllene epoxide (6.4%), β-selinene (6.2%) and α-humulene (5%) (Onayade and Adebajo, 2000). Studies on the volatile oil in the leaves of *M. koenigii* grown in Sri Lanka, isolated by steam distillation, identified 36 out of 53 compounds detected. The major constituents were β-thujene (5.8%), β-phellandrene (18.9%), (*E*)-β-ocimene (12.7%), β-caryophyllene (23.3%), α-humulene (4.3%) and β-bisabolene (3.14%) (Paranagama *et al.*, 2002). The major volatile compounds from *M. koenigii* are given in Fig. 23.1.

Wong and Chee (1996) analysed the volatile constituents of *M. koenigii* flowers by capillary GC and GC-MS following isolation by solvent extraction. Forty-eight compounds were identified, monoterpenoids and sesquiterpenoids accounting for 34.4 and 43.9% of the total volatiles, respectively. The major components were β-caryophyllene (24.2%), (*E*)-β-ocimene (18%) and linalool (8%). In a later study, Walde *et al.* (2006) isolated the volatiles of fresh leaf stalks and flowers by a simultaneous distillation and extraction method, followed by GC-MS analysis. Thirty-one components were identified in the leaf stalk oil, constituting 88.1% of the volatile oil. The major components were the mono- and sesquiterpene hydrocarbons (66.7%), the major ones of which were α-pinene (24.2%), β-pinene (6.9%), α-phellandrene (7.3%) and α-copaene (8.9%). In addition, the oil had nine oxygenated monoterpenes (14.2%) and four sesquiterpene alcohols (8.1%). In the flower oil, 24 components were identified (constituting 91.8% of the volatile oil), which constituted 87% mono- and sesquiterpenes. The major compounds in this class were *cis*-ocimene (34.1%), α-pinene (19.1%), γ-terpinene (6.7%) and β-caryophyllene (9.5%). It also contained seven oxygenated monoterpenes and three oxygenated sesquiterpenes,

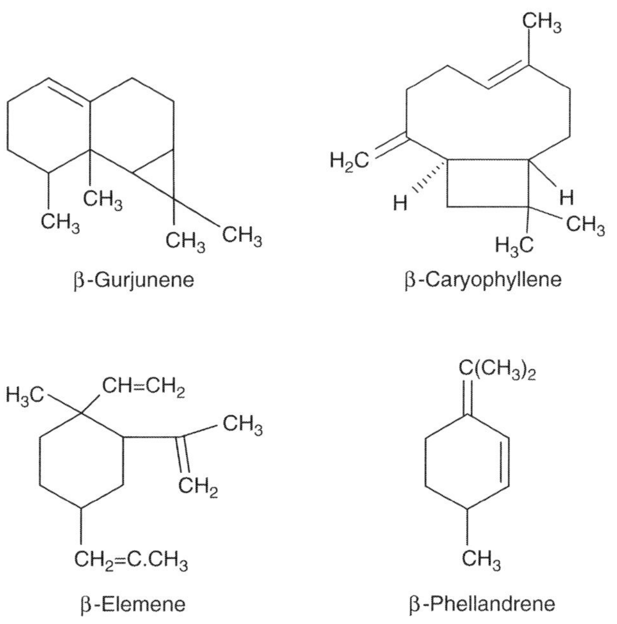

Fig. 23.1. Volatile compounds from *Murraya koenigii*.

constituting 4.7% of the oil. The larger number of oxygenated mono- and sesquiterpenes present appeared to be responsible for the intense odour associated with the stalk and flower parts of *M. koenigii*, as compared with the leaf. Table 23.2 describes the volatile component of seeds, fruit pulp, raw fruit and leaves and Table 23.3 illustrates the leaf oil composition in plants grown in Nigeria. β-Caryophyllene, bicyclogermacrene, caryophyllene epoxide, α-cadinol and α-humulene are some of the prominent compounds in Nigerian oil.

Carbazole alkaloids

Interestingly, the first discovery of carbazoles from plant sources, girinimbine, was reported in *M. Koenigii* (Chakraborty *et al.*, 1964). Since then, this species has proved to be a major source of carbazole alkaloids (Chakraborty, 1993).

From the stem bark, a new carbazole derivative, named murrayanine, has been isolated. It has been formulated as 1-methoxy-3-formylcarbazole (I) (Chakraborty *et al.*, 1965). Roy and Chakraborty (1974) reported the isolation of mahanimbine. The structure of a new hexacyclic carbazole alkaloid, isomurrayazoline, from stem bark has been shown to be 9a,10,11,12,13,13a-hexahydro-5,9,9,12-tetramethyl-1 (Bhattacharya *et al.*, 1982). The isolation of mahanimbine, girinimbine and two new carbazole alkaloids, isomahanimbine and koenimbidine, from the leaves and roots of the curry leaf plant has been reported. Spectroscopic and degradative evidence supports structures for girinimbine, mahanimbine, isomahanimbine, koenimbidine and murrayacine (Joshi *et al.*, 1970). The ethanolic extract of the leaves afforded a new carbazole alkaloid, which was identified as bismurrayafoline E on the basis of spectroscopic analysis (Nutan *et al.*, 1999). DL-*O*-Methylmahanine, 8-hydroxymahanimbine and pyranylcarbazole were synthesized from carabazole (Anwer *et al.*, 1972). 3-Methylcarbazoles were oxidized to the corresponding derivatives with excess 2,3-dichloro-5-6-dicynobezoquinonine. Murraycine was prepared similarly by oxidation of girinimbine (Anwer *et al.*, 1973). New biogenetically significant constituents, namely, 3-methyl carbazole and glycozoline, were reported by Adesina *et al.* (1988). A biogenetically important carbazole alkaloid from curry leaf was reported by Bhattacharyya *et al.* (1986). Its isolation supports the proposal that the pyranocarbazoles are formed from 2-hydroxy-3-methylcarbazole by the incorporation of C_5 and C_{10} units (Bhattacharyya and Chakraborty, 1984; Reisch *et al.*, 1994b).

Koenoline, a further cytotoxic carbazole alkaloid from *M. koenigii*, was described by

Table 23.2. Volatile components of seeds, fruit pulp, raw fruits and leaves of *Murraya koenigii*.

Compound	Seed oil (%)	Fruit pulp oil (%)	Raw fruit oil (%)	Leaf oil (%)
α-Pinene	51.60	21.8	46.32	16.79
Camphene	0.95	–	0.72	1.31
β-Pinene	12.56	4.6	10.10	4.72
β-Myrcene	7.29	3.34	5.92	2.65
α-Phellandrene	0.45	4.95	1.97	2.96
cis-β-Ocimene	0.31	1.27	13.94	28.49
trans-β-Ocimene	2.45	39.17	7.41	6.42
γ-Terpinene	9.28	5.49	0.41	6.63
cis-Linalool oxide	0.22	1.39	3.96	–
trans-Linalool oxide	5.82	–	0.82	–
Linalool	1.24	–	0.20	–
α-Copaene	6.00	10.32	5.04	20.16
β-Caryophellene	1.07	1.82	0.83	3.49
Isocaryophyllene	0.24	1.22	–	2.38

Table 23.3. Composition of leaf essential oil of *Murraya koenigii* grown in Nigeria.

Constituent[1]	Content (% of total oil)	Constituent	Content (% of total oil)
α-Pinene	1.5	β-Selinene	6.2
Sabinene	t	Zingiberene*	t
β-Pinene	t	Bicyclogermacrene	9.9
Myrcene	t	α-Muurolene*	t
α-Phellandrene	t	γ-Cadinene	0.2
p-Cymene	t	β-Sesquiphellandrene*	t
Limonene	2.2	δ-Cadinene	0.4
β-Phellandrene*	t	Cadina-1,4-diene	0.2
1,8-Cineole*	t	ε-Nerolidol	1.7
cis-β-Ocimene	t	Un (O-seq)	0.9
Dihydrotagetone	1.3	Ledol	0.4
Myrcene epoxide	t	Spathulenol	2.0
Linalool	0.2	Un (O-seq)	1.3
Camphor	t	caryophyllene Epoxide	6.4
Terpinen-4-ol	0.4	Globulol*	t
α-Terpineol	0.2	Un (O-seq)	2.0
Thymol	0.4	Widdrol	1.1
α-Copaene	t	Cedrol	1.5
β-Bourbounene	t	Humulene epoxide	2.7
β-Elemene	2.2	*epi*-Cubenol	2.1
β-Cubebene*	t	Selina-1,3,7(11)-trien-8-one	3.2
β-Caryophellene	20.5	Ledol isomer*	t
β-Cedrene*	t	α-Eudesmol*	t
Bicycloelemene	2.4	α-Cadinol	7.3
epi-Santalene	1.0	Selin-11-en-4α-ol	1.6
trans-β-Farnesene	1.8	Un (O-seq)	0.6
α-Humulene	5.0	α-Santalol	0.3
Alloaromadendrene	0.2	*epi*-Santalol	0.3
Curcumene	0.5	β-Santalol	0.4
γ-Muurolene	1.4	Un (O-seq)	0.8
Ledene	0.6		
Germacrene-B	t		

Source: Onayade and Adebajo (2000).

Note: [1]Listed in order of elution on Durabond-DBI GC column; marked (*) constituents were determined on CP-wax 52cb GC column; t = trace amount (< 0.2%), Un (O-seq) = unidentified oxygen-containing sesquiterpene. All Un (O-seq) < 0.5% are omitted from above list.

Fiebig *et al.* (1985). It exhibited cytotoxic activity against certain cell cultures. Its structure was established as 1-methoxy-3-hydroxymethylcarbazole by analysis of spectroscopic data and was confirmed by partial synthesis from murrayanine isolated from *M. siamensis* roots. Koenoline exhibited cytotoxic activity against the KB cell-culture test system. Bhattacharyya and Chakraborty (1984) described mukonal, a probable biogenetic intermediate of pyranocarbazole alkaloids, from stem bark of *M. koenigii*. The structure of the compound has been established as 2-hydroxy-3-formyl carbazole based on physical (UV, IR, ^{1}H NMR, ^{13}C NMR and mass spectrometry) and chemical transformations. Mukherjee *et al.* (1983) described mukonicine, another carbazole alkaloid from the leaves of *M. koenigii*. From physical methods coupled with chemical evidence, its structure was determined as 1,2-[2:2-dimethyl-Δ^3-pyrano]-3-methyl-6,

8 dimethoxycarbazole. A new C_{23}-carbazole alkaloid, mahanimbinol, was isolated from the stem of the Indian curry leaf plant, *M. koenigii*. It is a key precursor in the biosynthesis of some 20 other carbazole alkaloids reported previously from this plant (Rao *et al.*, 1980). Structure and synthesis of mukonine, a new carbazole alkaloid from *M. koenigii* [stem bark], was described by Chakraborty *et al.* (1978). The structure of murrayacinine, a carbazole alkaloid from *M. koenigii*, was described by Chakraborty *et al.* (1974). Apoptosis of curry leaf carbazole alkaloids was observed by Ito *et al.* (2006). The structures of carbazole alkaloids from *M. koenigii* are illustrated in Fig. 23.2.

Minor furocoumarins reported from seeds include xanthotoxin, isobyakangelicol, phellopterin, gosferol, neobyakangelicol, byakangelicol, byakangelicin and isogosferol (Adebajo and Reisch, 2000).

The chemical constituents of stem bark were reported by Sukari *et al.* (2001). The carbazole alkaloids, e.g. mahanimbine, girinimbine and murrayanine, were isolated and characterized from the petroleum ether extract of bark. A minor alkaloid, mahanine, was isolated from *M. koenigii* leaves (Rahman *et al.*, 1988). Based on silica gel column chromatography (CC) and preparative thin-layer chromatography (TLC) analyses, Ito *et al.* (1993) isolated 16 known carbazoles and carbazolequinones, three new monomeric and five new binary carbazole alkaloids (mukoenine-A, -B and -C, murrastifoline-F, *bis*-2-hydroxy-3-methylcarbazole, bismahanine,

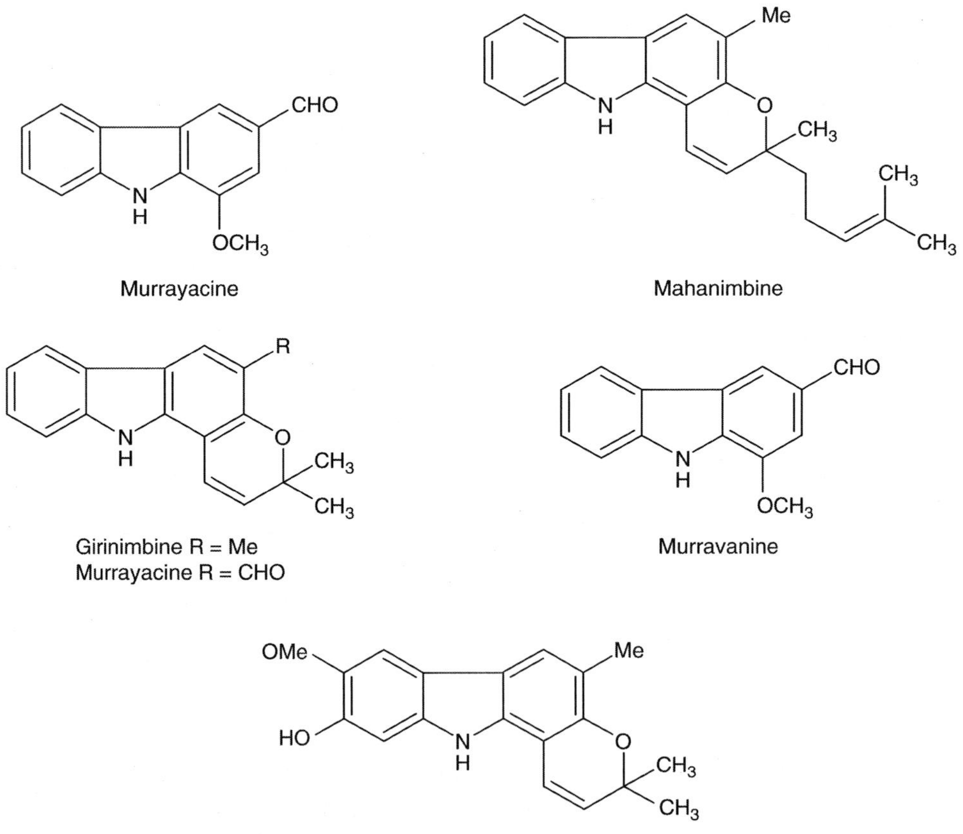

Fig. 23.2. Carbazole alkaloids from *Murraya koenigii*.

bikoeniquinone-A and bismurrayaquinone-A, respectively) from the acetone extracts of root and stem bark. Bikoeniquinone-A and bismurrayaquinone-A were found to contain a carbazole-1,4-quinone skeleton as a basic structural unit. Bhattacharyya *et al.* (1994) isolated two carbazole alkaloids from the stem bark. They were identified as 2-methoxy carbazole-3-methyl carboxylate and 1-hydroxy-3-methyl carbazole from spectral and chemical evidence. Wang *et al.* (2003) isolated two new carbazole alkaloids named murrayanine (1) and 8,8′-biskoenigine (2) from *M. koenigii*. The structure elucidations for 1 and 2 were carried out on the basis of 1D and 2D NMR experiments. Compound 1 was a novel carbazole alkaloid with a rare phenylpropanyl substitution and compound 2 was a symmetrical dimer of the carbazole alkaloid, koenigine. The synthesis of 2 from koenigine was carried out through oxidative coupling using a solid-state reaction. Chakraborty *et al.* (1997) isolated two new alkaloids, 9-carbethoxy-3-methylcarbazole and 9-formyl-3-methylcarbazole, and a known metabolite, 3-methyl-carbazole, from the roots of *M. koenigii*. All three compounds were identified by detailed spectral analyses, including 2D NMR studies, and their structures confirmed by synthesis. Of the two metabolites, the 9-formyl compound showed antitumour properties.

Lipids

Lipids from dry seed were extracted and the quantity of availability was around 4.4%, consisting of 85.4% neutral lipids, 5.1% glycolipids and 9.5% phospholipids. The fatty acid composition of total lipids indicated oleic, linoleic and palmitic acids to be the major components (Hemavathy, 1991).

23.5. Medicinal and Pharmacological Uses

Curry leaf has been used in folk medicine in China and other Asian countries as an analgesic, astringent, antidysenteric, antioxidant, febrifuge, hypolipidaemic, hypoglycaemic, for improvement of vision, to treat night-blindness and for regulation of fertility (Palaniswamy *et al.*, 2003). The cortex, roots and leaves of *M. koenigii*, a plant endemic to southern India, are used in Ayurvedic and Yunani (Unani) medicines for the treatment of febrile disorders, dysentery, diarrhoea and inflammation of the gums. The essential oil of this plant is thought to have antibacterial and antifungal properties. The essential oil obtained from the leaves by steam distillation was found to contain 11-selinen-4α,5α,7β,10β-4-ol (kongol or 11-eudesmen-4-α-ol) and the azulene derivative, 10-aromadendranol (globulol) (Wagner *et al.*, 1995). It is used in traditional medicine to treat constipation, colic, diarrhoea and hiccups. An aqueous extract of the leaves (200–800 mg/kg), collected from a garden in Sri Lanka, was administered to ethanol-treated rats. The extracts inhibited the development of ethanol-induced lesions in the corpus of the stomach (Ratnasooriya *et al.*, 1995). Differentiation in the pharmaceutical potential of the *Murraya* species was described by Kong *et al.* (1986).

The fresh juice of curry leaves, mixed with limejuice and sugar, cures morning sickness, nausea and vomiting due to indigestion. Chewing tender leaves helps to control diarrhoea, whereas matured leaves are beneficial in controlling diabetes and weight loss. Leaves ground with turmeric and taken daily are an effective remedy for allergic reactions. Curry leaves and black pepper beaten with sour curd are beneficial for indigestion (Irani, 2005).

Antioxidant

Five carbazole alkaloids isolated from the CH_2Cl_2 extract and their structures, namely, euchrestine B (compound 1), bismurrayafoline E (compound 2), mahanine (compound 3), mahanimbicine (compound 4) and mahanimbine (compound 5), exhibited antioxidative properties (Nakatani, 2000; Tachibana *et al.*, 2001). The levels of antioxidant vitamins in fresh curry leaves were

9744 ng of lutein, 212 ng of α-tocopherol and 183 ng of β-carotene per gram fresh weight (Palaniswamy *et al.*, 2003). The carbazole, mahanimbine, from curry leaf exhibits antioxidant activity (Ramsewak *et al.*, 1999).

Antimicrobial

Curry leaf has been shown to possess more antibacterial activity and less antifungal activity (Aqil and Ahmad, 2003). The compounds such as isomahanine and murrayanol isolated from the CCl_4 fraction exhibited good activity against 11 of 14 tested bacterial species (Nutan *et al.*, 1998). Chowdhury *et al.* (2001) isolated two new alkaloids, 1-formyl-3-methoxy-6-methylcarbazole (compound 1) and 6,7-dimethoxy-1-hydroxy-3-methylcarbazole (compound 2), from the leaves of *M. koenigii*. Compound 2 was found to be active against Gram-positive and Gram-negative bacteria and fungi. Rahman and Gray (2005) isolated a benzoisofuranone derivative, 3ε(1ε-hydroxyethyl)-7-hydroxy-1-isobenzofuranone, and a dimeric carbazole alkaloid, 3,3'-[oxybis (methylene)] *bis*-(9-methoxy-9*H*-carbazole), along with six known carbazole alkaloids and three known steroids from the stem bark of *M. koenigii*. The minimum inhibitory concentrations (MIC) of these compounds were found to be in the range 3.13–100 μg/ml when tested on *Staphylococcus aureus*, *Bacillus subtilis*, *Escherichia coli*, *Proteus vulgaris*, *Aspergillus niger and Candida albicans*. Based on ethnomedicine, *M. koenigii* (L.) Spreng. is used as a stimulant, antidysentery and for the management of diabetes mellitus (Adebajo *et al.*, 2004). The petroleum ether and crude chloroform extracts of curry leaf exhibited weak antibacterial activity against *B. cereus* (Sukari *et al.*, 2001).

The three bioactive carbazole alkaloids, e.g. mahanimbine, murrayanol and mahanine, were found to be mosquitocidal and antimicrobial. The antioxidant activity of carbazoles (1,1-diphenyl-2-picryl hydrazyl: DPH) from *M. koenigii* was found to be in order ascorbic acid > bismurraya foline E > euchestine B, mahanine and α-tocopherol > BHT > mahanimbirine and mahanimbine (Walde *et al.*, 2005).

Anti-inflammatory

Bioassay-guided fractionation of the acetone extract of the fresh leaves of *M. koenigii* resulted in the isolation of three bioactive carbazole alkaloids (mahanimbine, murrayanol and mahanine). Murrayanol showed activity against human prostaglandin H synthase (hPGHS-1) and hPGHS-2 (IC50 values of 109 and 208 μg/ml, respectively) in an anti-inflammatory assay (Ramsewak *et al.*, 1999). Essential oils (1.4% v/w) from *M. koenigii* leaves showed significant anti-inflammatory and analgesic activities in mice (Maiti *et al.*, 2004).

Antidiabetic

It is reported that curry leaves slow down the starch to glucose breakdown in diabetic patients. It has been recommended to promote them as preferable food adjuvants for diabetic patients (Grover *et al.*, 2002). In mild and moderate diabetic rats, feeding of 5, 10 and 15% diets caused a maximum reduction in blood sugar by 13.1, 16.3 and 21.4% (NS, $P < 0.05$ and 0.005) and 3.2, 5.58 and 8.21% (NS), respectively (Yadav *et al.*, 2002). The methanol extract of curry leaf was more effective in lowering the blood glucose levels in diabetic rats compared with the aqueous extract (Vinuthan *et al.*, 2007). There are a large number of reports on the use of curry leaf for diabetes. However, Iyer and Mani (1990), Adebajo *et al.* (2004) and Yadav *et al.* (2004) did not find any effect on diabetes parameters. This may be due to the quantity of curry leaf required to bring down the glucose level.

Pesticidal properties

The essential oil of curry leaf is composed of many chemicals with a multispectral activity. The monoterpene hydrocarbons and sesquiterpene hydrocarbons play a major role in this activity (Ray and Srivastava, 2006). Its leaves have been studied for their antifungal activity against three plant pathogenic fungi, i.e. *Rhizoctonia solani, R. bataticola* (*Macrophomina phaseolina*) and *Helminthosporium oryzae* (*Cochliobolus miyabeanus*). The essential oils had shown fungitoxity of more than 50% at 2000 ppm concentration. Benzene extract was found to be more antifungal than hexane extract. The antifungal activity of pure carbazoles, e.g. koenimbine, was improved by their derivatization into corresponding *N*-methyl koenimbine and *N*-methyl koenidine. However, hydrogenation of the pure carbazoles into dihydro derivatives caused reduction in fungitoxicity (Ray and Srivastava, 2006). It has also been proved to be effective against *Rhizopus stolonifer* (*R. stolonifer* var. *stolonifer*) and *Gloeosporium psidii* (*Colletotrichum coccodes*) infecting guava (Dwivedi *et al.*, 2002). The leaves and their extract were found to have nematicidal action (Padhi and Behera, 2000; Pandey, 2000, 2005).

The oil was toxic at concentrations of 340 ppm, and the varying effects of some Sri Lankan essential oils on oviposition and progeny production was reported by Pathak *et al.* (1997) and Abeywickrama *et al.* (2003). It has also been reported to be useful against the cabbage pests, *Pieris brassicae* and *Plutella xylostella* (Mehta *et al.*, 2005; Facknath, 2006), against the mustard aphid, *Lipaphis erysimi* (Srivastava and Kumar, 1999), and against *Callosobruchus chinensis* developing in green gram (*Vigna radiata*) and chickpea seeds.

It has great potential for use in mosquito control programmes. The three carbazole compounds (mahanimbine, murrayanol and mahanine) exhibited activity against *Aedes aegyptii* and against a battery of microorganisms (Ramsewak *et al.*, 1999). The *Murraya* extract was most toxic against *A. aegypti*, followed by *Culex quinquefasciatus* and *Anopheles stephensi* (Pathak *et al.*, 2000).

Other uses

Industrial uses

The major constituents of the oil are *cis*-caryophyllene (11.74%), dipentene (11.3%), α-eudesmol (9.61%), isocaryophyllene (8.41%) and β-elemene (7.09%). The composition of the oil suggests that it may find application as a fixative in soap and detergent perfumes (Chowdhury, 2000).

Plant growth inhibitors

Curry leaf has been reported to contain plant growth inhibitors. Bhattacharya *et al.* (1989) described the isolation and characterization of two coumarins present in *M. koenigii* bark. The growth-inhibitory properties are comparable to those of other coumarins, such as psoralen and xanthotoxin.

Effect on female sex hormones

The plant is reported to have an effect on the ovarian hormone profile and reproductive performance of animals (Mehrotra *et al.*, 2003, 2005).

23.6. Conclusion

M. koenigii (Linn), commonly known as the curry leaf plant, is highly valued for its characteristic aroma and medicinal properties. Its leaves are used extensively for culinary purposes, especially in curries and chutneys, but also in vegetable, fish and meat dishes, pickles, buttermilk preparations, curry powder blends, etc. The major volatile components in curry leaf are α-pinene, β-caryophyllene, (*E*)-β-ocimene, linalool and β-phellandrene. *M. koenigii* is a rich source of carbazole alkaloids. Its leaves, roots and bark are a tonic, stomachic and carminative. It is shown to possess a hypocholesterol effect and many other health benefits. The crop promises great scope in various biochemical and industrial applications in the future.

References

Abeywickrama, K., Paranagama, P., Adhikari, C. and Bandara, P. (2003) Deterrent effects of some Sri Lankan essential oils on oviposition and progeny production of the cowpea bruchid; *Callosobruchus maculatus* (F.) (Coleoptera; Bruchidae). *Journal of Food, Agriculture and Environment* 1(2), 254–257.

Adebajo, A.C. and Reisch, J. (2000) Minor furocoumarins of *Murraya koenigii*. *Fitoterapia* 71, 334–337.

Adebajo, A.C., Olayiwola, G., Verspohl, J.E., Iwalewa, E.O., Omisore, N.O.A., Bergenthal, D., Kumar, V. and Adesina, S.K. (2004) Evaluation of the ethno medical claims of *Murraya koenigii*. *Pharmaceutical Biology* 42, 610–620.

Adesina, S.K., Olatunji, O.A., Bergenthal, D. and Reisch, J. (1988) New biogenetically significant constituents of *Clausena anisata* and *Murraya koenigii*. *Pharmazie* 43, 221–222.

Ananthasamy, T.S., Kamat, V.N. and Pandya, H.G. (1960) Oxalate contents of leafy vegetables. *Current Science* 29, 133.

Anon. (1962) *The Wealth of India (Raw Materials)*, Volume 6. Publications and Information Directorate (CSIR), New Delhi.

Anwer, F., Kapil, R.S. and Popli, S.P. (1972) Terpenoid alkaloids from *Murraya koenigii* VII. Synthesis of DL-*O*-methylmahanine and related carbazoles. *Experientia* 28, 769.

Anwer, F., Masaldan, A.S., Kapil, R.S. and Popli, S.P. (1973) Synthesis of murraycine oxidation with DDQ (2,3-dichloro-5,6-dicynobenzoquinone) of the activated aromatic methyl group of the alkaloids of *Murraya koenigii*. *Indian Journal of Chemistry* 11, 1314–1315.

Aqil, F. and Ahmad, I. (2003) Broad-spectrum antibacterial and antifungal properties of certain traditionally used Indian medicinal plants. *World Journal of Microbiology and Biotechnology* 19(6), 653–657.

Bhattacharya, L., Roy, S.K. and Chakraborty, D.P. (1982) Structure of the carbazole alkaloid isomurrayazoline from *Murraya koenigii* [stem bark]. *Phytochemistry* 21, 2432–2433.

Bhattacharya, L., Mukhopadhyay, M. and Chakraborty, D.P. (1989) 3(1′,1′-Dimethyl allyl) xanthylentin and scopoletin, two plant growth inhibitors from *Murraya koenigii* Spreng. *Plant Physiology and Biochemistry* 16, 23–26.

Bhattacharyya, P. and Chakraborty, A. (1984) Mukonal, a probable biogenetic intermediate of pyranocarbazole alkaloids from [stem bark of] *Murraya koenigii*. *Phytochemistry* 23, 471–472.

Bhattacharyya, P., Jash, S.S. and Chowdhury, B.K. (1986) A biogenetically important carbazole alkaloid from *Murraya koenigii* Spreng. *Chemistry and Industry* 7, 246.

Bhattacharyya, P., Maiti, A.K., Basu, K. and Chowdhury, B.K. (1994) Carbazole alkaloids from *Murraya koenigii*. *Phytochemistry* 35, 1085–1086.

Chakraborty, D.P. (1993) Chemistry and biology of carbazole alkaloids. In: Cordell, G.A. (ed.) *Alkaloids*, Volume 44. Academic Press, New York, pp. 257–282.

Chakraborty, D.P., Barman, B.K. and Bose, P.K. (1964) Structure of girinimbine, a pyranocarbazole derivative isolated from *Murraya koenigii* Spreng. *Scientific Culture* 30, 445.

Chakraborty, D.P., Barman, B.K. and Bose, P.K. (1965) On the constitution of murrayanine, a carbazole derivative isolated from *Murraya koenigii* Spreng. *Tetrahedron* 21, 681–685.

Chakraborty, D.P., Bhattacharyya, P., Islam, A. and Roy, S. (1974) Structure of murrayacinine: a new carbazole alkaloid from *Murraya koenigii* Spreng. *Chemistry and Industry* 4, 165–166.

Chakraborty, D.P., Bhattacharyya, P., Roy, S., Bhattacharyya, S.P. and Biswas, A.K. (1978) Structure and synthesis of mukonine, a new carbazole alkaloid from *Murraya koenigii* [stem bark]. *Phytochemistry* 17, 834–835.

Chakraborty, M., Nath, A.C., Khasnobis, S., Chakrabarty, M., Konda, Y., Harigaya, Y. and Komiyama, K. (1997) Carbazole alkaloids from *Murraya koenigii*. *Phytochemistry* 46(4), 751–755.

Choudhury, R.P. and Garg, A.N. (2007) Variation in essential, trace and toxic elemental contents in *Murraya koenigii* – a spice and medicinal herb from different Indian states. *Food Chemistry* 104(4), 1454–1463.

Chowdhury, A.R. (2000) Essential oil from the leaves of *Murraya koenigii* (Linn.) Spreng. *Journal of Medicinal and Aromatic Plant Sciences* 22(1B), 643–645.

Chowdhury, B.K., Jha, S., Bhattacharyya, P. and Mukherjee, J. (2001) Two new carbazole alkaloids from *Murraya koenigii*. *Indian Journal of Chemistry Section-B* 40(6), 490–494.

Dwivedi, B.K., Pandey, R.C., Pandey, G., Pant, H.L. and Logani, R. (2002) Evaluation of angiospermic plant extracts against *Rhizopus stolonifer* and *Gloeosporium psidii* fungi of guava. *Bioved* 13(1/2), 129–134.

Facknath, S. (2006) Effects of phytoextracts and natural enemy to control *Plutella xylostella* L. (*Lepidoptera: Plutellidae*) in cabbage. *Allelopathy Journal* 17(2), 207–222

Fiebig, M., Pezzuto, J.M., Soejarto, D.D. and Kinghorn, A.D. (1985) Koenoline, a further cytotoxic carbazole alkaloid from *Murraya koenigii*. *Phytochemistry* 24, 3041–3043.

Grover, J.K., Yadav, S. and Vats, V. (2002) Medicinal plants of India with anti-diabetic potential. *Journal of Ethnopharmacology* 81(1), 81–100.

Hemavathy, T. (1991) Lipid composition of *Murraya koenigii* seed. *Journal of American Oil Chemical Society* 68, 651–652.

Hiremath, S.M. and Madalageri, B.B. (1997) Volatile aroma constituents of curry leaf cv. Suwasini. *Karnataka Journal of Agricultural Science* 10, 602–604.

Hiremath, S.M., Madalageri, B.B. and Basarkar, P.W. (1997) Variation for volatile oil content in curry leaf cultivars. *Karnataka Journal of Agricultural Science* 10, 913–915.

Irani, F. (2005) An encounter with ayurveda aromatherapy. *Aromatherapy Journal* (3/4), 29–35.

Ito, C., Thoyama, Y., Omura, M., Kajiura, I. and Furukawa, H. (1993) Alkaloidal constituents of *Murraya koenigii*. Isolation and structural elucidation of novel binary carbazolequinones and carbazole alkaloids. *Chemical and Pharmaceutical Bulletin* 41, 2096–2100.

Ito, C., Itoigawa, M., Nakao, K., Murata, T., Tsuboi, M., Kaneda, N. and Furukawa, H. (2006) Induction of apoptosis by carbazole alkaloids isolated from *Murraya koenigii*. *Phytomedicine* 13, 359–365.

Iyer, U.M. and Mani, U.V. (1990) Studies on the effect of curry leaves supplementation (*Murraya koenigii*) on lipid profile, glycated proteins and amino acids in non-insulin-dependent diabetic patients. *Plant Foods for Human Nutrition* (formerly *Qualitas Plantarum*) 40(4), 275–282.

Joshi, B.S., Kamat, V.N. and Gawad, D.H. (1970) On the structures of girinimbine, mahanimbine, isomahanimbine, koenimbidine and murrayacine. *Tetrahedron* 26, 1475–1482.

Kong, Y.C., Ng, G.K.H., Wat, C.K.H. and But, P.P.H. (1986) Pharmacognostic differentiation between *Murraya paniculata* (L.) Jack and *Murraya koenigii* (L.) Spreng. *International Journal of Crude Drug Research* 24, 167–170.

Kumar, V.S., Sharma, A., Tiwari, R. and Sushil, K. (1999) *Murraya koenigii* (curry leaf): a review. *Journal of Medicinal and Aromatic Plant Sciences* 21, 1139–1144.

Lal, R.K., Sharma, J.R., Singh, N., Misra, H.O. and Naqvi, A. (2000) Genetic associations and diversity in the genetic resources of curry leaf *Murraya koenigii*. *Journal of Medicinal and Aromatic Plant Sciences* 22–23(4A-1A), 216–220.

Lalitha, S., Thamburaj, S., Thangaraj, T. and Vijayakumar, M. (1997) Evaluation of curry leaf (*Murraya koenigii* Spreng) ecotypes. *South Indian Horticulture* 45, 78–80.

MacLeod, A.J. and Pieris, N.M. (1982) Analysis of the volatile essential oils of *Murraya koenigii* and *Pandanus latifolius*. *Phytochemistry* 21, 1653–1657.

Madalageri, B.B., Mahadev, and Hiremath, S.M. (1996) Dehydration methods, oil extraction and flavour components detection in curry leaf (*Murraya koenigii* Spreng) and detection of flavour components. *Karnataka Journal of Agricultural Sciences* 9, 284–288.

Maiti, A.K., Dash, G.K. and Patro, C.P. (2004) Anti-inflammatory and analgesic activity of leaf essential oil from *Murraya koenigii* Spreng. *Hamdard Medicus* 47(2), 22–26.

Mallavarapu, G.R., Ramesh, S., Syamasundar, K.V. and Chandrasekhara, R.S. (1999) Composition of Indian curry leaf oil. *Journal of Essential Oil Research* 11, 176–178.

Mallavarapu, G.R., Laxmi, R., Srinivasaiyer, R., Rao, L. and Ramesh, S. (2000) Volatile constituents of the leaf and fruit oils of *Murraya koenigii* Spreng. *Journal of Essential Oil Research* 12(6), 766–768.

Mehrotra, S., Umashanker, Jawaharlal, Majumdar, A.C. and Agarwal, S.K. (2003) Effect of indigenous medicinal plants on onset of puberty in immature female rats. *Indian Journal of Animal Reproduction* 24(2) 131–133.

Mehrotra, S., Singh, S.K., Umashankar and Agarwal, S.K. (2005) Effect of certain indigenous medicinal plants on ovarian hormone profile and reproductive performance in anestrus cattle. *Indian Journal of Animal Reproduction* 26, 20–23.

Mehta, P.K., Sood, A.K., Patial, A. and Lal, R. (2005) Evaluation of toxic and antifeedant properties of some plant extracts against major insect pests of cabbage. *Pesticide Research Journal* 17(2), 30–33.

Mukherjee, M., Mukherjee, S., Shaw, A.K. and Ganguly, S.N. (1983) Mukonicine, a carbazole alkaloid from leaves of *Murraya koenigii*. *Phytochemistry* 22, 2328–2329.

Nakatani, N. (2000) Phenolic antioxidants from herbs and spices. Proceedings 2nd International Conference on Food Factors (ICoFF '99), 12–17 December 1999, Kyoto, Japan. *BioFactors* 2000, 13(1–4), 141–146.

Nutan, M.T.H., Hasnat, A. and Rashid, M.A. (1998) Antibacterial and cytotoxic activities of *Murraya koenigii*. *Fitoterapia* 69, 173–175.

Nutan, M.T.H., Hasan, C.M. and Rashid, M.A. (1999) Bismurrayafoline E: a new dimeric carbazole alkaloid from *Murraya koenigii*. *Fitoterapia* 70(2), 130–133.

Onayade, O.A. and Adebajo, A.C. (2000) Composition of the leaf volatile oil of *Murraya koenigii* growing in Nigeria. *Journal of Herbs, Spices and Medicinal Plants* 7(4), 59–66.

Padhi, N.N. and Behera, G. (2000) Evaluation of nematicidal potential in ten indigenous plant species against *Meloidogyne incognita. Indian Phytopathology* 53(1), 28–31.

Palaniswamy, U.R., Stuart, J.D. and Caporuscio, C.A. (2002) Effects of storage temperature on the nutritive value of curry leaf. In: Janick, J. and Whipkey, A. (eds) *Trends in New Crops and New Uses*. Proceedings V National Symposium, Atlanta, Georgia, 10–13 November 2001. ASHA Press, Alexandria, Virginia, pp. 567–569.

Palaniswamy, U.R., Caporuscio, C. and Stuart, J.D. (2003) A chemical analysis of antioxidant vitamins in fresh curry leaf (*Murraya koenigii*) by reversed phase HPLC with UV detection. *Acta Horticulturae* 620, 475–478.

Pandey, R. (2000) Additive effect of three organic materials and nematicides on the reproduction of *Meloidogyne incognita* and yield of *Mentha arvensis. Nematropica* 30(2), 155–160.

Pandey, R. (2005) Management of *Meloidogyne incognita* in *Artemisia pallens* with bio-organics. *Phytoparasitica* 33(3), 304–308.

Paranagama, P.A., Wimalasena, S., Jayatilake, G.S., Jayawardena, A.L., Senanayake, U.M. and Mubarak, A.M. (2002) Analysis of volatile oil in the leaves of *Murraya koenigii. Journal of Tropical Medicinal Plants* 3, 187–189.

Parmar, C. and Kaushal, M.K. (1982) *Murraya koenigii.* In: Parmar, C. and Kaushal, M.K. (eds) *Wild Fruits.* Kalyani Publishers, New Delhi, pp. 45–48.

Pathak, N., Yadav, T.D., Jha, A.N. and Vasudevan, P. (1997) Contact and fumigant action of volatile essential oil of *Murraya koenigii* against *Callosobruchus chinensis. Indian Journal of Nematology* 59, 198–202.

Pathak, N., Mittal, P.K., Singh, O.P., Vidya Sagar, D. and Vasudevan, P. (2000) Larvicidal action of essential oils from plants against the vector mosquitoes *Anopheles stephensi* (Liston), *Culex quinquefasciatus* (Say) and *Aedes aegypti* (L.). *International Pest Control* 42(2), 53–55.

Philip, J. (1981) Curry leaf and its uses. *World Crops* 33, 125–127.

Rahman, A., Zaidi, R. and Firdous, S. (1988) NMR studies on mahanine. *Fitoterapia* 59, 494–495.

Rahman, M.M. and Gray, A.I. (2005) A benzoisofuranone derivative and carbazole alkaloids from *Murraya koenigii* and their antimicrobial activity. *Phytochemistry* 66, 1601–1606.

Raina, V.K., Lal, R.K., Savita, T., Khan, M., Syamasundar, K.V. and Srivastava, S.K. (2002) Essential oil composition of genetically diverse stocks of *Murraya koenigii* from India. *Flavour and Fragrance Journal* 17, 144–146.

Ramalakshmi, K., Sulochanamma, G. and Raghvan, B. (2000) Physico-chemical changes on processing of curry leaf (*Murraya koenigii* Spreng.). *Journal of Medicinal and Aromatic Plant Science* 22(1B), 510–516.

Ramsewak, R.S., Nair, M.G., Strasburg, G.M., DeWitt, D.L. and Nitiss, J.L. (1999) Biologically active carbazole alkaloids from *Murraya koenigii. Journal of Agricultural and Food Chemistry* 47, 444–447.

Rana, V.S., Juyal, J.P., Rashmi, M.S. and Blazquez, M.A. (2004) Chemical constituents of the volatile oil of *Murraya koenigii* leaves. *International Journal of Aromatherapy* 14, 23–25.

Rao, A.V.R., Bhide, K.S., Mujumdar, R.B. and Rama Rao, A.V. (1980) Mahanimbinol. *Chemistry and Industry* 17, 697–698.

Ratnasooriya, W.D., Premakumara, G.A.S. and Ananda, U.V.D.S. (1995) Protection by *Murraya koenigii* leaf extract against ethanol-induced gastric lesions in rats. *Medical Science Research* 23(1), 11–13.

Ray, D.P. and Srivastava, S. (2006) Curry leaf (*Murraya koenigii*): the aromatic biopesticide. *Journal of Interacademicia* 10, 231–235.

Reisch, J., Bergenthal, D., Adebajo, A.C. and Aladesanmi, A.J. (1994a) Furocoumarins of *Murraya koenigii* seeds [collected from Sri Lanka]. *Fitoterapia* 65(4), 380–381.

Reisch, J., Adebajo, A.C., Kumar, V. and Aladesanmi, A.J. (1994b) Two carbazole alkaloids from *Murraya koenigii. Phytochemistry* 36, 1073–1076.

Reisch, J., Adebajo, A.C., Aladesanmi, A.J., Adesina, K.S., Bergenthal, D. and Meve, U. (1994c) Chemotypes of *Murraya koenigii* growing in Sri Lanka. *Planta Medica* 60(3), 295–296.

Roy, S. and Chakraborty, D.P. (1974) Mahanimbine from *Murraya koenigii* Spreng. *Phytochemistry* 13, 2893.

Shanthala, M. and Prakash, J. (2005) Acceptability of curry leaf (*Murraya koenigii*) incorporated products and attitude toward consumption. *Journal of Food Processing and Preservation* 29, 33–44.

Srivastava, A. and Kumar, A. (1999) Bioefficacy of some plant extracts against *Lipaphis erysimi* (Kalt.) under laboratory conditions. *Journal of Insect Science* 12(1), 90–91.

Sukari, M.A., Ahmad, K., Aimi, N., Kitajima, M., Rahmani, M., Lian, G.E.C. and Ahmad, F.H. (2001) Carbazole alkaloids from stem bark of *Murraya koenigii. ACGA Chemical Research Communications* 13, 2–8.

Tachibana, Y., Kikuzaki, H., Lajis, N.H. and Nakatani, N. (2001) Antioxidative activity of carbazoles from *Murraya koenigii* leaves. *Journal of Agricultural and Food Chemistry* 49(11), 5589–5594.

Vinuthan, M.K., Girish Kumara, V., Narayanaswamya, M. and Veena, T. (2007) Lipid lowering effect of aqueous leaves extract of *Murraya koenigii* (curry leaf) on alloxan-induced male diabetic rats. *Pharmacognosy Magazine* 3(10), 112–115.

Wagner, I.W., Kalinowski, H.O. and Jork, H. (1995) Isolation and identification of 11-selinen-4alpha, 7beta-ol and 10-aromadendranol in the essential oil of *Murraya koenigii*. *Planta Medica* 61, 196–197.

Walde, S.G., Jyothirmayi, T., Rao, P.G.P., Shivaswamy, R. and Srinivas, P. (2005) Flavour volatiles of leaves, fruits and seed cotyledons of *Murraya koenigii* L. *Flavour and Fragrance Journal* 20, 169–172.

Walde, S.G., Jyothirmayi, T., Rao, P.G.P. and Srinivas, P. (2006) Flavour volatiles of flowers and stalks of *Murraya koenigii* L. *Flavour and Fragrance Journal* 21, 581–584.

Wang, Y.S., He, H.P., Shen, Y.M., Hong, X. and Hao, X.J. (2003) Two new carbazole alkaloids from *Murraya koenigii*. *Journal of Natural Products* 66, 416–418.

Wong, K.C. and Chee, S.G. (1996) Volatile constituents of *Murraya koenigii* Spreng flowers. *Journal of Essential Oil Research* 8, 545–547.

Wong, K.C. and Tie, D.Y. (1993) The essential oil of the leaves of *Murraya koenigii* Spreng. *Journal of Essential Oil Research* 5, 371–374.

Yadav, S., Vats, V., Dhunnoo, Y. and. Grover, J.K. (2002) Hypoglycemic and antihyperglycemic activity of *Murraya koenigii* leaves in diabetic rats. *Journal of Ethnopharmacology* 82(2–3), 111–116.

Yadav, S.P., Vats, V., Ammini, A.C. and. Grover, J.K. (2004) *Brassica juncea* (Rai) significantly prevented the development of insulin resistance in rats fed fructose-enriched diet. *Journal of Ethnopharmacology* 93(1), 113–116.

24 Bay Leaf

V.A. Parthasarathy, T. John Zachariah and B. Chempakam

24.1. Introduction

Bay (*Laurus nobilis* L.) leaf, or Turkish laurel, is an industrial plant used in foods, drugs and cosmetics. The dried leaves and essential oils are used extensively in the food industry for seasoning of meat products, soups and fish (Kilic *et al.*, 2004). Bay leaves are also known as laurel leaves and are native to the Mediterranean countries. They are large, light to olive green elliptical leaves about 8 cm long and 3–4 cm wide. In 2000, Turkey exported 3600 t of *L. nobilis* leaves worth US$7.5 million (Tainter and Grenis, 1993; Kilic *et al.*, 2004). The plant is widely cultivated in Europe, America and in the Arabian countries, from Libya to Morocco (Kumar *et al.*, 2001). Extracts from these plants have great potential in protecting stored spices from *Aspergillus flavus* (Geeta and Reddy, 1990).

24.2. Botany

Laurus nobilis is an evergreen shrub or, more rarely, a tree attaining a height of 15–20 m. The smooth bark may be olive green or reddish-blue. The luxurious, evergreen leaves are alternate with short stalks, lanceolate or lanceolate oblong, acuminate, 5–8 cm or longer and 3–4 cm wide, coriaceous pellucid-punctate and with revolute, entire wavy margins (Kumar *et al.*, 2001). The leaves of *L. nobilis* are plucked and dried under shade for use as a flavouring material in a variety of culinary preparations, especially in French cuisine. The plants are important for aromatic oils and spices, edible fruits and timber (e.g. from species of the largest genus, *Ocotea*). The true laurel – that of history and classical literature – is *L. nobilis,* called also bay and sweet bay. It is native to the Mediterranean, where to the ancients it symbolized victory and merit and was sacred to Apollo. The fragrant leaves are sold commercially as bay leaf, a seasoning. American laurel is the evergreen California laurel (*Umbellularia californica*), also called pepperwood, bay tree and Oregon myrtle (Anon., 2005). Its aromatic bark is used occasionally for medicinal tea and its pulverized leaves for soup and condiments. Many of the evergreen laurels are grown as hedges and, because of their handsome foliage, are used by florists. Table 24.1 gives the nutritional composition of bay leaves per 100 g.

24.3. Chemistry

Extraction of volatiles

There are various methods of extraction of essential oils, such as steam distillation,

Table 24.1. Nutritional composition of bay leaves per 100 g.

Composition	USDA Handbook (crumbled)	ASTA
Water (g)	5.44	4.50
Food energy (Kcal)	313	410
Protein (g)	7.61	7.50
Fat (g)	8.36	8.80
Carbohydrates (g)	74.96	75.40
Ash (g)	3.62	3.70
Calcium (g)	0.83	1.00
Phosphorus (mg)	113	110
Sodium (mg)	23	20
Potassium (mg)	529	600
Iron (mg)	43.0	53.3
Thiamine (mg)	0.009	0.100
Riboflavin (mg)	0.421	0.420
Niacin (mg)	2.005	2.000
Ascorbic acid (mg)	46.53	47.00
Vitamin A activity (RE)	618	618

Source: Tainter and Grenis (1993).

supercritical extraction and microwave hydrodistillation; 83–96% recovery can be obtained by steam distillation (Borges *et al.*, 2003). A microwave-assisted hydrodistillation protocol was modified to extract essential oils from laurel leaves. The essential oils of this plant generally are obtained by hydrodistillation or steam distillation. The volatile compounds obtained by microwave-assisted hydrodistillation and hydrodistillation methods were analysed by GC and GC-MS. Both distillation methods and analytical results were compared. 1,8-Cineole (46.8–54.2%) was the main component in the leaf oil. Although the distillation was accomplished in a shorter time, the oil yields and 1,8-cineole contents were slightly higher in the microwave-assisted hydrodistillation compared with the usual hydrodistillation. Thus, microwave-assisted hydrodistillation appears to be an effective method for the production of essential oils (Kosar *et al.*, 2005).

Caredda *et al.* (2002) described the extraction conditions for leaf oil by supercritical carbon dioxide extraction as follows: pressure, 90 bar; temperature, 50°C; and carbon dioxide flow, Phi = 1.0 kg/h. Waxes were entrapped in the first separator set at 90 bar and 10°C. The oil was recovered in the second separator working at 15 bar and 10°C. The main components were 1,8-cineole (22.8%), linalool (12.5%), α-terpinyl acetate (11.4%) and methyl eugenol (8.1%). Comparison with the hydrodistilled oil did not reveal any significant difference. Collection of samples at different extraction times during supercritical extraction allowed the change of the oil composition to be monitored. Lighter compounds, such as hydrocarbon and oxygenated monoterpenes, were extracted in shorter times than the heavier hydrocarbon and oxygenated sesquiterpenes. Beis and Dunford (2006) described supercritical fluid extraction of seed oil. The oil yield of ground seeds varied from 14 to 28%, depending on the method and particle size used for oil recovery. Yields were similar for both petroleum ether and SC-CO_2 extraction. The extraction yield decreased significantly with increasing particle size. The amount of extract collected increased exponentially with increasing SC-CO_2 pressure. The highest extraction yield was obtained at the highest temperature studied, 75°C. More than 45% of the oil was lauric acid. SC-CO_2 is a viable technique to obtain high-purity *L. nobilis* L. seed oil, which is a potential ingredient for the cosmetic industry (Vasudevan *et al.*, 1997).

Constituents of essential oil

The volatiles of fresh leaves, buds, flowers and fruits were isolated by solvent extraction and analysed by capillary gas chromatography-mass spectrometry. Their odour quality was characterized by gas chromatography-olfactometry–mass spectrometry (HRGC-O-MS) and aroma extract dilution analysis (AEDA). In fresh bay leaves, 1,8-cineole was the major component, together with α-terpinyl acetate, sabinene, α-pinene, β-pinene, β-elemene, α-terpineol, linalool and eugenol. Besides 1,8-cineole and the pinenes, the main components in the flowers were α-eudesmol, β-elemene and β-caryophyllene, in the fruits (*E*)-β-ocimene and bicyclogermacrene, and in the buds (*E*)-β-ocimene and germacrene D. The aliphatic ocimenes and farnesenes were absent in leaves. By using HRGC-O-MS, 21 odour compounds were identified in fresh leaves. Application of AEDA revealed (*Z*)-3-hexenal (fresh green), 1,8-cineole (eucalyptus), linalool (flowery), eugenol (clove), (*E*)-isoeugenol (flowery) and an unidentified compound (black pepper) with the highest flavour dilution factors (Kilic *et al.*, 2004).

Analysis of plant material from France indicated that the composition of the essential oil from the flowers showed differences compared with the essential oil from the leaves, having a high content of β-caryophyllene (10%), viridiflorene (12.2%), germacradienol (10.1%), β-elemene (9.7%) and (*E*)-ocimene (8%) (Fiorini *et al.*, 1997). Nhat *et al.* (1999) found a total of 49 compounds in the essential oil, the main constituents being 1,8-cineole [eucalyptol] (45.7%), α-terpinyl acetate (14.8%), sabinene (12.7%), α-pinene (4.8%), terpinene-4-ol (3.7%), α-terpineol (2.8%) and caryophyllene oxide (1.3%). The concentration of eugenol was lower than previously reported in the literature, which can affect the antioxidative and antimicrobial effect of the essential oil. Table 24.2 describes the percentage composition of the volatiles of the buds, flowers and fruits of *L. nobilis*. Fig. 24.1 illustrates the structures of the major volatiles of the bay leaf.

Table 24.2. Volatile compounds of buds, flowers and fruits of *L. nobilis* (%).

Compound	Bud	Flower	Fruit
Tricyclene	0.1	t	t
Thujone	0.1	0.2	0.1
α-Pinene	7.0	5.1	3.3
Camphene	3.4	2.4	1.7
Sabinene	2.4	1.7	1.7
β-Pinene	4.6	3.7	2.1
Myrcene	0.7	0.6	0.5
α-Phellandrene	t	0.1	t
Δ^3-Carene	–	0.4	–
p-Cymene	–	–	0.1
Limonene	t	t	t
1,8-Cineole	16.8	8.8	9.5
(*Z*)-β-Ocimene	0.1	0.3	–
Phenylacetaldehyde	0.1	–	–
(*E*)-β-Ocimene	8.1	2.7	22.1
(*E*)-Sabinene hydrate	0.1	t	t
Linalool	0.8	–	–
Pinocarvone	–	–	0.1
Borneol	0.7	0.4	0.3
α-Terpineol	–	t	0.4
Linalyl acetate	0.7	–	0.2
Bornyl acetate	2.0	2.1	1.1
2-Undecanone	–	–	0.1
δ-Terpinyl acetate	0.2	0.3	0.1
α-Terpinyl acetate	1.6	1.8	1.2
Eugenol	0.3	–	t
α-Ylangene	0.5	0.9	0.2
α-Copaene	0.2	0.3	0.1
iso-β-Elemene	0.1	0.4	0.1
β-Cubebene	–	–	t
β-Elemene	2.6	5.4	2.0
Eugenol methyl ether	0.3	–	0.1
(*E*)-β-Caryophyllene	0.9	5.1	0.3
(*E*)-Isoeugenol	0.3	0.5	0.2
α-Humulene	0.2	0.5	0.1
Alloaromadendrene	0.1	–	0.1
(*E*)-β-Farnesene	0.2	0.1	0.1
γ-Muurolene	–	–	t
GermacreneD	6.6	2.4	1.5
β-Selinene	0.1	0.3	0.1
Bicyclogermacrene	1.2	2.2	4.5
α-Farnesene	0.8	1.3	0.3
Germacrene A	0.8	1.1	0.6
γ-Cadinene	–	–	0.3
δ-Cadinene	t		0.1
Ni(sesquiterpene)	5.5	3.4	0.9
Elemol	0.4	–	–
Germacrene D-4-ol	0.7	–	–
α-Eudesmol	2.7	11.8	–
Costunolide	–	–	2.9

t = trace.
Source: Kilic *et al.* (2004).

Seasonal variation

Seasonal variations have also been observed in chemical composition. Anac (1986) studied the variation between fresh (March and June harvests) and dried (June harvest) leaves collected from mature trees in Istanbul. The respective essential oil yields were 0.86, 0.99 and 1.29%. Drying and harvest date also affected the qualitative composition; the content of the main component, 1,8-cineole, was 28.08, 40.62 and 42.70%, respectively, and the contents of the highly volatile components were somewhat higher in June than in March. A comparison with reported analyses indicated no major differ-

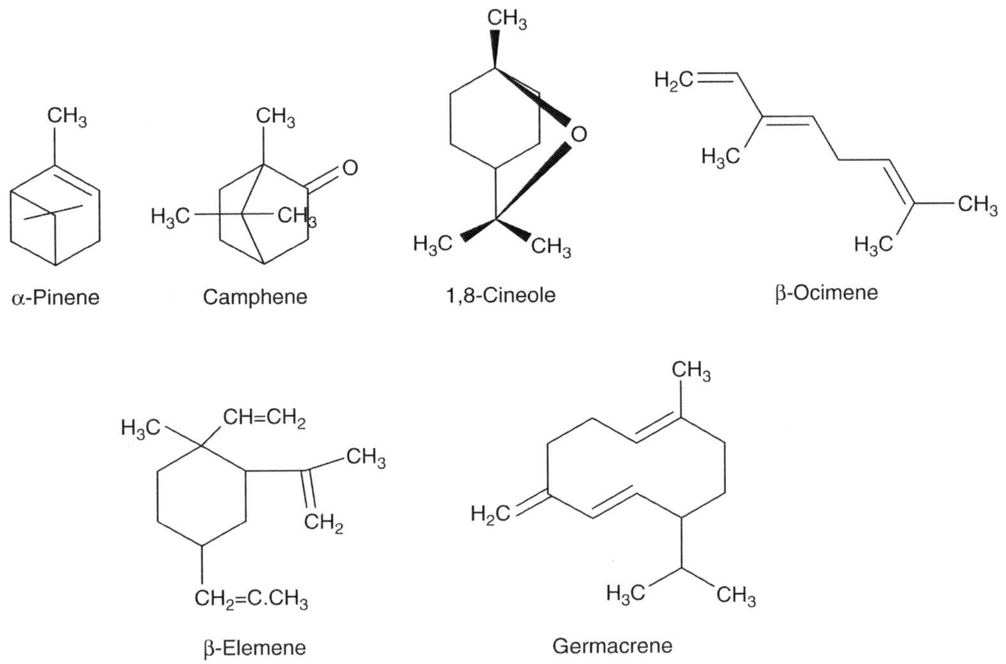

Fig. 24.1. Major volatiles in bay leaf.

ences between the Turkish oils and those normally found in commerce. Riaz *et al.* (1989) investigated the physical properties and chemical composition of essential oil extracted from *L. nobilis* leaves collected in March, July, September and November. Oil yield was lowest (0.13%) in March and highest in September (0.36%). GLC analysis showed 70 peaks corresponding to different terpenes, but only 27 of these were identified; data are reported for March, July and September samples. Cineole and eugenol were the major components in all three samples and there was very little difference in the amounts of all the components between the samples, except that linalyl acetate (a very minor component) was absent in the March samples. Kevseroglu *et al.* (2003) studied the ontogenetic and diurnal variability of laurel leaves. Leaves were collected from trees growing in Samsun (Turkey) every month from May to October, during the first 3 days of the month, in the morning (08.00 h), at noon (12.00 h) and in the evening (18.00 h). The leaves were dried in the shade for 3–4 weeks and then ground prior to essential oil analysis. Essential oil content in the leaves did not vary significantly during the day, but seasonal variation was significant. The highest essential oil percentage was found for leaves harvested in August (1.46%) and July (1.33%) and it was lowest for leaves collected during May and September (0.59 and 0.74%, respectively). There was a positive correlation between essential oil percentages in the leaves and atmospheric temperature.

The oil obtained from the dried leaves of *L. nobilis* (origin Tunisia; oil yield 1.5%) was analysed by capillary GC and GC-MS. The main components of the 24 constituents identified were 1,8-cineole [eucalyptol] (42.3%) and α-terpinyl acetate (11.2%) (Bouzouita *et al.*, 2001). Braun *et al.* (2001) identified 155 constituents: 76 monoterpenes, 46 sequiterpenoids, 10 phenylpropanoids and 23 others. δ-Terpinyl acetate was reported for the first time and was characterized using ^{1}H-, ^{13}C-NMR, GC-FTIR and GC-MS analysis.

24.4. Medicinal and Pharmacological Uses

Antimicrobial activity

Escherichia coli and many pathogens were inhibited by the essential oil of laurel (Baratta *et al.*, 1998; Friedman *et al.*, 2002; Bouzouita *et al.*, 2003; Mandeel *et al.*, 2003; Dadalioglu and Evrendilek, 2004). Trypanocidal constituents of the methanol extract of the dried leaves of *L. nobilis* L. resulted in the isolation of two guaianolides, dehydrocostus lactone (1) and zaluzanin D (2), and a new *p*-menthane hydroperoxide, (1*R*,4*S*)-1-hydroperoxy-*p*-menth-2-en-8-ol acetate (3). The minimum lethal concentrations of these compounds against epimastigotes of *Trypanosoma cruzi* were 6.3, 2.5 and 1.4 μM, respectively (Uchiyama *et al.*, 2002; Nakatani, 2003).

Inhibitors of nitric oxide production

Marino *et al.* (2005) isolated two new metabolites, 5α*H*,7α*H*-eudesman-4α, 6α,11, 12-tetraol (1) and 1*β*,15-dihydroxy-5α*H*, 7α*H*-eudesma-3,11(13)-dien-12,6 α-olide (2), from the methanolic extract of *L. nobilis* leaves (collected from spontaneous plants grown in Avellino, Campania, Italy). Their structures were determined through analysis of their one- and two-dimensional NMR spectral data (^{1}H- and ^{13}C-NMR, DEPT, COSY, HMQC, HMBC and ROESY). The relative stereochemistry was proposed based on the combined J-based configuration analysis and ROESY data. In addition, three known sesquiterpene lactones, santamarine (3), reynosin (4) and costunolide (5), along with blumenol C (6), were isolated and identified by spectral means. The isolated compounds 1–6 inhibited nitric oxide (NO) production in lipopolysaccharide (LPS)-activated murine macrophages. The most active, compound 2, potently inhibited NO_2 release with an IC50 value of 0.8 μM. The methanolic extract of leaves of *L. nobilis* (bay leaf) inhibited nitric oxide (NO) production in lipopolysaccharide (LPS)-activated mouse peritoneal macrophages (Matsuda *et al.*, 2000). Through bioassay-guided separation, 14 known sesquiterpenes were isolated from the active fraction and were examined for ability to inhibit NO production. Seven sesquiterpene lactones (costunolide, dehydrocostus lactone, eremanthine, zaluzanin C, magnolialide, santamarine and spirafolide) potently inhibited LPS-induced NO production (IC50 values of 1.2–3.8 μM). Other sesquiterpene constituents also showed inhibitory activity (IC50 < or = > 21 μM). α-Methylene-γ-butyrolactone also showed inhibitory activity (IC50 = 9.6 μM), while mokko lactone and watsonol A, reductants of the α-methylene-γ-butyrolactone moiety by $NaBH_4$ or DIBAL, and a 2-mercaptoethanol adduct of dehydrocostus lactone, showed little activity (IC50 < or = > 18 μM). These results indicated that the α-methylene–γ-butyrolactone moiety was important for activity. Furthermore, costunolide and dehydrocostus lactone inhibited inducible nitric oxide synthase (iNOS) induction in accordance.

Almost complete inhibition of 3-nitrotyrosine formation (91%) was achieved with the essential oil obtained from the leaves of *L. nobilis* (at 300 μg/ml). 1,8-Cineole (eucalyptol), accounting for 50% of this essential oil, was inactive in this model, thus evidencing a major role for the minor volatile compounds present in the leaves (Matsuda *et al.*, 2000; Chericoni *et al.*, 2005).

Antifungal activity

The potential of bay leaf essential oils against species belonging to *Eurotium*, *Aspergillus* and *Penicillium* genus has been demonstrated (Geeta and Reddy, 1990; Guynot *et al.*, 2003). Biological assays showed that fungitoxicity against *Fusarium moniliforme* (*Gibberella fujikuroi*), *Rhizoctonia solani*, *Sclerotinia sclerotiorum* and *Phytophthora capsici* was due to different concentrations of the phenolic fraction in the essential oils (Muller Riebau *et al.*, 1995; Pandey, 1997; Pandey and Dubey, 1997).

Anticonvulsant

The leaf essential oil of *L. nobilis*, which has been used as an antiepileptic remedy in Iranian traditional medicine, was evaluated for anticonvulsant activity against experimental seizures (Sayyah *et al.*, 2002). The essential oil protected mice against tonic seizures induced by maximal electroshock and especially by pentylenetetrazole. Components responsible for this effect may be methyleugenol, eugenol and pinene present in the essential oil. At anticonvulsant doses, the essential oil produced sedation and motor impairment. This effect seems to be related in part to cineol, eugenol and methyleugenol (Sayyah *et al.*, 2002).

Insecticidal

Essential oils from laurel were evaluated for fumigant toxicity against all developmental stages of the confused flour beetle (*Tribolium confusum*). GC-MS analysis showed that 1,8-cineole was the major component of the essential oils. The vapours of laurel essential oil were toxic to all the stages of *T. confusum* (Isikber *et al.*, 2006). Repellency and toxicity of essential oil from *L. nobilis* (Lauraceae) against the rust-red flour beetle (*T. castaneum* Herbst) were also reported by Andronikashvili and Reichmuth (2003). The toxicity of ethanol extracts from *L. nobilis* on the large diamondback moth, *Plutella xylostella*, was 55% (Erturk *et al.*, 2004).

The behavioural responses of adult female western flower thrips, *Frankliniella occidentalis*, to volatiles from meadowsweet (*Filipendula ulmaria*), bay laurel and sage (*Salvia officinalis*) were investigated in laboratory bioassays by Chermenskaya *et al.* (2001). Volatiles collected by entrainment of a solvent extract of *F. ulmaria* were more attractive than was the original extract. *F. occidentalis* also was attracted significantly to volatiles from *L. nobilis* and *S. officinalis*. Analysis by gas chromatography and mass spectrometry identified 1,8-cineole (eucalyptol) as one of the main volatile components of all three plant species. In coupled gas chromatography–electroantennography studies with *F. ulmaria*, both 1,8-cineole and methyl salicylate elicited responses from *F. occidentalis*. Eucarvone was identified as the major component of *F. ulmaria* volatiles, but showed no electrophysiological activity. The behavioural responses of thrips to a range of concentrations of 1,8-cineole and methyl salicylate were tested using a modified Pettersson 'star' olfactometer. 1,8-Cineole showed some attractant activity for the thrips at 0.01 mg, but methyl salicylate was repellent at all the concentrations tested.

The bruchid, *Acanthoscelides obtectus*, is one of the most damaging pests of kidney beans (*Phaseolus vulgaris*) worldwide. However, aromatic plants from the families Lamiaceae, Lauraceae, Myrtaceae and Poaceae can protect *P. vulgaris* by a direct or delayed insecticidal effect, through increased adult mortality and inhibition of reproduction (both oviposition and adult emergence). The insecticidal effect is due to the presence of factors other than those in the essential oils as there is no significant difference between the efficacy of distilled and intact plant extracts. Inhibition of reproduction is particularly important. The results suggest that lipid, as well as non-lipid allelochemicals, such as phenolics, or non-protein amino acids or flavonoids may be involved in the toxicity of extracts of aromatic plants to *A. obtectus* (Regnault Roger and Hamraoui, 1995; Mackeen *et al.*, 1997).

24.5. International Standards

Tables 24.3 and 24.4 describe the physical and chemical specifications for whole bay leaves and ground leaves. The minimum volatile oil content required for whole leaves is 1.5% and 1.0% for ground leaves.

24.6. Conclusion

The bay leaf belongs to the family Lauraceae and is one of the most popular culinary spices in the West. The bay leaf has been used as

Table 24.3. Whole bay (laurel leaves): chemical and physical specifications.

Specification	Suggested limits
ASTA cleanliness specifications	
Whole dead insects, by count	2
Mammalian excreta (mg/lb)	1
Other excreta (mg/lb)	10.0
Mould, % by weight	2.00
Insect defiled/infested, % by weight	2.50
Extraneous, % by weight	0.50
Detectable action levels	
Mouldy pieces by weight (av. %)	5
Insect-infested pieces by weight (av. %)	5
Mammalian excreta, after processing (mg/lb, av.)	1
Volatile oil (% min.)	1.5
Moisture (% max.)	9.0
Ash (% max.)	4.0
Acid-insoluble ash (% max.)	0.8

Source: Tainter and Grenis (1993).

Table 24.4. Ground bay: chemical and physical specifications.

Specification	Suggested limits
Detectable action levels	None
Volatile oil (% min.)	1.0
Moisture (% max.)	9.0
Total ash (% max.)	4.0
Acid-insoluble ash (% max.)	0.8
Military specifications	
Volatile oil (ml/100g) (% min.)	1.0
Moisture (% max.)	7.0
Total ash (% max.)	4.5
Acid-insoluble ash (% max.)	0.5
Granulation (% min. through a USS No. 30)	95
Bulk index (mg/100g)	220

Source: Tainter and Grenis (1993).

a herbal medicine and has pharmaceutical activity which includes antibacterial, antifungal, antidiabetic and anti-inflammatory effects. In fresh bay leaves, 1,8 cineole is the major aroma constituent. Other compounds of interest are α-terpinyl acetate, sabinene, α-pinene, β-pinene, β-elemene, α-terpineol, linalool and eugenol. The flowers and fruits also possess aroma. Season of harvest and time of harvest influence the aroma constituents. Considering the wide-ranging medicinal property, bay leaves need more attention in future.

References

Anac, O. (1986) Essential oil contents and chemical composition of Turkish laurel leaves. *Perfumer and Flavorist* 11(5), 73–75.

Andronikashvili, M. and Reichmuth, C. (2003) Repellency and toxicity of essential oils from *Ocimum gratissimum* (Lamiaceae) and *Laurus nobilis* (Lauraceae) from Georgia against the rust-red flour beetle (*Tribolium castaneum* Herbst) (Coleoptera: Tenebrionidae). *Proceedings of 8th International Working Conference on Stored Product Protection*, York, UK, 22–26 July 2002, pp. 749–762.

Anon. (2005) *The Columbia Encyclopedia, 2001–2005*, 6th edn. Columbia University Press, New York.

Baratta, M.T., Dorman, H.J.D., Deans, S.G., Biondi, D.M. and Ruberto, G. (1998) Chemical composition, antimicrobial and antioxidative activity of laurel, sage, rosemary, oregano and coriander essential oils. *Journal of Essential Oil Research* 10, 618–627.

Beis, S.H. and Dunford, N.T. (2006) Supercritical fluid extraction of daphne (*Laurus nobilis* L.) seed oil. *Journal of the American Chemical Society* 83, 953–957.

Borges, P., Fernandez, N. and Roncal, E. (2003). Isolation and characterization of essential oils of spice plants using steam distillation. *Alimentaria* 40(340), 125–127.

Bouzouita, N., Nafti, A., Chaabouni, M.M., Lognay, G.C., Marlier, M., Zghoulli, S. and Thonart, P. (2001) Chemical composition of *Laurus nobilis* oil from Tunisia. *Journal of Essential Oil Research* 13(2), 116–117.

Bouzouita, N., Kachouri, F., Hamdi, M. and Chaabouni, M.M. (2003) Antimicrobial activity of essential oils from Tunisian aromatic plants. *Flavour and Fragrance Journal* 18(5), 380–383.

Braun, N.A., Meier, M., Kohlenberg, B. and Hammerschmidt, F.J. (2001) delta-Terpinyl acetate. A new natural component from the essential leaf oil of *Laurus nobilis* L. (Lauraceae). *Journal of Essential Oil Research* 13(2), 95–97.

Caredda, A., Marongiu, B., Porcedda, S. and Soro, C. (2002) Supercritical carbon dioxide extraction and characterization of Laurus nobilis essential oil. *Journal of Agricultural and Food Chemistry* 50, 1492–1496.

Chericoni, S., Prieto, J.M., Iacopini, P. and Morelli, I. (2005) Essential oils of commonly used plants as inhibitors of peroxynitrite-induced tyrosine nitration. *Fitoterapia* 76(5), 481–483.

Chermenskaya, T.D., Burov, V.N., Maniar, S.P., Pow, E.M., Roditakis, N., Selytskaya, O.G., Shamshev, I.V., Wadhams, L.J. and Woodcock, C.M. (2001) Behavioural responses of western flower thrips, *Frankliniella occidentalis* (Pergande), to volatiles from three aromatic plants. *Insect Science and its Application* 21(1), 67–72.

Dadalioglu, I. and Evrendilek, G.A. (2004) Chemical compositions and antibacterial effects of essential oils of Turkish oregano (*Origanum minutiflorum*), bay laurel (*Laurus nobilis*), Spanish lavender (*Lavandula stoechas* L.), and fennel (*Foeniculum vulgare*) on common foodborne pathogens. *Journal of Agricultural and Food Chemistry* 52, 8255–8260.

Erturk, O., Kara, O., Sezer, E. and San, G. (2004) Toxicity effect of some plant extracts on development of larvae of *Plutella xylostella* (L.) (Lepidoptera; Plutellidae). *Ekoloji Cevre Dergisi* 13(50), 18–22.

Fiorini, C., Fouraste, I., David, B. and Bessiere, J.M. (1997) Composition of the flower, leaf and stem essential oils from *Laurus nobilis*. *Flavour and Fragrance Journal* 12, 91–93.

Friedman, M., Henika, P.R. and Mandrell, R.E. (2002) Bactericidal activities of plant essential oils and some of their isolated constituents against *Campylobacter jejuni, Escherichia coli, Listeria monocytogenes*, and *Salmonella enterica*. *Journal of Food Protection* 65(10), 1545–1560.

Geeta, G.S. and Reddy, T.K.R. (1990) *Aspergillus flavus* Link and its occurrence in relation to other *mycoflora* on stored spices. *Journal of Stored Products Research* 26, 211–213.

Guynot, M.E., Ramos, A.J., Seto, L., Purroy, P., Sanchis, V. and Marin, S. (2003) Antifungal activity of volatile compounds generated by essential oils against fungi commonly causing deterioration of bakery products. *Journal of Applied Microbiology* 94(5), 893–899.

Isikber, A.A., Alma, M.H., Kanat, M. and Karci, A. (2006) Fumigant toxicity of essential oils from *Laurus nobilis* and *Rosmarinus officinalis* against all life stages of *Tribolium confusum*. *Phytoparasitica* 34, 167–177.

Kevseroglu, K., Crak, C. and Ozyazc, G. (2003) A study on ontogenetic and diurnal variability of laurel (*Laurus nobilis* L.) leaves. *Turkish Journal of Field Crops* 8(1), 29–32.

Kilic, A., Hafizoglu, H., Kollmannsberger, H. and Nitz, S. (2004) Volatile constituents and key odorants in leaves, buds, flowers, and fruits of *Laurus nobilis* L. *Journal of Agricultural and Food Chemistry* 52, 1601–1606.

Kosar, M., Tunalier, Z., Ozek, T., Kurkcuoglu, M. and Baser, K.H.C. (2005) A simple method to obtain essential oils from *Salvia triloba* L. and *Laurus nobilis* L. by using microwave assisted hydrodistillation. *Zeitschiften fur Naturforschung Section C, Biosciences* 60, 501–504.

Kumar, S., Singh, J. and Sharma, A. (2001) Bay leaves. In: Peter, K.V. (ed.) *Handbook of Herbs and Spices*. Woodhead Publishing Limited, Cambridge, UK, pp. 52–61.

Mackeen, M.M., Ali, A.M., Abdullah, M.A., Nasir, R.M., Mat, N.B., Razak, A.R. and Kawazu, K. (1997) Antinematodal activity of some Malaysian plant extracts against the pine wood nematode, *Bursaphelenchus xylophilus*. *Pesticide Science* 51, 165–170.

Mandeel, Q., Hassan, A. and Isa, Z. (2003) Antibacterial activity of certain spice extracts. *Journal of Spices and Aromatic Crops* 12, 146–153.

Marino, S. de, Borbone, N., Zollo, F., Ianaro, A., Meglio, P. di and Iorizzi, M. (2005) New sesquiterpene lactones from *Laurus nobilis* leaves as inhibitors of nitric oxide production. *Planta Medica* 71, 706–710.

Matsuda, H., Kagerura, T., Toguchida, I., Ueda, H., Morikawa, T. and Yoshikawa, M. (2000) Inhibitory effects of sesquiterpenes from bay leaf on nitric oxide production in lipopolysaccharide activated macrophages: structure requirement and role of heat shock protein induction. *Life Sciences* 66(22), 2151–2157.

Muller Riebau, F., Berger, B. and Yegen, O. (1995) Chemical composition and fungitoxic properties to phytopathogenic fungi of essential oils of selected aromatic plants growing wild in Turkey. *Journal of Agricultural and Food Chemistry* 43, 2262–2266.

Nakatani, N. (2003) Biologically functional constituents of spices and herbs (2002's JSNFS award for excellence in research). *Journal of Japanese Society for Nutrition and Food Science* 56, 389–395.

Nhat, D.M., Takacsova, M., Jakubik, T. and DacVinh, N. (1999) The composition of essential oil from laurel leaf. *Czechoslovakian Journal of Food Science* 17(6), 201–203.

Pandey, R. (1997) Ecofriendly management of *Meloidogyne incognita* on *Hyoscyamus niger*. *Indian Journal of Nematology* 27, 175–178.

Pandey, V.N. and Dubey, N.K. (1997) Synergistic activity of extracted plant oils against *Pythium aphanidermatum* and *P. debaryanum*. *Tropical Agriculture* 74, 164–167.

Regnault Roger, C. and Hamraoui, A. (1995) Comparison of the insecticidal effects of water extracted and intact aromatic plants on *Acanthoscelides obtectus*, a bruchid beetle pest of kidney beans. *Chemoecology* 5(6), 1–5.

Riaz, M., Ashraf, C.M. and Chaudhary, F.M. (1989) Studies on the essential oil of the Pakistani *Laurus nobilis* Linn. in different seasons. *Pakistan Journal of Scientific and Industrial Research* 32, 33–35.

Sayyah, M., Valizadeh, J. and Kamalinejad, M. (2002) Anticonvulsant activity of the leaf essential oil of *Laurus nobilis* against pentylenetetrazole- and maximal electroshock–induced seizures. *Phytomedicine* 9(3), 212–216.

Tainter, D.R and Grenis, A.T. (1993) *Spices and Seasonings*. VCH Publishers, Inc., New York, pp. 40–43.

Uchiyama, N., Matsunaga, K., Kiuchi, F., Honda, G., Tsubouchi, A., Nakajima-Shimada, J. and Aoki, T. (2002) Trypanocidal terpenoids from *Laurus nobilis* L. *Chemical and Pharmaceutical Bulletin* 50(11), 1514–1516.

Vasudevan, P., Tandon, M. and Pathak, N. (1997) Fluid CO_2 extraction and hydrodistillation of certain biocidal essential oils and their constituents. *Journal of Scientific and Industrial Research* 56, 662–672.

Index